世界の昆虫 英名辞典
vol. 2 M-Z

矢野　宏二　編

A dictionary of English names of the world insects

Edited by Koji Yano

櫂歌書房

A dictionary of English names of the world insects

Date of publication 12. May. 2018
ISBN 978-4-434-24028-7
Edited by Koji Yano
Published by Touka Shobo
Printed and distributed by Touka Shobo
Sarayama 4-14-2, Minami-ku, Fukuoka-shi, 811-1365, Japan
Tel:+81-92-511-8111 e-mail: e@touka.com

目次

はじめに	iii
著者略歴	v
凡例	ix
参考文献	xiii
世界の昆虫英名	
A	1
M	【別冊 vol.2】 669
和名索引	【別冊 索引 vol.3】 1
学名索引	【別冊 索引 vol.3】 75

凡例

1．英名のアルファベット順に掲載したが、ハイフンの有無や語の結合があっても原則として同一タクサ（分類単位）は同じ所に掲載した。

2．英名が亜種で異なる場合は原則として別にし、原名亜種の英名を参照と記して関係が分かるようにした。

3．属名や種小名が英名に使われる場合、大文字か小文字かで統一せず、使用実態のままとした。固有名詞以外の英名は小文字で記したが、levant と Levant のように両方ある語もある。

4．日本以外の分布は動物地理区で記したが、全ての分布区を記したわけではなく、当該英名が使用される地域を示した方が多い。日本には分布しないが汎世界的に分布する種も汎世界と記した。

5．同じ英名が複数ある場合、下記のように整理してある。

 例。beet leaf miner (1) *Pegomya betae* (Curtis)(ハエ目、ハナバエ科)。分布。
   beet leafminer (2) *Liriomyza chenopodii* (Watt)(ハエ目、ハモグリバエ科)。分布。
   beet leafminer (3) テンサイモグリハナバエ *Pegomya cunicularia* (Rondani)
     (ハエ目、ハナバエ科)。分布。
  beet leafminer beet fly を見よ。

beet leafminer は上記のように4つあり、4種の英名であることが分かる。4つ目の beet fly を見ると、

  beet fly *Pegomya hyoscyami* (Panzer)(ハエ目、ハナバエ科)。分布。
  beet fly beet leaf miner (1) を見よ。

の2つあり、学名などを記述してある上段の beet fly が beet leafminer の別英名でもあることを示す。下段の beet fly は beet leaf miner (1) の種が beet fly ともいわれることを示す。

beet leafminer (1)(2)(3) は、他の英名の項で beet leafminer を見よ、と記す場合、(1)(2)(3) の3つのどれを見るかを区別するためである。例えば mangold fly を見ると beet leaf miner (1) を見よ、と記している。

このように、(XX を見よ) と記したのは、単に参照せよとの意味ではなく、XX の種の英名でもあることを示す。

また、同種で複数英名がある場合、例えば

  American cotton bollworm corn earworm (1)(2) を見よ。

と記しているが、American cotton bollworm は corn earworm (1) と同 (2) の2種の英名でもあることを示す。

6．同種で複数の異なる英名がある場合、分類順に掲載していないので一個所ではわからない。例えば *Agrius convolvuli* (Linnaeus) エビガラスズメは

  A. sweetpotato hawk-moth
  B. sweetpotato horn worm
  C. convolvulus hawkmoth
  D. morning-glory sphinx

の4英名がある。このうち、使用例が多く、定着していると思われる A をえらび、和名、学名、分類上の所属と分布、若干の解説を記した。そして B, C, D の英名の項では、これらの記述を省き、A を見よ、と記した。

A を選ぶ基準は上記したが、複数英名間に明らかな使用差がなかったり、使用地域により異なる場合もあり、多くの場合は編者の判断によった。

7．6の例にあげたエビガラスズメの英名の A と B、シャクガの *Nepytia canosaria* の英名の false hemlock moth と false hemlock looper などは成虫と幼虫の英名であるが、いずれも種の英名として成育期に関係なく使用され、幼虫名の方が定着している場合も多い。従って成虫は XX というとか、幼虫は XX というとかと記した場合もあるが、同様である。

8．同種で複数の英名がある場合、語尾などが少し異なる場合は同じところで記したが、かなり違う場合は別記した。例えば half-yellow moth のところに half-yellow ともいう、と記したが、yellow-cloaked midget は別記した。分類順（学名）のリストを用意すればよいが、紙数の関係で断念した。

9．種の解説は紙数の関係で最小限とし、日本産の主要害虫を主に記述した。幼虫が加害する場合が多いが、単に X の害虫と記した。また、加害植物全てを記さず、代表的な植物に限った。

10．解説の部分で、日本亜種の学名、和名を示した場合がある。例えば

  buff footman *Eilema deplana* (Esper) （チョウ目、ヒトリガ科）。旧北区。
   日本亜種は *E. d. puvescens* (Butler) ムジホソバ。

とあるのは、日本亜種の英名使用例はないが、旧北区の原名亜種は buff footman の英名であることを示す。

11. 英名に含まれるハイフンは採録した文献に準拠したので leaf miner, leaf-miner, leafminer のように統一していない。検索する場合は留意をお願いしたい。例えば armyworm と army worm、blow fly と blowfly、many-dotted appleworm moth と manydotted apple worm のように出典に準拠して掲載してある。これが英名であり、本書では採録文献の使用例に準拠したが、別の使用例（ハイフンの有無）もあるので注意をお願いしたい。このため、アルファベット順の掲載も leaf miner, leaf-miner, leafminer のあとに leafhopper がくるので注意いただきたい。また、通常はされない単語の結合もみられる。

12. 属以上のタクサの英名は下記のように複数形で記されるのが慣例である。
    stink bugs　　カメムシ科 Pentatomidae（カメムシ目）の昆虫の総称。
    cockchafers　　*Melolontha*（コウチュウ目、コガネムシ科）の昆虫の総称。
これはカメムシ科の種は stink bug、*Melolontha* 属の種は cockchafer といわれることを示すが、これらタクサの英名として使用されている。

13. 最近の分類学的研究の結果、掲載種の所属科名の変更が多く生じているので留意いただきたい（ジャノメチョウ、ドクチョウなどを含むタテハチョウ科やメイガ科など）。

14. 採録文献の英名に併記された学名は、現行学名に変更した場合がある。

世界の昆虫
英名辞典
vol. 2 M-Z

世界の民族
英名辞典
vol. 2 M-Z

英名	和名	学名	所属、分布、ほか
M			
maaky fly			blue blowfly を見よ
Mabashe river buff		*Deloneura immaculata* Trimen	(チョウ目、シジミチョウ科) エチオピア区
Mabille's awlet		*Bibasis phul* (Mabille)	(チョウ目、セセリチョウ科) 東洋区
Mabille's bent-skipper		*Camptopleura theramenes* Mabille	(チョウ目、セセリチョウ科) 新熱帯区
Mabille's Mylon		*Mylon illineatus* Mabille et Boullet	(チョウ目、セセリチョウ科) 新熱帯区
Mabille's red-eye			Mabille's red-eyed nightfighter を見よ
Mabille's red-eyed nightfighter		*Dyscophellus ethyras* Mabille	(チョウ目、セセリチョウ科) 新熱帯区
Mabille's sabre-wing		*Jemadia menechmus* (Mabille)	(チョウ目、セセリチョウ科) 新熱帯区
Mabille's skipper		*Polites puxillius* (Mabille)	(チョウ目、セセリチョウ科) 新熱帯区
Mabille's sootywing		*Bolla giselus* (Mabille)	(チョウ目、セセリチョウ科) 新熱帯区
macadamia cup moth		*Mecytha fasciata* (Walker)	(チョウ目、イラガ科) 豪州区
macadamia felted coccid			macadamia felted scale を見よ
macadamia felted scale		*Eriococcus ironsidei* Williams	(カメムシ目、フクロカイガラムシ科) 豪州区
macadamia flower caterpillar		*Cryptoblabes hemigypsa* Turner	(チョウ目、メイガ科) 豪州区
macadamia lace bug		*Ulonemia concava* Drake	(カメムシ目、グンバイムシ科) 豪州区
macadamia leafminer		*Acrocercops chionosema* Turner	(チョウ目、ホソガ科) 豪州区
macadamia mussel scale		*Lepidosaphes macadamiae* Williams	(カメムシ目、マルカイガラムシ科) 豪州区
macadamia nutborer	アシブトヒメハマキ	*Cryptophlebia ombrodelta* (Lower)	(チョウ目、ハマキガ科) 日本、東洋区、豪州区、大洋区
macadamia twig-girdler		*Xylorycta luteotactella* (Walker)	(チョウ目、マルハキバガ科) 豪州区
macadamia white scale		*Pseudaulacaspis brimblecombei* Williams	(カメムシ目、マルカイガラムシ科) 豪州区
Macao paper wasp		*Polistes macaensis* (Fabricius)	(ハチ目、スズメバチ科) 東洋区、新北区
macaque louse	サルジラミ	*Pedicinus obtusus* (Rudow)	(シラミ目、サルジラミ科) 日本、旧北区、東洋区
Macarius tufted skipper			marbled tufted skipper を見よ
macaw worm-fly			human bot fly を見よ
Macchia blue		*Tarucus thespis* (Linnaeus)	(チョウ目、シジミチョウ科) エチオピア区
MacClounie's Charaxes			red coast Charaxes を見よ
MacCulloch's forester moth		*Androloma maccullochii* (Kirby)	(チョウ目、ヤガ科) 新北区。MacCulloch's forester ともいう

英名	和名	学名	所属、分布、ほか
Macedonian blue		*Polyommatus philippi* (Lederer)	（チョウ目、シジミチョウ科）旧北区
Macedonian grayling		*Pseudochazara cingovskii* Gross	（チョウ目、タテハチョウ科）旧北区
Machacha brown		*Pseudonympha machacha* Riley	（チョウ目、タテハチョウ科）エチオピア区
Machacha opal		*Chrysoritis pelion* (Pennington)	（チョウ目、シジミチョウ科）エチオピア区
Machado's skimmer		*Orthetrum machadoi* Longfield	（トンボ目、トンボ科）エチオピア区
Machequena Acraea		*Acraea machequena* Grose-Smith	（チョウ目、タテハチョウ科）エチオピア区
machilids	イシノミ科	Machilidae	（イシノミ目）の昆虫の総称　世界から約300種、日本から14種
machilus bark beetle	アオガシノキクイムシ	*Xyleborus machili* Niijima	（コウチュウ目、キクイムシ科）日本
machilus cottony aphid	タブノキハアブラムシ	*Marhilaphis machili* (Takahashi)	（カメムシ目、アブラムシ科）日本、東洋区
machilus leaf gall midge	タブウスフシタマバエ	*Daphnephila machilicola* Yukawa	（ハエ目、タマバエ科）日本、東洋区
machilus oystershell scale			cymbidium scale を見よ
Macken's dart		*Acleros mackenii* (Trimen)	（チョウ目、セセリチョウ科）エチオピア区
Macken's skipper			Macken's dart を見よ
Mackinnon's swallowtail	マッキンノニオオアゲハ	*Papilio mackinnoni* Sharpe	（チョウ目、アゲハチョウ科）エチオピア区
Mackwood's hairstreak		*Satyrium mackwoodi* (Evans)	（チョウ目、シジミチョウ科）東洋区
Macleannan's skipper		*Parelbella macleannani* (Godman et Salvin)	（チョウ目、セセリチョウ科）新熱帯区
Macleay's grass yellow		*Eurema herla* (Macleay)	（チョウ目、シロチョウ科）豪州区
Macleay's spectre			giant prickly stick insect を見よ
Macleay's spectre stick insect			giant prickly stick insect を見よ
Macleay's swallowtail	キミドリタイマイ	*Graphium macleayanus* (Leach)	（チョウ目、アゲハチョウ科）豪州区
Macleay's swordtail			Macleay's swallowtail を見よ
MacNeill's shieldback		*Neduba macneilli* Rentz et Birchim	（バッタ目、キリギリス科）新北区
Macomo ranger		*Kedestes macomo* (Trimen)	（チョウ目、セセリチョウ科）エチオピア区
Macoun's arctic	カナダオオタカネヒカゲ	*Oeneis macounii* (Edwards)	（チョウ目、タテハチョウ科）新北区
Macqueen's hairstreak		*Jalmenus pseudictinus* Kerr et Macqueen	（チョウ目、シジミチョウ科）豪州区
Macrame moth		*Phaecasiophora confixana* (Walker)	（チョウ目、ハマキガ科）新北区
macraner	大雄型		異常に大きい雄アリ
macrergate	大職蟻型		異常に大きい職アリ

英名	和名	学名	所属、分布、ほか
macro-caddisflies			large caddisflies を見よ
macrocosma moth		*Niceteria macrocosma* Lower	（チョウ目、シャクガ科）豪州区
macrogyne	大雌型		異常に大きい雌アリ～女王アリ
macrolepidopterans			macro-moths を見よ
macro-moths	大蛾類	Macrolepidoptera	（チョウ目）の昆虫の総称
macros			macro-moths を見よ
macrozamia borer		*Tranes internatus* Pascoe	（コウチュウ目、ゾウムシ科）豪州区
macular Epitola		*Cephetola maculata* (Hawker-Smith)	（チョウ目、シジミチョウ科）エチオピア区
maculate curculio			spotted vegetable weevil を見よ
maculate ladybird	ヤホシテントウ	*Harmonia octomaculata* (Fabricius)	（コウチュウ目、テントウムシ科）日本、東洋区
maculate lancer		*Salanoemia sala* (Hewitson)	（チョウ目、セセリチョウ科）東洋区
maculated broad-winged planthopper	アミガサハゴロモ	*Pochazia albomaculata* (Uhler)	（カメムシ目、ハゴロモ科）日本
maculated lyonetid	ホシボシホソガ	*Callisto multimaculata* (Matsumura)	（チョウ目、ハモグリガ科）日本
maculated rice katydid	ホシササキリ	*Conocephalus maculatus* (Le Guillou)	（バッタ目、キリギリス科）、日本、東洋区、豪州区、エチオピア区
maculatus skipper	チャマダラセセリ	*Pyrgus maculatus maculatus* (Bremer et Grey)	（チョウ目、セセリチョウ科）日本、旧北区
Madagascan emperor		*Bunaea aslauga* Kirby	（チョウ目、ヤママユガ科）エチオピア区
Madagascan emperor swallowtail		*Papilio morondavana* Grose-Smith	（チョウ目、アゲハチョウ科）エチオピア区
Madagascan friar		*Amauris nossima* (Ward)	（チョウ目、タテハチョウ科）エチオピア区
Madagascan marbled praying mantis		*Polyspilota aeruginosa* (Goeze)	（カマキリ目、カマキリ科）エチオピア区。Madagascan marbled mantis ともいう
Madagascan moon moth			moon moth (1) を見よ
Madagascan sunset moth	ニシキオオツバメガ	*Chrysiridia rhipheus* (Drury)	（チョウ目、ツバメガ科）エチオピア区。マダガスカルの美麗種
Madagascan sunset moth			sunset moth を見よ
Madagascar beetle	サツマゴミムシダマシ	*Leichenum canaliculatum* (Fabricius)	（コウチュウ目、ゴミムシダマシ科）日本、東洋区、豪州区
Madagascar black beetle		*Heteronychus plebejus* (Klug)	（コウチュウ目、コガネムシ科）エチオピア区。イネ害虫
Madagascar diadem	ハガタムラサキ	*Hypolimnas dexithea* (Hewitson)	（チョウ目、タテハチョウ科）エチオピア区
Madagascar giant swallowtail	ホシボシジャコウアゲハ（アンテノールオオジャコウ）	*Pharmacophagus antenor* (Drury)	（チョウ目、アゲハチョウ科）エチオピア区
Madagascar green-veined Charaxes		*Charaxes antamboulou* Lucas	（チョウ目、タテハチョウ科）エチオピア区

英名	和名	学名	所属、分布、ほか
Madagascar hissing cockroach		*Gromphadorhina portentosa* (Schaum)	(ゴキブリ目、ブラベルスゴキブリ科) エチオピア区。マダガスカルの巨大種。hissing cockroach, hisser ともいう
Madagascar locust		*Locusta migratoria capito* (Saussure)	(バッタ目、バッタ科) エチオピア区
Madagascaran cockroach		*Gromphadorhina laevigata* Saussure et Zehntner	(ゴキブリ目、ブラベルスゴキブリ科) 新北区
madder hawk moth			bedstraw hawk を見よ　米国での英名
Madeira ant			big-headed ant (1) を見よ
Madeira cockroach	マデイラゴキブリ	*Leucophaea maderae* (Fabricius)	(ゴキブリ目、ハイイロゴキブリ科) 日本、全北区、熱帯圏。米国での英名。大西洋の Madeira 島に由来
Madeira grayling		*Hipparchia aristaeus maderensis* (Bethune-Baker)	(チョウ目、タテハチョウ科) 旧北区。southern grayling を参照
Madeira mealybug	マデイラコナカイガラムシ	*Phenacoccus madeirensis* Green	(カメムシ目、コナカイガラムシ科) 日本。マンゴー、ハイビスカスの害虫
Madeira roach			Madeira cockroach を見よ
Madeiran brimstone		*Gonepteryx maderensis* Felder	(チョウ目、シロチョウ科) 旧北区
Madeiran large white		*Pieris wollastoni* Butler	(チョウ目、シロチョウ科) 旧北区
Madeiran speckled wood		*Pararge xiphia* (Fabricius)	(チョウ目、タテハチョウ科) 旧北区
Madras ace		*Halpe honorei* (de Nicéville)	(チョウ目、セセリチョウ科) 東洋区
madrone shield-bearer		*Coptodisca arbutiella* Busck	(チョウ目、ツヤコガ科) 新北区
madrone skin miner		*Marmara arbutiella* Busck	(チョウ目、ホソガ科) 新北区
Maeonis hairstreak		*Apuecla maeonis* (Godman et Salvin)	(チョウ目、シジミチョウ科) 新熱帯区
maesitis hairstreak			amethyst hairstreak を見よ
Mafa grizzled skipper		*Spialia mafa* (Trimen)	(チョウ目、セセリチョウ科) エチオピア区
Mafa sandman			Mafa grizzled skipper を見よ
Magdalen underwing moth		*Catocala illecta* Walker	(チョウ目、ヤガ科) 新北区
Magdalena alpine	オオクロベニヒカゲ	*Erebia magdalena* Strecker	(チョウ目、タテハチョウ科) 新北区
Magdalena valley ringlet		*Splendeuptychia ackeryi* Huertas, Rios et Le Crom	(チョウ目、タテハチョウ科) 新熱帯区
Magellan birdwing	コウトウキシタアゲハ	*Troides magellanus* (Felder)	(チョウ目、アゲハチョウ科) 東洋区
magenta tip	ケリメネツマアカシロチョウ	*Colotis celimene* (Lucas)	(チョウ目、シロチョウ科) エチオピア区
magnetic termite	ジシャクシロアリ	*Amitermes meridionalis* (Froggatt)	(シロアリ目、シロアリ科) 豪州区。太陽に対応した扁平な巣を作る
magnifera long-horned beetle			artocarpus long-horned beetle を見よ
magnificent emperor			black emperor を見よ

英名	和名	学名	所属、分布、ほか
magnificent leafwing		*Coenophlebia archidona* Hewitson	(チョウ目、タテハチョウ科) 新熱帯区
magnificent oakblue		*Arhopala anarte* (Hewitson)	(チョウ目、シジミチョウ科) 東洋区
magnificent scalloped hawk		*Odontosida magnificum* (Rothschild)	(チョウ目、スズメガ科) エチオピア区
magnificent skimmer		*Macromia magnifica* McLachlan	(トンボ目、ヤマトンボ科) 新北区
magnificent swallowtail	ガラマスアゲハ	*Papilio garamas* (Geyer)	(チョウ目、アゲハチョウ科) 新北区、豪州区
magnificient shieldback		*Idiostatus magnificus* Hebard	(バッタ目、キリギリス科) 新北区
magnolia borer moth		*Euzophera magnolialis* Capps	(チョウ目、メイガ科) 新北区。magnolia root borer ともいう
magnolia marmorated aphid			Japanese elder aphid を見よ
magnolia scale		*Neolecanium cornuparvum* (Thro)	(カメムシ目、カタカイガラムシ科) 新北区
magnolia serpentine leafminer moth		*Phyllocnistis magnoliella* Chambers	(チョウ目、コハモグリガ科) 新北区
magnolia white scale			false oleander scale を見よ
Magou crow	マゴウルリマダラ	*Euploea magou* Martin	(チョウ目、タテハチョウ科) 東洋区。Magoue ともいう
magpie		*Abraxas grossulariata* (Linnaeus)	(チョウ目、シャクガ科) 旧北区。日本亜種は *A. g. conspurata* Butler スグリシロエダシャク
magpie China-mark			small magpie moth を見よ
magpie crow	デイオクレテイアヌスルリマダラ	*Euploea diocletianus diocletianus* (Fabricius)	(チョウ目、タテハチョウ科) 日本、東洋区
magpie crow	シロモンルリマダラ	*Euploea radamanthus* (Fabricius)	(チョウ目、タテハチョウ科) 東洋区
magpie flat		*Abraximorpha davidii ermasis* Fruhstorfer	(チョウ目、セセリチョウ科) 東洋区
magpie moth (1)	ユウマダラエダシャク	*Abraxas miranda* Butler	(チョウ目、シャクガ科) 日本、旧北区。シキミ、ニシキギの害虫
magpie moth		*Abraxas plumbeata* Cockerell	(チョウ目、シャクガ科) 東洋区
magpie moth		*Nyctemera secundiana* Lucas	(チョウ目、ヒトリガ科) 豪州区
magpie moth			senecio moth を見よ
magpie moth			magpie を見よ
magpie moth			black diamond moth を見よ
magpie moths		*Nyctemera*	(チョウ目、ヒトリガ科) の昆虫の総称　豪州での英名
maharaja Apollo		*Parnassius maharaja* Avinoff	(チョウ目、アゲハチョウ科) 東洋区
Maheta skipper		*Trapezites maheta* (Hewitson)	(チョウ目、セセリチョウ科) 豪州区
Mahoenui giant weta		*Deinacrida mahoenui* Gibbs	(バッタ目、Stenopelmatidae) 豪州区

英名	和名	学名	所属、分布、ほか
mahogany bark weevil		*Macrocopturus floridanus* (Fall)	（コウチュウ目、ゾウムシ科）新北区
mahogany flat			bed bug (1) を見よ
mahogany leafminer		*Phyllocnistis meliacella* Becker	（チョウ目、コハモグリガ科）新北区。mahogany leafminer moth ともいう
mahogany shoot borer		*Hypsipyla grandella* Zeller	（チョウ目、メイガ科）新北区、新熱帯区。成虫は mahogany shootborer moth
mahogany shoot borer (1)		*Hypsipyla robusta* (Moore)	（チョウ目、メイガ科）東洋区、エチオピア区
mahogany webworm		*Macalla thyrsisalis* Walker	（チョウ目、メイガ科）新北区。成虫は mahogany webworm moth
maiden moths		Thyretidae	（チョウ目）の昆虫の総称
maidenhair fern aphid			fern aphid (2) を見よ
maidenhair fern weevil		*Neosyagrius cordipennis* Lea	（コウチュウ目、ゾウムシ科）豪州区
maiden's blush		*Cyclophora punctaria* (Linnaeus)	（チョウ目、シャクガ科）旧北区
Maine copper		*Lycaena dorcas claytoni* Brower	（チョウ目、シジミチョウ科）新北区。
Maine snaketail		*Ophiogomphus mainensis* Packard	（トンボ目、サナエトンボ科）新北区
maize aphid			corn leaf aphid を見よ　南アフリカでの英名
maize billbug		*Calendra maidis* (Chittenden)	（コウチュウ目、ゾウムシ科）新北区。トウモロコシ害虫
maize borer			Neotropical corn borer を見よ
maize cob borer			maize ear borer を見よ
maize cob worm			corn earworm (2) を見よ　幼虫の英名
maize ear borer		*Mussidia nigrivenella* Ragonot	（チョウ目、メイガ科）エチオピア区
maize grain weevil			maize weevil を見よ
maize leafhopper		*Cicadulina bimaculata* (Evans)	（カメムシ目、オオヨコバイ科）豪州区
maize leafhopper		*Cicadulina mbila* (Naude)	（カメムシ目、オオヨコバイ科）エチオピア区
maize leafhopper			corn leafhopper を見よ
maize leafhopper			corn hopper を見よ
maize moth			peach moth を見よ
maize orange leafhopper		*Cicadulina bipunctella* (Matsumura)	（カメムシ目、オオヨコバイ科）日本、旧北区
maize planthopper			corn hopper を見よ　豪州での英名
maize seed beetle		*Clivina vagans* Putzys	（コウチュウ目、オサムシ科）豪州区
maize stalk borer (1)		*Busseola sorghicida* Thurau	（チョウ目、ヤガ科）エチオピア区。イネ、トウモロコシなどの害虫
maize stalk borer (2)		*Busseola fusca* Hampson	（チョウ目、ヤガ科）新北区、エチオピア区。トウモロコシ害虫。成虫は maize stalk borer moth
maize stalk borer			pink borer (2) を見よ

英名	和名	学名	所属、分布、ほか
maize tassel beetle		*Megalognatha rufiventris* Baly	（コウチュウ目、ハムシ科）エチオピア区
maize webworm		*Marasmia trapezalis* (Guenée)	（チョウ目、メイガ科）東洋区、エチオピア区、豪州区、新熱帯区
maize weevil	コクゾウムシ	*Sitophilus zeamais* Motschulsky	（コウチュウ目、オサゾウムシ科）日本、汎世界。貯穀害虫
major Datana moth			azalea caterpillar を見よ
major sallow moth		*Feralia major* J. B. Smith	（チョウ目、ヤガ科）新北区。major sallow ともいう
Makabusi sprite		*Pseudagrion makabusiense* Pinhey	（トンボ目、イトトンボ科）エチオピア区
Malabar banded peacock		*Papilio buddha* Westwood	（チョウ目、アゲハチョウ科）東洋区
Malabar banded swallowtail		*Papilio liomedon* Moore	（チョウ目、アゲハチョウ科）東洋区
Malabar brown		*Papilio dravidarum* Wood-Mason	（チョウ目、アゲハチョウ科）東洋区
Malabar flash		*Vadebra lankana* (Moore)	（チョウ目、シジミチョウ科）東洋区
Malabar raven			Malabar brown を見よ
Malabar rose		*Pachliopta pandiyana* Moore	（チョウ目、アゲハチョウ科）東洋区
Malabar spotted flat		*Celaenorrhinus ambareesa* (Moore)	（チョウ目、セセリチョウ科）東洋区
Malabar spreadwing		*Lestes malabaricus* Fraser	（トンボ目、アオイトトンボ科）東洋区
Malabar sprite		*Pseudagrion malabaricum* Fraser	（トンボ目、イトトンボ科）東洋区
Malabar tree nymph		*Idea malabarica* Moore	（チョウ目、タテハチョウ科）東洋区
malachiid beetles		Malachiidae	（コウチュウ目）の昆虫の総称　ジョウカイモドキ科の異名
malachite	ミドリタテハ	*Siproeta stelenes* (Linnaeus)	（チョウ目、タテハチョウ科）新北区、新熱帯区。斑紋がクジャク石に似ることから。malachite butterfly ともいう
malachite beetle			red-tipped flower beetle を見よ
malachites		Synlestidae	（トンボ目）の昆虫の総称
malacodermatous beetles		Malacodermata	（コウチュウ目）の昆虫の総称
malacopsyllid fleas		Malacopsyllidae	（ノミ目）の昆虫の総称
Malagasy grass yellow		*Eurema floricola* (Boisduval)	（チョウ目、シロチョウ科）エチオピア区
Malagasy sailer	コミスジモドキ	*Neptidopsis fulgurata* (Boisduval)	（チョウ目、タテハチョウ科）エチオピア区
Malana leafroller moth		*Olethreutes malana* (Fernald)	（チョウ目、ハマキガ科）新北区
malaria-carrying mosquitoes			malaria mosquitoes を見よ
malaria mosquito	シナハマダラカ	*Anopheles sinensis* Wiedemann	（ハエ目、カ科）日本、旧北区、東洋区。水田、池沼などに発生。三日熱マラリアを媒介
malaria mosquito (1)		*Anopheles quadrimaculatus* Say	（ハエ目、カ科）新北区

英名	和名	学名	所属、分布、ほか
malaria mosquito			spotted Anopheles を見よ
malaria mosquitoes	ハマダラカ属	*Anopheles*	（ハエ目、カ科）の昆虫の総称
malarial mosquitoes			malaria mosquitoes を見よ
Malay baron	モニナイナズマ	*Euthalia monina monina* (Moore)	（チョウ目、タテハチョウ科）東洋区
Malay count		*Tanaecia godartii asoka* Felder	（チョウ目、タテハチョウ科）東洋区
Malay forest bob		*Scobura phiditia* (Hewitson)	（チョウ目、セセリチョウ科）東洋区
Malay lacewing	ハレギチョウ	*Cethosia hypsea* Doubleday	（チョウ目、タテハチョウ科）東洋区。*C. h. hypsina* C. et R. Felder も同英名
Malay lilysquatter		*Paracercion malayanum* (Selys)	（トンボ目、イトトンボ科）東洋区
Malay red harlequin		*Paralaxita damajanti* (C. et R. Felder)	（チョウ目、シジミタテハ科）東洋区
Malay red harlequin (1)		*Paralaxita telesia lyclene* de Nicéville	（チョウ目、シジミタテハ科）東洋区
Malay staff sergeant	レタオオミスジ	*Athyma reta* Moore	（チョウ目、タテハチョウ科）東洋区。*A. r. moorei* (Fruhstorfer) も同英名
Malay tailed Judy		*Abisara savitri* C. et R. Felder	（チョウ目、シジミタテハ科）東洋区
Malay tiger	アフィニスカバマダラ	*Danaus affinis* (Fabricius)	（チョウ目、タテハチョウ科）東洋区、豪州区。*D. a. malayanus* Fruhstorfer も同英名
Malay viscount		*Tanaecia pelea pelea* Fabricius	（チョウ目、タテハチョウ科）東洋区
Malay Yeoman		*Cirrochroa emalea emalea* Guérin	（チョウ目、タテハチョウ科）東洋区
Malayan	タイワンクロギンシジミ（タイワンクロホシシジミ）	*Megisba malaya* (Horsfield)	（チョウ目、シジミチョウ科）東洋区。*M. m. sikkima* Moore も同英名
Malayan (1)		*Megisba strongyle nigra* (Miskin)	（チョウ目、シジミチョウ科）豪州区
Malayan Assyrian		*Terinos clarissa malayanus* Fruhstorfer	（チョウ目、タテハチョウ科）東洋区。Assyrian を参照
Malayan birdwing	キマダラマルバネアゲハ	*Troides amphrysus ruficollis* (Butler)	（チョウ目、アゲハチョウ科）東洋区。Malay birdwing ともいう
Malayan black rice bug			black paddy bug を見よ
Malayan bush brown		*Mycalesis fusca fusca* (C. et R. Felder)	（チョウ目、タテハチョウ科）東洋区
Malayan bushblue		*Arhopala ammon* (Hewitson)	（チョウ目、シジミチョウ科）東洋区
Malayan crow	シロオビルリマダラ	*Euploea camaralzeman malayica* (Butler)	（チョウ目、タテハチョウ科）東洋区
Malayan crow (1)		*Euploea rodtenbacheri* C. et R. Felder	（チョウ目、タテハチョウ科）東洋区
Malayan egg-fly	ヤエヤマムラサキ	*Hypolimnas anomala* (Wallace)	（チョウ目、タテハチョウ科）東洋区、豪州区
Malayan four-line blue		*Nacaduba pendleburyi pendleburyi* Corbet	（チョウ目、シジミチョウ科）東洋区

英名	和名	学名	所属、分布、ほか
Malayan fruit fly			Oriental fruit fly を見よ
Malayan jezebel		*Delias ninus ninus* (Wallace)	(チョウ目、シロチョウ科) 東洋区
Malayan jungle nymph			spiny stick insect を見よ
Malayan lascar		*Lasippa tiga siaka* Moore	(チョウ目、タテハチョウ科) 東洋区。Burmese lascar (1) を参照
Malayan moth			Malayan (1) を見よ
Malayan nawab	モオリフタオチョウ	*Polyura moori* (Distant)	(チョウ目、タテハチョウ科) 東洋区
Malayan owl	ムカシヒカゲ	*Neorina lowii* Doubleday	(チョウ目、タテハチョウ科) 東洋区。*N. l. neophyta* (Fruhstorfer) も同英名
Malayan plum Judy		*Abisara saturata kausambioides* de Nicéville et Martin	(チョウ目、シジミタテハ科) 東洋区
Malayan powderpost beetle			hairy powderpost beetle (1) を見よ
Malayan rice seedling fly			rice stem fly (2) を見よ
Malayan ringlet		*Ragadia makuta siponta* (Fruhstorfer)	(チョウ目、タテハチョウ科) 東洋区
Malayan rose			Ceylon rose を見よ
Malayan six ring			common six ring を見よ
Malayan sunbeam		*Curetis santana malayica* (Felder et Felder)	(チョウ目、シジミチョウ科) 東洋区
Malayan wood nymph			spiny stick insect を見よ
Malayan zebra	ゴマダラタイマイ（アサギタイマイ）	*Graphium delessertii* (Guérin-Méneville)	(チョウ目、アゲハチョウ科) 東洋区
Malaysian albatross		*Saletara liberia distanti* (Butler)	(チョウ目、シロチョウ科) 東洋区
Malaysian dead leaf mantis		*Deroplatys lobata* Guérin-Méneville	(カマキリ目、カマキリ科) 東洋区
Malaysian dead leaf mantis			giant dead leaf mantis を見よ
Malaysian fruit fly			solanum fruit fly (1) を見よ
Malaysian stick insect			spiny stick insect を見よ
Malibu shieldback		*Aglaothorax longipennis* Rentz et Weissman	(バッタ目、キリギリス科) 新北区
Malindeva skipper		*Hesperilla malindeva* Lower	(チョウ目、セセリチョウ科) 豪州区
malitious skipper		*Synapte malitiosa* (Herrich-Schäffer)	(チョウ目、セセリチョウ科) 新北区
mallee moths			concealer moths を見よ
mallee witchetygrub		*Cnemoplites blackburni* Lameere	(コウチュウ目、カミキリムシ科) 豪州区
mallee witchetygrub		*Cnemoplites edulis* Newman	(コウチュウ目、カミキリムシ科) 豪州区
mallophagans			chewing lice を見よ

英名	和名	学名	所属、分布、ほか
mallow		*Larentia clavaria* (Haworth)	（チョウ目、シャクガ科）旧北区。mallow フヨウに由来
mallow carpet			mallow を見よ
mallow moth			mallow を見よ
mallow scrub-hairstreak		*Strymon istapa* (Reakirt)	（チョウ目、シジミチョウ科）新北区、新熱帯区
mallow skipper	ウスイロチャマダラセセリ	*Carcharodus alceae* (Esper)	（チョウ目、セセリチョウ科）旧北区、東洋区
Malta groundstreak		*Calycopis malta* (Schaus)	（チョウ目、シジミチョウ科）新熱帯区
Maluti copper		*Aloeides maluti* Pringle	（チョウ目、シジミチョウ科）エチオピア区
Malvina purplewing		*Eunica malvina* Bates	（チョウ目、タテハチョウ科）新熱帯区
mamba	コロンナオナガタイマイ	*Graphium colonna* (Ward)	（チョウ目、アゲハチョウ科）エチオピア区。mamba は大型毒蛇の呼称
mamba swordtail			mamba を見よ
mammal chewing lice	ケモノハジラミ科	Trichodectidae	（ハジラミ目）の昆虫の総称　哺乳類に寄生
mammal lice			mammal chewing lice を見よ
mammal nest beetles		Leptinidae	（コウチュウ目）の昆虫の総称　ビーバーなど哺乳類に外部寄生
mammal sucking lice			wrinkled sucking lice を見よ
man faced bug	オオアカメムシ	*Catacanthus incarnatus* (Drury)	（カメムシ目、カメムシ科）日本、東洋区
Manado tiger	メナデンシスアサギマダラ	*Parantica menadensis* (Moore)	（チョウ目、タテハチョウ科）東洋区
Manado yellow tiger			Manado tiger を見よ
Manchester Argus			large heath を見よ
Manchester treble-bar		*Carsia sororiata anglica* Prout	（チョウ目、シャクガ科）旧北区
Manchurian ash elongate scolytid			Japanese platypodid を見よ
Manchurian fruit moth	リンゴコシンクイ	*Grapholita inopinata* (Heinrich)	（チョウ目、ハマキガ科）日本、旧北区。リンゴ害虫
mandarin grass yellow		*Eurema mandarinula* (Holland)	（チョウ目、シロチョウ科）エチオピア区
mandarin Tinea moth		*Tinea mandarinella* (Dietz)	（チョウ目、ヒロズコガ科）新北区
mandarin blue		*Charana mandarinus* (Hewitson)	（チョウ目、シジミチョウ科）東洋区
mandarin blues		*Charana*	（チョウ目、シジミチョウ科）の昆虫の総称
mandarin swallowtail		*Graphium mandarinus* Oberthür	（チョウ目、アゲハチョウ科）旧北区、東洋区
mandibulate archaic moths			mandibulate moths を見よ
mandibulate moths	コバネガ科	Micropterigidae	（チョウ目）の昆虫の総称　原始的なガで、成虫口器には大腮があって花粉、菌などを食べる。
Mandinga forester		*Bebearia mandinga* (Felder et Felder)	（チョウ目、タテハチョウ科）エチオピア区

英名	和名	学名	所属、分布、ほか
Manfreda giant skipper		*Stallingsia maculosa* (Freeman)	（チョウ目、セセリチョウ科）新北区、新熱帯区
mangel flea beetle			brassy flea beetle (1) をみよ
mangel fly			beet leaf miner (1) を見よ
mangifera long-horned beetle			artocarpus long-horned beetle を見よ
mango aphid			udo aphid を見よ
mango bark beetle		*Hypocryphalus mangiferae* Stebbing	（コウチュウ目、キクイムシ科）新北区
mango blossom midge		*Dasineura mangiferae* Felt	（ハエ目、タマバエ科）新熱帯区
mango flower beetle	ハイイロハナムグリ	*Protaetia fusca* (Herbst)	（コウチュウ目、コガネムシ科）日本、新北区、東洋区、大洋区、豪州区
mango fly		*Idiocerus niveosparsus* Lethierry	（カメムシ目、オオヨコバイ科）東洋区
mango fly		*Dacus frauenfeldi* Schiner	（ハエ目、ミバエ科）豪州区
mango fly			Oriental fruit fly を見よ
mango fruit fly		*Ceratitis cosyra* (Walker)	（ハエ目、ミバエ科）エチオピア区、
mango fruit fly (1)	ミナミアメリカミバエ	*Anastrepha fraterculus* (Wiedemann)	（ハエ目、ミバエ科）新熱帯区
mango gall fly		*Procontarinia matteiana* Kieffer et Cecconi	（ハエ目、タマバエ科）東洋区、エチオピア区
mango gall midge	マンゴーハフクレタマバエ	*Procontarinia mangicola* (Shi)	（ハエ目、タマバエ科）日本。マンゴー害虫
mango grey scale			vanda orchid scale を見よ
mango leaf miner			pear barkminer を見よ
mango leafhopper		*Anuritodus atlinsoni* (Lethierry)	（カメムシ目、オオヨコバイ科）東洋区
mango leaf-webber		*Achaea exea* Cramer	（チョウ目、ヤガ科）エチオピア区
mango leaf-webber		*Orthaga exvinacea* Hampson	（チョウ目、メイガ科）東洋区
mango nut weevil			mango weevil (2) を見よ
mango planthopper		*Colgaroides acuminata* (Walker)	（カメムシ目、アオバハゴロモ科）豪州区
mango psylla	マンゴーキジラミ	*Calophya mangiferae* Burckhardt et Basset	（カメムシ目、ヒメキジラミ科）日本、東洋区。マンゴー害虫
mango pulp weevil		*Sternochetus frigidus* (Fabricius)	（コウチュウ目、ゾウムシ科）東洋区
mango scale		*Leucaspis indica* Marlatt	（カメムシ目、マルカイガラムシ科）東洋区。マンゴ害虫
mango scale (1)		*Aulacaspis tubercularis* Newstead	（カメムシ目、マルカイガラムシ科）豪州区
mango scale (2)	アオキシロカイガラムシ	*Phenacaspis cockerelli* (Cooley)	（カメムシ目、マルカイガラムシ科）日本、旧北区、豪州区、エチオピア区。バナナ、マンゴー、サクラなどの害虫
mango scale			pulvinaria scale (2) を見よ
mango seed weevil			mango weevil (1) を見よ

英名	和名	学名	所属、分布、ほか
mango shield scale		*Coccus acuminatus* Signoret	(カメムシ目、カタカイガラムシ科) 新北区。マンゴ害虫
mango shield scale			mango soft scale を見よ
mango shoot borer		*Chlumetia transversa* Walker	(チョウ目、ヤガ科) 東洋区、豪州区
mango shoot caterpillar		*Bocchoris fatualis* (Lederer)	(チョウ目、メイガ科) 大洋区
mango shoot caterpillar (1)	ナカジロフサヤガ	*Penicillaria jocosatrix* Guenée	(チョウ目、ヤガ科) 日本、東洋区、大洋区、新北区、豪州区
mango soft scale	マンゴーカタカイガラムシ	*Milviscutulus mangiferae* (Green)	(カメムシ目、カタカイガラムシ科) 日本、新北区。マンゴー、カンキツ、クチナシなどの害虫
mango stone weevil			mango weevil (1) を見よ
mango tip borer			mango shoot caterpillar (1) を見よ
mango tip borer		*Chlumetia euthysticha* (Turner)	(チョウ目、ヤガ科) 豪州区
mango weevil (1)		*Cryptorhynchus mangiferae* (Fabricius)	(コウチュウ目、ゾウムシ科) 東洋区、新北区、新熱帯区、エチオピア区、大洋区
mango weevil (2)		*Sternochetus gravis* Fabricius	(コウチュウ目、ゾウムシ科) 東洋区
mango white scale		*Parlatoria crypta* McKenzie	(カメムシ目、マルカイガラムシ科) 東洋区
mangold aphid		*Rhopalosiphoninus staphyleae* (Koch)	(カメムシ目、アブラムシ科) 旧北区、豪州区
mangold aphid		*Rhopalosiphoninus tulipaellus* (Theobald)	(カメムシ目、アブラムシ科) 全北区、豪州区、エチオピア区、新熱帯区
mangold flea beetle			brassy flea beetle (1) を見よ
mangold flea beetle			beet flea beetle を見よ
mangold fly			beet leaf miner (1) を見よ
mangosteen leafminer		*Acrocercops zygonoma* Meyrick	(チョウ目、ホソガ科) 東洋区
Mangoura swallowtail		*Papilio mangoura* Hewitson	(チョウ目、アゲハチョウ科) エチオピア区
mangrove trig		*Anaxipha scia* Hebard	(バッタ目、クサヒバリ科) 新北区
mangrove ant-blue			Illidge's ant-blue を見よ
mangrove borer			Australian pine borer を見よ
mangrove buckeye			West Indian buckeye を見よ
mangrove buckeye			black buckeye を見よ
mangrove cicada		*Arunta interclusa* (Walker)	(カメムシ目、セミ科) 豪州区
mangrove flies			horse flies (1) を見よ
mangrove green trig		*Cyrtoxipha nola* Walker	(バッタ目、クサヒバリ科) 新北区
mangrove ground cricket		*Hygronemobius alleni* (Morse)	(バッタ目、コオロギ科) 新北区
mangrove jewel			dull jewel を見よ
mangrove skipper		*Phocides okeechobee* (Worthington)	(チョウ目、セセリチョウ科) 新北区
mangrove skipper (1)		*Phocides polybius* (Fabricius)	(チョウ目、セセリチョウ科) 新北区

英名	和名	学名	所属、分布、ほか
mangrove skipper (2)		*Phocides pigmalion* (Cramer)	(チョウ目、セセリチョウ科) 新北区
mangrove tree nymph	オオゴマダラ	*Idaea leuconoe chersonesia* (Fruhstorfer)	(チョウ目、タテハチョウ科) 東洋区。Siam tree nymph を参照
manidiid moths		Manidiidae	(チョウ目) の昆虫の総称
Manipur forester			Manipur woodbrown を見よ
Manipur golden fork		*Lethe kabrua* (Tytler)	(チョウ目、タテハチョウ科) 東洋区
Manipur jungle queen		*Stichophthalma sparta* de Nicéville	(チョウ目、タテハチョウ科) 東洋区
Manipur woodbrown		*Lethe violaceopicta* (Poujade)	(チョウ目、タテハチョウ科) 東洋区
Manipur yellow sailer		*Neptis namba* Tytler	(チョウ目、タテハチョウ科) 東洋区
Manitoba azure		*Celastrina ladon argentata* Fletcher	(チョウ目、シジミチョウ科) 新北区。spring azure を参照
Manitoba oakworm moth		*Anisota manitobensis* McDunnough	(チョウ目、ヤママユガ科) 新北区
Manley prominent			oak caterpillar を見よ
Mannerheim weevil	ルリイクビチョッキリ	*Deporaus mannerheimi* (Hummel)	(コウチュウ目、オトシブミ科) 日本、旧北区
manniferous cicada		*Cicada orni* Linnaeus	(カメムシ目、セミ科) 旧北区
Mann's ant cricket		*Myrmecophilus manni* Schimmer	(バッタ目、アリヅカコオロギ科) 新北区
manroot borer			vine borer を見よ
manroot borer moth			vine borer を見よ
Mansfield's hairstreak		*Callophrys gryneus mansfieldi* (Tilden)	(チョウ目、シジミチョウ科) 新北区。juniper hairstreak (1) を参照
mantid	ウスバカマキリ	*Mantis religiosa* (Linnaeus)	(カマキリ目、カマキリ科) 日本、全北区。19世紀末に欧州から米国に入ったため European mantid といわれる
mantid flies			false mantids を見よ
mantids (1)	カマキリ目	Mantodea	の昆虫の総称　nuns 他多くの英名がある
mantids (2)	カマキリ科	Mantidae	(カマキリ目) の昆虫の総称
mantids (3)		Galepsus	(カマキリ目、カマキリ科) の昆虫の総称
mantis			複数形　mantises でカマキリ目、カマキリ科の別英名。ギリシャ語の預言者の意
mantis flies			false mantids を見よ　英国での英名
mantis fly		*Mantispa styriaca* Poda	(アミメカゲロウ目、カマキリモドキ科) 旧北区
mantisflies		Mantispa	(アミメカゲロウ目、カマキリモドキ科) の昆虫の総称
mantisflies			false mantids を見よ
mantislike lacewings			false mantids を見よ
mantispids			false mantids を見よ
mantled Acrobasis moth		*Acrobasis palliolella* Ragonot	(チョウ目、メイガ科) 新北区
mantled baskettail		*Epitheca semiaquea* (Burmeister)	(トンボ目、エゾトンボ科) 新北区

英名	和名	学名	所属、分布、ほか
Manto tussock moth		*Dasychira manto* (Strecker)	（チョウ目、ドクガ科）新北区
Manu Oressinoma		*Oressinoma sorata* Salvin et Godman	（チョウ目、タテハチョウ科）新熱帯区
Manu spreadwing		*Antigonus decens* Bulter	（チョウ目、セセリチョウ科）新熱帯区
Manuel's skipper		*Polygonus manueli* Bell et Comstock	（チョウ目、セセリチョウ科）新北区
Manuel's skipper		*Polygonus savigny* (Latreille)	（チョウ目、セセリチョウ科）新北区、新熱帯区
manuka beetle		*Pyronota festiva* (Fabricius)	（コウチュウ目、コガネムシ科）豪州区
manuka moth			common manuka moth を見よ
manure flies			small dung flies を見よ
Manx Charaxes		*Charaxes nichetes* Grose-Smith	（チョウ目、タテハチョウ科）エチオピア区
manx robber fly		*Machimus cowini* (Hobby)	（ハエ目、ムシヒキアブ科）旧北区
many-banded daggerwing			road page を見よ 米国での英名
many-banded grey		*Exoplisia hypochalbe* Felder et Felder	（チョウ目、シジミタテハ科）新熱帯区
many-banded metalmark			many-banded grey を見よ
many-banded skipper		*Timochares trifasciata* (Hewitson)	（チョウ目、セセリチョウ科）新熱帯区
many-belted clear-wing			raspberry clearwing moth を見よ
many-combed bugs			bat bugs を見よ
manydotted apple worm		*Balsa malana* Fitch	（チョウ目、ヤガ科）新北区。幼虫はリンゴ害虫。many-dotted appleworm moth ともいう
many-lined			many-lined moth をみよ
many-lined angle moth		*Macaria multilineata* Packard	（チョウ目、シャクガ科）新北区。many-lined angle ともいう
many-lined carpet moth		*Anticlea multiferata* (Walker)	（チョウ目、シャクガ科）新北区。many-lined carpet ともいう
many-lined moth		*Costaconvexa polygrammata* (Borkhausen)	（チョウ目、シャクガ科）旧北区
many-lined prominent			angle-lined prominent moth を見よ
many-lined wainscot moth		*Leucania multilinea* Walker	（チョウ目、ヤガ科）新北区
many plume moth			twenty-plume moth を見よ
many plume moths	ニジュウシトリバガ科	Alucitidae	（チョウ目）の昆虫の総称 many plumed moths, many-plumes moths ともいう
many-plumed moth			twenty-plume moth を見よ
many-spotted blue-skipper		*Pythonides grandis assecla* Mabille	（チョウ目、セセリチョウ科）新熱帯区
many-spotted Dichomeris moth		*Dichomeris punctipennella* (Clemens)	（チョウ目、キバガ科）新北区

英名	和名	学名	所属、分布、ほか
many-spotted Ridens		*Ridens crison crison* (Godman et Salvin)	（チョウ目、セセリチョウ科）新熱帯区
many-spotted Scoparia moth		*Scoparia basalis* Walker	（チョウ目、メイガ科）新北区
many-spotted skipperling		*Piruna aea* (Dyar)	（チョウ目、セセリチョウ科）新北区、新熱帯区
many-spotted sootywing		*Bolla litus* (Dyar)	（チョウ目、セセリチョウ科）新熱帯区
many-spotted tiger-moth		*Hypercompe permaculata* (Packard)	（チョウ目、ヒトリガ科）新北区、新熱帯区
many-tailed oakblue		*Thaduka multicaudata* Moore	（チョウ目、シジミチョウ科）東洋区
many-tufted bushbrown		*Mycalesis mystes* de Nicéville	（チョウ目、タテハチョウ科）東洋区
maori		*Graphania dives* (Philpott)	（チョウ目、ヤガ科）旧北区
map			map butterfly を見よ
map butterfly		*Araschnia levana* (Linnaeus)	（チョウ目、タテハチョウ科）旧北区。日本亜種は *A. l. obscura* Fenton アカマダラ
map moth		*Clidia geographica* Fabricius	（チョウ目、ヤガ科）旧北区
map wing butterfly			common map を見よ
map-winged swift	キタコウモリ	*Hepialus fusconebulosa* De Geer	（チョウ目、コウモリガ科）日本、旧北区
maple aphid			California maple aphid を見よ
maple bark beetle			maple timber beetle を見よ
maple-basswood leafroller			maple leafroller moth を見よ
maple-basswood leafroller moth		*Cenopis pettitana* (Robinson)	（チョウ目、ハマキガ科）新北区
maple broad-mouth weevil	イタヤクチブトキクイゾウムシ	*Stenoscelis aceri* Konishi	（コウチュウ目、ゾウムシ科）日本
maple bud borer moth		*Proteoteras moffatiana* Fernald	（チョウ目、ハマキガ科）新北区
maple callus borer		*Synanthedon aceri* (Clemens)	（チョウ目、スカシバガ科）新北区。maple callus borer moth ともいう
maple case-bearer			maple leaf cutter を見よ
maple clearwing moth		*Synanthedon acerrubri* Engelhardt	（チョウ目、スカシバガ科）新北区
maple cutworm	イタヤキリガ	*Cosmia exigua* (Butler)	（チョウ目、ヤガ科）日本
maple gouty vein midge		*Dasineura communis* Felt	（ハエ目、タマバエ科）新北区
maple leaf beetle	イタヤハムシ	*Pyrrhalta fuscipennis* (Jacoby)	（コウチュウ目、ハムシ科）日本、旧北区
maple leaf beetle	カエデハムシ	*Pyrrhalta seminigra* (Jacoby)	（コウチュウ目、ハムシ科）日本。カエデ害虫
maple leaf blotch miner moth		*Cameraria aceriella* (Clemens)	（チョウ目、ホソガ科）新北区
maple leaf cutter		*Paraclemensia acerifoliella* Fitch	（チョウ目、マガリガ科）新北区。maple leafcutter moth ともいう

英名	和名	学名	所属、分布、ほか
maple leaf louse			apple mealybug を見よ
maple leaf mealybug			apple mealybug を見よ
maple leaf roller			larch webworm を見よ
maple leaf roller weevil		*Deporaus tristis* (Fabricius)	（コウチュウ目、オトシブミ科）旧北区
maple leaf roller weevil	イタヤハマキチョッキリ	*Byctiscus venustus* (Pascoe)	（コウチュウ目、オトシブミ科）日本
maple leafcutter		*Poraclemensia acerifoliella* (Fitch)	（チョウ目、マガリガ科）新北区
maple leafhopper	イタヤトガリヨコバイ	*Japananus aceri* (Matsumura)	（カメムシ目、オオヨコバイ科）日本、旧北区。カエデ害虫
maple leafroller moth		*Sparganothis pettitana* (Robinson)	（チョウ目、ハマキガ科）新北区
maple leaftier moth			Forsakal's button を見よ
maple longicorn			maple longicorn beetle を見よ
maple longicorn beetle	イタヤカミキリ	*Mecynippus pubicornis* Bates	（コウチュウ目、カミキリムシ科）日本。カエデ、ポプラなどの害虫
maple looper moth		*Parallelia bistriaris* Hübner	（チョウ目、ヤガ科）新北区
maple mocha			mocha を見よ
maple petiole borer		*Caulocampus acericaulis* (McGillivray)	（ハチ目、ハバチ科）新北区
maple prominent		*Ptilodontella cucullina* (Denis et Schiffermüller)	（チョウ目、シャチホコガ科）旧北区
maple prominent			saddled prominent を見よ
maple psylla	カエデキジラミ	*Psylla japonica* Kuwayama Jr.	（カメムシ目、キジラミ科）日本。カエデ害虫
maple pug		*Eupithecia inturbata* (Hübner)	（チョウ目、シャクガ科）旧北区
maple spanworm		*Ennomus margnarius* (Guenée)	（チョウ目、シャクガ科）新北区。maple spanworm moth ともいう
maple sucker		*Rhinocola aceris* (Linnaeus)	（カメムシ目、キジラミ科）旧北区
maple timber beetle	イタヤノキクイムシ	*Indocryphalus aceris* (Niijima)	（コウチュウ目、キクイムシ科）日本。カエデ害虫
maple tip borer moth		*Episimus tyrius* Heinrich	（チョウ目、ハマキガ科）新北区。maple leaftier moth ともいう
maple trumpet skeletonizer		*Epinotia aceriella* (Clemens)	（チョウ目、ハマキガ科）新北区。成虫は maple trumpet skeletonizer moth
maple twig borer	アオカミキリ	*Schwarzerium quadricolle* (Bates)	（コウチュウ目、カミキリムシ科）日本、旧北区。カエデ害虫
maple twig borer moth		*Proteoteras aesculana* Riley	（チョウ目、ハマキガ科）新北区
maple twist	コスジオビハマキ	*Choristoneura diversana* (Hübner)	（チョウ目、ハマキガ科）日本、旧北区。リンゴ、オウトウ、カキなどの害虫
maple webworm moth		*Pococera asperatella* (Clemens)	（チョウ目、メイガ科）新北区
maple Zale moth		*Zale galbanata* (Morrison)	（チョウ目、ヤガ科）新北区

英名	和名	学名	所属、分布、ほか
maplets	チビイシガキチョウ属	*Chersonesia*	（チョウ目、タテハチョウ科）の昆虫の総称
maps	イシガケチョウ属	*Cyrestis*	（チョウ目、タテハチョウ科）の昆虫の総称
Marasmia moth		*Marasmia cochrusalis* (Walker)	（チョウ目、メイガ科）新北区
Marathon Pixie		*Melanis marathon* (C. et R. Felder)	（チョウ目、シジミタテハ科）新熱帯区
marble			thrift tortrix moth を見よ
marble clover		*Heliothis viriplaca* (Hufnagel)	（チョウ目、ヤガ科）旧北区
marble dagger moth		*Acronicta marmorata* Smith	（チョウ目、ヤガ科）新北区
marble fritillary		*Brenthis daphne* (Bergsträsser)	（チョウ目、タテハチョウ科）旧北区。日本亜種は *B. d. rabdia* (Butler) ヒョウモンチョウ。marbled fritillary ともいう
marble gall wasp		*Andricus kollari* (Hartig)	（ハチ目、タマバチ科）旧北区。本種がカシ類の若い枝につくる大型の虫えいを marble gall や Devonshire woody gall, oak nut という
marble-yellow straw pearl			rape worm を見よ
marbled Argus			Italian marbled white を見よ
marbled beauty		*Cryphia peria* Schiffermüller	（チョウ目、ヤガ科）旧北区
marbled beauty		*Cryphia domestica* (Hufnagel)	（チョウ目、ヤガ科）旧北区
marbled bent-skipper		*Helias phalaenoides* Fabricius	（チョウ目、セセリチョウ科）新熱帯区
marbled blue		*Erysichton palmyra* (Felder)	（チョウ目、シジミチョウ科）豪州区
marbled blue butterfly		*Erysichton palmyra tasmanicus* (Miskin)	（チョウ目、シジミチョウ科）豪州区
marbled brown			marbled prominent を見よ
marbled button			mottled button を見よ
marbled carpet moth	ウチジロナミシャク	*Chloroclysta truncata* (Hufnagel)	（チョウ目、シャクガ科）日本、旧北区
marbled clover			marble clover を見よ
marbled coronet		*Hadena confusa* (Hufnagel)	（チョウ目、ヤガ科）旧北区
marbled diving beetle		*Thermonectus marmoratus* (Hope)	（コウチュウ目、ゲンゴロウ科）新北区
marbled dog's-tooth tortrix			mottled button を見よ
marbled duskywing		*Ebrietas anacreon* (Staudinger)	（チョウ目、セセリチョウ科）新熱帯区
marbled elf		*Eretis djaelaelae* (Wallengren)	（チョウ目、セセリチョウ科）エチオピア区。marbled elfin ともいう
marbled elfin		*Sarangesa lucidella* (Mabille)	（チョウ目、セセリチョウ科）エチオピア区
marbled flat		*Lobocla liliana* (Atkinson)	（チョウ目、セセリチョウ科）東洋区

英名	和名	学名	所属、分布、ほか
marbled fungus weevil			eastern fungus weevil を見よ
marbled grass skipper			Pompeius skipper を見よ
marbled green		*Cryphia muralis* (Forster)	（チョウ目、ヤガ科）旧北区
marbled-green Leuconycta moth		*Leuconycta lepidula* (Grote)	（チョウ目、ヤガ科）新北区。marbled-green Jaspidia ともいう
marbled grey		*Cryphia raptricula* (Denis et Schiffermüller)	（チョウ目、ヤガ科）旧北区
marbled gris			marbled grey を見よ
marbled knot-horn		*Euphodope marmorea* Haworth	（チョウ目、メイガ科）旧北区
marbled leafwing		*Hypna clytemnestra* (Cramer)	（チョウ目、タテハチョウ科）新熱帯区
marbled line-blue			marbled blue を見よ
marbled map	コクレスイシガケチョウ	*Cyrestis cocles* (Fabricius)	（チョウ目、タテハチョウ科）東洋区
marbled minor		*Oligia strigilis* (Linnaeus)	（チョウ目、ヤガ科）旧北区
marbled minor			streaked sphinx を見よ
marbled minor moth			marbled minor を見よ
marbled Morpho			Deidamia Morpho を見よ
marbled orchard tortrix		*Hedya dimidioalba* Retzius	（チョウ目、ハマキガ科）全北区
marbled orchard tortrix			grey bud moth を見よ
marbled prominent		*Drymonia dodonaea* (Denis et Schiffermüller)	（チョウ目、シャチホコガ科）旧北区
marbled pug		*Eupithecia irriguata* (Hübner)	（チョウ目、シャクガ科）旧北区
marbled ringlet	マーブルベニヒカゲ	*Erebia montana* (de Prunner)	（チョウ目、タテハチョウ科）旧北区
marbled rose chafer		*Liocola lugubris* (Herbst)	（コウチュウ目、コガネムシ科）旧北区
marbled sawyer		*Lochmaeocles marmoratus* Casey	（コウチュウ目、カミキリムシ科）新北区
marbled single-dot bell		*Chrysoesthia sexguttella* (Thunberg)	（チョウ目、キバガ科）旧北区
marbled single-dot bell			black-headed fireworm を見よ
marbled skipper		*Carcharodus lavatherae* (Esper)	（チョウ目、セセリチョウ科）旧北区
marbled skipper			African mallow を見よ
marbled spreadwing		*Potamanaxas andraemon* (Mabille)	（チョウ目、セセリチョウ科）新熱帯区
marbled spurwing		*Antigonus nearchus* (Latreille)	（チョウ目、セセリチョウ科）新熱帯区
marbled tooth roller		*Ancylis achatana* Denis et Schiffermüller	（チョウ目、ハマキガ科）旧北区
marbled tree-lichen			tree-lichen beauty をみよ

英名	和名	学名	所属、分布、ほか
marbled tuffet moth			laugher moth を見よ
marbled tuft			laugher moth を見よ
marbled tufted skipper		*Nisoniades macarius* (Herrich-Schäffer)	（チョウ目、セセリチョウ科）新熱帯区
marbled underwing moth		*Catocala marmorata* Edwards	（チョウ目、ヤガ科）新北区。marbled underwing ともいう
marbled vert			marbled green を見よ
marbled white	セイヨウシロジャノメ（シロジャノメ）	*Melanargia galathea* (Linnaeus)	（チョウ目、タテハチョウ科）旧北区
marbled white		*Hesperocharis graphites graphites* H. W. Bates	（チョウ目、シロチョウ科）新熱帯区。*H. g. avivolans* (Butler) も同英名。marbled white butterfly ともいう
marbled white-spot	シロフコヤガ	*Protodeltote pygarga* (Hufnagel)	（チョウ目、ヤガ科）日本、旧北区。イネ科牧草害虫
marbled Xenica		*Geitoneura klugii* (Guérin-Méneville)	（チョウ目、タテハチョウ科）豪州区
marbled yellow straw pearl			rape worm を見よ　marble-yellow straw pearl とも記す
marbleds		Jaspidiinae	（チョウ目、ヤガ科）の昆虫の総称
March brown		*Rhithrogena germanica* Eaton	（カゲロウ目、ヒラタカゲロウ科）旧北区
March browns			brown stream mayflies を見よ
March cranefly			common cranefly (1) を見よ
March day		*Diurnea fagella* (Denis et Schiffermüller)	（チョウ目、マルハキバガ科）旧北区
march flies	ケバエ科	Bibionidae	（ハエ目）の昆虫の総称
march flies			horse flies (1) を見よ　豪州での英名
march fly		*Bibio hortulanus* (Linnaeus)	（ハエ目、ケバエ科）旧北区
March moth		*Alsophila aescularia* (Denis et Schiffermüller)	（チョウ目、シャクガ科）旧北区
March usher			March moth を見よ
Marchall's Andean white	アミモンシロチョウ	*Hesperocharis marchallii* Guérin-Méneville	（チョウ目、シロチョウ科）新熱帯区
marching termites			rotten-wood termites を見よ
Marcia sister		*Adelpha cytherea marcia* Fruhstorfer	（チョウ目、タテハチョウ科）新熱帯区
Marco Polo's clouded yellow	マルコポーロモンキチョウ	*Colias marcopolo* (Grum Grshimailo)	（チョウ目、シロチョウ科）旧北区
Marcus skipper		*Vettius marcus* (Fabricius)	（チョウ目、セセリチョウ科）新熱帯区
Mardon skipper		*Polites mardon* (Edwards)	（チョウ目、セセリチョウ科）新北区
Margarita's blue			trident pencil-blue を見よ
Margarita's caper white		*Belenois margaritacea* Sharpe	（チョウ目、シロチョウ科）エチオピア区
margarodid scales			giant coccids を見よ
margarodids			giant coccids を見よ

英名	和名	学名	所属、分布、ほか
marginate Temnora		*Temnora marginata* (Walker)	（チョウ目、スズメガ科）エチオピア区
margined blister beetle		*Epicauta pestifera* Werner	（コウチュウ目、ツチハンミョウ科）新北区
margined blister beetle			old-fashioned potato beetle を見よ
margined diving beetle	フチトリゲンゴロウ	*Cybister limbatus* (Fabricius)	（コウチュウ目、ゲンゴロウ科）日本、東洋区
margined hedge blue		*Celatoxia marginata* (de Nicéville)	（チョウ目、シジミチョウ科）東洋区。*C. m. splendens* (Butler) も同英名
margined leaf chafer			margined vine chafer を見よ
margined lineblue		*Prosotas pia* Toxopeus	（チョウ目、シジミチョウ科）東洋区
margined vine chafer		*Anomala dubia* (Scopoli)	（コウチュウ目、コガネムシ科）旧北区
margined white		*Pieris marginalis* Scudder	（チョウ目、シロチョウ科）新北区
Marguarite's copper		*Aloeides margaretae* Tite et Dickson	（チョウ目、シジミチョウ科）エチオピア区
marguerite fly		*Phytomyza affinis* Fallén	（ハエ目、ハモグリバエ科）旧北区
Mariana albatross	マリアナトガリシロチョウ	*Appias mariana* Yata, Chainey et Vane-Wright	（チョウ目、シロチョウ科）東洋区
Mariana coconut beetle		*Brontispa mariana* Spatch	（コウチュウ目、ハムシ科）東洋区、豪州区
Maria's satyr		*Eretris maria* (Schaus)	（チョウ目、タテハチョウ科）新熱帯区
Marica blue ringlet		*Chloreuptychia marica* Weymer	（チョウ目、タテハチョウ科）新熱帯区
Maricopa harvester ant		*Pogonomyrmex maricopa* Wheeler	（ハチ目、アリ科）新北区
Marieps emperor		*Charaxes marieps* van Someren et Jackson	（チョウ目、タテハチョウ科）エチオピア区
marigold aphid		*Neotoxoptera oliveri* (Essig)	（カメムシ目、アブラムシ科）豪州区
marigold thrips		*Neohydatothrips samayunkur* Kudo	（アザミウマ目、アザミウマ科）日本、新北区、大洋区、豪州区
Marina			red-spotted patch を見よ
Marina checkerspot			red-spotted patch を見よ
marine blue	マリンカクモンシジミ	*Leptotes marina* (Reakirt)	（チョウ目、シジミチョウ科）新北区、新熱帯区
marine bug		*Aepophilus bonnairei* (Southwood et Leston)	（カメムシ目、ミズギワカメムシ科）旧北区
marine midge		*Telmatogeton macswaini* Wirth	（ハエ目、ユスリカ科）新北区
Mariposa copper	マリポーサベニシジミ	*Lycaena mariposa* (Reakirt)	（チョウ目、シジミチョウ科）新北区
maritime earwig	ハマベハサミムシ	*Anisolabis maritima* (Gene)	（ハサミムシ目、オオハサミムシ科）日本、新北区
maritime earwigs			long-horned earwigs を見よ
maritime sunflower borer moth		*Papaipema maritima* Bird	（チョウ目、ヤガ科）新北区
maritime swallowtail			short-tailed swallowtail を見よ

英名	和名	学名	所属、分布、ほか
Marius hairstreak		*Rekoa marius* (Lucas)	(チョウ目、シジミチョウ科) 新北区
marked spittlebug		*Sinopia signata* Sakakibara	(カメムシ目、コガシラアワフキ科) 新熱帯区
Marlatt whitefly	マーラットコナジラミ	*Aleurolobus marlatti* (Quaintance)	(カメムシ目、コナジラミ科) 日本、東洋区。カンキツ、クワ、ツツジなどの害虫
marmalade hoverfly			hover-fly (1) を見よ
marmorate broad-head leafhopper	ヒロズマダラヨコバイ	*Recilia latifrons* (Matsumura)	(カメムシ目、オオヨコバイ科) 日本、旧北区
marmorate thick-legged longicorn			marmorate thick-legged longicorn beetle を見よ
marmorate thick-legged longicorn beetle	ヒゲナガモモブトカミキリ	*Acanthocinus griseus* (Fabricius)	(コウチュウ目、カミキリムシ科) 日本、旧北区
marmorate-veined leafhopper	ブチミャクヨコバイ	*Drabescus nigrifemoratus* (Matsumura)	(カメムシ目、オオヨコバイ科) 日本、旧北区
marmorated aphid	トゲブチアブラムシ	*Tuberculatus capitatus* (Essig et Kuwana)	(カメムシ目、アブラムシ科) 日本、旧北区、東洋区
marmorated drepanid	マダラカギバ	*Callicilix abraxata* Butler	(チョウ目、カギバガ科) 日本
marmorated grasshopper	クルマバッタ	*Gastrimargus marmoratus* (Thunberg)	(バッタ目、バッタ科) 日本、旧北区、東洋区、エチオピア区
marmorated leaf bug	マダラカスミカメムシ	*Lygus saundersi* Reuter	(カメムシ目、カスミカメムシ科) 日本、旧北区
marmorated noctuid	マダラエグリバ	*Plusiodonta casta* (Butler)	(チョウ目、ヤガ科) 日本、旧北区
marmorated oak leafhopper	ゴマフハトムネヨコバイ	*Macropsis irrorata* (Matsumura)	(カメムシ目、オオヨコバイ科) 日本
Marmoress			Italian marbled white を見よ
marmot warble fly		*Oestromyia marmotae* Gedoelst	(ハエ目、ヒツジバエ科) 旧北区
maroonwing moth		*Sideridis maryx* (Guenée)	(チョウ目、ヤガ科) 新北区
marram weevil			sand weevil を見よ
married underwing moth		*Catocala nuptialis* Walker	(チョウ目、ヤガ科) 新北区。married underwing ともいう
marsh Acraea		*Telchinia rahira* (Boisduval)	(チョウ目、タテハチョウ科) エチオピア区
marsh beetles	マルハナノミ科	Helodidae	(コウチュウ目) の昆虫の総称 2〜4mm の微小種
marsh blue		*Harpendyreus noquasa* (Trimen et Bowker)	(チョウ目、シジミチョウ科) エチオピア区
marsh bluet			common bluet を見よ
marsh bluetail			common bluetail (1) を見よ
marsh buff			marsh moth を見よ
marsh carpet		*Perizoma sagittata* (Fabricius)	(チョウ目、シャクガ科) 旧北区。日本亜種は *P. s. albiflua* (Prout) ヤハズナミシャク
marsh commodore		*Precis ceryne* (Boisduval)	(チョウ目、タテハチョウ科) エチオピア区
marsh conehead		*Neoconocephalus palustris* (Blatchley)	(バッタ目、キリギリス科) 新北区
marsh cosmet			Schaffer's neb (1) を見よ
marsh crane fly		*Tipula paludosa* Meigen	(ハエ目、ガガンボ科) 日本

英名	和名	学名	所属、分布、ほか
marsh crane fly			common cranefly (1) を見よ
marsh cricket		*Gryllomorpha dalmatina* (Ocskay)	(バッタ目、コオロギ科) 旧北区
marsh cricket		*Pteronemobius heydenii* (Fischer)	(バッタ目、コオロギ科) 旧北区
marsh dagger		*Acronicta strigosa* (Denis et Schiffermüller)	(チョウ目、ヤガ科) 旧北区
marsh damsel bug		*Dolichonabis limbatus* (Dahlbom)	(カメムシ目、マキバサシガメ科) 旧北区
marsh dancer		*Onychargia atrocyana* Selys	(トンボ目、イトトンボ科) 東洋区
marsh eyed brown			eyed brown を見よ
marsh fern moth		*Fagitana littera* (Guenée)	(チョウ目、ヤガ科) 新北区
marsh firetail		*Telebasis digiticollis* Calvert	(トンボ目、イトトンボ科) 新北区
marsh five-spot burnet			five-spot burnet を見よ
marsh flies	ヤチバエ科	Sciomyzidae	(ハエ目) の昆虫の総称 幼虫が貝を捕食し、谷地に生息することに由来。世界に500種
marsh flies			shore flies を見よ 豪州での英名
marsh fly		*Poecilographa decora* (Loew)	(ハエ目、ヤチバエ科) 新北区
marsh-fly		*Tetanocera plebeja* Loew	(ハエ目、ヤチバエ科) 新北区
marsh fritillary	チョウセンヒョウモンモドキ	*Euphydryas aurinia* (Rottemburg)	(チョウ目、タテハチョウ科) 旧北区
marsh grass yellow			pale grass yellow を見よ
marsh grey		*Scoparia pallida* Stephens	(チョウ目、メイガ科) 旧北区
marsh ground beetles			bog ground beetles を見よ
marsh mallow moth		*Hydraecia osseola* Hucherardi	(チョウ目、ヤガ科) 旧北区
marsh marigold looper	ウラナミヒメシャク	*Scopula corrivalaria eccletica* Prout	(チョウ目、シャクガ科) 日本、旧北区
marsh moth		*Athetis pallustris* (Hübner)	(チョウ目、ヤガ科) 旧北区
marsh oblique-barred		*Hypenodes humidalis* Doubleday	(チョウ目、ヤガ科) 旧北区
marsh oblique-barred snout			marsh oblique-barred を見よ
marsh patroller			eyed bush brown を見よ
marsh pug		*Eupithecia pygmaeata* (Hübner)	(チョウ目、シャクガ科) 旧北区
marsh pug		*Eupithecia palustraria* Doubleday	(チョウ目、シャクガ科) 旧北区
marsh ringlet			large heath を見よ
marsh skimmer		*Orthetrum luzonicum* (Brauer)	(トンボ目、トンボ科) 東洋区
marsh springtail	ニセフシトビムシ	*Isotomurus palustris* (Müller)	(トビムシ目、ツチトビムシ科) 日本、全北区
marsh swift		*Borbo micans* (Holland)	(チョウ目、セセリチョウ科) エチオピア区

英名	和名	学名	所属、分布、ほか
marsh sylph		*Metisella meninx* (Trimen)	（チョウ目、セセリチョウ科）エチオピア区
marsh treader bugs			water measures を見よ
marsh treaders			water measures を見よ　marsh-treaders とも記す
marsh weevils	イネゾウムシ亜科	Erirhininae	（コウチュウ目、ゾウムシ科）の昆虫の総称
Marshall's Acraea mimic			Marshall's false monarch を見よ
Marshall's false monarch		*Mimacraea marshalli* Trimen	（チョウ目、シジミチョウ科）エチオピア区
Marshall's hawk		*Praedora marshalli* Rothschild et Jordan	（チョウ目、スズメガ科）エチオピア区
Marshall's highflyer		*Aphaneus marshalli* Neave	（チョウ目、シジミチョウ科）エチオピア区
Marshall's polyptychus		*Polyptychopsis marshalli* (Rothschild et Jordan)	（チョウ目、スズメガ科）エチオピア区
Marshall's striped hawk		*Rhodafra marshalli* Rothschild et Jordan	（チョウ目、スズメガ科）エチオピア区
marsupial chewing lice	ミナミケモノハジラミ科	Boopidae	（ハジラミ目）の昆虫の総称
marsupial coccid			Jacarand bug を見よ
marsupial coccids			giant coccids を見よ
marsupial lice	ケモノタンカクハジラミ科	Trimenoponidae	（ハジラミ目）の昆虫の総称　齧歯類、有袋類に寄生
Marsyas hairstreak	オオルリフタオシジミ	*Pseudolycaena marsyas* (Linnaeus)	（チョウ目、シジミチョウ科）新北区、新熱帯区
Martial hairstreak		*Strymon martialis* (Herrich-Schäffer)	（チョウ目、シジミチョウ科）新北区
Martial scrub-hairstreak			Martial hairstreak を見よ
martin bug			swallow bug (1) を見よ
Martinell's shieldback		*Idiostatus martinellii* Rentz	（バッタ目、キリギリス科）新北区
Martin's Bernardino blue		*Euphilotes bernardino martini* (Mattoni)	（チョウ目、シジミチョウ科）新熱帯区
Martin's blue		*Plebejus martini* (Allard)	（チョウ目、シジミチョウ科）旧北区
Marti's metallic tiger beetle		*Tetracha martii* (Perty)	（コウチュウ目、オサムシ科）新熱帯区
Mary's giant skipper			pecos giant skipper を見よ
Masai cupid		*Eicochrysops masai* (Bethue-Baker)	（チョウ目、シジミチョウ科）エチオピア区
Masai sprite		*Pseudagrion massaicum* Sjöstedt	（トンボ目、イトトンボ科）エチオピア区
masarid wasps			shining wasps を見よ
mascal			butterflies and moths を見よ　幼虫の英名
Mashuna ringlet		*Mashuna mashuna* Trimen	（チョウ目、タテハチョウ科）エチオピア区
masked bedbug hunter			masked hunter bug を見よ　米国での英名
masked bee	ヨーロッパチビムカシハナバチ	*Hylaeus pectoralis* Förster	（ハチ目、ムカシハナバチ科）日本、旧北区

英名	和名	学名	所属、分布、ほか
masked bees			obtuse-tongued bees を見よ
masked bug			masked hunter bug を見よ
masked chafer beetles		*Cyclocephala*	(コウチュウ目、コガネムシ科) の昆虫の総称
masked devil			greengrocer を見よ
masked hunter			masked hunter bug を見よ　北米での英名
masked hunter bug		*Reduvius personatus* (Linnaeus)	(カメムシ目、サシガメ科) 全北区。屋内でナンキンムシなどを捕食
masked Parahypenodes moth		*Parahypenodes quadralis* Barnes et McDunnough	(チョウ目、ヤガ科) 新北区
masked scale		*Mycetaspis personata* (Comstock)	(カメムシ目、マルカイガラムシ科) 旧北区
Maskell scale		*Morganella maskelli* Cockerell	(カメムシ目、マルカイガラムシ科) 新熱帯区。カンキツ害虫
Maskell scale (1)	ヒメナガカキカイガラムシ	*Lepidosaphes pallida* (Maskell)	(カメムシ目、マルカイガラムシ科) 日本、全北区、東洋区、大洋区。イヌマキ、スギ、カンキツなどの害虫
mason bee		*Hoplitis producta* (Cresson)	(ハチ目、ハキリバチ科) 新北区
mason bee		*Osmia cobaltina* Cresson	(ハチ目、ハキリバチ科) 新北区
mason bee			red Osmia を見よ
mason bee			wall bee を見よ
mason bees			leaf-cutting bees (1) を見よ
mason bees (1)		*Anthidium*	(ハチ目、ハキリバチ科) の昆虫の総称
mason bees (2)		*Dianthidium*	(ハチ目、ハキリバチ科) の昆虫の総称
mason bees (3)		*Osmia*	(ハチ目、ハキリバチ科) の昆虫の総称
mason wasp		*Pison spinolae* Shuckard	(ハチ目、アナバチ科) 豪州区
mason wasp (1)		*Odynerus spinipes* (Linnaeus)	(ハチ目、スズメバチ科) 旧北区
mason wasps		*Odynerus*	(ハチ目、スズメバチ科) の昆虫の総称
mason wasps			wasps (1) を見よ
mason wasps			potter wasps を見よ
Massilia sister		*Adelpha paraena* (Bates)	(チョウ目、タテハチョウ科) 新熱帯区。*A. p. massilia* (C. et R. Felder) も同英名
master blister beetle			desert blister beetle (1) を見よ
master's skipper		*Hesperilla mastersi* Waterhouse	(チョウ目、セセリチョウ科) 豪州区
Masui bamboo scale	マスイフサカイガラムシ	*Asterolecanium masuii* Kuwana	(カメムシ目、フサカイガラムシ科) 日本、旧北区
mat grass pyralid	イツトガ	*Calamotropha shichito* (Marumo)	(チョウ目、メイガ科) 日本、旧北区。イグサ害虫
mat rush leafminer	スズメノヤリハモグリバエ	*Cerodontha bimaculata* (Meigen)	(ハエ目、ハモグリバエ科) 日本。イグサ害虫
mat-rush pyralid	ムモンシロオオメイガ	*Scirpophaga praelata* (Scopoli)	(チョウ目、メイガ科) 日本、旧北区
mat rush sawfly	イハバチ	*Eutomostethus apicalis* (Matsumura)	(ハチ目、ハバチ科) 日本、旧北区。イグサ害虫

英名	和名	学名	所属、分布、ほか
mat rush worm		*Spilonota honesta* Meyrick	（チョウ目、ハマキガ科）日本
mat rush worm			mottled bran marble を見よ
Mather's Calephelis		*Calephelis matheri* McAlpine	（チョウ目、シジミタテハ科）新熱帯区
Mather's skipper		*Enosis matheri* Freeman	（チョウ目、セセリチョウ科）新熱帯区
Mathew's blue		*Neolucia mathewi* (Miskin)	（チョウ目、シジミチョウ科）豪州区
Mathew's ghost moth		*Gazoryctra mathewi* (Edwards)	（チョウ目、コウモリガ科）新北区
Mathew's groundstreak		*Electrostrymon mathewi* (Hewitson)	（チョウ目、シジミチョウ科）新熱帯区
Mathew's wainscot		*Mythimna flavicolor* (Barret)	（チョウ目、ヤガ科）旧北区
matriarcha katydid		*Saga pedo* (Pallas)	（バッタ目、キリギリス科）新北区
matriarchal katydid			matriarcha katydid を見よ
Matsumoto mealybug	マツモトコナカイガラムシ	*Crisicoccus serurata* (Kanda)	（カメムシ目、コナカイガラムシ科）日本。ナシなど果樹、プラタナスなど樹木害虫
Matsumoto snow flea	マツモトシロトビムシ	*Onychiurus matsumotoi* Kinoshita	（トビムシ目、シロトビムシ科）日本
Matsumura horntail	トドマツノキバチ	*Xoanon matsumurae* (Rohwer)	（ハチ目、キバチ科）日本、旧北区
Matsumura pine scale			pine scale (1) を見よ
Mau cupid		*Euchrysops mauensis* Bethune-Baker	（チョウ目、シジミチョウ科）エチオピア区
Maui banana Hedyleptan moth		*Omiodes musicola* Swezey	（チョウ目、メイガ科）大洋区。ハワイの Maui 島に由来
Maui upland damselfly		*Megalagrion jugorum* (Perkins)	（トンボ目、イトトンボ科）大洋区
Mauna Loa bean beetle		*Araeocorynus cunningi* Jekel	（コウチュウ目、ヒゲナガゾウムシ科）新北区。ハワイ島の火山に由来
Mauna Loa bean weevil			Mauna Loa bean beetle を見よ
Mauritius pink borer			African pink borer を見よ
Mauritius spotted cane borer			spotted stem borer を見よ
mauve line-blue			eastern dingy lineblue を見よ
mauve pittosporum scale		*Parlatoria pittospori* Maskell	（カメムシ目、マルカイガラムシ科）豪州区
mauve scallopwing		*Staphylus ascalaphus* (Staudinger)	（チョウ目、セセリチョウ科）新熱帯区
Mavors hairstreak			deep-green hairstreak を見よ
May beech piercer		*Pammene herrichiana* (Heinemann)	（チョウ目、ハマキガ科）旧北区
May beetles (1)		*Fruhstorferia*	（コウチュウ目、コガネムシ科）の昆虫の総称
May beetles (2)		*Phyllophaga*	（コウチュウ目、コガネムシ科）の昆虫の総称
May beetles (3)	コフキコガネ亜科	Melolonthinae	（コウチュウ目、コガネムシ科）の昆虫の総称 欧州で cockchafers といわれる

英名	和名	学名	所属、分布、ほか
May beetles			chafer beetles を見よ
May bug			June beetle (1) を見よ
May frit			small pearl-bordered fritillary を見よ
May highflier		*Hydriomena impluviata* (Denis et Schiffermüller)	（チョウ目、シャクガ科）旧北区。日本亜種は *H. i. insulata* Inoue ヒロオビナミシャク。May highflyer とも記される
May highflier		*Hydriomena coerulata* Fabricius	（チョウ目、シャクガ科）旧北区
May maid			dragonflies (1) を見よ
Mayan Calephelis		*Calephelis maya* McAlpine	（チョウ目、シジミタテハ科）新熱帯区
Mayan crescent		*Castilia myia* (Hewitson)	（チョウ目、タテハチョウ科）新熱帯区
Mayapple borer moth		*Papaipema rutila* (Guenée)	（チョウ目、ヤガ科）新北区
mayate		*Euphoria basilis* Brumeister	（コウチュウ目、コガネムシ科）新熱帯区。カンキツ害虫
Maybug			June beetle (1) を見よ　May bug の使用例もあり
Mayer's walkingstick		*Aplopus mayeri* Caudell	（ナナフシ目、ナナフシ科）新北区
mayflies	カゲロウ目	Ephemeroptera	の昆虫の総称　有翅昆虫で最も原始的な昆虫。世界に 2,100 種。mayfly, dayfly といわれる
mayflies			burrowing mayflies を見よ
mayflies			greendrakes を見よ
mayfly			green dark mayfly を見よ
mayfly			brown mayfly を見よ
Mazans scallopwing			Mazans sootywing を見よ
Mazans sootywing		*Staphylus mazans* (Reakirt)	（チョウ目、セセリチョウ科）新熱帯区
Mazarine blue		*Polyommatus semiargus* (Rottemburg)	（チョウ目、シジミチョウ科）旧北区。mazarine blue butterfly ともいう
Mbulu copper		*Aloeides mbuluensis* Pringle	（チョウ目、シジミチョウ科）エチオピア区
McAlpine's skipper		*Anthoptus macalpinei* Freeman	（チョウ目、セセリチョウ科）新熱帯区
McDunnough's leafwing		*Pero macdunnoughi* Cassino et Swett	（チョウ目、シャクガ科）新北区
McGregor's blue		*Lepidochrysops macgregori* Pennington	（チョウ目、シジミチョウ科）エチオピア区
McGuire's skipper		*Euphyes mcguirei* Freeman	（チョウ目、セセリチョウ科）新北区
McMaster's copper		*Aloeides macmasteri* Tite et Dickson	（チョウ目、シジミチョウ科）エチオピア区
McMaster's silver-spotted copper		*Trimenia macmasteri* (Dickson)	（チョウ目、シジミチョウ科）エチオピア区
McNeill's sootywing		*Pholisora gracielae* (McNeill)	（チョウ目、セセリチョウ科）新北区
McNeil's saltbush sootywing		*Hesperopsis gracielae* (MacNeil)	（チョウ目、セセリチョウ科）新北区。McNeil's sootywing ともいう
meadow Argus	サイパンタテハモドキ	*Junonia villida* (Fabricius)	（チョウ目、タテハチョウ科）豪州区

英名	和名	学名	所属、分布、ほか
meadow blue		*Cupidopsis cissus* (Godart)	(チョウ目、シジミチョウ科) エチオピア区
meadow brown	マキバジャノメ	*Maniola jurtina* (Linnaeus)	(チョウ目、タテハチョウ科) 旧北区。meadow brown butterfly ともいう
meadow brown Argus			meadow brown を見よ
meadowbrowns	マキバジャノメ属	*Maniola*	(チョウ目、タテハチョウ科) の昆虫の総称
meadow caterpillar			Virginia ctenucha moth を見よ
meadow copper			ruddy copper を見よ
meadow fritillary		*Mellicta parthenoides* (Keferstein)	(チョウ目、タテハチョウ科) 旧北区
meadow fritillary (1)		*Clossiana bellona* (Fabricius)	(チョウ目、タテハチョウ科) 新北区
meadow froghopper			cuckoo spit insect を見よ
meadow grasshopper		*Chorthippus longicornis* (Latreille)	(バッタ目、バッタ科) 旧北区
meadow grasshopper		*Chorthippus parallelus* (Zetterstedt)	(バッタ目、バッタ科) 旧北区
meadow grasshopper		*Chorthippus curtipennis* (Harris)	(バッタ目、バッタ科) 新北区
meadow grasshopper (1)		*Conocephalus brevipennis* (Scudder)	(バッタ目、キリギリス科) 新北区
meadow grasshopper			great green bush-cricket を見よ
meadow grasshoppers (1)	ササキリ亜科	Conocephalinae	(バッタ目、キリギリス科) の昆虫の総称
meadow grasshoppers			cone-headed grasshoppers を見よ　南アフリカでの英名
meadow grasshoppers			long-horned grasshoppers (1) を見よ
meadowhawks			darter dragonflies を見よ
meadow katydids			meadow grasshoppers (1) を見よ
meadow katydids			long-horned grasshoppers (1) を見よ
meadow leafhopper		*Paramesus nervosus* (Fallén)	(カメムシ目、オオヨコバイ科) 旧北区
meadow miller		*Axenus arvalis* Grote	(チョウ目、ヤガ科) 新北区
meadow moth			beet webworm を見よ
meadow plant bug		*Leptopterna dolabrata* (Linnaeus)	(カメムシ目、カスミカメムシ科) 全北区
meadow rue borer moth		*Papaipema unimoda* (Smith)	(チョウ目、ヤガ科) 新北区
meadow shade		*Cnephasia pascuana* (Hübner)	(チョウ目、ハマキガ科) 旧北区
meadow spittlebug		*Philaenus leucophthalmus* Linnaeus	(カメムシ目、アワフキムシ科) 全北区
meadow spittlebug			cuckoo spit insect を見よ　米国での英名
meadow white		*Pontia helice* (Linnaeus)	(チョウ目、シロチョウ科) エチオピア区
Mead's sulphur	アメリカダイダイモンキチョウ	*Colias meadii* Edwards	(チョウ目、シロチョウ科) 新北区

英名	和名	学名	所属、分布、ほか
Mead's wood nymph		*Cercyonis meadii* (Edwards)	（チョウ目、タテハチョウ科）新北区
Mead's wood nymph			red-eyed ringlet を見よ
Meakan bark beetle	メアカンキクイムシ	*Polygraphus meakanensis* Niijima	（コウチュウ目、キクイムシ科）日本
mealbeetle			yellow mealworm を見よ
meal moth	カシノシマメイガ	*Pyralis farinalis* (Linnaeus)	（チョウ目、メイガ科）日本、汎世界。貯穀害虫
meal sap beetle			dried fruit beetle (1) を見よ
meal snout moth			meal moth を見よ
meal worm moth			Indian meal moth (1) を見よ
mealie stalk borer			maize stalk borer (1)(2) を見よ
mealworm			yellow mealworm を見よ
mealworm			dark mealworm beetle を見よ
mealworm beetle			yellow mealworm を見よ
mealworm beetles			mealworms を見よ
mealworms		*Tenebrio*	（コウチュウ目、ゴミムシダマシ科）の昆虫の総称
mealy apple aphid			rosy apple aphid (1) を見よ
mealy bug			long-tailed mealybug (1) を見よ
mealy cabbage aphid			cabbage aphid を見よ
mealy fly			bluebottle fly を見よ
mealy knot-horn		*Nephopterix palumbella* Fabricius	（チョウ目、メイガ科）旧北区
mealy peach aphid		*Hyalopterus amygdali* (Blanchard)	（カメムシ目、アブラムシ科）旧北区
mealy plum aphid	モモコフキアブラムシ	*Hyalopterus pruni* (Geoffroy)	（カメムシ目、アブラムシ科）日本、汎世界。モモ、ウメの害虫
mealy plum louse			mealy plum aphid を見よ
mealy shield scale			pyriform scale を見よ　米国での英名
mealy wing			citrus whitefly (1) を見よ
mealybug		*Drosicha contrahen* Walker	（カメムシ目、ワタフキカイガラムシ科）旧北区。カンキツ害虫
mealybug		*Drosicha stebbingi* Green	（カメムシ目、ワタフキカイガラムシ科）東洋区。カンキツ害虫
mealybug		*Monophlebus dalbergiae* Green	（カメムシ目、ワタフキカイガラムシ科）東洋区。カンキツ害虫
mealybug		*Pseudococcus cryptus* Hemphill	（カメムシ目、コナカイガラムシ科）新熱帯区。カンキツ害虫
mealybug		*Pseudococcus filamentosus* Cockerell	（カメムシ目、コナカイガラムシ科）大洋区
mealybug		*Pseudococcus pseudofilamentosus* Betrem	（カメムシ目、コナカイガラムシ科）東洋区。カンキツ害虫
mealybug		*Pseudococcus virgatus* (Cockerell)	（カメムシ目、コナカイガラムシ科）東洋区、新熱帯区。カンキツ害虫

英名	和名	学名	所属、分布、ほか
mealybug		*Phenacoccus iceryoides* Green	(カメムシ目、コナカイガラムシ科) 東洋区
mealybug			citrus mealybug (2) を見よ
mealybug destroyer			mealybug ladybird を見よ
mealybug destructor			mealybug ladybird を見よ
mealybug ladybird	ツマアカオオヒメテントウ	*Cryptolaemus montrouzieri* Mulsant	(コウチュウ目、テントウムシ科) 日本、東洋区、新北区、豪州区
mealybug parasite	フジコナヒゲナガトビコバチ	*Leptomastix dactylopii* Howard	(ハチ目、トビコバチ科) 日本、全北区、大洋区、新熱帯区
mealybug parasitoid		*Tetracnemoidea sydneyensis* (Timberlake)	(ハチ目、トビコバチ科) 豪州区
mealybug predator fly		*Gitonides perspicax* Knab	(ハエ目、ショウジョウバエ科) 大洋区、新北区
mealybugs		*Pseudococcus*	(カメムシ目、コナカイガラムシ科) の昆虫の総称
mealybugs (1)	コナカイガラムシ科	Pseudococcidae	(カメムシ目) の昆虫の総称 世界に1100種
mealybugs (2)		Pseudococcinae	(カメムシ目、コナカイガラムシ科) の昆虫の総称
mealybugs (3)	フクロカイガラムシ科	Eriococcidae	(カメムシ目) の昆虫の総称
mealybugs			whiteflies を見よ
mealybugs			scale insects を見よ
mealywings			whiteflies を見よ
measuring worm			geometer moths を見よ
measuring worm moths			geometer moths を見よ
measuring worms			geometer moths を見よ
meat ant		*Iridomyrmex purpureus* (F. Smith)	(ハチ目、アリ科) 豪州区
meat ants		*Iridomyrmex*	(ハチ目、アリ科) の昆虫の総称
meat fly			flesh fly (1) を見よ
meat maggot			ニクバエの幼虫や、いわゆるウジを指す
meat skipper			cheese skipper を見よ 幼虫の英名
meatfly			bluebottle fly を見よ
mechanitis mimic	ホシキオビマダラ	*Melinaea lilis* Doubleday	(チョウ目、タテハチョウ科) 新熱帯区
mecopterans			scorpion flies (2) を見よ
medfly			Mediterranean fruit fly を見よ
median-spotted stink bug	ナカボシカメムシ	*Menida musiva* (Jakovlev)	(カメムシ目、カメムシ科) 日本、旧北区
median wasp	キオビホオナガスズメバチ	*Dolichovespula media* (Retzius)	(ハチ目、スズメバチ科) 日本、旧北区
Mediocre skipper		*Inglorius mediocris* Austin	(チョウ目、セセリチョウ科) 新熱帯区
Mediterranean black scale			black scale を見よ
Mediterranean blue			Mediterranean tiger blue を見よ

英名	和名	学名	所属、分布、ほか
Mediterranean brocade			Egyptian cotton leafworm を見よ
Mediterranean brocade			common cutworm を見よ
Mediterranean burnet moths		Heterogynidae	（チョウ目）の昆虫の総称
Mediterranean bush cricket		*Rhacocleis germanica* (Herrich-Schäffer)	（バッタ目、キリギリス科）旧北区
Mediterranean carnation leaf roller		*Cacoecimorpha pronubana* (Hübner)	（チョウ目、ハマキガ科）全北区、エチオピア区
Mediterranean carnatioon leaf miner		*Paraphytomyza dianthicola* (Venturi)	（ハエ目、ハモグリバエ科）旧北区
Mediterranean climbing cutworm			Egyptian cotton leafworm を見よ
Mediterranean cockroach			tawny cockroach (1) を見よ
Mediterranean earwig		*Forficula deciipiens* Gene	（ハサミムシ目、クギヌキハサミムシ科）旧北区
Mediterranean fig scale		*Lepidosaphes ficus* (Signoret)	（カメムシ目、マルカイガラムシ科）全北区。イチジク害虫
Mediterranean fireflies		*Luciola*	（コウチュウ目、ホタル科）の昆虫の総称
Mediterranean flannel moths		Somabrachyidae	（チョウ目）の昆虫の総称
Mediterranean flour moth	スジコナマダラメイガ	*Anagasta kuehniella* (Zeller)	（チョウ目、メイガ科）日本、汎世界。貯穀害虫
Mediterranean flour moth			raisin moth を見よ
Mediterranean fruit fly	チチュウカイミバエ	*Ceratitis capitata* (Wiedemann)	（ハエ目、ミバエ科）全北区、豪州区。著名な果実害虫
Mediterranean grape leaf beetle			vine flea beetle を見よ
Mediterranean hawk			Mediterranean hawkmoth を見よ
Mediterranean hawkmoth		*Hyles nicaea* (de Prunner)	（チョウ目、スズメガ科）旧北区
Mediterranean hawthorn weevil		*Otiorhynchus crataegi* Germar	（コウチュウ目、ゾウムシ科）旧北区
Mediterranean katydid			four-spot bush-cricket を見よ
Mediterranean mantid		*Iris oratoria* (Linnaeus)	（カマキリ目、カマキリ科）全北区
Mediterranean mantis			Mediterranean mantid を見よ
Mediterranean meal knot-horn			Mediterranean flour moth を見よ
Mediterranean pierrot			Mediterranean tiger blue を見よ
Mediterranean pine engraver beetle		*Orthotomicus erosus* (Wollaston)	（コウチュウ目、キクイムシ科）旧北区
Mediterranean scale			citrus scale を見よ

英名	和名	学名	所属、分布、ほか
Mediterranean skipper		*Gegenes nostradamus* (Fabricius)	（チョウ目、セセリチョウ科）旧北区
Mediterranean stick-insect		*Bacillus rossius* Rossi	（ナナフシ目、コブナナフシ科）旧北区
Mediterranean tiger blue	チビトラフシジミ	*Tarucus rosaceus* (Austaut)	（チョウ目、シジミチョウ科）旧北区
medium dagger moth		*Acronicta modica* Walker	（チョウ目、ヤガ科）新北区
medium green-banded swallowtail		*Papilio sosia* Rothschild et Jordan	（チョウ目、アゲハチョウ科）エチオピア区
medium olive		*Baetis vernus* Curtis	（カゲロウ目、コカゲロウ科）旧北区
medium-sized elongate scolytid	チュウガタナガキクイムシ	*Platypus modestus* Blandford	（コウチュウ目、ナガキクイムシ科）日本、東洋区
medium-sized pasture flood-water mosquito		*Ochlerotatus nigromaculis* (Ludlow)	（ハエ目、カ科）新北区
medium-sized saltmarsh mosquito		*Ochlerotatus melanimon* (Dyar)	（ハエ目、カ科）新北区
medium-sized semi-arid mosquito		*Ochlerotatus dorsalis* (Meigen)	（ハエ目、カ科）全北区、東洋区
Medora mimic white		*Dismorphia medora* (Doubleday)	（チョウ目、シロチョウ科）新熱帯区
megachilid bees			leef-cutting bees (2) を見よ
megalops skipper		*Cynea megalops* (Godman)	（チョウ目、セセリチョウ科）新熱帯区
megalopterans			alderflies (2) を見よ
megalopterous flies			alderflies (2) を見よ
megamerind flies	ホソバエ科	Megamerinidae	（ハエ目）の昆虫の総称
Megarus scrub-hairstreak		*Strymon megarus* (Godart)	（チョウ目、シジミチョウ科）新熱帯区
megaspilid wasps	オオモンクロバチ科	Megaspilidae	（ハチ目）の昆虫の総称
Megerle's silver-barred dwarf		*Elachista megerlella* (Hübner)	（チョウ目、クサモグリガ科）旧北区
Megerle's silver-barred dwarf		*Elachista adscitella* Stainton	（チョウ目、クサモグリガ科）旧北区
Megopis long-horned beetle		*Megopis reflexa* (Karsch)	（コウチュウ目、カミキリムシ科）大洋区
melaleuca leaf weevil			melaleuca snout beetle を見よ
melaleuca psyllid		*Boreioglycaspis melaleucae* Moore	（カメムシ目、キジラミ科）豪州区
melaleuca sawfly		*Lophyrotoma zonalis* Rohwer	（ハチ目、Pergidae）豪州区
melaleuca snout beetle		*Oxyops vitiosa* Pascoe	（コウチュウ目、ゾウムシ科）豪州区
melancholy metalmark		*Pheles melanchroia* (C. et R. Felder)	（チョウ目、シジミタテハ科）新熱帯区
melandryids		*Penthe*	（コウチュウ目、ナガクチキムシ科）の昆虫の総称

英名	和名	学名	所属、分布、ほか
Melantho tigerwing		*Thyridia psidii* (Linnaeus)	(チョウ目、タテハチョウ科) 新熱帯区。*T. p. melantho* Bates も同英名
melastoma borer		*Selca brunella* (Hampson)	(チョウ目、ヤガ科) 新北区
melastoma leafroller		*Bocchoris adipalis* (Lederer)	(チョウ目、メイガ科) 大洋区
Meleager's blue	ダフニスヒメシジミ	*Meleageria daphnis* (Denis et Schiffermüller)	(チョウ目、シジミチョウ科) 旧北区
Melicertes spreadwing		*Potamanaxas melicertes* (Godman et Salvin)	(チョウ目、セセリチョウ科) 新熱帯区
melicope long-horned beetle			Hawaiian alani long-horned beetle を見よ
Melino metalmark		*Caria melino* Dyar	(チョウ目、シジミタテハ科) 新熱帯区
meliosoma aphid			slender hairy aphid を見よ
Melissa arctic	ハイイロタカネヒカゲ	*Oeneis melissa* (Fabricius)	(チョウ目、タテハチョウ科) 全北区。日本亜種は *O. m. daisetsuzana* Matsumura ダイセツタカネヒカゲ
Melissa blue			orange-bordered blue を見よ
Melitaeoides checkerspot		*Chlosyne melitaeoides* (C. et R. Felder)	(チョウ目、タテハチョウ科) 新熱帯区
Melite mimic white		*Enantia melite* (Linnaeus)	(チョウ目、シロチョウ科) 新熱帯区
melittid bees	ケアシハナバチ科	Melittidae	(ハチ目) の昆虫の総称
Mella skipper		*Anatrytone mella* (Godman)	(チョウ目、セセリチョウ科) 新熱帯区
melodious ground cricket		*Eunemobius melodius* (Thomas et Alexander)	(バッタ目、コオロギ科) 新北区
meloid			blister beetle などツチハンミョウ科の種
meloids			blister beetles (1) を見よ　英国での英名
melon and cotton aphid			cotton aphid を見よ
melon aphid			cotton aphid を見よ
melon aphis			cotton aphid を見よ
melon borer			melon worm を見よ
melon caterpillar			melon worm を見よ
melon fly	ウリミバエ	*Zeugodacus cucurbitae* (Coquillett)	(ハエ目、ミバエ科) 日本、東洋区、大洋区。ウリ類、ナス、トマトなどの害虫
melon fruit fly			melon fly を見よ
melon ladybird		*Epilachna chrysomelina* (Fabricius)	(コウチュウ目、テントウムシ科) 旧北区。南アフリカでの英名
melon longicorn beetle	メロンサビカミキリ	*Apomecyna excavaticeps* Pic	(コウチュウ目、カミキリムシ科) 東洋区
melon moth			melon worm を見よ
melon moth			cotton caterpillar (1) を見よ
melon mottled-skipper		*Codatractus melon* (Godman et Salvin)	(チョウ目、セセリチョウ科) 新熱帯区
melon thrips	ミナミキイロアザミウマ	*Thrips palmi* Karny	(アザミウマ目、アザミウマ科) 日本、汎世界。多くの作物、果樹の害虫
melon thrips			honeysuckle thrips を見よ

英名	和名	学名	所属、分布、ほか
melon weevil		*Baris traegardhi* Aurivillius	(コウチュウ目、ゾウムシ科) エチオピア区
melon white-spotted longicorn			cucurbit stem borer を見よ
melon worm		*Diaphania hyalinata* (Linnaeus)	(チョウ目、メイガ科) 新北区、新熱帯区。成虫は melonworm moth
Melona sister		*Adelpha melona* (Hewitson)	(チョウ目、タテハチョウ科) 新熱帯区
Melscheimer's sack-bearer moth		*Cicinnus melscheimeri* (Harris)	(チョウ目、Mimallonidae) 新北区
Menander metalmark		*Menander menander purpurata* (Godman et Salvin)	(チョウ目、シジミタテハ科) 新熱帯区
Menapis tigerwing		*Mechanitis menapis* Hewitson	(チョウ目、タテハチョウ科) 新熱帯区。*M. m. doryssus* Bates も同英名
mendicants			mantids (2) を見よ
mendocino saturnid		*Saturnia mendocino* Behrens	(チョウ目、ヤママユガ科) 旧北区
Menelaus blue Morpho			blue Morpho (2) を見よ
Menelaus Morpho			blue Morpho (2) を見よ
mengeids		Mengeidae	(ネジレバネ目) の昆虫の総称　絶滅科
meniscus midges			dixid midges を見よ
mentha leaf webber			ocimum leaf folder を見よ
merchanized dandruff			body louse を見よ
merchant grain beetle	オオメノコギリヒラタムシ	*Oryzaephilus mercator* (Fauvel)	(コウチュウ目、ホソヒラタムシ科) 日本、汎世界。貯穀害虫
merchant grain beetle			saw-toothed grain beetle を見よ
Mercurial skipper		*Proteides mercurius* Fabricius	(チョウ目、セセリチョウ科) 新北区
Mercurial skipper		*Theagenes albiplaga* (Felder et Felder)	(チョウ目、セセリチョウ科) 新熱帯区
Mercury Island tusked weta			elephant weta を見よ
mere wainscot		*Photedes fluxa* (Hübner)	(チョウ目、ヤガ科) 旧北区。日本亜種は *P. f. rufata* (Kardakoff) ウスキモンヨトウ
meridian duskywing			southern duskywing を見よ
meridional termite			magnetic termite を見よ
Mermeria wood nymph		*Taygetis mermeria* (Cramer)	(チョウ目、タテハチョウ科) 新熱帯区
Merops sphinx moth		*Lintneria merops* (Boisduval)	(チョウ目、スズメガ科) 新熱帯区
merothripids			jumping thrips を見よ
Merrick's pyralid moth		*Loxostegopsis merrickalis* (Barnes et McDunnough)	(チョウ目、メイガ科) 新北区
merry may			dragonflies (1) を見よ

英名	和名	学名	所属、分布、ほか
merry Melipotis moth		*Melipotis jucunda* (Hübner)	（チョウ目、ヤガ科）新北区
merry shadowdamsel		*Drepanosticta hilaris* (Hagen)	（トンボ目、Platystictidae）東洋区
Meru paradise skipper		*Abantis meru* Evans	（チョウ目、セセリチョウ科）エチオピア区
Merveille-du-jour		*Dichonia aprilina* (Linnaeus)	（チョウ目、ヤガ科）旧北区
Mescalero shieldback		*Plagiostira mescaleroensis* Tinkham	（バッタ目、キリギリス科）新北区
Mescalero thread-legged katydid		*Arethaea mescalero* Hebard	（バッタ目、キリギリス科）新北区
mesem scale		*Pulvinaria mesembrianthemi* (Vallot)	（カメムシ目、カタカイガラムシ科）旧北区、エチオピア区。南アフリカでの英名
Mesentina sister		*Adelpha mesentina* Cramer	（チョウ目、タテハチョウ科）新熱帯区
Meske's Pero moth		*Pero meskaria* (Packard)	（チョウ目、シャクガ科）新北区
Meske's skipper			dixie skipper を見よ
Meske's underwing moth		*Catocala meskei* Grote	（チョウ目、ヤガ科）新北区
mesopsocid barklice			middle barklice を見よ
mesquite clearwing moth		*Carmenta prosopis* (Edwards)	（チョウ目、スカシバガ科）新北区
mesquite cutworm			indomitable Melipotis moth を見よ
mesquite katydid			elegant bush katydid を見よ
mesquite leaf tier moth		*Pococera euphemella* (Hulst)	（チョウ目、メイガ科）新北区
mesquite longhorn		*Callona rimosa* Buquet	（コウチュウ目、カミキリムシ科）新北区
mesquite looper moth		*Rindgea cyda* (Druce)	（チョウ目、シャクガ科）新北区
mesquite mealybug		*Spirococcus prosopides* (Cockerell)	（カメムシ目、コナカイガラムシ科）新北区
mesquite stinger moth		*Norape tenera* (Druce)	（チョウ目、Megalopygidae）新北区
mesquite webworm moth		*Friseria cockerelli* (Busck)	（チョウ目、キバガ科）新北区
Messalina underwing moth		*Catocala messalina* Guenée	（チョウ目、ヤガ科）新北区。Messalina underwing ともいう
Messana sister	キマエイチモンジ	*Adelpha messana* (C. et R. Felder)	（チョウ目、タテハチョウ科）新熱帯区
Mestra glasswing		*Hyalyris mestra* Hopffer	（チョウ目、タテハチョウ科）新熱帯区
Metaleuca ringlet		*Pareuptychia metaleuca* Boisduval	（チョウ目、タテハチョウ科）新北区、新熱帯区
metallescens skipper		*Polythrix metallescens* (Mabille)	（チョウ目、セセリチョウ科）新熱帯区
metallic-backed reedling		*Indolestes divisus* (Hagen)	（トンボ目、アオイトトンボ科）東洋区
metallic beetle	ウバタマムシ	*Chalcophora japonica* (Gory)	（コウチュウ目、タマムシ科）日本、旧北区、東洋区

英名	和名	学名	所属、分布、ほか
metallic blue hover fly		*Helophilus hochstetteri* Nowicki	(ハエ目、ハナアブ科) 豪州区
metallic Cerulean		*Jamides alecto* (Felder)	(チョウ目、シジミチョウ科) 東洋区。日本亜種は *J. a. dromicus* (Fruhstorfer) シロウラナミシジミ。*J. a. ageladas* (Fruhstorfer) も同英名
metallic Coleophora moth		*Coleophora mayrella* (Hübner)	(チョウ目、ツツミノガ科) 全北区
metallic colored geometrid	ハスオビキエダシャク	*Scardamia aurantiacaria* Bremer	(チョウ目、シャクガ科) 日本、旧北区
metallic flea beetle		*Clitea metallica* Chen	(コウチュウ目、ハムシ科) 旧北区。カンキツ害虫
metallic flea beetles		*Altica*	(コウチュウ目、ハムシ科) の昆虫の総称
metallic green hairstreak	インドオオミドリシジミ	*Chrysozephyrus duma* (Hewitson)	(チョウ目、シジミチョウ科) 東洋区
metallic green tomato fly		*Lamprolonchaea brouniana* (Bezzi)	(ハエ目、クロツヤバエ科) 豪州区
metallic hedge blue		*Celastrina melaena* (Doherty)	(チョウ目、シジミチョウ科) 東洋区
metallic leafcutter bees			mason bees (3) を見よ
metallic longhorn		*Nemophora metallica* (Poda)	(チョウ目、マガリガ科) 旧北区
metallic shield bug		*Scutiphora pedicellata* (Kirby)	(カメムシ目、キンカメムシ科) 豪州区
metallic small-headed flies		*Eulonchus*	(ハエ目、コガシラアブ科) の昆虫の総称
metallic sweat bees		*Agapostemon*	(ハチ目、コハナバチ科) の昆虫の総称
metallic wood-borers			flat-headed beetles を見よ
metallic wood-boring beetle		*Hippomelas sphenica* LeConte	(コウチュウ目、タマムシ科) 新北区
metallic wood-boring beetle			flatheaded apple tree borer を見よ
metallic woodboring beetles			flat-headed beetles を見よ
metallic wood-boring beetles			flat-headed beetles を見よ
metalmark butterflies			metalmarks (1) を見よ　米国での英名
metalmark moths	ハマキモドキガ科	Choreutidae	(チョウ目) の昆虫の総称
metalmarks		Riodininae	(チョウ目、シジミタテハ科) の昆虫の総称
metalmarks (1)	シジミタテハ科	Riodinidae	(チョウ目) の昆虫の総称
Metaponpneumata moth		*Metaponpneumata rogenhoferi* Möschler	(チョウ目、ヤガ科) 新北区
Metcalf's carpet			Devon carpet を見よ
methocid wasps	ツヤアリバチ科	Methocidae	(ハチ目) の昆虫の総称　コツチバチ科の亜科とされることあり
Meton hairstreak			tiger-eye hairstreak を見よ
metrosideros long-horned beetle			Hawaiian ohia long-horned beetle を見よ

英名	和名	学名	所属、分布、ほか
Metzner's tipped neb		*Metzneria metzneriella* (Stainton)	(チョウ目、キバガ科) 旧北区
Mexican corn rootworm		*Diabrotica virgifera zeae* Krysan et Smith	(コウチュウ目、ハムシ科) 新北区
Mexican Agapema moth		*Agapema anona* (Ottolengui)	(チョウ目、ヤママユガ科) 新北区
Mexican amberwing		*Perithemis intensa* Kirby	(トンボ目、トンボ科) 新北区、新熱帯区
Mexican anglewing		*Polygonia g-argenteum* (Doubleday)	(チョウ目、タテハチョウ科) 新熱帯区
Mexican Arcas	キプリアクジャクシジミ	*Arcas cypria* (Geyer)	(チョウ目、シジミチョウ科) 新熱帯区
Mexican azure		*Celastrina gozora* (Boisduval)	(チョウ目、シジミチョウ科) 新熱帯区
Mexican azure			holly blue を見よ
Mexican bean beetle		*Zabrotes pectoralis* (Sharp)	(コウチュウ目、ハムシ科) 新熱帯区
Mexican bean beetle (1)	インゲンテントウ	*Epilachna varivestris* Mulsant	(コウチュウ目、テントウムシ科) 日本、新北区、新熱帯区。ダイズ、インゲンなどの害虫
Mexican bean weevil			Brazil bean weevil を見よ
Mexican black scale		*Saissetia miranda* (Cockerell et Parrott)	(カメムシ目、カタカイガラムシ科) 新北区
Mexican bluewing			blue wing を見よ
Mexican bush katydid		*Scudderia mexicana* (Saussure)	(バッタ目、キリギリス科) 新北区
Mexican cactus fly		*Copestylum mexicanum* (Macquart)	(ハエ目、ハナアブ科) 新北区、新熱帯区
Mexican Calephelis		*Calephelis mexicana* McAlpine	(チョウ目、シジミタテハ科) 新熱帯区
Mexican cloudywing		*Thorybes mexicana* (Herrich-Schäffer)	(チョウ目、セセリチョウ科) 新北区
Mexican corn leafhopper		*Dalbulus elimatus* (Ball)	(カメムシ目、オオヨコバイ科) 新熱帯区
Mexican cotton boll weevil			boll weevil を見よ
Mexican crescent		*Phyciodes pallescens* (R. Felder)	(チョウ目、タテハチョウ科) 新北区、新熱帯区
Mexican cutworm tachinid		*Archytas cirphis* Curran	(ハエ目、ヤドリバエ科) 大洋区
Mexican cycadian	メキシコマルバネカラスシジミ	*Eumaeus toxea* (Godart)	(チョウ目、シジミチョウ科) 新北区、新熱帯区
Mexican dartwhite	アメリカキボシシロチョウ	*Catasticta nimbice* (Boisduval)	(チョウ目、シロチョウ科) 新北区
Mexican duskywing		*Erynnis mercurius* (Dyar)	(チョウ目、セセリチョウ科) 新熱帯区
Mexican eighty-eight			Astala eighty-eight を見よ
Mexican emperor	アメリカコムラサキ	*Doxocopa cyane mexicana* Bryk	(チョウ目、タテハチョウ科) 新熱帯区。Cyan emperor を参照
Mexican Entheus		*Entheus crux* Steinhauser	(チョウ目、セセリチョウ科) 新熱帯区
Mexican forktail		*Ischnura demorsa* (Hagen)	(トンボ目、イトトンボ科) 新北区

英名	和名	学名	所属、分布、ほか
Mexican fritillary		*Euptoieta hegesia* (Cramer)	(チョウ目、タテハチョウ科) 新北区、新熱帯区。*E. h. meridiana* Stichel も同英名
Mexican fruit fly	メキシコミバエ	*Anastrepha ludens* (Loew)	(ハエ目、ミバエ科) 新北区、新熱帯区
Mexican grass yellow			Mexican yellow を見よ
Mexican gray hairstreak		*Strymon bebrycia* (Hewitson)	(チョウ目、シジミチョウ科) 新北区
Mexican hoary edge		*Achalarus casica* (Herrich-Schäffer)	(チョウ目、セセリチョウ科) 新北区
Mexican jumping bean borer			jumping bean moth (1) を見よ
Mexican jumping bean moth			jumping bean moth (1) を見よ
Mexican jumping bean moth (1)		*Laspeyresia saltitans* (Westwood)	(チョウ目、ハマキガ科) 新北区。幼虫はトウダイグサ科の種子内部を食べた後、動作につれて種子がはねる
Mexican katydid			Mexican bush katydid を見よ
Mexican kite-swallowtail	エピダウスオナガタイマイ	*Eurytides epidaus epidaus* (Doubleday)	(チョウ目、アゲハチョウ科) 新熱帯区。*E. e. fenochionis* (Salvin et Godman), *E. e. tepicus* (Rothschild et Jordan) も同英名
Mexican leaf roller		*Amorbia emigratella* Busck	(チョウ目、ハマキガ科) 新北区、新熱帯区、大洋区
Mexican longtail		*Polythrix mexicanus* Freeman	(チョウ目、セセリチョウ科) 新北区、新熱帯区
Mexican longwing		*Heliconius hortense* Guérin-Méneville	(チョウ目、タテハチョウ科) 新熱帯区
Mexican-M hairstreak		*Parrhasius moctezuma* (Clench)	(チョウ目、シジミチョウ科) 新北区、新熱帯区
Mexican mealybug		*Phenacoccus gossypii* Townsend et Cockerell	(カメムシ目、コナカイガラムシ科) 新北区。温室害虫
Mexican Mellana		*Quasimellana mexicana* (Bell)	(チョウ目、セセリチョウ科) 新熱帯区
Mexican Mellana			common Mellana を見よ
Mexican metalmark		*Apodemia mejicanus mejicanus* (Behr)	(チョウ目、シジミタテハ科) 新熱帯区
Mexican moth borer			southern corn stalk borer (1) を見よ
Mexican mournful duskywing		*Erynnis tristis taitus* (Edwards)	(チョウ目、セセリチョウ科) 新熱帯区。mournful duskywing を参照
Mexican orangetip		*Anthocharis limonea* (Butler)	(チョウ目、シロチョウ科) 新北区
Mexican pine beetle		*Dendroctonus approximatus* Dietz	(コウチュウ目、キクイムシ科) 新北区
Mexican pine white			Chiricahua pine white を見よ
Mexican pine-satyr			Xicaque satyr を見よ
Mexican pygmy grasshopper		*Paratettix mexicanus* (Saussure)	(バッタ目、ヒシバッタ科) 新北区
Mexican queen		*Danaus gilippus strigosus* (Bates)	(チョウ目、タテハチョウ科) 新北区。queen を参照
Mexican rice borer		*Eoreuma loftini* (Dyar)	(チョウ目、メイガ科) 新北区

英名	和名	学名	所属、分布、ほか
Mexican Ridens		*Ridens miltas* (Godman et Salvin)	（チョウ目、セセリチョウ科）新熱帯区
Mexican ruby-eye		*Carystoides mexicana* Freeman	（チョウ目、セセリチョウ科）新熱帯区
Mexican sailor		*Dynamine postverta mexicana* d'Almeida	（チョウ目、タテハチョウ科）新熱帯区
Mexican satyr		*Taygetis weymeri* Draudt	（チョウ目、タテハチョウ科）新熱帯区
Mexican scarlet-eye		*Cephise mexicanus* Austin et Mielke	（チョウ目、セセリチョウ科）新熱帯区
Mexican silk moth		*Eutachyptera psidii* (Salle)	（チョウ目、ドクガ科）新熱帯区
Mexican silver-spot	オオウラギンドクチョウ	*Dione moneta* Hübner	（チョウ目、タテハチョウ科）新北区、新熱帯区。*D. m. poeyii* Butler も同英名
Mexican sister (1)		*Adelpha fessonia* (Hewitson)	（チョウ目、タテハチョウ科）新北区、新熱帯区
Mexican sister			Diaz's sister を見よ
Mexican sister			California sister (1) を見よ
Mexican sootywing		*Pholisora mejicanus* (Reakirt)	（チョウ目、セセリチョウ科）新北区
Mexican squash vine borer			squash vine borer を見よ　米国での英名
Mexican Telemiades		*Telemiades choricus* (Schaus)	（チョウ目、セセリチョウ科）新熱帯区
Mexican tiger moth		*Apantesis proxima* (Guérin-Méneville)	（チョウ目、ヒトリガ科）新北区
Mexican tiger swallowtail		*Papilio alexiares* Hopffer	（チョウ目、アゲハチョウ科）新熱帯区
Mexican tortoiseshell		*Nymphalis cyanomelas* Doubleday	（チョウ目、タテハチョウ科）新熱帯区
Mexican umber skipper		*Poanes melane vitellina* (Herrich-Schäffer)	（チョウ目、セセリチョウ科）新熱帯区。umber skipper を参照
Mexican wedgetail		*Acanthagrion quadratum* Selys	（トンボ目、イトトンボ科）新北区
Mexican yellow		*Eurema mexicana* (Boisduval)	（チョウ目、シロチョウ科）新北区、新熱帯区。Mexican grass yellow ともいう
Mexican Zestusa		*Zestusa elwesi* (Godman et Salvin)	（チョウ目、セセリチョウ科）新熱帯区
Miaba grass skipper		*Cobalopsis miaba* Schaus	（チョウ目、セセリチョウ科）新熱帯区
Miadora hairstreak			West-Mexican Ipidecla を見よ
Miahuatlan Emesis		*Emesis arnacis* Stichel	（チョウ目、シジミタテハ科）新熱帯区。
Miami blue		*Cyclargus thomasi* (Clench)	（チョウ目、シジミチョウ科）新北区
Miami blue (1)		*Cyclargus thomasi bethunebakeri* Comstock et Huntington	（チョウ目、シジミチョウ科）新北区。Miami blue butterfly ともいう
Miami eyed blue			Miami blue (1) を見よ
Mibu wormwood grained moth	セダカモクメ	*Cucullia perforata* Bremer	（チョウ目、ヤガ科）日本
Mibu wormwood looper	ムラサキキンウワバ	*Autographa buractica* (Staudinger)	（チョウ目、ヤガ科）日本

英名	和名	学名	所属、分布、ほか
Micandriana groundstreak		*Ziegleria micandriana* (Johnson)	（チョウ目、シジミチョウ科）新熱帯区
Michabo moth			Doris tiger moth を見よ
Michelle's desert goundstreak		*Strymon michelle* Nicolay et Robbins	（チョウ目、シジミチョウ科）新熱帯区
Michener's giant-skipper		*Agathymus micheneri* Stallings, Turner et Stallings	（チョウ目、セセリチョウ科）新熱帯区
micraner	小雄型		小型の雄アリ
micrergate	小職蟻型		小型の職アリ。microergate ともいう
micro-caddisflies	ヒメトビケラ科	Hydroptilidae	（トビケラ目）の昆虫の総称
micrographer bark-beetle		*Pityophthorus micrographus* (Linnaeus)	（コウチュウ目、キクイムシ科）旧北区
microgyne	小雌型		雌に2型あるアリの小型の雌
micro-lepidopterans			micro-moths を見よ
micromalthid beetle	チビナガヒラタムシ	*Micromalthus debilis* LeConte	（コウチュウ目、チビナガヒラタムシ科）日本、東洋区、大洋区、新北区、新熱帯区
micromalthid beetles			telephone pile beetles を見よ
micro-moths	小蛾類		（チョウ目）の小型の蛾で、microlepidoptera という
microphysid bugs			minute bladder bugs を見よ
microphysids			minute bladder bugs を見よ
micros			micro-moths を見よ
Micythus skipper		*Morys micythus* (Godman)	（チョウ目、セセリチョウ科）新熱帯区
mid-summer chafer			summer chafer を見よ
Midas opal		*Chrysoritis midas* (Pennington)	（チョウ目、シジミチョウ科）エチオピア区
Midas skipper		*Bungalotis midas* (Cramer)	（チョウ目、セセリチョウ科）新熱帯区
midboreal mayfly		*Ephemerella subvaria* (Hendrickson)	（カゲロウ目、マダラカゲロウ科）新北区
middle barklice	マルチャタテ科	Mesopsocidae	（チャタテムシ目）の昆虫の総称
middle-barred minor		*Oligia fasciuncula* (Haworth)	（チョウ目、ヤガ科）旧北区
middle feather clothes		*Ochsenheimeria mediopectinellus* (Haworth)	（チョウ目、Ochsenheimeriidae）旧北区
Middle Island tusked weta			elephant weta を見よ
middle lace border		*Scopula decorata* (Denis et Schiffermüller)	（チョウ目、シャクガ科）旧北区
Middlekauff's shieldback		*Idiostatus middlekauffi* Rentz	（バッタ目、キリギリス科）新北区
middle-palened geometrid	ナカウスエダシャク	*Alcis angulifera* (Butler)	（チョウ目、シャクガ科）日本、旧北区
midge		*Chironomus annularis* Meigen	（ハエ目、ユスリカ科）旧北区
midge			小昆虫、ユスリカなど Nematocera の種

英名	和名	学名	所属、分布、ほか
midge			plumed gnat を見よ
midge			house fly を見よ
midges	ユスリカ科	Chironomidae	（ハエ目）の昆虫の総称　世界に5,000種
midget moths	モグリチビガ科	Nepticulidae	（チョウ目）の昆虫の総称
midget moths			leafblotch miners (1) を見よ
midgets		*Mortonagrion*	（トンボ目、イトトンボ科）の昆虫の総称
midgets			midget moths を見よ
midgets			leafblotch miners (1) を見よ　Gracillariidae の異名 Lithocolletidae の英名
midland clubtail		*Gomphus fraternus* (Say)	（トンボ目、サナエトンボ科）新北区
midrib gall moth		*Sorhagenia nimbosa* (Braun)	（チョウ目、カザリバガ科）新北区
migrant hawker	マダラヤンマ	*Aeschna mixta* Latreille	（トンボ目、ヤンマ科）日本、旧北区、東洋区、エチオピア区
migrant hawker dragonfly			migrant hawker を見よ
migrant skipper			rice skipper を見よ
migrant spreadwing			southern emerald damselfly を見よ
migrants	移住型		アブラムシ類で第1次宿主上で発育した幹母の次世代を指し、有翅胎生雌で第2次宿主に移動する。migrante, migrantes ともいう
migratory bird grasshoppers			spur-throated grasshoppers (1) を見よ
migratory bird locusts			spur-throated grasshoppers (1) を見よ
migratory glider	ウスキタテハ	*Cymothoe caenis* Drury	（チョウ目、タテハチョウ科）エチオピア区
migratory grasshopper		*Melanoplus sanguinipes* (Fabricius)	（バッタ目、バッタ科）新北区
migratory grasshopper		*Melanoplus bilituratus* (Walker)	（バッタ目、バッタ科）新北区。トウモロコシ害虫
migratory locust			Asiatic migratory locust を見よ
migratory locust			desert locust を見よ
Mijburgh's blue		*Orachrysops mijburghi* Henning et Henning	（チョウ目、シジミチョウ科）エチオピア区
mikado ambrosia beetle	ミカドキクイムシ	*Scolytoplatypus mikado* Blandford	（コウチュウ目、キクイムシ科）日本、旧北区、東洋区。カキ、クリ害虫
mikado bark beetle			mikado ambrosia beetle を見よ
mikado fruit fly	ミカドハマダラミバエ	*Staurella mikado* (Matsumura)	（ハエ目、ミバエ科）日本
mikado minute garden cricket	シバスズ	*Pteronemobius mikado* (Shiraki)	（バッタ目、コオロギ科）日本
mikado okame cricket	ツヅレサセコオロギ	*Velarifictorus mikado* (Saussure)	（バッタ目、コオロギ科）日本、新北区
mikado pyralid	ミカドマダラメイガ	*Nephopterix mikadella* (Ragonot)	（チョウ目、メイガ科）日本
Milbert's tortoiseshell	アメリカコヒオドシ	*Aglais milberti* (Godart)	（チョウ目、タテハチョウ科）新北区、新熱帯区

英名	和名	学名	所属、分布、ほか
milfoil pug		*Eupithecia millefoliata* Rössler	（チョウ目、シャクガ科）旧北区
milichiid flies			jackal flies を見よ
military bug		*Spilostethus pandarus* (Scopoli)	（カメムシ目、ナガカメムシ科）旧北区
military meadow katydid		*Orchelimum militare* Rehn et Hebard	（バッタ目、キリギリス科）新北区
milk-vetch piercer	フタスジヒメハマキ	*Grapholita pallifrontana* Lienig et Zeller	（チョウ目、ハマキガ科）日本、旧北区
milkweed			monarch butterfly を見よ
milkweed aphid			oleander aphid を見よ
milkweed beetles			milkweed longhorns を見よ
milkweed bug			large milkweed bug を見よ
milkweed bugs		*Spilostethus*	（カメムシ目、ナガカメムシ科）の昆虫の総称
milkweed bugs		*Oncopeltus*	（カメムシ目、ナガカメムシ科）の昆虫の総称
milk-weed bugs			chinch bugs (1) を見よ
milkweed butterflies			monarchs を見よ
milkweed butterfly			monarch butterfly を見よ　英国での英名
milkweed longhorns		*Tetraopes*	（コウチュウ目、カミキリムシ科）の昆虫の総称
milk-weed moth			milkweed tussock moth を見よ
milkweed tigermoth			milkweed tussock moth を見よ　北米での英名
milkweed tussock			milkweed tussock moth を見よ
milkweed tussock moth		*Euchaetes egle* (Drury)	（チョウ目、ヒトリガ科）新北区。幼虫はトウワタにつく
milkweeds			tigers (1) を見よ
milk-white termite		*Coptotermes lacteus* (Froggatt)	（シロアリ目、ミゾガシラシロアリ科）豪州区
milky Cerulean		*Lampides lacteata* de Nicéville	（チョウ目、シジミチョウ科）東洋区
milky scarce flat		*Calleagris lacteus* (Mabille)	（チョウ目、セセリチョウ科）エチオピア区
milky urola moth		*Argyria lacteella* (Fabricius)	（チョウ目、メイガ科）新北区
mill moth			Mediterranean flour moth を見よ　英国での英名
Millar's buff		*Deloneura millari* Trimen	（チョウ目、シジミチョウ科）エチオピア区
Millar's hairtail		*Anthene millari* (Trimen)	（チョウ目、シジミチョウ科）エチオピア区
Millar's tiger		*Callioratis millari* Hampson	（チョウ目、シャクガ科）エチオピア区
miller		*Acronicta leporina* (Linnaeus)	（チョウ目、ヤガ科）旧北区。日本亜種は *A. l. leporella* Staudinger シロケンモン
Miller dagger moth		*Acronicta vulpina* Grote	（チョウ目、ヤガ科）新北区
miller moth			miller を見よ
miller whitish cutworm	シラナミアツバ	*Herminia innocens* Butler	（チョウ目、ヤガ科）日本、旧北区
millers			owlet moths を見よ　豪州での英名

英名	和名	学名	所属、分布、ほか
millers			leafroller moths を見よ　北米での英名
millers			butterflies and moths を見よ　米国での幼虫の英名
Miller's leaf sitter		*Gorgyra bule* Miller	（チョウ目、セセリチョウ科）エチオピア区
Miller's nightfighter		*Artitropa milleri* Riley	（チョウ目、セセリチョウ科）エチオピア区
Miller's playboy		*Deudorix cleora* Miller et Miller	（チョウ目、シジミチョウ科）東洋区
Miller's scarlet-eye		*Bungalotis milleri* Freeman	（チョウ目、セセリチョウ科）新熱帯区
Miller's sister		*Adelpha milleri* Beutelspacher	（チョウ目、タテハチョウ科）新熱帯区
Miller's skipperling		*Piruna millerorum* Steinhauser	（チョウ目、セセリチョウ科）新熱帯区
millet skipper			banana skipper (1) を見よ
millet head miner		*Heliocheilus albipunctella* de Joannis	（チョウ目、ヤガ科）東洋区、エチオピア区
millet stalk worm			European corn borer を見よ　幼虫の英名
millet stalk worm			Oriental corn borer を見よ
millet stem borer		*Coniesta ignefusalis* Hampson	（チョウ目、メイガ科）エチオピア区。maize 他の害虫
Milliere's December moth		*Poecilocampa alpina* (Frey et Wullschlevel)	（チョウ目、カレハガ科）旧北区
mimetic gumleaf grasshopper		*Goniaea opomaloides* (Walker)	（バッタ目、バッタ科）豪州区
mimetic swallowtail		*Papilio cynorta* Fabricius	（チョウ目、アゲハチョウ科）エチオピア区
mimic			danaid eggfly を見よ　米国での英名
mimic			viceroy を見よ
mimic Acraea		*Acraea mima* Neave	（チョウ目、タテハチョウ科）エチオピア区
mimic butterfly			danaid eggfly を見よ
mimic crescent	マネシミカズキタテハ	*Eresia pelonia* Hewitson	（チョウ目、タテハチョウ科）新熱帯区
mimic crescent		*Phyciodes incognitus* Gatrelle	（チョウ目、タテハチョウ科）新北区
mimic crescent (1)		*Castilia eranites* (Hewitson)	（チョウ目、タテハチョウ科）新熱帯区
mimic Obania		*Obania subvariegata* (Grose-Smith et Kirby)	（チョウ目、シジミチョウ科）エチオピア区
mimic phantom		*Pseudohaetera mimica* (Rosenberg et Talbot)	（チョウ目、タテハチョウ科）新熱帯区
mimic tigerwing		*Melinaea lilis flavicans* Hoffmann	（チョウ目、タテハチョウ科）新熱帯区。*M. l. imitata* Bates も同英名
mimic tigerwing			mechanitis mimic を見よ
mimic whites	トンボシロチョウ亜科	Dismorphiinae	（チョウ目、シロチョウ科）の昆虫の総称
mimosa flowerbud weevil		*Coelocephalapion aculeatum* (Fall)	（コウチュウ目、ミツギリゾウムシ科）豪州区。オジギソウの属名 *Mimosa* に由来
mimosa leaf beetle		*Chlamisus mimosae* Karren	（コウチュウ目、ハムシ科）豪州区

英名	和名	学名	所属、分布、ほか
mimosa sapphire		*Iolaus mimosae* Trimen	(チョウ目、シジミチョウ科) エチオピア区
mimosa seed weevil		*Acanthoscelides puniceus* Johnson	(コウチュウ目、ハムシ科) 東洋区
mimosa skipper		*Cogia calchas* (Herrich-Schäffer)	(チョウ目、セセリチョウ科) 新北区
mimosa stemboring moth		*Carmenta mimosa* Eichlin	(チョウ目、スカシバガ科) 豪州区
mimosa webmoth	ネムスガ	*Homadaula anisocentra* Meyrick	(チョウ目、ネムスガ科) 日本、新北区。ネムノキ害虫。mimosa webworm moth ともいう
mimosa yellow		*Eurema nise* (Cramer)	(チョウ目、シロチョウ科) 新北区、新熱帯区
miner moths			grass miner moths を見よ
miner pith borer			lesser pine shoot beetle を見よ
Minerva's owl-butterfly		*Opsiphanes blythekitzmillerae* Austin et Warren	(チョウ目、タテハチョウ科) 新熱帯区
miniature ghost moths		Palaeosetidae	(チョウ目) の昆虫の総称
minimus mealybug		*Phenacoccus minimus* Tinsley	(カメムシ目、コナカイガラムシ科) 新北区
mining bee		*Andrena haemorrhoa* (Fabricius)	(ハチ目、ヒメハナバチ科) 旧北区
mining bee		*Andrena vicina* Smith	(ハチ目、ヒメハナバチ科) 新北区
mining bee		*Andrena carlini* (Cockerell)	(ハチ目、ヒメハナバチ科) 新北区
mining bees			andrenid bees を見よ
mining bees			sweat bees (1) を見よ
mining bees			digger bees (1) を見よ
mining scale	カワムグリカイガラムシ	*Howardia biclavis* (Comstock)	(カメムシ目、マルカイガラムシ科) 日本 (小笠原)、全北区、豪州区。カンキツ、ツバキなどの害虫
Minnion's groundling		*Pulicalvaria piceaella* Kearfott	(チョウ目、キバガ科) 全北区
minor angle moth		*Macaria minorata* Packard	(チョウ目、シャクガ科) 新北区
minor ground mantid			ground mantis (1) を見よ
minor pine weevil		*Pissodes notatus* Fabricius	(コウチュウ目、ゾウムシ科) 旧北区
minor pith borer			pine shoot beetle (1) を見よ
minor Sakhalin fir webworm	マツチビヒメハマキ	*Coenobioides abietiella* (Matsumura)	(チョウ目、ハマキガ科) 日本、旧北区
minor shoulder-knot	ヌカビラネジロキリガ	*Brachylomia viminalis* (Fabricius)	(チョウ目、ヤガ科) 日本、旧北区
Minos skipper		*Zenis minos* (Latreille)	(チョウ目、セセリチョウ科) 新熱帯区
Minotaur			Minotaur beetle を見よ　ギリシャ伝説ミノタウルスに由来
Minotaur beetle		*Typhaeus typhoeus* (Linnaeus)	(コウチュウ目、センチコガネ科) 旧北区
mint aphid	ハッカイボアブラムシ	*Ovatus crataegarius* (Walker)	(カメムシ目、アブラムシ科) 日本、汎世界

英名	和名	学名	所属、分布、ほか
mint aphid		*Ovatus menthae* Walker	(カメムシ目、アブラムシ科) 旧北区
mint bentwing		*Opostega crepusculella* (Zeller)	(チョウ目、ヒラタモグリガ科) 旧北区
mint flea beetle		*Longitarsus menthaphagus* (Gentner)	(コウチュウ目、ハムシ科) 新北区。野菜害虫
mint flea beetle		*Longitarsus lycopi* (Foudras)	(コウチュウ目、ハムシ科) 旧北区
mint flea beetle		*Longitarsus waterhousei* Kutschera	(コウチュウ目、ハムシ科) 旧北区
mint leaf beetle		*Chrysolina menthastri* (Suffrian)	(コウチュウ目、ハムシ科) 旧北区
mint leaf miner		*Phytomyza petoei* Hering	(ハエ目、ハモグリバエ科) 旧北区
mint looper			gray looper moth を見よ
mint-loving Pyrausta moth		*Pyrausta acrionalis* (Walker)	(チョウ目、メイガ科) 新北区
mint plume moth		*Anstenoptilia marmarodactyla* (Dyar)	(チョウ目、トリバガ科) 新北区
mint rhizome worm	ハッカネムシガ	*Endothenia menthivora* (Oku)	(チョウ目、ハマキガ科) 日本、旧北区。ハッカ害虫
mint root borer		*Fumibotys fumalis* Guenée	(チョウ目、メイガ科) 新北区。成虫は mint root borer moth
Mintha widow		*Torynesis mintha* (Geyer)	(チョウ目、タテハチョウ科) エチオピア区
minute backswimmers			pygmy backswimmer bugs を見よ
minute bamboo scale	タケヒトフサカイガラムシ	*Asterolecanium minutum* Takahashi	(カメムシ目、フサカイガラムシ科) 日本、東洋区
minute bark beetles	カクホソカタムシ科	Cerylonidae	(コウチュウ目) の昆虫の総称
minute beetles			fringe-winged beetles を見よ
minute black scavenger flies	ニセケバエ科	Scatopsidae	(ハエ目) の昆虫の総称
minute black scavengers			minute bark scavenger flies を見よ
minute bladder bugs	フタガタカメムシ科	Microphysidae	(カメムシ目) の昆虫の総称　体長 1〜1.5 mm
minute bog beetles	マルケシムシ科	Sphaeriidae	(コウチュウ目) の昆虫の総称　1 mm 以下の微小種
minute bog beetles			false clown beetles を見よ
minute brown scavenger			plaster beetles を見よ
minute brown scavenger beetles			plaster beetles を見よ
minute bugs			minute bladder bugs を見よ
minute burdock weevil			burdock weevil を見よ
minute citrus leafhopper			smaller citrus leafhopper を見よ
minute cypress scale		*Carulaspis minima* (Targioni-Tozzetti)	(カメムシ目、マルカイガラムシ科) 新北区

英名	和名	学名	所属、分布、ほか
minute egg parasite		*Trichogramma minutum* Riley	(ハチ目、タマゴコバチ科) 新北区
minute egg parasites	タマゴコバチ科	Trichogrammatidae	(ハチ目) の昆虫の総称　世界に450種
minute fungus beetles	ミジンムシ科	Corylophidae	(コウチュウ目) の昆虫の総称
minute fungus beetles (1)		Orthoperidae	(コウチュウ目) の昆虫の総称
minute garden cricket	チビスズ	*Nemobius chibae* Shiraki	(バッタ目、コオロギ科) 日本
minute green weevil	チビアオゾウムシ	*Hyperstylus pallipes* Roelofs	(コウチュウ目、ゾウムシ科) 日本
minute ground beetles		Bembidiini	(コウチュウ目、オサムシ科) の昆虫の総称
minute house ant			vagrant ant を見よ
minute jumping bug			white-marked fleahopper を見よ
minute lady beetles		Microweisini	(コウチュウ目、テントウムシ科) の昆虫の総称
minute ladybird beetles			minute lady beetles を見よ
minute leaf miner moths			midget moths を見よ
minute leaf-roller weevil	ヒメシギゾウムシ	*Curculio hime* (Kono)	(コウチュウ目、ゾウムシ科) 日本
minute marsh-loving beetles	チビドロムシ科	Limnichidae	(コウチュウ目) の昆虫の総称
minute moss beetles	ダルマガムシ科	Hydraenidae	(コウチュウ目) の昆虫の総称　微小種
minute mould beetles		*Cortinicara*	(コウチュウ目、ヒメマキムシ科) の昆虫の総称
minute mould beetles			plaster beetles を見よ
minute mud-loving beetles	マルドロムシ科	Georyssidae	(コウチュウ目) の昆虫の総称
minute oval Abrostola moth		*Abrostola microvalis* Ottolengui	(チョウ目、ヤガ科) 新北区
minute pine bark beetle			pine shoot beetle (1) を見よ
minute pine weevil	チビマツアナアキゾウムシ	*Hylobitelus pinastri* (Gyllenhal)	(コウチュウ目、ゾウムシ科) 日本、旧北区
minute pine-bark beetle			pine bark beetle (1) を見よ
minute pirate bug		*Orius tristicolor* (White)	(カメムシ目、ハナカメムシ科) 新北区
minute pirate bug		*Coccivora californica* McAtee et Malloch	(カメムシ目、ハナカメムシ科) 新北区
minute pirate bug			mulberry flower bug を見よ
minute pirate bugs		*Anthocoris*	(カメムシ目、ハナカメムシ科) の昆虫の総称
minute pirate bugs			flower bugs を見よ　米国での英名
minute pitch-blister moth		*Retina houseri* (Miller)	(チョウ目、ハマキガ科) 新北区

英名	和名	学名	所属、分布、ほか
minute pubescent skin beetle	チビケカツオブシムシ	*Trinodes rufescens* Reitter	（コウチュウ目、カツオブシムシ科）日本
minute riffle bugs		*Microvelia*	（カメムシ目、カタビロアメンボ科）の昆虫の総称
minute scavenger beetle		*Cortinicara histalis* (Brown)	（コウチュウ目、ヒメマキムシ科）豪州区
minute scavenger beetles	ニセマキムシ科	Dasyceridae	（コウチュウ目）の昆虫の総称
minute seed weevils	サルゾウムシ亜科	Ceutorhynchinae	（コウチュウ目、ゾウムシ科）の昆虫の総称
minute shieldback		*Clinopleura minuta* Caudell	（バッタ目、キリギリス科）新北区
minute tree-fungus beetles	ツツキノコムシ科	Ciidae	（コウチュウ目）の昆虫の総称
minute two-spotted ladybird		*Diomus notescens* (Blackburn)	（コウチュウ目、テントウムシ科）豪州区
Miranda moth		*Proxenus miranda* (Grote)	（チョウ目、ヤガ科）新北区
Miranda underwing moth		*Catocala miranda* H. Edwards	（チョウ目、ヤガ科）新北区。Miranda underwing ともいう
mire pill beetle		*Curimopsis nigrita* (Palm)	（コウチュウ目、マルトゲムシ科）旧北区
mirid bug		*Miris striatus* (Linnaeus)	（カメムシ目、カスミカメムシ科）旧北区
mirid bugs			plant bugs (1) を見よ
mirids			plant bugs (1) を見よ
mirror turtle ant		*Cephalotes specularis* Brandao, Feitosa, Powell et Del-Claro	（ハチ目、アリ科）新熱帯区
miscanthus false mealybug	カヤノコナカイガラムシ	*Neoripersia japonica* (Kuwana)	（カメムシ目、コナカイガラムシ科）日本
miscanthus scurfy scale	ススキナガカキカイガラムシ	*Acanthomytilus miscanthi* Takahashi	（カメムシ目、マルカイガラムシ科）日本
mischievous bird grasshopper		*Schistocerca damnifica* (Saussure)	（バッタ目、バッタ科）新北区
miserabilis hairstreak		*Callophrys miserabilis* (Clench)	（チョウ目、シジミチョウ科）新熱帯区
Miskin's jewel		*Hypochrysops miskini* (Waterhouse)	（チョウ目、シジミチョウ科）豪州区
Miskin's swift		*Sabera dobboe* (Plötz)	（チョウ目、セセリチョウ科）豪州区
mission blue		*Aricia icarioides missionensis* (Hovanitz)	（チョウ目、シジミチョウ科）新北区
mistletoe borer moth		*Carmenta phoradendri* Engelhardt	（チョウ目、スカシバガ科）新北区。幼虫は mistletoe borer
mistletoe browntail moth		*Euproctis edwardsii* Newman	（チョウ目、ドクガ科）豪州区
mistletoe flower bug		*Anthocoris visci* Douglas	（カメムシ目、ハナカメムシ科）旧北区
mistletoe hairstreak			Johnson's hairstreak を見よ
mistletoe marble		*Celypha woodiana* (Barrett)	（チョウ目、ハマキガ科）旧北区
mistletoe sucker		*Psylla visci* Curtis	（カメムシ目、キジラミ科）旧北区

英名	和名	学名	所属、分布、ほか
Mitchell's marsh satyr		*Neonympha mitchellii* French	（チョウ目、タテハチョウ科）新北区。Mitchell's satyr ともいう
Mitchell's shieldback		*Pediodectes mitchelli* (Caudell)	（バッタ目、キリギリス科）新北区
mite beetles			minute bog beetles を見よ
mite eating ladybirds		*Stethorus*	（コウチュウ目、テントウムシ科）の昆虫の総称　豪州での英名
mite sandman		*Spialia paula* (Higgins)	（チョウ目、セセリチョウ科）エチオピア区
Mithrax duskywing			slaty skipper を見よ
mitred cricket		*Gryllus mitratus* Burmeister	（バッタ目、コオロギ科）旧北区。中国で鳴声鑑賞のため飼育
Mitsukuri sawfly			alder sawfly (1) を見よ
Mitterbach's red roller		*Ancylis mitterbachariana* (Denis et Schiffermüller)	（チョウ目、ハマキガ科）旧北区
mixed punch	メスシロシジミタテハ	*Dodona ouida* (Hewitson)	（チョウ目、シジミタテハ科）東洋区
Miyake lasiocampid	ミヤケカレハ	*Takanea miyakei* (Wileman)	（チョウ目、カレハガ科）日本
miyama stag beetle	ミヤマクワガタ	*Lucanus maculifemoratus* Motschulsky	（コウチュウ目、クワガタムシ科）日本
Miyazaki bark beetle	ミヤザキキクイムシ	*Xyleborus miyazakiensis* Murayama	（コウチュウ目、キクイムシ科）日本
Mnestra's ringlet	ムモンベニヒカゲ	*Erebia mnestra* (Hübner)	（チョウ目、タテハチョウ科）旧北区
mocha		*Cyclophora annulata* (Schulze)	（チョウ目、シャクガ科）旧北区
mocker bronze			eastern bush blue を見よ
mocker swallowtail	オスジロアゲハ	*Papilio dardanus* Brown	（チョウ目、アゲハチョウ科）エチオピア区。*P. d. cenea* (Stoll) も同英名
Mocquery's leaf sitter		*Gorgyra mocquerysii* Holland	（チョウ目、セセリチョウ科）エチオピア区
modest false dot		*Liptena modesta* (Kirby)	（チョウ目、シジミチョウ科）エチオピア区
modest Furcula moth		*Furcula modesta* (Hudson)	（チョウ目、シャチホコガ科）新北区
modest katydid		*Montezumina modesta* (Brunner)	（バッタ目、キリギリス科）新北区
modest Phtheochroa moth		*Phtheochroa modestana* (Busck)	（チョウ目、ハマキガ科）新北区
modest sphinx			poplar hawkmoth (2) を見よ
modest sylph		*Astictopterus inornatus* (Trimen)	（チョウ目、セセリチョウ科）エチオピア区
Moeller's silverfork		*Lethe moelleri* (Elwes)	（チョウ目、タテハチョウ科）東洋区
Mohave sootywing			Mojave sootywing を見よ
Mojave blue		*Euphilotes enoptes mojave* (Watson et Comstock)	（チョウ目、シジミチョウ科）新北区。
Mojave giant skipper		*Agathymus alliae* (Stallings et Turner)	（チョウ目、セセリチョウ科）新北区
Mojave sootywing		*Hesperopsis libya* (Scudder)	（チョウ目、セセリチョウ科）新北区
molannids			hood casemake caddisflies を見よ

英名	和名	学名	所属、分布、ほか
mold beetles			silken fungus beetles を見よ
mold lice			common barklice を見よ
mold-louse			booklouse (2) を見よ
mole cricket		*Gryllotalpa brachyptera* Tindale	（バッタ目、ケラ科）豪州区
mole cricket (1)	ケラ	*Gryllotalpa orientalis* Burmeister	（バッタ目、ケラ科）日本、東洋区、豪州区、大洋区、エチオピア区
mole cricket (2)		*Triamescaptor aotea* Tindale	（バッタ目、ケラ科）豪州区
mole cricket (3)		*Gryllotalpa gryllotalpa* (Linnaeus)	（バッタ目、ケラ科）旧北区
mole cricket wasp		*Larra luzonensis* Rohwer	（ハチ目、アナバチ科）大洋区
mole crickets (1)	ケラ科	Gryllotalpidae	（バッタ目）の昆虫の総称　世界に 2,300 種
mole crickets		Gryllotalpinae	（バッタ目、ケラ科）の昆虫の総称
mole crickets			two-clawed mole crickets を見よ
mole crickets			long-horned grasshoppers (3) を見よ
mole flea			giant flea を見よ
Molochina Eurybia			Molochina underleaf を見よ
Molochina underleaf		*Eurybia molochina* Stichel	（チョウ目、シジミタテハ科）新熱帯区
Molokai damselfly		*Megalagrion molokaiense* (Perkins)	（トンボ目、イトトンボ科）大洋区。ハワイの Molokai 島に由来
Molomo copper		*Aloeides molomo* (Trimen)	（チョウ目、シジミチョウ科）エチオピア区
momi sawfly	ハイマツハバチ	*Gilpinia abieticola* (Dalla Torre)	（ハチ目、マツハバチ科）日本、旧北区
mompha moth		*Mompha murtfeldtella* (Chambers)	（チョウ目、カザリバガ科）新北区
mompha moths			momphid moths を見よ
momphid moths		Momphidae	（チョウ目）の昆虫の総称
monacantha cochineal			mountain cochineal を見よ
Monachella satyr		*Steremnia monachella* Thieme	（チョウ目、タテハチョウ科）新熱帯区
monarch butterfly	オオカバマダラ	*Danaus plexippus plexippus* (Linnaeus)	（チョウ目、タテハチョウ科）日本、東洋区、大洋区、新北区。英国では monarch という。北米の個体群は秋に南部、メキシコに移動、集団で越冬
monarchs	マダラチョウ亜科	Danainae	（チョウ目、タテハチョウ科）の昆虫の総称 monarch butterflies ともいう
monarchs			tigers (1) を見よ
Monard's dropwing		*Trithemis monardi* Ris	（トンボ目、トンボ科）エチオピア区
Moneta longwing			Mexican silver-spot を見よ
Mongol		*Araschnia prorsoides* (Blanchard)	（チョウ目、タテハチョウ科）東洋区
Mongolian hornet			yellow hornet (1) を見よ
Mongolian map butterfly			Mongol を見よ
Mongolian oak-feeding silkworm			Chinese oak silkmoth を見よ

英名	和名	学名	所属、分布、ほか
Monica hairstreak		*Theritas monica* (Hewitson)	（チョウ目、シジミチョウ科）新熱帯区
monk		*Amauris tartarea* Mabille	（チョウ目、タテハチョウ科）エチオピア区
monk			monk skipper を見よ
monk moth			monk skipper を見よ
monk skipper		*Asholis capucinus* (Lucas)	（チョウ目、セセリチョウ科）新北区、新熱帯区
monkey blue		*Lepidochrysops methymna* (Trimen)	（チョウ目、シジミチョウ科）エチオピア区
monkey grasshoppers (1)		Eumastacidae	（バッタ目）の昆虫の総称
monkey grasshoppers (2)		Tanaoceridae	（バッタ目）の昆虫の総称　無翅、夜行性。北米に数種
monkey puzzle	インドミツオシジミ	*Rathinda amor* (Fabricius)	（チョウ目、シジミチョウ科）東洋区
monkey swordtail	セレベスオナガタイマイ（レサスオナガタイマイ）	*Graphium rhesus* (Boisduval)	（チョウ目、アゲハチョウ科）東洋区
monkeypod-kiwae caterpillar			indomitable Melipotis moth を見よ
monkeypod moth		*Polydesma umbricola* Boisduval	（チョウ目、ヤガ科）新北区
monkeypod psyllid		*Heteropsylla huasachae* Caldwell	（カメムシ目、キジラミ科）大洋区
monkeypod round-headed borer			two-lined albizzia longhorn を見よ
monkeys			giant lappet moths を見よ
monkshood aphid		*Delphinobium junackianum* (Karsch)	（カメムシ目、アブラムシ科）旧北区
monkshood dropwing		*Trithemis aconita* Lieftinck	（トンボ目、トンボ科）エチオピア区
Mono Lake Hygrotus diving beetle		*Hygrotus artus* Fall	（コウチュウ目、ゲンゴロウ科）新北区
monoedid beetles		Monoedidae	（コウチュウ目）の昆虫の総称
monolepta beetle			red-shouldered leaf beetle (1) を見よ
monommid beetles	タマムシモドキ科	Monommidae	（コウチュウ目）の昆虫の総称
monommids			monommid beetles を見よ
monopis moths		*Monopis*	（チョウ目、ヒロズコガ科）の昆虫の総称
Montagna mountain satyr		*Pedaliodes montagna* Adams et Bernard	（チョウ目、タテハチョウ科）新熱帯区
Montana six-plume moth			six-plume moth を見よ
montane crescent		*Anthanassa sitacles sitacles* (Godman et Salvin)	（チョウ目、タテハチョウ科）新熱帯区。*A. s. cortes* (Hall) も同英名
montane Euselasia		*Euselasia corduena* (Hewitson)	（チョウ目、シジミタテハ科）新熱帯区
montane grass-skipper			mountain skipper を見よ
montane heath-blue			mountain blue (1) を見よ
montane longtail		*Urbanus prodicus* Bell	（チョウ目、セセリチョウ科）新熱帯区

英名	和名	学名	所属、分布、ほか
montane longwing		*Heliconius clysonymus* Latreille	(チョウ目、タテハチョウ科) 新熱帯区
montane ochre			Phigalioides skipper を見よ
montane phantom		*Pseudohaetera hypaesia* (Hewitson)	(チョウ目、タテハチョウ科) 新熱帯区
montane sister			Donysa sister を見よ
montane skipper			Nevada skipper を見よ
montane tree nymph		*Sevenia garega* (Karsch)	(チョウ目、タテハチョウ科) エチオピア区
Monterey pine cone beetle			Monterey pine weevil を見よ
Monterey-pine resin midge		*Cecidomyia resinicoloides* Williams	(ハエ目、タマバエ科) 新北区
Monterey pine weevil		*Pissodes radiatae* Hopkins	(コウチュウ目、ゾウムシ科) 新北区
Montezuma's Calephelis		*Calephelis montezuma* McAlpine	(チョウ目、シジミタテハ科) 新熱帯区
Montezuma's cattleheart		*Parides montezuma* (Westwood)	(チョウ目、アゲハチョウ科) 新北区、新熱帯区
monticola blue		*Plebejus lupini monticola* (Clemence)	(チョウ目、シジミチョウ科) 新熱帯区
moody skipper		*Thoon modius* (Mabille)	(チョウ目、セセリチョウ科) 新熱帯区
Mooi River opal			Natal copper を見よ
moon moth		*Actias selene ningpoana* Felder et Felder	(チョウ目、ヤママユガ科) 東洋区。Indian moon moth も参照
moon moth (1)		*Argema mittrei* (Guérin-Méneville)	(チョウ目、ヤママユガ科) エチオピア区
moon moth			lunar moth (1)(2) を見よ
moon moths		Cercphanidae	(チョウ目) の昆虫の総称
moon satyr		*Pierella luna* Fabricius	(チョウ目、タテハチョウ科) 新北区、新熱帯区。*P. l. rubecula* Salvin et Godman も同英名
moon umber		*Zale lunata* (Drury)	(チョウ目、ヤガ科) 新北区、大洋区
moonlight jewel			blue jewel を見よ
moon-lined moth		*Spiloloma lunilinea* Grote	(チョウ目、ヤガ科) 新北区
moon-marked skipper		*Atrytonopsis lunus* Edwards	(チョウ目、セセリチョウ科) 新北区、新熱帯区
Moon's horse			rose chafer (1) を見よ
moonseed moth		*Plusiodonta compressipalpis* Guenée	(チョウ目、ヤガ科) 新北区
Moore's ace		*Halpe porus* (Mabille)	(チョウ目、セセリチョウ科) 東洋区
Moore's bush brown		*Mycalesis heri* Moore	(チョウ目、タテハチョウ科) 東洋区
Moore's cupid	ゴイシツバメシジミ	*Shijimia moorei* (Leech)	(チョウ目、シジミチョウ科) 旧北区、東洋区
Moore's freak			freak を見よ
Moorhen flea		*Dasypsyllus gallinulae* (Dale)	(ノミ目、ナガノミ科) 日本、旧北区、東洋区
Moorland clouded yellow		*Colias palaeno* (Linnaeus)	(チョウ目、シロチョウ科) 全北区。日本亜種は *C. p. aias* Fruhstorfer ミヤマモンキチョウ
Moorland emerald			northern emerald を見よ

英名	和名	学名	所属、分布、ほか
Moorland hawker			common hawker を見よ
Moorland Olethreutes moth	タカネナガバヒメハマキ	*Olethreutes schulziana* (Fabricius)	（チョウ目、ハマキガ科）日本、全北区
mopane worm		*Gonimbrasia belina* (Linnaeus)	（チョウ目、ヤママユガ科）エチオピア区
Mopsalis Diacme moth		*Diacme mopsalis* (Walker)	（チョウ目、メイガ科）新北区
Morant's blue		*Lepidochrysops hypopolia* (Trimen et Bowker)	（チョウ目、シジミチョウ科）エチオピア区
Morant's orange			Morant's skipper を見よ
Morant's skipper		*Parosmodes morantii* (Trimen)	（チョウ目、セセリチョウ科）エチオピア区
Morant's tree nymph		*Sevenia morantii* (Trimen)	（チョウ目、タテハチョウ科）エチオピア区
morbid owlet moth		*Chytolita morbidalis* (Guenée)	（チョウ目、ヤガ科）新北区。morbid owlet ともいう
mordellids			tumbling flower beetles (1) を見よ
Morelos Calephelis		*Calephelis yautepequensis* Maza et Turrent	（チョウ目、シジミタテハ科）新熱帯区
Morelos skipper		*Paratrytone decepta* Miller et Miller	（チョウ目、セセリチョウ科）新熱帯区
Moreton Bay fig psyllid		*Mecopsylla fici* (Tryon)	（カメムシ目、Homotomidae）豪州区。豪州東部の湾名に由来
Moreton Bay fig wasp		*Pleistodontes froggatti* Mayr	（ハチ目、イチジクコバチ科）豪州区
Morgan's sphinx		*Xanthopan morgani* (Walker)	（チョウ目、スズメガ科）旧北区
moringa hairy caterpillar		*Eupterote mollifera* Walker	（チョウ目、オビガ科）東洋区。moringa は植物名
Morishita's tiger	ヒポワッタンアサギマダラ	*Parantica hypowattan* Morishita	（チョウ目、タテハチョウ科）東洋区
Mormon creeping water bug		*Ambrysus mormon* Montandon	（カメムシ目、コバンムシ科）新北区
Mormon cricket		*Anabrus simplex* Haldeman	（バッタ目、キリギリス科）新北区
Mormon dark metalmark		*Apodemia mormo mormo* (C. et R. Felder)	（チョウ目、シジミタテハ科）新北区
Mormon fritillary	モルモンギンボシヒョウモン	*Speyeria mormonia* (Boisduval)	（チョウ目、タテハチョウ科）新北区
Mormon ground cricket		*Neonemobius mormonius* (Scudder)	（バッタ目、コオロギ科）新北区
Mormon light metalmark		*Apodemia mormo mejicanus* (Behr)	（チョウ目、シジミタテハ科）新北区
Mormon metalmark	カリフォルニアシジミタテハ	*Apodemia mormo langei* Comstock	（チョウ目、シジミタテハ科）新北区
morning-glory leaf miner	ヒルガオハモグリガ	*Bedellia somnulentella* (Zeller)	（チョウ目、ヒルガオハモグリガ科）日本、全北区、東洋区、豪州区、エチオピア区。サツマイモ、アサガオの害虫。morning-glory leafminer moth ともいう
morning glory leafminer			peach leafminer を見よ

英名	和名	学名	所属、分布、ほか
morning glory Pellicia			morning-glory skipper を見よ
morning glory plume moth		*Oidaematophorus monodactylus* (Linnaeus)	（チョウ目、トリバガ科）新北区
morning-glory prominent			oak-and-maple humpted caterpillar を見よ morning-glory prominent moth ともいう
morning-glory skipper		*Pellicia dimidiata* Herrich-Schäffer	（チョウ目、セセリチョウ科）新北区、新熱帯区
morning-glory sphinx			sweetpotato hawk-moth を見よ 米国での英名
morning glory sphinx moth			sweetpotato hawk-moth を見よ 米国での英名
morning glory tufted-skipper			morning-glory skipper を見よ
Moroccan copper		*Lycaena phoebus* (Blachier)	（チョウ目、シジミチョウ科）旧北区
Moroccan grayling		*Pseudochazara atlantis* (Austaut)	（チョウ目、タテハチョウ科）旧北区
Moroccan hairstreak		*Tomares mauretanicus* (Lucas)	（チョウ目、シジミチョウ科）旧北区
Moroccan locust		*Dociostaurus moroccanus* Thunberg	（バッタ目、バッタ科）旧北区、エチオピア区
Moroccan meadow brown		*Hyponephele maroccana* (Blachier)	（チョウ目、タテハチョウ科）旧北区
Moroccan migratory locust			Moroccan locust を見よ
Moroccan orange-tip		*Anthocharis euphenoides* Staudinger	（チョウ目、シロチョウ科）旧北区
Moroccan pearly heath		*Coenonympha arcanioides* (Pierret)	（チョウ目、タテハチョウ科）旧北区
Moroccan small skipper		*Thymelicus hamza* (Oberthür)	（チョウ目、セセリチョウ科）旧北区
Morocco orange-tip	ベリアクモマツマキチョウ	*Anthocharis belia* (Linnaeus)	（チョウ目、シロチョウ科）旧北区。Moroccan orange-tip ともいう
Morona sootywing		*Bolla morona* (Bell)	（チョウ目、セセリチョウ科）新熱帯区
Morosa sailer		*Neptis morosa* Overlaet	（チョウ目、タテハチョウ科）エチオピア区
Morpho	デイデイウスモルフォ	*Morpho didius* Hopffer	（チョウ目、タテハチョウ科）新熱帯区
Morpho		*Morpho glanadendis* Felder	（チョウ目、タテハチョウ科）新熱帯区
Morpho			Brazilian Morpho を見よ
Morpho			Cypris Morpho を見よ
Morpho			Deidamia Morpho を見よ
Morphos (1)		Morphinae	（チョウ目、タテハチョウ科）の昆虫の総称
Morphos	モルフォチョウ属	*Morpho*	（チョウ目、タテハチョウ科）の昆虫の総称
Morrill lace bug		*Corythucha morrilli* Osborn et Drake	（カメムシ目、グンバイムシ科）新北区
Morrison's goatweed		*Anaea aidea morrisonii* Holland	（チョウ目、タテハチョウ科）新北区

英名	和名	学名	所属、分布、ほか
Morrison's parlatoria		*Parlatoria morrisoni* McKenzie	（カメムシ目、マルカイガラムシ科）旧北区、東洋区。カンキツ害虫
Morrison's Pero moth		*Pero morrisonaria* (H. Edwards)	（チョウ目、シャクガ科）新北区
Morrison's silver spike		*Stinga morrisoni* (Edwards)	（チョウ目、セセリチョウ科）新北区、新熱帯区
Morrison's skipper			Morrison's silver spike を見よ
Morrison's sooty dart moth		*Pseudohermonassa tenuicula* (Morrison)	（チョウ目、ヤガ科）新北区
Morris's wainscot		*Photedes morrisii* (Dale)	（チョウ目、ヤガ科）旧北区
Morse's shieldback		*Aglaothorax morsei* (Caudell)	（バッタ目、キリギリス科）新北区
mortarjoint casemaker caddisflies	フトヒゲトビケラ科	Odontoceridae	（トビケラ目）の昆虫の総称
Moruus leafwing		*Memphis moruus* Fabricius	（チョウ目、タテハチョウ科）新熱帯区
mosaic			zebra (2) を見よ
mosaic Sparganothis moth		*Sparganothis xanthoides* (Walker)	（チョウ目、ハマキガ科）新北区
mosaic sylph			forest malachite を見よ
Moschler's bent-skipper		*Cycloglypha enega* (Möschler)	（チョウ目、セセリチョウ科）新熱帯区
Moschler's Liptena		*Liptena simplicia* Möschler	（チョウ目、シジミチョウ科）エチオピア区
Moschler's ruby-eye		*Moeros moeros* (Möschler)	（チョウ目、セセリチョウ科）新熱帯区
Moschler's tufted skipper			purplish-black skipper を見よ
Moselle fly			wheat midge を見よ　フランスの地名に由来
moso-bamboo eurytomid			bamboo gall chacid を見よ
mosquito			common mosquito を見よ
mosquito bees			stingless bees (1) を見よ
mosquito destroyer		*Eucorethra underwoodi* (Underwood)	（ハエ目、ケヨソイカ科）新北区
mosquito hawk			dragonflies (1) を見よ
mosquito hawks			dragonflies (1) を見よ
mosquitoes		*Psorophora*	（ハエ目、カ科）の昆虫の総称
mosquitoes (1)	カ科	Culicidae	（ハエ目）の昆虫の総称。mosquito の複数形は mosquitoes, mosquitos。世界に 3,000 種
mosquitoes			house mosquitoes を見よ
moss bugs		Peloridiidae	（カメムシ目）の昆虫の総称
moss carder bee			large carder bee を見よ
moss elfin		*Incisalia mossii* (H. Edwards)	（チョウ目、シジミチョウ科）新北区
moss fly		*Sciara pectoralis* Staeger	（ハエ目、クロバネキノコバエ科）旧北区
moss-green lappet		*Trabala viridana* Joicey et Talbot	（チョウ目、カレハガ科）東洋区
moss peacock	オビクジャクアゲハ	*Papilio palinurus* Fabricius	（チョウ目、アゲハチョウ科）東洋区

英名	和名	学名	所属、分布、ほか
Moss's elfin		*Callophrys mossii* (E. Edwards)	（チョウ目、シジミチョウ科）新北区
mossy Pseudobryomima moth		*Pseudobryomima muscosa* (Hampson)	（チョウ目、ヤガ科）新北区
mossy rose gall wasp		*Diplolepis rosae* (Linnaeus)	（ハチ目、タマバチ科）全北区。虫えいをmossy rose gall、bedeguar gall, bedeguar といい、緑から赤色になる
moth			蛾。イガの俗称
moth			cream-bordered green pea を見よ
moth borer			citrus fruit borer を見よ
moth bugs			flatid planthoppers を見よ
moth butterfly	アリノスシジミ	*Liphyra brassolis* Westwood	（チョウ目、シジミチョウ科）豪州区。*L. b. major* Rothschild、*L. b. abbreviata* Strand も同英名
moth flies	チョウバエ科	Psychodidae	（ハエ目）の昆虫の総称　世界に500種以上
moth fly			trickling filter fly を見よ
moth-fly			owl-midge を見よ
moth-fly			sprinkler sewage filter fly を見よ
moth lacewing		*Oliarces clara* Banks	（アミメカゲロウ目、Ithonidae）新北区
moth lacewings	オオアミメカゲロウ科	Ithonidae	（アミメカゲロウ目）の昆虫の総称
moth midges			moth flies を見よ
moth parasite			spiny tachina fly を見よ
Mother Shipton		*Euclidia mi* (Clerck)	（チョウ目、ヤガ科）旧北区
mother underwing moth		*Catocala parta* Guenée	（チョウ目、ヤガ科）新北区。mother underwing ともいう
mothering bug			parent bug を見よ
mother-of-pearl		*Salamis parhassus aethiops* de Beauvais	（チョウ目、タテハチョウ科）エチオピア区
mother-of-pearl (1)		*Haritala ruralis* Scopoli	（チョウ目、メイガ科）旧北区
mother-of-pearl			bean webworm を見よ
mother-of- pearl blue		*Polyommatus nivescens* Keferstein	（チョウ目、シジミチョウ科）旧北区
mother-of-pearl butterfly			forest mother-of-pearl (1) を見よ
mother-of-pearl Morpho	ウスアオモルフォ	*Morpho laertes* (Hübner)	（チョウ目、タテハチョウ科）新熱帯区
mother-of-pearl moth		*Argyria rufisignella* Zeller	（チョウ目、メイガ科）新北区
mother-of-pearl moth			mother-of-pearl (1) を見よ
mother-of-pearl Scoparia moth		*Scoparia diphtheralis* Walker	（チョウ目、メイガ科）豪州区
mother-of-pearls	シンジュタテハ属	*Salamis*	（チョウ目、タテハチョウ科）の昆虫の総称
mothlike lacewings			false mantids を見よ
mothlike netwings			moth lacewings を見よ
moths	蛾類	Heterocera	（チョウ目）の昆虫の総称

英名	和名	学名	所属、分布、ほか
mottled Argus		*Loxerebia narasingha* (Moore)	（チョウ目、タテハチョウ科）東洋区
mottled arum aphid	シクラメンコブアブラムシ	*Aulacorthum circumflexum* (Buckton)	（カメムシ目、アブラムシ科）日本、汎世界。シクラメン、セロリーの害虫
mottled beauty		*Alcis repandata* (Linnaeus)	（チョウ目、シャクガ科）旧北区
mottled beauty moth			mottled beauty を見よ
mottled bee-fly		*Thyridanthrax fenestratus* (Fallén)	（ハエ目、ツリアブ科）旧北区
mottled Bolla		*Bolla clytius* (Godman et Salvin)	（チョウ目、セセリチョウ科）新北区、新熱帯区
mottled Bomolocha moth		*Hypena palparia* Walker	（チョウ目、ヤガ科）新北区。mottled Bomolocha ともいう
mottled bran marble	イグサヒメハマキ	*Bactra furfurana* (Haworth)	（チョウ目、ハマキガ科）日本、旧北区。イグサ害虫
mottled brown cutworm moth		*Melanchra stipata* Walker	（チョウ目、ヤガ科）豪州区
mottled button		*Acleris maccana* Fristrup	（チョウ目、ハマキガ科）旧北区
mottled clover moth		*Heliothis dipsacea* (Linnaeus)	（チョウ目、ヤガ科）旧北区
mottled coast shade		*Eana penziana* (Thunberg)	（チョウ目、ハマキガ科）旧北区
mottled corn clothes moth			European grain moth を見よ
mottled cup moth		*Doratifera vulnerans* (Lewin)	（チョウ目、イラガ科）豪州区
mottled Cupes		*Cupes concolor* Westwood	（コウチュウ目、ナガヒラタムシ科）新北区
mottled darner		*Aeschna clepsydra* Say	（トンボ目、ヤンマ科）新北区
mottled duskywing		*Erynnis martialis* (Scudder)	（チョウ目、セセリチョウ科）新北区
mottled emigrant			mottled migrant を見よ
mottled Euchlaena moth		*Euchlaena tigrinaria* (Guenée)	（チョウ目、シャクガ科）新北区
mottled flower scarab			mango flower beetle を見よ　豪州での英名
mottled forest carpet moth		*Chloroclystis lichenodes* Purdie	（チョウ目、シャクガ科）豪州区
mottled grain moth			European grain moth を見よ
mottled grasshopper		*Myrmeleotettix maculatus* (Thunberg)	（バッタ目、バッタ科）旧北区
mottled grass-skipper			cynone skipper を見よ
mottled grass-veneer moth		*Neodactria luteolellus* (Clemens)	（チョウ目、メイガ科）新北区。mottled grass-veneer ともいう
mottled gray carpet moth		*Cladara limitaria* (Walker)	（チョウ目、シャクガ科）新北区
mottled gray cutworm			greater red dart moth を見よ
mottled green			Baltimore butterfly を見よ
mottled grey		*Scoparia basistrigalis* Knaggs	（チョウ目、メイガ科）旧北区

英名	和名	学名	所属、分布、ほか
mottled grey		*Colostygia multistrigaria* (Haworth)	(チョウ目、シャクガ科) 旧北区
mottled grey cutworm			greater red dart moth を見よ
mottled hairstreak		*Arawacus dolylas* (Cramer)	(チョウ目、シジミチョウ科) 新熱帯区
mottled katydid		*Ephippityha trigintiduoguttata* (Serville)	(バッタ目、キリギリス科) 豪州区
mottled leafwing		*Memphis arginussa eubaena* (Boisduval)	(チョウ目、タテハチョウ科) 新熱帯区
mottled leafwing		*Polygrapha cyanea* (Godman et Salvin)	(チョウ目、タテハチョウ科) 新熱帯区
mottled longtail		*Typhedanus undulatus* (Hewitson)	(チョウ目、セセリチョウ科) 新北区、新熱帯区
mottled migrant	ウラナミシロチョウ	*Catopsilia pyranthe* (Linnaeus)	(チョウ目、シロチョウ科) 日本、東洋区、豪州区。common migrant (2) を参照
mottled Prepona		*Prepona pylene philetas* Fruhstorfer	(チョウ目、タテハチョウ科) 新熱帯区
mottled prominent moth		*Macrurocampa marthesia* (Cramer)	(チョウ目、シャチホコガ科) 新北区
mottled pug		*Eupithecia exiguata* (Hübner)	(チョウ目、シャクガ科) 旧北区
mottled purplewing		*Eunica caelina* Godart	(チョウ目、タテハチョウ科) 新熱帯区
mottled roadside skipper		*Amblyscirtes nysa* Edwards	(チョウ目、セセリチョウ科) 新北区、新熱帯区
mottled rustic		*Caradrina morpheus* (Hufnagel)	(チョウ目、ヤガ科) 旧北区。mottled rustic moth ともいう
mottled sand grasshopper		*Spharagemon collare* (Scudder)	(バッタ目、バッタ科) 新北区。後翅は黄色で黒の縁
mottled satyr		*Steroma modesta* Weymer	(チョウ目、タテハチョウ科) 新熱帯区
mottled scrub-hairstreak		*Strymon mulucha* (Hewitson)	(チョウ目、シジミチョウ科) 新熱帯区
mottled sootywing			mottled Bolla を見よ
mottled tortoise beetle		*Deloyala guttata* (Olivier)	(コウチュウ目、ハムシ科) 新北区、大洋区。サツマイモ害虫
mottled umber		*Erannis defoliaria* (Clerck)	(チョウ目、シャクガ科) 旧北区。日本亜種は *E. d. gigantes* Inoue オオチャバネフユエダシャク
mottled umber moth			mottled umber を見よ
mottled-winged bush-cricket		*Jamaicana subguttata* (Walker)	(バッタ目、キリギリス科) 旧北区
mottles		*Logania*	(チョウ目、シジミチョウ科) の昆虫の総称
mouise-colored Euchaetias		*Pygarctia murina* (Stretch)	(チョウ目、ヒトリガ科) 新北区
mould beetles			silken fungus beetles を見よ
mound ant			yellow meadow ant を見よ
mound building ant		*Formica opaciventris* Emery	(ハチ目、アリ科) 新北区

英名	和名	学名	所属、分布、ほか
mound building termite		*Odontotermes obesus* (Rambur)	(シロアリ目、シロアリ科) 東洋区
mound building termite (1)		*Odontotermes redemanni* (Wasmann)	(シロアリ目、シロアリ科) 東洋区
mound building termites			soldierless termites (1) を見よ
Mount Donna Buang wingless stonefly		*Riekoperla darlingtoni* (Illies)	(カワゲラ目、Gripopterygidae) 豪州区。絶滅危惧種
Mount Stirling stonefly		*Thaumatoperla flaveola* Burns et Neboiss	(カワゲラ目、Eustheniidae) 豪州区。絶滅危惧種
Mountain Alcon blue		*Phengaris rebeli* (Hirschke)	(チョウ目、シジミチョウ科) 旧北区
mountain alder longicorn	タテスジゴマフカミキリ	*Mesosa senilis* Bates	(コウチュウ目、カミキリムシ科) 日本、旧北区
mountain alder narrow bark beetle	ハンノキホソキクイムシ	*Xyleborus schaufussi* Blandford	(コウチュウ目、キクイムシ科) 日本、東洋区
mountain alder striped xyleborus			bark ambrosia beetle を見よ
mountain Argus		*Paroeneis pumilus* (C. et R. Felder)	(チョウ目、タテハチョウ科) 東洋区
mountain Argus (1)		*Callerebia shallada* (Lang)	(チョウ目、タテハチョウ科) 東洋区
mountain Argus			northern brown Argus を見よ
mountain-ash bentwing		*Leucoptera scitella* Zeller	(チョウ目、ハモグリガ科) 旧北区
mountain-ash midget		*Phyllonorycter sorbi* (Stainton)	(チョウ目、ホソガ科) 旧北区
mountain ash sawfly		*Pristiphora geniculata* (Hartig)	(ハチ目、ハバチ科) 新北区
mountain ash seed chalcid		*Megastigmus brevicaudis* Ratzeburg	(ハチ目、オナガコバチ科) 旧北区
mountain beauty moth		*Syngrapha ignea* (Grote)	(チョウ目、ヤガ科) 新北区。mountain beauty ともいう
mountain blue		*Albulina pheretes* (Hübner)	(チョウ目、シジミチョウ科) 東洋区。
mountain blue (1)		*Neolucia hobartensis* (Miskin)	(チョウ目、シジミチョウ科) 豪州区
mountain blue			Ulysses butterfly を見よ
mountain blues		*Harpendyreus*	(チョウ目、シジミチョウ科) の昆虫の総称
mountain bumblebee		*Bombus monticola* Smith	(ハチ目、ミツバチ科) 旧北区
mountain burnet			Scotch burnet を見よ
mountain carpenter bee		*Xylocopa tabaniformis* Smith	(ハチ目、ミツバチ科) 新北区
mountain carpenter bee		*Xylocopa orpifex* Smith	(ハチ目、ミツバチ科) 新北区
mountain checkered skipper			southern checkered skipper を見よ
mountain cicada		*Cicadetta montana* (Scopoli)	(カメムシ目、セミ科) 旧北区
mountain clouded yellow	ヘリモンキチョウ	*Colias phicomone* (Esper)	(チョウ目、シロチョウ科) 旧北区

英名	和名	学名	所属、分布、ほか
mountain cloudy wing			Mexican cloudywing を見よ
mountain cochineal		*Dactylopius ceylonicus* (Green)	（カメムシ目、コチニールカイガラムシ科）東洋区、新熱帯区
mountain copper		*Aloeides thyra* (Linnaeus)	（チョウ目、シジミチョウ科）エチオピア区
mountain dappled white			dappled marble を見よ
mountain dappled white			dappled white を見よ
mountain darner		*Austroaeschna atrata* Martin	（トンボ目、ヤンマ科）豪州区
mountain-dwelling short-winged katydid		*Dichopetala oreoeca* Rehn et Hebard	（バッタ目、キリギリス科）新北区
mountain emperor		*Asterocampa montis* (Edwards)	（チョウ目、タテハチョウ科）新北区
mountain flattened leaf beetle	ミヤマヒラタハムシ	*Gastvalina peltoidea* (Gebler)	（コウチュウ目、ハムシ科）日本、旧北区
mountain fritillary		*Clossiana alaskensis* Lehmann	（チョウ目、タテハチョウ科）新北区
mountain fritillary (1)	キタヒメヒョウモン	*Boloria napaea* (Hoffmannsegg)	（チョウ目、タテハチョウ科）旧北区
mountain froghopper	ミヤマアワフキ	*Peuceptyelus nigroscutellatus* Matsumura	（カメムシ目、アワフキムシ科）日本、旧北区
mountain frosted weevil	ミヤマヒゲボソゾウムシ	*Phyllobius annectens* Sharp	（コウチュウ目、ゾウムシ科）日本、旧北区
mountain girdle moth		*Enypia griseata* Grossbeck	（チョウ目、シャクガ科）新北区
mountain greenstreak		*Cyanophrys longula* (Hewitson)	（チョウ目、シジミチョウ科）新北区、大洋区、新熱帯区
mountain green-veined white			dark-veined white を見よ
mountain hooded owlet moth		*Cucullia montanae* Grote	（チョウ目、ヤガ科）新北区
mountain humble bee		*Bombus lapponicus* Fabricius	（ハチ目、ミツバチ科）旧北区
mountain katydid		*Acripeza reticulata* Guérin-Méneville	（バッタ目、キリギリス科）豪州区
mountain leafhopper		*Colladonus montanus* (Van Duzee)	（カメムシ目、オオヨコバイ科）新北区
mountain longicorn beetle			chestnut trunk borer を見よ
mountain longwing			Mexican longwing を見よ
mountain mahogany hairstreak		*Satyrium tetra* (Edwards)	（チョウ目、シジミチョウ科）新北区
mountain mahogany looper		*Anacamptodes clivinaria* Guenée	（チョウ目、シャクガ科）新北区。成虫は mountain mahogany looper moth
mountain mahogany moth		*Ethmia discostrigella* (Chambers)	（チョウ目、スエヒロキバガ科）新北区、新熱帯区

英名	和名	学名	所属、分布、ほか
mountain Malachite		*Chlorolestes fasciatus* (Burmeister)	(トンボ目、Synlestidae) エチオピア区
mountain mantis		*Stagmomantis montana* (Rehn)	(カマキリ目、カマキリ科) 新熱帯区
mountain midges	アミカモドキ科	Deuterophlebidae	(ハエ目) の昆虫の総称 山地渓流に発生する
mountain midges			net-winged midges を見よ
mountain mimetic swallowtail		*Papilio plagiatus* Aurivillius	(チョウ目、アゲハチョウ科) エチオピア区
mountain moth		*Psodos quadrifaria* Sulzer	(チョウ目、シャクガ科) 旧北区
mountain mugwort gall midge	ヨモギクキコブタマバエ	*Rhopalomyia struma* Monzen	(ハエ目、タマバエ科) 日本、旧北区
mountain Parnassian		*Parnassius sminthheus* Doubleday	(チョウ目、アゲハチョウ科) 新北区
mountain pearl Charaxes		*Charaxes acuminatus* Thurau	(チョウ目、タテハチョウ科) エチオピア区
mountain pine beetle		*Dendroctonus jeffreyi* Hopkins	(コウチュウ目、キクイムシ科) 新北区
mountain pine beetle (1)	アメリカマツノキクイムシ	*Dendroctonus ponderosae* Hopkins	(コウチュウ目、キクイムシ科) 新北区
mountain pine coneworm		*Dioryctria yatesi* Mutuura et Munroe	(チョウ目、メイガ科) 新北区。成虫は mountain pine coneworm moth
mountain pinhole borer		*Platypus subgranosus* Schedl	(コウチュウ目、ナガキクイムシ科) 豪州区
mountain pride			Table Mountain beauty を見よ
mountain reedling		*Indolestes gracilis* (Hagen)	(トンボ目、アオイトトンボ科) 東洋区
mountain ringlet	マエトガリベニヒカゲ	*Erebia epiphron* Knoch	(チョウ目、タテハチョウ科) 旧北区
mountain river cruiser		*Macromia margarita* Westfall	(トンボ目、ヤマトンボ科) 新北区
mountain sandman		*Spialia spio* (Linnaeus)	(チョウ目、セセリチョウ科) エチオピア区
mountain shutwing		*Cordulephya montana* Tillyard	(トンボ目、エゾトンボ科) 豪州区
mountain skipper		*Anisynta monticolae* Olliff	(チョウ目、セセリチョウ科) 豪州区
mountain skolly		*Thestor montanus* van Son	(チョウ目、シジミチョウ科) エチオピア区
mountain small white		*Pieris ergane* (Geyer)	(チョウ目、シロチョウ科) 旧北区
mountain sootywing		*Staphylus vincula* (Plötz)	(チョウ目、セセリチョウ科) 新熱帯区
mountain spotted skipper		*Oreisplanus peronatus* (Kirby)	(チョウ目、セセリチョウ科) 豪州区
mountain sprite		*Pseudagrion draconis* Barnard	(トンボ目、イトトンボ科) エチオピア区
mountain stone weta		*Hemideina maori* (Pictet et Saussure)	(バッタ目、Stenopelmatidae) 豪州区
mountain sylph		*Metisella aegipan* (Trimen)	(チョウ目、セセリチョウ科) エチオピア区
mountain tiger		*Parantica sita* (Kollar)	(チョウ目、タテハチョウ科) 旧北区
mountain tortoiseshell			small tortoiseshell を見よ

英名	和名	学名	所属、分布、ほか
mountain white		*Leptophobia aripa elodia* (Boisduval)	（チョウ目、シロチョウ科）新熱帯区
mournful crow			long-branded crow を見よ
mournful Desmia moth		*Desmia ploralis* (Guenée)	（チョウ目、メイガ科）新北区
mournful duskywing		*Erynnis tristis* Boisduval	（チョウ目、セセリチョウ科）新北区
mournful sphinx moth		*Enyo lugubris* (Linnaeus)	（チョウ目、スズメガ科）新北区、新熱帯区。mournful sphinx ともいう
mournful Thyris moth		*Thyris sepulchralis* Guérin	（チョウ目、マドガ科）新北区
mourning bee		*Bombus funebris* Smith	（ハチ目、ミツバチ科）新熱帯区
mourning cloak			Camberwell beauty を見よ　米国での英名。mourning cloak butterfly ともいう
mourning cloak grasshopper			Carolina grasshopper を見よ
mourning underwing moth		*Catocala flebilis* Grote	（チョウ目、ヤガ科）新北区。mournful underwing ともいう
mouse			mouse moth を見よ
mouse beetle		*Leptinus testaceus* Müller	（コウチュウ目、Leptinidae）全北区
mouse blue		*Lepidochrysops puncticilia* (Trimen)	（チョウ目、シジミチョウ科）エチオピア区
mouse-colored lichen moth		*Pagara simplex* Walker	（チョウ目、ヒトリガ科）新北区
mouse flea			European mouse flea を見よ
mouse fleas			scaled fleas を見よ
mouse fleas			rodent fleas (1) を見よ
mouse moth		*Amphipyra tragopoginis* (Clerck)	（チョウ目、ヤガ科）旧北区
mouselike barklice	ホシチャタテ科	Myopsocidae	（チャタテムシ目）の昆虫の総称
mouse-nest beetle			mouse beetle を見よ
mousy angle moth		*Speranza argillacearia* (Packard)	（チョウ目、シャクガ科）新北区
Mozambique barred-blue			Mozambique silverline を見よ
Mozambique hawk			magnificent scalloped hawk を見よ
Mozambique silverline		*Cigaritis mozambica* (Bertolini)	（チョウ目、シジミチョウ科）エチオピア区
Mrs Raven flat		*Calleagris kobela* (Trimen)	（チョウ目、セセリチョウ科）エチオピア区
Mrs Raven skipper			Mrs Raven flat を見よ
Mt Arthur giant weta			Nelson alpine weta を見よ
Mt Cook giant weta		*Deinacrida pluvialis* Gibbs	（バッタ目、Stenopelmatidae）豪州区
muck beetle		*Ligyrus gibbosus* (De Geer)	（コウチュウ目、コガネムシ科）新北区
muckworm			くそ虫、ウジ
mud-backed beetles			minute mud-loving beetles を見よ
mud dauber	アメリカジガバチ	*Sceliphron caementarium* (Drury)	（ハチ目、アナバチ科）日本、全北区、新熱帯区、豪州区

英名	和名	学名	所属、分布、ほか
mud-dauber wasp		*Sceliphron fistularium* (Dahlbom)	（ハチ目、アナバチ科）新熱帯区
mud-dauber wasp			mud dauber を見よ
mud dauber wasps			mud daubers and spider wasps (1) を見よ
mud daubers			cicada killers を見よ
mud daubers			slender mudnest builders を見よ
mud daubers			wasps (1) を見よ
mud daubers			potter wasps を見よ
mud daubers and spider wasps		Trypoxylidae	（ハチ目）の昆虫の総称　アナバチ科 Sphecidae の亜科とされることあり
mud daubers and spider wasps (1)		Trypoxyloninae	（ハチ目、アナバチ科）の昆虫の総称
mud-eyes			トンボ目の一部の幼虫の呼称
mud-feather case		*Coleophora limosipennella* (Duponchel)	（チョウ目、ツツミノガ科）全北区
mudflat tiger beetle		*Cicindela trifasciata sigmoidea* LeConte	（コウチュウ目、ハンミョウ科）新北区
mudpot wasps	トックリバチ属	*Eumenes*	（ハチ目、スズメバチ科）の昆虫の総称
mud-pot wasps			potter wasps を見よ
mud-wasp		*Pseudomasaris vespoides* (Cresson)	（ハチ目、Massaridae）新北区
mud wasps			potter wasps を見よ
Muga silkmoth		*Antheraea assamensis* Westwood	（チョウ目、ヤママユガ科）東洋区
Muga silkworm			Muga silkmoth を見よ
mugwort case	ヨモギハナツツミノガ	*Coleophora artemisicolella* Bruand	（チョウ目、ツツミノガ科）日本、旧北区
mugwort cryptosiphum aphid	ヨモギフシアブラムシ	*Cryptosiphum artemisiae* Buckton	（カメムシ目、アブラムシ科）日本、旧北区、東洋区
mugwort cutworm	ヨモギガ	*Protoschinia scutosa* (Denis et Schiffermüller)	（チョウ目、ヤガ科）日本、旧北区
mugwort flat-body		*Exaeretia allisella* Stainton	（チョウ目、マルハキバガ科）旧北区
mugwort fruit fly	ヨモギハマダラミバエ	*Trypeta artemisiae* (Fabricius)	（ハエ目、ミバエ科）日本、旧北区
mugwort gall midge	ヨモギタマバエ	*Rhopalomyia yomogicola* (Matsumura)	（ハエ目、タマバエ科）日本、旧北区
mugwort hairy gall midge	ヨモギシロケフシタマバエ	*Rhopalomyia cinearius* Monzen	（ハエ目、タマバエ科）日本、旧北区
mugwort leaf beetle	ヨモギハムシ	*Chrysolina aurichalcea* (Mannerheim)	（コウチュウ目、ハムシ科）日本、旧北区、東洋区
mugwort leaf caterpillar	ヨモギネムシガ	*Epiblema foenella* (Linnaeus)	（チョウ目、ハマキガ科）日本、旧北区
mugwort leaf roller	ヨモギウストビホソハマキ	*Cochylidia richteriana* (Fischer von Röslerstamm)	（チョウ目、ホソハマキガ科）日本、旧北区
mugwort leafhopper	ヨモギヨコバイ	*Macrosteles artemisiae* (Matsumura)	（カメムシ目、オオヨコバイ科）日本

英名	和名	学名	所属、分布、ほか
mugwort long-horned aphid	ヨモギヒゲナガアブラムシ	*Macrosiphoniella yomogicola* (Matsumura)	(カメムシ目、アブラムシ科) 日本、旧北区
mugwort looper	ヨモギエダシャク	*Ascotis selenaria* (Denis et Schiffermüller)	(チョウ目、シャクガ科) 日本。ダイズ、カンキツ、チャ、バラなど多くの植物害虫
mugwort plume moth			sluggish plume を見よ
mugwort powdery aphid	ヨモギハアブラムシ	*Aphis kurosawai* Takahashi	(カメムシ目、アブラムシ科) 日本、東洋区
mugwort Rhopalomyia midge	ヨモギイボタマバエ	*Rhopalomyia japonica* Monzen	(ハエ目、タマバエ科) 日本
mugwort slender	ヨモギホソガ	*Leucospilapteryx omissella* (Stainton)	(チョウ目、ホソガ科) 日本、旧北区
mugwort woolly gall midge	ヨモギワタタマバエ	*Rhopalomyia giraldi* Kieffer et Trotter	(ハエ目、タマバエ科) 日本、旧北区
Muirhead's labyrinth		*Neope muirheadi nagasawae* Matsumura	(チョウ目、タテハチョウ科) 東洋区
Muir's cedar hairstreak			Muir's hairstreak を見よ
Muir's hairstreak		*Callophrys muiri* (H. Edwards)	(チョウ目、シジミチョウ科) 新北区
mulberry ambrosia beetle	クワノキクイムシ	*Xyleborus atratus* Eichhoff	(コウチュウ目、キクイムシ科) 日本、旧北区、東洋区。クワ害虫
mulberry bagworm		*Bambalina* sp.	(チョウ目、ミノガ科) 日本
mulberry banded gall midge	クワハマダラタマバエ	*Resseliella quadrifasciata* (Niwa)	(ハエ目、タマバエ科) 日本、旧北区
mulberry bark beetle	クワノコキクイムシ	*Cryphalus exiguus* Blandford	(コウチュウ目、キクイムシ科) 日本、旧北区、東洋区。クワ、カキなどの害虫
mulberry bark beetle			mulberry ambrosia beetle を見よ
mulberry black gall midge			mulberry bud gall midge を見よ
mulberry borer	クワカミキリ	*Apriona japonica* Thomson	(コウチュウ目、カミキリムシ科) 日本。クワ、カキ、リンゴなど果樹、プラタナスなどの樹木害虫
mulberry bud gall midge	クワクロタマバエ	*Asphondylia morivorella* (Naito)	(ハエ目、タマバエ科) 日本。クワ害虫
mulberry bug	オオメダカナガカメムシ	*Malcus japonicus* Ishihara et Hasegawa	(カメムシ目、メダカナガカメムシ科) 日本。クワ害虫
mulberry caterpillar	シロシタヨトウ	*Sarcopolia illoba* (Butler)	(チョウ目、ヤガ科) 日本、旧北区。ダイズ、アブラナ科野菜、カンキツ、クワ、サクラなど多くの植物害虫
mulberry crane fly	クリハマダラガガンボ	*Limonia nohirai* (Alexander)	(ハエ目、ガガンボ科) 日本、旧北区
mulberry flea beetle	クワノミハムシ	*Luperomorpha funesta* (Baly)	(コウチュウ目、ハムシ科) 日本。クワ、カンキツ、テンサイなどの害虫
mulberry flower bug	ヒメハナカメムシ	*Orius sauteri* (Poppius)	(カメムシ目、ハナカメムシ科) 日本、旧北区
mulberry frosted longicorn	クワサビカミキリ	*Mesosella simiola* Bates	(コウチュウ目、カミキリムシ科) 日本
mulberry gall midge			mulberry bud gall midge を見よ
mulberry geometrid	クワエダシャク	*Menophra atrilineata* Butler	(チョウ目、シャクガ科) 日本、旧北区、東洋区

英名	和名	学名	所属、分布、ほか
mulberry hawk		*Pseudoclanis postica* (Walker)	（チョウ目、スズメガ科）エチオピア区
mulberry leaf beetle	クワハムシ	*Fleutiauxia armata* (Baly)	（コウチュウ目、ハムシ科）日本、旧北区
mulberry leaf bug	モンキクロカスミカメムシ	*Deraeocoris ater* Jakovlev	（カメムシ目、カスミカメムシ科）日本、旧北区
mulberry leafminer	クワハモグリバエ	*Agromyza morivora* Sasakawa et Fukuhara	（ハエ目、ハモグリバエ科）日本。クワ害虫
mulberry leafroller	クワハマキ	*Olethreutes mori* (Matsumura)	（チョウ目、ハマキガ科）日本、旧北区。クワ害虫
mulberry leafstalk midge			mulberry bud gall midge を見よ
mulberry lecanium	ニシガハラワタカイガラムシ	*Pulvinaria nishigaharae* (Kuwana)	（カメムシ目、カタカイガラムシ科）日本
mulberry longicorn			mulberry borer を見よ
mulberry looper	クワトゲエダシャク	*Apochima excavata* (Dyar)	（チョウ目、シャクガ科）日本、旧北区。カンキツ、モモ害虫
mulberry looper	クワエダシャク	*Phthonondria atrilineata* (Butler)	（チョウ目、シャクガ科）日本、旧北区、東洋区。クワ害虫
mulberry minute bark beetle			mulberry bark beetle を見よ
mulberry moth			mulberry pyralid を見よ
mulberry oystershell scale	クワカキカイガラムシ	*Lepidosaphes kuwacola* Kuwana	（ケメムシ目、マルカイガラムシ科）日本
mulberry pyralid	クワノメイガ	*Glyphodes pyloalis* Walker	（チョウ目、メイガ科）日本、東洋区。クワ害虫
mulberry scale			white peach scale を見よ
mulberry shoot gall midge	クワシントメタマバエ	*Diplosis mori* Sasaki	（ハエ目、タマバエ科）日本。クワ害虫
mulberry silkmoth			silkworm を見よ　英国での英名
mulberry silkworm			silkworm を見よ
mulberry small bark beetle			mulberry bark beetle を見よ
mulberry small tiger longicorn	ヨツスジトラカミキリ	*Chlorophorus quinquefasciatus* (Castelnau et Gory)	（コウチュウ目、カミキリムシ科）日本、旧北区
mulberry small weevil	クワヒメゾウムシ	*Baris deplanata* Roelofs	（コウチュウ目、ゾウムシ科）日本
mulberry sucker	クワキジラミ	*Anomoneura mori* Schwarz	（カメムシ目、キジラミ科）日本、旧北区、クワ害虫
mulberry thrips	クワアザミウマ	*Pseudodendrothrips mori* (Niwa)	（アザミウマ目、アザミウマ科）日本、全北区、東洋区。クワ害虫
mulberry tiger moth	クワゴマダラヒトリ	*Thanatarctia imparilis* (Butler)	（チョウ目、ヒトリガ科）日本、旧北区、東洋区。クワ、カンキツ、チャ、サクラなど多くの植物害虫
mulberry tortrix	クロカクモンハマキ	*Archips endoi* Yasuda	（チョウ目、ハマキガ科）日本
mulberry tortrix			brown oak tortrix を見よ
mulberry urosema midge			mulberry shoot gall midge を見よ
mulberry white weevil	クワゾウムシ	*Episomus mundus* Sharp	（コウチュウ目、ゾウムシ科）日本

英名	和名	学名	所属、分布、ほか
mulberry whitefly		*Tetraleurodes mori* (Quaintance)	(カメムシ目、コナジラミ科) 新北区
mulberry whitefly	シナノコナジラミ	*Bemisia shinanoensis* Kuwana	(カメムシ目、コナジラミ科)日本。クワ、ユキヤナギ、ツツジの害虫
mulberry white-spotted longicorn			white-spotted longicorn beetle (1) を見よ
mulberry wild silkworm			wild mulberry silkmoth を見よ
mulberry wing		*Poanes massasoit* Scudder	(チョウ目、セセリチョウ科) 新北区
mulberry yellow-tail moth			browntail moth (2) を見よ
mule killer			dragonflies (1) を見よ
mule killers			mantids (2) を見よ
mule-killers			velvet ants (1) を見よ
mulga ant		*Polyrhachis macropus* Wheeler	(ハチ目、アリ科) 豪州区。mulga は植物名
mullberry leaftier moth		*Glyphodes sibillalis* Walker	(チョウ目、メイガ科) 新北区、新熱帯区
mullein			mullein moth (1) を見よ　モウズイカ (ゴマノハグサ科) に由来
mullein aphid		*Aphis verbasci* Schrank	(カメムシ目、アブラムシ科) 旧北区
mullein leaf bug		*Campylomma verbasci* (Meyer)	(カメムシ目、カスミカメムシ科) 旧北区
mullein moth (1)		*Cucullia verbasci* (Linnaeus)	(チョウ目、ヤガ科) 旧北区
mullein moth			shark moth を見よ
mullein shark			mullein moth (1) を見よ
mullein thrips		*Neoheegeria verbasci* (Osborn)	(アザミウマ目、クダアザミウマ科) 全北区
mullein wave		*Scopula marginepunctata* (Goeze)	(チョウ目、シャクガ科) 旧北区
mullein weevil		*Cionus hortulanus* (Fourcroy)	(コウチュウ目、ゾウムシ科) 旧北区
Muller's Mellana		*Quasimellana mulleri* (Bell)	(チョウ目、セセリチョウ科) 新熱帯区
Mullins' skipperling		*Piruna mullinsi* Freeman	(チョウ目、セセリチョウ科) 新熱帯区
multicolored Asian lady beetle			Asian multicolored lady beetle を見よ
multicolored sedgeminer moth		*Meropleon diversicolor* (Morrison)	(チョウ目、ヤガ科) 新北区
multiform leafroller moth		*Acleris flavivittana* (Clemens)	(チョウ目、ハマキガ科) 新北区
multinellus grass-veneer moth		*Crambus multilinellus* Fernald	(チョウ目、メイガ科) 新北区
multi-spotted darner		*Austroaeschna multipunctata* (Martin)	(トンボ目、ヤンマ科) 豪州区
multispotted flat		*Celaenorrhinus pulomaya* (Moore)	(チョウ目、セセリチョウ科) 東洋区

英名	和名	学名	所属、分布、ほか
multivoltine silkworm uji-fly			multivoltine tachina fly を見よ
multivoltine tachina fly	クワゴヤドリバエ	*Exorista sorbillans* (Wiedemann)	(ハエ目、ヤドリバエ科) 日本、旧北区、東洋区、豪州区、エチオピア区
Munakata nymphula	ムナカタミズメイガ	*Paraponyx ussuriensis* (Rebel)	(チョウ目、メイガ科) 日本、旧北区
mung moth			bean pod borer (1)(2) を見よ　mung bean 緑豆に由来。
Muridosca hairstreak		*Erora muridosca* (Dyar)	(チョウ目、シジミチョウ科) 新熱帯区
murky-legged black legionnaire		*Beris chalybata* (Forster)	(ハエ目、ミズアブ科) 旧北区
murkey meal caterpillar			small tabby を見よ
murmidiid beetles		Murmidiidae	(コウチュウ目) の昆虫の総称
Murphy's crow		*Euploea caespes* Ackery et Vane-Wright	(チョウ目、タテハチョウ科) 東洋区
Murphy's metalmark		*Apodemia murphyi* Austin	(チョウ目、シジミタテハ科) 新熱帯区
Murray River hunter		*Austrogomphus angelorum* Tillyard	(トンボ目、サナエトンボ科) 豪州区
Murray's skolly		*Thestor murrayi* Swanepoel	(チョウ目、シジミチョウ科) エチオピア区
muscid			house fly を見よ
muscid flies			house flies (1) を見よ
muscoid flies	裂額亜目	Schizophora	(ハエ目) の昆虫の総称
Muscosa sphinx		*Manduca muscosa* (Rothschild et Jordan)	(チョウ目、スズメガ科) 新北区、新熱帯区。Muscosa sphinx moth ともいう
museum beetle (1)	セマルヒョウホンムシ	*Gibbium psylloides* (Czenpinski)	(コウチュウ目、ヒョウホンムシ科) 日本、汎世界。標本の害虫
museum beetle (2)		*Anthrenus museorum* (Linnaeus)	(コウチュウ目、カツオブシムシ科) 豪州区
Musgrave's Psednura		*Psednura musgravei* Rehn	(バッタ目、オンブバッタ科) 豪州区
mushroom camel crickets		*Pristoceuthophilus*	(バッタ目、コロギス科) の昆虫の総称
mushroom cecid			mushroom white cecid を見よ
mushroom flies	キノコバエ科	Mycetophilidae	(ハエ目) の昆虫の総称
mushroom flies			dark-winged fungus flies を見よ　豪州での英名
mushroom fly	ツクリタケクロバネキノコバエ	*Lycoriella ingenua* (Dufour)	(ハエ目、クロバネキノコバエ科) 日本。ショウガ、シイタケなどの害虫
mushroom fly			mushroom phorid を見よ
mushroom gnat		*Mycetophila fungorum* (De Geer)	(ハエ目、キノコバエ科) 新北区
mushroom midge		*Lestremia cinerea* Macquart	(ハエ目、タマバエ科) 旧北区
mushroom midge		*Mycetophila speyeri* Barnes	(ハエ目、キノコバエ科) 旧北区
mushroom phorid		*Megaselia halterata* (Wood)	(ハエ目、ノミバエ科) 豪州区

英名	和名	学名	所属、分布、ほか
mushroom phorids		*Megaselia*	(ハエ目、ノミバエ科) の昆虫の総称
mushroom sciarid		*Lycoriella auripila* (Fitch)	(ハエ目、クロバネキノコバエ科) 豪州区
mushroom sciarid		*Lycoriella mali* (Fitch)	(ハエ目、クロバネキノコバエ科) 豪州区
mushroom springtail		*Ceratophysella denticulata* (Bagnall)	(トビムシ目、ヒメトビムシ科) 豪州区
mushroom springtail (1)		*Ceratophysella armata* (Nicolet)	(トビムシ目、ヒメトビムシ科) 旧北区
mushroom springtails		*Hypogastrura*	(トビムシ目、ヒメトビムシ科) の昆虫の総称
mushroom white cecid		*Heteropeza pygmaea* Winnertz	(ハエ目、タマバエ科) 豪州区、旧北区
mushroom yellow cecid		*Mycophila barnesi* Edwards	(ハエ目、タマバエ科) 豪州区
musical bush cricket		*Hapithus melodius* Walker	(バッタ目、コオロギ科) 新北区
musk			musk beetle を見よ
musk beetle		*Aromia moschata* (Linnaeus)	(コウチュウ目、カミキリムシ科) 旧北区。日本亜種は *A. m. ambrosiaca* (Steven) ジャコウカミキリ
musk mare			two-striped walking-stick を見よ
muskel			butterflies and moths を見よ 幼虫の英名
musky meal caterpillar			small tabby を見よ
muslin bombyx		*Penicillifera apicalis* Walker	(チョウ目、カイコガ科) 東洋区
muslin ermine		*Diaphora mendica* (Clerck)	(チョウ目、ヒトリガ科) 旧北区
muslin footman		*Nudaria mundana* (Linnaeus)	(チョウ目、ヒトリガ科) 旧北区
muslin moth			muslin ermine を見よ
muslin moth			muslin footman を見よ
mussel purple scale			purple scale を見よ
mussel scale			purple scale を見よ
mussel scale			oystershell scale (1) を見よ 英国での英名
mussel scale parasite		*Aphytis lepidosaphes* Compere	(ハチ目、ツヤコバチ科) 豪州区
Mussoorie pied flat		*Celaenorrhinus pero* Leech	(チョウ目、セセリチョウ科) 東洋区
mustached clubtail		*Gomphus adelphus* Selys	(トンボ目、サナエトンボ科) 新北区
mustard aphid			turnip aphid (1) を見よ
mustard beetle (1)		*Phaedon veronicae* Bedel	(コウチュウ目、ハムシ科) 旧北区
mustard beetle (2)		*Phaedon cochleariae* (Fabricius)	(コウチュウ目、ハムシ科) 旧北区
mustard leaf beetle			mustard beetle (2) を見よ
mustard sallow moth		*Pyreferra hesperidago* (Guenée)	(チョウ目、ヤガ科) 新北区
mustard sawfly		*Athalia lugens proxima* Klug	(ハチ目、ハバチ科) 東洋区
mustard white		*Pieris oleracea* (Harris)	(チョウ目、シロチョウ科) 新北区

英名	和名	学名	所属、分布、ほか
mustard white			green-veined white を見よ　*P. napi* Linnaeus の米国での英名
muste skipper		*Decinea mustea* Freeman	（チョウ目、セセリチョウ科）新熱帯区
mute scots-fir argent		*Ocnerostoma piniariella* Zeller	（チョウ目、スガ科）旧北区
muted dart			old man dart を見よ
muted hairstreak		*Electrostrymon canus* (Druce)	（チョウ目、シジミチョウ科）新北区
muted hairstreak		*Electrostrymon joya* (Dognin)	（チョウ目、シジミチョウ科）新北区、新熱帯区
mutillid wasps			velvet ants (1) を見よ
mutillids			velvet ants (1) を見よ
Mutina midistreak		*Tmolus mutina* (Hewitson)	（チョウ目、シジミチョウ科）新熱帯区
Muzaria Euchlaena moth		*Euchlaena muzaria* (Walker)	（チョウ目、シャクガ科）新北区
Mycerina untailed Charaxes		*Charaxes mycerina* (Godart)	（チョウ目、タテハチョウ科）エチオピア区
mycetaeid fungus beetles		Mycetaeidae	（コウチュウ目）の昆虫の総称　テントウムシダマシ科に含まれる
mycetophilids			mushroom flies を見よ
Mycon hairstreak		*Theclopsis mycon* (Godman et Salvin)	（チョウ目、シジミチョウ科）新熱帯区
mydas flies	ムシヒキアブモドキ科	Mydidae	（ハエ目）の昆虫の総称
mydas fly		*Mydas clavatus* Drury	（ハエ目、ムシヒキアブモドキ科）新北区
Mylitta crescent			thistle crescent を見よ
Mylitta crescentspot			thistle crescent を見よ
Mylitta greenwing	アカガネタテハ	*Dynamine mylitta* Cramer	（チョウ目、タテハチョウ科）新北区
Myller's nettle-tap	シロヘリハマキモドキ	*Prochoreutis myllerana* (Fabricius)	（チョウ目、ハマキモドキガ科）日本、旧北区
Mylotes cattleheart			cattle heart を見よ
Mylothrid Pentila		*Pentila tachyroides* Dewitz	（チョウ目、シジミチョウ科）エチオピア区
mymarid egg parasites			fairy flies を見よ
myopsocid barklice			mouselike barklice を見よ
myrica whitefly	ヤマモモコナジラミ	*Parabemisia myricae* (Kuwana)	（カメムシ目、コナジラミ科）日本、東洋区。ヤマモモ、モモ、クワ、チャ、サクラなどの害虫
Myris skipper		*Myrinia myris* (Mabille)	（チョウ目、セセリチョウ科）新熱帯区
myrmecophilic beetle		*Atemeles pubicollis* Brisout	（コウチュウ目、ハネカクシ科）旧北区
myrmecophilous beetle		*Claviger testaceus* Preyssler	（コウチュウ目、アリヅカムシ科）旧北区
myrmeleotids			antlions (1) を見よ
myrmicine ants			agricultural ants を見よ
myrsine-leaved oak bark beetle			oak bark beetle (2) を見よ
Myrtea metalmark			Godman's Sarota を見よ

英名	和名	学名	所属、分布、ほか
Mys skipper		*Zariaspes mys* (Hübner)	（チョウ目、セセリチョウ科）新熱帯区
mysterious Olethreutes moth		*Olethreutes mysteriana* Miller	（チョウ目、ハマキガ科）新北区
mystic skipper			long dash を見よ
mystical Euselasia		*Euselasia mystica* (Schaus)	（チョウ目、シジミタテハ科）新熱帯区
Mythra sister		*Adelpha mythra* (Godart)	（チョウ目、タテハチョウ科）新熱帯区

英名	和名	学名	所属、分布、ほか
N			
Nabakov's satyr	ナボコフジャノメ	*Cyllopsis pyracmon* (Butler)	(チョウ目、タテハチョウ科) 新北区
nabids			damsel bugs を見よ
Nabokov's lycid		*Nabokovia cuzquenha* Balint et Lamas	(チョウ目、シジミチョウ科) 新熱帯区
Naga duke		*Euthalia khama* Alpheraky	(チョウ目、タテハチョウ科) 東洋区
Naga hedge blue		*Oreolyce quadriplaga* (Snellen)	(チョウ目、シジミチョウ科) 東洋区
Naga sapphire		*Heliophorus kohimensis* (Tytler)	(チョウ目、シジミチョウ科) 東洋区
Naga treebrown		*Lethe naga* Doherty	(チョウ目、タテハチョウ科) 東洋区
Naga white		*Talbozia naganum* (Moore)	(チョウ目、シロチョウ科) 東洋区。*T. n. karumii* (Ikeda) カルミモンシロチョウ も同英名
Nagao xyleborus	ナガオキクイムシ	*Xyleborus nagaoensis* Murayama	(コウチュウ目、キクイムシ科) 日本
Nairobi eye		*Paederus eximius* Reiche	(コウチュウ目、ハネカクシ科) エチオピア区。Nairobi fly, Kenya fly ともいう
Nais metalmark		*Apodemia nais* (Edwards)	(チョウ目、シジミタテハ科) 新北区
Nais tiger moth		*Apantesis nais* (Drury)	(チョウ目、ヒトリガ科) 新北区
Nakano long-nosed planthopper	ナカノテングスケバ	*Dictyophara nakanonis* Matsumura	(カメムシ目、テングスケバ科) 日本、旧北区、大洋区
naked scales			soft scales を見よ
Namagualand Temnora		*Temnora namagua* Rothschild et Jordan	(チョウ目、スズメガ科) エチオピア区。南アの地名に由来
Namaqua arrowhead		*Phasis clavum* Murray	(チョウ目、シジミチョウ科) エチオピア区
Namaqua bar		*Cigaritis namaquus* (Trimen)	(チョウ目、シジミチョウ科) エチオピア区
Namaqua dancer		*Alenia namaqua* Vári	(チョウ目、セセリチョウ科) エチオピア区
Namaqua opal		*Chrysoritis aridus* (Pennington)	(チョウ目、シジミチョウ科) エチオピア区
Namaqua sandman			Namaqua dancer を見よ
Namaqua widow		*Tarsocera namaquensis* Vári	(チョウ目、タテハチョウ科) エチオピア区
Nambour canegrub			southern one-year canegrub を見よ 豪州東部の地名に由来
Namib desert beetle		*Stenocara gracilipes* Solier	(コウチュウ目、ゴミムシダマシ科) エチオピア区。Namib 砂漠に由来
Nancy's ranger		*Kedestes nancy* Collins et Larsen	(チョウ目、セセリチョウ科) エチオピア区
Nanda Devi Apollo		*Parnassius nandadevinensis* Weiss	(チョウ目、アゲハチョウ科) 東洋区
Nanina oak-slug moth		*Euclea nanina* Dyar	(チョウ目、イラガ科) 新北区
Nantucket pine tip moth		*Rhyacionia frustrana* (Comstock)	(チョウ目、ハマキガ科) 新北区、新熱帯区。米国東部の島名に由来
Napaea fritillary			mountain fritillary (1) を見よ
Napaea satyr		*Pedaliodes napaea* (Bates)	(チョウ目、タテハチョウ科) 新熱帯区

英名	和名	学名	所属、分布、ほか
Narbal hairstreak		*Olynthus narbal* (Stoll)	（チョウ目、シジミチョウ科）新熱帯区
narcissus bulb fly	スイセンハナアブ	*Merodon equestris* (Fabricius)	（ハエ目、ハナアブ科）日本、全北区。スイセン害虫
narcissus flies		*Eumerus*	（ハエ目、ハナアブ科）の昆虫の総称
narcissus fly			narcissus bulb fly を見よ
narcissus jewel	リュウグウニシキシジミ	*Hypochrysops narcissus* (Fabricius)	（チョウ目、シジミチョウ科）豪州区
narrow apex click beetle	コクロクシコメツキ	*Melanotus invectitius* Candèze	（コウチュウ目、コメツキムシ科）日本
narrow apple tortrix			Asiatic leafroller を見よ
narrow-banded crescent		*Telenassa berenice* (Felder)	（チョウ目、タテハチョウ科）新熱帯区
narrow-banded dartwhite		*Catasticta flisa flisa* (Herrich-Schäffer)	（チョウ目、シロチョウ科）新熱帯区。*C. f. flisandra* Reissinger, *C. f. flisella* Reissinger も同英名
narrow-banded demon		*Chamaelimnas tircis* C. et R. Felder	（チョウ目、シジミタテハ科）新熱帯区
narrow-banded froghopper	ホソオビトドアワフキ	*Aphrophora scutellata* Matsumura	（カメムシ目、アワフキムシ科）日本
narrow-banded noctuid	ツマオビアツバ	*Zanclognatha griselda* (Butler)	（チョウ目、ヤガ科）日本、旧北区
narrow-banded owl-butterfly			tamarindi owlet を見よ
narrow-banded red-eye		*Pteroteinon concaenira* Belcastro et Larsen	（チョウ目、セセリチョウ科）エチオピア区
narrow-banded satyr		*Aulocera brahminus* (Blanchard)	（チョウ目、タテハチョウ科）東洋区
narrow-banded shoemaker		*Prepona pylene* Hewitson	（チョウ目、タテハチョウ科）新熱帯区
narrow-banded swallowtail	ガリエナスアゲハ	*Papilio gallienus* Distant	（チョウ目、アゲハチョウ科）エチオピア区
narrow-banded thick-legged moth	ホソオビアシブトクチバ	*Parallelia arctotaenia* (Guenée)	（チョウ目、ヤガ科）日本、東洋区、豪州区、大洋区
narrow-banded velvet bob		*Koruthaialos rubecula* (Plötz)	（チョウ目、セセリチョウ科）東洋区
narrow-banded widow		*Dingana angusta* Henning et Henning	（チョウ目、タテハチョウ科）エチオピア区
narrow-barred carpet			gem (1) を見よ
narrow barklice	ホソチャタテ科	Stenopsocidae	（チャタテムシ目）の昆虫の総称　多くの種が汎世界
narrow-beaked katydid		*Turpilia rostrata* (Rehn et Hebard)	（バッタ目、キリギリス科）新北区
narrow blue-banded swallowtail			green-banded swallowtail を見よ
narrow-bordered bee			narrow-bordered bee hawkmoth を見よ
narrow-bordered bee hawk			narrow-bordered bee hawkmoth を見よ
narrow-bordered bee hawkmoth		*Hemaris tityus* (Linnaeus)	（チョウ目、スズメガ科）旧北区

英名	和名	学名	所属、分布、ほか
narrow-bordered five-spot burnet		*Zygaena lonicerae* (Scheven)	(チョウ目、マダラガ科) 旧北区
narrow-bordered six-spot burnet			burnet moth を見よ
narrow brown pine aphid		*Eulachnus rileyi* (Williams)	(カメムシ目、アブラムシ科) 旧北区
narrow brown scale	ホソカタカイガラムシ	*Coccus acutissimus* (Green)	(カメムシ目、カタカイガラムシ科) 日本、大洋区。マンゴー害虫
narrow clouded knot-horn		*Phycitodes saxicola* (Vaughan)	(チョウ目、メイガ科) 旧北区
narrow green-banded swallowtail	ルリアゲハ	*Papilio nireus lyaeus* Doubleday	(チョウ目、アゲハチョウ科) エチオピア区
narrow green pine aphid		*Eulachnus brevipilosus* Börner	(カメムシ目、アブラムシ科) 旧北区
narrow-headed ant		*Formica exsecta* Nyal	(ハチ目、アリ科) 旧北区
narrow-horned flour beetle			slender-horned flour beetle を見よ
narrow-lined beauty		*Baeotus japetus* (Staudinger)	(チョウ目、タテハチョウ科) 新熱帯区
narrow-lined hairstreak		*Arawacus leucogyna* (C. et R. Felder)	(チョウ目、シジミチョウ科) 新熱帯区
narrownecked grain beetle	アトグロホソアリモドキ	*Anthicus floralis* (Linnaeus)	(コウチュウ目、アリモドキ科) 日本、新北区
narrow-patch Carolella moth			two-spotted Carolella moth を見よ
narrow red-barred twist		*Batodes angustioranus* (Haworth)	(チョウ目、ハマキガ科) 旧北区
narrow rice bug			rice bug (1)(2) を見よ
narrow short-winged longicorn	ホソコバネカミキリ	*Necydalis pennata* Lewis	(コウチュウ目、カミキリムシ科) 日本、旧北区
narrow silver Y			epigaea looper moth を見よ
narrow snail eater		*Scaphinotus angusticollis* Mannerheim	(コウチュウ目、オサムシ科) 新北区。貝、ナメクジを捕食
narrow spark		*Sinthusa nasaka* (Horsfield)	(チョウ目、シジミチョウ科) 東洋区。*S. n. amba* Kirby も同英名
narrow spotted pine aphid		*Eulachnus agilis* (Kaltenbach)	(カメムシ目、アブラムシ科) 旧北区
narrow-striped forceptail		*Aphylla protracta* (Hagen)	(トンボ目、サナエトンボ科) 新北区、新熱帯区
narrow-waisted bark beetles	チビキカワムシ科	Salpingidae	(コウチュウ目) の昆虫の総称
narrow-winged awl			brown awl を見よ
narrow-winged blue-skipper		*Pythonides pteras* (Godman et Salvin)	(チョウ目、セセリチョウ科) 新熱帯区
narrow-winged bush grasshopper		*Aeolophides tenuipennis* (Scudder)	(バッタ目、バッタ科) 新北区
narrow-winged cymatophorid	ホソトガリバ	*Tethea octogesima* (Butler)	(チョウ目、トガリバガ科) 日本

英名	和名	学名	所属、分布、ほか
narrow-winged damselflies	イトトンボ科	Agrionidae	(トンボ目) の昆虫の総称 世界に1,000種以上
narrow-winged damselfly			lyre-tipped spreadwing を見よ
narrow-winged false feather	タテジマホソマイコガ	*Schreckensteinia festaliella* (Hübner)	(チョウ目、ホソマイコガ科) 日本、全北区、大洋区
narrow-winged grey		*Eudonia angustea* (Curtis)	(チョウ目、メイガ科) 旧北区
narrow-winged horsefly		*Tabanus maculicornis* Zetterstedt	(ハエ目、アブ科) 旧北区
narrow-winged katydid		*Elimaea punctifera* (Walker)	(バッタ目、キリギリス科) 大洋区
narrow-winged knot-horn		*Nephopteryx angustella* (Hübner)	(チョウ目、メイガ科) 旧北区
narrow-winged leaf miners			leafblotch miners (1) を見よ
narrow-winged mantid	チョウセンカマキリ	*Tenodera angustipennis* Saussure	(カマキリ目、カマキリ科) 日本、全北区
narrow-winged mantis			narrow-winged mantid を見よ
narrow-winged metalmark		*Apodemia multiplaga* Schaus	(チョウ目、シジミタテハ科) 新北区
narrow-winged midget moth		*Tarache augustipennis* Grote	(チョウ目、ヤガ科) 新北区。narrow-winged midget ともいう
narrow-winged Mylon		*Mylon ander* Evans	(チョウ目、セセリチョウ科) 新熱帯区
narrow-winged pearl white		*Elodina padusa* (Hewitson)	(チョウ目、アゲハチョウ科) 豪州区
narrow-winged prominent	ホソバシャチホコ	*Fentonia ocypete* (Bremer)	(チョウ目、シャチホコガ科) 日本、旧北区
narrow-winged pug		*Eupithecia nanata* (Hübner)	(チョウ目、シャクガ科) 旧北区
narrow-winged pug moth			narrow-winged pug を見よ
narrow-winged sand grasshopper		*Melanoplus angustipennis* (Dudge)	(バッタ目、バッタ科) 新北区
narrow-winged tree cricket			snowy tree cricket (1) を見よ
narrow-wriggle			common earwig を見よ
narrow yellow-tipped prominent			quercus caterpillar を見よ
narrowly green-banded swallowtail			green-banded swallowtail を見よ
Narses metalmark		*Lasaia narses* Staudinger	(チョウ目、シジミタテハ科) 新北区
Narva checkerspot	ホソバヒョウモンモドキ	*Chlosyne narva* Fabricius	(チョウ目、タテハチョウ科) 新熱帯区。Narva patch ともいう
Narva patch			Narva checkerspot を見よ
nasal bots of deer			nostril flies を見よ
nasal worm			sheep bot fly を見よ 幼虫の英名
Nasisi sapphire		*Iolaus nasisii* (Riley)	(チョウ目、シジミチョウ科) エチオピア区

英名	和名	学名	所属、分布、ほか
Nason's slug-moth		*Natada nasoni* (Herrich-Schäffer)	（チョウ目、イラガ科）新北区。Nason's slug ともいう
nasutiform termites		*Nasutitermes, Tenuirostritermes*	（シロアリ目、シロアリ科）の昆虫の総称
nasutiform termites			soldierless termites (1) を見よ
Natal Acraea	チャマダラホソチョウ	*Acraea natalica* Boisduval	（チョウ目、タテハチョウ科）エチオピア区
Natal Babul blue		*Azanus natalensis* (Trimen et Bowker)	（チョウ目、シジミチョウ科）エチオピア区
Natal Babul blue			topaz blue を見よ
Natal barred blue	ナンアフタオツバメ	*Cigaritis natalensis* (Westwood)	（チョウ目、シジミチョウ科）エチオピア区
Natal brown		*Coenyropsis natalii* (Boisduval)	（チョウ目、タテハチョウ科）エチオピア区
Natal codling moth			false codling moth (1) を見よ
Natal copper		*Chrysoritis lycegenes* Trimen	（チョウ目、シジミチョウ科）エチオピア区
Natal fruit fly	ナタールミバエ	*Ceratitis rosa* Karsh	（ハエ目、ミバエ科）エチオピア区。カンキツ害虫
Natal opal		*Chrysoritis natalensis* (van Son)	（チョウ目、シジミチョウ科）エチオピア区
Natal pansy			brown commodore を見よ
natal rocksitter		*Durbania limbata* Trimen	（チョウ目、シジミチョウ科）エチオピア区
Natal spotted blue			Natal Babul blue を見よ
Natal sprite		*Pseudagrion spernatum* Hagen	（トンボ目、イトトンボ科）エチオピア区
Natal Temnora		*Temnora natalis* Walker	（チョウ目、スズメガ科）エチオピア区
Natal tree nymph		*Sevenia natalensis* (Boisduval)	（チョウ目、タテハチョウ科）エチオピア区
Natal yellow-banded sapphire		*Iolaus diametra natalica* Vári	（チョウ目、シジミチョウ科）エチオピア区。yellow-banded sapphire (1) を参照
native bees			hairy colletid bees を見よ
native budworm		*Helicoverpa punctigera* (Wallengren)	（チョウ目、ヤガ科）豪州区
native bush cockroaches		*Celatoblatta*	（ゴキブリ目、ゴキブリ科）の昆虫の総称　ニュージーランドに分布
native currant sawfly			small gooseberry sawfly (1) を見よ　米国での英名
native drone fly		*Eristalinus punctulatus* (Macquart)	（ハエ目、ハナアブ科）豪州区
native drywood termite		*Cryptotermes primus* (Hill)	（シロアリ目、レイビシロアリ科）豪州区
native elm bark beetle		*Hylurgopinus rufipes* (Eichhoff)	（コウチュウ目、キクイムシ科）新北区。ニレ害虫
native fig moth		*Lactura caminaea* Meyrick	（チョウ目、マダラガ科）豪州区
native fire ant			fire ant (2) を見よ
native flower ant		*Thynnus zonatus* Guérin-Méneville	（ハチ目、コツチバチ科）豪州区

英名	和名	学名	所属、分布、ほか
native gooseberry sawfly			small gooseberry sawfly (1) を見よ
native holly leafminer			holly leafminer を見よ
native honey bees			stingless bees (2) を見よ　豪州での英名
Natterer's longwing		*Heliconius nattereri* Felder	（チョウ目、タテハチョウ科）新熱帯区
Nauplius crescent			Peruvian crescent を見よ
Nautes hairstreak		*Mithras nautes* (Cramer)	（チョウ目、シジミチョウ科）新熱帯区
Nautes Mithras			Nautes hairstreak を見よ
nautical borer		*Xylotrechus nauticus* (Mannerheim)	（コウチュウ目、カミキリムシ科）新北区
naval commodore		*Precis touhilimasa* Vieillot	（チョウ目、タテハチョウ科）エチオピア区
navel orangeworm		*Amyelois transitella* (Walker)	（チョウ目、メイガ科）新北区。navel orangeworm moth ともいう
navel orangeworm (1)		*Myelois venipars* Dyar	（チョウ目、メイガ科）新北区。カンキツ、クルミ害虫
navy blue burnet moth		*Gynautocera papilionaria* Guérin-Méneville	（チョウ目、マダラガ科）東洋区
navy dropwing		*Trithemis furva* Karsch	（トンボ目、トンボ科）エチオピア区
navy eighty-eight			Astala eighty-eight を見よ
Nawa froghopper	マダラアワフキ	*Awafukia nawae* (Matsumura)	（カメムシ目、アワフキムシ科）日本、旧北区
Nawa gall wasp			quercus gall wasp (2) を見よ
Nawa globular scale	ナワタマカイガラムシ	*Kermes nawae* Kuwana	（カメムシ目、タマカイガラムシ科）日本
nawabs	フタオチョウ属	*Polyura*	（チョウ目、タテハチョウ科）の昆虫の総称
Naxia sister		*Adelpha naxia naxia* (C. et R. Felder)	（チョウ目、タテハチョウ科）新熱帯区
Nayarit gemmed-satyr		*Cyllopsis nayarit* (Chermock)	（チョウ目、タテハチョウ科）新熱帯区
Nayarit Mellana		*Quasimellana nayana* (Bell)	（チョウ目、セセリチョウ科）新熱帯区
Nayarit skipper		*Parphorus* sp.	（チョウ目、セセリチョウ科）新熱帯区
Nazareus flasher		*Narcosius nazaraeus* Steinhauser	（チョウ目、セセリチョウ科）新熱帯区
Nea sister			scarce sister を見よ
Neamathla skipper	ヒメナストラセセリ	*Nastra neamathla* (Skinner et Williams)	（チョウ目、セセリチョウ科）新北区
neat cosmet		*Mompha divisella* Herrich-Schäffer	（チョウ目、カザリバガ科）旧北区
neatfly			blue blowfly を見よ
Neave's Judy		*Abisara neavei* Riley	（チョウ目、シジミタテハ科）エチオピア区
Neave's polyptychus		*Afroclanis neavi* (Hampson)	（チョウ目、スズメガ科）エチオピア区
Neave's silver spot		*Aphnaeus neavei* Bethune-Baker	（チョウ目、シジミチョウ科）エチオピア区

英名	和名	学名	所属、分布、ほか
Nebraska ant cricket		*Myrmecophilus nebrascensis* Lugger	（バッタ目、アリヅカコオロギ科）新北区
Nebraska clearwing moth		*Euhagena nebraskae* Edwards	（チョウ目、スカシバガ科）新北区
Nebraska cone-head		*Neoconocephalus nebrascensis* (Brunner)	（バッタ目、キリギリス科）新北区
nebs			twirler moths を見よ
Nebulosa crescent		*Anthanassa nebulosa nebulosa* (Godman et Salvin)	（チョウ目、タテハチョウ科）新熱帯区
nebulous lady beetle		*Scymnus nebulosus* LeConte	（コウチュウ目、テントウムシ科）新北区
nebulous Munroessa moth		*Elophila nebulosalis* (Fernald)	（チョウ目、メイガ科）新北区
nectar beetles		*Phyllotocus*	（コウチュウ目、コガネムシ科）の昆虫の総称
nectar scarab		*Phyllotocus apicalis* Macleay	（コウチュウ目、コガネムシ科）豪州区
nectar scarabs			nectar beetles を見よ
nectar-sucking blister beetle		*Nemognatha lurida* LeConte	（コウチュウ目、ツチハンミョウ科）新北区
needle-bending pine gall midge		*Contarinia baeri* (Prell)	（ハエ目、タマバエ科）旧北区
needle bug		*Ranatra dispar* Montandon	（カメムシ目、タイコウチ科）豪州区
needle bug			strawberry blossom weevil (2) を見よ
needle-bug			saucer bug (1) を見よ
needle-bugs			creeping water bugs を見よ
needle flies			rolled-winged stoneflies を見よ
needle fly		*Leuctra fusca* Linnaeus	（カワゲラ目、ハラジロオナシカワゲラ科）旧北区
needle miner moths			shiny head-standing moths を見よ
needle miners			yucca moths (1) を見よ　米国での英名
needle-nosed hop bug			hop capsid を見よ
needle-nosed sharpshooter		*Raphirhinus phosphoreus* (Linnaeus)	（カメムシ目、オオヨコバイ科）新熱帯区
needle-shortening gall midge			needle-shortening pine gall midge を見よ
needle-shortening pine gall midge		*Thecodiplosis brachyntera* (Schwagrichen)	（ハエ目、タマバエ科）旧北区
neelid springtails			short-horned springtails を見よ
negatoria canegrub		*Lepidiota negatoria* Blackburn	（コウチュウ目、コガネムシ科）豪州区
neglected clay			grey rustic を見よ
neglected jewelmark		*Sarota neglecta* Stichel	（チョウ目、シジミタテハ科）新熱帯区
neglected rustic			grey rustic を見よ
negro ant			silky ant を見よ　米国での英名

英名	和名	学名	所属、分布、ほか
negro bug		*Corimelaena pulicaria* (Germar)	(カメムシ目、ツチカメムシ科) 新北区。北米の普通種
negro bug		*Thyreocoris scarabaeioides* (Linnaeus)	(カメムシ目、Thyreocoridae) 旧北区
negro bugs		Corimelaeninae	(カメムシ目、ツチカメムシ科) の昆虫の英名
negro bugs			ground bugs を見よ
negundo twig-borer			boxelder twig borer を見よ
neighbour		*Haploa contigua* Walker	(チョウ目、ヒトリガ科) 新北区
neighbour moth			neighbour を見よ
Neita brown		*Neita neita* (Wallengren)	(チョウ目、タテハチョウ科) エチオピア区
Nelson alpine weta		*Deinacrida tibiospina* Salmon	(バッタ目、Stenopelmatidae) 豪州区
Nelson's hairstreak		*Callophrys gryneus nelsoni* (Boisduval)	(チョウ目、シジミチョウ科) 新北区。juniper hairstreak (1) を参照
nematocerans			thread horned flies を見よ
Nemesis mimic white			falcate Dismorphia を見よ
nemestrinids			tangle-veined flies を見よ
nemopterid flies			thread-winged lacewings を見よ
Neo-Guinean termite		*Nasutitermes princeps* (Desneux)	(シロアリ目、シロアリ科) 東洋区
neobule Acraea			wandering donkey Acraea を見よ
neolitsea leaf gall midge		*Pseudasphondylia neolitseae* Yukawa	(ハエ目、タマバエ科) 日本
neolitsea stem borer			litsea longicorn beetle を見よ
neon cuckoo bee		*Thyreus nitidulus* (Fabricius)	(ハチ目、ミツバチ科) 豪州区
Neora hairstreak			common Brangas を見よ
Neotropical bluet		*Enallgama novaehispaniae* Calvert	(トンボ目、イトトンボ科) 新北区
Neotropical corn borer		*Zeadiatraea lineolata* (Walker)	(チョウ目、メイガ科) 新北区、新熱帯区。イネ害虫
Neotropical cornstalk borer			Neotropical corn borer を見よ
Neotropical deer ked		*Lipotena mazamae* (Rondani)	(ハエ目、シラミバエ科) 新北区
neottiophilid flies			nest flies を見よ
neps			midget moths を見よ
nepticulid moths			midget moths を見よ
Nereina white		*Hesperocharis nereina* Hopffer	(チョウ目、シロチョウ科) 新熱帯区
neriid flies			cactus flies を見よ
Nero clearwing		*Godyris nero* (Hewitson)	(チョウ目、タテハチョウ科) 新熱帯区
Nero skipper		*Cobalopsis nero* (Herrich-Schäffer)	(チョウ目、セセリチョウ科) 新熱帯区
nerve-winged insects			lacewings (1) を見よ

英名	和名	学名	所属、分布、ほか
nervous skipper		*Udranomia kikkawai* (Weeks)	(チョウ目、セセリチョウ科) 新熱帯区
Nesaea metalmark		*Baeotis nesaea* Godman et Salvin	(チョウ目、シジミタテハ科) 新熱帯区
nesiotes megalagrion damselfly			flying earwig Hawaiian damselfly を見よ
nessus sphinx		*Amphion nessus* Cramer	(チョウ目、スズメガ科) 新北区
nessus sphinx moth		*Amphion floridensis* Clark	(チョウ目、スズメガ科) 新北区、新熱帯区。nessus sphinx ともいう
nest			鳥、虫などの群は swarm であるが、nest も使用される
nest building termite		*Nasutitermes costalis* (Holmgren)	(シロアリ目、シロアリ科) 新北区、新熱帯区
nest flies		Neottiophilidae	(ハエ目) の昆虫の総称
nest-fly		*Neottiophilum praeustum* (Meigen)	(ハエ目、Neottiophilidae) 旧北区
net-spinning caddis		*Philopotamus montanus* (Donovan)	(トビケラ目、カワトビケラ科) 旧北区
net-spinning caddisflies	シマトビケラ科	Hydropsychidae	(トビケラ目) の昆虫の総称
net-spinning caddisfly		*Macrostemum arcuatum* (Erichson)	(トビケラ目、シマトビケラ科) 新熱帯区
net-spinning caddisfly		*Macronema hageni* Banks	(トビケラ目、シマトビケラ科) 新熱帯区
net-veined bee fly		*Conophorus fenestratus* (Osten Sacken)	(ハエ目、ツリアブ科) 新北区
net-veined midges			net-winged midges を見よ
net-winged beetles		*Lycus*	(コウチュウ目、ベニボタル科) の昆虫の総称
net-winged beetles (1)	ベニボタル科	Lycidae	(コウチュウ目) の昆虫の総称
net-winged insects			lacewings (1) を見よ
net-winged midges	アミカ科	Blephariceridae	(ハエ目) の昆虫の総称
net-winged planthopper		*Biolleyana pictifrons* (Stål)	(カメムシ目、Nogodinidae) 新熱帯区
netted carpet		*Eustroma reticulata* (Denis et Schiffermüller)	(チョウ目、シャクガ科) 旧北区。日本亜種は *E. r. obsoletum* Djakonov アミメナミシャク。netted carpet moth ともいう
netted mountain			netted mountain moth を見よ
netted mountain moth		*Macaria carbonaria* (Clerck)	(チョウ目、シャクガ科) 旧北区
netted pug		*Eupithecia venosata* (Fabricius)	(チョウ目、シャクガ科) 旧北区
netted smudge		*Ypsolopha mucronella* (Scopoli)	(チョウ目、スガ科) 旧北区
netted smudge			scarce water-veneer を見よ
netted sweep		*Whittleia retiella* Newman	(チョウ目、ミノガ科) 日本、旧北区
netted sylph		*Metisella willemi* (Wallengren)	(チョウ目、セセリチョウ科) エチオピア区
nettle butterfly			small tortoiseshell を見よ

英名	和名	学名	所属、分布、ほか
nettle caterpillar		*Setora nitens* (Walker)	（チョウ目、イラガ科）東洋区、大洋区
nettle caterpillar		*Darna pallivitta* (Moore)	（チョウ目、イラガ科）日本、東洋区、大洋区。stinging nettle caterpillar ともいう
nettle groundbug		*Heterogaster urticae* (Fabricius)	（カメムシ目、ナガカメムシ科）旧北区
nettle grub		*Spatulifimbria castaneiceps* Hampson	（チョウ目、イラガ科）東洋区
nettle grub		*Hindothosea cervina* Moore	（チョウ目、イラガ科）東洋区
nettle leafhopper		*Eupteryx aurata* (Linnaeus)	（カメムシ目、オオヨコバイ科）旧北区
nettle sucker		*Trioza urticae* (Linnaeus)	（カメムシ目、トガリキジラミ科）旧北区
nettle-tree butterfly		*Libythea celtis* (Laicharting)	（チョウ目、タテハチョウ科）旧北区、東洋区。日本亜種は *L. c. celtoides* Fruhstorfer テングチョウ。
nettle tree leaf beetle		*Hoplostines laporteae* (Weise)	（コウチュウ目、ハムシ科）豪州区
nettle weevil		*Phyllobius pomaceus* Gyllenhal	（コウチュウ目、ゾウムシ科）旧北区
network-marked froghopper	アミメホソアワフキ	*Aphilaenus guttatus* (Matsumura)	（カメムシ目、アワフキムシ科）日本
network-marked leafhopper	チャイロヨコバイ	*Matsumurella praesul* (Horvath)	（カメムシ目、オオヨコバイ科）日本、旧北区
network-marked yellow tortrix			Japanese oak leafroller を見よ
Neumoegen's giant-skipper		*Agathymus neumoegeni judithae* (Stallings et Turner)	（チョウ目、セセリチョウ科）新北区
Neumoegen's quaker moth		*Oligia egens* (Walker)	（チョウ目、ヤガ科）新北区
neuropterans			lacewings (1) を見よ
neuropteroid insects		Neuropteroidea	の昆虫の総称
neuropteroids			neuropteroid insects を見よ
Nevada arctic	ネバダタカネヒカゲ	*Oeneis nevadensis* (C. et R. Felder)	（チョウ目、タテハチョウ科）新北区
Nevada blue		*Polyommatus golgus* (Hübner)	（チョウ目、シジミチョウ科）全北区
Nevada buck moth		*Hemileuca nevadensis* Stretch	（チョウ目、ヤママユガ科）新北区
Nevada cloudy wing			Mexican cloudywing を見よ
Nevada dampwood termite		*Zootermopsis nevadensis* Hagen	（シロアリ目、オオシロアリ科）新北区
Nevada fritillary		*Speyeria callipe nevadensis* (Edwards)	（チョウ目、タテハチョウ科）新北区。Callipe fritillary を参照
Nevada grayling		*Pseudochazara hippolyte* (Esper)	（チョウ目、タテハチョウ科）旧北区
Nevada sage grasshopper		*Melanoplus rugglesi* Gurney	（バッタ目、バッタ科）新北区
Nevada skipper		*Hesperia nevada* Scudder	（チョウ目、セセリチョウ科）新北区

英名	和名	学名	所属、分布、ほか
Nevada tiger-moth		*Grammia nevadensis* (Grote et Robinson)	（チョウ目、ヒトリガ科）新北区
Neville's windmill		*Byasa nevilli* (Wood-Mason)	（チョウ目、アゲハチョウ科）東洋区
new beauty			annulate mosaic を見よ
new celebean		*Bletogona inexspectata* Uemura	（チョウ目、タテハチョウ科）東洋区
new-clouded yellow	ミナミモンキチョウ	*Colias australis* Verity	（チョウ目、シロチョウ科）旧北区。new-clouded yellow butterfly ともいう
New England bluet		*Enallagma laterale* Morse	（トンボ目、イトトンボ科）新北区
New England buck moth		*Hemiteuca lucina* H. Edwards	（チョウ目、ヤママユガ科）新北区
new English skipper			European skipper を見よ
new forest burnet		*Zygaena viciae* (Denis et Schiffermüller)	（チョウ目、マダラガ科）旧北区
new forest burnet		*Zygaena meliloti* (Esper)	（チョウ目、マダラガ科）旧北区
new forest cicada			mountain cicada を見よ
new forest fly			forest fly を見よ
new forest piercer		*Laspeyresia conicolana* Heylaerts	（チョウ目、ハマキガ科）旧北区
new garden bumblebee			tree bumblebee を見よ
New Guinea birdwing			Priam's birdwing を見よ
New Guinea spiny stick insect		*Eurycantha calcarata* (Lucas)	（ナナフシ目、ナナフシ科）豪州区（ニューギニア）
New Guinea sugarcane weevil			sugarcane weevil borer を見よ
new-house borer		*Arhopalus productus* (LeConte)	（コウチュウ目、カミキリムシ科）新北区
New Jersey mosquito			golden saltmarsh mosquito を見よ
New Mexico range caterpillar		*Hemileuca oliviae* Cockerell	（チョウ目、ヤママユガ科）新北区、新熱帯区
New Orleans ant			Argentine ant を見よ
new small skipper			European skipper を見よ
new thistle case		*Coleophora peribenanderi* (Toll)	（チョウ目、ツツミノガ科）旧北区
new thistle nettle-tap			burdock leaf roller を見よ
new world screw worm			screwworm を見よ　成虫は new world screw-worm fly
new world yucca moth			yucca moth を見よ
New York weevil		*Ithycerus noveboracensis* (Forster)	（コウチュウ目、ミツギリゾウムシ科）新北区
New Zealand armyworm			southern armyworm (2) を見よ
New Zealand blue blowfly		*Calliphora quadrimaculata* (Swederus)	（ハエ目、クロバエ科）豪州区

英名	和名	学名	所属、分布、ほか
New Zealand drywood termite		*Kalotermes brouni* Froggatt	（シロアリ目、レイビシロアリ科）豪州区
New Zealand flax mealybug		*Trionymus diminutus* Leonardi	（カメムシ目、コナカイガラムシ科）旧北区
New Zealand flower thrips		*Thrips obscuratus* (Crawford)	（アザミウマ目、アザミウマ科）豪州区
New Zealand giant stick insect			giant stick insect を見よ
New Zealand glow-worm			glow-worm (1) を見よ
New Zealand grasshopper		*Phaulacridium marginale* (Walker)	（バッタ目、バッタ科）豪州区
New Zealand praying mantis		*Orthodera novaezealandiae* (Colenso)	（カマキリ目、カマキリ科）豪州区
New Zealand primitive moths		Minesarchaeidae	（チョウ目）の昆虫の総称
New Zealand vegetable bug		*Glaucias amyoti* (Dallas)	（カメムシ目、カメムシ科）豪州区
Newman fly		*Dacus newmani* (Perkins)	（ハエ目、ミバエ科）豪州区
Newman's Mathildana moth		*Mathildana newmanella* (Clemens)	（チョウ目、マルハキバガ科）新北区
Newton's sprite		*Pseudagrion newtoni* Pinhey	（トンボ目、イトトンボ科）エチオピア区
ni moth			cabbage looper (1) を見よ
Nicaeus Eurybia		*Eurybia nicaeus* (Fabricius)	（チョウ目、シジミタテハ科）新熱帯区
Nicaraguan emperor		*Doxocopa callianira* (Ménétriès)	（チョウ目、タテハチョウ科）新熱帯区
Nicelli's alder midget		*Phyllonorycter stettinensis* (Nicelli)	（チョウ目、ホソガ科）旧北区
Nicelli's hazel midget		*Phyllonorycter nicellii* (Stainton)	（チョウ目、ホソガ科）旧北区
Nicholl's ringlet	ロードスベニヒカゲ	*Erebia rhodopensis* Nicholl	（チョウ目、タテハチョウ科）旧北区
nickerbean blue		*Hemiargus ammon* Lucas	（チョウ目、シジミチョウ科）新北区
Nickerl			Nickerl's fritillary を見よ
Nickerl's fritillary		*Mellicta aurelia* (Nickerl)	（チョウ目、タテハチョウ科）旧北区
Nicobar albatross		*Appias panda chrysea* Fruhstorfer	（チョウ目、シロチョウ科）東洋区
Nicobar crow		*Euploea scherzeri* (Felder)	（チョウ目、タテハチョウ科）東洋区
Nicobar map		*Cyrestis tabula* de Nicéville	（チョウ目、タテハチョウ科）東洋区
Nicobar Yeoman		*Cirrochroa nicobarica* Wood-Mason	（チョウ目、タテハチョウ科）東洋区
nicoletiids			subterranean silverfishes を見よ
nidularia scale	ニズラリアカイガラムシ	*Nidularia japonica* Kuwana	（カメムシ目、フクロカイガラムシ科）日本
Nielsen geometrid	オビベニヒメシャク	*Idaea nielseni* (Hedemann)	（チョウ目、シャクガ科）日本、旧北区

英名	和名	学名	所属、分布、ほか
Nietner's grappletail		*Heliogomphus nietneri* (Selys)	(トンボ目、サナエトンボ科) 東洋区
Nietner's shadowdamsel		*Drepanosticta nietneri* (Fraser)	(トンボ目、Platystictidae) 東洋区
nigger	メドウスニセコジャノメ (シロオビイチモンジジャノメ)	*Orsotriaena medus* (Fabricius)	(チョウ目、タテハチョウ科) 東洋区、豪州区。*O. m. cinerea* (Butler), *O. m. moira* Waterhouse et Lyell も同英名。この英名は不適切であると、最近は dark grass-brown, smooth-eyed bushbrown といわれる
nigger			nigger-worm を見よ 幼虫の英名
nigger-worm		*Athalia colibri* Christ	(ハチ目、ハバチ科) 新北区
niggerhead termite		*Nasutitermes graveolus* (Hill)	(シロアリ目、シロアリ科) 豪州区
niggerhead termite (1)		*Nasutitermes walkeri* (Hill)	(シロアリ目、シロアリ科) 豪州区
niggerhead termite			nest building termite を見よ
niggers			ladybird beetles を見よ 幼虫の英名
night-biting mosquito		*Aedes nocturnus* (Theobald)	(ハエ目、カ科) 大洋区
nightfeeding sugarcane armyworm		*Leucania loreyi* (Duponchel)	(チョウ目、ヤガ科) 豪州区
night-flying dung beetle	オオツヤマグソコガネ	*Aphodius rufipes* (Linnaeus)	(コウチュウ目、コガネムシ科) 日本、全北区、新熱帯区
night moths			owlet moths を見よ
night-wandering dagger moth		*Acronicta noctivaga* Grote	(チョウ目、ヤガ科) 新北区
night wasp		*Provespa anomala* (Saussure)	(ハチ目、スズメバチ科) 東洋区
Nigidius Jezebel		*Delias ennia nigidius* Miskin	(チョウ目、シロチョウ科) 豪州区。yellow-banded Jezebel を参照
nigra scale	クロカタカイガラムシ	*Parasaissetia nigra* (Nietner)	(カメムシ目、カタカイガラムシ科) 日本、東洋区、新北区、汎熱帯。カンキツ、ハイビスカス害虫
Nigrella crescent		*Tegosa nigrella nigrella* (Bates)	(チョウ目、タテハチョウ科) 新熱帯区
nigricola skipper		*Cynea nigricola* Freeman	(チョウ目、セセリチョウ科) 新熱帯区
Nigrita bagworm moth		*Cryptothelea nigrita* (Barnes et McDunnough)	(チョウ目、ミノガ科) 新北区
nigrofasciate prominent	ワイルマンネグロシャチホコ	*Disparia nigrofasciata* (Wileman)	(チョウ目、シャチホコガ科) 日本、東洋区
nigrohamatum damselfly		*Megalagrion nigrohamatum* (Blackburn)	(トンボ目、イトトンボ科) 大洋区
Niisima small-headed planthopper	ニイジマナガウンカ	*Stenocranus niisimai* Matsumura	(カメムシ目、ウンカ科) 日本
Nikko skipper		*Niconiades nikko* Hayward	(チョウ目、セセリチョウ科) 新熱帯区
Nilgiri clouded yellow		*Colias nilgiriensis* C. et R. Felder	(チョウ目、シロチョウ科) 東洋区
Nilgiri fourring		*Ypthima chenui* (Guérin-Méneville)	(チョウ目、タテハチョウ科) 東洋区

英名	和名	学名	所属、分布、ほか
Nilgiri grass yellow		*Eurema nilgiriensis* Yata	（チョウ目、シロチョウ科）東洋区
Nilgiri tiger	ニルギリアサギマダラ	*Parantica nilgiriensis* (Moore)	（チョウ目、タテハチョウ科）東洋区
Nilgiri tit		*Hypolycaena nilgirica* Moore	（チョウ目、シジミチョウ科）東洋区
Nilgiri tit		*Chliaria nilgirica* (Moore)	（チョウ目、シジミチョウ科）東洋区
nimble meadow katydid		*Orchelimum volantum* McNeill	（バッタ目、キリギリス科）新北区
nine-spotted		*Amata phegea* Linnaeus	（チョウ目、カノコガ科）旧北区
nine-spotted lady beetle			nine-spotted ladybird を見よ
nine-spotted ladybird		*Coccinella novemnotata* Herbst	（コウチュウ目、テントウムシ科）新北区
nine-spotted ladybird beetle			nine-spotted ladybird を見よ
Niobe fritillary	ニオベウラギンヒョウモン	*Fabriciana niobe* (Linnaeus)	（チョウ目、タテハチョウ科）旧北区
Nipisquit ringlet		*Coenonympha nipisquit* (McDunnough)	（チョウ目、タテハチョウ科）新北区
Nippon hairy aphid	ニホンケクダアブラムシ	*Greenidia nipponica* Suenaga	（カメムシ目、アブラムシ科）日本
Nippon lithocolletis	クヌギキンモンホソガ	*Phyllonorycter nipponicella* (Issiki)	（チョウ目、ハモグリガ科）日本
Nippon pine bark beetle	ニホンカバイロキクイムシ	*Hylurgops niponicus* Murayama	（コウチュウ目、キクイムシ科）日本
Nippon sawfly			rose argid sawfly (2) を見よ
Nippon vegetable sawfly			cabbage sawfly (2) を見よ
Nisaee groundstreak		*Arumecla nisaee* (Godman et Salvin)	（チョウ目、シジミチョウ科）新熱帯区
Nitetis hairstreak		*Erora nitetis* (Godman et Salvin)	（チョウ目、シジミチョウ科）新熱帯区
nitidulid beetles			sap beetles (1) を見よ
nitidulids			sap beetles (1) を見よ
Nitobe bagworm			aurea bagworm moth を見よ
Nitobe looper moth	ニトベエダシャク	*Wilemania nitobei* (Nitobe)	（チョウ目、シャクガ科）日本、旧北区
Nitra skipper		*Mnasitheus nitra* Evans	（チョウ目、セセリチョウ科）新熱帯区
nits			sucking lice, human lice を見よ　シラミ目、ハジラミ目の卵のほか、一般に卵、幼虫を nit と俗称
nits			damson-hop aphid を見よ
nits			phthirapterous insects を見よ
nits			common fleas を見よ
Nivalis copper	ロッキーベニシジミ	*Lycaena nivalis* (Boisduval)	（チョウ目、シジミチョウ科）新北区
no-brand crow			striped black crow (1) を見よ

英名	和名	学名	所属、分布、ほか
no-brand grass yellow			small grass yellow (1) を見よ　豪州での英名
no patch white		*Colotis venosa* (Staudinger)	（チョウ目、シロチョウ科）エチオピア区
noble chafer		*Gnorimus nobilis* (Linnaeus)	（コウチュウ目、コガネムシ科）旧北区
noble commander		*Euryphurana nobilis* (Staudinger)	（チョウ目、タテハチョウ科）エチオピア区
noble Emesis		*Emesis fatimella* Westwood	（チョウ目、シジミタテハ科）新熱帯区
noble glory		*Myscelus nobilis* (Cramer)	（チョウ目、セセリチョウ科）新熱帯区
noble leafwing		*Fountainea nobilis* (Bates)	（チョウ目、タテハチョウ科）新熱帯区
noble shadowdamsel		*Drepanosticta digna* (Selys)	（トンボ目、Platystictidae）東洋区
noble swallowtail	チャバネアゲハ	*Papilio nobilis* Rogenhofer	（チョウ目、アゲハチョウ科）エチオピア区
noble tiger beetle		*Cicindela (Cicindela) formosa generosa* Dejean	（コウチュウ目、ハンミョウ科）新北区
noble white Charaxes		*Charaxes nobilis* Druce	（チョウ目、タテハチョウ科）エチオピア区
Nobunaga tussock moth	マエグロマイマイ	*Lymantria nobunaga* Nagano	（チョウ目、ドクガ科）日本
noctuid moths			owlet moths を見よ
noctuids			owlet moths を見よ
nocturnal ground beetles			darkling beetles (1) を見よ
nocturnal paper wasp		*Apoica pallens* (Fabricius)	（ハチ目、スズメバチ科）新熱帯区
nocturnal tiphiids		*Brachycistis*	（ハチ目、コツチバチ科）の昆虫の総称
Nogales metalmark		*Calephelis dreisbachi* McAlpine	（チョウ目、シジミタテハ科）新北区
Nogel's hairstreak		*Tomares nogelii* (Herrich-Schäffer)	（チョウ目、シジミチョウ科）旧北区
Noguchi citrus thrips	ノグチクダアザミウマモドキ	*Acallurothrips nogutii* (Kurosawa)	（アザミウマ目、クダアザミウマ科）日本
noisy shieldback		*Steiroxys strepens* Fulton	（バッタ目、キリギリス科）新北区
Nokomis fritillary	メスキギンボシヒョウモン	*Speyeria nokomis* (Edwards)	（チョウ目、タテハチョウ科）新北区。*S. n. melaena* Mooser et Gracia, *S. n. wenona* dos Passos et Grey, *S. n. coerulescens* (Holland)，*S. n. wenona* Passos et Grey も同英名
Nolcken's spreadwing		*Zea nolckeni* (Mabille)	（チョウ目、セセリチョウ科）新熱帯区
nolid moths	コブガ科	Nolidae	（チョウ目）の昆虫の総称
Nolloth's copper		*Aloeides nollothi* Tite et Dickson	（チョウ目、シジミチョウ科）エチオピア区
nomad			red-veined darter を見よ
nomad dart			common dart (1) を見よ
non-biting midges			midges を見よ
non conformist	クモガタキリガ	*Lithophane lamda* (Fabricius)	（チョウ目、ヤガ科）日本、旧北区

英名	和名	学名	所属、分布、ほか
nonconformist moth			non conformist を見よ
nondescript dagger moth		*Acronicta spinigera* Guenée	(チョウ目、ヤガ科) 新北区
nonpareil cosmet		*Cosmopteryx eximia* Haworth	(チョウ目、カザリバガ科) 旧北区
nonpareil cosmet			Schaffer's neb (1) を見よ
non-polished bark beetle	ツヤナシキクイムシ	*Xyleborus adumbratus* Blandford	(コウチュウ目、キクイムシ科) 日本、旧北区
non-spotted eucosmid	グミオオウスツマキヒメハマキ	*Hedya auricristana* (Walsingham)	(チョウ目、ハマキガ科) 日本、旧北区
nonsuch palmer		*Creteus cyrina* Hewitson	(チョウ目、セセリチョウ科) 東洋区
Noogoora burr longicorn		*Nupserpha vexator* (Pascoe)	(コウチュウ目、カミキリムシ科) 豪州区
Noogoora burr seed fly		*Euaresta aequalis* (Loew)	(ハエ目、ミバエ科) 豪州区
noon-fly		*Mesembrina meridiana* (Linnaeus)	(ハエ目、イエバエ科) 旧北区
nopal moth			cactus moth を見よ
Nopporo leafhopper	アカシヒメヨコバイ	*Paracyba akashiensis* (Takahashi)	(カメムシ目、オオヨコバイ科) 日本、旧北区
Norax hairstreak		*Ignata norax* (Godman et Salvin)	(チョウ目、シジミチョウ科) 新熱帯区
nordic blue			Reverdin's blue を見よ　米国での英名
Norfolk aeschna		*Anaciaeschna isosceles* (Müller)	(トンボ目、ヤンマ科) 旧北区
Norfolk coenagrion			Norfolk damselfly を見よ
Norfolk damselfly		*Coenagrion armatum* (Charpentier)	(トンボ目、イトトンボ科) 旧北区
Norfolk hawker			Norfolk aeschna を見よ
Norfolk Howard			bed bug (1) を見よ　Norfolk-Howards ともいう。英国での俗称でトコジラミ、ナンキンムシをいう
Norfolk Island cicada		*Kikihia convicta* (Distant)	(カメムシ目、セミ科) 豪州区
Norfolk Island pine bark-louse			araucaria scale を見よ
Norfolk Island pine felt scale			araucaria scale を見よ
Norfolk Island pine scale			araucaria mealybug を見よ
Norfolk swallowtail		*Papilio amynthor* Boisduval	(チョウ目、アゲハチョウ科) 豪州区
Norfolk wood-rush dwarf		*Elachista geminatella* (Herrich-Schäffer)	(チョウ目、クサモグリガ科) 旧北区
Norman's lineblue		*Nacaduba normani* Eliot	(チョウ目、シジミチョウ科) 東洋区
Norse grayling	ホッキョクタカネヒカゲ	*Oeneis norna* (Thunberg)	(チョウ目、タテハチョウ科) 旧北区。日本亜種は *O. n. asamana* Matsumura タカネヒカゲ
North African camel oestrid		*Cephalopina titillator* (Clark)	(ハエ目、ヒツジバエ科) 旧北区

英名	和名	学名	所属、分布、ほか
North African grass mantis			African grass mantis を見よ
North American cactus moth		*Melitara dentata* (Grote)	(チョウ目、メイガ科) 新北区、大洋区
North American cow-killer ant			common eastern velvet ant を見よ
North American grass webworm		*Nomophila nearctica* Munroe	(チョウ目、メイガ科) 新北区、新熱帯区
North Island zebra moth			lichen moth (1) を見よ
North Queensland day moth			zodiac moth を見よ
north saddleback moth			saddle back を見よ　北米での英名
Northants bentwing	リンゴハモグリガ	*Lyonetia prunifoliella* (Hübner)	(チョウ目、ハモグリガ科) 日本、旧北区。リンゴ、スモモ害虫
northeastern pine Zale			black-eyed Zale moth を見よ
north-eastern sawyer		*Monochamus notatus* (Drury)	(コウチュウ目、カミキリムシ科) 新北区
northern Antirrhea		*Antirrhea philoctetes casta* Bates	(チョウ目、タテハチョウ科) 新熱帯区
northern apple sphinx moth		*Sphinx poecila* Stephens	(チョウ目、スズメガ科) 新北区
northern arches		*Apamea exulis* (Lefebvre)	(チョウ目、ヤガ科) 旧北区
northern arches		*Apamea zeta assimilis* (Doubleday)	(チョウ目、ヤガ科) 旧北区
northern arches moth		*Drasteria hudsonica* (Grote et Robinson)	(チョウ目、ヤガ科) 新北区。northern arches ともいう
northern Argus	チャイロタテハモドキ	*Junonia erigone* (Cramer)	(チョウ目、タテハチョウ科) 東洋区、豪州区
northern armyworm		*Leucania separata* (Walker)	(チョウ目、ヤガ科) 豪州区
northern ash sphinx			Canadian sphinx を見よ
northern auger beetle		*Xylothrips religiosus* (Boisduval)	(コウチュウ目、ナガシンクイムシ科) 豪州区
northern balsam marble	キシタヒメハマキ	*Pristerognatha penthinana* (Guenée)	(チョウ目、ハマキガ科) 日本、全北区
northern barley leaf miner	キタムギハモグリバエ	*Chromatomyia fuscula* (Zetterstedt)	(ハエ目、ハモグリバエ科) 日本。ムギ害虫
northern Barrens tiger beetle		*Cicindela patruela* Dejean	(コウチュウ目、ハンミョウ科) 新北区
northern blow fly			blackbottle fly を見よ　カナダでの英名
northern blue		*Plebejus anna* (Edwards)	(チョウ目、シジミチョウ科) 新北区
northern blue (1)		*Lycaeides idas* Linnaeus	(チョウ目、シジミチョウ科) 新北区
northern blue			Reverdin's blue を見よ　米国での英名
northern bluet		*Enallagma annexum* (Hagen)	(トンボ目、イトトンボ科) 新北区

英名	和名	学名	所属、分布、ほか
northern bluet			common blue damselfaly を見よ
northern boxelder twig borer moth		*Proteoteras crescentana* Kearfott	(チョウ目、ハマキガ科) 新北区
northern broken dash		*Wallengrenia egeremet* (Scudder)	(チョウ目、セセリチョウ科) 新北区
northern brown			Scotch Argus を見よ
northern brown Argus		*Aricia artaxerxes* (Fabricius)	(チョウ目、シジミチョウ科) 旧北区
northern brown house moth		*Dasypodia cymatodes* Guenée	(チョウ目、ヤガ科) 豪州区
northern brown house moth (1)		*Speiredonia spectans* (Guenée)	(チョウ目、ヤガ科) 豪州区
northern buckeye			buckeye を見よ
northern Burdock borer moth		*Papaipema arctivorens* Hampson	(チョウ目、ヤガ科) 新北区
northern bush katydid		*Scudderia septentrionalis* (Serville)	(バッタ目、キリギリス科) 新北区
northern caddisflies	エグリトビケラ科	Limnophilidae	(トビケラ目) の昆虫の総称
northern caper white		*Belenois subeida* (Felder et Felder)	(チョウ目、シロチョウ科) エチオピア区
northern carpinus aphid	イヌシデクロマダラアブラムシ	*Neochromaphis carpinicola* (Takahashi)	(カメムシ目、アブラムシ科) 日本
northern casemaker caddisflies			northern caddisflies を見よ
northern cattle grub			cattle warble-fly (1) を見よ　米国での英名
northern cattle maggot			cattle warble fly (1) を見よ　北米での幼虫の英名
northern checkerspot		*Chlosyne palla* (Boisduval)	(チョウ目、タテハチョウ科) 新北区、新熱帯区
northern chequered skipper	カラフトタカネキマダラセセリ	*Carterocephalus sylvicola* (Meigen)	(チョウ目、セセリチョウ科) 日本、旧北区
northern cherrynose		*Macrotristria sylvara* (Distant)	(カメムシ目、セミ科) 豪州区
northern citrus butterfly		*Princeps fuscus canopus* (Westwood)	(チョウ目、アゲハチョウ科) 豪州区
northern Clito		*Clito aberrans* (Draudt)	(チョウ目、セセリチョウ科) 新熱帯区
northern clouded yellow			Greenland sulphur を見よ
northern cloudywing		*Thorybes pylades* Scudder	(チョウ目、セセリチョウ科) 新北区
northern coenagrion			spearhead bluet を見よ
northern conifer dart			northern variable dart moth を見よ
northern conifer tussock			pine tussock moth (1) を見よ
northern cordgrass borer moth		*Photedes panatela* (Smith)	(チョウ目、ヤガ科) 新北区
northern corn rootworm		*Diabrotica longicornis* (Say)	(コウチュウ目、ハムシ科) 新北区
northern corn rootworm		*Diabrotica barberi* Smith et Lawrence	(コウチュウ目、ハムシ科) 新北区

英名	和名	学名	所属、分布、ほか
northern crescent		*Phyciodes cocyta* (Cramer)	（チョウ目、タテハチョウ科）新北区
northern crescent		*Phyciodes selenis* (Kirby)	（チョウ目、タテハチョウ科）新北区
northern crescent			pearly crescentspot を見よ
northern currant aphid		*Hyperomyzus rhinanthi* (Schouteden)	（カメムシ目、アブラムシ科）旧北区
northern damselfly			spearhead bluet を見よ
northern dance fly		*Empis borealis* (Linnaeus)	（ハエ目、オドリバエ科）旧北区
northern dart		*Xestia alpicola* (Zetterstedt)	（チョウ目、ヤガ科）旧北区。亜種 *X. a. alpina* (Humphreys et Westwood) も同英名
northern deep-brown dart		*Aporophyla lueneburgensis* (Freyer)	（チョウ目、ヤガ科）旧北区
northern dingy dart			dingy dart を見よ
northern double drummer		*Thopha sessiliba* Distant	（カメムシ目、セミ科）豪州区
northern drab		*Orthosia opima* (Hübner)	（チョウ目、ヤガ科）旧北区
northern dragon		*Antipodogomphus neophytus* Fraser	（トンボ目、サナエトンボ科）豪州区
northern Ectima		*Ectima erycinoides* Felder	（チョウ目、タテハチョウ科）新熱帯区
northern eggar		*Lasiocampa quercus callunae* Palmer	（チョウ目、カレハガ科）旧北区
northern emerald	ホソミモリトンボ	*Somatochlora arctica* (Zetterstedt)	（トンボ目、エゾトンボ科）日本、旧北区
northern Eudeilinia moth		*Eudeilinia herminiata* (Guenée)	（チョウ目、カギバガ科）新北区
northern eyed-skipper		*Cyclosemia anastomosis* Mabille	（チョウ目、セセリチョウ科）新熱帯区
northern faceted-skipper		*Synapte pecta* Evans	（チョウ目、セセリチョウ科）新熱帯区
northern false wireworm		*Gonocephalum carpentariae* (Blackburn)	（コウチュウ目、ゴミムシダマシ科）豪州区、大洋区
northern February red		*Brachyptera putata* (Newman)	（カワゲラ目、ミジカオカワゲラ科）旧北区
northern field cricket		*Gryllus pennsylvanicus* Burmeister	（バッタ目、コオロギ科）新北区
northern finned prominent moth			large dark prominent を見よ
northern footman		*Eilema sericea* Gregson	（チョウ目、ヒトリガ科）旧北区
northern giant flag moth		*Dysschema howardi* (Edwards)	（チョウ目、ヒトリガ科）新北区
northern grass dart		*Taractrocera ilia* Waterhouse	（チョウ目、セセリチョウ科）豪州区
northern grasshopper		*Melanoplus borealis* (Fieber)	（バッタ目、バッタ科）新北区
northern grass-veneer		*Crambus furcatellus* Zetterstedt	（チョウ目、メイガ科）旧北区
northern green longwing		*Philaethria diatonica* (Fruhstorfer)	（チョウ目、タテハチョウ科）新熱帯区

英名	和名	学名	所属、分布、ほか
northern green looper moth		*Xanthorhoe benedicta* Meyrick	（チョウ目、シャクガ科）豪州区
northern greengrocer		*Cyclochila virens* Distant	（カメムシ目、セミ科）豪州区
northern grizzled skipper		*Pyrgus centaureae* (Rambur)	（チョウ目、セセリチョウ科）新北区、新熱帯区
northern hairstreak	オンタリオカラスシジミ	*Euristrymon ontario* (Edwards)	（チョウ目、シジミチョウ科）新北区
northern hairstreak		*Satyrium favonius ontario* (Edwards)	（チョウ目、シジミチョウ科）新北区
northern hairstreak (1)		*Jalmenus eichhorni* (Staudinger)	（チョウ目、シジミチョウ科）豪州区
northern house mosquito			common mosquito を見よ
northern imperial blue			northern hairstreak (1) を見よ　northern imperial blue butterfly ともいう
northern Jezebel			scarlet Jezebel を見よ
northern jungle queen	ムラサキワモンチョウ	*Stichophthalma camadeva* (Westwood)	（チョウ目、タテハチョウ科）東洋区
northern katydid			true katydid を見よ
northern marble			northern marblewing を見よ
northern marblewing		*Euchloe creusa* (Doubleday)	（チョウ目、シロチョウ科）新北区
northern masked chafer		*Cyclocephala borealis* Arrow	（コウチュウ目、コガネムシ科）新北区
northern Mestra			Amymone を見よ
northern metalmark		*Calephelis borealis* (Grote et Robinson)	（チョウ目、シジミタテハ科）新北区
northern mole cricket		*Neocurtilia hexadactyla* (Perty)	（バッタ目、ケラ科）新北区。夜行性で池や流れ近くに生息
northern mottled carpet			white-banded black moth を見よ
northern Nessaea	ベニモンミズイロタテハ	*Nessaea aglaura aglaura* (Doubleday)	（チョウ目、タテハチョウ科）新熱帯区
northern paper wasp			golden paper wasp を見よ
northern pearl		*Mutuuraia terrealis* (Treitschke)	（チョウ目、メイガ科）旧北区
northern pearl white		*Elodina perdita* Miskin	（チョウ目、シロチョウ科）豪州区
northern pearly eye			pearly eye (1) を見よ
northern pencil-blue			Gilbert's blue を見よ
northern Petrophora moth		*Petrophora subaequaria* (Walker)	（チョウ目、シャクガ科）新北区
northern pine looper moth		*Caripeta piniata* Packard	（チョウ目、シャクガ科）新北区
northern pine processionary moth			pine processionary を見よ
northern pine sphinx			pine tree sphinx を見よ　northern pine sphinx moth ともいう

英名	和名	学名	所属、分布、ほか
northern pine tussock moth			pine tussock moth (1) を見よ
northern pine weevil		*Pissodes approximatus* Hopkins	(コウチュウ目、ゾウムシ科) 新北区
northern pitch twig moth		*Petrova albicapitana* (Busck)	(チョウ目、ハマキガ科) 新北区。亜種 *P. a. arizonensis* (Heinrich) は pinyon pitch nodule moth
northern purple azure			purple azure を見よ
northern pygmy clubtail		*Lanthus parvulus* (Selys)	(トンボ目、サナエトンボ科) 新北区
northern Pyrausta moth		*Pyrausta borealis* Packard	(チョウ目、メイガ科) 新北区
northern rat flea	ヨーロッパネズミノミ	*Nosopsyllus fasciatus* (Bosc)	(ノミ目、ナガノミ科) 日本、全北区。北米での英名
northern rice cricket			southern rice katydid を見よ
northern rice cricket			northern rice katydid を見よ
northern rice katydid	ヒメクサキリ	*Homorocoryphus jezoensis* (Matsumura et Shiraki)	(バッタ目、キリギリス科) 日本、旧北区
northern ringlet		*Hypocysta irius* (Fabricius)	(チョウ目、タテハチョウ科) 豪州区
northern river hunter		*Austrogomphus doddi* Tillyard	(トンボ目、サナエトンボ科) 豪州区
northern robberfly		*Rhadiurgus variabilis* (Zetterstedt)	(ハエ目、ムシヒキアブ科) 旧北区
northern rock crawler		*Grylloblatta campodeiformis* Walker	(ガロアムシ目、ガロアムシ科) 新北区。体長3cm。氷河近くの岩に生息
northern rough bollworm			spotted bollworm (1) を見よ　豪州での英名
northern rowan pigmy		*Stigmella magdalenae* (Klimesch)	(チョウ目、モグリチビガ科) 旧北区
northern rubus aphid	イチゴトゲアブラムシ	*Matsumuraja rubi* (Matsumura)	(カメムシ目、アブラムシ科) 日本、旧北区、東洋区
northern rustic		*Standfussiana lucernea* (Linnaeus)	(チョウ目、ヤガ科) 旧北区
northern sandy dung beetle		*Euoniticellus intermedius* (Reiche)	(コウチュウ目、コガネムシ科) 豪州区
northern Selenia moth		*Selenia alciphearia* Walker	(チョウ目、シャクガ科) 新北区
northern Setabis		*Setabis lagus jansoni* (Butler)	(チョウ目、シジミタテハ科) 新熱帯区
northern short-tailed admiral			dimorphic admiral を見よ
northern sicklewing		*Eantis tamenund* (W. H. Edwards)	(チョウ目、セセリチョウ科) 新熱帯区
northern silver ochre			Maheta skipper を見よ
northern silver-stiletto		*Spiriverpa lunulata* Zetterstedt	(ハエ目、ツルギアブ科) 旧北区
northern silver-striped marble		*Olethreutes palustrana* (Liebig et Zeller)	(チョウ目、ハマキガ科) 旧北区

英名	和名	学名	所属、分布、ほか
northern snailshell caddisfly		*Helicopsyche borealis* (Hagen)	（トビケラ目、カタツムリトビケラ科）新北区
northern snout-skipper		*Anisochoria bacchus* Evans	（チョウ目、セセリチョウ科）新熱帯区
northern spinach		*Eulithis populata* (Linnaeus)	（チョウ目、シャクガ科）旧北区
northern spinach moth			northern spinach を見よ
northern spreadwing		*Lestes disjunctus* Selys	（トンボ目、アオイトトンボ科）新北区
northern summer mayfly		*Siphlomurus alternatus* (Say)	（カゲロウ目、フタオカゲロウ科）旧北区
northern tent caterpillar		*Malacosoma californicum pluviale* Dyar	（チョウ目、カレハガ科）新北区
Northern Terriotry fruit fly		*Dacus aquilonis* (May)	（ハエ目、ミバエ科）豪州区
northern true katydid			true katydid を見よ
northern variable dart moth		*Xestia badicollis* Grote	（チョウ目、ヤガ科）新北区。northern variable dart ともいう
northern vegetable grasshopper	サッポロフキバッタ	*Miramella sapporensis* Shiraki	（バッタ目、バッタ科）日本。ダイズ害虫
northern walkingstick			walking stick を見よ
northern wall brown		*Lasiommata petropolitana* (Fabricius)	（チョウ目、タテハチョウ科）旧北区
northern white skipper			large white skipper を見よ
northern white-faced darter			ruby whiteface を見よ
northern willow clearwing moth		*Synanthedon bolteri* (Edwards)	（チョウ目、スカシバガ科）新北区
northern winter			northern winter moth (1) を見よ
northern winter moth		*Cheimatobia boreata* Hübner	（チョウ目、シャクガ科）旧北区
northern winter moth (1)		*Operophtera fagata* (Scharfenberg)	（チョウ目、シャクガ科）旧北区
northern wood ant		*Formica aquilonia* Yarrow	（ハチ目、アリ科）旧北区
northern wood cricket		*Gryllus vernalis* Blatchley	（バッタ目、コオロギ科）新北区
northland tusked weta		*Anisoura nicobaria* Ander	（バッタ目、Stenopelmatidae）豪州区
northwest alpine	ロッキーベニヒカゲ	*Erebia vidleri* Elwes	（チョウ目、タテハチョウ科）新北区
northwest coast mosquito		*Aedes (Ochlerotatus) aboriginis* Dyar	（ハエ目、カ科）新北区
northwest ringlet		*Coenonympha tullia ampelos* W. H. Edwards	（チョウ目、タテハチョウ科）新北区。large heath を参照
northwestern fritillary		*Speyeria hesperis* (W. H. Edwards)	（チョウ目、タテハチョウ科）新北区

英名	和名	学名	所属、分布、ほか
northwestern leaf-cutting bee		*Megachile (Megachiloides) pascoensis* Mitchell	（ハチ目、ハキリバチ科）新北区
Norway maple aphid		*Periphyllus lyropictus* (Kessler)	（カメムシ目、アブラムシ科）新北区
Norway maple aphid		*Periphyllus aceris* (Linnaeus)	（カメムシ目、アブラムシ科）旧北区
Norway maple leaf-stalk borer			old maple-seed pigmy を見よ
Norway maple seedminer		*Ectoedemia sericopeza* (Zeller)	（チョウ目、モグリチビガ科）新北区
Norway spruce tortrix	トドマツチビハマキ	*Pseudohermenias clausthaliana* Saxen	（チョウ目、ハマキガ科）日本、旧北区
Norway spruce weevil			silver fir weevil を見よ
Norway wasp			Norwegian wasp を見よ
Norwegian wasp		*Dolichovespula norwegica* (Fabricius)	（ハチ目、スズメバチ科）旧北区
nose bot fly	アトアカウマバエ	*Gasterophilus haemorrhoidalis* (Linnaeus)	（ハエ目、ウマバエ科）日本、汎世界
nose flies			bot flies (1) (2) を見よ
no-see-ums			punkies を見よ　ヌカカは小さくて見難いことから You don't see them の北米先住民の英語に由来
nosodendrid beetles			wounded-tree beetles を見よ
nosodendrids			wounded-tree beetles を見よ
nostril flies		Cephenomyiinae	（ハエ目、ヒツジバエ科）の昆虫の総称
nostril fly of sheep			sheep bot fly を見よ
nosy blue		*Orachrysops nasutus* Henning et Henning	（チョウ目、シジミチョウ科）エチオピア区
notable apote		*Apote notabilis* Scudder	（バッタ目、キリギリス科）新北区
notch-mouthed ground beetles		Licinini	（コウチュウ目、オサムシ科）の昆虫の総称
notch-tipped flower longhorn beetle		*Typocerus sinuatus* (Newman)	（コウチュウ目、カミキリムシ科）新北区
notch-wing			maple spanworm を見よ
notch-winged geometer			maple spanworm を見よ　notch-winged geometer moth ともいう
notch-winged wave moth		*Idaea furciferata* (Packard)	（チョウ目、シャクガ科）新北区
notched crescent		*Anthanassa dracaena phlegias* (Godman)	（チョウ目、タテハチョウ科）新熱帯区
notched single-dot bell		*Epinotia crenana* Duponchel	（チョウ目、ハマキガ科）旧北区
noted shieldback		*Ateloplus notatus* Scudder	（バッタ目、キリギリス科）新北区
nothopanax looper moth		*Epirrhanthis alectoraria* Walker	（チョウ目、シャクガ科）豪州区
Notioplusia moth		*Notioplusia illustrata* (Guenée)	（チョウ目、ヤガ科）新北区、新熱帯区、豪州区、エチオピア区

英名	和名	学名	所属、分布、ほか
notodontid moths			prominents を見よ
notonectids			backswimmers を見よ
noun moth			nun moth を見よ
November carpet		*Oporinia dilutata* (Denis et Schiffermüller)	(チョウ目、シャクガ科) 旧北区
November day		*Diurnea phryganella* Hübner	(チョウ目、マルハキバガ科) 旧北区
November moth		*Epirrita dilutata* (Denis et Schiffermüller)	(チョウ目、シャクガ科) 旧北区
November moth (1)	カバエダシャク	*Colotois pennaria ussuriensis* Bang-Haas	(チョウ目、シャクガ科) 日本。リンゴ害虫
Novice		*Amauris ochlea* (Boisduval)	(チョウ目、タテハチョウ科) エチオピア区
noxia canegrub		*Lepidiota noxia* Britton	(コウチュウ目、コガネムシ科) 豪州区
Nubis skipper		*Onespa nubis* Steinhauser	(チョウ目、セセリチョウ科) 新熱帯区
Nuevo Leon checkerspot			Kendall's checkerspot を見よ
Nullabar caves cockroach		*Trogloblattella nullarborensis* Mackerras	(ゴキブリ目、チャバネゴキブリ科) 豪州区
Numata longwing		*Heliconius numata* (Cramer)	(チョウ目、タテハチョウ科) 新熱帯区
nun		*Tarache luctuosa* Denis et Schiffermüller	(チョウ目、ヤガ科) 旧北区
nun (1)		*Tarache aprica* (Smith)	(チョウ目、ヤガ科) 全北区
nun			nun moth を見よ
nun moth	ノンネマイマイ	*Lymantria monacha* (Linnaeus)	(チョウ目、ドクガ科) 日本、旧北区。リンゴ、カラマツ、ブナノキ害虫。成虫の止まった形が修道女の後姿に似ることから
nuns			mantids (1)(2) を見よ
nurscia sabre-wing		*Mimoniades nurscia* (Swainson)	(チョウ目、セセリチョウ科) 新熱帯区
nurse			保母虫 (幼虫を保護する昆虫。働きバチ、働きアリなど)
nursery pine sawfly		*Gilpinia frutetorum* Fabricius	(ハチ目、マツハバチ科) 旧北区
nut bud moth			variable red bell を見よ 米国での英名
nut bud tortrix			variable red bell を見よ 米国での英名
nut fruit moth	クリミガ	*Cydia kurokoi* (Amsel)	(チョウ目、ハマキガ科) 日本。クリ害虫
nut fruit tortrix			splended piercer を見よ
nut leaf blister moth		*Phyllonorycter coryli* Nicelli	(チョウ目、ホソガ科) 旧北区
nut leaf weevil		*Strophosomus melanogrammus* (Forster)	(コウチュウ目、ゾウムシ科) 旧北区
nut sawfly			hazel sawfly を見よ
nut scale		*Eulecanium tiliae* (Linnaeus)	(カメムシ目、カタカイガラムシ科) 旧北区
nut skipper			holly blue を見よ

英名	和名	学名	所属、分布、ほか
nut tortrix		*Laspeyresia amplana* Hübner	（チョウ目、ハマキガ科）旧北区
nut tortrix			chequered fruit-tree tortrix を見よ
nut-tree tuffet			nut-tree tussock を見よ
nut-tree tussock		*Colocasia coryli* (Linnaeus)	（チョウ目、ヤガ科）旧北区
nut weevil			acorn weevil (1) を見よ
nut weevils			acorn weevils (1) を見よ
nutgrass armyworm			African armyworm を見よ　米国での英名
nutgrass billbug			southern corn billbug を見よ
nutgrass borer moth	シロテントガリバヒメハマキ	*Bactra venosoma* (Zeller)	（チョウ目、ハマキガ科）日本、東洋区、大洋区、豪州区、全北区、エチオピア区。nutsedge borer ともいう
nutgrass weevil		*Athesapeuta cyperi* Marshall	（コウチュウ目、ゾウムシ科）新北区
nutmeg	タイリクウスイロヨトウ	*Discestra trifolii* (Hufnagel)	（チョウ目、ヤガ科）日本、全北区。幼虫はキャベツ、クローバーなど多くの作物害虫。nutmeg moth ともいう
nutmeg weevil			coffee weevil を見よ
Nuttall's blister beetle		*Lytta nuttalli* Say	（コウチュウ目、ツチハンミョウ科）新北区。Nuttall blister beetle とも記される
Nuttall's sheep moth		*Hemileuca nuttalli* (Strecker)	（チョウ目、ヤママユガ科）新北区
Nyasa silverline		*Cigaritis nyassae* (Butler)	（チョウ目、シジミチョウ科）エチオピア区
Nyctelius skipper			violet-banded skipper を見よ
nymph			昆虫、とくに不完全変態類の幼虫の英名
nymphalids			brush-footed butterflies を見よ
nymphomyiid flies	ハネカ科	Nymphomyiidae	（ハエ目）の昆虫の総称
nymphs			brush-footed butterflies を見よ　南米での英名
nymphula moth		*Nymphula ekthlipsis* (Grote)	（チョウ目、メイガ科）新北区
Nysa Jezebel		*Delias nysa* (Fabricius)	（チョウ目、シロチョウ科）豪州区
Nysa roadside skipper			mottled roadside skipper を見よ

英名	和名	学名	所属、分布、ほか
O			
Oahu damselfly		*Megalagrion oahuense* (Blackburn)	(トンボ目、イトトンボ科) 大洋区。ハワイのOahu島に由来
oak-and-maple humpted caterpillar		*Schizura ipomoeae* (Doubleday)	(チョウ目、シャチホコガ科) 新北区
oak aphid		*Myzocallis annulata* (Hartig)	(カメムシ目、アブラムシ科) 旧北区、豪州区
oak aphid (1)	カシワホシブチアブラムシ	*Tuberculatus quercicola* (Matsumura)	(カメムシ目、アブラムシ科) 日本、旧北区
oak aphid			oak leaf aphid (1) を見よ　豪州での英名
oak-apple			oak apple gall wasp を見よ　虫えいの英名
oak apple gall wasp		*Biorhiza pallida* (Olivier)	(ハチ目、タマバチ科) 旧北区。本種がカシにつくる虫えいは oak apple といい、40 mm になる
oak bark beetle (1)		*Scolytus intricatus* (Ratzeburg)	(コウチュウ目、キクイムシ科) 旧北区
oak bark beetle (2)	シラカシノキクイムシ	*Acanthotomicus spinosus* Blandford	(コウチュウ目、キクイムシ科) 日本。カシ類害虫
oak bark mining moth		*Spulerina simploniella* Fischer von Röslerstamm	(チョウ目、ホソガ科) 旧北区
oak bark scaler		*Encyclops coerulea* Jurecek	(コウチュウ目、カミキリムシ科) 新北区
oak beauty		*Biston stratara* (Hufnagel)	(チョウ目、シャクガ科) 旧北区。日本亜種は *B. s. hasegawai* Inoue チャオビトビモンエダシャク
oak beauty moth		*Phaeoura quernaria* (Smith)	(チョウ目、シャクガ科) 新北区。oak beauty ともいう
oak beauty moth			oak beauty を見よ
oak beetle			oak bark beetle (1) を見よ
oak beetle			great capricorn beetle を見よ
oak Besma moth		*Besma quercivoraria* (Guenée)	(チョウ目、シャクガ科) 新北区。oak Besma ともいう
oak blotch miner moth		*Tischeria quercitella* Clemens	(チョウ目、ムモンハモグリガ科) 新北区
oakblues		*Arhopala*	(チョウ目、シジミチョウ科) の昆虫の総称
oak borer (1)		*Buprestis gibbsi* LeConte	(コウチュウ目、タマムシ科) 新北区
oak borer (2)	カシノナガキクイムシ	*Platypus quercivorus* (Murayama)	(コウチュウ目、ナガキクイムシ科) 日本、東洋区。カシ、シイノキ類害虫
oak borer (3)		*Agrilus angustulus* (Illiger)	(コウチュウ目、タマムシ科) 旧北区
oak brindled-beauty			oak beauty を見よ
oak bud collared-gall cynipid		*Andricus curvator* Hartig	(ハチ目、タマバチ科) 旧北区
oak bud gall wasp		*Cynips quercifolii* Linnaeus	(ハチ目、タマバチ科) 旧北区
oak bud red-gall cynipid		*Cynips divisa* Hartig	(ハチ目、タマバチ科) 旧北区
oak burncow		*Coraebus florentinus* (Herbst)	(コウチュウ目、タマムシ科) 旧北区。幼虫の英名
oak bush-cricket		*Meconema thalassinum* (De Geer)	(バッタ目、キリギリス科) 全北区

英名	和名	学名	所属、分布、ほか
oak bush-crickets		Meconemidae	(バッタ目) の昆虫の総称　キリギリス科の異名
oak button gall			common spangle gall wasp によって作られた虫えい
oak caterpillar	オオトビモンシャチホコ	*Phalerodonta manleyi* (Leech)	(チョウ目、シャチホコガ科) 日本。クリ、カシ類害虫
oak cerambyx			great capricorn beetle を見よ
oak clearwing borer			red oak clearwing moth を見よ
oak clearwing moth		*Paranthrene asilipennis* (Boisduval)	(チョウ目、スカシバガ科) 新北区
oak curculio	クヌギシギゾウムシ	*Curculio robustus* (Roelofs)	(コウチュウ目、ゾウムシ科) 日本、旧北区
oak egger		*Lasiocampa quercus* (Linnaeus)	(チョウ目、カレハガ科) 旧北区。亜種 *L. q. callunae* Palmer は northern eggar という
oak egger moth			カレハガの類。tent caterpillar moths (1) を見よ
oak-feeding silkworm			Japanese silkmoth を見よ
oak felt scale		*Eriococcus quercus* (Comstock)	(カメムシ目、フクロカイガラムシ科) 新北区
oak flea beetle		*Haltica quercetorum* Foudras	(コウチュウ目、ハムシ科) 旧北区
oak flea weevil		*Rhynchaenus quercus* (Linnaeus)	(コウチュウ目、ゾウムシ科) 旧北区
oak fringed scale	ナラフサカイガラムシ	*Asterolecanium japonicum* Cockerell	(カメムシ目、フサカイガラムシ科) 日本、東洋区
oak gall		*Loxaulus maculipennis* Ashmead	(ハチ目、タマバチ科) の虫えい。新北区
oak gall		*Neuroterus clavensis* Weld	(ハチ目、タマバチ科) の虫えい。新北区
oak-gall			gall-nut wasp の虫えい
oak-gall			oak apple gall wasp を見よ　虫えいの英名
oakgall borer moth		*Synanthedon decipiens* Edwards	(チョウ目、スカシバガ科) 新北区。幼虫は oak gall borer
oak gall chalcid		*Torymus californicus* (Ashmead)	(ハチ目、オナガコバチ科) 新北区
oakgall clearwing			oakgall borer moth を見よ
oak gall snout moth			orange-tufted Oneida moth を見よ
oak gall wasp	ナライガフシバチ	*Andricus japonicus* Ashmead	(ハチ目、タマバチ科) 日本
oak gall wasp			marble gall wasp を見よ
oak gall weevil		*Curculio villosus* Fabricius	(コウチュウ目、ゾウムシ科) 旧北区
oak globular			kermes scale を見よ
oak groundling		*Adrasteia sedulitella* (Busck)	(チョウ目、キバガ科) 新北区
oak hairstreak			southern hairstreak を見よ
oak hawk-moth		*Marumba quercus* (Denis et Schiffermüller)	(チョウ目、スズメガ科) 旧北区
oak hooktip		*Drepana glaucata* (Scopoli)	(チョウ目、カギバガ科) 旧北区

763

英名	和名	学名	所属、分布、ほか
oak hook-tip		*Drepana binaria* (Hufnagel)	(チョウ目、カギバガ科) 旧北区
oak jumping weevil	カシワノミゾウムシ	*Orchestes japonicus* (Hustache)	(コウチュウ目、ゾウムシ科) 日本。カシ類害虫
oak lace bug		*Corythucha arcuata* (Say)	(カメムシ目、グンバイムシ科) 新北区。カシ類害虫
oak lappet			lappet を見よ　原名亜種の英名
oak large spined aphid			oak aphid (1) を見よ
oak lasiocampid			oak egger を見よ
oak leaf aphid		*Tuberculoides annulatus* (Hartig)	(カメムシ目、Callaphididae) 旧北区
oak leaf aphid (1)		*Myzocallis castanicola* Baker	(カメムシ目、アブラムシ科) 旧北区
oak leaf beetle			cereal leaf beetle (1) を見よ
oak leaf blister-gall cynipid			spangle gall wasp を見よ
oak leaf blotchminer		*Cameraria agrifoliella* Braun	(チョウ目、ホソガ科) 新北区
oakleaf butterflies			leaf butterflies を見よ
oak leaf cherry-gall cynipid			cherry gall wasp を見よ
oak leaf cupped-gall cynipid		*Neuroterus tricolor* (Hartig)	(ハチ目、タマバチ科) 旧北区
oak leaf miner			oak flea weevil を見よ
oak leaf phylloxera		*Phylloxera glabra* (von Heyden)	(カメムシ目、ネアブラムシ科) 旧北区
oak leaf roller			green oak roller を見よ
oak leaf roller			rusty button を見よ
oak leaf roller weevil		*Attelabus nitens* (Scopoli)	(コウチュウ目、オトシブミ科) 旧北区
oak leaf smooth-gall cynipid		*Neuroterus albipes* (Schenck)	(ハチ目、タマバチ科) 旧北区。本種がカシ類の葉につくる虫えいを smooth spangle という
oak leaf spangle-gall cynipid			common spangle gall wasp を見よ
oak leaf striped-gall cynipid			green velvet bud gall wasp を見よ
oak leaf sucker		*Trioza remota* Forster	(カメムシ目、トガリキジラミ科) 旧北区
oak leafhopper (1)		*Typhlocyba quercus* (Fabricius)	(カメムシ目、オオヨコバイ科) 旧北区
oak leafhopper (2)	カシヒメヨコバイ	*Aguriahana quercus* (Matsumura)	(カメムシ目、オオヨコバイ科) 日本。クリ、カシ害虫
oak leafminer			Zeller's midget moth を見よ
oak leaf-mining sawfly		*Protenusa pygmaea* (Klug)	(ハチ目、ハバチ科) 旧北区
oak leaf-mining weevil			oak flea weevil を見よ
oak leafroller		*Archips semiferana* (Walker)	(チョウ目、ハマキガ科) 新北区。oak leafroller moth ともいう

英名	和名	学名	所属、分布、ほか
oak leafroller			green oak roller を見よ
oak leafroller moth		*Ancylis burgessiana* (Zeller)	(チョウ目、ハマキガ科) 新北区
oak leaf-rolling weevil			oak leaf roller weevil を見よ
oakleafs			leaf butterflies を見よ
oak leaftier		*Croesia semipurpurana* (Kearfott)	(チョウ目、ハマキガ科) 新北区、新熱帯区。oak leaf roller ともいう。成虫は oak leaftier moth
oak leaftier moth		*Psilocorsis quercicella* Clemens	(チョウ目、マルハキバガ科) 新北区
oak leaf-tying Psilocorsis moth			oak leaftier moth を見よ
oakleaves (oakleaf)		*Mallika*	(チョウ目、タテハチョウ科) の昆虫の総称
oak lecanium		*Parthenolecanium quercifex* (Fitch)	(カメムシ目、カタカイガラムシ科) 新北区
oak longicorn			oak longicorn beetle を見よ
oak longicorn beetle	ハラアカコブカミキリ	*Moechotypa diphysis* (Pascoe)	(コウチュウ目、カミキリムシ科) 日本、旧北区。シイタケ、材木害虫
oak looper			oak-worm looper を見よ
oak louse		*Lachnus iliciphilus* (Del Guercio)	(カメムシ目、アブラムシ科) 旧北区
oak lutestring		*Cymatophorima diluta* (Denis et Schiffermüller)	(チョウ目、トガリバガ科) 旧北区
oak lyctid	ナラヒラタキクイムシ	*Lyctus linearis* Goeze	(コウチュウ目、ナガシンクイムシ科) 日本、全北区
oak moth	ウスイロクチブサガ	*Ypsolopha parenthesellus* (Linnaeus)	(チョウ目、スガ科) 日本、旧北区
oak moth			oak egger を見よ
oak moths		Dioptidae	(チョウ目) の昆虫の総称
oak noctuid	オニベニシタバ	*Catocala dula* Bremer	(チョウ目、ヤガ科) 日本、旧北区
oak nut			marble gall wasp を見よ
oak Nycteoline			oak Nycteoline moth を見よ
oak Nycteoline moth		*Nycteola revayana* (Scopoli)	(チョウ目、ヤガ科) 旧北区
oak phylloxera		*Phylloxera coccinea* (Von Heyden)	(カメムシ目、ネアブラムシ科) 旧北区
oak phylloxera		*Phylloxera quercus* Fonscolombe	(カメムシ目、ネアブラムシ科) 旧北区
oak pinhole borer		*Platypus cylindrus* (Fabricius)	(コウチュウ目、ナガキクイムシ科) 旧北区
oak planthopper			oak leafhopper (2) を見よ
oak platypodid			oak borer (2) を見よ
oak platypodid borer			oak borer (2) を見よ
oak porcelain midget		*Phyllonorycter saportella* (Duponchel)	(チョウ目、ホソガ科) 旧北区

英名	和名	学名	所属、分布、ほか
oak procession moth		*Thaumetopoea processionea* (Linnaeus)	（チョウ目、シャチホコガ科）旧北区。oak processionary, oak processionary moth ともいう。幼虫は oak processionary caterpillar
oak processionary			oak procession moth を見よ
oak processionary caterpillar			oak procession moth を見よ
oak processionary moth			oak procession moth を見よ
oak red-barred twist			brown oak tortrix を見よ
oak-ribbed skeletonizer moth			live oak ribbed casemaker を見よ
oak-roller moth			green oak roller を見よ
oak roller worm			green oak roller を見よ
oak root gall			oak apple gall wasp によって作られた虫えい
oak rustic		*Dryobota labecula* (Esper)	（チョウ目、ヤガ科）旧北区
oak sapling borer		*Goes tessulatus* (Haldemann)	（コウチュウ目、カミキリムシ科）新北区
oak sapwood borer			oak bark beetle (1) を見よ
oak scale	クサマルカイガラムシ	*Quadraspidiotus cryptoxanthus* (Cockerell)	（カメムシ目、マルカイガラムシ科）日本、旧北区
oak scale (1)	カシノアカカイガラムシ	*Kuwania quercus* (Kuwana)	（カメムシ目、ワタフキカイガラムシ科）日本、東洋区。カシ、シイノキ類害虫
oak shoot sawfly	アカガシクキバチ	*Janus kashivorus* Yano et Sato	（ハチ目、クキバチ科）日本。カシ類害虫
oak silkworm			Chinese oak silkmoth を見よ
oak skeletonizer		*Bucculatrix ainsliella* Murtfeldt	（チョウ目、チビガ科）新北区。oak skeletonizer moth ともいう
oak skeletonizer moth		*Carcina quercana* Fabricius	（チョウ目、マルハキバガ科）全北区
oak slug sawfly		*Caliroa annulipes* (Klug)	（ハチ目、ハバチ科）旧北区
oak-spangle			lenticular gall wasp を見よ　虫えいの英名
oak spined aphid	カシワトゲマダラアブラムシ	*Tuberculatus kashiwae* (Matsumura)	（カメムシ目、アブラムシ科）日本
oak-sponge			oak apple gall wasp を見よ　虫えいの英名
oak stag beetle		*Platyceroides agassizi* (LeConte)	（コウチュウ目、クワガタムシ科）新北区
oak-strobile			artichoke gall を見よ
oak sucker	カシトガリキジラミ	*Heterotrioza remota* (Forster)	（カメムシ目、トガリキジラミ科）日本、旧北区。カシ害虫
oak terminal-shoot gall midge		*Arnoldiola quercus* (Binnic)	（ハエ目、タマバエ科）旧北区
oak tiger longicorn			oak tiger longicorn beetle を見よ
oak tiger longicorn beetle	クリストフコトラカミキリ	*Plagionotus christophi* (Kraatz)	（コウチュウ目、カミキリムシ科）日本、旧北区
oak timberworm		*Arrhenodes minutus* (Drury)	（コウチュウ目、ミツギリゾウムシ科）新北区
oak timberworm beetle			oak timberworm を見よ

英名	和名	学名	所属、分布、ほか
oak tortrix			green oak roller を見よ
oak treehopper		*Platycotis vittata* (Fabricius)	（カメムシ目、ツノゼミ科）新北区
oak-tree pug		*Eupithecia dodoneata* Guenée	（チョウ目、シャクガ科）旧北区
oak trumpet skeletonizer moth		*Catastega timidella* Clemens	（チョウ目、ハマキガ科）新北区
oak tussock caterpillar			yellow-spotted tiger mogh を見よ
oak tussock moth	カシワマイマイ	*Lymantria mathura aurora* Butler	（チョウ目、ドクガ科）日本。リンゴ、ナシ、クリ、カシ類害虫
oak twig pruner			twig pruner を見よ
oak vein pocket gall		*Macrodiplosis erubescens* (Osten Sacken)	（ハエ目、タマバエ科）の虫えい。新北区
oak wasp beetle		*Plagionotus arcuatus* (Linnaeus)	（コウチュウ目、カミキリムシ科）旧北区
oak wax scale		*Cerococcus quercus* Comstock	（カメムシ目、フサカイガラムシ科）新北区
oak webworm		*Archips fervidanus* (Clemens)	（チョウ目、ハマキガ科）新北区。oak webworm moth ともいう
oak winter highflier		*Hydriomena nubilofasciata* (Packard)	（チョウ目、シャクガ科）新北区。oak winter highflier moth ともいう
oak wood wasp		*Xiphydria longicollis* (Geoffroy)	（ハチ目、クビナガキバチ科）旧北区
oak-worm looper		*Lagoa fiscellaria somniaria* Hulst	（チョウ目、Megalopygidae）新北区
oakworm moths			oak moths を見よ
oakworms			oak moths を見よ
oak xylococcus	カブラカイガラムシ	*Beesonia napiformis* (Kuwana)	（カメムシ目、カブラカイガラムシ科）日本
oak yellow underwing		*Catocala nymphagoga* (Esper)	（チョウ目、ヤガ科）旧北区
oasis bluetail		*Ischnura fountaineae* Morton	（トンボ目、イトトンボ科）旧北区
oasis firetail		*Telebasis incolumis* Williamson et Williamson	（トンボ目、イトトンボ科）新熱帯区
oat aphid			oat bird-cherry aphid を見よ
oat aphid			apple aphid (2) を見よ
oat aphid			grain aphid (1) を見よ
oat aphis			greenbug を見よ
oat-apple aphid			apple grain aphid (1) を見よ
oat-apple aphid			grain aphid (1)(2) を見よ
oat bird-cherry aphid	ムギクビレアブラムシ	*Rhopalosiphum padi* (Linnaeus)	（カメムシ目、アブラムシ科）日本、全北区、東洋区。イネ、ムギ、トウモロコシ、リンゴ、ナシなどの害虫
oat fly			frit fly を見よ
oat gall midge			oat stem midge を見よ
oat leaf beetle			cereal leaf beetle (1) を見よ

英名	和名	学名	所属、分布、ほか
oat stem midge		*Mayetiola avenae* (Marchal)	（ハエ目、タマバエ科）旧北区
oat thrips		*Frankliniella tenuicornis* (Uzel)	（アザミウマ目、アザミウマ科）旧北区
Oaxacan bent-skipper		*Camptopleura oaxaca* Freeman	（チョウ目、セセリチョウ科）新熱帯区
Oaxacan checkerspot		*Chlosyne cynisca* (Godman et Salvin)	（チョウ目、タテハチョウ科）新熱帯区
Oaxacan crescent		*Anthanassa otanes oaxaca* Beutelspacher	（チョウ目、タテハチョウ科）新熱帯区
Oaxacan Emesis		*Emesis saturata* Godman et Salvin	（チョウ目、シジミタテハ科）新熱帯区
Oaxacan hairstreak		*Nicolaea* sp.	（チョウ目、シジミチョウ科）新熱帯区
Oaxacan pine-satyr		*Paramacera chinanteca* Miller	（チョウ目、タテハチョウ科）新熱帯区
Oaxacan skipperling		*Piruna jonka* Steinhauser	（チョウ目、セセリチョウ科）新熱帯区
Oaxacan sootywing		*Bolla fenestra* Steinhauser	（チョウ目、セセリチョウ科）新熱帯区
Oaxacan Zobera		*Zobera oaxaquena* Steinhauser	（チョウ目、セセリチョウ科）新熱帯区
Oberthur's admiral		*Chalinga elwesi* Oberthür	（チョウ目、タテハチョウ科）旧北区
Oberthur's anomalouse blue		*Polyommatus fabressei* (Oberthür)	（チョウ目、シジミチョウ科）旧北区
Oberthur's black-veined white		*Aporia bieti* (Oberthür)	（チョウ目、シロチョウ科）旧北区
Oberthur's freak		*Calinaga davidis* Oberthür	（チョウ目、タテハチョウ科）旧北区
Oberthur's grizzled skipper		*Pyrgus armoricanus* (Oberthür)	（チョウ目、セセリチョウ科）旧北区
Oberthur's pathfinder		*Catuna oberthueri* Karsch	（チョウ目、タテハチョウ科）エチオピア区
Oberthur's sister		*Adelpha boeotia oberthuri* (Boisduval)	（チョウ目、タテハチョウ科）新熱帯区
Oberthur's white		*Sinopieris dubernardi* Oberthür	（チョウ目、シロチョウ科）旧北区
obese pygmy grasshopper		*Paxilla obesa* (Scudder)	（バッタ目、ヒシバッタ科）新北区
obese springtail		*Morulina multatuberculata* (Coleman)	（トビムシ目、イボトビムシ科）新北区
Obi Island birdwing			golden birdwing (2) を見よ
objurgatella moth		*Alucita objurgatella* (Walsingham)	（チョウ目、ニジュウシトリバガ科）大洋区
oblique-banded geometrid	ハスオビエダシャク	*Descoreba simplex* Butler	（チョウ目、シャクガ科）日本、旧北区
oblique banded leafroller		*Choristoneura rosaceana* (Harris)	（チョウ目、ハマキガ科）新北区。温室、園芸害虫。oblique-banded leafroller moth ともいう
oblique-banded roller			oblique banded leafroller を見よ
oblique-banded tiger moth	フトスジモンヒトリ	*Spilarctia obliquizonata* (Miyake)	（チョウ目、ヒトリガ科）日本

英名	和名	学名	所属、分布、ほか
oblique-barred grey conch		*Piercea alismana* (Ragonot)	(チョウ目、ハマキガ科) 旧北区
oblique brocade moth		*Xylomoia indirecta* (Grote)	(チョウ目、ヤガ科) 新北区
oblique carpet		*Orthonama vittata* (Borkhausen)	(チョウ目、シャクガ科) 旧北区
oblique carpet		*Orthonama lignata* (Hübner)	(チョウ目、シャクガ科) 旧北区
oblique Eumorpha		*Eumorpha obliquus* (Rothschild et Jordan)	(チョウ目、スズメガ科) 新熱帯区
oblique grass moth		*Amolita obliqua* Smith	(チョウ目、ヤガ科) 新北区
oblique Heterocampa moth		*Heterocampa obliqua* Packard	(チョウ目、シャチホコガ科) 新北区
oblique-lined tea geometrid	エグリヅマエダシャク	*Odontopera arida* (Butler)	(チョウ目、シャクガ科) 日本
oblique looper moth		*Iridopsis obliquaria* (Grote)	(チョウ目、シャクガ科) 新北区
oblique striped		*Phibalapteryx virgata* (Hufnagel)	(チョウ目、シャクガ科) 旧北区
oblique-striped elongate weevil	ハスジカツオゾウムシ	*Lixus acutipennis* (Roelofs)	(コウチュウ目、ゾウムシ科) 日本、旧北区
oblique-striped emerald moth		*Synchlora bistriaria* (Packard)	(チョウ目、シャクガ科) 新北区
oblique-striped fern moth		*Azelina galleria* Walker	(チョウ目、シャクガ科) 豪州区
oblique tortrix		*Ctenopseutis obliquana* Walker	(チョウ目、ハマキガ科) 豪州区
oblique-waved fern moth		*Azelina variabilis* Warren	(チョウ目、シャクガ科) 豪州区
oblique Zale moth		*Zale obliqua* (Guenée)	(チョウ目、ヤガ科) 新北区。oblique Zale ともいう
oblong scale			European fruit lecanium を見よ
oblong sedge borer moth		*Capsula oblonga* Grote	(チョウ目、ヤガ科) 新北区。oblong sedge borer ともいう
oblong-winged katydid		*Amblycorypha oblongifolia* (De Geer)	(バッタ目、キリギリス科) 新北区。体長 25 mm。花上に普通
obscure-barred coast groundling		*Gnorimoschema ocellatellum* Boyd	(チョウ目、キバガ科) 旧北区
obscure blotch-back bell		*Gypsonoma oppressana* (Treitschke)	(チョウ目、ハマキガ科) 日本、旧北区
obscure blotched piercer		*Pammene albuginana* (Guenée)	(チョウ目、ハマキガ科) 旧北区
obscure Bolla		*Bolla brennus* (Godman et Salvin)	(チョウ目、セセリチョウ科) 新北区、新熱帯区
obscure click beetle		*Agriotes obscurus* (Linnaeus)	(コウチュウ目、コメツキムシ科) 旧北区
obscure-dash flat-body		*Depressaria pimpinellae* Zeller	(チョウ目、マルハキバガ科) 旧北区
obscure grasshopper		*Opeia obscura* (Thomas)	(バッタ目、バッタ科) 新熱帯区

英名	和名	学名	所属、分布、ほか
obscure ground mantid		*Litaneutria obscura* Scudder	（カマキリ目、カマキリ科）新北区
obscure lotus pigmy		*Trifurcula cryptella* (Stainton)	（チョウ目、モグリチビガ科）旧北区
obscure mealybug		*Pseudococcus obscurus* Essig	（カメムシ目、コナカイガラムシ科）新北区
obscure mealybug			tuber mealybug を見よ
obscure noctuid	クロクモヤガ	*Hermonassa cecilia* Butler	（チョウ目、ヤガ科）日本、旧北区
obscure pondweed moth		*Paraponynx obscuralis* (Grote)	（チョウ目、メイガ科）新北区
obscure root weevil		*Sciopithes obscurus* Horn	（コウチュウ目、ゾウムシ科）新北区
obscure rove beetles	ヒゲブトハネカクシ亜科	Aleocharinae	（コウチュウ目、ハネカクシ科）の昆虫の総称
obscure scale	カシクロマルカイガラムシ	*Melanaspis obscure* (Comstock)	（カメムシ目、マルカイガラムシ科）日本、新北区。クルミ害虫
obscure scaly cricket		*Cycloptilum ainiktos* Love et Walker	（バッタ目、コオロギ科）新北区
obscure sealed bell	シロズスソモンヒメハマキ	*Eucosma aemulana* (Schlager)	（チョウ目、ハマキガ科）日本、旧北区
obscure silver-striped piercer		*Cydia cosmophorana* (Treitschke)	（チョウ目、ハマキガ科）旧北区
obscure skipper		*Panoquina panoquinoides* (Skinner)	（チョウ目、セセリチョウ科）新北区、新熱帯区
obscure sootywing			obscure Bolla を見よ
obscure sphinx moth		*Erinnyis obscura* (Fabricius)	（チョウ目、スズメガ科）新北区、新熱帯区。obscure sphinx ともいう
obscure tree nymph		*Sallya morantii* (Trimen)	（チョウ目、タテハチョウ科）エチオピア区
obscure tree nymph			Morant's tree nymph を見よ
obscure underwing moth		*Catocala obscura* Strecker	（チョウ目、ヤガ科）新北区。obscure underwing ともいう
obscure wainscot	ノヒラキヨトウ	*Mythimna obsoleta* (Hübner)	（チョウ目、ヤガ科）日本、旧北区
obscure-wedged midget		*Phyllonorycter viminiella* (Stainton)	（チョウ目、ホソガ科）旧北区
obsolete white-spots		*Osmodes omar* Swinhoe	（チョウ目、セセリチョウ科）エチオピア区
obstinate furniture beetle			furniture beetle (2) を見よ
obtuse Euchlaena moth		*Euchlaena obtusaria* (Hübner)	（チョウ目、シャクガ科）新北区
obtuse-tongued bees		Prosopidae	（ハチ目）の昆虫の総称
obtuse yellow moth		*Azenia obtusa* (Herrich-Schäffer)	（チョウ目、ヤガ科）新北区
occelated burnet		*Zygaena carliolica* (Scopoli)	（チョウ目、マダラガ科）旧北区
occidental metalmark		*Exoplisia azuleja* Callaghan, Llorente et Luis	（チョウ目、シジミタテハ科）新熱帯区
occult Drasteria moth		*Drasteria occulta* (Hy. Edwards)	（チョウ目、ヤガ科）新北区

英名	和名	学名	所属、分布、ほか
occult sphinx		*Manduca occulta* (Rothschild et Jordan)	（チョウ目、スズメガ科）新北区、新熱帯区。occult sphinx moth ともいう
ocean skaters			sea skaters を見よ
oceanic burrower bug		*Geotomus pygmaeus* (Dallas)	（ケメムシ目、ツチカメムシ科）新北区
oceanic cricket		*Gryllus oceanicus* Le Guillou	（バッタ目、コオロギ科）東洋区
oceanic embiid		*Aposthonia oceania* (Ross)	（シロアリモドキ目、シロアリモドキ科）新北区
oceanic field cricket		*Teleogryllus oceanicus* (Le Guillou)	（バッタ目、コオロギ科）新北区、豪州区、大洋区
oceanic Hawaiian damselfly		*Megalagrion oceanicum* McLachlan	（トンボ目、イトトンボ科）大洋区
oceanic megalagrion damselfly			oceanic Hawaiian damselfly を見よ
Oceia hairstreak		*Laothus oceia* (Godman et Salvin)	（チョウ目、シジミチョウ科）新熱帯区
ocellate bog fritillary			bog fritillary を見よ
ocellate gall midge			ocellate maple gall midge を見よ
ocellate maple gall midge		*Acericecis ocellaris* (Osten Sacken)	（ハエ目、タマバエ科）新北区
ocellate rove beetles	ヨツメハネカクシ亜科	Omaliinae	（コウチュウ目、ハネカクシ科）の昆虫の総称
ocellated chafer		*Isonychus ocellatus* Burmeister	（コウチュウ目、コガネムシ科）新熱帯区
ocellated commodore		*Precis coelestina* Dewitz	（チョウ目、タテハチョウ科）エチオピア区
ocellated owlet			brown mort bleu を見よ
ocher ringlet		*Coenonympha ochracea* Edwards	（チョウ目、タテハチョウ科）新北区
ocher ringlet			large heath を見よ　米国での英名
ochra metalmark		*Synargis ochra* (H. Bates)	（チョウ目、シジミタテハ科）新熱帯区
ochraceous bombyx		*Gunda ochracea* Walker	（チョウ目、カイコガ科）東洋区
ochraceous wave		*Idaea serpeniata* (Hufnagel)	（チョウ目、シャクガ科）旧北区
ochre dagger moth		*Acronicta morula* Grote et Robinson	（チョウ目、ヤガ科）新北区
ochre Euchlaena moth		*Euchlaena marginaria* (Minot)	（チョウ目、シャクガ科）新北区
ochre-fringed neb			poplar leafminer (2) を見よ
ochre ringlet		*Coenonympha tullia ochracea* W. H. Edwards	（チョウ目、タテハチョウ科）新北区。large heath を参照
ochre-tinged slender		*Phyllocnistis suffusella* Zeller	（チョウ目、コハモグリガ科）旧北区
ochre-tipped darner		*Austroaeschna weiskei* Förster	（トンボ目、ヤンマ科）豪州区
ochreous dart			common dart (1) を見よ

英名	和名	学名	所属、分布、ほか
ochreous Emesis		*Emesis lupina* Godman et Salvin	（チョウ目、シジミタテハ科）新熱帯区
ochreous flat-body		*Agonopterix propinquella* (Treitschke)	（チョウ目、マルハキバガ科）旧北区
ochreous juniper argent	ネズミサシミモグリガ	*Argyresthia praecocella* Zeller	（チョウ目、メムシガ科）日本、旧北区
ochreous plume		*Platyptilia ochrodactyla* Denis et Schiffermüller	（チョウ目、トリバガ科）旧北区
ochreous pug		*Eupithecia indigata* (Hübner)	（チョウ目、シャクガ科）旧北区
ochreous saltern groundling		*Scrobipalpa suadella* (Richardson)	（チョウ目、キバガ科）旧北区
ochreous small case		*Coleophora caespitiella* Zeller	（チョウ目、ツツミノガ科）旧北区
ochreous-spotted patch		*Bucculatrix nigricomella* (Zeller)	（チョウ目、チビガ科）旧北区
ochreous twist	ハイイロウスモンハマキ	*Capua vulgana* (Frölich)	（チョウ目、ハマキガ科）日本、旧北区、東洋区
ochsenheimeriid moths			cereal stem moths を見よ
Ochus hairstreak		*Panthiades ochus* (Godman et Salvin)	（チョウ目、シジミチョウ科）新熱帯区
Ochus skipper		*Eutychide subcordata ochus* Godman	（チョウ目、セセリチョウ科）新熱帯区
Ocimum leaf folder		*Syngamia abruptalis* Walker	（チョウ目、メイガ科）東洋区、エチオピア区
Ocnus ringlet		*Magneuptychia ocnus* Butler	（チョウ目、タテハチョウ科）新熱帯区
Ocola skipper		*Panoquina ocola* (Edwards)	（チョウ目、セセリチョウ科）新北区、新熱帯区
Ocotillo slug moth		*Cryptophobetron oropeso* Barnes	（チョウ目、イラガ科）新北区
octagonal casemaker moth		*Coleophora octagonella* Walsingham	（チョウ目、ツツミノガ科）新北区
October thorn moth		*Tetracis jubararia* Hulst	（チョウ目、シャクガ科）新北区
odd beetle	マサカカツブシムシ	*Thylodrias contractus* Motschulsky	（コウチュウ目、カツオブシムシ科）日本、全北区。昆虫標本を加害
odd-spot blue		*Turanana endymion* (Freyer)	（チョウ目、シジミチョウ科）旧北区
odd-spot blue	ヘリブトルリシジミ	*Turanana panagaea* (Herrich-Schäffer)	（チョウ目、シジミチョウ科）旧北区
odiniid flies	トゲアシモグリバエ科	Odiniidae	（ハエ目）の昆虫の総称
odiniids			odiniid flies を見よ
Odites metalmark		*Juditha odites* (Cramer)	（チョウ目、シジミタテハ科）新熱帯区
odontocerids			mortarjoint casemaker caddisflies を見よ
odoriferous house ant			odorous house ant を見よ
odorous house ant		*Tapinoma sessile* (Say)	（ハチ目、アリ科）新北区

英名	和名	学名	所属、分布、ほか
oecophorid moths			concealer moths を見よ
oedemerid beetles			false blister beetles を見よ
oedipodine grasshoppers			band-winged grasshoppers を見よ
Oedipus owl butterfly		*Caligo oedipus* Stichel	（チョウ目、タテハチョウ科）新熱帯区
Oehmig's grayling		*Hipparchia caldeirensis* (Oehmig)	（チョウ目、タテハチョウ科）旧北区
Oenomais leafwing		*Memphis oenomais* (Boisduval)	（チョウ目、タテハチョウ科）新熱帯区
Oenotrus sphinx			oleander sphinx moth を見よ
off-white Hedya moth			common long-cloaked marble を見よ
Offa leafwing		*Memphis offa* Druce	（チョウ目、タテハチョウ科）新熱帯区
Oguma leafhopper	オグマブチミャクヨコバイ	*Drabescus ogumae* Matsumura	（カメムシ目、オオヨコバイ科）日本、旧北区
Ohio emerald dragonfly		*Somatochlora hineana* Williamson	（トンボ目、エゾトンボ科）新北区。絶滅危惧種
oil beetle	オオツチハンミョウ	*Meloe proscarabaeus* Linnaeus	（コウチュウ目、ツチハンミョウ科）日本、旧北区
oil beetle			short-winged blister beetle を見よ
oil beetles		*Meloe*	（コウチュウ目、ツチハンミョウ科）の昆虫の総称
oil beetles (1)	ツチハンミョウ亜科	Meloinae	（コウチュウ目、ツチハンミョウ科）の昆虫の総称
oil beetles			blister beetles (1) を見よ　英国での英名
oil cicada			large brown cicada を見よ
oil clock			oil beetle を見よ
oil clocks			oil beetles (1) を見よ
oil palm bag worm		*Pteroma pendula* de Joannis	（チョウ目、ミノガ科）東洋区
oil palm bunch moth		*Tirathaba mundella* Walker	（チョウ目、メイガ科）東洋区
Oileus giant owl			placid giant owl を見よ
Oileus owl butterfly		*Caligo oileus* C. et R. Felder	（チョウ目、タテハチョウ科）新熱帯区。Oileus giant owl ともいう
oinophilid moths		Oinophilidae	（チョウ目）の昆虫の総称
Oishi sawfly	コナラナメクジハバチ	*Caliroa oishii* (Takeuchi)	（ハチ目、ハバチ科）日本
Oithona tiger moth		*Grammia oithona* (Strecker)	（チョウ目、ヒトリガ科）新北区。Oithona moth ともいう
Okabe small longicorn	ヒメナガサビカミキリ	*Pterolophia leiopodina* (Bates)	（コウチュウ目、カミキリムシ科）日本
Okajima hairy aphid	オカジマケクダアブラムシ	*Greenidia okajimai* Suenaga	（カメムシ目、アブラムシ科）日本
Okame cricket	ハラオカメコオロギ	*Loxoblemmus arietulus* Saussure	（バッタ目、コオロギ科）日本

英名	和名	学名	所属、分布、ほか
Okamoto looper moth	オカモトトゲエダシャク	*Apochima juglansiaria* (Graeser)	(チョウ目、シャクガ科) 日本、旧北区
Okefenokee Zale moth		*Zale perculta* Franclemont	(チョウ目、ヤガ科) 新北区
Okitsu citrus cottony scale			cottony citrus scale (3) を見よ
Oklahoma clubtail		*Gomphus oklahomensis* Pritchard	(トンボ目、サナエトンボ科) 新北区
okra leafhopper	フタテンミドリヒメヨコバイ	*Amrasca biguttula* (Ishida)	(カメムシ目、ヒメヨコバイ科) 日本、東洋区、大洋区
okra leafworm moth		*Anomis illita* Guenée	(チョウ目、ヤガ科) 新北区、東洋区、豪州区、エチオピア区
Olane azure		*Ogyris olane* Hewitson	(チョウ目、シジミチョウ科) 豪州区
old-fashioned potato beetle		*Epicauta marginata* (Fabricius)	(コウチュウ目、ツチハンミョウ科) 新北区
old-fashioned potato beetle			striped blister beetle (1) を見よ
old-fashioned potato bugs			blister beetles (1) を見よ　英国での英名
old fathers long-legs			crane flies (1) を見よ
old gold Isocorypha moth		*Isocorypha mediostriatella* (Clemens)	(チョウ目、ヒロズコガ科) 新北区
old house borer			house longhorn を見よ　北米での英名。old-house borer とも記す
old lady		*Mormo maura* (Linnaeus)	(チョウ目、ヤガ科) 旧北区
old lady moth			old lady を見よ
old ladymoth			southern old lady moth を見よ
old maid			bay underwing moth を見よ
old man dart		*Agrotis vetusta* Walker	(チョウ目、ヤガ科) 新北区
old maple-seed pigmy		*Nepticula sericopeza* Zeller	(チョウ目、モグリチビガ科) 全北区
old Rocky Mountain locust			Rocky Mountain grasshopper を見よ
old wife underwing moth		*Catocala palaeogama* Guenée	(チョウ目、ヤガ科) 新北区。old wife underwing ともいう
old world bollworm			corn earworm (2) を見よ
old world bollworm moth			corn earworm (2) を見よ　豪州での英名
old world brimstones	ヤマキチョウ属	*Gonepteryx*	(チョウ目、シロチョウ科) の昆虫の総称
old world butterfly moth	イカリモンガ科	Callidulidae	(チョウ目) の昆虫の総称
old world cabbage webworm			cabbage webworm (1) を見よ
old world mouse louse			solenophagous louse を見よ
old world screw-worm		*Chrysomya bezziana* (Villeneuve)	(ハエ目、クロバエ科) 東洋区、エチオピア区
old world spiny-winged moths			spiny-winged moths を見よ

英名	和名	学名	所属、分布、ほか
old world swallowtail		*Papilio machaon* Linnaeus	（チョウ目、アゲハチョウ科）全北区。*P. m. hudsonianus* Clark も同英名
old world webworm			cabbage webworm (1) を見よ
oleander aphid	キョウチクトウアブラムシ	*Aphis nerii* Boyer et Fonscolombe	（カメムシ目、アブラムシ科）日本、汎世界。キョウチクトウ害虫
oleander butterfly			common Indian crow を見よ　豪州での使用
oleander butterfly			common crow を見よ
oleander hawk			oleander hawk-moth を見よ
oleander hawk-moth	キョウチクトウスズメ	*Daphnis nerii* (Linnaeus)	（チョウ目、スズメガ科）日本、東洋区、旧北区。キョウチクトウ害虫
oleander pit scale		*Asterolecanium pustulans* (Cockerell)	（カメムシ目、フサカイガラムシ科）新北区、大洋区
oleander scale	シロマルカイガラムシ	*Aspidiotus hederae* (Vallot)	（カメムシ目、マルカイガラムシ科）日本、旧北区。オリーブ、ラン害虫
oleander scale			mango scale (2) を見よ
oleander scale			white peach scale を見よ　中米での英名
oleander scale			ivy scale を見よ
oleander sphinx moth		*Erinnyis oenotrus* (Cramer)	（チョウ目、スズメガ科）新北区、新熱帯区
olethreutid moths		Olethreutidae	（チョウ目）の昆虫の総称　Tortricidae の異名
olethreutine moths	ヒメハマキガ亜科	Olethreutinae	（チョウ目、ハマキガ科）の昆虫の総称
olinda bug			Fuller rose beetle を見よ
olivaceous Olethreutes moth		*Olethreutes olivaceana* (Fernald)	（チョウ目、ハマキガ科）新北区
olivaceous Phaneta moth		*Phaneta olivaceana* (Riley)	（チョウ目、ハマキガ科）新北区
olive	ドロキリガ	*Ipimorpha subtusa* (Denis et Schiffermüller)	（チョウ目、ヤガ科）日本、旧北区
olive-and-black carpet moth		*Acasis viridata* (Packard)	（チョウ目、シャクガ科）新北区。olive-and-black carpet ともいう
olive angle shades		*Phlogophora iris* Guenée	（チョウ目、ヤガ科）新北区。olive angle shades moth ともいう
olive arches moth		*Lacinipolia olivacea* (Morrison)	（チョウ目、ヤガ科）新北区
olive Arta moth		*Arta olivalis* Grote	（チョウ目、メイガ科）新北区
olive bark beetle		*Luperisinus californicus* Swaine	（コウチュウ目、キクイムシ科）新北区。オリーブ害虫
olive bark beetle		*Phloeotribus scarabeoides* Bern	（コウチュウ目、キクイムシ科）新北区。オリーブ害虫
olive beetles		*Eleodes*	（コウチュウ目、ゴミムシダマシ科）の昆虫の総称
olive brindled pearl		*Pyrausta olivalis* (Denis et Schiffermüller)	（チョウ目、メイガ科）旧北区
olive clouded skipper		*Lerodea arabus* (Edwards)	（チョウ目、セセリチョウ科）新北区、新熱帯区
olive-clouded skipper		*Lerodea dysaules* Godman	（チョウ目、セセリチョウ科）新北区
olive clubtail		*Stylurus olivaceus* (Selys)	（トンボ目、サナエトンボ科）新北区
olive crescent	シロオビクルマコヤガ	*Trisateles emortualis* (Denis et Schiffermüller)	（チョウ目、ヤガ科）日本、旧北区

英名	和名	学名	所属、分布、ほか
olive crescent snout			olive crescent を見よ
olive flat		*Chamunda chamunda* (Moore)	(チョウ目、セセリチョウ科) 東洋区
olive fly			olive fruit fly を見よ
olive fruit fly		*Dacus oleae* (Gmelin)	(ハエ目、ミバエ科) 全北区。オリーブ害虫
olive-green cutworm		*Dargida procincta* Grote	(チョウ目、ヤガ科) 新北区。成虫は olive green cutworm moth
olive haired swift			Borbo skipper を見よ
olive hairstreak		*Mitoura gryneus* (Hübner)	(チョウ目、シジミチョウ科) 新北区
olive hairstreak			juniper hairstreak (1) を見よ
olive kernel borer			olive moth (1) を見よ
olive kidney		*Zenobia subtusa* Denis et Schiffermüller	(チョウ目、ヤガ科) 旧北区
olive lace bug		*Frogattia olivinia* Froggatt	(カメムシ目、グンバイムシ科) 豪州区
olive Metron		*Metron zimra* (Hewitson)	(チョウ目、セセリチョウ科) 新熱帯区
olive moth		*Prays oleellus* Fabricius	(チョウ目、スガ科) 旧北区。オリーブ害虫
olive moth (1)		*Prays oleae* (Bernard)	(チョウ目、スガ科) 旧北区
olive moth			olive を見よ
olive owlet		*Selepa celtis* Moore	(チョウ目、ヤガ科) 東洋区
olive parlatoria			olive scale を見よ
olive parlatoria scale			olive scale を見よ　豪州での英名
olive prominent		*Desmeocraera latex* Druce	(チョウ目、シャチホコガ科) エチオピア区
olive psylla		*Euphyllura phillyreae* Forster	(カメムシ目、Aphalaridae) 旧北区
olive psyllid		*Euphyflura olivina* Costa	(カメムシ目、キジラミ科) 旧北区。オリーブ害虫
olive purplewing	ウラキスジアイイロタテハ	*Eunica chlororhoa* Salvin	(チョウ目、タテハチョウ科) 新熱帯区
olive satin-flower case		*Coleophora lithargyrinella* Zeller	(チョウ目、ツツミノガ科) 旧北区
olive scale	オリーブクロホシカイガラムシ	*Parlatoria oleae* (Colvee)	(カメムシ目、マルカイガラムシ科) 汎世界。オリーブ害虫
olive scale			black scale を見よ
olive scolytid		*Phloeotribus oleae* Fabricius	(コウチュウ目　キクイムシ科) 旧北区。オリーブ害虫
olive-shaded bird-dropping moth		*Ponometia candefacta* (Hübner)	(チョウ目、ヤガ科) 全北区
olive shot-hole borer			fig tree bark borer を見よ
olive skipper		*Pyrgus serratulae* (Rambur)	(チョウ目、セセリチョウ科) 旧北区
olive thrips		*Liothrips oleae* (Costa)	(アザミウマ目、クダアザミウマ科) 旧北区。オリーブ害虫
olive upright		*Rhithrogena semicolorata* (Curtis)	(カゲロウ目、ヒラタカゲロウ科) 旧北区
olive weevil	オリーブアナアキゾウムシ	*Pimelocerus perforatus* (Roelofs)	(コウチュウ目、ゾウムシ科) 日本、旧北区

英名	和名	学名	所属、分布、ほか
olive whitefly		*Aleurolobus olivinus* Silvestri	（カメムシ目、コナジラミ科）旧北区
Olivencia tigerwing		*Forbestra olivencia* (Bates)	（チョウ目、タテハチョウ科）新熱帯区
Olivier's tubic		*Dasycera oliviella* (Fabricius)	（チョウ目、マルハキバガ科）旧北区
Olympia		*Euchloe olympia* (Edwards)	（チョウ目、シロチョウ科）新北区
Olympia marble			Olympia を見よ
Olympia marblewing			Olympia を見よ
Olynthia sister		*Adelpha olynthia* Felder et Felder	（チョウ目、タテハチョウ科）新熱帯区
Omango scale			vanda orchid scale を見よ
Omiltemi skipper		*Paratrytone omiltemensis* Steinhauser	（チョウ目、セセリチョウ科）新熱帯区
omnivorous beetles			polyphagous beetles を見よ
omnivorous leaf roller			platynota を見よ
omnivorous leafroller moth		*Archips purpurana* (Clemens)	（チョウ目、ハマキガ科）新北区
omnivorous leaftier		*Cnephasia longana* (Haworth)	（チョウ目、ハマキガ科）全北区。omnivorous leaftier moth ともいう
omnivorous looper		*Sabulodes caberata* Guenée	（チョウ目、シャクガ科）新北区。多食性
omnivorous looper			avocado looper を見よ
omnivorous pinhole borer		*Crossotarsus omnivorus* Lea	（コウチュウ目、ゾウムシ科）豪州区
omnivorous Platynota moth		*Platynota rostrana* (Walker)	（チョウ目、ハマキガ科）全北区、新熱帯区
omnivorous tussock moth		*Acyphas leucomelas* (Walker)	（チョウ目、ドクガ科）豪州区
Onaca skipper		*Vettius onaca* Evans	（チョウ目、セセリチョウ科）新熱帯区
once-married underwing moth		*Catocala unijuga* Walker	（チョウ目、ヤガ科）新北区。once-married underwing ともいう
one-banded geometrid			lilac beauty (1) を見よ
one-dotted cymatophorid	ヒトモントガリバ	*Tetheella fluctuosa isshikii* (Matsumura)	（チョウ目、トガリバガ科）日本、旧北区。satin luterstring も参照
one-eyed sphinx			willow hawk-moth (2) を見よ　one-eyed sphinx moth ともいう
one-lined Sparganothis moth		*Sparganothis unifasciana* (Clemens)	（チョウ目、ハマキガ科）新北区
one-lined Zale moth		*Zale unilineata* (Grote)	（チョウ目、ヤガ科）新北区
one pip policeman	キモンアオバセセリ	*Coeliades anchises* (Gerstaecker)	（チョウ目、セセリチョウ科）エチオピア区
one-pupil satyr		*Cissia similis* (Butler)	（チョウ目、タテハチョウ科）新熱帯区
one-spot grass yellow	ウスイロキチョウ	*Eurema andersoni* (Moore)	（チョウ目、シロチョウ科）東洋区
one spot redwing		*Phylloxiphia punctum* (Rothschild)	（チョウ目、スズメガ科）エチオピア区

英名	和名	学名	所属、分布、ほか
one-spotted Prepona		*Archaioprepona demophon* (Linnaeus)	（チョウ目、タテハチョウ科）新熱帯区。*A. d. occidentalis* (Stoffel et Descimon), *A. d. centralis* (Fruhstorfer) も同英名
one-spotted stink bug		*Euschistus variolarius* (Palisot et Beauvois)	（カメムシ目、カメムシ科）新北区
one-spotted variant moth		*Hypagyrtis unipunctata* (Haworth)	（チョウ目、シャクガ科）新北区
one-striped ringtail		*Erpetogomphus bothrops* Garrison	（トンボ目、サナエトンボ科）新熱帯区
one-tooth longwing		*Capnobotes unodontus* Rentz et Birchim	（バッタ目、キリギリス科）新北区
Onega glasswing		*Oleria onega* (Hewitson)	（チョウ目、タテハチョウ科）新熱帯区
Oneil's hawk		*Pantophaea oneili* (Clark)	（チョウ目、スズメガ科）エチオピア区
onion aphid	ネギアブラムシ	*Neotoxoptera formosana* (Takahashi)	（カメムシ目、アブラムシ科）日本、全北区、東洋区、豪州区。ネギ害虫
onion bulb fly	ハイジマハナアブ	*Eumerus strigatus* (Fallén)	（ハエ目、ハナアブ科）日本、全北区。タマネギ、スイセンなどの害虫
onion fly			onion maggot を見よ
onion fly			onion bulb fly を見よ
onion leafminer			leek moth を見よ
onion louse			onion thrips を見よ
onion maggot	タマネギバエ	*Hylemya (Delia) antiqua* (Meigen)	（ハエ目、ハナバエ科）日本、全北区。タマネギ害虫。米国での幼虫の英名
onion maggot			bean seed fly を見よ　豪州での幼虫の英名
onion moth			leek moth を見よ
onion plant bug		*Labopidicola altii* Knight	（カメムシ目、カスミカメムシ科）新北区。タマネギ害虫。長翅型短翅型あり
onion sucker		*Trioza tremblayi* Wagner	（カメムシ目、トガリキジラミ科）旧北区
onion thrips	ネギアザミウマ	*Thrips tabaci* Lindeman	（アザミウマ目、アザミウマ科）日本、汎世界。タバコ、ウリ類、タマネギ、カンキツなど多くの植物害虫
onion thrips			grape thrips を見よ
Onohara xyleborus	オノハラキクイムシ	*Xyleborus onoharensis* Murayama	（コウチュウ目、キクイムシ科）日本
onosmella			streaked silver case を見よ
Ontake bark beetle	オンタケキクイムシ	*Pityogenes foveolatus* Eggers	（コウチュウ目、キクイムシ科）日本、旧北区
Onuki bamboo scale			bamboo scale (1) を見よ
Oopsis long-horned beetle		*Oopsis nutator* (Fabricius)	（コウチュウ目、カミキリムシ科）大洋区
Oosthuizen's blue		*Lepidochrysops oosthuizeni* Swanepoel et Vári	（チョウ目、シジミチョウ科）エチオピア区
opal ciliate blue		*Anthene opalina* Stempffer	（チョウ目、シジミチョウ科）エチオピア区
opal copper		*Chrysoritis thyshe* (Linnaeus)	（チョウ目、シジミチョウ科）エチオピア区
opal oakblue		*Arhopala opalina* (Moore)	（チョウ目、シジミチョウ科）東洋区
opal slim		*Aciagrion dondoense* Dijkstra	（トンボ目、イトトンボ科）エチオピア区

英名	和名	学名	所属、分布、ほか
opaque sawyer	オオマルクビヒラタカミキリ	*Asemum striatum* (Linnaeus)	(コウチュウ目、カミキリムシ科) 日本、全北区。トドマツなどマツ類害虫
opaque 6-lineblue		*Erora opisena* (H. Druce)	(チョウ目、シジミチョウ科) 東洋区
opaque sweep			bagworm moth (1) を見よ
opaque wood white		*Leptosia wigginsi* (Dixey)	(チョウ目、シロチョウ科) エチオピア区
open-winged skippers		Pyrginae	(チョウ目、セセリチョウ科) の昆虫の総称
Ophia hairstreak		*Nicolaea ophia* (Hewitson)	(チョウ目、シジミチョウ科) 新熱帯区
Opisena hairstreak		*Erora opisena* (H. Druce)	(チョウ目、シジミチョウ科) 新熱帯区
opomyzid flies	ヒメコバエ科	Opomyzidae	(ハエ目) の昆虫の総称
opossum beetles		Monommatidae	(コウチュウ目) の昆虫の総称
opostegid moths			eye-cap moths を見よ
Oppia hairstreak		*Thereus oppia* (Godman et Salvin)	(チョウ目、シジミチョウ科) 新熱帯区
opuntia biocontrol beetle		*Archlagocheirus funestus* (Thomson)	(コウチュウ目、カミキリムシ科) 大洋区、豪州区。ウチワサボテンの防除に使用
orache brocade			orache moth を見よ
orache gelechid	ヘルマンアカザキバガ	*Chrysoesthia hermannella* (Fabricius)	(チョウ目、キバガ科) 日本、全北区、エチオピア区
orache groundling		*Scrobipalpa atriplicella* (Fischer von Röslerstamm)	(チョウ目、キバガ科) 旧北区
orache leafminer moth			marbled single-dot bell を見よ
orache moth		*Trachea atriplicis* (Linnaeus)	(チョウ目、ヤガ科) 旧北区。日本亜種は *T. a. gnoma* (Butler) シロスジアオヨトウで buckwheat cutworm という
orange Acraea			small orange Acraea (1) を見よ
orange Acraea			large orange Acraea を見よ
orange admiral	アフリカオナガタテハ	*Antanartia delius* (Drury)	(チョウ目、タテハチョウ科) エチオピア区
orange admiral (1)	レテキオビアカタテハ	*Hypanartia lethe* Fabricius	(チョウ目、タテハチョウ科) 新北区、新熱帯区
orange aeroplane	フトオビキンミスジ	*Pantoporia consimilis* (Boisduval)	(チョウ目、タテハチョウ科) 豪州区。orange plane ともいう
orange albatross	ベニシロチョウ	*Appias nero* (Fabricius)	(チョウ目、シロチョウ科) 東洋区。*A. n. figulina* Butler も同英名
orange albatross			rare albatross を見よ
orange amberwing		*Perithemis cornelia* Ris	(トンボ目、トンボ科) 新熱帯区
orange-and-black cinnabar		*Callimorpha jacobaea* Linnaeus	(チョウ目、ヒトリガ科) 旧北区
orange and black royal		*Tajuria megistia* (Hewitson)	(チョウ目、シジミチョウ科) 東洋区
orange-and-lemon		*Eronia leda* (Boisduval)	(チョウ目、シロチョウ科) エチオピア区
orange Antinephele		*Antinephele lunulata* Rothschild et Jordan	(チョウ目、スズメガ科) エチオピア区
orange awlet		*Burana harisa* (Moore)	(チョウ目、セセリチョウ科) 東洋区。*B. h. consobrina* Plötz も同英名

英名	和名	学名	所属、分布、ほか
orange awlet			orange-striped awl を見よ
orange bagworm		*Platoeceticus gloverii* Packard	(チョウ目、ミノガ科) 旧北区。カンキツ害虫
orange-banded Andrena		*Andrena prunorum* Cockerell	(ハチ目、ヒメハナバチ科) 新北区
orange-banded daggerwing		*Marpesia corita corita* (Westwood)	(チョウ目、タテハチョウ科) 新熱帯区。*M. c. phiale* (Godman et Salvin) も同英名
orange-banded 88		*Diaethria pandama* (Doubleday)	(チョウ目、タテハチョウ科) 新北区、新熱帯区
orange-banded hairstreak		*Satyrium ledereri* (Boisduval)	(チョウ目、シジミチョウ科) 旧北区
orange-banded hairstreak			striped hairstreak を見よ
orange banded lancer		*Pseudokerana fulgur* (de Nicéville)	(チョウ目、セセリチョウ科) 東洋区
orange-banded metalmark		*Crocozona coecias* (Hewitson)	(チョウ目、シジミタテハ科) 新熱帯区
orange-banded plane		*Lexias aeropa* (Linnaeus)	(チョウ目、タテハチョウ科) 豪州区
orange-banded Protea			Protea scarlet を見よ
orange-banded sister	チャオビイチモンジ	*Adelpha saundersii* (Hewitson)	(チョウ目、タテハチョウ科) 新熱帯区
orange banner	ツマグロタテハ	*Temenis laothoe hondurensis* Fruhstorfer	(チョウ目、タテハチョウ科) 新熱帯区。*T. l. quilapayunia* Maza et Turrent も同英名。tomato を参照
orange-barred carpet moth		*Dysstroma hersiliata* (Guenée)	(チョウ目、シャクガ科) 新北区
orange-barred conch		*Aethes margaritana* (Duponchel)	(チョウ目、ホソハマキガ科) 旧北区
orange-barred Emesis		*Emesis cypria* C. et R. Felder	(チョウ目、シジミタテハ科) 新熱帯区
orange-barred giant sulphur			orange-barred sulphur を見よ
orange-barred playboy		*Deudorix diocles* Hewitson	(チョウ目、シジミチョウ科) エチオピア区
orange-barred sulphur	ベニモンオオキチョウ	*Phoebis philea* (Linnaeus)	(チョウ目、シロチョウ科) 新北区、新熱帯区
orange barred tiger			banded orange heliconian を見よ
orange-barred velvet		*Panara phereclus* (Linnaeus)	(チョウ目、シジミタテハ科) 新熱帯区
orange basket-worm		*Cryotithelea gloveri* Packard	(チョウ目、ミノガ科) 新北区
orange beetle-borer		*Stromatium barbatum* Fabricius	(コウチュウ目、カミキリムシ科) 東洋区。カンキツ害虫
orange-bellied metalmark		*Brachyglenis dodone* (Godman et Salvin)	(チョウ目、シジミタテハ科) 新熱帯区
orange-belted bumblebee			red-tailed bumble bee (2) を見よ
orange Bematistes	カバイロホソチョウ	*Acraea tellus* (Aurivillius)	(チョウ目、タテハチョウ科) エチオピア区

英名	和名	学名	所属、分布、ほか
orange-black Hawaiian damselfly			orange-black megalagrion damselfly を見よ
orange-black megalagrion damselfly		*Megalagrion xanthomelas* (Selys)	(トンボ目、イトトンボ科) 大洋区
orange blackboy		*Gudanga browni* (Distant)	(カメムシ目、セミ科) 豪州区
orange blossom bug		*Dionconotus cruentatus* (Brulle)	(カメムシ目、カスミカメムシ科) 旧北区。カンキツ害虫
orange bluet		*Enallagma signatum* (Hagen)	(トンボ目、イトトンボ科) 新北区
orange-bodied Altinote		*Altinote alcione sodalis* Butler	(チョウ目、タテハチョウ科) 新熱帯区
orange-bodied Pygarctia			yellow-edged Pygarctia を見よ
orange-bordered Argus		*Polyommatus astrarche* Bergsträsser	(チョウ目、シジミチョウ科) 東洋区
orange-bordered blue	メリッサミヤマシジミ	*Plebejus melissa* (Edwards)	(チョウ目、シジミチョウ科) 新北区
orange-bordered Pixie		*Melanis cinaron* (C. et R. Felder)	(チョウ目、シジミタテハ科) 新熱帯区
orange-bordered satyr		*Pseudomaniola gigas* (Godman et Salvin)	(チョウ目、タテハチョウ科) 新熱帯区
orange bush brown		*Mycalesis terminus* (Fabricius)	(チョウ目、タテハチョウ科) 豪州区
orange case moth		*Hyalarcta hubneri* (Westwood)	(チョウ目、ミノガ科) 豪州区。カンキツ害虫
orange caterpillar parasite		*Netalia producta* (Brulle)	(ハチ目、ヒメバチ科) 豪州区
orange chewing louse of sheep			Angora goat biting louse (1) を見よ
orange chewing louse of sheep and goats			Angora goat biting louse (1) を見よ
orange clouded yellow	ヒメベニモンキチョウ	*Colias stoliczkana* Moore	(チョウ目、シロチョウ科) 東洋区
orange-coloured wheat midge			wheat midge を見よ　米国での英名
orange-costa Euselasia		*Euselasia procula* (Godman et Salvin)	(チョウ目、シジミタテハ科) 新熱帯区
orange-crescent groundstreak		*Ziegleria guzanta* (Schaus)	(チョウ目、シジミチョウ科) 新熱帯区
orange dart		*Suniana sunias rectivitta* (Mabille)	(チョウ目、セセリチョウ科) 豪州区
orange-disked Altinote		*Altinote stratonice* (Latreille)	(チョウ目、タテハチョウ科) 新熱帯区。*A. s. oaxaca* (Miller et Miller) も同英名
orange dog			citrus swallowtail (2) を見よ
orange dog			giant swallowtail を見よ
orange dogs			*Papilio* 属 (チョウ目、アゲハチョウ科) の幼虫
orange-dotted shoot moth		*Rhyacionia pinicolana* Zeller	(チョウ目、ハマキガ科) 旧北区

英名	和名	学名	所属、分布、ほか
orange ear			frosted orange (1) を見よ
orange-edged roadside skipper		*Amblyscirtes fimbriata* (Plötz)	(チョウ目、セセリチョウ科) 新北区、新熱帯区
orange emperor		*Anax speratus* Hagen	(トンボ目、ヤンマ科) エチオピア区
orange emperor	チャイロミナミフタオチョウ	*Charaxes latona* Butler	(チョウ目、タテハチョウ科) 東洋区、豪州区
orange fairy		*Cicadetta sulcata* (Distant)	(カメムシ目、セミ科) 豪州区
orange-flash crow	マルバネルリマダラ	*Euploea leucostictos* (Gmelin)	(チョウ目、タテハチョウ科) 東洋区、豪州区
orange flat		*Sarangesa phidyle* (Walker)	(チョウ目、セセリチョウ科) エチオピア区
orange-flushed eyemark			Lasus metalmark を見よ
orange footman		*Eilema sororcula* (Hufnagel)	(チョウ目、ヒトリガ科) 旧北区
orange-fronted threadtail		*Prodasineura vittata* (Selys)	(トンボ目、モノサシトンボ科) エチオピア区
orange fruitborer		*Isotenes miserana* (Walker)	(チョウ目、ハマキガ科) 豪州区。カンキツ害虫
orange fruit fly			Mediterranean fruit fly を見よ
orange gaster			red ant (1) を見よ
orange giant skipper			California giant skipper (1) を見よ
orange giant sulphur	アメリカオオキチョウ (アルガンテオオキチョウ)	*Phoebis agarithe* (Boisduval)	(チョウ目、シロチョウ科) 新北区
orange grass dart		*Taractrocera anisomorpha* (Lower)	(チョウ目、セセリチョウ科) 豪州区
orange gull	キシタシロチョウ	*Cepora judith malaya* (Fruhstorfer)	(チョウ目、シロチョウ科) 東洋区
orange harlequin			harlequin (1) を見よ
orange-headed Epicallima moth		*Epicallima argenticinctella* (Clemens)	(チョウ目、マルハキバガ科) 新北区
orange headed leafhopper			yellow rice leafhopper を見よ
orange-headed Metron		*Metron chrysogastra chrysogastra* (Butler)	(チョウ目、セセリチョウ科) 新熱帯区
orange-headed roadside skipper			red-headed roadside skipper を見よ
orange Holomelina moth		*Virbia aurantiaca* (Hübner)	(チョウ目、ヒトリガ科) 新北区。orange Holomelina ともいう
orange-horned green colonel		*Odontomyia angulata* (Panzer)	(ハエ目、ミズアブ科) 旧北区
orange-humped mapleworm			orange-humped mapleworm moth を見よ
orange-humped mapleworm moth		*Symmerista leucitys* Franclemont	(チョウ目、シャチホコガ科) 新北区
orange-humped oakworm			white-headed prominent moth を見よ
orange Jezebel		*Delias aruna inferna* Butler	(チョウ目、シロチョウ科) 豪州区。golden Jezebel を参照

英名	和名	学名	所属、分布、ほか
orange kite-swallowtail	キイロオナガタイマイ	*Eurytides thyastes marchandii* (Boisduval)	（チョウ目、アゲハチョウ科）新熱帯区。*E. t. occidentalis* (Maza) も同英名
orange lacewing		*Cethosia penthesilea* (Cramer)	（チョウ目、タテハチョウ科）東洋区、豪州区
orange ladybird	シロジュウロクホシテントウ	*Halyzia sedecimguttata* Linnaeus	（コウチュウ目、テントウムシ科）日本、旧北区
orange larch tubemaker moth		*Coleotechnites laricis* (Freeman)	（チョウ目、キバガ科）新北区。幼虫は orange larch tubemaker
orange-legged grasshopper		*Spharagemon equale* (Say)	（バッタ目、バッタ科）新北区
orange-legged robberfly		*Dioctria oelandica* (Linnaeus)	（ハエ目、ムシヒキアブ科）旧北区
orange-lobed flash			blue Cornelian を見よ
orange lurid glicer		*Cymothoe hesiodotus* Staudinger	（チョウ、タテハチョウ科）エチオピア区
orange maggot			Mexican fruit fly を見よ
orange mapwing			orange admiral (1) を見よ
orange-marked moth		*Melanchra paracausta* Meyrick	（チョウ目、ヤガ科）豪州区
orange migrant	キシタウスキシロチョウ	*Catopsilia scylla* (Linnaeus)	（チョウ目、シロチョウ科）豪州区。東洋区の *C. s. cornelia* Fabricius も同英名
orange mint moth		*Pyrausta orphisalis* Walker	（チョウ目、メイガ科）新北区
orange moth		*Gymnandrosoma aurantianum* Costa	（チョウ目、ハマキガ科）新熱帯区。カンキツ害虫
orange moth		*Melanchra rubescens* Walker	（チョウ目、ヤガ科）豪州区
orange moth			orange thorn を見よ
orange-necked blue Xenandra		*Xenandra poliotactis* (Stichel)	（チョウ目、シジミタテハ科）新熱帯区
orange oakleaf	コノハチョウ	*Kallima inachus* (Boisduval)	（チョウ目、タテハチョウ科）旧北区、東洋区。日本亜種 *K. i. eucerea* Fruhstorfer、台湾亜種 *K. i. formosana* Fruhstorfer も同英名
orange ochre			Eliena skipper を見よ
orange owl-butterfly		*Opsiphanes boisduvallii* Doubleday	（チョウ目、タテハチョウ科）新北区、新熱帯区。orange owlet ともいう
orange palm dart		*Cephrenes augiades sperthias* (Felder)	（チョウ目、セセリチョウ科）豪州区
orange Panopoda moth		*Panopoda repanda* (Walker)	（チョウ目、ヤガ科）新北区
orange-patch ciliate blue		*Anthene rufoplagata* (Bethune-Baker)	（チョウ目、シジミチョウ科）エチオピア区
orange patch white		*Colotis pleione* (Klug)	（チョウ目、シロチョウ科）エチオピア区
orange patch white			yellow patch white を見よ
orange-patched crescent		*Anthanassa drusilla* (C. et R. Felder)	（チョウ目、タテハチョウ科）新北区、新熱帯区
orange-patched satyr		*Euptychia fetna* Butler	（チョウ目、タテハチョウ科）新熱帯区
orange-patched smoky moth		*Pyromorpha dimidiata* (Herrich-Schäffer)	（チョウ目、マダラガ科）新北区

英名	和名	学名	所属、分布、ほか
orange piercing moth		*Achaea obvia* Hampson	(チョウ目、ヤガ科) エチオピア区
orange piercing moth			fruit-piercing moth (2) を見よ
orange playboy			apricot playboy を見よ
orange pulvinaria			California red scale を見よ
orange pulvinaria scale			California red scale を見よ
orange punch	ネパールトラフシジミタテハ	*Dodona egeon* (Westwood)	(チョウ目、シジミタテハ科) 旧北区、東洋区。*D. e. confluens* Corbet も同英名
orange puppies			orange dogs を見よ
orange puppy			giant swallowtail を見よ
orange-rayed pearl	スジモンカバノメイガ	*Nascia cilialis* (Christoph)	(チョウ目、メイガ科) 日本、旧北区
orange-red glider		*Cymothoe aramis* (Hewitson)	(チョウ、タテハチョウ科) エチオピア区
orange red skirt		*Choaspes hemixanthus furcatus* Evans	(チョウ目、セセリチョウ科) 東洋区
orange rice leafhopper		*Thaia oryzivora* Ghauri	(カメムシ目、オオヨコバイ科) 東洋区
orange-rimmed skipper		*Chalypyge chalybea chalybea* (Scudder)	(チョウ目、セセリチョウ科) 新熱帯区。*C. c. chloris* (Evans) も同英名
orange ringlet		*Hypocysta adiante* (Hübner)	(チョウ目、タテハチョウ科) 豪州区
orange roadside skipper		*Amblyscirtes simius* Edwards	(チョウ目、セセリチョウ科) 新北区
orange sallow		*Xanthia citrago* (Linnaeus)	(チョウ目、ヤガ科) 旧北区
orange sawyer		*Anoplium inerme* (Newman)	(コウチュウ目、カミキリムシ科) 新北区。カンキツ害虫
orange scale			California red scale を見よ
orange shadowdragon		*Neurocordulia xanthosoma* (Williamson)	(トンボ目、エゾトンボ科) 新北区
orange skipperling		*Copaeodes aurantiaca* (Hewitson)	(チョウ目、セセリチョウ科) 新北区、新熱帯区
orange spiny whitefly			citrus spiny whitefly を見よ　米国での英名
orange-spot duke		*Siseme neurodes* (Ménétriès)	(チョウ目、シジミタテハ科) 新熱帯区
orange spot hairtail		*Neurellipes lusones* Hewitson	(チョウ目、シジミチョウ科) エチオピア区
orange-spot Scoparia moth		*Scoparia submarginalis* Walker	(チョウ目、メイガ科) 豪州区
orange-spotted carpet moth		*Dysstroma walkerata* (Pearsall)	(チョウ目、シャクガ科) 新北区
orange-spotted castniid		*Synemon parthenoides* Felder	(チョウ目、カストニアガ科) 豪州区
orange-spotted cockroach			South American cockroach を見よ
orange-spotted emerald		*Oxygastra curtisii* (Dale)	(トンボ目、エゾトンボ科) 旧北区

英名	和名	学名	所属、分布、ほか
orange-spotted Euselasia		*Euselasia argentea* (Hewitson)	(チョウ目、シジミタテハ科) 新熱帯区
orange-spotted flower moth		*Syngamia florella* (Stoll)	(チョウ目、メイガ科) 新北区
orange-spotted hopper			short-tailed Arizona skipper を見よ
orange-spotted Idia moth		*Idia diminuendis* (Barnes et McDunnough)	(チョウ目、ヤガ科) 新北区。orange-spotted Idia ともいう
orange-spotted ladybird		*Coccinella leonina* (Fabricius)	(コウチュウ目、テントウムシ科) 東洋区、豪州区
orange spotted Nephele		*Nephele discifera* Karsch	(チョウ目、スズメガ科) エチオピア区
orange-spotted Prepona		*Prepona deiphile brooksiana* Godman et Salvin	(チョウ目、タテハチョウ科) 新熱帯区。*P. d. diaziana* Miller et Miller, *P. d. escalantiana* Stoffel et Mast, *P. d. ibarra* Beutelspacher, *P. d. lambertoana* Llorente, Luis et Gonzalez も同英名
orange-spotted Pyrausta			orange mint moth を見よ
orange-spotted shoot			spotted shoot を見よ
orange-spotted skipper		*Zenonia zeno* (Trimen)	(チョウ目、セセリチョウ科) エチオピア区
orange-spotted skipper			short-tailed Arizona skipper を見よ
orange-spotted skipper		*Atarnes sallei* (Felder et Felder)	(チョウ目、セセリチョウ科) 新熱帯区
orange-spotted sun moth			orange-spotted castniid を見よ
orange-spotted tachinid		*Phasiomyia splendida* (Coquillett)	(ハエ目、ヤドリバエ科) 新北区
orange-spotted tiger clearwing			common tiger (2) を見よ
orange sprite		*Celaenorrhinus galenus* Fabricius	(チョウ目、セセリチョウ科) エチオピア区
orange spruce needleminer moth		*Coleotechnites piceaella* (Kearfott)	(チョウ目、キバガ科) 新北区。幼虫は orange spruce needleminer
orange staff sergeant	タイワンイチモンジ	*Athyma cama* Moore	(チョウ目、タテハチョウ科) 東洋区
orange stem borer (1)		*Indarbela quadrinotata* Walker	(チョウ目、Metarbelidae) 東洋区
orange stem borer (2)		*Dirphya nigricornis* Olivier	(コウチュウ目、カミキリムシ科) エチオピア区。コーヒー害虫
orange streamcruiser		*Hesperocordulia berthoudi* Tillyard	(トンボ目、エゾトンボ科) 豪州区
orange-striped awl		*Burana jaina* (Moore)	(チョウ目、セセリチョウ科) 東洋区
orange-striped eighty-eight			orange-banded 88 を見よ
orange-striped Emesis		*Emesis cypria paphia* R. Felder	(チョウ目、シジミタテハ科) 新熱帯区。orange-barred Emesis を参照
orange-striped leafwing		*Memphis philumena xenica* (Bates)	(チョウ目、タテハチョウ科) 新熱帯区。streaky leafwing を参照

英名	和名	学名	所属、分布、ほか
orange-striped oakworm			orange-tipped oakworm moth を見よ
orange-striped sister			yellow-striped sister を見よ
orange-striped stonefly			February red を見よ
orange-striped threadtail		*Protoneura cara* Calvert	(トンボ目、ミナミイトトンボ科) 新北区、新熱帯区
orange sulfur			alfalfa butterfly を見よ
orange sun moth	キゴマダラヒトリ	*Argina astrea* (Drury)	(チョウ目、ヒトリガ科) 日本、東洋区、大洋区、豪州区、エチオピア区
orange swallowtail			King page を見よ
orange swallow-tailed moth		*Thinopteryx crocopterata* (Kollar)	(チョウ目、シャクガ科) 旧北区、東洋区
orange swift		*Hepialus sylvinus* (Linnaeus)	(チョウ目、コウモリガ科) 旧北区
orange-tail awl		*Bibasis sena* (Moore)	(チョウ目、セセリチョウ科) 東洋区。*B. s. uniformis* Elwes et Edwards も同英名
orange-tailed clearwing		*Synanthedon andrenaeformis* (Laspeyres)	(チョウ目、スカシバガ科) 旧北区
orangetailed potter wasp		*Delta latreillei petiolaris* (Schulz)	(ハチ目、スズメバチ科) 新北区
orange Telemiades		*Telemiades megallus* Mabille	(チョウ目、セセリチョウ科) 新熱帯区
orange Theope butterfly	カカオシジミモドキ	*Theope eudocia* Westwood	(チョウ目、シジミチョウ科) 新熱帯区
orange thorn	スモモエダシャク	*Angerona prunaria* (Linnaeus)	(チョウ目、シャクガ科) 旧北区。スモモ、ウメ、サクラ害虫
orange tiger			banded orange heliconian を見よ
orange tiger			common tiger (3) を見よ
orange tiger butterfly		*Danaus affinis alexis* (Waterhouse et Lyell)	(チョウ目、タテハチョウ科) 豪州区
orange tiger moth			tiger moth (1) を見よ
orange tip	ナカスジツマアカシロチョウ	*Colatis evenina* (Wallengren)	(チョウ目、シロチョウ科) エチオピア区
orange tip (1)		*Anthocharis cardamines* (Linnaeus)	(チョウ目、シロチョウ科) 全北区。日本亜種は *A. c. isshikii* Matsumura クモマツマキチョウ。orange tip butterfly ともいう
orange tip			desert orange tip (2) を見よ
orange-tip			Sara orange-tip を見よ
orange tip moth		*Adoxophyes fasciculana* (Walker)	(チョウ目、ハマキガ科) 東洋区、大洋区
orange-tipped azure			Dodd's azure を見よ
orange-tipped grasshopper		*Omocestus haemorrhoidalis* (Charpentier)	(バッタ目、バッタ科) 旧北区
orange-tipped oakworm moth		*Anisota senatoria* (Smith)	(チョウ目、ヤママユガ科) 新北区。幼虫は orange-tipped oakworm

英名	和名	学名	所属、分布、ほか
orange-tipped pea-blue		*Everes lacturnus* (Godart)	（チョウ目、シジミチョウ科）豪州区、東洋区
orange tips	ツマアカシロチョウ属	*Colotis*	（チョウ目、シロチョウ科）の昆虫の総称
orange-toned Mecyna moth		*Mecyna submedialis* Grote	（チョウ目、メイガ科）新北区
orange tortrix		*Argyrotaenia citrana* (Fernald)	（チョウ目、ハマキガ科）新北区
orange tortrix (1)		*Argyrotaenia franciscana* (Walsingham)	（チョウ目、ハマキガ科）新北区。orange tortrix moth ともいう
orange tree bug			bronze orange bug を見よ
orange-tufted Oneida moth		*Oneida lunulalis* Hulst	（チョウ目、メイガ科）新北区
orange underwing		*Archiearis parthenias* (Linnaeus)	（チョウ目、シャクガ科）新北区。日本亜種は *A. p. bella* (Inoue)　カバシャク
orange underwing (1)		*Jodia croceago* (Denis et Schiffermüller)	（チョウ目、ヤガ科）旧北区
orange underwing moth		*Notoreas brephos* Walker	（チョウ目、シャクガ科）豪州区
orange-underwings		Archieariinae	（チョウ目、ヤガ科）の昆虫の総称
orange upperwing			orange underwing (1) を見よ
orange-veined blue		*Aricia neurona* (Skinner)	（チョウ目、シジミチョウ科）新北区
orange-washed sister			Cocala sister を見よ
orange waxtail			common citril を見よ
orange webworm moth		*Dichomeris citrifoliella* (Chambers)	（チョウ目、キバガ科）新北区
orange wheat blossom midge			wheat midge を見よ
orange whisp		*Agriocnemis ruberrima* Balinsky	（トンボ目、イトトンボ科）エチオピア区
orange white		*Hesperocharis crocea crocea* Bates	（チョウ目、シロチョウ科）新熱帯区。*H. c. jaliscana* Schaus も同英名
orange white-spot skipper		*Trapezites heteromacula* Meyrick et Lower	（チョウ目、セセリチョウ科）豪州区
orange whitefly		*Paraleyrodes naranjae* Dozier	（カメムシ目、コナジラミ科）大洋区、新熱帯区。カンキツ害虫
orangewing moth		*Mellilla xanthometata* Walker	（チョウ目、シャクガ科）新北区
orange-winged dropwing			Kirby's dropwing を見よ
orange-winged grasshopper		*Arphia conspersa ramona* Rehn	（バッタ目、バッタ科）新北区
orangeworm			orange fruitborer を見よ
orb underwing moth		*Catocala orba* Kuznezov	（チョウ目、ヤガ科）新北区。orb underwing ともいう
orbed narrow-wing moth		*Magusa divaricata* Grote	（チョウ目、ヤガ科）新北区。orbed narrow-wing ともいう
orbed red-underwing skipper			Hungarian skipper を見よ

英名	和名	学名	所属、分布、ほか
orbed sulphur		*Phoebis orbis* (Poey)	(チョウ目、シロチョウ科) 新北区
orchard butterfly			large citrus butterfly を見よ
orchard cicada		*Platypedia areolata* Uhler	(カメムシ目、セミ科) 新北区
orchard ermine		*Yponomeuta padella* (Linnaeus)	(チョウ目、スガ科) 旧北区。orchard ermine moth ともいう
orchard grass aphid			cocksfoot aphid を見よ
orchard midget			silver-spotted midget を見よ
orchard piercer		*Laspeyresia prunivorana* Ragonot	(チョウ目、ハマキガ科) 旧北区
orchard swallowtail			large citrus butterfly を見よ
orchid aphid (1)		*Cerataphis lataniae* (Boisduval)	(カメムシ目、アブラムシ科) 旧北区、豪州区
orchid aphid (2)		*Macrosiphum luteum* (Buckton)	(カメムシ目、アブラムシ科) 全北区、豪州区、大洋区
orchid aphid			fringed orchid aphid を見よ
orchid bees	シタバチ亜科	Euglossinae	(ハチ目、ミツバチ科) の昆虫の総称
orchid beetle		*Stethopachys formosa* Baly	(コウチュウ目、ハムシ科) 豪州区
orchid crane fly	ベッコウガガンボ	*Ctenophora fasciata* Coquillett	(ハエ目、ガガンボ科) 日本、旧北区
orchid dupe		*Lissopimpla excelsa* (Costa)	(ハチ目、ヒメバチ科) 豪州区。orchid dupe wasp ともいう
orchid embiid	シロアリモドキ	*Oligotoma saundersii* (Westwood)	(シロアリモドキ目、シロアリモドキ科) 日本、新北区、汎熱帯
orchid flash			black-and-white tit を見よ
orchid fly		*Eurytoma orchidearum* (Westwood)	(ハチ目、カタビロコバチ科) 全北区
orchid hooktip			arched hooktip を見よ
orchid mantis			flower mantis を見よ
orchid mealybug		*Pseudococcus microcirculus* McKenzie	(カメムシ目、コナカイガラムシ科) 新北区
orchid oystershell scale			cymbidium scale を見よ
orchid parlatoria scale			cattleya scale を見よ　豪州での英名
orchid praying mantis			flower mantis を見よ
orchid scale	ランシロカイガラムシ	*Diaspis boisduvalii* Signoret	(カメムシ目、マルカイガラムシ科) 日本、全北区、豪州区、汎熱帯。ラン害虫
orchid scale			cattleya scale を見よ
orchid soft scale		*Coccus pseudohesperidum* (Cockerell)	(カメムシ目、カタカイガラムシ科) 新北区、大洋区
orchid springtail		*Orchesella cincta* (Linnaeus)	(トビムシ目、アヤトビムシ科) 全北区
orchid springtails		*Orchesella*	(トビムシ目、アヤトビムシ科) の昆虫の総称
orchid stem miner	ランミモグリバエ	*Japanagromyza tokunagai* (Sasakawa)	(ハエ目、ハモグリバエ科) 日本。ラン害虫

英名	和名	学名	所属、分布、ほか
orchid thrips	ランノオビアザミウマ	*Chaetanaphothrips orchidii* (Moulton)	(アザミウマ目、アザミウマ科) 日本、旧北区。ラン、シクラメン、カエデ類害虫
orchid thrips			dendrobium thrips を見よ
orchid thrips			yellow orchid thrips を見よ
orchid tit	サイトウフタオルリシジミ	*Chliaria othona* (Hewitson)	(チョウ目、シジミチョウ科) 東洋区
orchid weevil	オオランヒメゾウムシ	*Orchidophilus aterrimus* (Waterhouse)	(コウチュウ目、ゾウムシ科) 日本、豪州区、大洋区。ラン害虫
orcinus skipper		*Udranomia orcinus* (C. et R. Felder)	(チョウ目、セセリチョウ科) 新熱帯区
Orcus checkered-skipper		*Pyrgus orcus* (Stoll)	(チョウ目、セセリチョウ科) 新熱帯区
Orcynia hairstreak			Imma hairstreak を見よ
Oreas anglewing		*Polygonia oreas* (Edwards)	(チョウ目、タテハチョウ科) 新北区
Oreas comma			Oreas anglewing を見よ
Oreas copper		*Aloeides oreas* Tite et Dickson	(チョウ目、シジミチョウ科) エチオピア区
Oregon ant cricket			ant cricket (2) を見よ
Oregon Catoptria moth		*Catoptria oregonica* (Grote)	(チョウ目、メイガ科) 新北区
Oregon Cycnia			Oregon Euchaetias を見よ　Oregon Cycnia moth ともいう
Oregon Euchaetias		*Euchaetias oregonensis* (Stretch)	(チョウ目、ヒトリガ科) 新北区
Oregon fir sawyer		*Monochamus scutellatus oregonensis* LeConte	(コウチュウ目、カミキリムシ科) 新北区
Oregon stag beetle		*Platycerus oregonensis* Westwood	(コウチュウ目、クワガタムシ科) 新北区
Oregon swallowtail		*Papilio machaon oregonius* Edwards	(チョウ目、アゲハチョウ科) 新北区
Oregon tiger beetle		*Cicindela oregona* LeConte	(コウチュウ目、ハンミョウ科) 新北区
Oregon wainscott		*Leucania oregona* Smith	(チョウ目、ヤガ科) 新北区
Oregon wireworm		*Melanotus longulus oregonensis* (LeConte)	(コウチュウ目、コメツキムシ科) 新北区
Orestessa metalmark		*Synargis orestessa* Hübner	(チョウ目、シジミタテハ科) 新熱帯区
Orestia glassy Acraea		*Acraea orestia* Hewitson	(チョウ目、タテハチョウ科) エチオピア区
organ-pipe mud-daubers			mud daubers and spider wasps (1) を見よ
Oria skipper		*Drephalys oria* Evans	(チョウ目、セセリチョウ科) 新熱帯区
oriander skipper		*Drephalys oriander* (Hewitson)	(チョウ目、セセリチョウ科) 新熱帯区
Orichora brown		*Oreixenica orichora* (Meyrick)	(チョウ目、タテハチョウ科) 豪州区
Oriens metalmark		*Mesene oriens* Butler	(チョウ目、シジミタテハ科) 新熱帯区
Oriental armyworm			armyworm (2) を見よ
Oriental asparagus beetle	カタボシクビナガハムシ	*Crioceris orientalis* Jacoby	(コウチュウ目、ハムシ科) 日本、東洋区

英名	和名	学名	所属、分布、ほか
Oriental bark moth	カラマツミキモグリガ	*Cydia laricicolana* (Kuznetzov)	(チョウ目、ハマキガ科) 日本。カラマツ害虫
Oriental bean beetle			bean leaf beetle (1) を見よ
Oriental bee hawk moth			larger pellucid hawk moth を見よ
Oriental beetle			Asiatic beetle を見よ
Oriental blow fly			Oriental latrine fly を見よ
Oriental broad-nosed weevils		Eremninae	(コウチュウ目、ゾウムシ科) の昆虫の総称
Oriental carpenter moth			goat moth (2) を見よ
Oriental chinch bug	カンシャコバネナガカメムシ	*Cavelerius saccharivorus* (Okajima)	(カメムシ目、ナガカメムシ科) 日本、東洋区。サトウキビ害虫
Oriental clouded yellow		*Colias erate* (Esper)	(チョウ目、シロチョウ科) 旧北区。日本亜種は *C. e. poliographus* Motschulsky モンキチョウ
Oriental cockroach			common cockroach を見よ
Oriental corn borer	アワノメイガ	*Ostrinia furnacalis* (Guenée)	(チョウ目、メイガ科) 日本、旧北区、東洋区、豪州区。トウモロコシ、ナス、トマト、イネ科牧草などの害虫
Oriental cotton stainer			cotton bug を見よ
Oriental cowpea bruchid			adzuki weevil を見よ
Oriental face fly	ノイエバエ	*Musca hervei* Villeneuve	(ハエ目、イエバエ科) 日本、旧北区、東洋区、大洋区
Oriental false chinch bug	ホソメダカナガカメムシ	*Ninomimus flavipes* (Matsumura)	(カメムシ目、ナガカメムシ科) 日本、旧北区
Oriental fir bark moth	トドマツミキモグリガ	*Cydia pactolana yasudai* (Oku)	(チョウ目、ハマキガ科) 日本。トドマツ害虫
Oriental fir budworm	トウヒオオハマキ	*Lozotaenia coniferana* (Issiki)	(チョウ目、ハマキガ科) 日本、旧北区。トウヒ、モミ、トドマツなどの害虫
Oriental fruit fly	ミカンコミバエ	*Bactrocera dorsalis* (Hendel)	(ハエ目、ミバエ科) 日本、東洋区、新北区。カンキツほかパパイア、バナナ、マンゴーも加害する著名害虫
Oriental fruit fly egg parasite		*Opius oophilus* Fullaway	(ハチ目、コマユバチ科) 大洋区
Oriental fruit fly egg parasite		*Biosteres arisanus* (Sonan)	(ハチ目、コマユバチ科) 大洋区
Oriental fruit moth	ナシヒメシンクイ	*Grapholita molesta* (Busck)	(チョウ目、ハマキガ科) 日本、世界の温暖地域。リンゴ、ナシ、モモ、サクラなどの大害虫。アジアから侵入したことに由来する北米での英名
Oriental garden fleahopper	クロトビカスミカメムシ	*Halticiellus insularis* (Usinger)	(カメムシ目、カスミカメムシ科) 日本、東洋区、大洋区。ダイズ、ウリ類、アブラナ科野菜の害虫
Oriental grapevine looper	ウストビモンナミシャク	*Eulithis lederi inurbana* (Prout)	(チョウ目、シャクガ科) 日本
Oriental grass root aphid	オカボノクロアブラムシ	*Tetraneura nigriabdominalis* (Sasaki)	(カメムシ目、アブラムシ科) 日本、旧北区、東洋区、豪州区、エチオピア区。イネ、ムギ、トウモロコシ害虫
Oriental green rice leafhopper			green rice leafhopper (2) を見よ

英名	和名	学名	所属、分布、ほか
Oriental green stink bug			green stink bug (1) を見よ
Oriental greenwing		*Neurobasis chinensis chinensis* (Linnaeus)	（トンボ目、イトトンボ科）東洋区
Oriental hive bee			Indian honey bee を見よ
Oriental honey bee	トウヨウミツバチ	*Apis cerana* Fabricius	（ハチ目、ミツバチ科）東洋区
Oriental hornet		*Vespa orientalis* Linnaeus	（ハチ目、スズメバチ科）旧北区、東洋区
Oriental horntail			Asian horntail を見よ
Oriental house fly		*Musca domestica vicina* Macquart	（ハエ目、イエバエ科）新北区
Oriental lappet	カレハガ	*Gastropacha orientalis* Sheljuzhko	（チョウ目、カレハガ科）日本、旧北区。リンゴ、など果樹、サクラの害虫
Oriental latrine fly	オビキンバエ	*Chrysomya megacephala* (Fabricius)	（ハエ目、クロバエ科）日本、旧北区、東洋区、大洋区、豪州区
Oriental leafworm moth			common cutworm を見よ
Oriental leopard moth	ゴマフボクトウ	*Zeuzera multistrigata leuconota* Butler	（チョウ目、ボクトウガ科）日本、旧北区。カキ、チャ、サクラ、ツバキ、カシ類などの害虫
Oriental longheaded grasshopper			Oriental long-headed locust (2) を見よ
Oriental long-headed locust (1)	オンブバッタ	*Atractomorpha lata* (Motschulsky)	（バッタ目、オンブバッタ科）日本、旧北区、東洋区
Oriental long-headed locust (2)	ショウリョウバッタ	*Acrida cinerea* (Thunberg)	（バッタ目、バッタ科）日本、旧北区、東洋区、豪州区、エチオピア区
Oriental marble		*Falcuna orientalis* (Bethune-Baker)	（チョウ目、シジミチョウ科）エチオピア区
Oriental marbled skipper		*Carcharodes orientalis* Reverdin	（チョウ目、セセリチョウ科）旧北区
Oriental meadow brown		*Hyponephele lupina* (Costa)	（チョウ目、タテハチョウ科）旧北区
Oriental mealybug parasite		*Pauridia peregrina* Timberlake	（ハチ目、トビコバチ科）大洋区
Oriental migratory locust	トウヨウトビバッタ	*Locusta migratoria manilensis* (Meyen)	（バッタ目、バッタ科）東洋区
Oriental mole cricket			mole cricket (1) を見よ
Oriental moth	イラガ	*Cnidocampa flavescens* (Walker)	（チョウ目、イラガ科）日本、旧北区。開帳30mm。多くの果樹、樹木、チャ、バラの害虫
Oriental Parnassian moths		Ratardidae	（チョウ目）の昆虫の総称
Oriental peach moth			Oriental fruit moth を見よ
Oriental rat flea	ケオプスネズミノミ	*Xenopsylla cheopis* (Rothschild)	（ノミ目、ヒトノミ科）日本、全北区、豪州区
Oriental rice borer			rice stem borer を見よ
Oriental rice seedling fly		*Atherigona orientalis* Schiner	（ハエ目、イエバエ科）エチオピア区
Oriental satin moth			satin moth (1) を見よ
Oriental scale	オスベッキーマルカイガラムシ	*Aonidiella orientalis* (Newstead)	（カメムシ目、マルカイガラムシ科）日本、東洋区、豪州区、新北区。カンキツ害虫

英名	和名	学名	所属、分布、ほか
Oriental scale parasite		*Comperiella lemniscata* Compere et Annecke	（ハチ目、トビコバチ科）日本
Oriental scale predator		*Telsimia* sp.	（コウチュウ目、テントウムシ科）豪州区
Oriental scale predator		*Chilocorus bailey* Blackburn	（コウチュウ目、テントウムシ科）豪州区
Oriental scarlet			scarlet skimmer を見よ
Oriental silverfish	ヤマトシミ	*Ctenolepisma villosa* (Fabricius)	（シミ目、シミ科）日本、東洋区
Oriental skipper		*Carcharodus orientalis* Reverdin	（チョウ目、セセリチョウ科）旧北区
Oriental skipper			Oriental marbled skipper を見よ
Oriental stink bug		*Plautia stali* Scott	（カメムシ目、カメムシ科）新北区
Oriental stink bug			brown-winged green bug を見よ
Oriental straight swift			Ceylon swift を見よ
Oriental strawberry leafroller			Atmore's twist を見よ
Oriental swallowtail moths	アゲハモドキガ科	Epicopeiidae	（チョウ目）の昆虫の総称
Oriental tea tortrix	チャハマキ	*Homona magnamima* Diakonoff	（チョウ目、ハマキガ科）日本、東洋区。チャ、果樹、園芸植物の害虫
Oriental tobacco budworm	タバコガ	*Helicoverpa assulta* (Guenée)	（チョウ目、ヤガ科）日本、旧北区、東洋区、大洋区、豪州区、エチオピア区。タバコ、ジャガイモ、ウリ類、キクなどの害虫
Oriental tobacco budworm moth			Oriental tobacco budworm を見よ
Oriental toon moth			mahogany shoot borer (1) を見よ
Oriental tussock moth	ドクガ	*Euproctis subflava* (Bremer)	（チョウ目、ドクガ科）日本、旧北区。ダイズ、カンキツ、チャ、カシ類などの害虫
Oriental wheat-stem sawfly			black grain stem sawfly を見よ　北アフリカでの英名
Oriental woolly aphid	ホンシュウマツカサアブラムシ	*Pineus orientalis* (Dreyfus)	（カメムシ目、カサアブラムシ科）日本、旧北区。マツ害虫
Oriental yellow scale			Oriental scale を見よ
original round-spot		*Pilodeudorix otraeda* (Hewitson)	（チョウ目、シジミチョウ科）エチオピア区
Origo groundstreak		*Calycopis origo* (Godman et Salvin)	（チョウ目、シジミチョウ科）新熱帯区
Orina Acraea	ベニオビホソチョウ	*Acraea orina* Hewitson	（チョウ目、タテハチョウ科）エチオピア区
Orion	オリオンタテハ	*Historis odius* (Fabricius)	（チョウ目、タテハチョウ科）新北区
Orion cercopian			Orion を見よ　*H. o. dious* Lamas の英名
Orius skipper		*Naevolus orius* (Mabille)	（チョウ目、セセリチョウ科）新熱帯区
orizaba silk moth		*Rothschildia orizaba* (Westwood)	（チョウ目、ヤママユガ科）全北区
Orizaba sootywing		*Bolla oriza* Evans	（チョウ目、セセリチョウ科）新熱帯区
orlflies			alderflies (1) を見よ

英名	和名	学名	所属、分布、ほか
orlflies			lacewings (1) を見よ　米国での英名
Orma		*Mopala orma* (Plötz)	(チョウ目、セセリチョウ科) エチオピア区
Ormiston's oakblue		*Arhopala ormistoni* Riley	(チョウ目、シジミチョウ科) 東洋区
ormyrids	タマヤドリコバチ科	Ormyridae	(ハチ目) の昆虫の総称
ornamental knot-horn		*Pempeliella ornatella* (Denis et Schiffermüller)	(チョウ目、メイガ科) 旧北区
ornamented Utetheisa			ornate bella moth を見よ
ornate aphid			violet aphid (1) を見よ
ornate Bella moth		*Utetheisa ornatrix* (Linnaeus)	(チョウ目、ヒトリガ科) 新北区、新熱帯区。Bella moth, ornate moth ともいう
ornate brigadier		*Odontomyia ornata* (Meigen)	(ハエ目、ミズアブ科) 旧北区
ornate checkered beetle		*Trichodes ornatus* Say	(コウチュウ目、カッコウムシ科) 新北区。幼虫はハチ類を捕食
ornate dusk-flat			rare red-eye を見よ
ornate green Charaxes	オオアオフタオチョウ	*Charaxes subornatus* Schultze	(チョウ目、タテハチョウ科) エチオピア区
ornate Hydriris moth		*Hydriris ornatalis* (Duponchel)	(チョウ目、メイガ科) 新北区
ornate Junea		*Junea doraete* (Hewitson)	(チョウ目、タテハチョウ科) 新熱帯区
ornate ochre		*Trapezites genevieveae* (Atkins)	(チョウ目、セセリチョウ科) 豪州区
ornate pit scale insects			ornate pit scales を見よ
ornate pit scales	フジツボカイガラムシ科	Cerococcidae	(カメムシ目) の昆虫の総称　小さな科
ornate plant bug		*Closterocoris ornatus* (Reuter)	(カメムシ目、カスミカメムシ科) 新北区
ornate Temnora		*Temnora inornata* (Rothschild)	(チョウ目、スズメガ科) エチオピア区
ornate tiger moth			tiger moth (1) を見よ
Ornythion swallowtail		*Papilio ornythion* Boisduval	(チョウ目、アゲハチョウ科) 新北区
Orseis crescentspot		*Phyciodes orseis* Edwards	(チョウ目、タテハチョウ科) 新北区
ortalid flies		Ortalidae	(ハエ目) の昆虫の総称　マダラバエ科の異名。
Ortalus hairstreak		*Thereus ortalus* (Godman et Salvin)	(チョウ目、シジミチョウ科) 新熱帯区
Orthesia leafwing		*Memphis mora orthesia* (Godman et Salvin)	(チョウ目、タテハチョウ科) 新熱帯区
orthezia lady beetle		*Hyperaspis jocosa* (Mulsant)	(コウチュウ目、テントウムシ科) 新北区
orthoperid beetles			minute fungus beetles (1) を見よ
orthopterans			locusts を見よ
orthopteroid insects		Orthopteroidea	の昆虫の総称　locusts を見よ
orthopteron			直翅類の昆虫。locusts を見よ

英名	和名	学名	所属、分布、ほか
orthopterous insects			locusts を見よ
orthorrhaphous flies	直縫群	Orthorrhapha	(ハエ目) の昆虫の総称
osage orange sphinx			Hagen's sphinx を見よ
Osborn scale		*Diaspidiotus osborni* (Newell et Cockerell)	(カメムシ目、マルカイガラムシ科) 新北区
Osca skipper		*Rhinthon osca* (Plötz)	(チョウ目、セセリチョウ科) 新北区
osier case-bearer		*Coleophora luscinaepennella* (Treitschke)	(チョウ目、ツツミノガ科) 旧北区
osier green moth			cream-bordered green pea を見よ
osier hornet clearwing		*Sphecia bembeciformis* Hübner	(チョウ目、スカシバガ科) 旧北区
osier leaf aphid		*Chaitophorus beuthami* (Börner)	(カメムシ目、アブラムシ科) 旧北区
osier leaf-folding midge		*Rhabdophaga marginemfirquens* (Bremi)	(ハエ目、タマバエ科) 旧北区
osier midget		*Phyllonorycter viminetorum* (Stainton)	(チョウ目、ホソガ科) 旧北区
osier weevil			willow beetle を見よ
Osiris blue		*Cupido osiris* (Meigen)	(チョウ目、シジミチョウ科) 旧北区
Osiris sooty blue			African cupid を見よ
Oslar's eacles			Oslar's imperial moth を見よ
Oslar's imperial moth		*Eacles oslari* Rothschild	(チョウ目、ヤママユガ科) 新北区
Oslar's oakworm moth		*Anisota oslari* Rothschild	(チョウ目、ヤママユガ科) 新北区、新熱帯区
Oslar's roadside skipper		*Amblyscirtes oslari* (Skinner)	(チョウ目、セセリチョウ科) 新北区、新熱帯区
osmanthus pinnaspis scale	サカキホソカイガラムシ	*Pinnaspis uniloba* (Kuwana)	(カメムシ目、マルカイガラムシ科) 日本、東洋区、大洋区
osmanthus woolly aphid	ヒイラギハマキワタムシ	*Prociphilus osmanthae* Essig et Kuwana	(カメムシ目、アブラムシ科) 日本
Osmunda borer moth		*Papaipema speciosissima* (Grote et Robinson)	(チョウ目、ヤガ科) 新北区
osmylid flies	ヒロバカゲロウ科	Osmylidae	(アミメカゲロウ目) の昆虫の総称
ostomatid beetles			gnawing beetles (1) を見よ 米国での英名。Trogositidae の異名である Ostomidae に由来
ostreiform scale			oystershell scale (1) を見よ
Ostrinia nubilaris parasitoid	メイガヒゲナガコマユバチ	*Macrocentrus grandii* Goidanich	(ハチ目、コマユバチ科) 日本、全北区
Osumi xyleborus	オオスミキクイムシ	*Xyleborus osumiensis* Murayama	(コウチュウ目、キクイムシ科) 日本
Otanes crescent			blackened crescent を見よ
othniids			false tiger beetles (1) を見よ
otitid flies			picture-winged flies を見よ
otitid fly		*Ceroxys latiuscula* (Loew)	(ハエ目、ハネフリバエ科) 新北区
otter			ghost moth を見よ
otter moth			ghost moth を見よ

英名	和名	学名	所属、分布、ほか
Ottoe skipper		*Hesperia ottoe* Edwards	(チョウ目、セセリチョウ科) 新北区
Ottoman brassy ringlet	オットマンベニヒカゲ	*Erebia ottomana* Herrich-Schäffer	(チョウ目、タテハチョウ科) 旧北区
Otway stonfly		*Eusthenia nothofagi* Zwick	(カワゲラ目、Eustheniidae) 豪州区。絶滅危惧種。豪州のオトウエイ国立公園に由来
Ouachita spiketail		*Cordulegaster talaria* Tennessen	(トンボ目、オニヤンマ科) 新北区。米国のウオシタ山地に由来
Outeniqua blue		*Lepidochrysops outeniqua* Swanepoel et Vári	(チョウ目、シジミチョウ科) エチオピア区
outer barklice		Ectopsocidae	(チャタテムシ目) の昆虫の総称
outhouse flies		*Sylvicola*	(ハエ目、カバエ科) の昆虫の総称
Outis skipper		*Cogia outis* (Skinner)	(チョウ目、セセリチョウ科) 新北区
outlet sword	カギバアゲハ	*Meandrusa payeni* (Boisduval)	(チョウ目、アゲハチョウ科) 東洋区。前翅端が尖っていることに因む
oval Abrostola moth		*Abrostola ovalis* Guenée	(チョウ目、ヤガ科) 新北区。oval Arostola ともいう
oval-based prominent moth		*Peridea basitriens* (Walker)	(チョウ目、シャチホコガ科) 新北区
oval froghopper	マルアワフキ	*Lepyronia coleoptrata* (Linnaeus)	(カメムシ目、アワフキムシ科) 日本、旧北区
oval Guinea pig louse	カビアマルハジラミ	*Gyropus ovalis* Burmeister	(ハジラミ目、ナガケモノハジラミ科) 日本、汎世界。モルモットに寄生
oval leaf beetles	サルハムシ亜科	Eumolpinae	(コウチュウ目、ハムシ科) の昆虫の総称
ovate dagger moth		*Acronicta ovata* Grote	(チョウ目、ヤガ科) 新北区
ovate shieldback		*Aglaothorax ovata* (Scudder)	(バッタ目、キリギリス科) 新北区
Overberg skolly		*Thestor overbergensis* Heath et Pringle	(チョウ目、シジミチョウ科) エチオピア区
overcast skipper		*Lerema lumina* (Herrich-Schäffer)	(チョウ目、セセリチョウ科) 新熱帯区
Overlaet's glasswing		*Ornipholidotos overlaeti* Stempffer	(チョウ目、シジミチョウ科) エチオピア区
Ovinia skipper			sheep skipper (1) を見よ
ovoviviparous cockroaches	オオゴキブリ科	Panesthiidae	(ゴキブリ目) の昆虫の総称
ovoviviparous cockroaches			cockroaches (4) を見よ
owl butterflies	フクロチョウ属	*Caligo*	(チョウ目、タテハチョウ科) の昆虫の総称
owl butterfly (1)	メムノーンフクロチョウ	*Caligo memnon* Felder	(チョウ目、タテハチョウ科) 新熱帯区
owl butterfly (2)	メキシコフクロチョウ	*Caligo uranus* Herrich-Schäffer	(チョウ目、タテハチョウ科) 新熱帯区
owl butterfly			purple owl を見よ
owl butterfly			Idomeneus giant owl をみよ
owl-eyed bird-dropping moth		*Cerma cora* Hübner	(チョウ目、ヤガ科) 新北区
owlflies	ツノトンボ科	Ascalaphidae	(アミメカゲロウ目) の昆虫の総称　旧世界に多い

英名	和名	学名	所属、分布、ほか
owl-midge		*Pericoma fuliginosa* (Meigen)	(ハエ目、チョウバエ科) 旧北区
owl-midges			sand flies を見よ
owl midges			moth flies を見よ
owl moth		*Thysania zenobia* (Cramer)	(チョウ目、ヤガ科) 新北区、新熱帯区
owl moth			ligustrum moth を見よ
owl moths			owlet moths を見よ
owlet moths	ヤガ科	Noctuidae	(チョウ目) の昆虫の総称 世界に 25,000 種
owlet moths and underwings			owlet moths を見よ
owlets			flower moths を見よ
owls	ムカシヒカゲ属	*Neorina*	(チョウ目、タテハチョウ科) の昆虫の総称
owls			Morphos (1) を見よ
Owyhee harvester ant			harvester ant (5) を見よ
ox beetle		*Strategus antaeus* Fabricius	(コウチュウ目、コガネムシ科) 新北区
ox beetle		*Strategus aloeus* (Linnaeus)	(コウチュウ目、コガネムシ科) 新北区、新熱帯区
ox beetle			unicorn beetle を見よ
ox bot			cattle warble fly (1)(2) を見よ 幼虫の英名。ox-bot とも記す
ox bot fly			cattle warble fly (1)(2) を見よ
ox breeze fly			great gad fly を見よ
ox-eyed pansy			eyed pansy を見よ
oxfly			cattle warble fly (1) (2) を見よ
ox warble			cattle warble fly (1) (2) を見よ
ox warble flies			warble flies (2) を見よ
ox warble fly			cattle warble-fly (1) を見よ
ox warbles			warble flies (2) を見よ
oxaeid bees		Oxaeidae	(ハチ目) の昆虫の総称
Oxalis whitefly	カタバミコナジラミ	*Aleyrodes shizuokensis* Kuwana	(カメムシ目、コナジラミ科) 日本、新北区、大洋区
oxycanus grassgrub		*Oxycanus antipoda* (Herrich-Schäffer)	(チョウ目、コウモリガ科) 豪州区
Oyamel skipper		*Poanes monticola* (Godman)	(チョウ目、セセリチョウ科) 新熱帯区
oyster scale			oystershell scale (1) を見よ
oystershell bark louse			oystershell scale (1) を見よ
oystershell bark scale			oystershell scale (2) を見よ
oystershell Metrea moth		*Metrea ostreonalis* Grote	(チョウ目、メイガ科) 新北区
oystershell scale (1)	リンゴカキカイガラムシ	*Lepidosaphes ulmi* (Linnaeus)	(カメムシ目、マルカイガラムシ科) 日本、汎世界。カキ殻の形をした園芸、果樹の著名害虫。北米での英名
oystershell scale (2)		*Quadraspidiotus ostraeformis* (Curtis)	(カメムシ目、マルカイガラムシ科) 新北区、豪州区

英名	和名	学名	所属、分布、ほか
oystershell scale			pear scale (1) を見よ
oyster-shell scales		Lepidosaphinae	(カメムシ目、マルカイガラムシ科) の昆虫の総称
Ozark clubtail		*Gomphus ozarkensis* Westfall	(トンボ目、サナエトンボ科) 新北区
Ozark swallowtail		*Papilio joanae* Heitzman	(チョウ目、アゲハチョウ科) 新北区

英名	和名	学名	所属、分布、ほか
P			
Pachinus longwing		*Heliconius pachinus* Salvin	(チョウ目、タテハチョウ科) 新熱帯区
pachyneurid flies	キノコバエモドキ科	Pachyneuridae	(ハエ目) の昆虫の総称
pachyneurid gnats			pachyneurid flies を見よ
pachytroctids			thick barklice を見よ
Pacific albatross		*Appias athama* (Blanchard)	(チョウ目、シロチョウ科) 大洋区
Pacific ambush bug		*Phymata pacifica* Evans	(カメムシ目、ヒゲブトサシガメ科) 新北区
Pacific azure		*Celastrina echo echo* (W. H. Edwards)	(チョウ目、シジミチョウ科) 新熱帯区
Pacific beetle cockroach		*Diploptera punctata* (Eschsholtz)	(ゴキブリ目、ブラベルスゴキブリ科) 新北区、大洋区
Pacific brown lacewing		*Hemerobius pacificus* Banks	(アミメカゲロウ目、ヒメカゲロウ科) 新北区
Pacific clubtail		*Gomphus kurilis* Hagen	(トンボ目、サナエトンボ科) 新北区
Pacific coast wireworm		*Limonius canus* (LeConte)	(コウチュウ目、コメツキムシ科) 新北区。ブドウ害虫
Pacific cockroach		*Euthyrrhapha pacifica* (Coquebert)	(ゴキブリ目、カブトゴキブリ科) 東洋区、新北区
Pacific cuckoo wasp		*Chrysis pacifica* (Say)	(ハチ目、セイボウ科) 新北区
Pacific dampwood termite			common dampwood termite を見よ
Pacific damsel bug	セスジマキバサシガメ	*Nabis kinbergii* Reuter	(カメムシ目、マキバサシガメ科) 日本、大洋区、豪州区
Pacific damselfly		*Megalagrion pacificum* McLachlan	(トンボ目、イトトンボ科) 大洋区。絶滅危惧種
Pacific flatheaded borer		*Chrysobothris mali* Horn	(コウチュウ目、タマムシ科) 新北区。リンゴ他多数の果樹害虫
Pacific forktail		*Ischnura cervula* Selys	(トンボ目、イトトンボ科) 新北区
Pacific fritillary			western meadow fritillary を見よ
Pacific green sphinx moth		*Proserpinus lucidus* (Boisduval)	(チョウ目、スズメガ科) 新北区、新熱帯区。Pacific green sphinx ともいう
Pacific Hawaiian damselfly			Pacific damselfly を見よ
Pacific kissing bug		*Oncocephalus pacificus* (Kirkaldy)	(カメムシ目、サシガメ科) 新北区
Pacific meadow katydid		*Conocephalus occidentalis* (Morse)	(バッタ目、キリギリス科) 新北区
Pacific mealybug			passionvine mealybug を見よ
Pacific megalagrion damselfly			Pacific damselfly を見よ
Pacific outer barklouse		*Ectopsocus californicus* Banks	(チャタテムシ目、Ectopsocidae) 新北区
Pacific peach tree borer		*Synanthedon exitiosa graefi* (Hy. Edwards)	(チョウ目、スカシバガ科) 新北区。アーモンド害虫
Pacific pelagic water strider	コガタウミアメンボ	*Halobates sericeus* Eschscholtz	(カメムシ目、アメンボ科) 日本、新北区、大洋区、大西洋の熱帯〜亜熱帯圏

英名	和名	学名	所属、分布、ほか
Pacific roach			Pacific cockroach を見よ
Pacific spiketail			yellow-backed biddie を見よ
Pacific spotted mayfly		*Callibaetis pacificus* Seeman	(カゲロウ目、コカゲロウ科) 新北区
Pacific stylops		*Stylops pacifica* Bohart	(ネジレバネ目、ハチネジレバネ科) 新北区
Pacific tent caterpillar		*Malacosoma constricta* (Edwards)	(チョウ目、カレハガ科) 新北区
Pacific tigerwing		*Aeria eurimedia pacifica* Godman et Salvin	(チョウ目、タテハチョウ科) 新熱帯区
Pacific willow leaf beetle			grey willow leaf beetle を見よ
Packard grasshopper		*Melanoplus packardi* Scudder	(バッタ目、バッタ科) 新北区
Packard's concealer moth		*Semioscopis packardella* Clemens	(チョウ目、マルハキバガ科) 新北区
Packard's Eusarca moth		*Eusarca packardaria* (McDunnough)	(チョウ目、シャクガ科) 新北区
Packard's flatbody moth			Packard's concealer moth を見よ
Packard's girdle moth		*Enypia packardata* Taylor	(チョウ目、シャクガ科) 新北区
Packard's lichen moth		*Cisthene packardii* (Grote)	(チョウ目、ヒトリガ科) 新北区
Packard's wave moth		*Cyclophora packardi* (Prout)	(チョウ目、シャクガ科) 新北区。Packard's wave ともいう
Packard's white flannel moth		*Alarodia slossoniae* (Packard)	(チョウ目、イラガ科) 新北区
pacuvius duskywing	ヒメクロミヤマセセリ	*Erynnis pacuvius* (Lintner)	(チョウ目、セセリチョウ科) 新北区
paddle caterpillar			funerary dagger moth を見よ
paddle-tailed darner		*Aeschna palmata* Hagen	(トンボ目、ヤンマ科) 新北区
paddy armyworm			rice armyworm を見よ
paddy armyworm			armyworm (2) を見よ
paddy borer			yellow rice borer を見よ
paddy bug			rice bug (1) (2) を見よ
paddy case-bearer			rice case worm (1) を見よ
paddy climbing cutworm			armyworm (1) を見よ
paddyfield parasol		*Neurothemis intermedia intermedia* (Rambur)	(トンボ目、トンボ科) 東洋区
paddy fly			rice bug (2) を見よ
paddy grasshopper			rice grasshopper (3) を見よ
paddy hispid			rice hispa を見よ
paddy leaf-roller			rice leafroller を見よ
paddy root caterpillar		*Acigona chrysographella* (Kollar)	(チョウ目、メイガ科) 東洋区。イネ害虫
paddy root weevil		*Hydronomidius molitor* Faust	(コウチュウ目、ゾウムシ科) 東洋区。イネ害虫

英名	和名	学名	所属、分布、ほか
paddy slug caterpillar		*Pasara bicolor* (Walker)	（チョウ目、イラガ科）東洋区。イネ害虫
paddy stem borer			dark-headed rice borer を見よ
paddy stem borer			yellow rice borer を見よ
paddy stem maggot			rice whorl maggot (2) を見よ
paddy swarming caterpillar			rice armyworm を見よ
paddy thrips			rice thrips を見よ
Padilla glasswing		*Oleria padilla* Hewitson	（チョウ目、タテハチョウ科）新熱帯区
paederine rove beetles	アリガタハネカクシ亜科	Paederinae	（コウチュウ目、ハネカクシ科）の昆虫の総称
Paetus hairstreak		*Aubergina paetus* (Godman et Salvin)	（チョウ目、シジミチョウ科）新熱帯区
Pagenstecher's castor		*Ariadne pagenstecheri* (Suffert)	（チョウ目、タテハチョウ科）エチオピア区
pagoda bagworm		*Pagodiella hekmeyeri* Heylaerts	（チョウ目、ミノガ科）東洋区
Pahaska skipper		*Hesperia pahaska* Leussler	（チョウ目、セセリチョウ科）新北区、新熱帯区
Paignton snout			stout snout を見よ
paintbrush checkerspot			Anicia checkerspot
paintbrush swift		*Baoris farri* (Moore)	（チョウ目、セセリチョウ科）東洋区
paintbrush swift		*Baoris oceia* (Hewitson)	（チョウ目、セセリチョウ科）東洋区
painted apple moth		*Teia anartoides* Walker	（チョウ目、ドクガ科）豪州区
painted Arachnis			painted tiger moth を見よ
painted beauty		*Batesia hypochlora* C. et R. Felder	（チョウ目、タテハチョウ科）新熱帯区
painted beauty			American painted lady を見よ
painted bug		*Bagrada hilaris* (Burmeister)	（カメムシ目、カメムシ科）新北区、エチオピア区
painted capparis bug		*Stenozygum personatum* Walker	（カメムシ目、カメムシ科）新北区。カンキツ害虫
painted courtesan	アカホシトガリゴマダラ	*Euripus consimilis* (Westwood)	（チョウ目、タテハチョウ科）東洋区
painted crescent		*Phyciodes pictus* (Edwards)	（チョウ目、タテハチョウ科）新北区
painted crescentspot			painted crescent を見よ
painted cup moth		*Doratifera oxleyi* (Newman)	（チョウ目、イラガ科）豪州区
painted damsel		*Hesperagrion heterodoxum* (Selys)	（トンボ目、イトトンボ科）新北区
painted empress		*Apaturopsis cleochares* (Hewitson)	（チョウ目、タテハチョウ科）エチオピア区
painted grasshawk		*Neurothemis stigmatizans* (Fabricius)	（トンボ目、トンボ科）東洋区、豪州区、新北区、エチオピア区
painted hickory borer		*Megacyllene caryae* (Gahan)	（コウチュウ目、カミキリムシ科）新北区

英名	和名	学名	所属、分布、ほか
painted Jezebel	ベニモンシロチョウ	*Delias hyparete* (Linnaeus)	（チョウ目、シロチョウ科）東洋区。*D. h. metarete* Butler も同英名
painted lady	ヒメアカタテハ	*Cynthia cardui* (Linnaeus)	（チョウ目、タテハチョウ科）日本、全北区、東洋区。painted lady butterfly ともいう
painted lady beetle		*Mulsantina picta* (Randall)	（コウチュウ目、テントウムシ科）新北区
painted leafhopper		*Endria inimica* (Say)	（カメムシ目、オオヨコバイ科）新北区
painted lichen moth		*Hypoprepia fucosa* (Hübner)	（チョウ目、ヒトリガ科）新北区
painted maple aphid		*Drepanaphis acerifoliae* Thomas	（カメムシ目、アブラムシ科）新北区
painted meal moth	ネグロシマメイガ	*Pyralis pictalis* (Curtis)	（チョウ目、メイガ科）日本、旧北区、東洋区
painted neb		*Eulamprotes wilkella* (Linnaeus)	（チョウ目、キバガ科）旧北区
painted pine moth		*Orgyia australis* Walker	（チョウ目、ドクガ科）豪州区
painted ringlet		*Aphysoneura pigmentaria* Karsch	（チョウ目、タテハチョウ科）エチオピア区
painted ringlets		*Aphysoneura*	（チョウ目、タテハチョウ科）の昆虫の総称
painted sawtooth		*Prioneris sita* (C. et R. Felder)	（チョウ目、シロチョウ科）東洋区
painted skipper		*Hesperilla picta* Leach	（チョウ目、セセリチョウ科）豪州区
painted tiger moth		*Arachnis picta* (Packard)	（チョウ目、ヒトリガ科）新北区
painted vine moth			Joseph's coat moth を見よ
painted waxtail		*Ceriagrion cerinorubellum* (Brauer)	（トンボ目、イトトンボ科）東洋区
Paiute dancer		*Argia alberta* Kennedy	（トンボ目、イトトンボ科）新北区。米国原住民族に由来
Pakistani skipper		*Eogenes lesliei* (Evans)	（チョウ目、セセリチョウ科）旧北区
Palaearctic migratory locust			Asiatic migratory locust を見よ
Palaearctic sweetpotato horn worm			sweetpotato hawk-moth を見よ
Palaeno sulphur		*Colias palaeno chippewa* Edwards	（チョウ目、シロチョウ科）新北区。Moorland clouded yellow を参照
Palamedes swallowtail	パラメデスアゲハ	*Papilio palamedes* (Drury)	（チョウ目、アゲハチョウ科）新北区。Palamedes ともいう
Palatka skipper			saw-grass skipper を見よ
pale Actinote		*Actinote lapitha lapitha* (Staudinger)	（チョウ目、タテハチョウ科）新熱帯区。*A. l. calderoni* Schaus も同英名
pale alder moth		*Tacparia detersata* (Guenée)	（チョウ目、シャクガ科）新北区
pale ant-aphid			galling aphid を見よ
pale apple leafroller		*Pseudexentera mali* Freeman	（チョウ目、ハマキガ科）新北区。pale apple leafroller moth ともいう
pale arctic			pale arctic clouded yellow を見よ
pale arctic clouded yellow	アムールモンキチョウ	*Colias tyche* (de Böber)	（チョウ目、シロチョウ科）旧北区

英名	和名	学名	所属、分布、ほか
pale arctic clouded yellow			Moorland clouded yellow を見よ
pale arctic clouded yellow			Labrador sulphur を見よ
pale Babul blue		*Azanus mirza* (Plötz)	（チョウ目、シジミチョウ科）エチオピア区
pale Baileya moth		*Baileya levitans* (Smith)	（チョウ目、ヤガ科）新北区
pale-banded crescent		*Anthanassa tulcis* (Bates)	（チョウ目、タテハチョウ科）新北区、新熱帯区
pale-banded dart moth		*Agnorisma badinodis* Grote	（チョウ目、ヤガ科）新北区。pale-banded dart ともいう
pale-banded drepanid			scarce hook-tip を見よ
pale beauty moth		*Campaea perlata* Guenée	（チョウ目、シャクガ科）新北区。pale beauty ともいう
pale black-waved brown moth		*Selidosema rudiata* Walker	（チョウ目、シャクガ科）豪州区
pale blue		*Euphilotes pallescens* (Tilden et Downey)	（チョウ目、シジミチョウ科）新北区
pale-blue eyed-metalmark		*Mesosemia coelestis* Godman et Salvin	（チョウ目、シジミタテハ科）新熱帯区
pale blue hairstreak	オナガミツオシジミ	*Hypolycaena liara plana* Talbot	（チョウ目、シジミチョウ科）エチオピア区
pale bluet		*Enallagma pallidum* Root	（トンボ目、イトトンボ科）新北区
pale-brand bushbrown			intermediate bushbrown を見よ
pale brindled beauty		*Apocheima pilosaria* (Denis et Schiffermüller)	（チョウ目、シャクガ科）旧北区
pale brindled beauty moth			pale brindled beauty を見よ
pale-brown long-horn		*Nematopogon pilella* (Denis et Schiffermüller)	（チョウ目、マガリガ科）旧北区
palebrown sawfly		*Pseudoperga lewisii* (Westwood)	（ハチ目、Pergidae）豪州区
pale brown waved moth		*Asthena subpurpureata* Walker	（チョウ目、シャクガ科）豪州区
pale burrower mayflies	オオシロカゲロウ科	Polymitarcidae	（カゲロウ目）の昆虫の総称
pale bushblue		*Arhopala aberrans* (de Nicéville)	（チョウ目、シジミチョウ科）東洋区
pale Cerulean		*Jamides cytus* (Boisduval)	（チョウ目、シジミチョウ科）豪州区
pale-chequered smoke		*Taleporia tubulosa* (Retzius)	（チョウ目、ミノガ科）旧北区
pale chrysanthemum aphid	キクメダカアブラムシ	*Coloradoa rufomaculata* (Wilson)	（カメムシ目、アブラムシ科）日本、汎世界。キク害虫
pale ciliate blue		*Anthene lycaenoides* (Felder)	（チョウ目、シジミチョウ科）豪州区
pale clouded yellow	モトモンキチョウ	*Colias hyale* (Linnaeus)	（チョウ目、シロチョウ科）旧北区、東洋区。pale clouded yellow butterfly ともいう
pale clouded yellow			Oriental clouded yellow を見よ

英名	和名	学名	所属、分布、ほか
pale-clubbed hairstreak			Hemon blue hairstreak を見よ
pale coast conch		*Cochylis pallidana* Zeller	(チョウ目、ホソハマキガ科) 旧北区
pale conch		*Phtheochroa inopiana* (Haworth)	(チョウ目、ハマキガ科) 旧北区
pale corn clothes		*Nemapogon personellus* (Pierce et Metcalf)	(チョウ目、ヒロズコガ科) 旧北区
pale cotton stainer			cotton stainer (1) を見よ
pale cracker		*Hamadryas amphichloe* (Boisduval)	(チョウ目、タテハチョウ科) 新北区、新熱帯区
pale crescent			pallid crescentspot を見よ
pale daggerwing		*Marpesia harmonia* (Klug)	(チョウ目、タテハチョウ科) 新熱帯区
pale damsel bug		*Nabis capsiformis* (Germar)	(カメムシ目、マキバサシガメ科) 新北区
pale darter			pale palm dart (1) を見よ
pale Doberes		*Doberes hewitsonius* (Reakirt)	(チョウ目、セセリチョウ科) 新熱帯区
pale drab bell		*Eucosma obumbratana* (Lienig et Zeller)	(チョウ目、ハマキガ科) 旧北区
pale drosophila			pomace fly を見よ
pale-edged Selenisa moth		*Selenisa sueroides* (Guenée)	(チョウ目、ヤガ科) 新北区
pale eggar		*Trichiura crataegi* (Linnaeus)	(チョウ目、カレハガ科) 旧北区
pale Emesis		*Emesis vulpina* Godman et Salvin	(チョウ目、シジミタテハ科) 新熱帯区
pale emperor		*Asterocampa subpallida* (Barnes et McDunnough)	(チョウ目、タテハチョウ科) 新北区
pale Enargia moth		*Enargia decolor* (Walker)	(チョウ目、ヤガ科) 新北区。pale Enargia ともいう
pale Epidelta			pale Phalaenostola moth を見よ
pale evening dun		*Procloeon*	(カゲロウ目、コカゲロウ科) の昆虫の総称
pale evening mayfly		*Procloeon bifidum* (Bengtsson)	(カゲロウ目、コカゲロウ科) 旧北区
pale-faced forestskimmer		*Cratilla lineata calverti* Foerster	(トンボ目、トンボ科) 東洋区
pale flea beetle		*Systena blanda* Melsheimer	(コウチュウ目、ハムシ科) 新北区
pale forester		*Lethe latiaris* (Hewitson)	(チョウ目、タテハチョウ科) 東洋区
pale four-line blue		*Nacaduba hermus* (Felder)	(チョウ目、シジミチョウ科) 東洋区
pale-fringed moss-snipefly		*Ptiolina nigra* Zetterstedt	(ハエ目、シギアブ科) 旧北区
pale garden moth		*Selidosema fenerata* Felder	(チョウ目、シャクガ科) 豪州区
pale giant horsefly			great gad fly を見よ
pale Glyph moth		*Protodeltote albidula* (Guenée)	(チョウ目、ヤガ科) 新北区
pale grand imperial		*Neocheritra fabronia* (Hewitson)	(チョウ目、シジミチョウ科) 東洋区

英名	和名	学名	所属、分布、ほか
pale grass blue		*Zizeeria maha* (Kollar)	（チョウ目、シジミチョウ科）旧北区、東洋区。日本亜種は *Z. m. argia* (Ménétriès) ヤマトシジミ
pale grass blue			pale green blue を見よ
pale grass yellow		*Eurema hapale* (Mabille)	（チョウ目、シロチョウ科）エチオピア区
pale gray bird-dropping moth		*Antaeotricha leucillana* (Zeller)	（チョウ目、マルハキバガ科）新北区
pale green awlet		*Burana gomata* (Moore)	（チョウ目、セセリチョウ科）東洋区。*B. g. lalita* Fruhstorfer も同英名
pale green blue			pale grass blue を見よ
pale green darner		*Triacanthagyna septima* (Selys)	（トンボ目、ヤンマ科）新北区、新熱帯区
pale green notodontid			common Gluphisia moth を見よ
pale green plant bug			green grape capsid を見よ
pale-green sailer	ユウヅキミスジ	*Neptis zaida* Doubleday	（チョウ目、タテハチョウ科）東洋区
pale green triangle			pale triangle を見よ
pale green triangle			blue Jay を見よ
palegreen triangle butterfly		*Graphium eurypylus lycaon* (C. et R. Felder)	（チョウ目、アゲハチョウ科）豪州区
pale green waved moth		*Asthena pulchraria* Doubleday	（チョウ目、シャクガ科）豪州区
pale grey carpet		*Lithostege griseata* (Denis et Schiffermüller)	（チョウ目、シャクガ科）旧北区
pale hairtail		*Anthene butleri* (Oberthür)	（チョウ目、シジミチョウ科）エチオピア区。*A. b. livida* (Trimen) も同英名
pale-headed aspen leafroller moth		*Anacampsis niveopulvella* (Chambers)	（チョウ目、マルハキバガ科）新北区
pale-headed Phaneta moth		*Phaneta ochrocephala* (Walsingham)	（チョウ目、ハマキガ科）新北区
pale-headed striped borer			rice stem borer を見よ
pale hedge blue		*Lycaenopsis cardia* Moore	（チョウ目、シジミチョウ科）東洋区
pale hedge blue		*Celastrina cardia* (Felder)	（チョウ目、シジミチョウ科）東洋区
pale hedge blue (1)	タッパンルリシジミ	*Udara dilecta* (Moore)	（チョウ目、シジミチョウ科）日本、東洋区
pale Himalayan oakblue		*Arhopala dodonaea* Moore	（チョウ目、シジミチョウ科）東洋区
pale hockeystick sailer		*Neptis manasa* Moore	（チョウ目、タテハチョウ科）旧北区、東洋区
pale Homochlodes moth		*Homochlodes fritillaria* (Guenée)	（チョウ目、シャクガ科）新北区
pale hunter		*Austrogomphus amphiclitus* (Selys)	（トンボ目、サナエトンボ科）豪州区。pale hunter dragonfly ともいう
pale Jezebel		*Delias sanaca* (Moore)	（チョウ目、シロチョウ科）旧北区、東洋区
pale juniper webworm			barred gold conch を見よ
pale knapweed flat-body	シナノシロホシマルハキバガ	*Agonopterix pallorella* (Zeller)	（チョウ目、マルハキバガ科）日本、旧北区
pale leaf sitter		*Gorgyra pali* Evans	（チョウ目、セセリチョウ科）エチオピア区

英名	和名	学名	所属、分布、ほか
pale leafcutting bee		*Megachile concinna* Smith	(ハチ目、ハキリバチ科) 新北区
pale leafwing		*Zaretis callidryas* (R. Felder)	(チョウ目、タテハチョウ科) 新熱帯区
pale-legged earwig	キアシハサミムシ	*Euborellia plebeja* (Dohrn)	(ハサミムシ目、ハサミムシ科) 日本、東洋区、エチオピア区
pale legume bug		*Lygus elisus* Van Duzee	(カメムシ目、カスミカメムシ科) 新北区
pale-lemon sallow		*Xanthia ocellaris* (Borkhausen)	(チョウ目、ヤガ科) 旧北区
pale lice			smooth sucking lice を見よ
pale lichen moth		*Crambidia pallida* (Packard)	(チョウ目、ヒトリガ科) 新北区
pale-lined tiger moth			white ermine を見よ
pale-lobed hairstreak		*Thereus cithonius* (Godart)	(チョウ目、シジミチョウ科) 新熱帯区
pale long scale			Maskell scale (1) を見よ
pale-marked angle moth		*Macaria signaria* (Hübner)	(チョウ目、シャクガ科) 全北区
pale Metanema moth		*Metanema inatomaria* Guenée	(チョウ目、シャクガ科) 新北区。pale Metanema ともいう
pale Metarranthis moth		*Metarranthis indeclinata* (Walker)	(チョウ目、シャクガ科) 新北区
pale ministreak		*Ministrymon una scopas* (Godman et Salvin)	(チョウ目、シジミチョウ科) 新熱帯区
pale mottle		*Logania marmorata* Moore	(チョウ目、シジミチョウ科) 東洋区
pale mottled willow		*Caradrina clavipalpis* Scopoli	(チョウ目、ヤガ科) 旧北区
pale mottled willow moth			pale mottled willow を見よ
pale mountain satyr		*Lymanopoda eubagioides* Butler	(チョウ目、タテハチョウ科) 新熱帯区
pale Mylon			pallid Mylon を見よ
pale November moth		*Epirrita christyi* (Allen)	(チョウ目、シャクガ科) 旧北区
pale oak beauty		*Serraca punctinalis* (Scopoli)	(チョウ目、シャクガ科) 旧北区
pale oak eggar			pale eggar を見よ
pale oak egger			nettle-tree butterfly を見よ
pale oblique-barred moth		*Eucymatoge gobiata* Felder	(チョウ目、シャクガ科) 豪州区
pale ochraceous wave		*Idaea ochrata* (Scopoli)	(チョウ目、シャクガ科) 旧北区
pale orange dart		*Ocybadistes hypomeloma* Lower	(チョウ目、セセリチョウ科) 豪州区
pale-orange darter			pale palm dart (1) を見よ
pale orange-spot Scoparia			pale orange-spot Scoparia moth を見よ
pale orange-spot Scoparia moth		*Scoparia chimeria* Meyrick	(チョウ目、メイガ科) 豪州区
pale orange-spot shoot			cone pitch moth を見よ

英名	和名	学名	所属、分布、ほか
pale orange spot shoot moth			cone pitch moth を見よ
pale orange spot tortrix			cone pitch moth を見よ
pale owl			owl butterfly (1) を見よ
pale owl-butterfly		*Caligo telamonius memnon* (C. et R. Felder)	(チョウ目、タテハチョウ科) 新熱帯区。yellow-fronted owl を参照
pale palm dart (1)	ネッタイアカセセリ	*Telicota colon* (Fabricius)	(チョウ目、セセリチョウ科) 東洋区、豪州区。日本亜種は *T. c. stinga* Evans。*T. c. argeus* (Plötz) も同英名
pale palm dart (2)		*Telicota augias* (Linnaeus)	(チョウ目、セセリチョウ科) 東洋区、豪州区
pale-palped shieldback		*Steiroxys pallidipalpus* (Thomas)	(バッタ目、キリギリス科) 新北区
pale-patched hairstreak		*Mithras* sp.	(チョウ目、シジミチョウ科) 新熱帯区
pale pea-blue			silver forget-me-not (1) を見よ
pale Phalaenostola moth		*Phalaenostola metonalis* (Walker)	(チョウ目、ヤガ科) 新北区。pale Phalaenostola ともいう
pale pinion	ナカグロホソキリガ	*Lithophane socia* (Hufnagel)	(チョウ目、ヤガ科) 日本、旧北区
pale pinion (1)		*Lithophane hepatica* (Clerck)	(チョウ目、ヤガ科) 旧北区
pale platon			pale pinion (1) を見よ
pale plume		*Platyptilia pallidactyla* Haworth	(チョウ目、トリバガ科) 旧北区
pale prominent			sallow kitten (2) を見よ
pale ranger		*Kedestes brunneostriga* (Plötz)	(チョウ目、セセリチョウ科) エチオピア区
pale ranger		*Kedestes callicles* Hewitson	(チョウ目、セセリチョウ科) エチオピア区
pale-rayed skipper		*Vidius perigenes* (Godman)	(チョウ目、セセリチョウ科) 新北区
pale-reddish small noctuid	ウスベニコヤガ	*Perynea subrosea* (Butler)	(チョウ目、ヤガ科) 日本、旧北区
pale rice-plant weevil	アカイネゾウモドキ	*Dorytomus roelofsi* Faust	(コウチュウ目、ゾウムシ科) 日本、旧北区
pale ringlets	イワヒバシャノメ属	*Acrophtalmia*	(チョウ目、タテハチョウ科) の昆虫の総称
pale rush case		*Coleophora glaucicolella* Wood	(チョウ目、ツツミノガ科) 旧北区
pale scaly cricket		*Cycloptilum exsanguis* Love et Walker	(バッタ目、コオロギ科) 新北区
pale shining arches			pale shining brown を見よ
pale shining brown		*Polia bombycina* (Hufnagel)	(チョウ目、ヤガ科) 旧北区。日本亜種は *P. b. grisea* (Butler) オオチャイロヨトウ
pale-shining clay case		*Coleophora lutipennella* (Zeller)	(チョウ目、ツツミノガ科) 旧北区
pale shoulder		*Acontia lucida* (Hufnagel)	(チョウ目、ヤガ科) 旧北区
pale shouldered brocade		*Lacanobia thalassina* (Hufnagel)	(チョウ目、ヤガ科) 旧北区。pale-shoulder's brocade ともいう。日本亜種は *L. t. contrastata* (Bryk) ミヤマヨトウ で rubus caterpillar

英名	和名	学名	所属、分布、ほか
pale-shouldered cloud		*Cloantha hyperici* (Denis et Schifermüller)	（チョウ目、ヤガ科）旧北区
pale-shouldered knot-horn		*Sciota hostilis* (Stephens)	（チョウ目、メイガ科）旧北区
pale sicklewing			pallid batwing を見よ
pale-sided cutworm			pale-sided cutworm moth を見よ
pale-sided cutworm moth		*Agrotis malefida* Guenée	（チョウ目、ヤガ科）新北区
pale small-branced swift			banana skipper (1) を見よ
pale snaketail		*Ophiogomphus severus* Hagen	（トンボ目、サナエトンボ科）新北区
pale snout butterfly			southern snout を見よ
pale southern broken-dash		*Wallengrenia otho clavus* (Erichson)	（チョウ目、セセリチョウ科）新熱帯区。broken dash を参照
pale spark		*Sinthusa virgo* (Elwes)	（チョウ目、シジミチョウ科）東洋区
pale-spotted emperor		*Anax guttatus* (Burmeister)	（トンボ目、ヤンマ科）日本、旧北区、東洋区、豪州区、エチオピア区
pale spotted gooseberry sawfly		*Nematus leucotrochus* Hartig	（ハチ目、ハバチ科）旧北区
pale-spotted rice leafhopper	シラホシスカシヨコバイ	*Scaphoideus festivus* Matsumura	（カメムシ目、オオヨコバイ科）日本、旧北区
pale stem grasshopper		*Adreppus* sp.	（バッタ目、バッタ科）豪州区
pale stigma		*Mesogona acetosellae* (Denis et Schiffermüller)	（チョウ目、ヤガ科）旧北区
pale straw pearl		*Pyrausta lutealis* Hübner	（チョウ目、メイガ科）旧北区
pale-streaked grass veneer		*Crambus selasellus* Hübner	（チョウ目、メイガ科）旧北区
pale striped dawnfly		*Capila zennara* (Moore)	（チョウ目、セセリチョウ科）東洋区
pale-striped flea beetle			pale flea beetle を見よ
pale sugarcane planthopper	ウスイロノウンカ	*Numata muiri* (Kirkaldy)	（カメムシ目、ウンカ科）日本、東洋区
pale sulphur	ウスキオオキチョウ（ナンベイウスキシロチョウ）	*Aphrissa statira* Cramer	（チョウ目、シロチョウ科）新北区、新熱帯区。Statira ともいう
pale swallowtail		*Papilio (Pterourus) eurymedon* Lucas	（チョウ目、アゲハチョウ科）新北区
pale tiger swallowtail			pale swallowtail を見よ
pale-tipped black moth	ウスヅマクチバ	*Dinumma deponens* Walker	（チョウ目、ヤガ科）日本、旧北区、東洋区
pale-tipped cycadian		*Theorema eumenia* Hewitson	（チョウ目、シジミチョウ科）新熱帯区
pale tortoise beetle		*Cassida flaveola* Thunberg	（コウチュウ目、ハムシ科）旧北区
pale triangle	ミナミミカドアゲハ	*Graphium eurypylus* (Linnaeus)	（チョウ目、アゲハチョウ科）豪州区

英名	和名	学名	所属、分布、ほか
pale tussock		*Calliteara pudibunda* (Linnaeus)	（チョウ目、ドクガ科）旧北区
pale tussock moth			pale tussock を見よ
pale tussock moth			checkered tussock moth を見よ
pale underwing purple		*Eriocrania subpurpurella* (Haworth)	（チョウ目、スイコバネガ科）旧北区
pale-veined Isturgia moth		*Isturgia dislocaria* (Packard)	（チョウ目、シャクガ科）新北区
pale wanderer		*Pareronia avatar* (Moore)	（チョウ目、シロチョウ科）東洋区
pale watery		*Baetis fuscatus* (Linnaeus)	（カゲロウ目、コカゲロウ科）旧北区
pale watery dun		*Centroptilum*	（カゲロウ目、コカゲロウ科）の昆虫の総称
pale western cutworm			pale western cutworm moth を見よ
pale western cutworm moth		*Agrotis orthogonia* Morrison	（チョウ目、ヤガ科）新北区。トウモロコシ害虫
pale wineberry moth		*Izatha attactella* Walker	（チョウ目、マルハキバガ科）豪州区
pale-winged Crocidiphora moth		*Crocidiphora tuberculatis* Lederer	（チョウ目、メイガ科）新北区
pale-winged gray moth		*Iridopsis ephyraria* (Walker)	（チョウ目、シャクガ科）新北区。pale-winged gray ともいう
pale wormwood shark			wormwood を見よ
pale Y moth		*Plusia oxygramma* Hübner	（チョウ目、ヤガ科）豪州区
pale yellow Acraea		*Hyalites obeira* (Hewitson)	（チョウ目、タテハチョウ科）エチオピア区
pale-yellow Acraea		*Hyalites burni* Butler	（チョウ目、タテハチョウ科）エチオピア区
pale-yellow leafroller			apple leaf-curling moth を見よ
pale yellowish geometrid	ウスアオエダシャク	*Parabapta clarissa* (Butler)	（チョウ目、シャクガ科）日本、旧北区
Palegon hairstreak		*Thereus palegon* Cramer	（チョウ目、シジミチョウ科）新熱帯区
paler commodore		*Precis cuama* (Hewitson)	（チョウ目、タテハチョウ科）エチオピア区
paler Diacme moth		*Diacme elealis* (Walker)	（チョウ目、メイガ科）新北区
paler Dolichomia moth			spruce needleworm moth を見よ
paler froghopper	ウスイロアワフキ	*Aphrophora major* Uhler	（カメムシ目、アワフキムシ科）日本、旧北区
paler greenish delicate geometrid			little emerald を見よ
paler Moodna moth		*Moodna pallidostrinella* Neunzig	（チョウ目、メイガ科）新北区
paler oystershell scale			Maskell scale (1) を見よ
paler prominent	ウスキシャチホコ	*Mimopydna pallida* (Butler)	（チョウ目、シャチホコガ科）日本、旧北区
paler tiger longicorn	ウスイロトラカミキリ	*Xylotrechus cuneipennis* (Kraatz)	（コウチュウ目、カミキリムシ科）日本、旧北区
palin scarlet-eye		*Dyscophellus ramusis ramon* Evans	（チョウ目、セセリチョウ科）新熱帯区
palingeniid mayflies			spinyheaded burrower mayflies を見よ

英名	和名	学名	所属、分布、ほか
Palla butterfly	ヒメシロオビオナガタテハ	*Palla ussheri* (Butler)	(チョウ目、タテハチョウ科) エチオピア区
Pallas duke		*Siseme pallas* (Latreille)	(チョウ目、シジミタテハ科) 新熱帯区
Pallas' fritillary	ウラギンスジヒョウモン	*Argyronome laodice* (Pallas)	(チョウ目、タテハチョウ科) 旧北区。日本のウラギンスジヒョウモン *A. l. japonica* (Ménétriès) も同英名
Pallas' sailer		*Neptis sappho* Pallas	(チョウ目、タテハチョウ科) 旧北区
pallid Argus		*Callerebia scanda* (Kollar)	(チョウ目、タテハチョウ科) 東洋区
pallid batwing		*Achlyodes pallida* Felder	(チョウ目、セセリチョウ科) 新熱帯区
pallid blue			pale blue を見よ
pallid crescentspot		*Phyciodes pallida* (Edwards)	(チョウ目、タテハチョウ科) 新北区
pallid dotted blue			pale blue を見よ
pallid faun		*Melanocyma faunula faunula* (Westwood)	(チョウ目、タテハチョウ科) 東洋区
pallid five-ring		*Ypthima savara* (Grose-Smith)	(チョウ目、タテハチョウ科) 東洋区
pallid forester		*Lethe satyavati* de Nicéville	(チョウ目、タテハチョウ科) 東洋区
pallid gemmed-satyr		*Cyllopsis pallens* Miller	(チョウ目、タテハチョウ科) 新熱帯区
pallid hairstreak		*Arawacus hypocrita* (Schaus)	(チョウ目、シジミチョウ科) 新熱帯区
pallid monkey moth		*Tagora pallida* (Walker)	(チョウ目、オビガ科) 東洋区
pallid Mylon		*Mylon pelopidas* (Fabricius)	(チョウ目、セセリチョウ科) 新熱帯区
pallid nawab	ミズイロオビモンフタオチョウ	*Polyura arja* (C. et R. Felder)	(チョウ目、タテハチョウ科) 東洋区
pallid oakblue		*Amblypodia alesia* Felder	(チョウ目、シジミチョウ科) 東洋区。
pallid ringlet		*Forsterinaria pallida* Pena et Lamas	(チョウ目、タテハチョウ科) 新熱帯区
Pallid royal		*Tajuria albiplaga* de Nicéville	(チョウ目、シジミチョウ科) 東洋区
pallid scarlet-eye		*Bungalotis quadratum quadratum* Sepp	(チョウ目、セセリチョウ科) 新熱帯区
pallid shieldback		*Inyodectes pallidus* Rentz et Birchim	(バッタ目、キリギリス科) 新北区
pallid spreadwing		*Lestes pallidus* Rambur	(トンボ目、アオイトトンボ科) エチオピア区
pallid tiger swallowtail			pale swallowtail を見よ
pallid-winged grasshopper			banded wing grasshopper を見よ
pallied dart		*Potanthus pallidus* (Evans)	(チョウ目、セセリチョウ科) 東洋区
pallisade sawfly			poplar sawfly (1) を見よ
pallopterid flies		Pallopteridae	(ハエ目) の昆虫の総称
pallopterids			pallopterid flies を見よ
palm aphid		*Cerataphis variabilis* Hille Ris Lambers	(カメムシ目、アブラムシ科) 豪州区

英名	和名	学名	所属、分布、ほか
palm aphid			orchid aphid (1) を見よ
palm bob			Indian palm bob を見よ
palm borer			California palm borer を見よ
palm bud moth			coconut moth (1) を見よ
palm dart			pale palm dart (2) を見よ
palm darts		*Telicota*	(チョウ目、セセリチョウ科) の昆虫の総称
palm date scale			date palm scale を見よ
palm flower moth		*Litoprosopus coachella* Hill	(チョウ目、ヤガ科) 新北区
palm king		*Amathusia phidippus phidippus* (Linnaeus)	(チョウ目、タテハチョウ科) 東洋区
palm leaf beetle		*Brontispa longissima* (Gestro)	(コウチュウ目、ハムシ科) 東洋区、大洋区、豪州区。ヤシ害虫
palm leaf housemaker			exclamation moth を見よ
palm leaf skeletonizer		*Homaderdra subalella* (Chambers)	(チョウ目、ツツミノガ科) 新北区。palm leaf skeletonizer moth ともいう
palm mealybug	ヤシコナカイガラムシ	*Palmicultor palmarum* (Ehrhorn)	(カメムシ目、コナカイガラムシ科) 日本、東洋区、大洋区、新北区
palm mealybug			coconut mealybug を見よ　南アフリカでの英名
palm moth			palm flower moth を見よ
palm moths		Agonoxenidae	(チョウ目) の昆虫の総称
palm redeye		*Erionota hiraca* (Moore)	(チョウ目、セセリチョウ科) 東洋区
palm redeye (1)	オオマエキセセリ	*Erionota thrax* (Linnaeus)	(チョウ目、セセリチョウ科) 旧北区、東洋区、大洋区
palm rhinoceros beetle			rhinoceros beetle (2) を見よ
palm scale	ジャワマルカイガラムシ	*Abgrallaspis palmae* (Cockerell)	(カメムシ目、マルカイガラムシ科) 日本、汎熱帯。バナナ、パイナップル害虫
palm scale			dictyospermum scale を見よ
palm scale			Latania scale を見よ　南アフリカでの英名
palm scale			red date scale を見よ
palm scale insects			palm scales を見よ
palm scales		Phoenicococcidae	(カメムシ目) の昆虫の総称
palm scales			lac insects を見よ
palm seed scolytid		*Coccotrypes carpophagus* Hornung	(コウチュウ目、キクイムシ科) 新熱帯区
palm seed weevil		*Caryobruchus gleditsiae* (Linnaeus)	(コウチュウ目、ハムシ科) 新北区
palm seedborer			button weevil を見よ
palm skipper		*Zophopetes dysmepila* (Trimen)	(チョウ目、セセリチョウ科) エチオピア区
palm thrips			banded-winged palm thrips を見よ
palm tree nightfighter			palm skipper を見よ

英名	和名	学名	所属、分布、ほか
palm weevil (1)	カバイクビチョッキリ	*Deporaus betulae* (Linnaeus)	(コウチュウ目、オトシブミ科) 日本、旧北区
palm weevil		*Paramasius distortus* (Gemminger et Harold)	(コウチュウ目、ゾウムシ科) 新熱帯区
palm weevil			Asiatic palm weevil を見よ
palm weevil			South American palm weevil を見よ
palm weevil borer	ヨツボシヤシコクゾウムシ	*Diocalandra frumenti* (Fabricius)	(コウチュウ目、オサゾウムシ科) 日本、東洋区、豪州区、エチオピア区
palm whitefly		*Aleurotrachelus atratus* Hempel	(カメムシ目、コナジラミ科) 全北区、新熱帯区
palmer worm		*Dichomeris ligulella* Hübner	(チョウ目、キバガ科) 新北区。palmerworm moth ともいう。時に大発生して幼虫が植物の葉を食いつくす
Palmer's metalmark		*Apodemia palmerii arizona* Austin	(チョウ目、シジミタテハ科) 新熱帯区
Palmer's metalmark			gray metalmark を見よ
palmetto borer moth		*Litoprosopus futilis* (Grote et Robinson)	(チョウ目、ヤガ科) 新北区
palmetto bug			Florida woods cockroach を見よ
palmetto conehead		*Belocephalus sabalis* Davis	(バッタ目、キリギリス科) 新北区
palmetto leaf miner			palm leaf skeletonizer を見よ
palmetto scale		*Comstockiella sabalis* (Comstock)	(カメムシ目、マルカイガラムシ科) 新北区
palmetto skipper		*Euphyes arpa* (Boisduval et LeConte)	(チョウ目、セセリチョウ科) 新北区
palmflies	ルリモンジャノメ属	*Elymnias*	(チョウ目、タテハチョウ科) の昆虫の総称
palmfly	メスジロルリモンジャノメ	*Elymnias agondas* (Boisduval)	(チョウ目、タテハチョウ科) 豪州区。palm fly とも記す
Palmiet sprite		*Pseudagrion furcigerum* (Rambur)	(トンボ目、イトトンボ科) エチオピア区
palmking butterflies			brush-footed butterflies を見よ
palni bushbrown		*Mycalesis mamerta davisoni* Moore	(チョウ目、タテハチョウ科) 東洋区。blind-eye bushbrown を参照
palni dart		*Potanthus palnia* (Evans)	(チョウ目、セセリチョウ科) 東洋区
palni fourring		*Ypthima ypthimoides* (Moore)	(チョウ目、タテハチョウ科) 東洋区
palo verde beetle		*Derobrachus geminatus* LeConte	(コウチュウ目、カミキリムシ科) 新北区。palo verde borer beetle, palo verde root borer ともいう
Palos Verdes blue		*Glaucopsyche lygdamus palosverdesensis* Perkins et Emmel	(チョウ目、シジミチョウ科) 新北区
Palu swallowtail			Sulawesi clubtail を見よ
Paludis brown		*Pseudonympha paludis* Riley	(チョウ目、タテハチョウ科) エチオピア区
Pama skipper		*Zonia zonia panamensis* Nicolay	(チョウ目、セセリチョウ科) 新熱帯区

英名	和名	学名	所属、分布、ほか
pamakani gall fly		*Procecidochares utilis* (Stone)	(ハエ目、ミバエ科) 大洋区、新熱帯区。雑草の *Eupatorium adenophorum* Sprengel (ハワイ名 Hamakua pamakani) に虫えいをつくる
Pamela	メスアカシロチョウ	*Perrhybris pamela* (Stoll)	(チョウ目、シロチョウ科) 新熱帯区
pamphiliids			web-spinning sawflies を見よ
pan opal		*Chrysoritis pan* (Pennington)	(チョウ目、シジミチョウ科) エチオピア区
Panama grass tubeworm moth		*Acrolophus panamae* Busck	(チョウ目、Acrolophidae) 新北区
Panama sharpshooter		*Paraulacizes panamensis* (Fowler)	(カメムシ目、オオヨコバイ科) 新熱帯区
Panama skipperling		*Dalla eryonas* (Hewitson)	(チョウ目、セセリチョウ科) 新熱帯区
Panamanian Sarota		*Sarota gamelia gamelia* Godman et Salvin	(チョウ目、シジミタテハ科) 新熱帯区
Panamanian Theope		*Theope barea* Godman et Salvin	(チョウ目、シジミタテハ科) 新熱帯区
pandanus mealybug		*Laminicoccus pandani* (Cockerell)	(カメムシ目、コナカイガラムシ科) 新北区
Pandemis leafroller		*Pandemis pyrusana* Kearfott	(チョウ目、ハマキガ科) 新北区。成虫は Pandemis leafroller moth
Pandora moth		*Coloradia pandora* Blake	(チョウ目、ヤママユガ科) 新北区
Pandora pinemoth			Pandora moth を見よ
Pandora sphinx		*Eumorpha pandorus* (Hübner)	(チョウ目、スズメガ科) 新北区。pandorous sphinx moth ともいう
pandorous sphinx			satellite sphinx moth を見よ
pangola grass aphid		*Schizaphis hypersiphonata* Basu	(カメムシ目、アブラムシ科) 豪州区
panicle thrips			cereal thrips を見よ
panicum planthopper	ヒエウンカ	*Sogatella vibix* (Haupt)	(カメムシ目、ウンカ科) 日本、旧北区。ヒエ害虫
pannaria wave moth		*Leptostales pannaria* (Guenée)	(チョウ目、シャクガ科) 新北区
Panoptes blue		*Pseudophilotes panoptes* (Hübner)	(チョウ目、シジミチョウ科) 旧北区
panorpians			scorpion flies (1) を見よ
pansies	アフリカタテハモドキ属	*Precis*	(チョウ目、タテハチョウ科) の昆虫の総称
pansies			buckeyes を見よ
pansy daggerwing		*Marpesia marcella* (Felder)	(チョウ目、タテハチョウ科) 新熱帯区
Pantala			globe trotter を見よ
panther			leopard を見よ
panther-spotted grasshopper		*Poecilotettix pantherina* (Walker)	(バッタ目、バッタ科) 新北区。乾燥地のキク科植物に多い
Pantiacolla blue ringlet		*Cepheuptychia* sp.	(チョウ目、タテハチョウ科) 新熱帯区
pantry moths		*Anagasta*	(チョウ目、メイガ科) の昆虫の総称
pantry moths		*Ephestia*	(チョウ目、メイガ科) の昆虫の総称

英名	和名	学名	所属、分布、ほか
Paona hairstreak	パオナメスルリシジミ	*Chrysozephyrus paona* (Tytler)	(チョウ目、シジミチョウ科) 東洋区
papago thread-legged katydid			thread-legged katydid を見よ
papataci sand fly		*Phlebotomus papatasi* Scopoli	(ハエ目、チョウバエ科) 旧北区
papaya fly			papaya fruit fly を見よ
papaya fruit fly		*Toxotrypana curvicauda* (Gerstaecker)	(ハエ目、ミバエ科) 新北区、新熱帯区
papaya hornworm			Alope sphinx moth を見よ
papaya mealybug		*Paracoccus marginatus* Williams et Granara de Willink	(カメムシ目、コナカイガラムシ科) 新北区、新熱帯区
papaya scale			white peach scale を見よ　米国での英名
papaya webworm		*Homolapalpia dalera* Dyar	(チョウ目、メイガ科) 新北区。パパイア害虫
papaya webworm moth		*Davara caricae* (Dyar)	(チョウ目、メイガ科) 新北区
paper birch leaftier			striped birch pyralid moth を見よ
paper-butterflies			tree nymphs (1) を見よ
paper wasp (1)		*Ropalidia revolutionalis* (Saussure)	(ハチ目、スズメバチ科) 豪州区
paper wasp (2)		*Polistes exclamens* Viereck	(ハチ目、スズメバチ科) 新北区、大洋区
paper wasp			French wasp を見よ
paper wasps			papernest wasps (1)(2)(3) を見よ
paperback cicada		*Cicadetta hackeri* (Distant)	(カメムシ目、セミ科) 豪州区
paperback sawfly			melaleuca sawfly を見よ
papernest wasp			paper wasp (1) を見よ
papernest wasps (1)		*Polistes*	(ハチ目、スズメバチ科) の昆虫の総称
papernest wasps (2)		*Ropalidia*	(ハチ目、スズメバチ科) の昆虫の総称
papernest wasps (3)		Polistinae	(ハチ目、スズメバチ科) の昆虫の総称
papernest wasps			hornets (1)(2) を見よ
Paphos blue		*Galucopsyche paphos* Chapman	(チョウ目、シジミチョウ科) 旧北区
Papilionider			swallowtails を見よ
papilios			swallowtails を見よ
Papuan army worm		*Tirocola plagiata* Walker	(チョウ目、ヤガ科) 日本
Papuan Cerulean		*Jamides nemophila* (Butler)	(チョウ目、シジミチョウ科) 豪州区
Papuan evening brown			Constantia evening brown を見よ
Papuan line-blue			Felder's lineblue (1) を見よ
Papuan snow flat		*Tagiades nestus* (C. Felder)	(チョウ目、セセリチョウ科) 豪州区
paradise birdwing	ゴクラクトリバネアゲハ	*Ornithoptera paradisea* Staudinger	(チョウ目、アゲハチョウ科) 東洋区
paradise jewel		*Hypochrysops hippuris* (Hewitson)	(チョウ目、シジミチョウ科) 豪州区

英名	和名	学名	所属、分布、ほか
paradise phantom		*Cithaerias phantoma* Fassl	（チョウ目、タテハチョウ科）新熱帯区
paradise skipper		*Abantis paradisea* (Butler)	（チョウ目、セセリチョウ科）エチオピア区
parallel-banded Euselasia		*Euselasia orfita* (Cramer)	（チョウ目、シジミタテハ科）新熱帯区
parallel-banded leafroller moth			spotted fireworm moth を見よ
parandra long-horned beetle		*Parandra puncticeps* Sharp	（コウチュウ目、カミキリムシ科）大洋区
parasite flies			tachina flies を見よ
parasite moths		Cyclotornidae	（チョウ目）の昆虫の総称
parasitic beetles			stylopids を見よ
parasitic flies			tachina flies を見よ
parasitic grain wasp		*Cephalonomia waterstoni* Gahan	（ハチ目、アリガタバチ科）新北区
parasitic leaf-cut weevil	ヤドカリチョッキリ	*Paradoporaus depressus* (Faust)	（コウチュウ目、オトシブミ科）日本、旧北区
parasitic lice			sucking lice を見よ　豪州での英名
parasitic sawflies			parasitic wood wasps を見よ
parasitic wasps	ツチバチ上科	Scolioidea	（ハチ目）の昆虫の総称
parasitic wasps			chalcid wasps (2) を見よ
parasitic wasps			sapygid wasps を見よ
parasitic wasps			common braconids を見よ
parasitic wasps			chalcid wasps (2) を見よ
parasitic wood wasps	ヤドリキバチ科	Orussidae	（ハチ目）の昆虫の総称
parasitic yellow jacket			cuckoo wasp (1) を見よ
parasol ant		*Atta fervens* Say	（ハチ目、アリ科）新北区
parasol ants			leafcutter ants を見よ　葉をくわえて歩く姿がパラソルに似ることに由来
parasol ants			agricultural ants を見よ
parasols		*Neurothemis*	（トンボ目、トンボ科）の昆虫の総称
parchment skipper		*Atrytonopsis pittacus* (Edwards)	（チョウ目、セセリチョウ科）新北区、新熱帯区
parent bug		*Elasmucha grisea* (Linnaeus)	（カメムシ目、ツノカメムシ科）旧北区
Parepa mountain satyr		*Parapedaliodes parepa* (Hewitson)	（チョウ目、タテハチョウ科）新熱帯区
Paris peacock	ルリモンアゲハ	*Achillides paris* (Linnaeus)	（チョウ目、アゲハチョウ科）旧北区、東洋区
parlatoria date scale			date palm scale を見よ　属名 *Parlatoria* に由来
parlatoria scale			cattleya scale を見よ
parlatoria scales		*Parlatoria*	（カメムシ目、マルカイガラムシ科）の昆虫の総称
parlour palm thrips			banded-winged palm thrips を見よ
Parnassian butterfly			large parnassian を見よ　米国での英名
Parnassian moths		Pterothysanidae	（チョウ目）の昆虫の総称

英名	和名	学名	所属、分布、ほか
Parnassians	ウスバアゲハ亜科	Parnassiinae	（チョウ目、アゲハチョウ科）の昆虫の総称
Parnassians			Apollos を見よ　米国での英名
Parnassius acdestis		*Parnassius acdestis* Oberthür	（チョウ目、アゲハチョウ科）東洋区
Parnassius actius	オオミヤマウスバアゲハ	*Parnassius actius* (Eversmann)	（チョウ目、アゲハチョウ科）東洋区
Paroeca sister		*Adelpha paroeca paroeca* (Bates)	（チョウ目、タテハチョウ科）新熱帯区
Paron hairstreak		*Temecla paron* (Godman et Salvin)	（チョウ目、シジミチョウ科）新熱帯区
parrot sticktight flea		*Hectopsylla psittaci* von Frauenfeld	（ノミ目、ヒトノミ科）旧北区
parsley aphid			hawthorn aphid (2) を見よ
parsley swallowtail			black swallowtail を見よ
parsley weevil			carrot weevil を見よ
parsleyworm			black swallowtail を見よ
parsnip and willow aphid			carrot aphid (1) を見よ
parsnip leaf miner		*Trypeta fratria* Loew	（ハエ目、ミバエ科）新北区
parsnip moth (1)		*Depressaria pastinacella* (Duponchel)	（チョウ目、マルハキバガ科）全北区
parsnip moth (2)		*Depressaria heracliana* (Linnaeus)	（チョウ目、マルハキバガ科）新北区。ボウフウを食べる
parsnip root aphid			hogweed aphid を見よ
parsnip seed wasp		*Systole* sp.	（ハチ目、カタビロコバチ科）豪州区
parsnip swallowtail		*Papilio ajax* Linnaeus	（チョウ目、アゲハチョウ科）新北区
parsnip webworm			parsnip moth (1) (2) を見よ　成虫は parsnip webworm moth
parsnip webworm moth			parsnip moth (1) を見よ
parsnip webworms			concealer moths を見よ　幼虫の英名
parthenice tiger moth		*Grammia parthenice* (Kirby)	（チョウ目、ヒトリガ科）新北区。parthenice moth ともいう
parthenium stemgalling moth		*Epiblema strenuana* (Walker)	（チョウ目、ハマキガ科）新北区、豪州区。カンキツ害虫
parthenocissus looper	ハガタナミシャク	*Eustroma melancholicum* (Butler)	（チョウ目、シャクガ科）日本
particoloured auger beetle		*Mesoxylion collaris* (Erichson)	（コウチュウ目、ナガシンクイムシ科）豪州区
Parva clearwing		*Pteronymia parva* (Salvin)	（チョウ目、タテハチョウ科）新熱帯区
Parvula skipper		*Toxidia parvula* (Plötz)	（チョウ目、セセリチョウ科）豪州区
Pasadena masked chafer		*Cyclocephala pasadenae* (Casey)	（コウチュウ目、コガネムシ科）新北区、大洋区
Pasania aphid	コケクダアブラムシ	*Eutrichosiphum pasaniae* (Okajima)	（カメムシ目、アブラムシ科）日本、東洋区
pasha		*Herona marathus* Doubleday	（チョウ目、タテハチョウ科）東洋区

英名	和名	学名	所属、分布、ほか
pasha			great nawab を見よ
pashford pot beetle	モモグロチビツツハムシ	*Cryptocephalus exiguus* Schneider	(コウチュウ目、ハムシ科) 日本、旧北区
paspalum whitegrub		*Lepidiota laevis* Arrow	(コウチュウ目、コガネムシ科) 豪州区
passalid beetles			peg beetles (1) を見よ
passalids			peg beetles (1) を見よ
passenger		*Dysgonia algira* (Linnaeus)	(チョウ目、ヤガ科) 旧北区
passenger moth			passenger を見よ
passion butterfly			gulf fritillary を見よ
passionvine bug			leaf-footed plant bug (1) を見よ 豪州での英名
passionvine hopper		*Scolypoda australis* (Walker)	(カメムシ目、ハゴロモ科) 豪州区
passionvine mealybug		*Planococcus minor* (Maskell)	(カメムシ目、コナカイガラムシ科) 豪州区
Passova firetip		*Passova passova* (Hewitson)	(チョウ目、セセリチョウ科) 新熱帯区
Pastazena crescent		*Tegosa pastazena* (Bates)	(チョウ目、タテハチョウ科) 新熱帯区
pastel skimmer		*Sympetrum corruptum* (Hagen)	(トンボ目、トンボ科) 新北区
pastoral ants			typical ants を見よ
pasture day moth		*Apina callisto* (Angas)	(チョウ目、ヤガ科) 豪州区
pasture Eumorpha		*Eumorpha phorbas* (Cramer)	(チョウ目、スズメガ科) 新熱帯区
pasture grass veneer moth		*Crambus saltuellus* Zeller	(チョウ目、メイガ科) 新北区
pasture grasshopper		*Melanoplus confusus* Scudder	(バッタ目、バッタ科) 新北区
pasture leaf bug	マキバカスミカメムシ	*Lygus rugulipennis* Poppius	(カメムシ目、カスミカメムシ科) 日本、旧北区
pasture pill beetle		*Amphicyrta dentipes* Erichson	(コウチュウ目、マルトゲムシ科) 新北区
pasture skipper		*Vehilius stictomenes* (Butler)	(チョウ目、セセリチョウ科) 新熱帯区
pasture tunnel moth			philobota を見よ
pasture webworm		*Hednota crypsichroa* Lower	(チョウ目、メイガ科) 豪州区
pasture webworm		*Hednota panteucha* (Meyrick)	(チョウ目、メイガ科) 豪州区
pasture webworm		*Hednota relatalis* (Walker)	(チョウ目、メイガ科) 豪州区
pasture webworm (1)		*Hednota longipalpella* (Meyrick)	(チョウ目、メイガ科) 豪州区
pasture webworm (2)		*Hednota pedionoma* (Meyrick)	(チョウ目、メイガ科) 豪州区
pasture whitegrubs		*Rhopaea*	(コウチュウ目、コガネムシ科) の昆虫の総称
pasture wireworm			common click beetle (2) を見よ

英名	和名	学名	所属、分布、ほか
pasturegrass armyworm parasite		*Apanteles militaris* (Walsh)	(ハチ目、コマユバチ科) 大洋区
Patagonian grayling		*Etcheverrius chiliensis* (Guérin-Méneville)	(チョウ目、タテハチョウ科) 新熱帯区
patched noctuid			sallow を見よ
patches			ribbed-cocoon maker moths を見よ
Patchmarhi bushbrown		*Mycalesis mercea* Evans	(チョウ目、タテハチョウ科) 東洋区
patchwork leafcutter			common leaf-cutter bee を見よ
patent-leather beetle			bess-beetle を見よ
patent-leather beetles			peg beetles (1) を見よ
Patricia blue	ランタナアフリカゴマシジミ	*Lepidochrysops patricia* (Trimen)	(チョウ目、シジミチョウ科) エチオピア区。Patrician blue ともいう
Patricia's roadside-skipper		*Amblyscirtes patriciae* (Bell)	(チョウ目、セセリチョウ科) 新熱帯区
Patrobas skipper		*Elbella patrobas mexicana* Mielke	(チョウ目、セセリチョウ科) 新熱帯区
Patton's tiger		*Hyphoraia testudinaria* (Geoffroy et Fourcroy)	(チョウ目、ヒトリガ科) 旧北区
Paua beetle		*Geotrupes spiniger* (Marsham)	(コウチュウ目、センチコガネ科) 旧北区
Paula's clearwing		*Oleria paula* (Weymer)	(チョウ目、タテハチョウ科) 新北区、新熱帯区
Paulina yellow		*Eurema paulina* (Bates)	(チョウ目、シロチョウ科) 新熱帯区
Paulinus map		*Cyrestis paulinus* C. et R. Felder	(チョウ目、タテハチョウ科) 東洋区
paulownia leaf weevil	クロタマゾウムシ	*Cionus helleri* Reitter	(コウチュウ目、ゾウムシ科) 日本。キリ害虫
Paul's buff		*Pentila pauli* Staudinger	(チョウ目、シジミチョウ科) エチオピア区
pauper pug			Fletcher's pug を見よ
pauper skipper		*Panoquina pauper pauper* (Mabille)	(チョウ目、セセリチョウ科) 新熱帯区
paussids			ant-beetles を見よ
pavement ant	トビイロシワアリ	*Tetramorium caespitum* (Linnaeus)	(ハチ目、アリ科) 日本、全北区、大洋区。米国での英名
Pavlovski's Monopis moth		*Monopis pavlovski* Zagulajev	(チョウ目、ヒロズコガ科) 日本、全北区
Pavon		*Doxocopa pavon* (Latreille)	(チョウ目、タテハチョウ科) 新北区。Pavon butterfly ともいう
Pavon emperor	ツマアカアメリカコムラサキ	*Doxocopa pavon theodora* (Lucas)	(チョウ目、タテハチョウ科) 新熱帯区。Pavon を参照
Pawnee skipper		*Hesperia pawnee* Dodge	(チョウ目、セセリチョウ科) 新北区
pawpaw sphinx			black alder sphinx を見よ　pawpaw sphinx moth ともいう
pawpaw webworm			papaya webworm を見よ
pea and bean thrips			pea thrips (1) を見よ
pea and bean weevil			pea leaf weevil を見よ
pea and bean weevils		*Sitona*	(コウチュウ目、ゾウムシ科) の昆虫の総称
pea and bean weevils			pea weevils を見よ

英名	和名	学名	所属、分布、ほか
pea aphid		*Macrosiphum pisi* Kaltenbach	(カメムシ目、アブラムシ科) 汎世界
pea aphid (1)	エンドウヒゲナガアブラムシ	*Acyrthosiphon pisum* (Harris)	(カメムシ目、アブラムシ科) 日本、汎世界。エンドウ、ダイズなどの害虫
pea aphid parasite		*Aphidius smithi* Sharma et Rao	(ハチ目、コマユバチ科) 大洋区
pea beetle		*Bruchus pisi* Linnaeus	(コウチュウ目、ハムシ科) 旧北区
pea beetle			pea weevil を見よ
pea blue	ウラナミシジミ	*Lampides boeticus* (Linnaeus)	(チョウ目、シジミチョウ科) 日本、旧北区、東洋区、豪州区、エチオピア区。pea blue butterfly ともいう
pea bug			pea weevil を見よ
pea fly		*Kleinschmidtimyia pisi* (Kleinshmidt)	(ハエ目、ハモグリバエ科) 豪州区
pea-fowl louse		*Myrsidea phaestoma* (Nitzsch)	(ハジラミ目、タンカクトリハジラミ科) 大洋区
pea gall			rose spiked pea-gall cynipid により作られた虫えい
pea green moth			green oak roller を見よ　pea-green moth とも記す
pea green oak curl			green oak roller を見よ　pea-green oak curl とも記す
pea green tortrix			green oak roller を見よ
pea leaf miner		*Liriomyza flaveola* (Fallén)	(ハエ目、ハモグリバエ科) 全北区
pea leaf miner		*Liriomyza strigata* (Meigen)	(ハエ目、ハモグリバエ科) 旧北区
pea leaf miner			South American leafminer を見よ
pea leaf weevil	アカアシチビコフキゾウムシ	*Sitona lineata* (Linnaeus)	(コウチュウ目、ゾウムシ科) 日本、全北区。ラッカセイ害虫。北米での英名
pea midge		*Contarinia pisi* (Winnertz)	(ハエ目、タマバエ科) 旧北区
pea moth	エンドウシンクイ	*Cydia nigricana* (Stephens)	(チョウ目、ハマキガ科) 日本、全北区。エンドウ害虫
pea moth			broom brocade を見よ
pea pear-shaped weevil		*Apion pisi* (Fabricius)	(コウチュウ目、ホソクチゾウムシ科) 旧北区
pea pod Argus			pea blue を見よ　pea-pod Argus とも記す
pea pod borer			lima-bean pod borer を見よ
pea seed beetle		*Bruchus pallidicornis* (Boheman)	(コウチュウ目、ハムシ科) 旧北区
pea seed beetle			pea weevil を見よ
pea stem fly			French bean fly を見よ
pea thrips		*Caliothrips indicus* (Bagnall)	(アザミウマ目、アザミウマ科) 東洋区
pea thrips (1)		*Kakothrips robustus* (Uzel)	(アザミウマ目、アザミウマ科) 全北区
pea tortricid			pea moth を見よ
pea weevil	エンドウゾウムシ	*Bruchus pisorum* Linnaeus	(コウチュウ目、ハムシ科) 日本、汎世界。エンドウ、貯穀害虫
pea weevil			broadbean weevil を見よ

英名	和名	学名	所属、分布、ほか
pea weevil			pea leaf weevil を見よ
pea weevils	マメゾウムシ亜科	Bruchinae	(コウチュウ目、ハムシ科) の昆虫の総称
peach aphid		*Brachycaudus schwartzi* (Börner)	(カメムシ目、アブラムシ科) 全北区、東洋区
peach aphid			green peach aphid を見よ
peach aphid			mealy peach aphid を見よ
peach bark beetle		*Phloeotribus liminaris* (Harris)	(コウチュウ目、キクイムシ科) 新北区
peach beauty		*Peria lamis* Cramer	(チョウ目、タテハチョウ科) 新熱帯区
peach blossom			peach blossom moth を見よ
peach blossom moth		*Thyatira batis* (Linnaeus)	(チョウ目、トガリバガ科) 旧北区、豪州区。日本亜種は *T. b. japonica* Werny モントガリバ
peach capnodis		*Capnodis tenebrionis* (Linnaeus)	(コウチュウ目、タマムシ科) 旧北区
peach curculio	モモチョッキリ	*Rhynchites heros* Roelofs	(コウチュウ目、オトシブミ科) 日本、旧北区。モモ、リンゴ、ウメなどの害虫
peach curl aphid			peach aphid を見よ
peach fall cankerworm	クロテンフユシャク	*Inurois membranaria* (Christoph)	(チョウ目、シャクガ科) 日本、旧北区
peach flower moth	ノコメトガリキリガ	*Tolerta divergens* (Butler)	(チョウ目、ヤガ科) 日本、旧北区。モモ、ナシ、ウメ害虫
peach fruit borer			peach moth を見よ
peach fruit fly		*Bactrocera zonata* (Saunders)	(ハエ目、ミバエ科) 太洋区
peach fruit moth	モモシンクイガ	*Carposina niponensis* Walsingham	(チョウ目、シンクイガ科) 日本、旧北区。モモ、ナシ、リンゴなどの害虫
peach geometrid	ヒロバツバメアオシャク	*Maxates illiturata* (Walker)	(チョウ目、シャクガ科) 日本、旧北区。モモ害虫
peach greenish geometrid			peach geometrid を見よ
peach horn worm	モモスズメ	*Marumba gaschkowitschii echephron* (Boisduval)	(チョウ目、スズメガ科) 日本、旧北区。モモ、リンゴ、サクラなどの害虫
peach leafminer	モモハモグリガ	*Lyonetia clerkella* (Linnaeus)	(チョウ目、ハモグリガ科) 日本、旧北区、東洋区、エチオピア区。モモ、リンゴ、サクラ害虫
peach leafminer moth			peach leafminer を見よ
peach moth	モモノゴマダラノメイガ	*Conogethes punctiferalis* (Guenée)	(チョウ目、メイガ科) 日本、東洋区。モモ、リンゴなど果樹やカシの害虫
peach moth			Oriental fruit moth を見よ
peach-potato aphid			green peach aphid を見よ
peach root borer			peach tree borer を見よ　北米での英名
peach scale			European fruit lecanium を見よ
peach slug	モモハバチ	*Caliroa matsumotonis* (Harukawa)	(ハチ目、ハバチ科) 日本。モモ、ナシ、ウメ害虫
peach sword stripe night moth			southern oak dagger moth を見よ
peach tip moth			Oriental fruit moth を見よ　peach tip-moth とも記す

英名	和名	学名	所属、分布、ほか
peach tree borer		*Synanthedon exitiosa* (Say)	（チョウ目、スカシバガ科）新北区。モモの害虫
peach tree borer moth			peach tree borer を見よ　北米での英名
peach trunk aphid			clouded peach stem aphid を見よ
peach twig borer	モモキバガ	*Anarsia lineatella* Zeller	（チョウ目、キバガ科）日本、全北区、大洋区。peach twig borer moth ともいう
peach twig moth			peach twig borer を見よ
peach white scale			white peach scale を見よ
peach worm			peach twig borer を見よ
peacock		*Inachis io* Linnaeus	（チョウ目、タテハチョウ科）全北区。日本亜種は *I. i. geisha* (Stichel) クジャクチョウ。斑紋がクジャクの目に似ることから。peacock butterfly ともいう
peacock			dragonflies (1) を見よ
peacock awl	トウヨウアオネセセリ	*Allora doleschallii* (Felder)	（チョウ目、セセリチョウ科）新北区、豪州区
peacock Brenthia moth		*Brenthia pavonacella* Clemens	（チョウ目、ハマキモドキガ科）新北区
peacock eye			peacock を見よ
peacock flies			fruit flies を見よ　交尾の時、雄はクジャクのように翅をすばやく動かすことに由来
peacock flies			picture-winged flies を見よ
peacock hairstreak	トモイロキリシマミドリシジミ	*Euaspa pavo* (de Nicéville)	（チョウ目、シジミチョウ科）東洋区
peacock jewel	クジャクニシキシジミ	*Hypochrysops pythias* (C. et R. Felder)	（チョウ目、シジミチョウ科）豪州区
peacock louse		*Goniodes parviceps* Piaget	（ハジラミ目、チョウカクハジラミ科）新北区
peacock louse		*Goniodes pavonis* (Linnaeus)	（ハジラミ目、チョウカクハジラミ科）全北区
peacock moth			southern old lady moth を見よ
peacock moth			giant emperor を見よ
peacock moth			birch angle moth を見よ
peacock pansy	タテハモドキ	*Junonia almana* (Linnaeus)	（チョウ目、タテハチョウ科）日本、東洋区。*J. a. javana* C. Felder も同英名
peacock royal	ヤドリギツバメ	*Tajuria cippus* Fabricius	（チョウ目、シジミチョウ科）東洋区。*T. c. maxentius* Fruhstorfer も同英名
peak white	ソーゲンシロチョウ	*Pontia callidice* (Hübner)	（チョウ目、シロチョウ科）旧北区、東洋区
Peakesia grasshopper		*Peakesia* sp.	（バッタ目、バッタ科）豪州区
Peal's palmfly		*Elymnias pealii* Wood-Mason	（チョウ目、タテハチョウ科）東洋区
peanut bug			South American lantern fly を見よ
peanut scarab		*Heteronyx piceus* Blanchard	（コウチュウ目、コガネムシ科）豪州区
peanut scarab		*Heteronyx rugosipennis* Macleay	（コウチュウ目、コガネムシ科）豪州区
peanut trash bug		*Elasmolomus sordidus* (Fabricius)	（カメムシ目、ナガカメムシ科）豪州区

英名	和名	学名	所属、分布、ほか
pear amatid	カノコガ	*Amata fortunei* (Orza)	(チョウ目、カノコガ科) 日本、旧北区
pear and cherry sawfly			pear sawfly (1) を見よ
pear and cherry slug			pear sawfly (1) を見よ
pear and cherry slugworm			pear sawfly (1) を見よ
pear Anuraphis	ナシオマルアブラムシ	*Anuraphis farfarae* (Koch)	(カメムシ目、アブラムシ科) 日本
pear aphid	ナシアブラムシ	*Schizaphis piricola* (Matsumura)	(カメムシ目、アブラムシ科) 日本、東洋区
pear aphid			pear Anuraphis を見よ
pear aphid			woolly pear aphid を見よ
pear aphid			bedstraw aphid を見よ
pear barkminer	ナシホソガ	*Spulerina astaurota* (Meyrick)	(チョウ目、ホソガ科) 日本、東洋区。ナシ、リンゴ、ボケ害虫
pear-bedstraw aphid			bedstraw aphid を見よ
pear black bark beetle	ナナカマドノキクイムシ	*Polygraphus nigriclytris* Niijima	(コウチュウ目、キクイムシ科) 日本、旧北区
pear-blight beetle			shot-hole borer (3) を見よ
pear blossom midge		*Contarinia pyri* Bouché	(ハエ目、タマバエ科) 旧北区
pear blossom weevil			apple bud weevil を見よ
pear borer (1)	ルリカミキリ	*Bacchisa fortunei japonica* (Gahan)	(コウチュウ目、カミキリムシ科) 日本。ナシ、リンゴ、サクラなどの害虫
pear borer (2)	ギンマダラメイガ	*Numonia heringii* (Ragonot)	(チョウ目、メイガ科) 日本、旧北区、東洋区
pear borer		*Aegeria pyri* Harris	(チョウ目、スカシバガ科) 新北区
pear borer			sinuate pear tree borer を見よ
pear bryobia			grass-pear bryobia を見よ
pear cimbex			pear cimbicid sawfly を見よ
pear cimbicid sawfly	ナシアシブトハバチ	*Palaeocimbex carinulatus* (Konow)	(ハチ目、コンボウハバチ科) 日本、旧北区。ナシ、サクラ害虫
pear codling moth		*Laspeyresia pyrivora* Danilevsky	(チョウ目、ハマキガ科) 旧北区
pear-coltsfoot aphid			pear Anuraphis を見よ
pear cutworm			lunar spotted pinion を見よ
pear diplosis			pear midge を見よ
pear driller			pear fruit moth を見よ
pear fall cankerworm	シロオビフユシャク	*Alsophila japonensis* (Warren)	(チョウ目、シャクガ科) 日本。リンゴ、ウメ、サクラなどの害虫
pear flat-headed borer			beech borer を見よ
pear flower bud weevil			apple blossom weevil を見よ
pear fruit borer			pear borer (2) を見よ
pear fruit chafer			bumble flower beetle を見よ

英名	和名	学名	所属、分布、ほか
pear fruit moth	ナシマダラメイガ	*Acrobasis pyrivorella* (Matsumura)	（チョウ目、メイガ科）日本、旧北区。ナシ、リンゴ害虫
pear fruit sawfly	ナシミハバチ	*Hoplocampa pyricola* Rohwer	（ハチ目、ハバチ科）日本。ナシ害虫
pear gnat			pear midge を見よ
pear-grass aphid		*Melanaphis pyraria* (Passerini)	（カメムシ目、アブラムシ科）旧北区
pear green aphid	ナシミドリオオアブラムシ	*Nippolachnus piri* Matsumura	（カメムシ目、アブラムシ科）日本、旧北区
pear-hogweed aphid			hogweed aphid を見よ
pear lace bug	ナシグンバイ	*Stephanitis nashi* Esaki et Takeya	（カメムシ目、グンバイムシ科）日本、旧北区。ナシ、リンゴ、モモ、サクラなどの害虫
pear lace bug (1)		*Stephanitis pyri* Fabricius	（カメムシ目、グンバイムシ科）旧北区
pear leaf blister moth		*Leucoptera malifoliella* Costa	（チョウ目、ハモグリガ科）旧北区
pear leaf blister moth			mountain-ash bentwing を見よ
pear leaf curling midge			pear leaf-roller (2) を見よ　pear leaf-curling midge とも記す
pear leaf eating weevil			common leaf weevil を見よ
pear leaf midge			pear leaf-roller (2) を見よ
pear leaf roller	セモンカギバヒメハマキ	*Ancylis mandarinana* Walsingham	（チョウ目、ハマキガ科）日本、旧北区
pear leaf roller (1)		*Byctiscus betulae* (Linnaeus)	（コウチュウ目、オトシブミ科）旧北区
pear leaf-roller (2)		*Dasineura pyri* (Bouché)	（ハエ目、タマバエ科）全北区、豪州区
pear leaf roller (3)		*Croesia holmiana* (Linnaeus)	（チョウ目、ハマキガ科）全北区
pear leaf roller			apple leafroller (1) を見よ
pear leaf roller pyralid	アカオビマダラメイガ	*Conobathra bifidella* (Leech)	（チョウ目、メイガ科）日本
pear leafminer	ナシチビガ	*Bucculatrix pyrivorella* Kuroko	（チョウ目、チビガ科）日本、旧北区。ナシ害虫
pearleaf worm	リンゴスカシクロバ	*Illiberis pruni* Dyar	（チョウ目、マダラガ科）日本、旧北区。リンゴ、ナシ、ウメ、サクラ害虫
pear mealybug	ナシコナカイガラムシ	*Dysmicoccus wistariae* (Green)	（カメムシ目、コナカイガラムシ科）日本、全北区。ナシ、サクラ、カエデなどの害虫
pear mealybug			grape mealybug を見よ
pear midge		*Contarinia pyrivora* (Riley)	（ハエ目、タマバエ科）全北区
pear midge			pear leaf-roller (2) を見よ
pear midges			fruit flies を見よ
pear moth			pear fruit moth を見よ
pear oystershell scale	ナシカキカイガラムシ	*Lepidosaphes conchiformioides* Borchsenius	（カメムシ目、マルカイガラムシ科）日本、旧北区
pear phylloxera	キナコネアブラムシ	*Aphanostigma iaksuiense* (Kishida)	（カメムシ目、ネアブラムシ科）日本

英名	和名	学名	所属、分布、ほか
pear plant bug		*Lygocoris communis* (Knight)	(カメムシ目、カスミカメムシ科) 新北区
pear psylla		*Psylla pyri* (Linnaeus)	(カメムシ目、キジラミ科) 旧北区
pear psylla (1)	フタホシナシキジラミ	*Psylla simulans* Foerster	(カメムシ目、キジラミ科) 全北区
pear psylla			pear sucker を見よ
pear psyllid			pear sucker を見よ
pear psyllid			pear psylla (1) を見よ
pear pyralid			pear fruit moth を見よ
pear red-striped pyralid	ナシアカスジマダラメイガ	*Nephopterix bicolorella* Leech	(チョウ目、メイガ科) 日本、旧北区、東洋区
pear root aphid		*Eriosoma pyricola* Baker et Davidson	(カメムシ目、アブラムシ科) 新北区、豪州区
pear sawfly (1)	オウトウナメクジハバチ	*Caliroa cerasi* (Linnaeus)	(ハチ目、ハバチ科) 日本、全北区、豪州区
pear sawfly (2)		*Hoplocampa brevis* (Klug)	(ハチ目、ハバチ科) 旧北区
pear scale (1)	ナシクロホシカイガラムシ	*Parlatoreopsis pyri* (Marlatt)	(カメムシ目、マルカイガラムシ科) 日本、全北区。ナシ、ウメ、クワ、サクラ、カエデなどの害虫
pear scale		*Quadraspidiotus pyri* (Lichtenstein)	(カメムシ目、マルカイガラムシ科) 豪州区
pear scale			Italian pear scale を見よ
pear scale			oystershell scale (1) を見よ
pear-shaped weevil		*Apion pomonae* (Fabricius)	(コウチュウ目、ホソクチゾウムシ科) 旧北区
pear-shaped weevils			seed weevils (2) を見よ
pear shortwing beetle		*Glahyra umbellatarum* (von Schreber)	(コウチュウ目、カミキリムシ科) 旧北区
pear slug			pear sawfly (1) を見よ
pear slug sawfly			pear sawfly (1) を見よ
pear slugworm			pear sawfly (1) を見よ
pear stinging caterpillar	ナシイラガ	*Narosoideus flavidorsalis* (Staudinger)	(チョウ目、イラガ科) 日本、旧北区。ナシ、リンゴ、ダイズなどの害虫
pear stink bug	ナシカメムシ	*Urochela luteovaria* Distant	(カメムシ目、クヌギカメムシ科) 日本、旧北区。ナシ、リンゴ、サクラなどの害虫
pear sucker	ナシキジラミ	*Psylla pyrisuga* Foerster	(カメムシ目、キジラミ科) 日本、旧北区。ナシ害虫
pear sucker			pear psylla (1) を見よ
pear thrips	ナシアザミウマ	*Taeniothrips inconsequens* (Uzel)	(アザミウマ目、アザミウマ科) 日本、全北区、新熱帯区
pear tingid			pear lace bug (1) を見よ
pear tortrix		*Cydia pyrivora* Danilevskii	(チョウ目、ハマキガ科) 旧北区
pear-tree oyster scale			oystershell scale (1) を見よ
pear-tree psylla			pear psylla (1) を見よ
pear web-spinning sawfly			social pear sawfly (1) を見よ
pear weevil		*Magdalis barbicornis* (Latreille)	(コウチュウ目、ゾウムシ科) 旧北区

英名	和名	学名	所属、分布、ほか
pear weevil			apple bud weevil を見よ
pear white scale	ナシシロナガカイガラムシ	*Lopholeucaspis japonica* (Cockerell)	(カメムシ目、マルカイガラムシ科) 日本。ナシ、チャ、バラ、カエデなどの害虫
pear white-banded geometrid	ナシモンエダシャク	*Garaeus mirandus* (Butler)	(チョウ目、シャクガ科) 日本、旧北区
pear-winged leafwing		*Anaea halice* Godart	(チョウ目、タテハチョウ科) 新熱帯区
pear yellow aphid	ナシマルアブラムシ	*Sapphaphis piri* Matsumura	(カメムシ目、アブラムシ科) 日本、旧北区
pearl-band grass veneer		*Catoptria margaritella* (Denis et Schiffermüller)	(チョウ目、メイガ科) 旧北区
pearl-bordered fritillary	ミヤマヒョウモン	*Boloria euphrosyne* (Linnaeus)	(チョウ目、タテハチョウ科) 全北区
pearl Charaxes	オオネシロフタオチョウ	*Charaxes varanes* (Cramer)	(チョウ目、タテハチョウ科) エチオピア区
pearl crescent			pearly crescentspot を見よ
pearl grass veneer		*Catoptria pinella* (Linnaeus)	(チョウ目、メイガ科) 旧北区
pearl-mussel grass-veneer		*Catoptria permutatella* (Herrich-Schäffer)	(チョウ目、メイガ科) 旧北区
pearl owl		*Taenaris artemis* Snellen van Vollenhoven	(チョウ目、タテハチョウ科) 豪州区
pearl sawfly			pear sawfly (1) を見よ
pearl skipper			silver-spotted skipper (1) を見よ
pearl-spotted Charaxes		*Charaxes jahlusa* (Trimen)	(チョウ目、タテハチョウ科) エチオピア区
pearl-spotted emperor			pearl-spotted Charaxes を見よ
pearl-spotted forest sylph		*Ceratrichia argyrosticta* (Plötz)	(チョウ目、セセリチョウ科) エチオピア区
pearl white		*Euchloe ausonia dephalis* Hübner	(チョウ目、シロチョウ科) 東洋区。dappled marble を参照
pearls			leaftiers を見よ
pearly crescentspot		*Phyciodes tharos* (Drury)	(チョウ目、タテハチョウ科) 新北区
pearly Euselasia		*Euselasia pusilla pusilla* (R. Felder)	(チョウ目、シジミタテハ科) 新熱帯区。*E. p. mazai* Beutelspacher も同英名
pearly eye (1)		*Enodia anthedon* Clark	(チョウ目、タテハチョウ科) 新北区
pearly eye (2)		*Enodia portlandia* (Fabricius)	(チョウ目、タテハチョウ科) 新北区
pearly-gray hairstreak		*Strephonota tephraeus* (Geyer)	(チョウ目、シジミチョウ科) 新熱帯区
pearly-gray hairstreak			Tephraeus hairstreak を見よ
pearly-grey wave			sweetfern geometer moth を見よ
pearly hairstreak		*Theritas theocritus* (Fabricius)	(チョウ目、シジミチョウ科) 新熱帯区
pearly heath	ベニヒメヒカゲ	*Coenonympha arcania* (Linnaeus)	(チョウ目、タテハチョウ科) 旧北区

英名	和名	学名	所属、分布、ほか
pearly leafwing		*Consul electra electra* (Westwood)	(チョウ目、タテハチョウ科) 新熱帯区。*C. e. adustus* (Lamas) も同英名
pearly marblewing	アメリカツマグロシロチョウ	*Euchloe hyantis* (Edwards)	(チョウ目、シロチョウ科) 新北区
pearly underwing	ニセタマナヤガ	*Peridroma saucia* (Hübner)	(チョウ目、ヤガ科) 日本、全北区、エチオピア区、新熱帯区。北米での英名。トウモロコシ、アルファルファ害虫。pearly underwing moth ともいう
pearly winged lichen moth		*Crambidia casta* (Packard)	(チョウ目、ヒトリガ科) 新北区
pearly wood nymph moth		*Endryas unio* (Hübner)	(チョウ目、ヤガ科) 新北区。pearly wood nymph ともいう
Pearsall's carpet moth		*Venusia pearsalli* (Dyar)	(チョウ目、シャクガ科) 新北区
pease blossom			pease blossom moth を見よ
pease-blossom moth		*Charidon delphinii* (Linnaeus)	(チョウ目、ヤガ科) 旧北区
pebble hooktip		*Drepana falcataria* (Linnaeus)	(チョウ目、カギバガ科) 旧北区
pebble prominent		*Notodonta ziczac* (Linnaeus)	(チョウ目、シャチホコガ科) 旧北区
pecan bark borer moth		*Synanthedon geliformis* (Walker)	(チョウ目、スカシバガ科) 新北区。幼虫は pecan bark borer
pecan blotchminer moth		*Phyllonorycter caryaealbella* (Chambers)	(チョウ目、ホソガ科) 新北区
pecan borer		*Conopia scitula* (Harris)	(チョウ目、スカシバガ科) 新北区。ペカン害虫
pecan bud moth		*Gretchena bolliana* (Slingerland)	(チョウ目、ハマキガ科) 新北区
pecan carpenterworm		*Cossula magnifica* (Strecker)	(チョウ目、ボクトウガ科) 新北区。pecan carpenterworm moth ともいう
pecan casebearer (1)		*Acrobasis hebescella* Hulst	(チョウ目、メイガ科) 新北区
pecan casebearer (2)		*Acrobasis grossbecki* Barnes et McDunnough	(チョウ目、メイガ科) 新北区
pecan cigar casebearer		*Coleophora laticomella* Clemens	(チョウ目、ツツミノガ科) 新北区。成虫は pecan cigar casebearer moth
pecan cigar casebearer		*Coleophora caryaefoliella* (Clemens)	(チョウ目、ツツミノガ科) 新北区。ペカン害虫.
pecan leaf casebearer		*Acrobasis juglandis* (LeBaron)	(チョウ目、メイガ科) 新北区。成虫は pecan leaf casebearer moth
pecan leaf casebearer			mantled Acrobasis moth を見よ
pecan leaf casebearer			pecan casebearer (2) を見よ
pecan leaf miner			pecan serpentine leaf miner を見よ
pecan leaf phylloxera		*Phylloxera notabilis* Pergande	(カメムシ目、ネアブラムシ科) 新北区
pecan leafminer moth		*Cameraria caryaefoliella* (Clemens)	(チョウ目、ホソガ科) 新北区
pecan nursery casemaker			pecan nut case-bearer (1) を見よ
pecan nut case-bearer (1)		*Acrobasis caryivorella* Ragonot	(チョウ目、メイガ科) 新北区

英名	和名	学名	所属、分布、ほか
pecan nut casebearer		*Acrobasis nuxvorella* Neunzig	（チョウ目、メイガ科）新北区。成虫は pecan nut casebearer moth
pecan nut casebearer		*Acrobasis caryae* (Grote)	（チョウ目、メイガ科）新北区。ペカン害虫
pecan nut casebearer			pecan casebearer (1) を見よ
pecan phylloxera		*Phylloxera devastatrix* Pergande	（カメムシ目、ネアブラムシ科）新北区
pecan serpentine leaf miner		*Stigmella juglandifoliella* (Clemens)	（チョウ目、モグリチビガ科）新北区
pecan shuckworm			hickory shuckworm を見よ
pecan spittle bug		*Clastoptera achatina* Germar	（カメムシ目、コガシラアワフキ科）新北区
pecan tree borer			dogwood moth を見よ
pecan weevil		*Curculio caryae* (Horn)	（コウチュウ目、ゾウムシ科）新北区
peccary lice	リスハジラミ科	Pecaroecidae	（シラミ目）の昆虫の総称
Pech's white		*Euchloe pechi* Staudinger	（チョウ目、シロチョウ科）旧北区
Peck's pug moth		*Eupithecia peckorum* Heitzman et Enns	（チョウ目、シャクガ科）新北区
Peck's skipper (1)		*Polites coras* (Cramer)	（チョウ目、セセリチョウ科）新北区
Peck's skipper (2)	ペックアカセセリ	*Polites peckius* (Kirby)	（チョウ目、セセリチョウ科）新北区
pecos giant skipper		*Agathymus mariae* (Barnes et Benjamin)	（チョウ目、セセリチョウ科）新北区、新熱帯区
pectinate-horned click beetle	クシコメツキ	*Melanotus legatus* Candèze	（コウチュウ目、コメツキムシ科）日本、旧北区
pectinate-horned prominent	クシヒゲシャチホコ	*Ptilophora nohirae* (Matsumura)	（チョウ目、シャチホコガ科）日本、旧北区
pectinate-horned pyralid	クシヒゲシマメイガ	*Sybrida approximans* (Leech)	（チョウ目、メイガ科）日本
peculiar giant cupid		*Lepidochrysops peculiaris* (Rogenhofer)	（チョウ目、シジミチョウ科）エチオピア区
peddlers			tortoise beetles (1) を見よ　米国での幼虫の英名
pedegral skipper		*Atrytonopsis frappenda* (Dyar)	（チョウ目、セセリチョウ科）新熱帯区
pediculosis	シラミ		sucking lice を見よ
pedilid beetles			cardinal beetles を見よ　アカハネムシ科 Pyrochroidae の亜科として Pedilinae が使用されたことに由来
pedilids			cardinal beetles を見よ
pedunculate ground beetles		Scaritini	（コウチュウ目、オサムシ科）の昆虫の総称
peg beetles		*Odontotaenius*	（コウチュウ目、クロツヤムシ科）の昆虫の総称
peg beetles (1)	クロツヤムシ科	Passalidae	（コウチュウ目）の昆虫の総称
peg bug			bess-beetle を見よ
peg-top coccids		Apiomorphidae	（カメムシ目）の昆虫の総称
Peigler's oakworm moth		*Anisota peigleri* Riotte	（チョウ目、ヤママユガ科）新北区

英名	和名	学名	所属、分布、ほか
Pelarge metalmark		*Calospila pelarge* (Godman et Salvin)	(チョウ目、シジミタテハ科) 新熱帯区
pelargonium aphid		*Acyrthosiphon malvae* (Mosley)	(カメムシ目、アブラムシ科) 旧北区
pelea long-horned beetle			Hawaiian alani long-horned beetle を見よ
pelecinid wasps		Pelecinidae	(ハチ目) の昆虫の総称
pelecinids			pelecinid wasps を見よ
pelecorhynchid flies		Pelecorhynchidae	(ハエ目) の昆虫の総称　アブ科の亜科 Scarphiinae の異名 Pelecorhynchinae に由来
Peleides blue Morpho		*Morpho helenor peleides* (Kollar)	(チョウ目、タテハチョウ科) 新熱帯区
pelican pouch louse		*Piageticella bursaepelecani* (Perry)	(ハジラミ目、チョウカクハジラミ科) 新北区
Pelidne			blueberry sulphur を見よ
pellucid broad-winged planthopper	スケバハゴロモ	*Euricania facialis* (Walker)	(カメムシ目、ハゴロモ科) 日本、旧北区
pellucid hawk moth	スキバホウジャク	*Hemaris radians* (Walker)	(チョウ目、スズメガ科) 日本、旧北区
pellucid hawk moth			larger pellucid hawk moth を見よ
pellucid silk moth	ウスタビガ	*Rhodinia fugax* (Butler)	(チョウ目、ヤママユガ科) 日本、旧北区。オウトウ、カエデの害虫
pellucid-spotted silk moth			pellucid silk moth を見よ
pellucid tortoise beetle	スキバジンガサハムシ	*Aspidomorpha transparipennis* (Motschulsky)	(コウチュウ目、ハムシ科) 日本、旧北区
pellucid zygaenid	ミノウスバ	*Pryeria sinica* Moore	(チョウ目、マダラガ科) 日本、旧北区。マサキ害虫
Pelon skipperling		*Dalla kemneri* Steinhauser	(チョウ目、セセリチョウ科) 新熱帯区
pelt moth			casemaking clothes moth (1) を見よ
Pemberton's windmill	クロヘリジャコウアゲハ	*Byasa plutonius* (Oberthür)	(チョウ目、アゲハチョウ科) 東洋区
pemphredon wasps			aphid wasps (1)(2) を見よ
pencilled blue	アブルリミナミシジミ	*Candalides absimilis* (C. Felder)	(チョウ目、シジミチョウ科) 豪州区。pencilled blue butterfly ともいう
Pendlebury's zebra		*Graphium ramaceus pendleburyi* (Corbet)	(チョウ目、アゲハチョウ科) 東洋区
Penelope's Acraea	ベニシタホソチョウ	*Acraea penelope* Staudinger	(チョウ目、タテハチョウ科) エチオピア区
Penelope's ringlet		*Cissia penelope* Fabricius	(チョウ目、タテハチョウ科) 新熱帯区
penicillate scale	タケマルカイガラムシ	*Odonaspis penicillata* Green	(カメムシ目、マルカイガラムシ科) 日本、全北区
peninsula blue		*Lepidochrysops oreas* Tite	(チョウ目、シジミチョウ科) エチオピア区
peninsula skolly		*Thestor yildizae* Kocak	(チョウ目、シジミチョウ科) エチオピア区
peninsular metalmark		*Apodemia virgulti peninsularis* Emel, Emel et Pratt	(チョウ目、シジミタテハ科) 新熱帯区
penitent underwing moth		*Catocala piatrix* (Grote)	(チョウ目、ヤガ科) 新北区

英名	和名	学名	所属、分布、ほか
pen-marked sphinx			great ash sphinx を見よ
pennant tussock		*Psalis africana* Kiriakoff	(チョウ目、ドクガ科) エチオピア区
pennants		*Macrodiplax*	(トンボ目、トンボ科) の昆虫の総称
Pennington's blue		*Lepidochrysops penningtoni* Dickson	(チョウ目、シジミチョウ科) エチオピア区
Pennington's brown		*Pseudonympha penningtoni* Riley	(チョウ目、タテハチョウ科) エチオピア区
Pennington's buff		*Cnodontes penningtoni* Bennet	(チョウ目、シジミチョウ科) エチオピア区
Pennington's copper		*Aloeides penningtoni* Tite et Dickson	(チョウ目、シジミチョウ科) エチオピア区
Pennington's opal		*Chrysoritis penningtoni* (Riley)	(チョウ目、シジミチョウ科) エチオピア区
Pennington's playboy		*Deudorix penningtoni* van Son	(チョウ目、シジミチョウ科) エチオピア区
Pennington's Protea		*Capys penningtoni* Riley	(チョウ目、シジミチョウ科) エチオピア区
Pennington's sailer		*Neptis penningtoni* van Son	(チョウ目、タテハチョウ科) エチオピア区
Pennington's skolly		*Thestor penningtoni* van Son	(チョウ目、シジミチョウ科) エチオピア区
Pennsylvania leatherwing beetle		*Chauliognathus pennsylvanicus* (De Geer)	(コウチュウ目、ジョウカイボン科) 新北区。Pennsylvania leatherwing ともいう
Pennsylvania wood cockroach			wood cockroach を見よ
Pennsylvania yellowjacket			western yellowjacket を見よ
penny adder			dragonflies (1) を見よ
penny doctors			tiger beetles (1) を見よ　ニュージーランドでの英名
Penstemon borer moth		*Penstemonia edwardsii* (Beutenmüller)	(チョウ目、スカシバガ科) 新北区
pentameren			pentamerous beetles を見よ
pentamerous beetles		*Pentamera*	(コウチュウ目) の昆虫の総称
pentatomid bugs			stink bugs を見よ
pentatomids			stink bugs を見よ
Pentila	マネシシジミ属	*Pentila*	(チョウ目、シジミチョウ科) の昆虫の総称
Pentz's shade			chestnut buttom を見よ
Pentz's shade			mottled coast shade をみよ
peony scale	チャノマルカイガラムシ	*Pseudaonidia paeonia* (Cockerell)	(カメムシ目、マルカイガラムシ科) 日本、全北区。チャ、カキ、ツバキ、ツツジなどの害虫
pepper-and-salt			peppered moth を見よ
pepper-and-salt moth		*Biston cognataria* (Guenée)	(チョウ目、シャクガ科) 新北区
pepper-and-salt moth			peppered moth を見よ
pepper-and-salt skipper		*Amblyscirtes hegon* Scudder	(チョウ目、セセリチョウ科) 新北区

英名	和名	学名	所属、分布、ほか
pepper flower-bud moth		*Symmetrischema capsicum* (Bradley et Povolny)	(チョウ目、キバガ科) 新北区、新熱帯区
pepper maggot	トウガラシミバエ	*Zonosemata electa* (Say)	(ハエ目、ミバエ科) 新北区。幼虫はナス他の害虫
pepper tree caterpillar		*Bombycomorpha bifascia* Walker	(チョウ目、カレハガ科) エチオピア区
pepper tree caterpillar		*Bombycomorpha pallida* Distant	(チョウ目、カレハガ科) エチオピア区
pepper tree psyllid		*Calophya rubra* (Blanchard)	(カメムシ目、ヒメキジラミ科) 大洋区、新熱帯区
pepper tree psyllid		*Calophya schini* Tuthill	(カメムシ目、ヒメキジラミ科) 新北区、新熱帯区、豪州区
pepper weevil	トウガラシゾウムシ	*Anthonomus eugenii* Cano	(コウチュウ目、ゾウムシ科) 新北区
peppered blue skipper			common blue-skipper を見よ
peppered cockroach		*Archimandrita tesselata* Rehn	(ゴキブリ目、ブラベルスゴキブリ科) 新熱帯区。peppered roach ともいう
peppered Haimbachia moth		*Haimbachia placidella* (Haimbach)	(チョウ目、メイガ科) 新北区
peppered hopper		*Platylesches ayresii* (Trimen)	(チョウ目、セセリチョウ科) エチオピア区
peppered moth		*Biston betularia* (Linnaeus)	(チョウ目、シャクガ科) 全北区。日本亜種は *B. b. parvus* Leech オオシモフリエダシャクで、large frosted geometrid という
peppered spurwing		*Antigonus mutilatus* Hopffer	(チョウ目、セセリチョウ科) 新熱帯区
peppergrass beetle		*Galeruca browni* Blake	(コウチュウ目、ハムシ科) 新北区
peppermint aphid			potato aphid (2) を見よ
peppermint flea beetle	ハッカトビハムシ	*Longitarsus nipponensis* Csiki	(コウチュウ目、ハムシ科) 日本。ハッカ害虫
peppermint leaf beetle	ハッカハムシ	*Chrysolina exanthematica* (Wiedemann)	(コウチュウ目、ハムシ科) 日本、旧北区。ハッカ害虫
peppermint leafminer	ハッカハモグリバエ	*Phytomyza tetrasticha* Hendel	(ハエ目、ハモグリバエ科) 日本。ハッカ害虫
peppermint looper		*Paralaea beggaria* (Guenée)	(チョウ目、シャクガ科) 豪州区
peppermint phytometra	コヒサゴキンウワバ	*Diachrysia nadeja* (Oberthür)	(チョウ目、ヤガ科) 日本、旧北区
peppermint Pyrausta	ハッカノメイガ	*Pyrausta aurata* (Scopoli)	(チョウ目、メイガ科) 日本、旧北区。ハッカ害虫
peppermint small weevil	ハッカヒメゾウムシ	*Baris menthae* Kono	(コウチュウ目、ゾウムシ科) 日本
peppermint stick insect		*Megacrania batesii* (Kirby)	(ナナフシ目、ナナフシ科) 東洋区、豪州区
Perak lascar	パラカキンミスジ	*Pantoporia paraka* (Butler)	(チョウ目、タテハチョウ科) 東洋区
perchers		*Diplacodes*	(トンボ目、トンボ科) の昆虫の総称
perching Saliana		*Saliana esperi esperi* Evans	(チョウ目、セセリチョウ科) 新北区、新熱帯区
Perenna leafwing		*Memphis perenna perenna* (Godman et Salvin)	(チョウ目、タテハチョウ科) 新熱帯区

英名	和名	学名	所属、分布、ほか
Perfida skipper		*Anatrytone perfida* Möschler	（チョウ目、セセリチョウ科）新熱帯区
pergid sawflies		Pergidae	（ハチ目）の昆虫の総称
pergids			pergid sawflies を見よ
Periander metalmark		*Rhetus periander naevianus* Stichel	（チョウ目、シジミタテハ科）新熱帯区。blue doctor を参照
pericopid moth		*Gnophaela vermiculata* (Grote)	（チョウ目、ヒトリガ科）新北区
pericopid moths		Pericopidae	（チョウ目）の昆虫の総称
perilampids	マルハラコバチ科	Perilampidae	（ハチ目）の昆虫の総称
Perilla aphid	シソヒゲナガアブラムシ	*Acyrthosiphon perillae* (Shinji)	（カメムシ目、アブラムシ科）日本
Perilla crescent			Acraea mimic (1) を見よ
Perilla leaf roller	ベニフキノメイガ	*Pyrausta panopealis* (Walker)	（チョウ目、メイガ科）日本、東洋区、豪州区、新熱帯区。シソ害虫
Perilla long-horned aphid			perilla aphid を見よ
Perilla looper			burnished brass を見よ
periodical cicada			seventeen-year locust (1) を見よ
periodical cicadas		*Magicicada*	（カメムシ目、セミ科）の昆虫の総称　周期セミの属で、17年セミ3種、13年セミ3種あり
periscelid flies		Periscelidae	（ハエ目）の昆虫の総称
perlid stoneflies			common stoneflies (1) を見よ
perlids			common stoneflies (1) を見よ
perlids			stoneflies を見よ　Perlidae (common stoneflies) の英名が目にも拡大して使用されたのだろう
perlodid stoneflies			predatory stoneflies を見よ
permanent apple aphid			apple aphid (1) を見よ
permanent blackberry aphid		*Aphis ruborum* (Börner)	（カメムシ目、アブラムシ科）旧北区、東洋区
permanent carrot aphid	ユリノキヒゲナガアブラムシ	*Aphis lambersi* (Börner)	（カメムシ目、アブラムシ科）旧北区
permanent currant aphid		*Aphis schneideri* (Börner)	（カメムシ目、アブラムシ科）旧北区
permanent dock aphid			dock aphid (1) を見よ
permanent parsnip aphid		*Dysaphis bonomii* (Hille Ris Lambers)	（カメムシ目、アブラムシ科）旧北区
perniciosus skipper		*Chrysoplectrum perniciosus* (Herrich-Schäffer)	（チョウ目、セセリチョウ科）新熱帯区
pernicious scale			San Jose scale を見よ
pernicious scale			San Jose scale を見よ
perny silk moth			Chinese oak silkmoth を見よ
perothopid beetles			beech-tree beetles を見よ

英名	和名	学名	所属、分布、ほか
perothopids			beech-tree beetles を見よ
perplexed arches moth		*Drasteria perplexa* (Edwards)	(チョウ目、ヤガ科) 新北区。perplexing arches ともいう
perplexed button			broad-barred marble (2) を見よ
perplexing gemmed-satyr		*Cyllopsis perplexa* Miller	(チョウ目、タテハチョウ科) 新熱帯区
Perseus opal		*Chrysoritis perseus* (Henning)	(チョウ目、シジミチョウ科) エチオピア区
Persian grass blue		*Chilades galba* (Lederer)	(チョウ目、シジミチョウ科) 旧北区
Persian skipper		*Spialia phlomidis* (Herrich et Schäffer)	(チョウ目、セセリチョウ科) 旧北区
persicaria aphid	タデヨツオヒゲナガアブラムシ	*Akkaia polygoni* Takahashi	(カメムシ目、アブラムシ科) 日本、旧北区
persimmon bagworm	ヒメミノガ	*Psyche niphonica* (Hori)	(チョウ目、ミノガ科) 日本
persimmon bark borer	フタモンマダラメイガ	*Euzophera batangensia* Caradja	(チョウ目、メイガ科) 日本。カキ、クリ、リンゴ害虫
persimmon borer		*Sannina uroceriformis* Walker	(チョウ目、スカシバガ科) 新北区。persimmon borer moth ともいう
persimmon cochlid	ヒメクロイラガ	*Scopelodes contracta* Walker	(チョウ目、イラガ科) 日本、旧北区。カキ、ヤマモモ、プラタナスなどの害虫
persimmon fruit moth	カキノヘタムシガ	*Stathmopoda masinissa* Meyrick	(チョウ目、ニセマイコガ科) 日本、旧北区、東洋区。カキ害虫
persimmon gelechid	ゴマフシロキバガ	*Odites leucostola* (Meyrick)	(チョウ目、ヒゲナガキバガ科) 日本
persimmon leaf beetle			claycolored leaf beetle を見よ
persimmon leafminer	カキホソガ	*Cuphodes diospyrosella* (Issiki)	(チョウ目、ホソガ科) 日本。カキ害虫
persimmon leafroller			Leche's twist を見よ
persimmon pseudaonidia scale			trilobe scale を見よ
persimmon psylla		*Trioza diospyri* (Ashmead)	(カメムシ目、トガリキジラミ科) 新北区。persimmon psyllid ともいう
persimmon root borer			persimmon borer を見よ
persimmon tiger longicorn	キスジトラカミキリ	*Cyrtoclytus caproides* Bates	(コウチュウ目、カミキリムシ科) 日本、旧北区
persimmon tree borer	ヒメスカシバ	*Synanthedon tenuis* (Butler)	(チョウ目、スカシバガ科) 日本、旧北区
persimmon weevil	カキゾウムシ	*Pseudocneorhinus obesus* Roelofs	(コウチュウ目、ゾウムシ科) 日本、旧北区
persistent Saliana		*Saliana antoninus* (Latreille)	(チョウ目、セセリチョウ科) 新熱帯区
persius duskywing		*Erynnis persius* (Scudder)	(チョウ目、セセリチョウ科) 新北区
Perth clothes moth		*Archinemapogon laterellus* (Thunberg)	(チョウ目、ヒロズコガ科) 旧北区
Peru Altinote		*Altinote momina* Jordan	(チョウ目、タテハチョウ科) 新熱帯区
Peru bird-dropping skipper		*Milanion cramba* Evans	(チョウ目、セセリチョウ科) 新熱帯区

英名	和名	学名	所属、分布、ほか
Peruvian cattleheart mimic	ホソオタイマイ	*Mimoides xynias* (Hewitson)	（チョウ目、アゲハチョウ科）新熱帯区
Peruvian crescent		*Eresia nauplius* (Linnaeus)	（チョウ目、タテハチョウ科）新熱帯区
Peruvian dubia cockroach			South American cockroach を見よ
Peruvian fly		*Paratheresia claripalpis* (Wulp)	（ハエ目、ヤドリバエ科）新北区、新熱帯区
Peruvian larder beetle		*Dermestes peruvianus* Laporte	（コウチュウ目、カツオブシムシ科）旧北区
Peruvian leafhopper		*Proconia marmorata* (Fabricius)	（カメムシ目、オオヨコバイ科）旧北区
Peruvian longwing		*Heliconius peruvianus* Felder	（チョウ目、タテハチョウ科）新熱帯区
Peruvian puna skipper		*Hylephila peruana* Draudt	（チョウ目、セセリチョウ科）新熱帯区
Peruvian shield mantis		*Choeradodis rhombicollis* Latreille	（カマキリ目、カマキリ科）新北区、新熱帯区
Peruvian stick mantis		*Pseudovates peruviana* (Rehn)	（カマキリ目、カマキリ科）新熱帯区
Peruvian twin-tailed satyr		*Daedalma vertex* Pyrcz	（チョウ目、タテハチョウ科）新熱帯区
Peruvian underleaf		*Eurybia* sp.	（チョウ目、シジミタテハ科）新熱帯区
pest fruit fly		*Bactrocera tau* (Walker)	（ハエ目、ミバエ科）日本、東洋区
Petelina hairstreak		*Ocaria petelina* (Hewitson)	（チョウ目、シジミチョウ科）新熱帯区
Peter's fairy playboy		*Paradeudorix petersi* (Stempffer et Bennett)	（チョウ目、シジミチョウ科）エチオピア区
petiolate hymenoptera	ハチ亜目（細腰亜目）	Apocrita	（ハチ目）の昆虫の総称
petite wood white			immaculate wood white を見よ
Petiver's drill	ホソキオビヘリホシヒメハマキ	*Dichrorampha petiverella* (Linnaeus)	（チョウ目、ハマキガ科）日本、旧北区
Peto skipper		*Anastrus petius peto* Evans	（チョウ目、セセリチョウ科）新熱帯区
petroleum fly		*Helaeomyia petrolei* (Coquillett)	（ハエ目、ミギワバエ科）新北区、新熱帯区
Petrovna skipper		*Adlerodea petrovna* (Schaus)	（チョウ目、セセリチョウ科）新熱帯区
petty-whin case		*Coleophora genistae* Stainton	（チョウ目、ツツミノガ科）旧北区
Phadka grasshopper		*Hieroglyphus nigrorepletus* (Bolivar)	（バッタ目、バッタ科）東洋区
Phaeton primrose sphinx moth		*Euproserpinus phaeton* Grote et Robinson	（チョウ目、スズメガ科）新北区、新熱帯区
phalacrid beetles			shining flower beetles を見よ
Phalakron blue		*Polyommatus andronicus* Coutsis et Ghavalas	（チョウ目、シジミチョウ科）旧北区
Phaleros hairstreak		*Panthiades phaleros* (Linnaeus)	（チョウ目、シジミチョウ科）新熱帯区
phaloniid moths	ホソハマキガ科	Cochylidae	（チョウ目）の昆虫の総称　本科の異名 Phaloniidae に由来

英名	和名	学名	所属、分布、ほか
phantom crane flies	コシボソガガンボ科	Ptychopteridae	（ハエ目）の昆虫の総称
phantom crane flies			false crane flies を見よ
phantom darner		*Triacanthagyna trifida* (Rambur)	（トンボ目、ヤンマ科）新熱帯区
phantom flutterer		*Rhyothemis semihyalina* (Desjardins)	（トンボ目、トンボ科）エチオピア区
phantom gnats			phantom midges (1) を見よ
phantom hemlock looper		*Nepytia phantasmaria* (Strecker)	（チョウ目、シャクガ科）新北区。成虫は phantom hemlock looper moth という
phantom knight		*Psaltoda brachypennis* (Moss et Moulds)	（カメムシ目、セミ科）豪州区
phantom larva		*Corethra plumicornis* Fabricius	（ハエ目、ケヨソイカ科）旧北区
phantom larva			ghost larva を見よ
phantom larva			Clear Lake gnat を見よ
phantom midges		*Chaoborus*	（ハエ目、ケヨソイカ科）の昆虫の総称
phantom midges (1)	ケヨソイカ科	Chaoboridae	（ハエ目）の昆虫の総称
Phaon crescent		*Phyciodes phaon* (Edwards)	（チョウ目、タテハチョウ科）新北区
Phaon crescentspot			Phaon crescent を見よ
Pharaoh ant			fire ant (2) を見よ
Pharaoh's ant	イエヒメアリ	*Monomorium pharaonis* (Linnaeus)	（ハチ目、アリ科）日本．汎世界。古代エジプト王の名に由来。屋内生息性。red ant (2) も参照。Pharaoh ant とも記す
Pharsalus Acraea			East African forest Acraea を見よ
phasmatids			stick insects を見よ
phasmids			walking sticks (2) を見よ
Phegeus hairstreak		*Theritas phegeus* (Hewitson)	（チョウ目、シジミチョウ科）新熱帯区
phellodendron scale			cymbidium scale を見よ
Phemiades skipper		*Phemiades* sp.	（チョウ目、セセリチョウ科）新熱帯区
Phenomoe glasswing		*Oleria phenomoe* Doubleday	（チョウ目、タテハチョウ科）新熱帯区
Phidias firetip		*Pyrrhopyge phidias* (Linnaeus)	（チョウ目、セセリチョウ科）新熱帯区
Phidyle skipper		*Mimia phidyle phidyle* (Godman et Salvin)	（チョウ目、セセリチョウ科）新熱帯区
Phigalia skipper		*Trapezites phigalia* (Hewitson)	（チョウ目、セセリチョウ科）豪州区
Phigalioides skipper		*Trapezites phigalioides* Waterhouse	（チョウ目、セセリチョウ科）豪州区
Philippine hedge blue		*Celastrina philippina* (Semper)	（チョウ目、シジミチョウ科）東洋区
Philippine katydid		*Phaneroptera furcifera* Stål	（バッタ目、キリギリス科）新北区
Philippine ladybeetle		*Epilachna philippinensis* (Dieke)	（コウチュウ目、テントウムシ科）東洋区

英名	和名	学名	所属、分布、ほか
Philippine Mormon		*Papilio polytes alphenor* Cramer	(チョウ目、アゲハチョウ科) 東洋区。common Mormon を参照
Philippine subterranean termite		*Coptotermes vastator* Light	(シロアリ目、ミゾガシラシロアリ科) 大洋区
Philippine swift		*Caltoris philippina* (Herrich-Schäffer)	(チョウ目、セセリチョウ科) 東洋区
Philip's arctic		*Oeneis philipi* Troubridge	(チョウ目、タテハチョウ科) 新北区
Philip's arctic			early arctic を見よ
phillyrea whitefly			ash whitefly を見よ
philobota		*Philobota productella* (Walker)	(チョウ目、マルハキバガ科) 豪州区
phlox moth		*Schinia indiana* Smith	(チョウ目、ヤガ科) 新北区
phlox plant bug		*Lopidea davisi* Knight	(カメムシ目、カスミカメムシ科) 新北区
Phoebus			large parnassian を見よ
Phoebus Parnassian			large parnassian を見よ
phoenicean Pyrausta moth		*Pyrausta phoenicealis* (Walker)	(チョウ目、メイガ科) 新北区
phoenicoccid scales			palm scales を見よ
phoenix		*Eulithis prunata* (Linnaeus)	(チョウ目、シャクガ科) 旧北区。日本亜種は *E. p. leucoptera* (Diakonov) チョウセンハガタナミシャク
phoenix emerald moth		*Dichordophora phoenix* (Prout)	(チョウ目、シャクガ科) 新北区
phoenix moth			phoenix を見よ
phorid flies			humpbacked flies を見よ
phorids			humpbacked flies を見よ
phragmites planthopper	ヨシウンカ	*Chloriona japonica* Matsumura	(カメムシ目、ウンカ科) 日本
phthirapterans			phthirapterous insects を見よ
phthirapterous insects		Phthiraptera	の昆虫の総称
phycitid moths			cereal moths を見よ
Phylaca sister		*Adelpha phylaca phylaca* (Bates)	(チョウ目、タテハチョウ科) 新熱帯区
Phyllira moth		*Apantesis phyllira* Hampson	(チョウ目、ヒトリガ科) 新北区
Phyllira tiger moth		*Grammia phyllira* (Drury)	(チョウ目、ヒトリガ科) 新北区
phylloxera			grapeleaf louse を見よ
phylloxeras			bark aphids を見よ
phylloxerids			bark aphids を見よ
physopods			thrips (2) を見よ
phytophaga destructor			Hessian fly を見よ
phytophagous beetles		Phytophaga	(コウチュウ目) の昆虫の総称
phytophagous hymenopterans			horntails (1) を見よ

英名	和名	学名	所属、分布、ほか
picea frosted aphid	コナフキトドイロアブラムシ	*Cinara costata* (Zetterstedt)	（カメムシ目、アブラムシ科）日本、全北区
pickerelweed borer		*Bellura densa* (Walker)	（チョウ目、ヤガ科）新北区。pickerelweed borer moth ともいう
pickle worm	アメリカウリノメイガ	*Diaphania nitidalis* (Stoll)	（チョウ目、メイガ科）新北区、新熱帯区。成虫は pickleworm moth
pictate noctuid			fruit-piercing moth (11) を見よ
Pictet's shieldback		*Idiostatus californicus* Pictet	（バッタ目、キリギリス科）新北区
picticollis canegrub		*Lepidiota picticollis* Lea	（コウチュウ目、コガネムシ科）豪州区
picture-winged flies	ハネフリバエ科	Otitidae	（ハエ目）の昆虫の総称　picture-wing flies とも記す
picture-winged flies			fruit flies を見よ
picture-winged fly		*Delphinia picta* (Fabricius)	（ハエ目、ハネフリバエ科）新北区
picture-winged leaf moths			window-winged moths を見よ
pictured grasshopper		*Dactylotum pictum* (Thomas)	（バッタ目、バッタ科）新北区
pictured rove beetle		*Thinopinus pictus* LeConte	（コウチュウ目、ハネカクシ科）新北区
pictured-wing flies			broadmouthed flies を見よ
pie-dish beetles		*Helea*	（コウチュウ目、ゴミムシダマシ科）の昆虫の総称
pied blue		*Phlyaria cyara* (Hewitson)	（チョウ目、シジミチョウ科）エチオピア区
pied blue		*Pithecops dionisius* (Boisduval)	（チョウ目、シジミチョウ科）豪州区
pied flat		*Celaenorrhinus morena* Evans	（チョウ目、セセリチョウ科）東洋区
pied flat			common snow flat を見よ
pied flats		*Coladenia*	（チョウ目、セセリチョウ科）の昆虫の総称
pied hoverfly			large syrphid fly を見よ
pied paddy skimmer			pied parasol を見よ
pied parasol		*Neurothemis tullia* (Drury)	（トンボ目、トンボ科）東洋区
pied pierrots		*Tuxentius*	（チョウ目、シジミチョウ科）の昆虫の総称
pied piper		*Eurytela hiarbas angustata* Lathy	（チョウ目、タテハチョウ科）エチオピア区
pied piper		*Spioniades abbreviata* (Mabille)	（チョウ目、セセリチョウ科）新熱帯区
pied shield bug		*Tritomegas bicolor* (Linnaeus)	（カメムシ目、ツチカメムシ科）旧北区
pied shieldbug		*Sehirus bicolor* (Linnaeus)	（カメムシ目、ツチカメムシ科）旧北区
pied spot		*Hemistigma albipuncta* (Rambur)	（トンボ目、トンボ科）エチオピア区
pied-winged robberfly		*Pamponerus germanicus* (Linnaeus)	（ハエ目、ムシヒキアブ科）旧北区
Piedmont anomalouse blue		*Polyommatus humedasae* (Toso et Balletto)	（チョウ目、シジミチョウ科）旧北区
Piedmont clubtail		*Gomphus parvidens* Currie	（トンボ目、サナエトンボ科）新北区

英名	和名	学名	所属、分布、ほか
Piedmont ringlet		*Erebia meolans* (Prunner)	(チョウ目、タテハチョウ科) 旧北区
piercers		*Laspeyresia*	(チョウ目、ハマキガ科) の昆虫の総称
pieris whitefly	イヌツゲシロコナジラミ	*Aleurotuberculatus similis* Takahashi	(カメムシ目、コナジラミ科) 日本。イヌツゲ、アセビ害虫
Pierre's Acraea		*Acraea encedana* Pierre	(チョウ目、タテハチョウ科) エチオピア区
pierrots		*Castalius*	(チョウ目、シジミチョウ科) の昆虫の総称
pies			pied pierrots を見よ
pig louse			cleg (1) を見よ
pig louse			hog louse を見よ
pigeon body louse		*Hohorstiella lata* (Piaget)	(ハジラミ目、タンカクトリハジラミ科) 旧北区
pigeon bug		*Cimex columbarius* Jenyns	(カメムシ目、トコジラミ科) 全北区。ニワトリに寄生
pigeon flea		*Ceratophyllus columbae* (Gervais)	(ノミ目、ナガノミ科) 旧北区
pigeon fly		*Pseudolynchia maura* Bigot	(ハエ目、シラミバエ科) 新北区
pigeon fly (1)		*Pseudolynchia canariensis* (Macquart)	(ハエ目、シラミバエ科) 日本、東洋区、全北区
pigeon horntail			pigeon tremex を見よ
pigeon louse		*Colpocephalum turbinatum* Denny	(ハジラミ目、タンカクハジラミ科) 日本、汎世界
pigeon louse			slender pigeon louse (1) を見よ
pigeon louse			small pigeon louse (2) を見よ
pigeon louse			large pigeon louse を見よ
pigeon louse fly		*Pseudolynchia brunnea* (Latreille)	(ハエ目、シラミバエ科) 新北区、新熱帯区
pigeon louse fly			pigeon fly (1) を見よ
pigeon tremex		*Tremex columba* (Linnaeus)	(ハチ目、キバチ科) 新北区
pigeonpea borer		*Ancylostomia stercorea* Zeller	(チョウ目、メイガ科) 新熱帯区
pigeonpea pod borer			pigeonpea borer を見よ
piger grass tubeworm moth		*Acrolophus piger* (Dyar)	(チョウ目、Acrolophidae) 新北区
pigface scale			cottony pigface scale を見よ
Pigmalion skipper			mangrove skipper (2) を見よ
pigmies		*Nepticula*	(チョウ目、モグリチビガ科) の昆虫の総称
pigmies			midget moths を見よ
pigmy backswimmer bugs			pygmy backswimmer bugs を見よ
pigmy footman		*Eilema pygmaeola* (Doubleday)	(チョウ目、ヒトリガ科) 旧北区。pygmy footman ともいう
pigmy footman		*Eilema lutarella* (Linnaeus)	(チョウ目、ヒトリガ科) 旧北区
pigmy grasshoppers			grouse locusts を見よ
pigmy locusts			grouse locusts を見よ

英名	和名	学名	所属、分布、ほか
pigmy mole cricket		*Tridactylus variegatus* (Latreille)	（バッタ目、ノミバッタ科）旧北区
pigmy mole cricket			smaller sand cricket を見よ
pigmy mole crickets			long-horned grasshoppers (3) を見よ
pigmy moths			midget moths を見よ
pigmy scrub hopper		*Acromachus pygmaeus* (Fabricius)	（チョウ目、セセリチョウ科）東洋区
pigmy skipper		*Gegenes pumilio* (Hoffmannsegg)	（チョウ目、セセリチョウ科）旧北区
pigmy Y piercer		*Pammene populana* (Fabricius)	（チョウ目、ハマキガ科）旧北区
pigweed caterpillar			beet armyworm を見よ
pilbara dragon		*Antipodogomphus hodgkini* Watson	（トンボ目、サナエトンボ科）豪州区
pilfer			white-marked spider beetle を見よ
pill beetle		*Byrrhus pilula* Linnaeus	（コウチュウ目、マルトゲムシ科）旧北区
pill beetles	マルトゲムシ科	Byrrhidae	（コウチュウ目）の昆虫の総称
pill-makers			chafer beetles を見よ
pilose biting horse louse		*Trichodectes pilosus* Giebel	（ハジラミ目、ケモノハジラミ科）全北区。米国での英名
Pilza skipper		*Paratrytone pilza* (Evans)	（チョウ目、セセリチョウ科）新熱帯区
Pima Apilocrocis moth		*Apilocrocis pimalis* Barnes et McDunnough	（チョウ目、メイガ科）新北区
Pima dancer		*Argia pima* Garrison	（トンボ目、イトトンボ科）新北区
Pima orange-tip	アリゾナツマキチョウ	*Anthocharis pima* Edwards	（チョウ目、シロチョウ科）新北区
pimelia cutworm moth		*Melanchra rhodopleura* Meyrick	（チョウ目、ヤガ科）豪州区
Pimpinel pug		*Eupithecia pimpinellata* (Hübner)	（チョウ目、シャクガ科）旧北区
pimple-headed hunter		*Austrogomphus mjobergi* Sjöstedt	（トンボ目、サナエトンボ科）豪州区
pimple psyllid			Eugenia psyllid を見よ
pimpline wasps		Pimplinae	（ハチ目、ヒメバチ科）の昆虫の総称
pinhead wisp		*Agriocnemis femina femina* (Brauer)	（トンボ目、イトトンボ科）東洋区
pinhole beetles	ナガキクイムシ科	Platypodidae	（コウチュウ目）の昆虫の総称
pinhole borer			oak pinhole borer を見よ
pin-hole borers			pinhole beetles を見よ
pin oak clearwing moth		*Paranthrene pellucida* Greenfield et Karandinos	（チョウ目、スカシバガ科）新北区
pin-striped vermilion slug moth		*Monoleuca semilascia* (Walker)	（チョウ目、イラガ科）新北区
pintail		*Acisoma panorpoides* Rambur	（トンボ目、トンボ科）東洋区、エチオピア区
pin-tailed pondhawk		*Erythemis plebeja* (Burmeister)	（トンボ目、トンボ科）新熱帯区

英名	和名	学名	所属、分布、ほか
pintails		*Acisoma*	（トンボ目、トンボ科）の昆虫の総称
pinara moth		*Pinara divisa* (Walker)	（チョウ目、カレハガ科）豪州区
pincate beetles			darkling beetles (1) を見よ　米国での英名
pincertails		*Onychogomphus*	（トンボ目、サナエトンボ科）の昆虫の総称
pinch beetles			stag beetles (1) を見よ
pinchbuck		*Rhagium sycophanta* (Schrank)	（コウチュウ目、カミキリムシ科）旧北区
pincher wasps	カマバチ科	Dryinidae	（ハチ目）の昆虫の総称
pincher-wig			common earwig を見よ
pinching beetle		*Pseudolucanus capreolus* Linnaeus	（コウチュウ目、クワガタムシ科）新北区
pinching bug			pinching beetle を見よ
pinching bugs			stag beetles (1) を見よ
pine adelgid	エゾマツカサアブラムシ	*Pineus pini* (Macquart)	（カメムシ目、カサアブラムシ科）日本、旧北区
pine adelgids			conifer aphids を見よ
pine agaric crane fly	キノコガガンボ	*Ula fungicola* Nobuchi	（ハエ目、ガガンボ科）日本。マツタケ害虫
pine agaric humpbacked fly	マツタケバエ	*Megaselia flava* (Fallén)	（ハエ目、ノミバエ科）日本。マツタケ害虫
pine agaric maggot	オオマツタケバエ	*Tetragoneura matsutakei* (Sasaki)	（ハエ目、キノコバエ科）日本。マツタケ害虫
pine agaric moth fly	マツタケチョウバエ	*Psychodocha fungicola* (Tokunaga)	（ハエ目、チョウバエ科）日本。マツタケ害虫
pine agaric phorid			pine agaric humpbacked fly を見よ
pine aphid	マツホソオオアブラムシ	*Eulachnus thunbergii* Wilson	（カメムシ目、アブラムシ科）日本、旧北区、東洋区、豪州区
pine aphid			sylvestris aphid を見よ
pine aphids			conifer aphids を見よ　豪州での英名
pine aphids			jumping plantlice を見よ
pine bagmoth			persimmon bagworm を見よ
pine bark adelges			pine bark adelgid を見よ
pine bark adelgid		*Pineus strobi* (Hartig)	（カメムシ目、カサアブラムシ科）全北区
pine bark anobiid	マツザイシバンムシ	*Ernobius mollis* (Fabricius)	（コウチュウ目、シバンムシ科）日本、全北区、豪州区、エチオピア区。乾材害虫
pine bark aphid			pine bark adelgid を見よ　北米での英名
pine bark beetle	マツノホソスジキクイムシ	*Hylastes parallelus* Chapuis	（コウチュウ目、キクイムシ科）日本、東洋区。マツ害虫
pine bark beetle	マツノヒロスジキクイムシ	*Hylastes plumbeus* Blandford	（コウチュウ目、キクイムシ科）日本、旧北区、東洋区。マツ害虫
pine bark beetle (1)	キイロコキクイムシ	*Cryphalus fulvus* Niijima	（コウチュウ目、キクイムシ科）日本、旧北区。マツ害虫
pine bark beetle (2)	マツノスジキクイムシ	*Hylurgops interstitialis* (Chapuis)	（コウチュウ目、キクイムシ科）日本、旧北区、豪州区。マツ害虫
pine bark beetle			black pine bark beetle を見よ
pine bark beetle			pine shoot beetle (2) を見よ
pine bark beetles			pine engraver beetles を見よ

英名	和名	学名	所属、分布、ほか
pine bark engraver	マツカワノキクイムシ	*Orthotomicus proximus* (Eichhoff)	（コウチュウ目、キクイムシ科）日本、旧北区
pine bark scale			pine scale (1) を見よ
pine bark weevil		*Aesiotes notabilis* Pascoe	（コウチュウ目、ゾウムシ科）豪州区
pine Barrens bluet		*Enallagma recurvatum* Davis	（トンボ目、イトトンボ科）新北区
pine beau			pine beauty を見よ
pine beauty		*Panolis flammea* (Denis et Schiffermüller)	（チョウ目、ヤガ科）旧北区。日本亜種 *P. f. japonica* Droudt マツキリガ は pine cutworm という
pine beauty moth			pine beauty を見よ
pine beetle			pine shoot beetle (1) を見よ
pine big weevil			pine weevil (2) を見よ
pine black		*Zabrachia tenella* (Jaennicke)	（ハエ目、ミズアブ科）旧北区
pine borer		*Buprestis adjecta* LeConte	（コウチュウ目、タマムシ科）新北区
pine borer			horntail (1) を見よ
pine bud gall midge	マツシントメタマバエ	*Contarinia matsusintome* Haraguti et Monzen	（ハエ目、タマバエ科）日本。マツ害虫
pine bud moth			small black-specked groundling を見よ
pine bud moth			cone pitch moth を見よ
pine bud tortricid			cone pitch moth を見よ
pine budgall midge		*Contarinia coloradensis* Felt	（ハエ目、タマバエ科）新北区
pine butterfly	マツノキシロチョウ	*Neophasia menapia* (C. et R. Felder)	（チョウ目、シロチョウ科）新北区
pine candle moth		*Exeteleia nepheos* Freeman	（チョウ目、キバガ科）新北区
pine carpenterworm moth		*Givira lotta* Barnes et McDunnough	（チョウ目、ボクトウガ科）新北区
pine carpet		*Thera firmata* (Hübner)	（チョウ目、シャクガ科）旧北区
pine caterpillar		*Dendrolimus punctatus* (Walker)	（チョウ目、カレハガ科）旧北区、東洋区
pine caterpillar			pine moth を見よ
pine caterpillar moth			pine caterpillar を見よ
pine chafer		*Anomala oblivia* Horn	（コウチュウ目、コガネムシ科）新北区
pine chafer			Fuller を見よ
pine chermid		*Pineus bornei* (Annand)	（カメムシ目、カサアブラムシ科）旧北区、豪州区
pine colaspis		*Colaspis pini* Barber	（コウチュウ目、ハムシ科）新北区
pine-cone bug	マツヒラタナガカメムシ	*Gastrodes grossipes japonicus* (Stål)	（カメムシ目、ナガカメムシ科）旧北区。イネ害虫
pine-cone moth			pine knot-horn を見よ
pine cone moth		*Epinotia nigricana* (Herrich-Schäffer)	（チョウ目、ハマキガ科）旧北区
pine cone weevil		*Pissodes validirostris* Gyllenhal	（コウチュウ目、ゾウムシ科）旧北区

英名	和名	学名	所属、分布、ほか
pine conelet bug		*Platylyus luridus* (Reuter)	（カメムシ目、カスミカメムシ科）新北区
pine conelet looper		*Nepytia semiclusaria* (Walker)	（チョウ目、シャクガ科）新北区。成虫は pine conelet looper moth
pine cosmet		*Batrachedra pinicolella* (Zeller)	（チョウ目、カザリバガ科）旧北区
pine crescent		*Phyciodes sitalces* Godman et Salvin	（チョウ目、タテハチョウ科）新北区
pine devil moth		*Citheronia sepulcralis* (Grote et Robinson)	（チョウ目、ヤママユガ科）新北区。幼虫はマツにつく。pine-devil moth ともいう
pine elfin		*Incisalia niphon clarki* Freeman	（チョウ目、シジミチョウ科）新北区。eastern pine elfin を参照
pine emperor		*Nudaurelia cytherea* (Fabricius)	（チョウ目、ヤママユガ科）エチオピア区
pine emperor moth			pine emperor を見よ
pine engrave			pine engraver (1) を見よ
pine engraver	ホンスンキクイムシ	*Orthotomicus suturalis* (Gyllenhal)	（コウチュウ目、キクイムシ科）日本、旧北区
pine engraver		*Ips oregoni* Swaine	（コウチュウ目、キクイムシ科）新北区
pine engraver	マツスジキクイムシ	*Hylurgus parallerus* (Chapuis)	（コウチュウ目、キクイムシ科）日本、旧北区、東洋区
pine engraver (1)		*Ips pini* (Say)	（コウチュウ目、キクイムシ科）新北区
pine engraver beetles		*Ips*	（コウチュウ目、キクイムシ科）の昆虫の総称
pine engravers			pine engraver beetles を見よ
pine eucosmid			pine tip moth (1) を見よ
pine false looper moth		*Zale duplicata* (Bethune)	（チョウ目、ヤガ科）新北区。pine false looper Zale, pine false looper ともいう
pine false webworm		*Acantholyda erythrocephala* (Linnaeus)	（ハチ目、ヒラタハバチ科）新北区。pine false web-worm とも記す
pine flatbug		*Aradus cinnamomeus* (Panzer)	（カメムシ目、ヒラタカメムシ科）旧北区
pine flattened sawfly	マツヒラタハバチ	*Cephalcia nigricoxae* (Matsumura)	（ハチ目、ヒラタハバチ科）日本、旧北区
pine-flower snout beetles		Nemonychidae	（コウチュウ目）の昆虫の総称
pine froghopper	コミヤマアワフキ	*Peureptyelus indentatus* (Uhler)	（カメムシ目、アワフキムシ科）日本
pine froghopper (1)	マツアワフキ	*Aphrophora flavipes* Uhler	（カメムシ目、アワフキムシ科）日本、旧北区。マツ害虫
pine gall weevil		*Podapion gallicola* Riley	（コウチュウ目、ゾウムシ科）新北区
pine geometrid moth			bordered white beauty を見よ
pine giant mealybug	マツワラジカイガラムシ	*Drosicha pinicola* (Kuwana)	（カメムシ目、ワタフキカイガラムシ科）日本、旧北区。マツ害虫
pine green sawfly	マツノミドリハバチ	*Nesodiprion japonicus* (Marlatt)	（ハチ目、マツハバチ科）日本、全北区。マツ害虫
pine hawk		*Hyloicus pinastri* Linnaeus	（チョウ目、スズメガ科）旧北区
pine hawk moth	クロスズメ	*Hyloicus caligineus* Butler	（チョウ目、スズメガ科）日本、旧北区。マツ害虫
pine hawk moth			pine hawk を見よ

英名	和名	学名	所属、分布、ほか
pine hook-tipped twist		*Archips operana* Wilkes	(チョウ目、ハマキガ科) 旧北区
pine hook-tipped twist		*Archips piceana* Linnaeus	(チョウ目、ハマキガ科) 旧北区
pine horntail	ヒゲジロキバチ	*Urocerus antennatus* (Marlatt)	(ハチ目、キバチ科) 日本、旧北区
pine horntails			steel-blue wood wasps を見よ
pine hornworm	マツクロスズメ	*Hyloicus morio morio* Rothschild et Jordan	(チョウ目、スズメガ科) 日本
pine ips	マツノムツバキクイムシ	*Ips acuminatus* (Gyllenhal)	(コウチュウ目、キクイムシ科) 日本、旧北区、東洋区。マツ害虫
pine jewel beetle			metallic beetle を見よ
pine katydid		*Hubbellia marginifera* (Walker)	(バッタ目、キリギリス科) 新北区
pine key conehead		*Belocephalus micanopy* Vavis	(バッタ目、キリギリス科) 新北区
pine knot-horn	マツマダラメイガ	*Dioryctria abietella* (Denis et Schiffermüller)	(チョウ目、メイガ科) 日本、旧北区。マツ害虫
pine ladybird		*Exochomus quadripustulatus* (Linnaeus)	(コウチュウ目、テントウムシ科) 旧北区
pine lappet			pine-tree lappet を見よ
pine lappet moth			pine-tree lappet を見よ
pine lasiocampid			pine-tree lappet を見よ
pine leaf adelgid		*Pineus pinifoliae* (Fitch)	(カメムシ目、カサアブラムシ科) 新北区
pine leaf gall midge			pine needle gall midge を見よ
pine leaf miner			pine needle miner (1) を見よ
pine leaf roller			pine tortricid を見よ
pine leaf scale			pine scale (2) を見よ
pine leaf scale			pine needle scale を見よ
pine leafhopper	マツヒメヨコバイ	*Empoasca abietis* (Matsumura)	(カメムシ目、オオヨコバイ科) 日本、旧北区。マツ害虫
pine leaf-mining moth		*Clavigesta purdeyi* (Durrant)	(チョウ目、ハマキガ科) 旧北区
pine long proboscis aphid	マツクチナガオオアブラムシ	*Stomaphis pini* Takahashi	(カメムシ目、アブラムシ科) 日本
pine looper			bordered white beauty を見よ　幼虫の英名
pine looper moth			bordered white beauty を見よ
pine loopers		*Chlenias*	(チョウ目、シャクガ科) の昆虫の総称
pine margarodid scale			pine giant mealybug を見よ
pine mealybug	マツコナカイガラムシ	*Crisicoccus pini* (Kuwana)	(カメムシ目、コナカイガラムシ科) 日本、旧北区。マツ害虫
pine measuringworm moth		*Hypagyrtis piniata* (Packard)	(チョウ目、シャクガ科) 新北区
pine moth	マツカレハ	*Dendrolimus spectabilis* (Butler)	(チョウ目、カレハガ科) 日本、旧北区。マツ害虫

英名	和名	学名	所属、分布、ほか
pine moth			bordered white beauty を見よ
pine moth			pine-tree lappet を見よ
pine needle gall midge	マツバノタマバエ	*Thecodiplosis japonensis* Uchida et Inouye	(ハエ目、タマバエ科) 日本、旧北区。マツ害虫
pine needle miner	アカマツハモグリスガ	*Ocnerostoma friesei* Svensson	(チョウ目、スガ科) 日本、旧北区。マツ害虫
pine needle miner (1)		*Exoteleia pinifoliella* (Chambers)	(チョウ目、キバガ科) 新北区。pine needle miner moth ともいう
pine needle miner			mute scots-fir argent を見よ
pine needle scale		*Phenacaspis pinifoliae* (Fitch)	(カメムシ目、マルカイガラムシ科) 新北区
pine needle scale			pine scale (2) を見よ
pine needle sheath miner		*Zelleria haimbachi* Busck	(チョウ目、スガ科) 新北区。成虫は pine needle sheathminer moth
pine noctuid			pine beauty を見よ
pine noctuid moth			pine beauty を見よ
pine phycita			pine salebria moth を見よ
pine procession moth			pine processionary moth を見よ
pine processionary		*Thaumetopoea pinivora* Treitschke	(チョウ目、シャチホコガ科) 旧北区
pine processionary caterpillar			pine processionary moth を見よ
pine processionary moth	マツノギョウレツケムシガ	*Thaumetopoea pityocampa* (Denis et Schiffermüller)	(チョウ目、シャチホコガ科) 旧北区。pine processionary ともいう
pine reproduction weevil		*Cylindrocopturus eatoni* Buchanan	(コウチュウ目、ゾウムシ科) 新北区
pine resin-gall moth		*Retinia resinella* (Linnaeus)	(チョウ目、ハマキガ科) 旧北区
pine resin-gall tortrix			pine resin-gall moth を見よ
pine root aphid		*Stagona pini* (Burmeister)	(カメムシ目、アブラムシ科) 旧北区
pine root bark beetle	マツノネキクイムシ	*Hylurgus ligniperda* (Fabricius)	(コウチュウ目、キクイムシ科) 日本、旧北区、豪州区。マツ害虫
pine root collar weevil		*Hylobius radicis* Buchanan	(コウチュウ目、ゾウムシ科) 新北区
pine root scolytid			pine bark beetle (2) を見よ
pine root tip weevil		*Hylobius assimilis* Boheman	(コウチュウ目、ゾウムシ科) 新北区
pine rugose bark beetle			pine bark beetle (2) を見よ
pine salebria moth	マツアカマダラメイガ	*Dioryctria preyeri* Ragonot	(チョウ目、メイガ科) 日本、旧北区、東洋区。マツ害虫
pine satyr		*Paramacera allyni* Miller	(チョウ目、タテハチョウ科) 新北区
pine sawflies		*Diprion*	(ハチ目、マツハバチ科) の昆虫の総称
pine sawflies			conifer sawflies を見よ
pine sawfly (1)	マツノクロホシハバチ	*Diprion nipponicus* Rohwer	(ハチ目、マツハバチ科) 日本。カラマツ、マツ害虫
pine sawfly		*Diprion pini* Linnaeus	(ハチ目、マツハバチ科) 旧北区
pine sawfly			fox colored sawfly を見よ

英名	和名	学名	所属、分布、ほか
pine sawfly			pine green sawfly を見よ
pine sawyer			Japanese pine sawyer を見よ
pine sawyer beetle		*Ergates spiculatus* (LeConte)	(コウチュウ目、カミキリムシ科) 新北区
pine sawyer beetle			Japanese pine sawyer を見よ
pine scale		*Chionaspis heterophyllae* Cooley	(カメムシ目、マルカイガラムシ科) 新北区
pine scale (1)	マツモグリカイガラムシ	*Matsucoccus matsumurae* (Kuwana)	(カメムシ目、ワタフキカイガラムシ科) 日本、全北区。マツ害虫
pine scale (2)	マツカキカイガラムシ	*Lepidosaphes pini* (Maskell)	(カメムシ目、マルカイガラムシ科) 日本、旧北区、東洋区、大洋区。マツ害虫
pine shoot		*Clavigesta sylvestrana* (Curtis)	(チョウ目、ハマキガ科) 旧北区
pine shoot beetle (1)	マツノコキクイムシ	*Tomicus minor* (Hartig)	(コウチュウ目、キクイムシ科) 日本、旧北区、東洋区。マツ害虫
pine shoot beetle (2)	マツノキクイムシ	*Tomicus piniperda* (Linnaeus)	(コウチュウ目、キクイムシ科) 日本、旧北区、東洋区。マツ害虫
pine shoot borer		*Eucosma gloriola* Heinrich	(チョウ目、ハマキガ科) 新北区
pine shoot borer			pine knot-horn を見よ
pine shoot moth	マツアカシンムシ	*Rhyacionia dativa* Heinrich	(チョウ目、ハマキガ科) 日本、東洋区
pine shoot moth (1)		*Rhyacionia buoliana* (Denis et Schiffermüller)	(チョウ目、ハマキガ科) 全北区
pine six-dentate bark beetle			pine ips を見よ
pine sphinx			Abbott's pine sphinx を見よ
pine spittlebug		*Aphrophora parallela* (Say)	(カメムシ目、アワフキムシ科) 新北区
pine spittlebug			pine froghopper (1) を見よ
pine-sprout tortrix			pine shoot moth (1) を見よ
pine stump weevil		*Mitrastethus australiae* Lea	(コウチュウ目、ゾウムシ科) 豪州区
pine thrips		*Taeniothrips pini* Uzel	(アザミウマ目、アザミウマ科) 旧北区
pine thrips			greenhouse thrips を見よ 南アフリカでの英名
pine tip moth	マツノシンマダラメイガ	*Dioryctria sylvestrella* (Ratzeburg)	(チョウ目、メイガ科) 日本、旧北区。マツ害虫
pine tip moth		*Xestia adela* Franclemont	(チョウ目、ヤガ科) 新北区、新熱帯区
pine tip moth (1)	マツズアカシンムシガ	*Retinia cristata* (Walsingham)	(チョウ目、ハマキガ科) 日本、旧北区。マツ害虫
pine tip moth (2)		*Rhyacionia adana* Heinrich	(チョウ目、ハマキガ科) 新北区
pine tip moth (3)		*Rhyacionia busnella* Busck	(チョウ目、ハマキガ科) 新北区
pine tip moth (4)		*Rhyacionia subtropica* Miller	(チョウ目、ハマキガ科) 新北区、新熱帯区
pine tip moth			Nantucket pine tip moth を見よ
pine tortoise scale		*Toumeyella parvicornis* (Cockerell)	(カメムシ目、カタカイガラムシ科) 新北区

英名	和名	学名	所属、分布、ほか
pine tortricid	マツアトキハマキ	*Archips oporanus* (Linnaeus)	(チョウ目、ハマキガ科) 日本。マツ、スギ、モミ類などの害虫
pine-tree beetle			six-toothed bark beetle (1) を見よ
pine tree cricket		*Oecanthus pini* Beutenmüller	(バッタ目、カンタン科) 新北区
pine tree emperor moth			pine emperor を見よ
pine tree katydids			crickets (3) を見よ
pine-tree lappet		*Dendrolimus pini* (Linnaeus)	(チョウ目、カレハガ科) 旧北区
pine tree lappet moth			pine moth を見よ
pine tree sphinx		*Lapara bombycoides* Walker	(チョウ目、スズメガ科) 新北区
pine tree sphinx			pine hawk を見よ
pine tree thrips			greenhouse thrips を見よ　南アフリカでの英名
pine tube moth		*Argyrotaenia pinatubana* (Kearfott)	(チョウ目、ハマキガ科) 新北区
pine tube moth (1)		*Argyrotaenia velutinana* (Walker)	(チョウ目、ハマキガ科) 新北区。リンゴ他果樹害虫
pine tussock moth		*Dasychira pinicola* (Dyar)	(チョウ目、ドクガ科) 新北区。マツ害虫
pine tussock moth		*Dasychira leucophaea* J. E. Smith	(チョウ目、ドクガ科) 新北区
pine tussock moth (1)		*Dasychira plagiata* Walker	(チョウ目、ドクガ科) 新北区
pine twig moth	マツトビヒメハマキ	*Gravitarmata margarotana* (Heinemann)	(チョウ目、ハマキガ科) 日本、旧北区。マツ、モミ、トドマツなどの害虫
pine twig moths		*Rhyacionia*	(チョウ目、ハマキガ科) の昆虫の総称
pine warajicoccus			pine giant mealybug を見よ
pine web sawfly			pine flattened sawfly を見よ
pine web-spinning sawflies		*Acantholyda*	(ハチ目、ヒラタハバチ科) の昆虫の総称
pine web-spinning sawfly			black-tipped sawfly を見よ
pine webworm		*Pococera robustella* (Zeller)	(チョウ目、メイガ科) 新北区。幼虫はマツ害虫。成虫は　pine webworm moth
pine weevil (1)		*Hylobius abietis* (Linnaeus)	(コウチュウ目、ゾウムシ科) 旧北区
pine weevil (2)	マツアナアキゾウムシ	*Callirus haroldi* (Faust)	(コウチュウ目、ゾウムシ科) 日本、旧北区。マツ害虫
pine weevils		*Pissodes*	(コウチュウ目、ゾウムシ科) の昆虫の総称
pine weevils	アナアキゾウムシ亜科	Hylobiinae	(コウチュウ目、ゾウムシ科) の昆虫の総称
pine white			pine butterfly を見よ
pine white-spotted weevil	マツシラホシゾウムシ	*Shirahoshizo insidiosus* (Roelofs)	(コウチュウ目、ゾウムシ科) 日本
pine witchetygrub		*Cacodacnus planicollis* (Blackburn)	(コウチュウ目、カミキリムシ科) 豪州区
pine wood wasp			horntail (1) を見よ

英名	和名	学名	所属、分布、ほか
pine wood weevil	マツコブキクイゾウムシ	*Xenomineites destructor* Wollaston	(コウチュウ目、ゾウムシ科) 日本
pine yellow-spotted weevil			yellow-spotted pine weevil を見よ
pineapple beetle			yellow-shouldered souring beetle を見よ
pineapple gall midges		*Rhabdophaga*	(ハエ目、タマバエ科) の昆虫の総称
pineapple long scale	クロイトカイガラムシ	*Gymnaspis aechmeae* Newstead	(カメムシ目、マルカイガラムシ科) 旧北区、大洋区
pineapple mealybug	パイナップルコナカイガラムシ	*Dysmicoccus brevipes* (Cockerell)	(カメムシ目、コナカイガラムシ科) 日本、東洋区、新北区、汎熱帯。パイナップル、マンゴー害虫
pineapple mealybug			pineapple scale を見よ
pineapple sap beetle			yellow-shouldered souring beetle を見よ
pineapple scale	アナナスシロカイガラムシ	*Diaspis bromeliae* (Kerner)	(カメムシ目、マルカイガラムシ科) 日本、東洋区、新北区、汎熱帯。パイナップル害虫
pineapple scurf scale			pineapple scale を見よ　南アフリカでの英名
pineapple weevil		*Metamasius ritchiei* Marshall	(コウチュウ目、ゾウムシ科) 新北区
pinebark borers			(コウチュウ目、キクイムシ科) の昆虫の総称
pineine plantlice		Pineinae	(カメムシ目、カサアブラムシ科) の昆虫の総称
Pinhey's whisp		*Agriocnemis pinheyi* Balinsky	(トンボ目、イトトンボ科) エチオピア区
pinion-spotted pug		*Eupithecia insigniata* (Hübner)	(チョウ目、シャクガ科) 旧北区。日本亜種は *E. i. insignioides* Wehrli アミモンカバナミシャク。pinion spotted pug とも記す
pinion-streaked snout	クロスジヒメアツバ	*Schrankia costaestrigalis* (Stephens)	(チョウ目、ヤガ科) 日本、旧北区、エチオピア区、豪州区。エンドウ害虫
pinions		Xyleninae	(チョウ目、ヤガ科) の昆虫の総称
pink Acraea	ホシボシホソチョウ	*Acraea caecilia* (Fabricius)	(チョウ目、タテハチョウ科) エチオピア区
pink and green potato aphid			potato aphid (1) を見よ
pink and green tomato and rose aphis		*Macrosiphum gei* (Koch)	(カメムシ目、アブラムシ科) 旧北区
pink and green tomato aphid			potato aphid (1) を見よ
pink-banded sister		*Adelpha lycorias* (Godart)	(チョウ目、タテハチョウ科) 新熱帯区。*A. l. melanthe* (Bates) は rayed sister
pink-barred Lithacodia moth		*Pseudeustrotia carneola* (Guenée)	(チョウ目、ヤガ科) 新北区。pink-barred Pseudeustrotia ともいう
pink-barred sallow	キイロキリガ	*Xanthia togata* (Esper)	(チョウ目、ヤガ科) 日本、旧北区
pink bellied moth		*Oenochroma vinaria* (Guenée)	(チョウ目、シャクガ科) 豪州区
pink-bodied Altinote		*Altinote neleus* (Latreille)	(チョウ目、タテハチョウ科) 新熱帯区
pink bollworm	ワタアカミムシ	*Pectinophora gossypiella* (Saunders)	(チョウ目、キバガ科) 日本、汎世界。著名なワタ害虫
pink bollworm			tobacco budworm を見よ
pink bollworm moth			pink bollworm を見よ

英名	和名	学名	所属、分布、ほか
pink-bordered yellow moth		*Phytometra rhodarialis* Walker	（チョウ目、ヤガ科）新北区。pink-bordered yellow ともいう
pink borer (1)	イネヨトウ	*Sesamia inferens* (Walker)	（チョウ目、ヤガ科）日本、東洋区、豪州区。イネ害虫
pink borer (2)		*Chilo partellus* (Swinhoe)	（チョウ目、メイガ科）東洋区、エチオピア区。イネ害虫
pink borer			durra stalk borer を見よ　北アフリカでの英名
pink borer			durra stem borer を見よ
pink-checked cattleheart			cattle heart を見よ
pink cornworm	トウモロコシトガリホソガ	*Pyroderces rileyi* (Walsingham)	（チョウ目、カザリバガ科）日本、全北区、豪州区、新熱帯区、大洋区。貯穀害虫。pink bud moth, pink scavenger ともいう
pink cutworm			brown cutworm (1) を見よ
pink-edged conch		*Falseuncaria degreyana* (McLachlan)	（チョウ目、ホソハマキガ科）旧北区
pink-edged sulphur	オンタリオモンキチョウ	*Colias interior interior* Scudder	（チョウ目、シロチョウ科）新北区
pink-fringed Dolichomia moth		*Dolichomia binodulalis* Zeller	（チョウ目、メイガ科）新北区
pink glowworm		*Microphotus angustus* LeConte	（コウチュウ目、ホタル科）新北区
pink grass-yellow			Macleay's grass yellow を見よ
pink ground pearl		*Eumargarodes laingi* Jakubski	（カメムシ目、ワタフキカイガラムシ科）豪州区
pink gum-lerp		*Candiaspina densitexta* Taylor	（カメムシ目、キジラミ科）豪州区
pink hawk moth			elephant hawk moth を見よ　米国での英名
pink hibiscus mealybug			hibiscus mealybug (1) を見よ
pink-legged tiger moth		*Spilosoma latipennis* Stretch	（チョウ目、ヒトリガ科）新北区
pink margined green		*Nemoria leptalea* Ferguson	（チョウ目、シャクガ科）新北区
pink-masked pyralid moth		*Aglossa disciferalis* (Dyar)	（チョウ目、メイガ科）新北区
pink mealybug			grey sugarcane mealybug を見よ
pink mealybug			pink sugarcane mealybug を見よ
pink mealybug of sugarcane			pink sugarcane mealybug を見よ
pink orchid mantis			flower mantis を見よ
pink-patched looper moth		*Eosphoropteryx thyatyroides* (Guenée)	（チョウ目、ヤガ科）新北区。pink-tinted beauty ともいう
pink pineapple mealybug			pineapple mealybug を見よ
pink polyptychus		*Rufoclanis fulgurans* (Rothschild et Jordan)	（チョウ目、スズメガ科）エチオピア区
pink porina moth		*Wiseana signata* Walker	（チョウ目、コウモリガ科）豪州区
pink prominent moth		*Hyparpax aurora* (Smith)	（チョウ目、シャチホコガ科）新北区

英名	和名	学名	所属、分布、ほか
pink scavenger			pink cornworm を見よ
pink scavenger caterpillar			pink cornworm を見よ　成虫は pink scavenger caterpillar moth
pink-shaded fern moth		*Callopistria mollissima* (Guenée)	（チョウ目、ヤガ科）新北区
pink shot blue		*Lemonia dumi* (Linnaeus)	（チョウ目、Lemoniidae）エチオピア区
pink skimmer		*Orthetrum pruinosum neglectum* (Rambur)	（トンボ目、トンボ科）東洋区
pink-spotted bollworm		*Pectinophora scutigera* (Holdaway)	（チョウ目、キバガ科）豪州区
pink-spotted cattleheart	フタエヘリボシジャコウアゲハ	*Parides photinus* (Doubleday)	（チョウ目、アゲハチョウ科）新北区
pink-spotted dart moth		*Pseudohermonassa bicarnea* (Guenée)	（チョウ目、ヤガ科）新北区。pink-spotted dart ともいう
pink spotted hawk-moth		*Agrius cingulatus* (Fabricius)	（チョウ目、スズメガ科）全北区。幼虫はサツマイモ害虫、成虫は sweet potato sphinx ともいう
pink spotted swallowtail (1)		*Papilio pharnaces* (Doubleday)	（チョウ目、アゲハチョウ科）新北区
pink-spotted swallowtail		*Papilio rogeri* Boisduval	（チョウ目、アゲハチョウ科）新北区。新熱帯区の *P. r. pharnaces* Doubleday も同英名
pink-spotted trig		*Anaxipha* sp.	（バッタ目、クサヒバリ科）新北区
pink stalk borer			African pink borer を見よ
pink stem borer			pink borer (1) を見よ
pink-striped oakworm moth		*Anisota virginiensis* (Drury)	（チョウ目、ヤママユガ科）新北区。幼虫は pink-striped oakworm
pink-striped wave		*Pellonia vibicaria* Clerck	（チョウ目、シャクガ科）旧北区
pink sugarcane mealybug	サトウキビコナカイガラムシ	*Saccharicoccus sacchari* (Cockerell)	（カメムシ目、コナカイガラムシ科）日本、東洋区、新北区、汎熱帯。サトウキビ、イネ害虫
pink-tipped yellow moth		*Gymnobathra flavidella* Meyrick	（チョウ目、マルハキバガ科）豪州区
pink underwing		*Catocala concumbena* Walker	（チョウ目、ヤガ科）新北区。幼虫はヤナギとポプラを食う。pink underwing moth ともいう
pink underwing			ornate Bella moth を見よ
pink-washed leafroller moth		*Metendothenia separatana* (Kearfott)	（チョウ目、ハマキガ科）新北区
pink-washed looper moth		*Enigmogramma basigera* (Walker)	（チョウ目、ヤガ科）新北区、新熱帯区。pink-washed looper ともいう
pink wax scale			red wax scale を見よ
pink wax scale parasite		*Aenasoidea varia* Girault	（ハチ目、トビコバチ科）豪州区
pink wax scale parasite	ルビーアカヤドリコバチ	*Anicetus beneficus* Ishii et Yasumatsu	（ハチ目、トビコバチ科）日本、旧北区、東洋区。ルビーロウムシの寄生蜂として著名
pink waxy scale			red wax scale を見よ
pink webspinner			webspinner を見よ
pink-winged grasshopper		*Atractomorpha sinensis* Bolivar	（バッタ目、オンブバッタ科）新北区、大洋区
pink-winged phasma		*Podacanthus typhon* Gray	（ナナフシ目、ナナフシ科）豪州区

英名	和名	学名	所属、分布、ほか
pinkish forester		*Bebearia eliensis* (Hewitson)	(チョウ目、タテハチョウ科) エチオピア区
pinnule scale		*Chrysomphalus pinnulifer* (Maskell)	(カメムシ目、マルカイガラムシ科) 旧北区
pinon cone beetle		*Conophthorus edulis* Hopkins	(コウチュウ目、キクイムシ科) 新北区
pint-spotted hawkmoth		*Herse cingulata* (Fabricius)	(チョウ目、スズメガ科) 全北区、新熱帯区
pintail		*Acisoma panorpoides* Rambur	(トンボ目、トンボ科) 東洋区、エチオピア区
Pintiati shieldback		*Eremopedes pintiati* Rentz	(バッタ目、キリギリス科) 新北区
pinworms			ship-timber beetles を見よ
pinyon cone moth		*Eucosma bobana* Kearfott	(チョウ目、ハマキガ科) 新北区
pinyon pine engraver		*Ips confusus* (LeConte)	(コウチュウ目、キクイムシ科) 新北区
pinyon pine scale		*Matsucoccus acalyptus* Herbert	(カメムシ目、ワタフキカイガラムシ科) 新北区
pinyon spindlegall midge		*Pinyonia edulicola* Gagné	(ハエ目、タマバエ科) 新北区
pioneer			brown-veined white を見よ
pioneers		*Anaphaeis*	(チョウ目、シロチョウ科) の昆虫の総称
piophilids			skipper flies を見よ
pipeline swallowtail			pipevine swallowtail を見よ
pipe-organ wasp		*Trypoxylon (Trypargilum) politum* (Say)	(ハチ目、アナバチ科) 新北区
pipers	ミスジモドキ属	*Eurytela*	(チョウ目、タテハチョウ科) の昆虫の総称
pipevine swallowtail	アオジャコウアゲハ	*Battus philenor* (Linnaeus)	(チョウ目、アゲハチョウ科) 東洋区
pipevine swallowtails	アオジャコウ属	*Battus*	(チョウ目、アゲハチョウ科) の昆虫の総称 *Papilio* 属の幼虫にも使われる
pirate		*Catacroptera cloanthe* (Stoll)	(チョウ目、タテハチョウ科) エチオピア区。pirate butterfly ともいう
pirate ant		*Cardiocondyla pirata* Seifert et Frohschammer	(ハチ目、アリ科) 東洋区
pirate bugs			ハナカメムシ科のように刺して食う虫
pirate looper moth			Epigaea looper moth を見よ
pirates	オオタテハモドキ属	*Catacroptera*	(チョウ目、タテハチョウ科) の昆虫の総称
Pirus skipperling			russet skipper を見よ
Pisis groundstreak		*Calycopis pisis* (Godman et Salvin)	(チョウ目、シジミチョウ科) 新熱帯区
Pisonis mimic			brown-bordered white を見よ
piss ant			fire ant (1) を見よ
pissmires			アリの一部の階級に対する英名
pistachio emerald moth		*Hethemia pistasciaria* (Guenée)	(チョウ目、シャクガ科) 新北区
pistachio thrips		*Liothrips pistaciae* Kreutzberg	(アザミウマ目、クダアザミウマ科) 旧北区

英名	和名	学名	所属、分布、ほか
pistachio-seed chalcid		*Megastigmus pistaciae* Walker	（ハチ目、オナガコバチ科）旧北区
pistol case bearer (1)		*Coleophora multipulvella* Chambers	（チョウ目、ツツミノガ科）新北区
pistol case bearer (2)		*Coleophora atromarginata* Braun	（チョウ目、ツツミノガ科）新北区
pistol case bearer			Stainton's catchfly case を見よ
pistol casebearer			apple pistol casebearer (1) を見よ
pistol case-bearer			goose-feather case（1）を見よ
pit-making oak scale			golden oak scale (1) を見よ
pit scale insects			pit scales を見よ
pit scales	フサカイガラムシ科	Asterolecanidae	（カメムシ目）の昆虫の総称 世界に200種
pitch-eating weevil		*Pachylobius picivorus* Germar	（コウチュウ目、ゾウムシ科）新北区
pitch mass borer		*Synanthedon pini* (Kellicott)	（チョウ目、スカシバガ科）新北区。pitch mass borer moth ともいう
pitch moth		*Blastodacna putripennella* Zeller	（チョウ目、カザリバガ科）全北区
pitch moth			sequoia pitch moth を見よ
pitch-pine looper			eastern pine looper を見よ
pitch pine tip moth		*Rhyacionia rigidana* (Fernald)	（チョウ目、ハマキガ科）新北区
pitch twig moth		*Petrova comstockiana* (Fernald)	（チョウ目、ハマキガ科）新北区
pitcher-plant borer moth		*Papaipema appassionata* (Harvey)	（チョウ目、ヤガ科）新北区
pitcher-plant mosquito		*Wyeomyia (Wyeomyia) smithii* (Coquillett)	（ハエ目、カ科）新北区
pitcher plant moth		*Exyra rolandiana* Grote	（チョウ目、ヤガ科）新北区。幼虫はウツボカズラ類を食う
pith moth			hawthorn cosmet を見よ
pith moth			apple black cosmet を見よ
pithworms			click beetles (1) を見よ
Pithy sister		*Adelpha pithys* (Bates)	（チョウ目、タテハチョウ科）新熱帯区
Pitman's ciliate blue		*Anthene pitmani* Stempffer	（チョウ目、シジミチョウ科）エチオピア区
pitted apple beetle		*Geloptera porosa* Lea	（コウチュウ目、ハムシ科）豪州区。カンキツ害虫
pittosporum beetle		*Lamprolina aeneipennis* (Boisduval)	（コウチュウ目、ハムシ科）豪州区
pittosporum bug		*Pseudapines geminata* (Van Duzee)	（カメムシ目、カメムシ科）豪州区
pittosporum leafminer		*Phytoliriomyza pittosporphylli* (Hering)	（ハエ目、ハモグリバエ科）豪州区
pittosporum longicorn		*Strongylurus thoracicus* (Pascoe)	（コウチュウ目、カミキリムシ科）豪州区
pittosporum looper moth			tarata looper moth を見よ

英名	和名	学名	所属、分布、ほか
pittosporum scale		*Asterolecanium arabidis* Signoret	（カメムシ目、フサカイガラムシ科）旧北区
Pixie		*Melanis pixe* (Boisduval)	（チョウ目、シジミタテハ科）新北区、新熱帯区
place bug			forest bug を見よ
Placenta tiger moth		*Grammia placentia* (Smith)	（チョウ目、ヒトリガ科）新北区
placid giant owl		*Caligo placidanus* Staudinger	（チョウ目、タテハチョウ科）新熱帯区
plague caterpillar			banana fruit caterpillar を見よ
plague flea			Oriental rat flea を見よ
plague locust			アフリカトビバッタのように時に大発生するバッタ。African migratory locust を参照
plague soldier beetle		*Chauliognathus lugubris* (Fabricius)	（コウチュウ目、ジョウカイボン科）豪州区
plague thrips			blossom thrips (1) を見よ
plain banded awl		*Hasora vitta* (Butler)	（チョウ目、セセリチョウ科）東洋区
plain bent-skipper		*Ebrietas elaudia livius* Mabille	（チョウ目、セセリチョウ科）新熱帯区
plain blackneck			blackneck を見よ
plain bushblue	ウラマダラムラサキシジミ	*Flos apidanus* (Cramer)	（チョウ目、シジミチョウ科）東洋区。*F. a. saturatus* Snellen も同英名
plain bushbrown		*Mycalesis malsarida* Butler	（チョウ目、タテハチョウ科）東洋区
plain clay		*Eugnorisma depuncta* (Linnaeus)	（チョウ目、ヤガ科）旧北区
plain earl		*Tanaecia jahnu* (Moore)	（チョウ目、タテハチョウ科）東洋区
plain-eyed brown horsefly		*Tabanus miki* Brauer	（ハエ目、アブ科）旧北区
plain-eyed grey horsefly		*Tabanus cordiger* Meigen	（ハエ目、アブ科）旧北区
plain-faced blueberry dart			blueberry dart moth を見よ
plain floodwater mosquito		*Ochlerotatus trivittatus* (Coquillett)	（ハエ目、カ科）新北区
plain golden Y			golden Y-moth を見よ
plain gold-fringed drill	コキオビヘリホシヒメハマキ	*Dichrorampha gueneeana* Obraztsov	（チョウ目、ハマキガ科）日本、旧北区
plain hedge blue	ホリシャルリシジミ	*Celastrina lavendularis himilcon* (Fruhstorfer)	（チョウ目、シジミチョウ科）日本、東洋区。*C. lavendularis* (Moore) の英名。*C. l. isabella* Corbet も同英名
plain lacewing		*Cethosia penthesilea methypsea* (Butler)	（チョウ目、タテハチョウ科）東洋区。orange lacewing を参照
plain longtail		*Urbanus simplicius* (Stoll)	（チョウ目、セセリチョウ科）新北区
plain nawab	オオオビモンフタオチョウ	*Polyura hebe plautus* Fruhstorfer	（チョウ目、タテハチョウ科）東洋区
plain orange awlet		*Burara anadi* de Nicéville	（チョウ目、セセリチョウ科）東洋区
plain orange tip		*Colotis eucharis* (Fabricius)	（チョウ目、シロチョウ科）東洋区
plain palm dart		*Cephrenes chrysozona* (Plötz)	（チョウ目、セセリチョウ科）豪州区

英名	和名	学名	所属、分布、ほか
plain plushblue		*Amblypodia apidanus* Doherty	(チョウ目、シジミチョウ科) 東洋区。
plain puffin	クモガタシロチョウ	*Appias indra* (Moore)	(チョウ目、シロチョウ科) 東洋区
plain pug		*Eupithecia simpliciata* (Haworth)	(チョウ目、シャクガ科) 旧北区
plain pug	ウスオビシロエダシャク	*Eupithecia subnotata* (Hübner)	(チョウ目、シャクガ科) 旧北区
plain pumpkin beetle		*Aulacophora abdominalis* (Fabricius)	(コウチュウ目、ハムシ科) 豪州区
plain purplewing	アオネアイイロタテハ	*Eunica sydonia caresa* (Hewitson)	(チョウ目、タテハチョウ科) 新熱帯区。Godart's purplewing を参照
plain ringlet		*Forsterinaria inornata* (C. et R. Felder)	(チョウ目、タテハチョウ科) 新熱帯区
plain ringlet			prairie ringlet を見よ
plain sailer		*Neptis cartica* Moore	(チョウ目、タテハチョウ科) 東洋区
plain satyr		*Cissia pompilia* (C. et R. Felder)	(チョウ目、タテハチョウ科) 新熱帯区
plain shortwing beetle		*Nathrius brevipennis* (Mulsant)	(コウチュウ目、カミキリムシ科) 旧北区
plain sulphur		*Dercas lycorias* (Doubleday)	(チョウ目、シロチョウ科) 旧北区、東洋区
plain tailless oakblue		*Arhopala asopia* (Hewitson)	(チョウ目、シジミチョウ科) 東洋区
plain three-ring		*Ypthima lycus* de Nicéville	(チョウ目、タテハチョウ科) 東洋区
plain tiger			African monarch を見よ
plain tufted lancer		*Isma iapis* (de Nicéville)	(チョウ目、セセリチョウ科) 東洋区
plain vagrant			Buquet's vagrant を見よ
plain wave		*Idaea straminata* (Borkhausen)	(チョウ目、シャクガ科) 旧北区
plain wave		*Sterrha inornata* (Haworth)	(チョウ目、シャクガ科) 旧北区
plain-winged Holomelina			immaculate Holomelina を見よ
plain yellow twist			timothy tortrix を見よ
plains blue royal		*Tajuria jehana* Moore	(チョウ目、シジミチョウ科) 東洋区
plains clubtail		*Gomphus externus* Hagen	(トンボ目、サナエトンボ科) 新北区
plains cupid	クロマダラソテツシジミ	*Chilades pandava* (Horsfield)	(チョウ目、シジミチョウ科) 東洋区
plains false wireworm			false wireworm (3) を見よ
plains forktail		*Ischnura damula* Calvert	(トンボ目、イトトンボ科) 新北区
plains gray skipper		*Polites rhesus* (Edwards)	(チョウ目、セセリチョウ科) 新北区
plains lubber grasshopper			lubber grasshopper を見よ
plains Schizura moth		*Schizura apicalis* (Grote et Robinson)	(チョウ目、シャチホコガ科) 新北区
plane			Australian plane を見よ　*B. p. yargama* Couchman も同英名。plane butterfly ともいう

英名	和名	学名	所属、分布、ほか
plane midget		*Lithocolletis platani* Staudinger	（チョウ目、ホソガ科）旧北区
plant ant		*Azteca ulei* Forel	（ハチ目、アリ科）新熱帯区
plant bug		*Helopeltis antonii* Waterhouse	（カメムシ目、カスミカメムシ科）。東洋区。カンキツ害虫
plant bug		*Helopeltis collaris* Stål	（カメムシ目、カスミカメムシ科）。東洋区。カンキツ害虫
plant bugs		*Closterocoris*	（カメムシ目、カスミカメムシ科）の昆虫の総称
plant bugs (1)	カスミカメムシ科	Miridae	（カメムシ目）の昆虫の総称　カメムシ目最大の科で世界に 10,000 種以上。カメムシ目にも同英名が使用される
plant bugs			lygus bugs を見よ
plant-eaters			horntails (1) を見よ
plant-eating beetles			phytophagous beetles を見よ
plant hoppers			lantern bugs を見よ
plant lice	腹吻群	Sternorrhyncha	（カメムシ目）の昆虫の総称
plant lice			aphids (1) を見よ
plant sucker		*Kikihia subalpina* (Hudson)	（カメムシ目、セミ科）豪州区
plantain frit			Glanville fritillary を見よ
plantain groundling		*Scrobipalpa samadensis plantaginella* (Stainton)	（チョウ目、キバガ科）旧北区
planthopper parasite moth		*Fulgoraecia exigua* (Hy. Edwards)	（チョウ目、イラガ科）新北区
planthopper parasite moths			planthopper parasites を見よ
planthopper parasites	セミヤドリガ科	Epipyropidae	（チョウ目）の昆虫の総称
planthoppers		*Metcalfa*	（カメムシ目、アオバハゴロモ科）の昆虫の総称
planthoppers	ハゴロモ科	Ricaniidae	（カメムシ目）の昆虫の総称
planthoppers (1)	ウンカ科	Delphacidae	（カメムシ目）の昆虫の総称　世界に 1,300 種
planthoppers			lantern bugs を見よ
planthoppers			lantern flies を見よ
plantlice			aphids (2) を見よ
plaster bee		*Colletes compactus* Cresson	（ハチ目、ムカシハナバチ科）新北区
plaster bees	ムカシハナバチ亜科	Colletinae	（ハチ目、ムカシハナバチ科）の昆虫の総称
plaster bees			yellow-faced bees (1) を見よ
plaster bees			obtuse-tongued bees を見よ
plaster beetle	クビレヒメマキムシ	*Cartodere constricta* (Gyllenhal)	（コウチュウ目、ヒメマキムシ科）日本、汎世界
plaster beetle		*Coninomus nodifer* Westwood	（コウチュウ目、ヒメマキムシ科）旧北区
plaster beetle			squarenosed fungus beetle を見よ
plaster beetles	ヒメマキムシ科	Lathridiidae	（コウチュウ目）の昆虫の総称
plasterer bee			plaster bee を見よ

英名	和名	学名	所属、分布、ほか
plasterer bees			obtuse-tongued bees を見よ
plataspid bugs	マルカメムシ科	Plataspidae	(カメムシ目) の昆虫の総称
plate-thigh beetles	マルハナノミダマシ科	Eucinetidae	(コウチュウ目) の昆虫の総称　3 mm 以下の微小虫
plateau dragonlet		*Erythrodiplax basifusca* (Calvert)	(トンボ目、トンボ科) 新北区、新熱帯区
plateau satyr		*Zischkaia pellonia* (Godman)	(チョウ目、タテハチョウ科) 新熱帯区
plateau skipperling		*Piruna cyclosticta* (Dyar)	(チョウ目、セセリチョウ科) 新熱帯区
plateau spreadwing		*Lestes alacer* Hagen	(トンボ目、アオイトトンボ科) 新北区
platycnemid damselflies	モノサシトンボ科	Platycnemidae	(トンボ目) の昆虫の総称
platygasterid parasites			platygasterid wasps を見よ
platygasterid wasps	ハラビロクロバチ科	Platygastridae	(ハチ目) の昆虫の総称
platygasterids			platygasterid wasps を見よ
platynota		*Platynota stultana* (Walsingham)	(チョウ目、ハマキガ科) 新北区、大洋区、新熱帯区。カンキツ害虫
platypodids			pinhole beetles を見よ
platypterus skipper		*Tosta platypterus* (Mabille)	(チョウ目、セセリチョウ科) 新熱帯区
platypus beetle			mountain pinhole borer を見よ
platystomatid flies			broadmouthed flies を見よ
playboys			Cornelians を見よ
playboys		*Pilodeudorix*	(チョウ目、シジミチョウ科) の昆虫の総称
pleasant dagger moth		*Acronicta laetifica* Smith	(チョウ目、ヤガ科) 新北区
pleasing fungus beetle		*Megalodacne heros* (Say)	(コウチュウ目、オオキノコムシ科) 新北区
pleasing fungus beetles		*Cypherotylus*	(コウチュウ目、オオキノコムシ科) の昆虫の総称
pleasing fungus beetles		*Ischyrus*	(コウチュウ目、オオキノコムシ科) の昆虫の総称
pleasing fungus beetles		*Triplax*	(コウチュウ目、オオキノコムシ科) の昆虫の総称
pleasing fungus beetles (1)	オオキノコムシ科	Erotylidae	(コウチュウ目) の昆虫の総称
pleasing lacewings	クシヒゲカゲロウ科	Dilaridae	(アミメカゲロウ目) の昆虫の総称　旧世界に少し分布。新世界には稀
pleasure sister			Hubner's sister を見よ
Plebeia skipper		*Hesperilla chrysotricha plebeia* Waterhouse	(チョウ目、セセリチョウ科) 豪州区
plebeian sphinx		*Paratrea plebeja* (Fabricius)	(チョウ目、スズメガ科) 新北区。plebian sphinx moth ともいう
plecopterans			stoneflies を見よ
pleid bugs			pygmy backswimmer bugs を見よ
pleid water bugs			pygmy backswimmer bugs を見よ

英名	和名	学名	所属、分布、ほか
pleioblastus gall midge	メダケタマバエ	*Geromyia nawai* (Monzen)	(ハエ目、タマバエ科) 日本。タケ害虫
Pleistocene fritillary		*Clossiana natazhati* (Gibson)	(チョウ目、タテハチョウ科) 新北区
pleocomids		Pleocomidae	(コウチュウ目) の昆虫の総称
plerergate	膨職蟻型		腹部が食物で満たされ球状となった職蟻
Ploetz's skipper		*Acleros ploetzi* Mabille	(チョウ目、セセリチョウ科) エチオピア区
Plotz sootywing		*Staphylus oeta* (Plötz)	(チョウ目、セセリチョウ科) 新熱帯区
plum and apple leafhopper			oak leafhopper (1) を見よ
plum and peach scale			globose scale を見よ
plum aphid			leaf-curl plum aphid を見よ
plum aphid			waterlily aphid を見よ
plum beetle		*Tetrops praeustus* (Linnaeus)	(コウチュウ目、カミキリムシ科) 旧北区
plum borer	ツツムネチョッキリ	*Involvulus cylindricollis* (Schilsky)	(コウチュウ目、オトシブミ科) 日本
plum borer (1)		*Rhynchites cupreus* (Linnaeus)	(コウチュウ目、オトシブミ科) 旧北区
plum bud moth			lesser cloaked marble を見よ
plum cambium midge			plum cambium miner を見よ
plum cambium miner		*Phytobia cerasiferae* Kangas	(ハエ目、ハモグリバエ科) 旧北区
plum cankerworm	ウメエダシャク	*Cystidia couaggaria* (Guenée)	(チョウ目、シャクガ科) 日本、旧北区。ウメ、モモ、リンゴ、サクラ害虫
plum curculio	スモモゾウムシ	*Conotrachelus nenuphar* (Herbst)	(コウチュウ目、ゾウムシ科) 新北区
plum cutworm	カタハリキリガ	*Lithophane rosinae* (Hufnagel)	(チョウ目、ヤガ科) 日本
plum fruit maggot			plum fruit moth を見よ
plum fruit moth		*Cydia funebrana* (Treitschke)	(チョウ目、ハマキガ科) 旧北区
plum fruit sawfly		*Hoplocampa minuta* (Christ)	(ハチ目、ハバチ科) 旧北区
plum gouger		*Coccotorus scutellaris* (LeConte)	(コウチュウ目、ゾウムシ科) 新北区
plum hairy caterpillar		*Euproctis fraterna* Moore	(チョウ目、ドクガ科) 東洋区
plum Judy	オキナワシジミタテハ	*Abisara echerius* (Stoll)	(チョウ目、シジミタテハ科) 旧北区、東洋区
plum leaf-curling aphid			oat bird-cherry aphid を見よ
plum leaf curling midge		*Dasineura tortrix* (Loew)	(ハエ目、タマバエ科) 旧北区
plum leaf gall midge		*Putoniella marsupialis* Loew	(ハエ目、タマバエ科) 旧北区
plum leaf roller			thicket twist を見よ

英名	和名	学名	所属、分布、ほか
plum leaf sawfly			gooseberry sawfly (2) を見よ
plum leaf sawfly			small gooseberry sawfly (1) を見よ
plum leaf worm			twin-spotted quaker を見よ
plum leafhopper		*Macropsis trimaculata* (Fitch)	(カメムシ目、オオヨコバイ科) 新北区
plum lecanium			globose scale を見よ
plum louse			mealy plum aphid を見よ
plum moth			plum fruit moth を見よ
plum piercer			plum fruit moth を見よ
plum plant-louse		*Myzus lythri* (Schrank)	(カメムシ目、アブラムシ科) 旧北区
plum pulvinaria			large cottony scale (1) を見よ
plum rhynchites			plum borer (1) を見よ
plum sawfly		*Hoplocampa flava* (Linnaeus)	(ハチ目、ハバチ科) 旧北区
plum sawfly			plum fruit sawfly を見よ
plum scale			olive scale を見よ
plum scales			parlatoria scales を見よ
plum slug caterpillar		*Parasa latistriga* Walker	(チョウ目、イラガ科) エチオピア区
plum sphinx			wild cherry sphinx を見よ
plum-stem piercer		*Magdalis armigera* (Fourcroy)	(コウチュウ目、ゾウムシ科) 旧北区
plum sucker		*Psylla pruni* Geoffroy	(カメムシ目、キジラミ科) 旧北区
plum-thistle aphid			thistle aphid を見よ
plum tortrix			lesser cloaked marble を見よ
plum tree borer			lesser peach tree borer を見よ
plum tree plant louse			oat bird-cherry aphid を見よ
plum tree underwing		*Catocala ultronia* Hübner	(チョウ目、ヤガ科) 新北区。幼虫はアンズ、リンゴ、サクラなどの害虫
plum web-spinning sawfly		*Neurotoma inconspicua* (Norton)	(ハチ目、ヒラタハバチ科) 新北区
plum weevil			red-legged weevil (1) を見よ
plum worm	ヒメトビネマダラメイガ	*Acrobasis rufilimbalis* (Wileman)	(チョウ目、メイガ科) 日本
Plumbago blue			zebra blue (1)(2) を見よ
plumbeous hairstreak		*Thecla chalybeia* (Leech)	(チョウ目、シジミチョウ科) 東洋区
plumbeous silverline		*Cigaritis schistacea* (Moore)	(チョウ目、シジミチョウ科) 東洋区
plume fruit moth		*Grapholita funebrana* (Treitschke)	(チョウ目、ハマキガ科) 旧北区。日本亜種は *G. f. cerasivora* Matsumura サクラヒメハマキ
plume moths	トリバガ科	Pterophoridae	(チョウ目) の昆虫の総称　世界に約 500 種
plumed bees			yellow-faced bees (1) を見よ
plumed fan-foot		*Pechipogo plumigeralis* (Hübner)	(チョウ目、ヤガ科) 旧北区

英名	和名	学名	所属、分布、ほか
plumed gnat	オオユスリカ	*Chironomus plumosus* (Linnaeus)	(ハエ目、ユスリカ科) 日本、旧北区
plumed gnats			midges を見よ
plumed prominent		*Ptilophora plumigera* (Denis et Schiffermüller)	(チョウ目、シャチホコガ科) 旧北区
plumed wasps			eulophid wasps を見よ
plumeria borer		*Lagocheirus undatus* (Voet)	(コウチュウ目、カミキリムシ科) 新北区、新熱帯区、大洋区
plumeria caterpillar			giant grey sphinx を見よ
plumeria whitefly			bay whitefly を見よ
plumes			plume moths を見よ
plumose scale	イチジクケマルカイガラムシ	*Morganella longispina* (Morgan)	(カメムシ目、マルカイガラムシ科) 日本、汎世界。イチジク、カンキツ、クワ、カエデなどの害虫
plumose sweep		*Oreopsyche muscella* (Denis et Schiffermüller)	(チョウ目、ミノガ科) 旧北区
plush		*Sithon nedymond nedymond* (Cramer)	(チョウ目、シジミチョウ科) 東洋区
plusia moths			gems (1) を見よ
plusiine moths			gems (1) を見よ
plutellid moths		Plutellidae	(チョウ目) の昆虫の総称　スガ科の異名
Pluto Euptera		*Euptera pluto* (Ward)	(チョウ目、タテハチョウ科) エチオピア区
Pluto sphinx moth		*Xylophanes pluto* (Fabricius)	(チョウ目、スズメガ科) 新北区、新熱帯区。pluto sphinx ともいう
Plutonia banner		*Epiphile iblis plutonia* H. Bates	(チョウ目、タテハチョウ科) 新熱帯区
Plutus opal		*Chrysoritis plutus* (Pennington)	(チョウ目、シジミチョウ科) エチオピア区
pod			イナゴの卵袋、カイコのマユ
pod lover		*Hadena perplexa* (Denis et Schiffermüller)	(チョウ目、ヤガ科) 旧北区、東洋区。*H. p. capsophila* (Duponchel) も同英名
pod midge			cabbage gall midge を見よ
pod pea borer			lima-bean pod borer を見よ
pod sucking bug			brown bean bug (1)(2) を見よ
podocarpus aphid	マキシンハアブラムシ	*Neophyllaphis podocarpi* Takahashi	(カメムシ目、アブラムシ科) 日本、東洋区、大洋区
podocarpus longicorn			podocarpus longicorn beetle を見よ
podocarpus longicorn beetle	ケブカトラカミキリ	*Hirticlytus conosus* (Matsushita)	(コウチュウ目、カミキリムシ科) 日本。イヌマキ害虫
podurid springtails		Hydropoduridae	(トビムシ目) の昆虫の総称
podurids			podurid springtails を見よ
Poecila sphinx			northern apple sphinx moth を見よ
poetry moth		*Neumoegenia poetica* Grote	(チョウ目、ヤガ科) 新北区
Pogge's mimic		*Pseudacraea poggei* Dewitz	(チョウ目、タテハチョウ科) エチオピア区

英名	和名	学名	所属、分布、ほか
Pogge's wanderer		*Bematistes poggei* (Dwitz)	(チョウ目、タテハチョウ科) エチオピア区
poinciana longicorn		*Agrianome spinicollis* (Macleay)	(コウチュウ目、カミキリムシ科) 豪州区
poinciana looper	ホウオウボククチバ	*Pericyma cruegeri* (Butler)	(チョウ目、ヤガ科) 日本、東洋区、新北区、豪州区、大洋区
pointed caper			African veined white を見よ
pointed ciliate blue		*Anthene lycaenina* (Felder)	(チョウ目、シジミチョウ科) 東洋区。*A. l. miya* Fruhstorfer も同英名
pointed copper		*Aloeides apicalis* Tite et Dickson	(チョウ目、シジミチョウ科) エチオピア区
pointed-footed flies			spear-winged flies を見よ
pointed gourd vine borer			cucurbit longicorn を見よ
pointed groundling		*Scrobipalpa acuminatella* (Sircom)	(チョウ目、キバガ科) 旧北区
pointed leafhopper	トガリヨコバイ	*Doratulina producta* (Matsumura)	(カメムシ目、オオヨコバイ科) 日本、東洋区
pointed leafwing		*Fountainea eurypyle confusa* (Hall)	(チョウ目、タテハチョウ科) 新熱帯区。*F. e. glanzi* (Roger, Escalante et Coronado) も同英名
pointed leafwing		*Anaea eurypyle* C. et R. Felder	(チョウ目、タテハチョウ科) 新北区、新熱帯区
pointed lineblue		*Ionolyce helicon* (Felder)	(チョウ目、シジミチョウ科) 東洋区、豪州区。*I. h. merguiana* (Moore) も同英名
pointed palmfly		*Elymnias penanga* (Westwood)	(チョウ目、タテハチョウ科) 東洋区
pointed pierrot		*Niphanda cymbia* de Nicéville	(チョウ目、シジミチョウ科) 東洋区
pointed pierrot			blue tiger (1) を見よ
pointed pierrot			common tiger blue を見よ
pointed sallow moth		*Epiglaea apiata* (Grote)	(チョウ目、ヤガ科) 新北区
pointed sister			Iphiclus sister を見よ
pointed-tailed wasps	クロバチ上科	Proctotrupoidea	(ハチ目) の昆虫の総称
pointed-tailed wasps			serphid wasps を見よ
pointed-winged flies			spear-winged flies を見よ
pointed-winged wave			angled wave moth を見よ
pointed-wingflies			spear-winged flies を見よ
poison hemlock moth			Alstroemer's flat-body を見よ
poison ivy leaf miner		*Caloptilia murtfeldtella* (Busck)	(チョウ目、ホソガ科) 新北区
polaris fritillary	ホソバホッキョクヒョウモン	*Clossiana polaris* (Boisduval)	(チョウ目、タテハチョウ科) 全北区。polar fritillary ともいう
police car moth			pericopid moth を見よ
policeman		*Coeliades chalybe* Westwood	(チョウ目、セセリチョウ科) エチオピア区
policeman flies	ドロバチモドキ亜科	Nyssoninae	(ハチ目、アナバチ科) の昆虫の総称

英名	和名	学名	所属、分布、ほか
Polina crescent		*Eresia polina* Hewitson	(チョウ目、タテハチョウ科) 新熱帯区
Poling's giant skipper		*Agathymus polingi* (Skinner)	(チョウ目、セセリチョウ科) 新北区
Poling's hairstreak		*Euristrymon polingi* (Capps)	(チョウ目、シジミチョウ科) 新北区
Poling's hairstreak		*Satyrium polingi* (Barnes et Benjamin)	(チョウ目、シジミチョウ科) 新北区
Poling's thread-legged katydid		*Arethaea polingi* Hebard	(バッタ目、キリギリス科) 新北区
polished chafer			shiny chafer を見よ
polished dart moth		*Euxoa perpolita* (Morrison)	(チョウ目、ヤガ科) 新北区
polished horntail			small blue horntail を見よ
polished lady beetles		Synonsychini	(コウチュウ目、テントウムシ科) の昆虫の総称
polished ladybird beetles			polished lady beetles を見よ
polistes wasps			papernest wasps (3) を見よ
Polixenes arctic	ファブリシャンタカネヒカゲ	*Oeneis polixenes* (Fabricius)	(チョウ目、タテハチョウ科) 全北区
polka dot		*Pardopsis punctatissima* (Boisduval)	(チョウ目、タテハチョウ科) エチオピア区
polka dot moth		*Syntomeida epilais* Walker	(チョウ目、ヒトリガ科) 新北区
polka dot moths			syntomid tiger moths を見よ　米国での英名
polka-dot wasp moth			polka dot moth を見よ
Polla blue-skipper		*Paches polla* (Mabille)	(チョウ目、セセリチョウ科) 新北区、新熱帯区
pollard weevil			confused flour beetle を見よ
pollen beetle		*Meligethes aeneus* (Fabricius)	(コウチュウ目、ケシキスイ科) 旧北区
pollen beetle			spotted maize beetle を見よ
pollen beetles (1)		*Meligethes*	(コウチュウ目、ケシキスイ科) の昆虫の総称
pollen beetles		*Dicranolaius*	(コウチュウ目、ジョウカイモドキ科) の昆虫の総称
pollen beetles		*Coryna*	(コウチュウ目、ツチハンミョウ科) の昆虫の総称
pollen beetles			sap beetles (1) を見よ
pollen feeding beetles			false blister beetles を見よ　pollen-feeding beetles とも記す
pollen rove beetle	ハナムグリハネカクシ	*Eusphalerum pollens* (Sharp)	(コウチュウ目、ハネカクシ科) 日本
pollen wasp		*Ceramius palaestinensis* (Gordani Soika)	(ハチ目、スズメバチ科) 旧北区
pollen wasps		Masarinae	(ハチ目、スズメバチ科) の昆虫の総称
pollinating fig insect			fig wasp を見よ
Polusca mountain satyr		*Pedaliodes polusca* (Hewitson)	(チョウ目、タテハチョウ科) 新熱帯区
Polyclea skipper		*Paratrytone polyclea* Godman	(チョウ目、セセリチョウ科) 新熱帯区

英名	和名	学名	所属、分布、ほか
polyctenids			bat bugs を見よ
Polyctor skipper		*Polyctor polyctor* Prittwitz	(チョウ目、セセリチョウ科) 新熱帯区
Polydamas swallowtail	キオビアオジャコウアゲハ	*Battus polydamas* (Linnaeus)	(チョウ目、アゲハチョウ科) 新北区、新熱帯区
polygynous ant		*Leptothorax acervorum* (Fabricius)	(ハチ目、アリ科) 旧北区
polygynous carpenter ant		*Camponotus planatus* Roger	(ハチ目、アリ科) 新熱帯区
Polyhymno moth		*Polyhymno luteostrigella* Chambers	(チョウ目、キバガ科) 新北区
Polymnia tigerwing		*Mechanitis polymnia lycidice* Bates	(チョウ目、タテハチョウ科) 新熱帯区
polymorphic longwing		*Heliconius hecale* (Fabricius)	(チョウ目、タテハチョウ科) 新熱帯区
polymorphic pondweed moth		*Parapoynx maculalis* (Clemens)	(チョウ目、メイガ科) 新北区
polyphagans			polyphagous beetles を見よ
polyphagous beetles	多食亜目	Polyphaga	(コウチュウ目) の昆虫の総称　カブトムシ亜目ともいう
polyphagous pinhole borer		*Platypus australis* Chapuis	(コウチュウ目、ナガキクイムシ科) 豪州区
Polyphemus moth		*Antheraea polyphemus* (Cramer)	(チョウ目、ヤママユガ科) 新北区
polypsocids			gnawing barklice を見よ
Polysema skipper		*Proeidosa polysema* (Lower)	(チョウ目、セセリチョウ科) 豪州区
Polyxo leafwing		*Memphis polyxo* (Druce)	(チョウ目、タテハチョウ科) 新熱帯区
pomace flies			vinegar flies (1) を見よ　米国での英名
pomace fly	キイロショウジョウバエ	*Drosophila melanogaster* Meigen	(ハエ目、ショウジョウバエ科) 日本、汎世界。遺伝学研究に寄与した著名種
pome looper		*Chloroclystis testulata* (Guenée)	(チョウ目、シャクガ科) 豪州区
pomegranate leaf roller			apple tortrix を見よ
pomegranate playboy		*Deudorix livia* (Klug)	(チョウ目、シジミチョウ科) エチオピア区
pomelo stink bug		*Cappaea taprobanensis* (Dallas)	(カメムシ目、カメムシ科) 東洋区。カンキツ害虫
pomice flies			vinegar flies (1) を見よ
Pompeius skipper		*Pompeius pompeius* (Latreille)	(チョウ目、セセリチョウ科) 新北区、新熱帯区
pompiloid wasps	ベッコウバチ上科	Pompiloidea	(ハチ目) の昆虫の総称
pond beetle		*Acilius sulcatus* (Linnaeus)	(コウチュウ目、ゲンゴロウ科) 旧北区
pond clubtails		*Arigomphus*	(トンボ目、サナエトンボ科) の昆虫の総称
pond leaf beetle		*Galerucella nympheae* (Linnaeus)	(コウチュウ目、ハムシ科) 全北区
pond-lily leaf-beetle			waterlily leaf beetle を見よ

英名	和名	学名	所属、分布、ほか
pond olive	フタバカゲロウ	*Cloeon dipterum* (Linnaeus)	(カゲロウ目、コカゲロウ科) 日本、旧北区
pond skater			common pond skater を見よ
pond skaters	両生カメムシ群	Amphibicorisa	(カメムシ目) の昆虫の総称
pond skaters			water striders (1) を見よ　英国での英名
pondapple leafroller moth		*Argyrotaenia amatana* (Dyar)	(チョウ目、ハマキガ科) 新北区
pondcruisers		*Epophthalmia*	(トンボ目、エゾトンボ科) の昆虫の総称
ponddamsel		*Megalagrion amaurodyctum waianaeanum* (Perkins)	(トンボ目、イトトンボ科) 大洋区。*M. a. pales* Perkins も同英名
ponddamsel			Adytum swamp damselfly を見よ
ponddamsel			Molokai damselfly を見よ
ponderosa pine bark borer		*Canonura princeps* (Walker)	(コウチュウ目、カミキリムシ科) 新北区
ponderosa pine cone beetle		*Conophthorus ponderosae* Hopkins	(コウチュウ目、キクイムシ科) 新北区
ponderosa pine cone worm		*Dioryctria auranticella* Grote	(チョウ目、メイガ科) 新北区。成虫は ponderosa pineconeworm moth
ponderosa pine needle miner		*Exoteleia anomala* Hodges	(チョウ目、キバガ科) 新熱帯区
ponderosa pine seedworm moth		*Cydia piperana* Kearfott	(チョウ目、ハマキガ科) 新北区
ponderosa pine tip moth		*Rhyacionia zozana* (Kearfott)	(チョウ目、ハマキガ科) 新北区
ponderous borer			pine sawyer beetle を見よ
ponderous borer beetle			pine sawyer beetle を見よ
ponderous spur-throated grasshopper			spur-throated grasshopper を見よ
Pondo emperor		*Charaxes pondoensis* van Someren	(チョウ目、タテハチョウ科) エチオピア区
Pondo shadefly		*Coenyra aurantiaca* Riley	(チョウ目、タテハチョウ科) エチオピア区
Pondoland widow		*Dira oxylus* (Trimen)	(チョウ目、タテハチョウ科) エチオピア区
pondside pyralid moth		*Elophila icciusalis* (Walker)	(チョウ目、メイガ科) 新北区
pondweed bug		*Mesovelia furcata* Mulsant et Rey	(カメムシ目、ミズカメムシ科) 旧北区
pondweed bugs			water treaders を見よ
pondweed leafhopper		*Macrosteles cyane* (Boheman)	(カメムシ目、オオヨコバイ科) 日本、旧北区
ponerine ants			keleps を見よ
ponerines			keleps を見よ
Pontasis Euselasia		*Euselasia pontasis* Callaghan, Llorente et Luis	(チョウ目、シジミタテハ科) 新熱帯区
pontic blue		*Polyommatus coelestinus* (Eversmann)	(チョウ目、シジミチョウ科) 旧北区

英名	和名	学名	所属、分布、ほか
Ponza grayling		*Hipparchia sbordonii* Kudma	（チョウ目、タテハチョウ科）旧北区
poor knights giant weta		*Deinacrida fallai* Salmon	（バッタ目、Stenopelmatidae）豪州区
popinjay		*Stibochiona nicea* (Gray)	（チョウ目、タテハチョウ科）東洋区
poplar admiral	オオイチモンジ	*Limenitis populi jezoensis* Matsumura	（チョウ目、タテハチョウ科）日本、旧北区。*L. populi* (Linnaeus) の英名
poplar and willow borer			willow beetle を見よ　米国での英名
poplar aphid			lettuce root aphid を見よ
poplar beetle			red poplar leaf beetle を見よ
poplar beetle			willow beetle を見よ
poplar borer		*Saperda calcarata* Say	（コウチュウ目、カミキリムシ科）新北区
poplar borer			poplar longhorn を見よ
poplar borer			small poplar longhorn を見よ
poplar branchlet borer moth			poplar grey bell を見よ
poplar broad-headed leafhopper	マダラヒロズヨコバイ	*Oncopsis tristis* (Zetterstedt)	（カメムシ目、オオヨコバイ科）日本、旧北区
poplar bud borer moth		*Meroptera cviatella* Dyar	（チョウ目、メイガ科）新北区
poplar-buttercup aphid		*Thecabius affinis* (Kaltenbach)	（カメムシ目、アブラムシ科）旧北区
poplar carpenterworm moth			aspen carpenterworm moth を見よ
poplar clearwing moth			dusky clearwing moth を見よ
poplar clearwing moth			Doll's clearwing moth を見よ
poplar cloaked bell		*Gypsonoma aceriana* (Duponchel)	（チョウ目、ハマキガ科）旧北区
poplar cloaked shoot			poplar cloaked bell を見よ
poplar-cudweed aphid		*Pemphigus populinigrae* (Schrank)	（カメムシ目、アブラムシ科）旧北区
poplar cutworm			olive を見よ
poplar dagger		*Apatele megacephala* Schiffermüller	（チョウ目、ヤガ科）旧北区
poplar flea beetle		*Chalcoides aurea* (Fourcroy)	（コウチュウ目、ハムシ科）旧北区
poplar flea weevil		*Rhynchaenus populi* (Fabricius)	（コウチュウ目、ゾウムシ科）旧北区
poplar gall aphid			lettuce root aphid を見よ　豪州での英名
poplar gall aphids			root aphids を見よ
poplar gall midge		*Contarinia petioli* (Kieffer)	（ハエ目、タマバエ科）旧北区
poplar gelechid			poplar sober を見よ
poplar gracilarid			triangle-marked slender を見よ

英名	和名	学名	所属、分布、ほか
poplar grey	シベチャケンモン	*Subacronicta concerpta* (Draudt)	（チョウ目、ヤガ科）日本
poplar grey			poplar grey moth を見よ
poplar grey bell		*Epinotia nisella* (Clerck)	（チョウ目、ハマキガ科）旧北区
poplar grey moth		*Acronicta megacephala* (Denis et Schiffermüller)	（チョウ目、ヤガ科）旧北区
poplar hawk			poplar hawkmoth (1) を見よ
poplar hawk moth	エゾシモフリスズメ	*Meganoton seribae* (Austaut)	（チョウ目、スズメガ科）日本、旧北区
poplar hawkmoth (1)		*Laothoe populi* (Linnaeus)	（チョウ目、スズメガ科）旧北区
poplar hawkmoth (2)		*Pachysphinx modesta* Harris	（チョウ目、スズメガ科）新北区
poplar horned aphid	ドロムネアブラムシ	*Doraphis populi* (Maskell)	（カメムシ目、アブラムシ科）日本、旧北区
poplar hornet clearwing			hornet moth を見よ
poplar kitten		*Furcula bifida* (Brahm)	（チョウ目、シャチホコガ科）旧北区
poplar kitten (1)		*Harpyia hermelina* Goeze	（チョウ目、シャチホコガ科）旧北区
poplar lasiocampid	ホシカレハ	*Gastropacha augustipennis* Walker	（チョウ目、カレハガ科）日本、旧北区
poplar leaf aphid		*Chaitophorus leucomelas* Koch	（カメムシ目、アブラムシ科）旧北区
poplar leaf beetle	ヤナギハムシ	*Chrysomela vigintipunctata* (Scopoli)	（コウチュウ目、ハムシ科）日本、旧北区。ヤナギ害虫
poplar leaf beetle			red poplar leaf beetle を見よ
poplar leaf gall aphid	ドロハタマワタムシ	*Epipemphigus niisimae* (Matsumura)	（カメムシ目、アブラムシ科）日本、旧北区
poplar leaf miner (1)		*Phyllocnistis unipunctella* (Stephens)	（チョウ目、コハモグリガ科）旧北区
poplar leafminer (2)		*Zeugophora scutellaris* Suffrian	（コウチュウ目、ハムシ科）旧北区
poplar leafminer moth		*Phyllonorycter populiella* (Chambers)	（チョウ目、ホソガ科）新北区
poplar leaf-mining weevil			poplar flea weevil を見よ
poplar leaf roller		*Byctiscus populi* (Linnaeus)	（コウチュウ目、オトシブ科）旧北区
poplar leaf roller			Brander's great marble を見よ
poplar leafroller moth		*Pseudosciaphila duplex* (Walsingham)	（チョウ目、ハマキガ科）新北区
poplar leaf-roller weevil	ドロハマキチョッキリ	*Byctiscus puberulus* (Motschulsky)	（コウチュウ目、オトシブミ科）日本、旧北区
poplar leaf roller weevil			pear leaf roller (1) を見よ
poplar leafcurl midge		*Prodiplosis morrisi* Gagné	（ハエ目、タマバエ科）新北区
poplar leaffolding sawfly		*Phyllocolpa bozemani* (Cooley)	（ハチ目、ハバチ科）新北区

英名	和名	学名	所属、分布、ほか
poplar leafhopper	ヒロズキンヨコバイ	*Idiocerus populi* (Linnaeus)	(カメムシ目、オオヨコバイ科) 日本、旧北区
poplar leaf-stalk aphid			lettuce root aphid を見よ
poplar leaf-stem gall aphid		*Pemphigus populitransversus* Riley	(カメムシ目、アブラムシ科) 新北区、エチオピア区
poplar leaf-tier			poplar tent-maker を見よ
poplar leopard			common leopard を見よ
poplar longhorn		*Saperda carcharias* (Linnaeus)	(コウチュウ目、カミキリムシ科) 全北区
poplar longhorns		*Saperda*	(コウチュウ目、カミキリムシ科) の昆虫の総称
poplar longicorn beetle			musk beetle を見よ
poplar lutestring		*Tethea or* (Denis et Schiffermüller)	(チョウ目、トガリバガ科) 旧北区。日本亜種は *T. or akannensis* (Matsumura) アカントガリバ
poplar midget		*Lithocolletis populifoliella* Treitschke	(チョウ目、ホソガ科) 旧北区
poplar petiole gall aphid			poplar leaf-stem gall aphid を見よ
poplar petiole gall moth			aspen petiole gall moth を見よ
poplar prominent (1)	ツマアカシャチホコ	*Clostera anachoreta* (Denis et Schiffermüller)	(チョウ目、シャチホコガ科) 日本。ポプラ害虫
poplar prominent (2)	セグロシャチホコ	*Clostera anastomosis* (Linnaeus)	(チョウ目、シャチホコガ科) 日本。ポプラ害虫
poplar pyralid	オオキノメイガ	*Botyodes principalis* Leech	(チョウ目、メイガ科) 日本、東洋区
poplar pyrausta	タイワンウスキノメイガ	*Botyodes diniasalis* (Walker)	(チョウ目、メイガ科) 日本、旧北区、東洋区
poplar sawfly	ポプラハバチ	*Trichiocampus populi* Okamoto	(ハチ目、ハバチ科) 日本、旧北区。ポプラ害虫
poplar sawfly (1)	サクツクリハバチ	*Stauronematus compressicornis* (Fabricius)	(ハチ目、ハバチ科) 日本、旧北区。ポプラ害虫
poplar sawfly (2)		*Trichiocampus viminalis* (Fallén)	(ハチ目、ハバチ科) 旧北区
poplar scale		*Quadraspidiotus gigas* (Thiem et Gerneck)	(カメムシ目、マルカイガラムシ科) 旧北区
poplar sober	シモフリツヅリガ	*Anacampsis populella* (Clerck)	(チョウ目、キバガ科) 日本、旧北区
poplar spanworm			small engrailed を見よ
poplar-spiral-gall aphid		*Pemphigus spyrothecae* Passerini	(カメムシ目、アブラムシ科) 旧北区
poplar stem leafhopper		*Macropsis virescens* (Gmelin)	(カメムシ目、オオヨコバイ科) 旧北区
poplar tent caterpillar			apical prominent moth を見よ
poplar tent-maker		*Ichthyura inclusa* (Hübner)	(チョウ目、シャチホコガ科) 新北区
poplar tentmaker			angle-lined prominent moth を見よ
poplar twig clearwing moth			dusky clearwing moth を見よ

英名	和名	学名	所属、分布、ほか
poplar twig gall aphid		*Pemphigus populiramulorum* Riley	（カメムシ目、アブラムシ科）新北区
poplar twig gall aphid	ドロタマワタムシ	*Pemphigus dorocola* Matsumura	（カメムシ目、アブラムシ科）日本、旧北区
poplar vagabond aphid		*Mordvilkoja vagabunda* (Walsh)	（カメムシ目、アブラムシ科）新北区
poplar weevil			willow beetle を見よ
poppy bee		*Osmia papaveris* Latreille	（ハチ目、ハキリバチ科）旧北区
poppy gall midge		*Dasineura papaveris* Winnertz	（ハエ目、タマバエ科）旧北区
poppy leaf roller			tobacco leaf worm を見よ
Porcelain gray dash-lined looper			Porcelain gray moth を見よ
Porcelain gray moth		*Protoboarmia porcelaria* (Guenée)	（チョウ目、シャクガ科）新北区
porchlight moth			filigreed moth を見よ
porcus sphinx moth		*Xylophanes porcus* (Hübner)	（チョウ目、スズメガ科）新北区、新熱帯区。porcus sphinx ともいう
porina moth			spring porina moth を見よ
porina moths		*Wiseana*	（チョウ目、コウモリガ科）の昆虫の総称　豪州での英名
porphyry knot-horn			porphyry knothorn moth を見よ
porphyry knothorn moth		*Numonia suavella* (Zincken)	（チョウ目、メイガ科）旧北区
porphyry pigmy		*Stigmella regiella* (Herrich-Schäffer)	（チョウ目、モグリチビガ科）旧北区
Port Arthur buck moth		*Hemileuca* sp.	（チョウ目、ヤママユガ科）新北区
porter's rustic		*Athetis hospes* (Freyer)	（チョウ目、ヤガ科）旧北区
Porthos untailed Charaxes		*Charaxes porthos* Grose-Smith	（チョウ目、タテハチョウ科）エチオピア区
Portia			Florida leafwing を見よ
Portia widow		*Palpopleura portia* (Drury)	（トンボ目、トンボ科）エチオピア区
Portland dart			Portland moth を見よ
Portland moth		*Ochropleura praecox* (Linnaeus)	（チョウ目、ヤガ科）旧北区。日本亜種は *O. p. flavomaculata* (Graeser) ホソアオバヤガ
Portland riband wave		*Idaea degeneraria* (Hübner)	（チョウ目、シャクガ科）旧北区。Portland ribbon wave ともいう
Porto Rica changa			changa を見よ
Porto Rican mole cricket			changa を見よ
Portorica changa			changa を見よ
Portuguese dappled white		*Euchloe tagis* (Hübner)	（チョウ目、シロチョウ科）旧北区
portulaca leafmining weevil		*Hypurus bertrandi* Perris	（コウチュウ目、ゾウムシ科）新北区、豪州区

英名	和名	学名	所属、分布、ほか
post-burn Datana moth		*Datana ranaeceps* (Guérin-Méneville)	（チョウ目、シャチホコガ科）新北区
postman	メルポメネーアカスジドクチョウ	*Heliconius melpomene* (Linnaeus)	（チョウ目、タテハチョウ科）新北区
postman butterfly			common longwing を見よ
postman moth			postman を見よ
postmans	ドクチョウ属	*Heliconius*	（チョウ目、タテハチョウ科）の昆虫の総称
posturing Arta moth		*Arta statalis* Grote	（チョウ目、メイガ科）新北区
pot-bellied emerald beetle			green dock beetle (1) を見よ
potato and peach aphis			green peach aphid を見よ
potato aphid (1)	チュウリップヒゲナガアブラムシ	*Macrosiphum euphorbiae* (Thomas)	（カメムシ目、アブラムシ科）日本、汎世界。ジャガイモ、チュウリップ、ナス、トマトなどの害虫
potato aphid (2)	ジャガイモヒゲナガアブラムシ	*Aulacorthum solani* (Kaltenbach)	（カメムシ目、アブラムシ科）日本、汎世界。ジャガイモ、チュウリップ、ナス、トマトなどの害虫
potato aphid			green peach aphid を見よ
potato aphid			bulb and potato aphid を見よ
potato beetle			Colorado potato beetle を見よ
potato bug			Colorado potato beetle を見よ
potato bug			Jerusalem cricket を見よ
potato bug			potato capsid を見よ
potato capsid		*Calocoris norvegicus* (Gmelin)	（カメムシ目、カスミカメムシ科）旧北区、豪州区
potato cutworm		*Pseudoleucania bilitura* (Guenée)	（チョウ目、ヤガ科）新熱帯区
potato flea beetle		*Epitrix cucumeris* (Harris)	（コウチュウ目、ハムシ科）新北区。カンキツ害虫
potato flea beetle		*Psylliodes affinis* (Paykull)	（コウチュウ目、ハムシ科）旧北区
potato flea beetles		*Psylliodes*	（コウチュウ目、ハムシ科）の昆虫の総称
potato ladybird	オオニジュウヤホシテントウ	*Epilachna vigintioctomaculata* Motschulsky	（コウチュウ目、テントウムシ科）日本、旧北区。ジャガイモ、ダイズ、ナス、タバコなどの害虫
potato leafhopper		*Cicadella aurata* Linnaeus	（カメムシ目、オオヨコバイ科）旧北区
potato leafhopper		*Typhlocyba jucunda* Uhler	（カメムシ目、オオヨコバイ科）旧北区
potato leafhopper (1)	ジャガイモヒメヨコバイ	*Empoasca fabae* Harris	（カメムシ目、オオヨコバイ科）新北区。ジャガイモ、マメ、クローバーなどの害虫
potato leafminer	ジャガイモモグリハナバエ	*Pegomya dulcamarae* Wood	（ハエ目、ハナバエ科）日本、旧北区。ジャガイモ害虫
potato maggot			seed potato maggot を見よ　幼虫の英名
potato moth		*Stoeberthinus testaceus* Butler	（チョウ目、キバガ科）豪州区、大洋区
potato moth			potato tuber moth (1) を見よ
potato myzus			potato aphid (2) を見よ
potato plant louse			potato aphid (1) を見よ
potato psylla			tomato psyllid を見よ

英名	和名	学名	所属、分布、ほか
potato psyllid			tomato psyllid を見よ
potato sawfly		*Pachyprotasis variegata* (Fallén)	(ハチ目、ハバチ科) 旧北区
potato scab gnat	ジャガイモクロバネキノコバエ	*Pnyxia scabiei* (Hopkins)	(ハエ目、クロバネキノコバエ科) 日本、全北区、新熱帯区。ジャガイモ害虫。雌は無翅
potato skin borer			rosy rustic を見よ
potato stalk borer		*Trichobaris trinotata* (Say)	(コウチュウ目、ゾウムシ科) 新北区
potato stalk-borer			burdock borer を見よ
potato stem borer			frosted orange (1) を見よ　米国での英名
potato stem borer			rosy rustic を見よ
potato sucker		*Trioza nigricornis* Foerster	(カメムシ目、キジラミ科) 旧北区
potato thrips			onion thrips を見よ
potato tuber moth (1)	ジャガイモガ	*Phthorimaea operculella* (Zeller)	(チョウ目、キバガ科) 日本、新北区、豪州区。著名なジャガイモ害虫
potato tuber moth (2)		*Scrobipalpopsis solanivora* Povolny	(チョウ目、キバガ科) 新熱帯区。中米のジャガイモ害虫
potato tuber moths		*Scrobipalpa*	(チョウ目、キバガ科) 北米での英名
potato tuberworm			potato tuber moth (1) を見よ　幼虫の英名
potato tuberworm			tobacco stem borer (1) を見よ
potato wireworm		*Hapatesus hirtus* Candèze	(コウチュウ目、コメツキムシ科) 豪州区
potato worm		*Sphinx quinquemaculata* (Haworth)	(チョウ目、スズメガ科) 旧北区
Potchefstroom blue		*Lepidochrysops procera* (Trimen)	(チョウ目、シジミチョウ科) エチオピア区
Potosi skipper		*Anatrytone potosiensis* (Freeman)	(チョウ目、セセリチョウ科) 新熱帯区
Potrillo skipper		*Cabares potrillo* (Lucas)	(チョウ目、セセリチョウ科) 新北区
potter flower bee		*Anthophora retusa* (Linnaeus)	(ハチ目、コシブトハナバチ科) 旧北区
potter flower bees			long-tongued bees を見よ
potter flower bees			hairy-footed bees を見よ
potter wasp		*Eumenes fraternus* Say	(ハチ目、スズメバチ科) 新北区
potter wasp		*Eumenes pemiformis* (Fabricius)	(ハチ目、スズメバチ科) 旧北区
potter wasp			heath potter を見よ
potter wasp			Australian hornet を見よ
potter wasps	ドロバチ亜科	Eumeninae	(ハチ目、スズメバチ科) の昆虫の総称
potter wasps			hornets (2) を見よ
Poulton's sapphire		*Iolaus poultoni* (Riley)	(チョウ目、シジミチョウ科) エチオピア区
poultry body louse			chicken body louse を見よ　豪州での英名
poultry bodylice			poultry lice を見よ
poultry bug		*Haematosiphon inodorus* (Dugès)	(カメムシ目、トコジラミ科) 新北区。ニワトリに寄生
poultry chewing lice			poultry lice を見よ　poultry-chewing lice とも記す

英名	和名	学名	所属、分布、ほか
poultry flea			chicken flea を見よ
poultry fluff louse			chicken flea を見よ
poultry head louse			chicken head louse を見よ　豪州での英名
poultry house moth		*Niditinea spretella* (Denis et Schiffermüller)	(チョウ目、ヒロズコガ科) 新北区
poultry house moth (1)		*Niditinea fuscipunctella* (Haworth)	(チョウ目、ヒロズコガ科) 新北区
poultry lice	タンカクハジラミ科 (トリハジラミ科)	Menoponidae	(ハジラミ目)の昆虫の総称　多くのトリに寄生
poultry shaft louse			chicken louse を見よ　豪州での英名
poultry stickfast flea			sticktight flea を見よ
poultry wing louse			chicken wing louse を見よ
pourthiaea cottony scale	ウシコロシワタカイガラムシ	*Pulvinaria photiniae* Kuwana	(カメムシ目、カタカイガラムシ科) 日本
pourthiaea leaf curling aphid	ウシコロシハマキワタムシ	*Prociphilus ushikoroshi* Shinji	(カメムシ目、アブラムシ科) 日本
powder blue Charaxes		*Charaxes pythodoris* Hewitson	(チョウ目、タテハチョウ科) エチオピア区
powder moth		*Eufidonia notataria* (Walker)	(チョウ目、シャクガ科) 新北区
powder post beetles			(コウチュウ目) ナガシンクイムシ科のヒラタキクイムシ亜科、ヒョウホンムシ科、シバンムシ科、ナガシンクイムシ科の多くの種を指す。シバンムシ科の *Anobium punctatum* De Geer が代表
powdered baron			Malay baron を見よ
powdered bigwing moth		*Lobophora nivigerata* Walker	(チョウ目、シャクガ科) 新北区
powdered brimstone	アラビアヤマキチョウ	*Gonepteryx farinosa* Zeller	(チョウ目、シロチョウ科) 旧北区
powdered dagger			reed dagger を見よ
powdered dancer		*Argia moesta* (Hagen)	(トンボ目、イトトンボ科) 新北区
powdered dusky skipper		*Acleros nigrapex* Strand	(チョウ目、セセリチョウ科) エチオピア区
powdered Epitolina		*Epitolina melissa* (Druce)	(チョウ目、シジミチョウ科) エチオピア区
powdered fulvous			powdered rustic を見よ
powdered green hairstreak	フカミドリシジミ	*Chrysozephyrus zoa* (de Nicéville)	(チョウ目、シジミチョウ科) 東洋区
powdered grey flat-body		*Agonopterix curvipunctosa* (Haworth)	(チョウ目、マルハキバガ科) 旧北区
powdered grey spurwing		*Antigonus erosus* (Hübner)	(チョウ目、セセリチョウ科) 新熱帯区
powdered knot-horn		*Pempeliella diluta* (Haworth)	(チョウ目、メイガ科) 旧北区
powdered oakblue		*Arhopala bazalus* (Hewitson)	(チョウ目、シジミチョウ科) 東洋区。日本亜種は *A. b. turbata* (Butler) ムラサキツバメ。*A. b. zalinda* Corbet も同英名
powdered pearl		*Psammotis pulveralis* Hübner	(チョウ目、メイガ科) 旧北区

英名	和名	学名	所属、分布、ほか
powdered quaker		*Orthosia gracilis* (Denis et Schiffermüller)	(チョウ目、ヤガ科) 旧北区
powdered rustic		*Hoplodrina superstes* (Ochsenheimer)	(チョウ目、ヤガ科) 旧北区
powdered snout			waved tabby を見よ
powdered-straw flat-body			greenweed flat-body moth を見よ
powdered wainscot		*Arsilonche albovenosa* Goeze	(チョウ目、ヤガ科) 旧北区
powderleg longicorn			basket longhorn beetle を見よ
powderpost beetle		*Lyctus parallelocollis* Blackburn	(コウチュウ目、ナガシンクイムシ科) 豪州区
powderpost beetle (1)	ヒラタキクイムシ	*Lyctus brunneus* (Stephens)	(コウチュウ目、ナガシンクイムシ科) 日本、汎世界。乾材害虫
powderpost beetles			false powderpost beetles を見よ
powderpost beetles			lyctids を見よ
powderpost bostrichid		*Amphicerus cornutus* Pallas	(コウチュウ目、ナガシンクイムシ科) 新北区
powderpost termite			West Indian drywood termite を見よ
powder-post termites			drywood termites (1)(2)(3) を見よ
powder-striped sprite			Kersten's sprite を見よ
powdery green sapphire	タムベニモンシジミ	*Heliophorus tamu* (Kollar)	(チョウ目、シジミチョウ科) 東洋区
powdery looper moth			barred umber を見よ
Powell's grayling		*Neohipparchia powelli* (Oberthür)	(チョウ目、タテハチョウ科) 旧北区
Poweshiek skipperling		*Oarisma poweshiek* (Parker)	(チョウ目、セセリチョウ科) 新北区
Praeclara underwing moth		*Catocala praeclara* Grote et Robinson	(チョウ目、ヤガ科) 新北区。Praeclara underwing ともいう
Praevia dart moth		*Xestia praevia* Lafontaine	(チョウ目、ヤガ科) 新北区
prairie bird-dropping moth		*Ponometia binocula* (Grote)	(チョウ目、ヤガ科) 新北区
prairie flea beetle		*Altica canadensis* Gentner	(コウチュウ目、ハムシ科) 新北区
prairie floodwater mosquito		*Ochlerotatus spencerii idahoensis* (Theobald)	(ハエ目、カ科) 新北区
prairie grain wireworm		*Ctenicera destructor* (Brown)	(コウチュウ目、コメツキムシ科) 新北区
prairie meadow katydid		*Conocephalus saltans* (Scudder)	(バッタ目、キリギリス科) 新北区
prairie mole cricket		*Gryllotalpa major* Saussure	(バッタ目、ケラ科) 新北区
prairie ringlet	アメリカヒメヒカゲ	*Coenonympha inornata* Edwards	(チョウ目、タテハチョウ科) 新北区
prairie tree cricket		*Oecanthus argentinus* (Saussure)	(バッタ目、カンタン科) 新北区
prairie wireworm			prairie grain wireworm を見よ

英名	和名	学名	所属、分布、ほか
Pratt's shieldback		*Pediodectes pratti* (Caudell)	(バッタ目、キリギリス科) 新北区
praying insects			mantids (2) を見よ
praying mantid			mantid を見よ
praying mantids			mantids (1)(2)(3) を見よ　祈る姿に似ることに由来。日本では「拝み虫」の俗称あり。preying manntids を参照
praying mantis			mantid を見よ　英国での英名
praying mantis			green mantid を見よ
preachers			mantids (2) を見よ
predaceous diving beetle		*Cybister fimbriolatus* (Say)	(コウチュウ目、ゲンゴロウ科) 新北区
predaceous diving beetles			diving beetles (1) を見よ
predaceous water beetles			diving beetles (1) を見よ
predacious capsid bug		*Cyrtorhinus fulvus* Knight	(カメムシ目、カスミカメムシ科) 大洋区
predacious corn planthopper bug			corn delphacid predator を見よ
predacious diving beetles			diving beetles (1) を見よ
predacious ground beetle		*Calosoma blaptoides tehuacanum* (Lapouge)	(コウチュウ目、オサムシ科) 大洋区
predacious ground beetle			Japanese ground beetle を見よ　米国での英名
predacious ground beetles			ground beetles を見よ
predacious mosquito	サキジロカクイカ	*Culex fuscanus* Wiedemann	(ハエ目、カ科) 日本、東洋区、大洋区
predacious thrips			banded thrips を見よ
predatory backswimmer		*Notonecta maculata* Fabricius	(カメムシ目、マツモムシ科) 旧北区
predatory bush cricket			matriarcha katydid を見よ
predatory gall midge		*Feltiella acarisuga* (Vallot)	(ハエ目、タマバエ科) 汎世界(新熱帯区を除く)　ハダニを捕食
predatory shield bug		*Oechalia schellembergii* (Guérin-Méneville)	(カメムシ目、カメムシ科) 豪州区
predatory shield bug (1)		*Cermatulus nasalis* (Westwood)	(カメムシ目、カメムシ科) 豪州区
predatory stink bug		*Alcaeorrhynchus grandis* Dallas	(カメムシ目、カメムシ科) 新北区。大豆害虫を捕食するほか、ナス他にもつく
predatory stoneflies	アミメカワゲラ科	Perlodidae	(カワゲラ科) の昆虫の総称　幼虫は雑食〜捕食性
predatory thrips		*Franklinothrips vespiformis* (Crawford)	(アザミウマ目、シマアザミウマ科) 日本、新北区、新熱帯区、豪州区、大洋区
predatory window fly			house windowfly を見よ

英名	和名	学名	所属、分布、ほか
Prenda roadside skipper		*Amblyscirtes prenda* Evans	（チョウ目、セセリチョウ科）新北区
prescott pine scale		*Matsucoccus vexillorum* Morrison	（カメムシ目、ワタフキカイガラムシ科）新北区
pretty chalk carpet	ナカジロナミシャク	*Melanthia procellata* (Denis et Schiffermüller)	（チョウ目、シャクガ科）日本、旧北区
pretty flat-body		*Depressaria pulcherrimella* Stainton	（チョウ目、マルハキバガ科）旧北区
pretty marbled		*Lithacodia deceptoria* Scopoli	（チョウ目、ヤガ科）旧北区
pretty mimic white		*Dismorphia thermesia* (Godart)	（チョウ目、シロチョウ科）新熱帯区
pretty pinion			pretty pinion carpet を見よ
pretty pinion carpet		*Perizoma blandiata* (Denis et Schiffermüller)	（チョウ目、シャクガ科）旧北区
pretty-pinion rivulet			pretty pinion carpet を見よ
pretty smidge		*Digitivalva perlepidella* (Stainton)	（チョウ目、アトヒゲコガ科）旧北区
pretty-thigh shieldback		*Idiostatus callimeris* Rehn et Hebard	（バッタ目、キリギリス科）新北区
preying flowers			mantids (2) を見よ
preying mantid			mantid を見よ
preying mantids			mantids (1)(2)(3) を見よ　他昆虫などを捕獲することに由来
Priam's birdwing	メガネトリバネアゲハ	*Ornithoptera priamus* (Linnaeus)	（チョウ目、アゲハチョウ科）東洋区、豪州区
Priapus batwing	キイロハゲタカアゲハ	*Atrophaneura priapus* (Boisduval)	（チョウ目、アゲハチョウ科）東洋区
prickly acacia seed beetle		*Bruchidius sahlbergi* Schilsky	（コウチュウ目、ハムシ科）豪州区
prickly-ash elfin	サンショウコツバメ	*Ahlbergia haradai* Igarashi	（チョウ目、シジミチョウ科）東洋区
prickly bush-cricket		*Cosmoderus maculatus* (Kirby)	（バッタ目、キリギリス科）旧北区
prickly pear bug		*Chelinidea tabulata* (Burmeister)	（カメムシ目、ヘリカメムシ科）豪州区
prickly pear cochineal		*Dactylopius opuntiae* (Cockerell)	（カメムシ目、コチニールカイガラムシ科）新北区、豪州区
prickly pear cochineal			cactus mealybug を見よ
prickly pear moth-borer		*Tucumania tapiacola* Dyar	（チョウ目、メイガ科）豪州区
prickly pear scale			cactus scale を見よ
prickly pear scale of the karroo			cactus scale を見よ
prickly stick insect		*Acanthoxyla geisovii* (Kaup)	（ナナフシ目、ナナフシ科）旧北区、豪州区
prickly stick insects		*Acanthoxyla*	（ナナフシ目、ナナフシ科）の昆虫の総称
prickly stick-insect		*Acanthoxyla prasina* (Westwood)	（ナナフシ目、ナナフシ科）旧北区

英名	和名	学名	所属、分布、ほか
primitive ant			Sri Lankan relict ant を見よ
primitive ants			keleps を見よ
primitive bristletails			forest silverfishes を見よ
primitive caddisflies	ナガレトビケラ科	Rhyacophilidae	（トビケラ目）の昆虫の総称
primitive coccoids	古カイガラムシ群	Archaecoccoidea	（カメムシ目）の昆虫の総称
primitive crane flies	ニセヒメガガンボ科	Tanyderidae	（ハエ目）の昆虫の総称
primitive ghost moths			African primitive ghost moths を見よ
primitive gnats	原蚊類	Protophthalmia	（ハエ目）の昆虫の総称
primitive katydids		Prophalangopsidae	（バッタ目）の昆虫の総称　世界に4種
primitive katydids			monkey grasshoppers (1) を見よ
primitive minnow mayflies	フタオカゲロウ科	Siphlonuridae	（カゲロウ目）の昆虫の総称
primitive moths		Mnesarchaeidae	（チョウ目）の昆虫の総称
primitive moths (1)	スイコバネガ科	Eriocraniidae	（チョウ目）の昆虫の総称
primitive moths			mandibulate moths を見よ
primitive sawflies		Megalodontoidea	（ハチ目）の昆虫の総称
primitive termites			Australian termites を見よ
primitive weevils			straight snouted weevils を見よ
primorye scale			willow oystershell scale を見よ
primrose cochylid moth		*Afroposia oenotherana* (Riley)	（チョウ目、ハマキガ科）新北区
primrose flag	ウラキシロチョウ	*Melete lycimnia* (Cramer)	（チョウ目、シロチョウ科）新熱帯区。
primrose moth		*Schinia florida* (Guenée)	（チョウ目、ヤガ科）新北区
primrose thrips		*Taeniothrips picipes* (Zetterstedt)	（アザミウマ目、アザミウマ科）旧北区
primula aphid		*Microlophium primulae* (Theobald)	（カメムシ目、アブラムシ科）旧北区、豪州区
primula aphid			mottled arum aphid を見よ
prince baskettail		*Epitheca princeps* Hagen	（トンボ目、エゾトンボ科）新北区
princess flash		*Deudorix smilis* Hewitson	（チョウ目、シジミチョウ科）豪州区
principalis jewelmark		*Anteros principalis* Hopffer	（チョウ目、シジミタテハ科）新熱帯区
Pringle's arrowhead		*Phasis pringlei* Dickson	（チョウ目、シジミチョウ科）エチオピア区
Pringle's blue		*Lepidochrysops pringlei* Dickson	（チョウ目、シジミチョウ科）エチオピア区
Pringle's copper		*Aloeides pringlei* Tite et Dickson	（チョウ目、シジミチョウ科）エチオピア区
Pringle's widow		*Torynesis pringlei* Dickson	（チョウ目、タテハチョウ科）エチオピア区
prionine longhorns	ノコギリカミキリ亜科	Prioninae	（コウチュウ目、カミキリムシ科）の昆虫の総称
prisoner sphinx			arrow sphinx を見よ
Pritchard ground mealybug		*Rhizoecus dianthi* Green	（カメムシ目、コナカイガラムシ科）新北区

英名	和名	学名	所属、分布、ほか
Pritchard mealybug		*Rhizoecus pritchardi* McKenzie	（カメムシ目、コナカイガラムシ科）新北区
privet aphid		*Myzus ligustri* (Mosley)	（カメムシ目、アブラムシ科）全北区
privet hawk		*Sphinx ligustri* Linnaeus	（チョウ目、スズメガ科）旧北区。日本亜種は *S. l. amurensis* Oberthür エゾコエビガラスズメ
privet hawk moth		*Psilogramma menephron* (Cramer)	（チョウ目、スズメガ科）東洋区、豪州区
privet hawk moth			privet hawk を見よ
privet leaf miner		*Caloptilia fraxinella* (Ely)	（チョウ目、ホソガ科）新北区
privet leafminer			lilac leaf miner を見よ
privet thrips		*Dendrothrips ornatus* (Jablonowski)	（アザミウマ目、アザミウマ科）全北区
prized hedge blue			pale hedge blue (1) を見よ
procampodeid entotrophs		Procampodeidae	（コムシ目）の昆虫の総称
processionary caterpillar		*Thaumetopoea wilkinsoni* Tams	（チョウ目、シャチホコガ科）旧北区
processionary caterpillar (1)		*Ochrogaster lunifer* Herrich-Schäffer	（チョウ目、Thaumetopoeidae）豪州区
processionary caterpillars			processionary moths を見よ　幼虫の英名
processionary moth			pine processionary moth を見よ
processionary moth			oak procession moth を見よ
processionary moths		Thaumetopoeidae	（チョウ目）の昆虫の総称
processionary worms			processionary moths を見よ　南米での英名
Prochyta satyr		*Mygona prochyta* (Hewitson)	（チョウ目、タテハチョウ科）新熱帯区
proctotrupid wasps	シリボソクロバチ科	Proctotrupidae	（ハチ目）の昆虫の総称
proctotrupid wasps			pointed-tailed wasps を見よ
proctotrupids			proctotrupid wasps を見よ
proctotrupids			pointed-tailed wasps を見よ
proctotrupoid wasps			pointed-tailed wasps を見よ
Procula longwing		*Eueides procula asidia* Schaus	（チョウ目、タテハチョウ科）新熱帯区
projapygids		Projapygidae	（コムシ目）の昆虫の総称
projecta gray moth		*Cleora projecta* (Walker)	（チョウ目、シャクガ科）新北区。projecta gray ともいう
Prolata Melipotis moth		*Melipotis prolata* (Walker)	（チョウ目、ヤガ科）新北区
promethea moth			spicebush silk moth を見よ
prominent moths			prominents を見よ
prominents	シャチホコガ科	Notodontidae	（チョウ目）の昆虫の総称　世界に 2,000 種以上
promiscuous angle moth		*Macaria promiscuata* (Ferguson)	（チョウ目、シャクガ科）新北区

英名	和名	学名	所属、分布、ほか
pronged clubtail		*Gomphus graslinii* Rambur	（トンボ目、サナエトンボ科）旧北区
pronggill mayflies	トビイロカゲロウ科	Leptophlebiidae	（カゲロウ目）の昆虫の総称
pronghorn clubtail		*Gomphus graslinellus* Walsh	（トンボ目、サナエトンボ科）新北区
Pronus longtail		*Urbanus pronus* Evans	（チョウ目、セセリチョウ科）新北区、新熱帯区
proof maggot			cattle warble fly (2) を見よ　幼虫の英名
prop beetles			hister beetles を見よ
propertius duskywing	オオクロミヤマセセリ	*Erynnis propertius* (Scudder et Burgess)	（チョウ目、セセリチョウ科）新北区
prophalangopsine grasshoppers		Prophalangopsinae	（バッタ目、キリギリス科）の昆虫の総称
Propst's shieldback		*Neduba propsti* Rentz et Weissmann	（バッタ目、キリギリス科）新北区
Proserpina leafwing		*Memphis proserpina proserpina* (Salvin)	（チョウ目、タテハチョウ科）新熱帯区
prosopis round-headed borer			Kiawe roundheaded borer を見よ
Prosoplus long-horned beetle		*Prosoplus bankii* (Fabricius)	（コウチュウ目、カミキリムシ科）大洋区
Protea butterflies		*Capys*	（チョウ目、シジミチョウ科）の昆虫の総称　幼虫が Protea の花を食う
Protea Charaxes			Protea emperor を見よ
Protea emperor		*Charaxes pelias* (Cramer)	（チョウ目、タテハチョウ科）エチオピア区
Protea scarlet		*Capys alphaeus* (Cramer)	（チョウ目、シジミチョウ科）エチオピア区
protean shieldback		*Atlanticus testaceus* (Scudder)	（バッタ目、キリギリス科）新北区
Proteus longtail			long-tailed skipper を見よ
Proteus scale			cattleya scale を見よ
protocalliphorine flies		Protocalliphorinae	（ハエ目、クロバエ科）の昆虫の総称
proturans	カマアシムシ目	Protura	の昆虫の総称　世界に約 650 種
proturids			proturans を見よ
proud sphinx moth		*Proserpinus gaurae* (Smith)	（チョウ目、スズメガ科）新北区、新熱帯区
Provencal short-tailed blue		*Cupido alcetas* (Hoffmannsegg)	（チョウ目、シジミチョウ科）旧北区
Provencal fritillary		*Mellicta dejone* (Geyer)	（チョウ目、タテハチョウ科）旧北区
Provence burnet moth		*Zygaena occitanica* (Villers)	（チョウ目、マダラガ科）旧北区
Provence Chalk hill blue		*Polyommatus hispana* (Herrich-Schäffer)	（チョウ目、シジミチョウ科）旧北区
Provence hairstreak		*Tomares ballus* (Fabricius)	（チョウ目、シジミチョウ科）旧北区、エチオピア区
Provence orange-tip			Moroccan orange-tip を見よ
Proxenus blue-skipper		*Pythonides proxenus* (Godman et Salvin)	（チョウ目、セセリチョウ科）新熱帯区

英名	和名	学名	所属、分布、ほか
pruinose bean weevil		*Stator pruininus* Horn	（コウチュウ目、ハムシ科）新北区
pruinose scarab		*Sericesthis geminata* Boisduval	（コウチュウ目、コガネムシ科）豪州区
pruinosed bloodtail		*Lathrecista asiatica asiatica* (Fabricius)	（トンボ目、トンボ科）東洋区
prune leafhopper		*Typhlocyba prunicola* (Edwards)	（カメムシ目、オオヨコバイ科）全北区
prune moth			plum fruit moth を見よ
prune thrips			pear thrips を見よ　米国での英名
prunus bud moth	ウメスカシクロバ	*Illiberis psychina* (Oberthür)	（チョウ目、マダラガ科）日本、旧北区。ウメ、モモ、ナシ、サクラ害虫
prunus cambium miner			plum cambium miner を見よ
prunus leafminer beetle	ウメチビタマムシ	*Trachys inconspicua* (E. Saunders)	（コウチュウ目、タマムシ科）日本、旧北区、東洋区。ウメ、モモ害虫
Prussian roach			German cockroach を見よ
Pryer mulberry leaf roller	スカシノメイガ	*Glyphodes pryeri* Butler	（チョウ目、メイガ科）日本、旧北区
Pryer tortris	プライヤハマキ	*Acleris affinitana* (Snellen)	（チョウ目、ハマキガ科）日本、旧北区
pselaphid beetles			antloving beetles (1) を見よ
pselaphid-like rove beetles		Euaesthetinae	（コウチュウ目、ハネカクシ科）の昆虫の総称
Pseudanthracia moth		*Pseudanthracia coracias* (Guenée)	（チョウ目、ヤガ科）新北区
pseudo click beetles			throscid beetles を見よ
pseudocaeciliid barklice			false lizard barklice を見よ
psocids			booklice (2) を見よ
psocids			common barklice を見よ
psocopterans			booklice (2) を見よ
psoid beetles		Psoidae	（コウチュウ目）の昆虫の総称　ナガシンクイムシ科に含まれることあり
psoid branch and twig beetles			psoid beetles を見よ
psyche	クロテンシロチョウ属	*Leptosia*	（チョウ目、シロチョウ科）の昆虫の総称
psyche			psyche butterfly を見よ
psyche butterfly	クロテンシロチョウ	*Leptosia nina* (Fabricius)	（チョウ目、シロチョウ科）東洋区、豪州区。*L. n. malayana* Fruhstorferと日本亜種　*L. n. niobe* (Wallace) クロテンシロチョウ　も同英名
psychedelic Jones moth		*Thaumatographa jonesi* (Brower)	（チョウ目、ハマキガ科）新北区
psyllas			jumping plantlice を見よ
psyllids			jumping plantlice を見よ
psyllines		Psyllina	（カメムシ目）の昆虫の総称
psyllipsocids			cave barklice を見よ
pterergate	有翅職蟻		痕跡的な翅をもつ職蟻、兵蟻

英名	和名	学名	所属、分布、ほか
pteromalid wasps			jewel wasps を見よ
pteromalids			jewel wasps を見よ
pteronarcids			giant stoneflies (1) を見よ
ptiliid beetles			feather-winged beetles を見よ
ptiliids			feather-winged beetles を見よ
ptilodactylid beetles			soft-bodied plant beetles を見よ
ptilodactylids			soft-bodied plant beetles を見よ
ptinid beetles			spider beetles を見よ
Ptolyca crescentspot		*Anthanassa ptolyca* (Bates)	（チョウ目、タテハチョウ科）新熱帯区
Ptunarra brown butterfly		*Oreixenica ptunarra* Couchman	（チョウ目、タテハチョウ科）豪州区
pubescent round bark beetle	ケブカマルキクイムシ	*Sphaerotrypes pila* Blandford	（コウチュウ目、キクイムシ科）日本、旧北区、東洋区
pubic lice	ケジラミ科	Phthiridae	（シラミ目）の昆虫の総称
pubic louse			crab louse を見よ
public louse			crab louse を見よ
Puente colosal hairstreak		*Strymon* sp.	（チョウ目、シジミチョウ科）新熱帯区
puffball beetle		*Caenocara*	（コウチュウ目、シバンムシ科）の昆虫の総称
puffins and albatrosses	トガリシロチョウ属	*Appias*	（チョウ目、シロチョウ科）の昆虫の総称
pug loopers		*Eupithecia*	（チョウ目、シャクガ科）の昆虫の総称
pug moths			geometer moths を見よ
puget sound wireworm		*Ctenicera aeripennis* (Kirby)	（コウチュウ目、コメツキムシ科）新北区。亜種 *C. a. destructor* (Brown) は prairie grain wireworm という
Pugnacious lancer		*Pemara pugnana* (de Nicéville)	（チョウ目、セセリチョウ科）東洋区
pugs		*Chloroclystis*	（チョウ目、シャクガ科）の昆虫の総称
pugs			pug loopers を見よ
pugs			geometer moths を見よ
pugs			pea weevils を見よ
pulicid fleas			common fleas を見よ
pullback moth		*Euscirrhopterus poeyi* Grote	（チョウ目、ヤガ科）新北区、新熱帯区
pulse beetle			adzuki weevil を見よ
pulse beetles			pea weevils を見よ
pulse beetles			seed weevils (1) を見よ
pulse-seed weevils			pea weevils を見よ
pulse weevils			pea weevils を見よ
pulvinaria scale	モミジワタカイガラムシ	*Pulvinaria horii* Kuwana	（カメムシ目、カタカイガラムシ科）日本。リンゴ、ナシ、カエデ、カシなどの害虫
pulvinaria scale		*Pulvinaria cellulosa* Green	（カメムシ目、カタカイガラムシ科）豪州区
pulvinaria scale (1)		*Pulvinaria flavescens* Brethes	（カメムシ目、カタカイガラムシ科）新熱帯区。カンキツ害虫

英名	和名	学名	所属、分布、ほか
pulvinaria scale (2)	タイワンワタカイガラムシ	*Pulvinaria polygonata* Cockerell	（カメムシ目、カタカイガラムシ科）日本、東洋区。カンキツ害虫
pulvinaria scale			camellia scale (1) を見よ
pulvinaria scale			cottony citrus scale (1) を見よ
Pumice dotted-blue		*Philotiella leona* Hammond et McCorkl	（チョウ目、シジミチョウ科）新北区
pumpkin beetle		*Aulacophora hilaris* (Boisduval)	（コウチュウ目、ハムシ科）豪州区
pumpkin bug			cucurbit shield bug を見よ
pumpkin caterpillar			cotton caterpillar (1) を見よ
pumpkin fly		*Dacus bivittatus* (Bigot)	（ハエ目、ミバエ科）エチオピア区
pumpkin fruit fly	カボチャミバエ	*Paradacus depressus* (Shiraki)	（ハエ目、ミバエ科）日本、旧北区、東洋区。カボチャ害虫
pumpkin gall wasp		*Dryocosmus minusculus* Weld	（ハチ目、タマバチ科）新北区
Puna chequered blue		*Madeleinea koa* Druce	（チョウ目、シジミチョウ科）新熱帯区
Puna clouded yellow		*Colias euxanthe* C. et R. Felder	（チョウ目、シロチョウ科）新熱帯区
Puna silverwing		*Punargentus lamna* Thieme	（チョウ目、タテハチョウ科）新熱帯区
punches	トラフシジミタテハ属	*Dodona*	（チョウ目、シジミタテハ科）の昆虫の総称
Punchinello		*Zemeros flegyas* (Cramer)	（チョウ目、シジミタテハ科）東洋区。*Z. f. albipunctatus* Butler も同英名
punctate blister beetle		*Epicauta puncticollis* Mannerheim	（コウチュウ目、ツチハンミョウ科）新北区
punctate flower chafer		*Polystigma punctata* (Donovan)	（コウチュウ目、コガネムシ科）豪州区
punctate leaf beetles		Orsodacninae	（コウチュウ目、クビナガハムシ科）の昆虫の総称
punctate leafhopper			red-maculated leafhopper を見よ
puncturevine seed weevil		*Microlarinus lareyniei* (Jacquelin)	（コウチュウ目、ゾウムシ科）新北区
puncturevine stem weevil		*Microlarinus lypriformis* Wollaston	（コウチュウ目、ゾウムシ科）新北区
punic scarlet		*Axiocerses punicea* (Grose-Smith)	（チョウ目、シジミチョウ科）エチオピア区
punkies		*Culicoides*	（ハエ目、ヌカカ科）の昆虫の総称　人血を吸う。単数形は punkie
punkies			biting midges (1) を見よ
punky			punkies を見よ　punkey ともいう
pupa			昆虫の蛹
pupilled skipper		*Polites pupillus* (Plötz)	（チョウ目、セセリチョウ科）新熱帯区
pupiparous flies	蛹生類	Pupipara	（ハエ目）の昆虫の総称
purbeck mason-wasp		*Pseudepipona herrichii* (Saussure)	（ハチ目、スズメバチ科）旧北区
pure-banded dartwhite		*Catasticta teutila teutila* (Doubleday)	（チョウ目、シロチョウ科）新熱帯区。*C. t. flavifasciata* (Beutelspacher) も同英名

英名	和名	学名	所属、分布、ほか
pure lichen moth		*Crambidia pura* Barnes et McDunnough	(チョウ目、ヒトリガ科) 新北区
Purepecha skipperling		*Piruna purepecha* Warren et Gonzalez	(チョウ目、セセリチョウ科) 新熱帯区
puriri moth		*Aenetus virescens* (Doubleday)	(チョウ目、コウモリガ科) 豪州区
puritan tiger beetle		*Cicindela puritana* Horn	(コウチュウ目、ハンミョウ科) 新北区
Puritana wood nymph		*Taygetomorpha puritana* (Weeks)	(チョウ目、タテハチョウ科) 新熱帯区
purple and gold flitter		*Zographetus satwa* (de Nicéville)	(チョウ目、セセリチョウ科) 東洋区
purple arches moth		*Polia purpurissata* Grote	(チョウ目、ヤガ科) 新北区
purple azure	ヒメミイロヤドリギシジミ	*Ogyris zosine* Hewitson	(チョウ目、シジミチョウ科) 豪州区
purple-backed cabbageworm			chequered straw pearl を見よ
purple bar	ナカクロモンシロナミシャク	*Cosmorhoe ocellata* (Linnaeus)	(チョウ目、シャクガ科) 日本、旧北区
purple bar carpet			purple bar を見よ
purple-barred carpet		*Larentia ocellata* (Linnaeus)	(チョウ目、シャクガ科) 旧北区
purple-barred yellow		*Lythria purpuraria* (Linnaeus)	(チョウ目、シャクガ科) 旧北区
purple-barred yellow carpet			purple-barred yellow を見よ
purple beak			Australian beak を見よ
purple-black drill		*Dichrorampha simpliciana* (Haworth)	(チョウ目、ハマキガ科) 旧北区
purple bluet		*Enallagma coecum* (Hagen)	(トンボ目、イトトンボ科) 新熱帯区
purple bog fritillary			Titania's fritillary を見よ
purple bordered gold	ベニヒメシャク	*Idaea muricata* (Hufnagel)	(チョウ目、シャクガ科) 日本、旧北区
purple-bordered wave			purple bordered gold を見よ
purple brown-eye		*Chaetocneme porphyropsis* (Meyrick et Lower)	(チョウ目、セセリチョウ科) 豪州区
purple-brown hairstreak		*Hypolycaena philippus* (Fabricius)	(チョウ目、シジミチョウ科) エチオピア区
purple-brown tailless oakblue		*Arhopala arvina* (Hewitson)	(チョウ目、シジミチョウ科) 東洋区
purple bushbrown		*Mycalesis orseis* Hewitson	(チョウ目、タテハチョウ科) 東洋区。*M. o. nautilus* Butler も同英名
purple carrot-seed moth			Blunt's flat-body を見よ
purple Cerulean			dark Cerulean (2) を見よ
purple clay	ミヤマアカヤガ	*Diarsia brunnea* (Denis et Schiffermüller)	(チョウ目、ヤガ科) 日本、旧北区
purple cloud	ヒメモクメヨトウ	*Actinotia polyodon* (Clerck)	(チョウ目、ヤガ科) 日本、旧北区

英名	和名	学名	所属、分布、ほか
purple-coloured rhynchites		*Rhynchites bacchus* (Linnaeus)	(コウチュウ目、オトシブミ科) 旧北区
purple copper		*Paralucia spinifera* (Edwards et Common)	(チョウ目、シジミチョウ科) 豪州区
purple-crested slug moth		*Adoneta spinuloides* (Herrich-Schäffer)	(チョウ目、イラガ科) 新北区
purple crow			dwarf crow を見よ
purple duke		*Eulaceura osteria kumana* Fruhstorfer	(チョウ目、タテハチョウ科) 東洋区
purple dun		*Paraleptophlebia cincta* (Retzius)	(カゲロウ目、トビイロカゲロウ科) 旧北区
purple dusk-flat			purple brown-eye を見よ
purple-edged copper	メスグロベニシジミ	*Lycaena hippothoe* (Linnaeus)	(チョウ目、シジミチョウ科) 旧北区
purple-edged ermel	ウスグロヒメスガ	*Swammerdamia pyrella* (Villers)	(チョウ目、スガ科) 旧北区。スモモ害虫
purple emperor (1)	コムラサキ	*Apatura metis substitula* Butler	(チョウ目、タテハチョウ科) 日本、旧北区
purple emperor (2)	チョウセンコムラサキ	*Apatura iris* (Linnaeus)	(チョウ目、タテハチョウ科) 旧北区。purple emperor butterfly ともいう
purple gem		*Chloroselas mazoensis* (Trimen)	(チョウ目、シジミチョウ科) エチオピア区
purple giant Epitola		*Epitola urania* Kirby	(チョウ目、シジミチョウ科) エチオピア区
purple-glazed oakblue		*Arhopala agaba* (Hewitson)	(チョウ目、シジミチョウ科) 東洋区
purple hairstreak	ケルクスミドリシジミ	*Quercusia quercus* (Linnaeus)	(チョウ目、シジミチョウ科) 旧北区。purple hairstreak butterfly ともいう
purple hindwinged catocala	ムラサキシタバ	*Catocala fraxini jezoensis* Matsumura	(チョウ目、ヤガ科) 日本、旧北区
purple Lamb's quarters cutworm			buckwheat cutworm を見よ
purple Lamb's quarters mealy aphid	アカザコフキアブラムシ	*Hayhurstia atriplicis* (Linnaeus)	(カメムシ目、アブラムシ科) 日本、全北区、東洋区
purple leaf blue		*Amblypodia anita* Hewitson	(チョウ目、シジミチョウ科) 東洋区
purple lesser fritillary		*Clossiana titania boisduvalii* Herrich-Schäffer	(チョウ目、タテハチョウ科) 新北区
purple line-blue			small purple line blue を見よ
purple-lined borer			rice stem borer を見よ
purple-lined borer of sugarcane			rice stem borer を見よ
purple marbled		*Eublemma ostrina* (Hübner)	(チョウ目、ヤガ科) 旧北区
purple moonbeam		*Philiris fulgens* (Grose-Smith et Kirby)	(チョウ目、シジミチョウ科) 豪州区
purple moonbeam			common moonbeam を見よ
purple mort bleu		*Eryphanis polyxena* (Meerburgh)	(チョウ目、タテハチョウ科) 新熱帯区

英名	和名	学名	所属、分布、ほか
purple oak-blue			dull oak blue を見よ
purple owl	ツマキフクロチョウ（ムラサキフクロチョウ）	*Caligo beltrao* Illiger	（チョウ目、タテハチョウ科）新熱帯区
purple Plagodis moth		*Plagodis kuetzingi* (Grote)	（チョウ目、シャクガ科）新北区。purple Plagodis ともいう
purple sapphire	ウラフチベニシジミ	*Heliophorus epicles* (Godart)	（チョウ目、シジミチョウ科）東洋区。*H. e. tweediei* Eliot も同英名
purple scale	ミカンカキカイガラムシ	*Lepidosaphes beckii* (Newman)	（カメムシ目、マルカイガラムシ科）日本、新北区、エチオピア区、汎熱帯。カンキツ害虫
purple scale destroyer			scale-eating ladybird を見よ
purple scum springtail		*Hypogastrura vernalis* (Carl)	（トビムシ目、ヒメトビムシ科）豪州区
purple shaded gem		*Euchalcia illustris* Fabricius	（チョウ目、ヤガ科）旧北区
purple-shaded gem		*Euchalcia variabilis* (Piller)	（チョウ目、ヤガ科）旧北区
purple-shaded skipper		*Eutocus matildae vinda* Evans	（チョウ目、セセリチョウ科）新熱帯区
purple-shot copper	ムラサキオオベニシジミ（アオベニシジミ）	*Lycaena alciphron* (Rottemburg)	（チョウ目、シジミチョウ科）旧北区
purple spotted flitter		*Zographetus ogygia* (Hewitson)	（チョウ目、セセリチョウ科）東洋区
purplespotted lily aphid		*Macrosiphum lilii* (Monell)	（カメムシ目、アブラムシ科）新北区
purple spotted swallowtail	ミイロタイマイ	*Graphium weiskei* (Ribbe)	（チョウ目、アゲハチョウ科）東洋区、豪州区
purple-stained skipper		*Zenis jebus hemizona* (Dyar)	（チョウ目、セセリチョウ科）新熱帯区
purple stem borer			pink borer (1) を見よ
purple stem borer moth			pink borer (1) を見よ
purple stink bug	ムラサキカメムシ	*Carpocoris purpureipennis* De Geer	（カメムシ目、カメムシ科）日本、旧北区。イネ、タマネギなどの害虫
purple-striped shootworm moth		*Zeiraphera unfortunana* Ferris et Kruse	（チョウ目、ハマキガ科）新北区。purple-striped shootworm ともいう
purple swift		*Caltoris tulsi* (de Nicéville)	（チョウ目、セセリチョウ科）東洋区
purple thorn	ムラサキエダシャク	*Selenia tetralunaria* (Hufnagel)	（チョウ目、シャクガ科）日本、旧北区
purple tip	イオネツマアカシロチョウ	*Colotis ione* (Godart)	（チョウ目、シロチョウ科）エチオピア区
purple-topped Euselasia		*Euselasia regipennis regipennis* (Butler et Druce)	（チョウ目、シジミタテハ科）新熱帯区
purple treble-bar		*Aplocera praeformata* Hübner	（チョウ目、シャクガ科）旧北区
purple-washed eyed-metalmark		*Mesosemia lamachus* Hewitson	（チョウ目、シジミタテハ科）新熱帯区

英名	和名	学名	所属、分布、ほか
purple-washed eyemark			purple-washed eyed-metalmark を見よ
purple-washed skipper		*Panoquina sylvicola* (Herrich-Schäffer)	（チョウ目、セセリチョウ科）新北区
purple-washed skipper		*Panoquina lucas* (Fabricius)	（チョウ目、セセリチョウ科）新北区、新熱帯区
purple-webbed ministreak		*Ministrymon phrutus* (Geyer)	（チョウ目、シジミチョウ科）新熱帯区
purple white-streak argent		*Argyresthia albistria* Haworth	（チョウ目、メムシガ科）旧北区
purplewinged mantid		*Tenodera australasiae* (Leach)	（カマキリ目、カマキリ科）豪州区、新北区。purple-winged mantis ともいう
purple-winged mantid			purplewinged mantid を見よ
purple yellow-barred clothes		*Monopis obviella* (Denis et Schiffermüller)	（チョウ目、ヒロズコガ科）旧北区
purples			primitive moths (1) を見よ
purplish bent-skipper		*Camptopleura termon* (Hopffer)	（チョウ目、セセリチョウ科）新熱帯区
purplish birch-miner moth			gold-sprinkled purple を見よ
purplish-black skipper		*Nisoniades rubescens* (Möschler)	（チョウ目、セセリチョウ科）新北区、新熱帯区
purplish-blue cricket hunter		*Chlorion cyaneum* Dahlbom	（ハチ目、アナバチ科）新北区
purplish chafer	コガネムシ	*Mimela splendens* (Gyllenhal)	（コウチュウ目、コガネムシ科）日本
purplish copper	ムラサキベニシジミ	*Lycaena helloides* (Boisduval)	（チョウ目、シジミチョウ科）新北区、新熱帯区
purplish double-lined gray			projecta gray moth を見よ
purplish fritillary		*Clossiana montinus* (Scudder)	（チョウ目、タテハチョウ科）新北区
purplish fritillary			arctic fritillary を見よ
purplish looper moth			purple thorn を見よ
purplish marble		*Celypha rosaceana* (Schlager)	（チョウ目、ハマキガ科）旧北区
purplish Metarranthis moth		*Metarranthis homuraria* (Grote et Robinson)	（チョウ目、シャクガ科）新北区
purplish prominent	ムラサキシャチホコ	*Uropyia meticulodina* (Oberthür)	（チョウ目、シャチホコガ科）日本、旧北区
purplish stem borer			pink borer (1) を見よ
purplish tufted skipper			purplish-black skipper を見よ
purse casemake caddisflies			micro-caddisflies を見よ
purse casemaker caddisflies			micro-caddisflies を見よ

英名	和名	学名	所属、分布、ほか
purslane moth		*Euscirrhopterus gloveri* Grote et Robinson	(チョウ目、ヤガ科) 新北区
pursuer		*Potamarcha*	(トンボ目、トンボ科) の昆虫の総称
Pusilla crescent		*Dagon pusilla* (Salvin)	(チョウ目、タテハチョウ科) 新熱帯区
Pusilla purplewing		*Eunica pusilla* Bates	(チョウ目、タテハチョウ科) 新熱帯区
Pusilla skipper		*Sostrata pusilla* Godman et Salvin	(チョウ目、セセリチョウ科) 新熱帯区
puss			puss moth を見よ
puss caterpillar			puss caterpillar moth を見よ
puss caterpillar moth		*Megalopyge opercularis* (Smith)	(チョウ目、Megalopygidae) 新北区。カンキツ害虫
puss moth		*Cerura vinula* (Linnaeus)	(チョウ目、シャチホコガ科) 旧北区。日本亜種は *C. v. felina* Butler モクメシャチホコ
puss moths			prominents を見よ
puss moths			American puss moths を見よ
Pussy's toes Pyrausta moth		*Pyrausta unifascialis* (Packard)	(チョウ目、メイガ科) 新北区
pustular oak scale			golden oak scale (1) を見よ
Putman scale		*Diaspidiotus ancylus* (Putman)	(カメムシ目、マルカイガラムシ科) 新北区
Putman's cicada		*Platypedia minor* Uhler	(カメムシ目、セミ科) 新北区、豪州区
Putnam's looper moth		*Plusia putnami* Grote	(チョウ目、ヤガ科) 新北区
putty-patched moth		*Olethreutes griseoalbana* (Walsingham)	(チョウ目、ハマキガ科) 新北区
puzzle-mark skipperling		*Dalla dognini* (Mabille)	(チョウ目、セセリチョウ科) 新熱帯区
Pygas eighty-eight			Godart's numberwing を見よ
pygmy ant		*Plagiolepis pygmaea* (Latreille)	(ハチ目、アリ科) 旧北区
pygmy backswimmer		*Plea minutissima* Leach	(カメムシ目、マルミズムシ科) 旧北区
pygmy backswimmer bugs	マルミズムシ科	Pleidae	(カメムシ目) の昆虫の総称
pygmy backswimmers			pygmy backswimmer bugs を見よ
pygmy basker		*Aethriamanta rezia* Kirby	(トンボ目、トンボ科) エチオピア区
pygmy bee			pygmy mangold beetle を見よ
pygmy blue			western pygmy blue を見よ
pygmy damselfly		*Nehalennia speciosa* (Charpentier)	(トンボ目、イトトンボ科) 旧北区
pygmy grass blue		*Zizula hylax pygmaea* (Snellen)	(チョウ目、シジミチョウ科) 東洋区。tiny grass blue (1) を参照
pygmy grasshoppers			grouse locusts を見よ
pygmy hog sucking louse		*Haematopinus oliveri* Mishra et Singh	(シラミ目、ケモノジラミ科) 東洋区。絶滅危惧種
pygmy leafhoppers			dwarf leafhoppers を見よ

英名	和名	学名	所属、分布、ほか
pygmy mangel beetle			pygmy mangold beetle を見よ
pygmy mangold beetle		*Atomaria linearis* Stephens	（コウチュウ目、キスイムシ科）旧北区
pygmy mole cricket	ノミバッタ	*Xya japonica* (de Haan)	（バッタ目、ノミバッタ科）日本、東洋区
pygmy mole crickets	ノミバッタ科	Tridactylidae	（バッタ目）の昆虫の総称
pygmy moths			midget moths を見よ
pygmy Paectes moth		*Paectes pygmaea* Hübner	（チョウ目、ヤガ科）新北区
pygmy posy		*Drupadia rufotaenia rufotaenia* Fruhstorfer	（チョウ目、シジミチョウ科）東洋区
pygmy Redectis moth		*Redectis pygmaea* Grote	（チョウ目、ヤガ科）新北区。pygmy Redectis ともいう
pygmy sandgrinder		*Arenopsaltria pygmaea* (Distant)	（カメムシ目、セミ科）豪州区
pygmy scrub hopper		*Aeromachus pygmaeus* (Fabricius)	（チョウ目、セセリチョウ科）東洋区
pygmy snaketail		*Ophiogomphus howei* Bromley	（トンボ目、サナエトンボ科）新北区
pygmy soldier		*Oxycera pygmaea* (Fallén)	（ハエ目、ミズアブ科）旧北区
pygmy wisp		*Agriocnemis pygmaea* (Rambur)	（トンボ目、イトトンボ科）東洋区
pygmy yellowtail		*Cicadetta murrayensis* (Distant)	（カメムシ目、セミ科）豪州区
pyralid moths			grass moths (2) を見よ
pyralids			grass moths (2) を見よ
pyralina spreadwing			variegated skipper（1）を見よ
pyramid ant	ピラミッドアリ	*Conomyrma insana* (Burkley)	（ハチ目、アリ科）新北区
pyramid ants		*Conomyrma*	（ハチ目、アリ科）の昆虫の総称
Pyramus opal		*Chrysoritis pyramus* (Pennington)	（チョウ目、シジミチョウ科）エチオピア区
pyraustid moths			leaftiers を見よ
Pyrenees brassy ringlet		*Erebia rondoui* Oberthür	（チョウ目、タテハチョウ科）旧北区
Pyrgines			open-winged skippers を見よ
pyrgota fly		*Pyrgota undata* Wiedemann	（ハエ目、デガシラバエ科）新北区
pyrgotid flies			light flies を見よ
pyriform scale	ナシガタカタカイガラムシ	*Protopulvinaria pyriformis* (Cockerell)	（カメムシ目、カタカイガラムシ科）日本、新北区、汎熱帯。カンキツ、クチナシ害虫
pyroderces			pink cornworm を見よ
Pyropina phantom		*Cithaerias pyropina* Godman et Salvin	（チョウ目、タテハチョウ科）新熱帯区
pyrrhocorid bugs			red bugs (1) を見よ
pythid beetles	キカワムシ科	Pythidae	（コウチュウ目）の昆虫の総称
Python skipper		*Atrytonopsis python* (Edwards)	（チョウ目、セセリチョウ科）新北区、新熱帯区
pyxis hairstreak		*Nicolaea pyxis* (Johnson)	（チョウ目、シジミチョウ科）新熱帯区

英名	和名	学名	所属、分布、ほか
Q			
quaker		*Neopithecops zalmora* (Butler)	(チョウ目、シジミチョウ科) 東洋区
quaker (1)		*Neopithecops lucifer* (Rober)	(チョウ目、シジミチョウ科) 豪州区。*N. l. heria* (Fruhstorfer) も同英名
Quarre's fingertail		*Gomphidia quarrei* (Schouteden)	(トンボ目、サナエトンボ科) エチオピア区
Quasigadira hairstreak		*Ignata* sp.	(チョウ目、シジミチョウ科) 新熱帯区
Quasilatagus hairstreak		*Lathecla* sp.	(チョウ目、シジミチョウ科) 新熱帯区
Quasileda ministreak		*Ministrymon* sp.	(チョウ目、シジミチョウ科) 新熱帯区
Quebec Phlyctaenia moth		*Anania quebecensis* (Munroe)	(チョウ目、メイガ科) 新北区
queen			ハチ、アリ、シロアリなど社会性昆虫で活発に産卵する雌虫、いわゆる女王
queen	クイーンマダラ(ジョオウマダラ)	*Danaus gilippus* (Cramer)	(チョウ目、タテハチョウ科) 新北区、新熱帯区
queen			queen butterfly を見よ
Queen Alexandra's birdwing	アレクサンドラトリバネアゲハ	*Ornithoptera alexandrae* (Rothschild)	(チョウ目、アゲハチョウ科) 東洋区、豪州区。世界最大のチョウ
Queen Alexandra's sulphur	アレクサンドラモンキチョウ	*Colias alexandra* Edwards	(チョウ目、シロチョウ科) 新北区
queen butterfly		*Danaus gilippus berenice* (Cramer)	(チョウ目、タテハチョウ科) 新北区
queen butterfly			common yellow swallowtail を見よ
queen cracker	カスリタテハ	*Hamadryas arethusa* Denis et Schiffermüller	(チョウ目、タテハチョウ科) 旧北区
queen flasher		*Panacea regina* Bates	(チョウ目、タテハチョウ科) 新熱帯区
queen malachite		*Chlorolestes nylephtha* Barnard	(トンボ目、Synlestidae) エチオピア区
Queen of England frit			dark green fritillary を見よ
Queen of Spain fritillary	スペインヒョウモン	*Issoria lathonia* (Linnaeus)	(チョウ目、タテハチョウ科) 旧北区、東洋区
queen purple-tip	ツマムラサキシロチョウ	*Colotis regina* (Trimen)	(チョウ目、シロチョウ科) エチオピア区
queen swallowtail			Androgeus を見よ
Queen Victoria's birdwing	ビクトリアトリバネアゲハ	*Ornithoptera victoriae* Gray	(チョウ目、アゲハチョウ科) 東洋区
queen wasp			女王ハチ、雌ハチ
queens			monarchs を見よ　南米での英名
queens			tigers (1) を見よ
Queensland bollworm			peach moth を見よ
Queensland fruit fly	クイーンスランドミバエ	*Bactrocera tryoni* (Froggatt)	(ハエ目、ミバエ科) 豪州区、大洋区

英名	和名	学名	所属、分布、ほか
Queensland pine beetle		*Calymmaderus incisus* Lea	(コウチュウ目、シバンムシ科) 豪州区
Queensland pink bollworm			pink-spotted bollworm を見よ
quercus caterpillar	ツマキシャチホコ	*Phalera assimilis* (Bremer et Grey)	(チョウ目、シャチホコガ科) 日本、旧北区。クリ、カシ類害虫
quercus gall midge	クヌギヒメコブタマバエ	*Ametrodiplosis acutissima* (Monzen)	(ハエ目、タマバエ科) 日本、旧北区。カシ類害虫
quercus gall wasp	ナライガタマバチ	*Andricus mukaigawae* (Mukaigawa)	(ハチ目、タマバチ科) 日本、旧北区。カシ類害虫
quercus gall wasp (1)	クヌギイガタマバチ	*Trichagalma serratae* (Ashmead)	(ハチ目、タマバチ科) 日本。カシ類害虫
quercus gall wasp (2)	ナラリンゴタマバチ	*Biorhiza nawai* (Ashmead)	(ハチ目、タマバチ科) 日本。カシ類害虫
quercus hornet moth	カシワスカシバ	*Sesia rhynchioides* (Butler)	(チョウ目、スカシバガ科) 日本。クリ害虫
quercus lasiocampid	クヌギカレハ	*Knugia undans flaveola* (Motschulsky)	(チョウ目、カレハガ科) 日本。リンゴ、クリ、カシ類害虫
quercus lasiocampid (1)	ヤマダカレハ	*Knugia yamadai* (Nagano)	(チョウ目、カレハガ科) 日本、旧北区。クリ、カシ類害虫
quercus leafminer	アラカシハモグリバエ	*Japanagromyza quercus* (Sasakawa)	(ハエ目、ハモグリバエ科) 日本。カシ類害虫
quercus scale	カシフサカイガラムシ	*Asterolecanium gerplexum* Russell	(カメムシ目、フサカイガラムシ科) 日本
quercus spined aphid			oak aphid (1) を見よ
quercus stem gall midge			quercus gall midge を見よ
quercus stink bug	クヌギカメムシ	*Urostylis westwoodi* Scott	(カメムシ目、クヌギカメムシ科) 日本、旧北区。カシ類害虫
quercus sucker	クリトガリキジラミ	*Trioza quercicola* Shinji	(カメムシ目、トガリキジラミ科) 日本。クリ、カシ類害虫
quercus thrips	カシアザミウマ	*Pseudanaphothrips querci* (Moulton)	(アザミウマ目、アザミウマ科) 東洋区
quercus tree borer	カシコスカシバ	*Synanthedon quercus* (Matsumura)	(チョウ目、スカシバガ科) 日本、旧北区
question mark	オナガシータテハ	*Polygonia interrogationis* Fabricius	(チョウ目、タテハチョウ科) 新北区、新熱帯区
question mark			angelwings を見よ
Quickelberge's blue		*Lepidochrysops quickelbergei* Swanepoel	(チョウ目、シジミチョウ科) エチオピア区
Quickelberge's copper		*Aloeides quickelbergei* Tite et Dickson	(チョウ目、シジミチョウ科) エチオピア区
quiet-calling katydids	ヒメツユムシ亜科	Meconematinae	(バッタ目、キリギリス科) の昆虫の総称
quiet underwing			sweet underwing moth を見よ
quilted metalmark		*Voltinia umbra* (Boisduval)	(チョウ目、シジミタテハ科) 新熱帯区
quince borer		*Coryphodema tristis* (Drury)	(チョウ目、ボクトウガ科) エチオピア区
quince cottony scale			Suwako cottony-cussion scale を見よ

英名	和名	学名	所属、分布、ほか
quince curculio		*Conotrachelus crataegi* Walsh	（コウチュウ目、ゾウムシ科）新北区
quince moth		*Euzophera bigella* Zeller	（チョウ目、メイガ科）旧北区
quince scale			Latania scale を見よ
quince treehopper		*Glossonotus crataegi* (Fitch)	（カメムシ目、ツノゼミ科）新北区
quino checkerspot		*Euphydryas editha quino* Behr	（チョウ目、タテハチョウ科）新北区、新熱帯区
Quintina glasswing		*Oleria quintina* (Felder et Felder)	（チョウ目、タテハチョウ科）新熱帯区

英名	和名	学名	所属、分布、ほか
R			
rabbit bot fly		*Cuterebra cuniculi* (Clark)	（ハエ目、ヒツジバエ科）新北区
rabbit bot fly		*Cuterebra princeps* (Austen)	（ハエ目、ヒツジバエ科）新北区
rabbit bots		Cuterebridae	（ハエ目）の昆虫の総称　ネズミ類に寄生
rabbit brush webbing moth		*Synnoma lynosyrana* Walsingham	（チョウ目、ハマキガ科）新北区
rabbitbush flower moth		*Schinia unimacula* Smith	（チョウ目、ヤガ科）新北区
rabbit flea			European rabbit flea を見よ
rabbit louse	ウサギジラミ	*Haemodipus ventricosus* (Denny)	（シラミ目、ホソケジラミ科）日本、全北区、豪州区。ウサギに寄生
rabbit stickfast flea		*Echidnophaga perilis* Jordan	（ノミ目、ヒトノミ科）豪州区
rabbit sucking louse			rabbit louse を見よ
racket-tailed emerald		*Dorocordulia libera* (Selys)	（トンボ目、エゾトンボ科）新北区
Radcliffe's dagger moth		*Acronicta radcliffei* Harvey	（チョウ目、ヤガ科）新北区
Radford's flame shoulder		*Ochropleura leucogaster* (Freyer)	（チョウ目、ヤガ科）旧北区
radiant skipper		*Choranthus radians* Lucas	（チョウ目、セセリチョウ科）新北区
radiant skipper		*Callimormus radiola radiola* (Mabille)	（チョウ目、セセリチョウ科）新熱帯区
radiata pine shoot weevil		*Merimnetes oblongus* Blackburn	（コウチュウ目、ゾウムシ科）豪州区
radicoles			bark aphids を見よ
radish fly		*Paregle radicum* (Linnaeus)	（ハエ目、イエバエ科）旧北区
radish fly			cabbage root fly (1) を見よ
radish maggot			cabbage root fly (3) を見よ
radish root maggot	ヒメダイコンバエ	*Delia planipalpis* (Stein)	（ハエ目、ハナバエ科）日本。全北区。アブラナ科野菜害虫
Raffles's oakblue		*Arhopala pseudomuta* (Staudinger)	（チョウ目、シジミチョウ科）東洋区
Raffray's white		*Belenois raffrayi* (Oberthür)	（チョウ目、シロチョウ科）エチオピア区
ragged skipper		*Caprona cassualalla* Bethune-Baker	（チョウ目、セセリチョウ科）エチオピア区
ragged skipper		*Caprona pillaana* Wallengren	（チョウ目、セセリチョウ科）エチオピア区
ragged Temnora		*Temnora zantus* (Herrich-Schäffer)	（チョウ目、スズメガ科）エチオピア区
raggedy skimmer			black saddlebags skimmer を見よ
ragi root aphid			Oriental grass root aphid を見よ
ragi stem borer			pink borer (1) を見よ
ragweed beetle	ブタクサハムシ	*Ophraella communa* LeSage	（コウチュウ目、ハムシ科）日本

英名	和名	学名	所属、分布、ほか
ragweed borer		*Eucosma strenuana* (Walker)	(チョウ目、ハマキガ科) 新北区
ragweed borer moth			parthenium stemgalling moth を見よ
ragweed flower moth		*Schinia rivulosa* Guenée	(チョウ目、ヤガ科) 新北区。ragweed flower ともいう
ragweed plant bug		*Chlamydatus associatus* (Uhler)	(カメムシ目、カスミカメムシ科) 新北区。ブタクサにつく
ragwort flea beetle		*Longitarsus flavicornis* (Stephens)	(コウチュウ目、ハムシ科) 豪州区
ragwort flea beetle		*Longitarsus jacobaeae* Waterhouse	(コウチュウ目、ハムシ科) 豪州区
ragwort stem borer moth		*Papaipema insulidens* (Bird)	(チョウ目、ヤガ科) 新北区。ragwort stem borer ともいう
rain beetle			violet ground beetle を見よ
rain beetles		*Pleocoma*	(コウチュウ目、コガネムシ科) の昆虫の総称 雌は無翅、小雨で現れる
rain breeze cleg			cleg (1) を見よ
rain breeze fly			cleg (1) を見よ
rainforest Acraea		*Acraea boopis* Wichgraf	(チョウ目、タテハチョウ科) エチオピア区
rain-forest Acraea		*Acraea admatha* Hewitson	(チョウ目、タテハチョウ科) エチオピア区
rainforest brown		*Cassionympha cassius* (Godart)	(チョウ目、タテハチョウ科) エチオピア区
rain-forest faceted-skipper		*Synapte silius* (Latreille)	(チョウ目、セセリチョウ科) 新熱帯区
rain-forest hoary-skipper		*Carrhenes calidius* Godman et Salvin	(チョウ目、セセリチョウ科) 新熱帯区
rainforest hunter		*Zephyrogomphus longipositor* (Watson)	(トンボ目、サナエトンボ科) 豪州区
rainforest vicetail		*Hemigomphus theischingeri* Watson	(トンボ目、サナエトンボ科) 豪州区
rain moth			bardee grub (1) を見よ
rainbow bluet		*Enallagma antennatum* (Say)	(トンボ目、イトトンボ科) 新北区
rainbow grasshopper		*Dactylotum bicolor variegatum* (Scudder)	(バッタ目、バッタ科) 新北区。翅は腹部より短い
rainbow leaf beetle		*Chrysolina cerealis* (Linnaeus)	(コウチュウ目、ハムシ科) 旧北区
rainbow metalmark		*Caria trochilus* Erichson	(チョウ目、シジミタテハ科) 新熱帯区
rainbow stag beetle			king stag beetle を見よ
rainpool spreadwing		*Lestes forficula* Rambur	(トンボ目、アオイトトンボ科) 新北区、新熱帯区
raisin bees		*Chalcodoma*	(ハチ目、ハキリバチ科) の昆虫の総称
raisin honey			rice moth を見よ
raisin moth	ホシブドウマダラメイガ	*Cadra figulilella* (Gregson)	(チョウ目、メイガ科) 日本、汎世界。乾果害虫
raisin moth			almond moth を見よ
Rajah Brooke's birdwing	アカエリトリバネアゲハ	*Trogonoptera brookiana* Wallace	(チョウ目、アゲハチョウ科) 東洋区。*T. b. albescens* Rothschild も同英名

英名	和名	学名	所属、分布、ほか
Rajahs Charaxes	フタオチョウ属	*Charaxes*	(チョウ目、タテハチョウ科) の昆虫の総称
Ramajana bambootail		*Disparoneura ramajana* Lieftinck	(トンボ目、Protoneuridae) 東洋区
Rambur's forktail		*Ischnura ramburii* (Selys)	(トンボ目、イトトンボ科) 大洋区、新北区、新熱帯区
ramie caterpillar			ramie moth を見よ
ramie longicorn	ラミーカミキリ	*Paraglenea fortunei* (Saunders)	(コウチュウ目、カミキリムシ科) 日本、旧北区
ramie longicorn beetle			ramie longicorn を見よ
ramie moth	フクラスズメ	*Arcte coerula* (Guenée)	(チョウ目、ヤガ科) 日本、旧北区、東洋区、豪州区、大洋区。ナシ、モモなど果樹、カラムシの害虫
ramie moth caterpillar			ramie moth を見よ
Ramon's blue		*Hemiargus ramon* (Dognin)	(チョウ目、シジミチョウ科) 新熱帯区
ranchman's tiger moth		*Platyprepia virginalis* (Boisduval)	(チョウ目、ヒトリガ科) 新北区
ranchman's tiger moth			rangeland tiger moth を見よ
range caterpillar			New Mexico range caterpillar を見よ
range crane fly		*Tipula simplex* Doane	(ハエ目、ガガンボ科) 新北区。芝害虫
range grass moth			clover looper を見よ
rangeland tiger moth		*Platyprepia guttata* Boisduval	(チョウ目、ヒトリガ科) 新北区
Rannoch brindled beauty		*Lycia lapponaria* (Boisduval)	(チョウ目、シャクガ科) 旧北区
Rannoch case	コケモモツツミノガ	*Coleophora glitzella* Hofmann	(チョウ目、ツツミノガ科) 旧北区
Rannoch looper	キトビエダシャク	*Itame brunneata* (Thunberg)	(チョウ目、シャクガ科) 日本、旧北区
Rannoch looper		*Itame fulvaria* Villers	(チョウ目、シャクガ科) 旧北区
Rannoch sprawler		*Brachionycha nubeculosa* (Esper)	(チョウ目、ヤガ科) 旧北区。日本亜種は *B. n. jezoensis* Matsumura エゾモクメキリガ
rapacious flangetail		*Ictinogomphus rapax* (Rambur)	(トンボ目、サナエトンボ科) 東洋区
rape beetle			pollen beetle を見よ
rape blossom beetle			pollen beetle を見よ
rape bug			cabbage bug (2) を見よ
rape caterpillar			red-belly grained moth を見よ
rape flea beetle			cabbage stem flea beetle を見よ
rape leaf beetle	ホタルハムシ	*Monolepta dichroa* Harold	(コウチュウ目、ハムシ科) 日本
rape stem weevil		*Ceutorhynchus napi* (Gyllenhal)	(コウチュウ目、ゾウムシ科) 旧北区
rape winter stem weevil			black winter weevil を見よ

英名	和名	学名	所属、分布、ほか
rape worm	ウスベニノメイガ	*Evergestis extimalis* (Scopoli)	（チョウ目、メイガ科）日本、旧北区。アブラナ科野菜の害虫
raphidians			snake-flies (1) を見よ
raphidiid snakeflies			snake-flies (3) を見よ
rapid plant bug		*Adelphocoris rapidus* (Say)	（カメムシ目、カスミカメムシ科）新北区。体長7〜8 mm。スイバ他につく
rapids clubtail		*Gomphus quadricolor* Walsh	（トンボ目、サナエトンボ科）新北区
rare albatross	アンガウルシロチョウ	*Appias ada* (Stoll)	（チョウ目、シロチョウ科）豪州区
rare bladder cicada		*Cystopsaltria immaculata* Goding et Froggatt	（カメムシ目、セミ科）豪州区
rare blue doctor			long-tailed metalmark を見よ
rare borer moth		*Papaipema dribi* Barnes et Benjamin	（チョウ目、ヤガ科）新北区。rare borer ともいう
rare click beetles			cerophytid beetles を見よ
rare longtail		*Urbanus viridis* Freeman	（チョウ目、セセリチョウ科）新熱帯区
rare Morant skipper		*Parosmodes lentiginosa* (Holland)	（チョウ目、セセリチョウ科）エチオピア区
rare Musanga Acraea		*Acraea vesperalis* Grose-Smith	（チョウ、タテハチョウ科）エチオピア区
rare pathfinder skipper		*Pardaleodes xanthopeplus* Holland	（チョウ目、セセリチョウ科）エチオピア区
rare red-eye		*Chaetocneme denitza* (Hewitson)	（チョウ目、セセリチョウ科）豪州区
rare sand quaker moth		*Caradrina meralis* Morrison	（チョウ目、ヤガ科）新北区
rare silver spot		*Aphnaeus argyrocyclus* Holland	（チョウ目、シジミチョウ科）エチオピア区
rare skipper		*Problema bulenta* (Boisduval et LeConte)	（チョウ目、セセリチョウ科）新北区
rare spring moth		*Heliomata infulata* (Grote)	（チョウ目、シャクガ科）新北区
rare spurwing		*Systasea microsticta* Dyar	（チョウ目、セセリチョウ科）新熱帯区
rare tufted skipper			confused skipper を見よ
rare white-spot skipper		*Trapezites luteus* (Tepper)	（チョウ目、セセリチョウ科）豪州区
rascal dart moth			pale-sided cutworm moth を見よ
Raschke's cosmet		*Mompha raschkiella* (Zeller)	（チョウ目、カザリバガ科）旧北区
rasin bees		*Chalicodoma*	（ハチ目、ハキリバチ科）の昆虫の総称
rasin moth			almond moth を見よ
Raspa skipper		*Paratrytone raspa* (Evans)	（チョウ目、セセリチョウ科）新熱帯区
raspberry and loganberry beetle			raspberry beetle (1) を見よ
raspberry aphid	マメクロアブラムシ	*Acyrthosiphon chelidonii* (Kaltenbach)	（カメムシ目、アブラムシ科）旧北区。エンドウ害虫
raspberry aphid (1)		*Aphis idaei* van der Goot	（カメムシ目、アブラムシ科）旧北区

英名	和名	学名	所属、分布、ほか
raspberry aphid			rubus aphid (2) を見よ
raspberry beetle (1)		*Byturus tomentosus* (De Geer)	(コウチュウ目、キスイモドキ科) 旧北区
raspberry beetle (2)		*Byturus fumatus* Fabricius	(コウチュウ目、キスイモドキ科) 旧北区
raspberry beetles			fruit worm beetles を見よ
raspberry borer			raspberry bright を見よ 幼虫の英名
raspberry bright		*Lampronia rubiella* (Bjerkander)	(チョウ目、マガリガ科) 旧北区。raspberry bud moth ともいう
raspberry bud dagger moth			southern oak dagger moth を見よ raspberry bud moth ともいう
raspberry budmoth		*Acronicta intermedia* Warren	(チョウ目、ヤガ科) 新北区
raspberry budmoth		*Heterocrossa rubophaga* Dugdale	(チョウ目、シンクイガ科) 豪州区
raspberry cane borer		*Oberea bimaculata* Olivier	(コウチュウ目、カミキリムシ科) 新北区
raspberry cane borer		*Phorbia rubivora* Coquillett	(ハエ目、ハナバエ科) 新北区
raspberry cane maggot			loganberry cane fly を見よ 北米での幼虫の英名
raspberry cane midge		*Thomasiniana theobaldi* Barnes	(ハエ目、タマバエ科) 旧北区
raspberry clearwing			raspberry clearwing moth を見よ
raspberry clearwing moth	カラフトスカシバ	*Bembecia hylaeiformis* Laspeyres	(チョウ目、スカシバガ科) 日本、旧北区
raspberry crown borer		*Pennisetia marginata* (Harris)	(チョウ目、スカシバガ科) 新北区、大洋区。raspberry crown borer moth ともいう
raspberry crown borer			raspberry clearwing moth を見よ
raspberry flea beetle		*Batophila aerata* (Marshall)	(コウチュウ目、ハムシ科) 旧北区
raspberry flea beetle (1)		*Batophila rubi* (Paykull)	(コウチュウ目、ハムシ科) 旧北区
raspberry flower midge			gooseberry flower midge を見よ
raspberry fruit weevil			raspberry flea beetle (1) を見よ
raspberry fruitworm		*Byturus univolor* Say	(コウチュウ目、キスイモドキ科) 新北区
raspberry fruitworms			fruit worm beetles を見よ
raspberry gall fly			blackberry stem gall midge を見よ
raspberry leaf sawfly			small raspberry sawfly (2) を見よ
raspberry leaf-mining sawfly		*Metallus albipes* (Cameron)	(ハチ目、ハバチ科) 旧北区
raspberry leaf-mining sawfly		*Metallus pumilus* (Klug)	(ハチ目、ハバチ科) 旧北区
raspberry leafroller		*Olethreutes permundana* (Clemens)	(チョウ目、ハマキガ科) 新北区。raspberry leafroller moth ともいう

英名	和名	学名	所属、分布、ほか
raspberry leaf-roller moth		*Epinotia medioviridana* (Kearfott)	(チョウ目、ハマキガ科) 新北区
raspberry looper			large looper moth を見よ
raspberry moth		*Lampronia corticella* (Linnaeus)	(チョウ目、マガリガ科) 旧北区
raspberry moth			raspberry bright を見よ
raspberry Pyrausta moth		*Pyrausta signatalis* (Walker)	(チョウ目、メイガ科) 新北区
raspberry root borer			raspberry crown borer を見よ
raspberry root borer			raspberry clearwing moth を見よ
raspberry sawfly		*Empria tridens* (Konow)	(ハチ目、ハバチ科) 旧北区
raspberry sawfly (1)		*Monophadnoides geniculatus* (Hartig)	(ハチ目、ハバチ科) 日本、全北区
raspberry sawfly			small raspberry sawfly (1) を見よ　豪州での英名
raspberry shoot borer			raspberry bright を見よ
raspberry stem bud moth			raspberry bright を見よ
raspberry stem gall midge			blackberry stem gall midge を見よ
raspberry wave moth		*Leptostales laevitaria* (Geyer)	(チョウ目、シャクガ科) 新北区
raspberry weevil			clay-coloured weevil を見よ
rastrococcus mealybug		*Rastrococcus truncatispinus* Williams	(カメムシ目、コナカイガラムシ科) 豪州区
rat flea		*Xenopsylla vexabilis* (Jordan)	(ノミ目、ヒトノミ科) 豪州区、大洋区、新北区
rat flea			northern rat flea を見よ
rat flea			Oriental rat flea を見よ
rat fleas		*Xenopsylla*	(ノミ目、ヒトノミ科) の昆虫の総称
rat fleas			rodent fleas (1) を見よ
rat louse			spined rat louse を見よ　英国での英名
rat-tailed maggot			drone fly を見よ　幼虫の尾端がネズミの尾状に長いことに由来
rat-tailed maggots			drone flies を見よ　幼虫の英名
Rathvon's scarab		*Lichnanthe rathvoni* LeConte	(コウチュウ目、コガネムシ科) 新北区
ratoon shootborer		*Ephysteris promptella* (Staudinger)	(チョウ目、キバガ科) 豪州区
rattle grasshopper		*Psophus stridulus* (Linnaeus)	(バッタ目、バッタ科) 旧北区。後翅が赤褐色
rattle moth			bella moth を見よ
rattlebox moth			ornate Bella moth を見よ
rattlebox moth			Bella moth を見よ　北米での英名
rattler conehead		*Neoconocephalus affinis* (Beauvois)	(バッタ目、キリギリス科) 新北区

英名	和名	学名	所属、分布、ほか
rattler round-winged katydid		*Amblycorypha rotundifolia* (Scudder)	（バッタ目、キリギリス科）新北区
rattlesnake-master borer moth		*Papaipema eryngii* Bird	（チョウ目、ヤガ科）新北区
Ratzeburg's bell		*Zeiraphera ratzeburgiana* (Ratzeburg)	（チョウ目、ハマキガ科）全北区
Ratzer's ringlet	クリステイベニヒカゲ	*Erebia christi* Ratzer	（チョウ目、タテハチョウ科）旧北区
Rauparaha's copper		*Lycaena rauparaha* (Fereday)	（チョウ目、シジミチョウ科）豪州区
raven-feather case		*Coleophora fuscedinella* Zeller	（チョウ目、ツツミノガ科）旧北区
raven-feather case			apple casebearer を見よ　英国での英名
raven-feather owlet		*Scythris grandipennis* (Haworth)	（チョウ目、キヌバコガ科）旧北区
Rawson's metalmark		*Calephelis rawsoni* McAlpine	（チョウ目、シジミタテハ科）新北区
rayed blue		*Actizera lucida* (Trimen)	（チョウ目、シジミチョウ科）エチオピア区
rayed blue	ウラジロミナミシジミ	*Candalides heathi* (Cox)	（チョウ目、シジミチョウ科）豪州区
rayed purplewing	マルビナアイイロタテハ	*Eunica malvina albida* Jenkins	（チョウ目、タテハチョウ科）新熱帯区。*E. m. almae* Vargas, Llorente et Luis も同英名。Malvina purplewing を参照
rayed sister		*Adelpha melanthe* (Bates)	（チョウ目、タテハチョウ科）新熱帯区
Raymond's bush cricket		*Yersinella raymondi* (Yersin)	（バッタ目、キリギリス科）旧北区
Raymundo's skipper		*Myrinia raymundo* Freeman	（チョウ目、セセリチョウ科）新熱帯区
Ray's midget		*Phyllonorycter schreberella* (Fabricius)	（チョウ目、ホソガ科）旧北区
razor grinder		*Henicopsaltria eydouxii* (Guérin-Méneville)	（カメムシ目、セミ科）豪州区
Razowski's Aethes moth		*Aethes razowskii* Sabourin et Miller	（チョウ目、ホソハマキガ科）新北区
Reakirt's blue	クロテンチビウラナミシジミ	*Hemiargus isola* (Reakirt)	（チョウ目、シジミチョウ科）新北区
real stink beetle			stink beetle を見よ
Real's wood white		*Leptidea reali* (Reissinger)	（チョウ目、シロチョウ科）旧北区
reaper dart moth			dark-sided cutworm を見よ
rear horses			mantids (2) を見よ　カマキリは驚くと馬のように棒立ちになることから
Reaumur's long-horn		*Adela rufimitrella* (Scopoli)	（チョウ目、マガリガ科）旧北区
Rebel's striped hawk		*Hippotion rebeli* Rothschild et Jordan	（チョウ目、スズメガ科）エチオピア区
recondite webworm moth		*Diathrausta reconditalis* (Walker)	（チョウ目、メイガ科）新北区
rectal horse bot fly			horse bot fly を見よ

英名	和名	学名	所属、分布、ほか
rectilinea slug moth		*Apoda rectilinea* (Grote et Robinson)	(チョウ目、イラガ科) 新北区
red admirable			red admiiral (1) を見よ
red admiral		*Bassaris gonerilla* (Fabricius)	(チョウ目、タテハチョウ科) 豪州区
red admiral (1)	アタランタアカタテハ	*Vanessa atalanta* (Linnaeus)	(チョウ目、タテハチョウ科) 全北区。*V. a. rubria* (Fruhstorfer) も同英名。red admiral butterfly ともいう
red and black burying beetle		*Nicrophorus marginatus* Weise	(コウチュウ目、シデムシ科) 新北区
red and black froghopper			black and red froghopper を見よ
red and black wasp mimic			Florida oakgoal moth を見よ
red and blue beetle			Australian rice beetle を見よ
red-and-blue firetail		*Telebasis rubricauda* Bick et Bick	(トンボ目、イトトンボ科) 新熱帯区
red-and-blue leafhopper			candy-striped leafhopper を見よ
red-and-green stink bug		*Banasa dimiata* (Say)	(カメムシ目、カメムシ科) 新北区
red-and-white wave moth		*Idaea basinta* (Schaus)	(チョウ目、シャクガ科) 新北区
red ant		*Dorylus orientalis* Westwood	(ハチ目、アリ科) 東洋区
red ant	ケズネアカヤマアリ	*Formica (Formica) truncorum* Fabricius	(ハチ目、アリ科) 日本、旧北区
red ant	シワクシケアリ	*Myrmica ruginodis* Nylander	(ハチ目、アリ科) 日本
red ant (1)	ツムギアリ	*Oecophilla smaragdina* Fabricius	(ハチ目、アリ科) 東洋区、豪州区。樹上で葉を紡ぎあわせ、手まり状の巣を作る。攻撃性が強い
red ant (2)	キイロクシケアリ	*Myrmica rubra* (Linnaeus)	(ハチ目、アリ科) 日本、旧北区。この英名はイエヒメアリをさすこともある
red ant			common red ant を見よ
red ant			slave-maker ant を見よ
red ants			red myrmicine ants を見よ
red apple capsid		*Psallus ambiguus* (Fallén)	(カメムシ目、カスミカメムシ科) 旧北区
red assassin bug		*Haematoloecha rubescens* Distant	(カメムシ目、サシガメ科) 新北区
red-back phytometra	セアカキンウワバ	*Erythroplusia pyropia* (Butler)	(チョウ目、ヤガ科) 日本、東洋区
red-back stink bug	セアカツノカメムシ	*Acanthosoma denticauda* Jakovlev	(カメムシ目、ツノカメムシ科) 日本、旧北区
red-backed cutworm moth		*Euxoa (Euxoa) ochrogaster* (Guenée)	(チョウ目、ヤガ科) 新北区。幼虫は red-backed cutworm
red-backed froghopper	ムネアカアワフキ	*Hindoloides bipunctatus* (Haupt)	(カメムシ目、トゲアワフキムシ科) 日本、東洋区
redbacked oedemerid		*Ananea bicolor* (Fairmaire)	(コウチュウ目、カミキリモドキ科) 新北区

英名	和名	学名	所属、分布、ほか
red bamboo longicorn beetle	ベニカミキリ	*Purpuricenus temminckii* (Guérin-Méneville)	(コウチュウ目、カミキリムシ科) 日本、旧北区、東洋区。タケ害虫
red-band fritillary			spotted fritillary を見よ
red-banded Aemilia		*Aemilia ambigua* (Strecker)	(チョウ目、ヒトリガ科) 新北区
red-banded Altinote		*Altinote dicaeus* Latreille	(チョウ目、タテハチョウ科) 新熱帯区。zea clearwing を参照
red-banded blister beetle		*Mylabris ligata* Marsham	(コウチュウ目、ツチハンミョウ科) エチオピア区
red-banded firetip		*Amysoria galgala* (Hewitson)	(チョウ目、セセリチョウ科) 新熱帯区
red-banded hairstreak		*Calycopis cecrops* (Fabricius)	(チョウ目、シジミチョウ科) 新北区
red-banded Jezebel			Union Jack butterfly を見よ
red-banded leafhopper			candy-striped leafhopper を見よ
red banded leaf roller			apple leaf roller (2) を見よ
red banded leaf roller			pine tube moth (1) を見よ
red-banded longicorn beetle	アカジマトラカミキリ	*Akajimatora bella* (Matsumura et Matsushita)	(コウチュウ目、カミキリムシ科) 日本。ケヤキ害虫。red-banded longicorn beetle ともいう
red-banded Pereute		*Pereute leucodrosime* (Kollar)	(チョウ目、シロチョウ科) 新熱帯区
red-banded sand wasp			sand wasp (1) を見よ
red-banded shield bug			unibanded stink bug を見よ
red-banded thrips	アカオビアザミウマ	*Selenothrips rubrocinctus* (Giard)	(アザミウマ目、アザミウマ科) 日本、豪州区、新北区、エチオピア区、東洋区、新熱帯区。カキ、マンゴー害虫
red-banded widow		*Dingana alticola* Henning et Henning	(チョウ目、タテハチョウ科) エチオピア区
red bandit		*Cicadetta melete* (Walker)	(カメムシ目、セミ科) 豪州区
red baron		*Urothemis aliena* Selys	(トンボ目、トンボ科) 豪州区
red-barred Amarynthis	ベニスジシジミタテハ	*Amarynthis meneria* (Cramer)	(チョウ目、シジミタテハ科) 新熱帯区
red-barred gold		*Micropteryx thunbergella* (Fabricius)	(チョウ目、コバネガ科) 旧北区
red-barred hairstreak		*Badecla lanckena* Schaus	(チョウ目、シジミチョウ科) 新熱帯区
red barred tortrix			fruit tree tortrix を見よ　red-barred tortrix moth ともいう
red-base Jezebel	アカネシロチョウ	*Delias pasithoe* (Linnaeus)	(チョウ目、シロチョウ科) 旧北区、東洋区。*D. p. parthenope* Wallace も同英名
red-based groundstreak		*Arumecla galliena* (Hewitson)	(チョウ目、シジミチョウ科) 新熱帯区
red-based longicorn beetle			red-shouldered longicorn beetle を見よ
red-based tiger longicorn			red-shouldered tiger longicorn beetle を見よ

英名	和名	学名	所属、分布、ほか
red basker		*Urothemis assignata* (Selys)	(トンボ目、トンボ科) エチオピア区
red-bellied aphid			rice root aphid を見よ
red belly		*Baeturia varicolor* Distant	(カメムシ目、セミ科) 豪州区
red-belly black-dotted arctiid	アカハラゴマダラヒトリ	*Spilosoma punctaria* (Stoll)	(チョウ目、ヒトリガ科) 日本、旧北区
red-belly grained moth	キバラモクメキリガ	*Xylena formosa* (Butler)	(チョウ目、ヤガ科) 日本、旧北区、東洋区
red-belly tussock moth			fir tussock moth (1) を見よ
red-belted awl-fly		*Xylophagus cinctus* (De Geer)	(ハエ目、キアブ科) 旧北区
red-belted clearwing			apple clearwing moth を見よ
red-black oedemerid		*Eobia bicolor* (Fairmaire)	(コウチュウ目、カミキリモドキ科) 新北区
red blackboy		*Gudanga boulayi* Distant	(カメムシ目、セミ科) 豪州区
red-blue checkered beetle		*Trichodes nuttalli* (Kirby)	(コウチュウ目、カッコウムシ科) 新北区
red-bodied swallowtail		*Pachliopta polydorus* (Linnaeus)	(チョウ目、アゲハチョウ科) 豪州区
red bollworm		*Diparopsis castanea* Hampson	(チョウ目、ヤガ科) エチオピア区、新熱帯区。南米での英名。ワタ害虫
red bollworm (1)		*Diparopsis watersi* Rothschild	(チョウ目、ヤガ科) エチオピア区
red bollworms		*Diparopsis*	(チョウ目、ヤガ科) の昆虫の総称
redbolts		*Rhodothemis*	(トンボ目、トンボ科) の昆虫の総称
red-bordered brown		*Gyrocheilus patrobas* (Hewitson)	(チョウ目、タテハチョウ科) 新北区
red-bordered emerald moth		*Nemoria lixaria* (Guenée)	(チョウ目、シャクガ科) 新北区
red-bordered metalmark		*Caria ino* Godman et Salvin	(チョウ目、シジミタテハ科) 新北区
red-bordered Pixie			Pixie を見よ
red-bordered satyr			red-bordered brown を見よ
red-bordered stink bug		*Edessa rufomarginata* (De Geer)	(カメムシ目、カメムシ科) 新熱帯区
red-bordered wave moth		*Idaea demissaria* (Hübner)	(チョウ目、シャクガ科) 新北区
red borer			red coffee borer を見よ
red branch borer			red coffee borer を見よ
redbreast		*Papilio alcmenor* C. et R. Felder	(チョウ目、アゲハチョウ科) 東洋区
red-breast Jezebel	ミヤマアカネシロチョウ	*Delias acalis* (Godart)	(チョウ目、シロチョウ科) 東洋区
red-brindled dwarf		*Elachista rufocinerea* (Haworth)	(チョウ目、クサモグリガ科) 旧北区
red brown paper wasp		*Polistes olivaceus* (De Geer)	(ハチ目、スズメバチ科) 大洋区

英名	和名	学名	所属、分布、ほか
red-brown rice-flour beetle			red flour beetle を見よ
red bud borer		*Thomasiniana oculiperda* (Rubsaamen)	（ハエ目、タマバエ科）旧北区
redbud bruchid		*Gibbobruchus mimus* (Say)	（コウチュウ目、ハムシ科）新北区
redbud leaf-folder		*Fascista cercerisella* (Chambers)	（チョウ目、キバガ科）新北区。成虫は redbud leaffolder moth
red bud maggot			red bud borer を見よ　幼虫の英名
red bug		*Largus succinetus* (Linnaeus)	（カメムシ目、オオホシカメムシ科）新北区。体長 17 mm
red bug (1)		*Pyrrhocoris apterus* (Linnaeus)	（カメムシ目、ホシカメムシ科）旧北区
red bugs (1)	オオホシカメムシ科	Largidae	（カメムシ目）の昆虫の総称
red bugs (2)	ホシカメムシ科	Pyrrhocoridae	（カメムシ目）の昆虫の総称
red bulldog ant			bulldog ant (1) を見よ　red bull ant ともいう
red caliph		*Enispe euthymius* (Doubleday)	（チョウ目、タテハチョウ科）東洋区
red carpenter ant (1)		*Camponotus (Camponotus) ferrugineus* (Fabricius)	（ハチ目、アリ科）新北区
red carpenter ant (2)		*Camponotus floridanus* Buckley	（ハチ目、アリ科）新北区、新熱帯区
red carpet		*Xanthorhoe munitata* (Hübner)	（チョウ目、シャクガ科）旧北区
red carpet		*Xanthorhoe decoloraria* (Esper)	（チョウ目、シャクガ科）全北区
red cat louse			cat biting louse を見よ　南アフリカでの英名
red-ceder tip moth			mahogany shoot borer (1) を見よ
red chestnut	ネムロウスモンヤガ	*Cerasatis rubricosa* (Denis et Schiffermüller)	（チョウ目、ヤガ科）日本、旧北区
red chestnut moth			red chestnut を見よ
red chestnut rustic			red chestnut を見よ
red click beetle		*Elater sanguineus* Linnaeus	（コウチュウ目、コメツキムシ科）旧北区
red clover gall gnat		*Campylomyza ormerodi* (Kieffer)	（ハエ目、タマバエ科）旧北区
red clover seed weevil		*Tychius stephensi* Schoenherr	（コウチュウ目、ゾウムシ科）新北区
red clover thrips	ツメクサクダアザミウマ	*Haplothrips niger* (Osborn)	（アザミウマ目、クダアザミウマ科）日本、全北区、豪州区。マメ科牧草害虫
red coast Charaxes		*Charaxes macclouni* Butler	（チョウ目、タテハチョウ科）エチオピア区
red coat			bed bug (1) を見よ
red coffe borer	コーヒーゴマフボクトウ	*Zeuzera coffeae* Nietner	（チョウ目、ボクトウガ科）日本、東洋区、豪州区。コーヒー害虫
red coffee stem borer			red coffee borer を見よ
red-collared firetip		*Aspitha aegenoria* (Hewitson)	（チョウ目、セセリチョウ科）新熱帯区

英名	和名	学名	所属、分布、ほか
red-collared firetip		*Mysoria affinis* (Herrich-Schäffer)	(チョウ目、セセリチョウ科) 新熱帯区
red collared longhorn beetle		*Dinoptera collaris* (Linnaeus)	(コウチュウ目、カミキリムシ科) 旧北区
red-costate green moth	アカマエアオリンガ	*Earias pudicana* Staudinger	(チョウ目、ヤガ科) 日本、東洋区
red-costate tiger moth	マエアカヒトリ	*Amsacta lactinea* (Cramer)	(チョウ目、ヒトリガ科) 日本、東洋区。カンキツ害虫
red-crescent scrub-hairstreak			reddish hairstreak を見よ
red-crossed button slug moth		*Tortricidia pallida* (Herrich-Schäffer)	(チョウ目、イラガ科) 新北区
red copper			mountain copper を見よ
red cotton bug			cotton bug を見よ
red cotton bug			cotton stainer (1) を見よ
red cotton bugs		*Dysdercus*	(カメムシ目、ホシカメムシ科) の昆虫の総称 *D. cardinalis* (Gerstaecker), *D. suturellus* (Herrich-Schäffer) など新北区の種が多い
red cracker	ウラベニカスリタテハ	*Hamadryas amphinome* (Linnaeus)	(チョウ目、タテハチョウ科) 新熱帯区。*H. a. mexicana* (Lucas), *H. a. mazai* Jenkins も同英名
red cuckoo bee		*Nomada angelarum* Cockerell	(ハチ目、コシブトハナバチ科) 新北区
red cucujid		*Cucujus clavipes* Fabricius	(コウチュウ目、ヒラタムシ科) 新北区
red currant aphid			currant aphid を見よ
red currant-arrow aphid			red currant-arrow grass aphid を見よ
red currant-arrow grass aphid		*Aphis triglochinis* Theobald	(カメムシ目、アブラムシ科) 旧北区
red currant blister aphid			currant aphid を見よ
red daggerwing			ruddy daggerwing を見よ
red damselflies			narrow-winged damselflies を見よ
red dart moth		*Diarsia rubifera* (Grote)	(チョウ目、ヤガ科) 新北区
red date palm scale			red date scale を見よ
red date scale		*Phoenicococcus marlatti* Cockerell	(カメムシ目、Phoenicococcidae) 新北区。ナツメヤシ害虫
red deer louse		*Solenopotes burmeisteri* (Fahrenholz)	(シラミ目、ケモノホソジラミ科) 旧北区
red demon		*Mesene epaphus* (Stoll)	(チョウ目、シジミタテハ科) 新熱帯区
red-disc bushbrown		*Mycalesis oculus* Marshall	(チョウ目、タテハチョウ科) 東洋区
red-disked alpine	ムモンチャイロベニヒカゲ	*Erebia discoidalis discoidalis* Kirby	(チョウ目、タテハチョウ科) 新北区
red-dotted carpet moth		*Euchoeca rubropunctaria* Doubleday	(チョウ目、シャクガ科) 豪州区
red-dotted skipper		*Melanopyge erythrosticta* (Godman et Salvin)	(チョウ目、セセリチョウ科) 新熱帯区

英名	和名	学名	所属、分布、ほか
red dry bamboo longicorn			red bamboo longicorn beetle を見よ
red dwarf honey bee			lesser honey bee を見よ
red-edged Acleris moth		*Acleris albicomana* (Clemens)	（チョウ目、ハマキガ科）新北区
red-edged Euselasia		*Euselasia eumedia* (Hewitson)	（チョウ目、シジミタテハ科）新熱帯区
red-edged white		*Belenois rubrosignata rubrosignata* (Weymer)	（チョウ目、シロチョウ科）エチオピア区
redeye		*Psaltoda moerens* (Germar)	（カメムシ目、セミ科）豪州区。redeye cicada ともいう
red-eye bushbrown		*Heteropsis adolphei* (Guérein-Méneville)	（チョウ目、タテハチョウ科）東洋区
red-eyed button slug moth		*Heterogenea shurtleffi* Packard	（チョウ目、イラガ科）新北区
red-eyed damselfly		*Erythromma najas* (Hansemann)	（トンボ目、イトトンボ科）旧北区。日本亜種は *E. n.. baicalense* Belyshev ゴトウアカメイトンボ
red-eyed devil			greater arid-land katydid を見よ
red-eyed ringlet		*Coenonympha meadii* Edwards	（チョウ目、タテハチョウ科）新北区
red-eyed underleaf		*Teratophthalma maenades* (Hewitson)	（チョウ目、シジミタテハ科）新熱帯区
red-eyed wood nymph			Mead's wood nymph を見よ
redeyes		*Erionota*	（チョウ目、セセリチョウ科）の昆虫の総称
redeyes		*Matapa*	（チョウ目、セセリチョウ科）の昆虫の総称
red Egyptian ant			Pharaoh's ant を見よ
red elm bark weevil		*Magdalis armicollis* (Say)	（コウチュウ目、ゾウムシ科）新北区
red-faced dragonlet		*Erythrodiplax fusca* (Rambur)	（トンボ目、トンボ科）新北区、新熱帯区
red fairy		*Cicadetta froggatti* (Distant)	（カメムシ目、セミ科）豪州区
red false click beetle	アカヒメコメツキモドキ	*Anadastus filiformis* (Fabricius)	（コウチュウ目、コメツキモドキ科）日本、東洋区
red-feather carl		*Tischeria ekebladella* (Bjerkander)	（チョウ目、ムモンハモグリガ科）旧北区
red flasher	ルリオビウラベニタテハ	*Panacea prola* (Doubleday)	（チョウ目、タテハチョウ科）新熱帯区
red flat bark beetle			red cucujid を見よ
red flea		*Echidnophaga myrmecobii* Rothschild	（ノミ目、ヒトノミ科）豪州区
red flour beetle	コクヌストモドキ	*Tribolium castaneum* (Herbst)	（コウチュウ目、ゴミムシダマシ科）日本、汎世界。貯穀害虫
red flower longicorn	アカハナカミキリ	*Corymbia succedanea* (Lewis)	（コウチュウ目、カミキリムシ科）日本、旧北区、東洋区
red fly			wheat midge を見よ　アジアでの英名
red-footed bean weevil			broadbean weevil を見よ

英名	和名	学名	所属、分布、ほか
red forest Charaxes		*Charaxes boueti* Feisthamel	(チョウ目、タテハチョウ科) エチオピア区
red-fringed emerald moth		*Nemoria bistriaria* Hübner	(チョウ目、シャクガ科) 新北区
red-fringed ermil		*Falseuncaria ruficiliana* (Haworth)	(チョウ目、ホソハマキガ科) 旧北区
red-fronted emerald moth		*Nemoria rubrifrontaria* (Packard)	(チョウ目、シャクガ科) 新北区
red fungus bug		*Achilus flammeus* Kirby	(カメムシ目、コガシラウンカ科) 豪州区
red girdle moth		*Caripeta aequaliaria* Grote	(チョウ目、シャクガ科) 新北区
red glider (1)	ベニイロタテハ	*Cymothoe sangaris* (Godart)	(チョウ目、タテハチョウ科) エチオピア区
red glider (2)		*Cymothoe excelsa* Neustetter	(チョウ目、タテハチョウ科) エチオピア区
red glider			common glider (1) を見よ
red-green carpet		*Chloroclysta siterata* (Hufnagel)	(チョウ目、シャクガ科) 旧北区
red-green carpet (1)		*Chloroclysta miata* (Linnaeus)	(チョウ目、シャクガ科) 旧北区
red groundling		*Brachythemis lacustris* (Kirby)	(トンボ目、トンボ科) エチオピア区
redgum basket lerp		*Cardiaspina retator* Taylor	(カメムシ目、キジラミ科) 豪州区
red gum lerp psyllid			redgum sugar lerp を見よ
red gum lerp psyllid		*Glycaspis brimlecombei* Moore	(カメムシ目、キジラミ科) 全北区、豪州区、エチオピア区
redgum pit scale		*Lachnodius eucalypti* (Maskell)	(カメムシ目、フクロカイガラムシ科) 豪州区
redgum sugar lerp		*Glycaspis blakei* (Moore)	(カメムシ目、キジラミ科) 豪州区
redhaired bark beetle			pine root bark beetle を見よ
red-haired velvet ant		*Dasymutilla coccineohirta* Blake	(ハチ目、アリバチ科) 新北区
red hairy caterpillar		*Amsacta albistriga* Walker	(チョウ目、ヒトリガ科) 東洋区
red harvester ant		*Pogonomyrmex (Pogonomyrmex) barbatus* (Smith)	(ハチ目、アリ科) 新北区
red-headed Ancylis moth		*Ancylis muricana* (Walsingham)	(チョウ目、ハマキガ科) 新北区
red-headed ash borer		*Neoclytus acuminatus* (Fabricius)	(コウチュウ目、カミキリムシ科) 全北区、大洋区
red-headed biting louse			sheep biting louse を見よ
red-headed bush cricket			handsome bush cricket を見よ
red-headed chestnut		*Conistra erythrocephala* (Denis et Schiffermüller)	(チョウ目、ヤガ科) 旧北区
red-headed firetip		*Pyrrhopyge zenodorus* Godman et Salvin	(チョウ目、セセリチョウ科) 新熱帯区

英名	和名	学名	所属、分布、ほか
red-headed flea beetle		*Systena frontalis* (Fabricius)	(コウチュウ目、ハムシ科) 新北区
red-headed flies			broadmouthed flies を見よ
red-headed inchworm moth		*Semiothisa bisignata* (Walker)	(チョウ目、シャクガ科) 新北区。幼虫は red-headed inchworm
redheaded Jack pine sawfly		*Neodiprion rugifrons* Middleton	(ハチ目、マツハバチ科) 新北区
red-headed louse			sheep biting louse を見よ
red-headed meadow katydid		*Orchelimum erythrocephalum* Davis	(バッタ目、キリギリス科) 新北区
redheaded pasture cockchafer		*Adorypharus coulonii* (Burmeister)	(コウチュウ目、コガネムシ科) 豪州区
red-headed pigmy		*Stigmella ruficapitella* (Haworth)	(チョウ目、モグリチビガ科) 旧北区
red-headed pine sawfly		*Neodiprion lecontei* (Fitch)	(ハチ目、マツハバチ科) 新北区
red-headed Polybia wasp		*Polybia ruficeps* Schrottky	(ハチ目、スズメバチ科) 新熱帯区
red-headed roadside skipper		*Amblyscirtes phylace* (Edwards)	(チョウ目、セセリチョウ科) 新北区、新熱帯区
red-headed sawfly	ニホンアカズヒラタハバチ	*Acantholyda nipponica* Yano et Sato	(ハチ目、ヒラタハバチ科) 日本
red-headed sheep louse			sheep biting louse を見よ
red-headed spruce	オオアカズヒラタハバチ	*Cephalcia issikii* Takeuchi	(ハチ目、ヒラタハバチ科) 日本
redheaded whitegrub		*Dasygnathus dejeani* Macleay	(コウチュウ目、コガネムシ科) 豪州区
red Helen		*Papilio helenus* Linnaeus	(チョウ目、アゲハチョウ科) 旧北区。日本亜種は *P. h. nicconicolens* Butler モンキアゲハ
red hill copper		*Aloeides egerides* Tite et Dickson	(チョウ目、シジミチョウ科) エチオピア区
red Himalayan flash		*Bidaspa micans* (Bremer et Grey)	(チョウ目、シジミチョウ科) 東洋区
red-hindwinged catocala			rosy underwing を見よ
red-horned cardinal click beetle		*Ampedus rufipennis* (Stephens)	(コウチュウ目、コメツキムシ科) 旧北区
red-horned grain beetle		*Platydema ruficorne* (Sturm)	(コウチュウ目、ゴミムシダマシ科) 新北区
red humped apple caterpillar			red-humped caterpillar を見よ
red humped apple worm			red-humped caterpillar を見よ
red-humped appleworm moth			red-humpted caterpillar を見よ 北米での英名
red-humped caterpillar		*Schizura concinna* (Smith et Abbott)	(チョウ目、シャチホコガ科) 新北区。リンゴ害虫
red-humped caterpillar moth			red-humpted caterpillar を見よ

英名	和名	学名	所属、分布、ほか
red-humped maple worm			white-headed prominent moth を見よ
red-humped oakworm moth		Symmerista canicosta Franclemont	（チョウ目、シャチホコガ科）新北区。幼虫は red-humped oakworm
red ichneumons	アメバチ亜科	Ophioninae	（ハチ目、ヒメバチ科）の昆虫の総称
red imperial		Suasa lisides (Hewitson)	（チョウ目、シジミチョウ科）東洋区
red imported fire ant			imported fire ant (1) を見よ
red jumping plant louse	ベニキジラミ	Psylla coccinea Kuwayama	（カメムシ目、キジラミ科）日本、旧北区、東洋区
red lacewing	ベニハレギチョウ	Cethosia biblis (Drury)	（チョウ目、タテハチョウ科）東洋区。C. b. pemanggilensis Eliot, C. b. perakana Fruhstorfer も同英名
red lacewing			eastern red lacewing を見よ
red leaf-beetle			red pumpkin beetle を見よ
red-legged bean weevil			broadbean weevil を見よ
red-legged beech moth		Proteodes carnifex Butler	（チョウ目、マルハキバガ科）豪州区
red-legged beetle			red-legged ham beetle を見よ
red-legged bug			forest bug を見よ
red-legged Diacrisia			pink-legged tiger moth を見よ
red-legged drepanid	アシベニカギバ	Oreta pulchripes Butler	（チョウ目、カギバガ科）日本
red-legged earwigs	ハサミムシ科	Carcinophoridae	（ハサミムシ目）の昆虫の総称
red-legged flea beetle		Derocrepis erythropus (Melsheimer)	（コウチュウ目、ハムシ科）新北区
red-legged flea beetle		Derocrepis rufipes (Linnaeus)	（コウチュウ目、ハムシ科）旧北区
red-legged grasshopper		Melanoplus femurrubrum (De Geer)	（バッタ目、バッタ科）新北区、大洋区。米国の普通種でトウモロコシ害虫。線虫の中間宿主
red-legged green longicorn			red-legged green longicorn beetle を見よ
red-legged green longicorn beetle	アカアシオオアオカミキリ	Chloridolum japonicum Harold	（コウチュウ目、カミキリムシ科）日本、旧北区。カシ類害虫
red-legged ham beetle	アカアシホシカムシ	Necrobia rufipes (De Geer)	（コウチュウ目、カッコウムシ科）日本、汎世界。貯穀害虫
red-legged ichneumon		Pimpla instigator (Fabricius)	（ハチ目、ヒメバチ科）旧北区
red-legged Methiola		Methiola picta Sjöstedt	（バッタ目、バッタ科）豪州区
red-legged plum weevil			red-legged weevil (1) を見よ
red-legged threadtail		Protoneura sanguinipes Westfall	（トンボ目、ミナミイトトンボ科）新熱帯区
red-legged velvet ant		Smicromyrme rufipes (Fabricius)	（ハチ目、アリバチ科）旧北区、東洋区。日本亜種は S. r. lewisi Mickel ヒトホシアリバチ
redlegged weevil		Catasarcus impressipennis (Boisduval)	（コウチュウ目、ゾウムシ科）豪州区
red-legged weevil (1)		Otiorhynchus clavipes (Bonsdorff)	（コウチュウ目、ゾウムシ科）旧北区

英名	和名	学名	所属、分布、ほか
red-letter flat-body		*Agonopterix ocellana* (Fabricius)	（チョウ目、マルハキバガ科）旧北区
red lily beetle			scarlet lily beetle を見よ
redline doctor		*Ancyluris meliboeus* (Fabricius)	（チョウ目、シジミタテハ科）新熱帯区
red-line quaker		*Agrochola lota* (Clerck)	（チョウ目、ヤガ科）旧北区
red-line sapphire		*Iolaus sidus* Trimen	（チョウ目、シジミチョウ科）エチオピア区
red-line sapphire blue			red-line sapphire を見よ
red-lined clearwing moth		*Synanthedon refulgens* (Edwards)	（チョウ目、スカシバガ科）新北区
red-lined geometrid		*Chrysiphona ocultaria* (Donovan)	（チョウ目、シャクガ科）旧北区
red-lined Panopoda moth		*Panopoda rufimargo* (Hübner)	（チョウ目、ヤガ科）新北区
red-lined scrub-hairstreak			Mexican gray hairstreak を見よ
red locust	アカトビバッタ	*Nemadacris septemfasciata* (Serville)	（バッタ目、バッタ科）エチオピア区
red longhorn beetle		*Stictoleptura rubra* (Linnaeus)	（コウチュウ目、カミキリムシ科）旧北区
red longicorn beetle			red bamboo longicorn beetle を見よ
red long-winged planthopper	アカハネナガウンカ	*Diostrombus politus* Uhler	（カメムシ目、ハネナガウンカ科）日本、旧北区
red louse			cattle chewing louse を見よ
red-maculated leafhopper	チマダラヒメヨコバイ	*Eurythroneura mori* (Matsumura)	（カメムシ目、オオヨコバイ科）日本、旧北区
red maggot			wheat midge を見よ　幼虫の英名
red-mantled dragonlet		*Erythrodiplax fervida* (Erichson)	（トンボ目、トンボ科）新熱帯区
red mapwing			Keferstein's mapwing を見よ
red-margined assassin bug		*Scadra rufidens* Stål	（カメムシ目、サシガメ科）新北区
red-margined miltochrista			rosy footman を見よ
red-marked tent-maker			willow nestmaker を見よ　北米での英名
red-marmorate dried-fish beetle	アカマダラカツオブシムシ	*Trogoderma varium* (Matsumura et Yokoyama)	（コウチュウ目、カツオブシムシ科）日本、旧北区
red-marmorate long-winged planthopper	アカフハネナガウンカ	*Epotiocerus flexuosus* (Uhler)	（カメムシ目、ハネナガウンカ科）日本、旧北区、東洋区
red marsh trotter			keyhole glider を見よ
red melon beetle			red pumpkin beetle を見よ
red milkweed beetle			asclepias reddish longicorn を見よ
red milkweed beetles			milkweed longhorns を見よ
red mound ants		*Formica*	（ハチ目、アリ科）の昆虫の総称
red myrmicine ant			red ant (2) を見よ

英名	和名	学名	所属、分布、ほか
red myrmicine ants		*Myrmica*	(ハチ目、アリ科) の昆虫の総称
redneck longhorned beetle		*Aromia bungii* (Faldemann)	(コウチュウ目、カミキリムシ科) 東洋区
red-necked apple caterpillar			yellow-necked caterpillar を見よ
red-necked bark beetle	アカクビキクイムシ	*Xyleborus rubricollis* Eichhoff	(コウチュウ目、キクイムシ科) 日本、全北区
red-necked cane borer		*Agrilus ruficollis* (Fabricius)	(コウチュウ目、タマムシ科) 新北区
red-necked footman		*Atolmis rubricollis* (Linnaeus)	(チョウ目、ヒトリガ科) 旧北区
red-necked peanut worm		*Stegasta bosquella* Chambers	(チョウ目、キバガ科) 新北区、新熱帯区。ピーナツ害虫。red-necked peanutworm moth ともいう
red-necked tiger longicorn	クビアカトラカミキリ	*Xylotrechus rufilius rufilius* Bates	(コウチュウ目、カミキリムシ科) 日本、旧北区
red oak borer		*Enaphalodes rufulus* (Haldeman)	(コウチュウ目、カミキリムシ科) 新北区、大洋区
red oak clearwing moth		*Paranthrene simulans* (Grote)	(チョウ目、スカシバガ科) 新北区。幼虫は red oak clearwing borer
red oak roller			oak leaf roller weevil を見よ
red orange scale			California red scale を見よ
red orchid scale		*Furcaspis biformis* (Cockerell)	(カメムシ目、マルカイガラムシ科) 新北区
red Osmia		*Osmia rufa* (Linnaeus)	(ハチ目、ハキリバチ科) 旧北区
red palm weevil			Asiatic palm weevil を見よ
red-patch Epitolina		*Epitolina catori* Bethune-Baker	(チョウ目、シジミチョウ科) エチオピア区
red-patched leafwing		*Siderone syntyche syntyche* Hewitson	(チョウ目、タテハチョウ科) 新熱帯区
red-patched stink bug	ベニモンツノカメムシ	*Elasmostethus humeralis* Jakovlev	(カメムシ目、ツノカメムシ科) 日本、旧北区
red pavement ant			pavement ant を見よ
red-pepper gnats			biting midges (1) を見よ
red pierrot		*Talicada nyseus* (Guérin-Méneville)	(チョウ目、シジミチョウ科) 東洋区
red pine cone beetle		*Conophthorus resinosae* Hopkins	(コウチュウ目、キクイムシ科) 新北区
red pine needle midge		*Thecodiplosis pini resinosae* Kearby	(ハエ目、タマバエ科) 新北区
red pine needleminer moth		*Coleotechnites resinosae* (Freeman)	(チョウ目、キバガ科) 新北区
red pine sawfly		*Neodiprion nanulus nanulus* Schedl	(ハチ目、マツハバチ科) 新北区
red pine scale		*Matsucoccus resinosae* Bean et Godwin	(カメムシ目、ワタフキカイガラムシ科) 新北区
red pine shoot moth		*Dioryctria resinosella* Mutuura	(チョウ目、メイガ科) 新北区

英名	和名	学名	所属、分布、ほか
red pine tip moth		*Rhyacionia busckana* Heinrich	（チョウ目、ハマキガ科）新北区
red pinecone borer moth		*Eucosma monitorana* Heinrich	（チョウ目、ハマキガ科）新北区
red playboy			apricot playboy を見よ
red plum maggot			plum fruit moth を見よ　幼虫の英名
red poplar leaf beetle	ドロノキハムシ	*Chrysomela populi* (Linnaeus)	（コウチュウ目、ハムシ科）日本、旧北区、東洋区。ポプラ害虫
red postman			common longwing を見よ
red pumpkin beetle		*Rhapidopalpa foveicollis* Lucas	（コウチュウ目、ハムシ科）．旧北区
red-rayed Euselasia		*Euselasia hieronymi hieronymi* (Salvin et Godman)	（チョウ目、シジミタテハ科）新熱帯区
red rim	アカヘリタテハ	*Biblis hyperia aganisa* Boisduval	（チョウ目、タテハチョウ科）新北区、新熱帯区。crimson-banded black を参照
red ring skirt			Circe (1) を見よ
red roarer		*Psaltoda aurora* Distant	（カメムシ目、セミ科）豪州区
red rock skimmer		*Paltothemis lineatipes* Karsch	（トンボ目、トンボ科）新北区、新熱帯区
red rose maggot			three-dotted rose bell を見よ　幼虫の英名
red Rumex weevil		*Apion frumentarium* (Linnaeus)	（コウチュウ目、ホソクチゾウムシ科）旧北区
red-rust flour beetle			red flour beetle を見よ
red rust grain beetle			rusty grain beetle を見よ
red rust thrips			orchid thrips を見よ
red saddlebags		*Tramea onusta* (Hagen)	（トンボ目、トンボ科）新北区、新熱帯区。red-mantled saddlebags ともいう
red satyr	ヒイロコジャノメ	*Megisto rubricata* (Edwards)	（チョウ目、タテハチョウ科）新北区
red scale			California red scale を見よ
red scale of citrus			California red scale を見よ
red scale of Florida			Florida red scale を見よ
red scale parasite		*Aphytis lingnanensis* Compere	（ハチ目、ツヤコバチ科）豪州区
red scale parasite		*Aphytis melinus* DeBach	（ハチ目、ツヤコバチ科）豪州区
red scale parasite		*Encarsia perniciosi* (Tower)	（ハチ目、ツヤコバチ科）豪州区
red scale parasite		*Comperiella bifasciata* Howard	（ハチ目、トビコバチ科）豪州区
red scale parasite (1)		*Aphytis chrysomphalis* (Mercet)	（ハチ目、ツヤコバチ科）新北区、豪州区
red-shanked carder-bee		*Bombus derhamellus* Kirby	（ハチ目、ミツバチ科）旧北区
red-shanked carder bee			knapweed carder bee を見よ
red-shanked grasshopper		*Xanthippus corallipes* (Haldeman)	（バッタ目、バッタ科）新北区

英名	和名	学名	所属、分布、ほか
red-shouldered beetle			red-shouldered ham beetle を見よ
red-shouldered bug		*Jadera haematoloma* (Herrich-Schäffer)	（カメムシ目、ヒメヘリカメムシ科）新北区、新熱帯区
red-shouldered ctenucha		*Ctenucha rubroscapus* (Ménétriès)	（チョウ目、カノコガ科）新北区
red-shouldered ham beetle	アカクビホシカムシ	*Necrobia ruficollis* (Fabricius)	（コウチュウ目、カッコウムシ科）日本、汎世界。貯穀害虫
red-shouldered leaf beetle		*Saxinus saucia* LeConte	（コウチュウ目、ハムシ科）新北区
red-shouldered leaf beetle (1)		*Monolepta australis* (Jacoby)	（コウチュウ目、ハムシ科）豪州区。カンキツ害虫
red-shouldered longicorn beetle	アカネカミキリ	*Phymatodes maaki* (Kraatz)	（コウチュウ目、カミキリムシ科）日本、旧北区。ブドウ害虫
red-shouldered shot-hole borer		*Xylobiops basilare* (Say)	（コウチュウ目、ハムシ科）新北区。ペカン害虫
red-shouldered stink bug		*Thyanta pallidovirens* (Stål)	（カメムシ目、カメムシ科）新北区
red-shouldered stink bug		*Thyanta accerra* McAtee	（カメムシ目、カメムシ科）新北区
red-shouldered tiger longicorn			red-shouldered tiger longicorn beetle を見よ
red-shouldered tiger longicorn beetle	アカネトラカミキリ	*Brachyclytus singularis* Kraatz	（コウチュウ目、カミキリムシ科）日本、旧北区。ブドウ害虫
red-shouldered vine longicorn			red-shouldered longicorn beetle を見よ
red skimmer		*Libellula saturata* Uhler	（トンボ目、トンボ科）新北区。全体赤色の種
red skimmers		*Libellula*	（トンボ目、トンボ科）の昆虫の総称
red-specked case		*Coleophora lineolea* (Haworth)	（チョウ目、ツツミノガ科）旧北区
red-splashed sulphur		*Phoebis avellaneda* (Herrich-Schäffer)	（チョウ目、シロチョウ科）新熱帯区
redspot		*Zesius chrysomallus* Hübner	（チョウ目、シジミチョウ科）東洋区
red spot ciliate blue			red-spot hairtail を見よ
red spot diadem	ベニモンシロムラサキ	*Hypolimnas usambara* (Ward)	（チョウ目、タテハチョウ科）エチオピア区
red-spot duke		*Dophla evelina* (Stoll)	（チョウ目、タテハチョウ科）東洋区
red-spot duke Latin		*Dophla evelina compta* (Fruhstorfer)	（チョウ目、タテハチョウ科）東洋区。red-spot duke を参照
red-spot hairtail		*Anthene lunulata lunulata* (Trimen)	（チョウ目、シジミチョウ科）エチオピア区
red-spot Jezebel	アカネキシタカザリシロチョウ	*Delias descombesi* (Boisduval)	（チョウ目、シロチョウ科）東洋区
red-spot Marquis		*Bassarona recta monilis* Moore	（チョウ目、タテハチョウ科）東洋区。redtail Marquis を参照
redspot sawtooth	スジグロマダラシロチョウ	*Prioneris philonome* Boisduval	（チョウ目、シロチョウ科）東洋区

英名	和名	学名	所属、分布、ほか
red-spot sawtooth		*Prioneris clemanthe* (Doubleday)	(チョウ目、シロチョウ科) 東洋区
red-spotted admiral			white admiral (1) を見よ
red-spotted crescent			red-spotted patch を見よ
red-spotted green moth			azalea rough bollworm を見よ
red-spotted hairstreak			large lantana butterfly を見よ
red-spotted lady beetles		Chilocorini	(コウチュウ目、テントウムシ科) の昆虫の総称
red-spotted ladybird beetles			red-spotted lady beetles を見よ
red-spotted leaf beetle	サクラケブカハムシ	*Pyrrhalta semifulva* (Jakoby)	(コウチュウ目、ハムシ科) 日本、旧北区、東洋区
red-spotted Lithacodia moth		*Maliattha concinnimacula* (Guenée)	(チョウ目、ヤガ科) 新北区
red-spotted longhorn beetle		*Batocera rufomaculata* (De Geer)	(コウチュウ目、カミキリムシ科) 東洋区。イチジク害虫
red-spotted longicorn			red-spotted longicorn beetle を見よ
red-spotted longicorn beetle	ホシベニカミキリ	*Eupromus ruber* (Dalman)	(コウチュウ目、カミキリムシ科) 日本、旧北区、東洋区。クスノキ害虫
red-spotted ministreak		*Ministrymon una* (Hewitson)	(チョウ目、シジミチョウ科) 新熱帯区
red-spotted patch		*Chlosyne marina* (Geyer)	(チョウ目、タテハチョウ科) 新北区
red-spotted patch			Melitaeoides checkerspot を見よ
red-spotted purple		*Limenitis arthemis astyanax* (Fabricius)	(チョウ目、タテハチョウ科) 新北区
red-spotted sap beetle		*Glischrochilus obtusus* (Say)	(コウチュウ目、ケシキスイ科) 新北区
red-spotted sap beetle		*Glischrochilus quadrisignatus* (Say)	(コウチュウ目、ケシキスイ科) 新北区
red-spotted scale			Florida red scale を見よ
red-spotted skipper		*Melanopyge mulleri* (Bell)	(チョウ目、セセリチョウ科) 新熱帯区
red-spotted stink bug		*Brachystethus rubromaculatus* Dallas	(カメムシ目、カメムシ科) 新熱帯区
red-spotted swallowtail			ruby-spotted swallowtail を見よ
red-spotted sweetpotato moth		*Polygrammodes elevata* (Fabricius)	(チョウ目、メイガ科) 新北区
red-spotted tussock moth	アカモンドクガ	*Orgyia recens approximans* Butler	(チョウ目、ドクガ科) 日本、旧北区
red spruce leaf-miner			Minnion's groundling を見よ 米国での英名
red stem-borer		*Zeuzera postexcisa* Hampson	(チョウ目、ボクトウガ科) 東洋区
red-streaked Mompha moth		*Mompha eloisella* (Clemens)	(チョウ目、カザリバガ科) 新北区
red stripe weevil			Asiatic palm weevil を見よ

英名	和名	学名	所属、分布、ほか
red-striped arge	ベニスジヒメシャク	*Timandra griseata prouti* (Inoue)	(チョウ目、シャクガ科) 日本
red-striped delicate geometrid		*Aroga trialbamaculella* (Chambers)	(チョウ目、キバガ科) 新北区
red-striped fireworm moth	アカスジキンカメムシ	*Poecilocoris lewisi* Distant	(カメムシ目、キンカメムシ科) 日本、旧北区
red-striped golden stink bug	アカスジアオリンガ	*Pseudoips sylpha* (Butler)	(チョウ目、ヤガ科) 日本、旧北区
red-striped green moth	アカスジヒゲブトカスミカメムシ	*Eolygus rubrolineatus* (Matsumura)	(カメムシ目、カスミカメムシ科) 日本、旧北区
red-striped leaf bug			scarlet leafwing を見よ
red-striped leafwing	フチベニクロカミキリ	*Asiates holodendri* Pallas	(コウチュウ目、カミキリムシ科) 旧北区
red-striped longicorn beetle	マダラハネナガウンカ	*Pamendanga matsumurae* (Muir)	(カメムシ目、ハネナガウンカ科) 日本、旧北区、東洋区
red-striped long-winged planthopper		*Epinotia radicana* (Heinrich)	(チョウ目、ハマキガ科) 新北区
red-striped needleworm moth			rose argid sawfly (1) を見よ
red-striped sawfly	アカスジカメムシ	*Grophosoma rubrolineatum* (Westwood)	(カメムシ目、カメムシ科) 日本、旧北区
red striped sugarcane scale		*Saccharipulvinaria iceryi* (Signoret)	(カメムシ目、カタカイガラムシ科) 東洋区
red-striped threadtail		*Elattoneura tenax* (Hagen)	(トンボ目、Protoneuridae) 東洋区
red-studded skipper		*Antigonus stator* Godman	(チョウ目、セセリチョウ科) 新北区
red-studded skipper		*Noctuana haematospila* (Felder et Felder)	(チョウ目、セセリチョウ科) 新熱帯区
red-studded skipper		*Noctuana stator* (Godman et Salvin)	(チョウ目、セセリチョウ科) 新熱帯区
red sunflower seed weevil		*Smicronyx fulvus* LeConte	(コウチュウ目、ゾウムシ科) 新北区
red sword-grass		*Xylena vetusta* (Hübner)	(チョウ目、ヤガ科) 旧北区
red swordgrass moth		*Xylena nupera* Lintner	(チョウ目、ヤガ科) 新北区
red sword-grass moth			red sword-grass を見よ
red tab policeman		*Coeliades keithloa* (Wallengren)	(チョウ目、セセリチョウ科) エチオピア区。red-tab policeman とも記す
redtail	アジアイトトンボ	*Ischnura asiatica* (Brauer)	(トンボ目、イトトンボ科) 日本、旧北区
redtail Marquis		*Bassarona recta* (de Nicéville)	(チョウ目、タテハチョウ科) 東洋区
red tail moth			pale tussock を見よ　red-tail moth とも記される。
red-tailed beech roller moth		*Dasychira pudibunda* (Linnaeus)	(チョウ目、ドクガ科) 旧北区
red-tailed bumble bee (1)		*Bombus lapidarius* (Linnaeus)	(ハチ目、ミツバチ科) 旧北区
red-tailed bumble bee (2)		*Bombus ternarius* Say	(ハチ目、ミツバチ科) 新北区、大洋区

英名	和名	学名	所属、分布、ほか
red-tailed flesh fly			flesh fly (2) を見よ
red tailed forester		*Lethe sinorix* (Hewitson)	(チョウ目、タテハチョウ科) 東洋区。tailed red forester ともいう
red-tailed pennant		*Brachymesia furcata* (Hagen)	(トンボ目、トンボ科) 新熱帯区
red-tailed spector		*Euerythra phasma* Harvey	(チョウ目、ヒトリガ科) 新北区。red-tailed specter moth ともいう
redtailed spider wasp	ツマアカベッコウ	*Tachypompilus analis* (Fabricius)	(ハチ目、ベッコウバチ科) 日本、東洋区、新北区
red-tailed tachina		*Winthemia quadripustulata* (Fabricius)	(ハエ目、ヤドリバエ科) 全北区
red thrips			common black vine thrips を見よ
red-tip		*Colotis antevippe gavisa* (Wallengren)	(チョウ目、シロチョウ科) エチオピア区
red-tipped clearwing		*Synanthedon formicaeformis* (Esper)	(チョウ目、スカシバガ科) 旧北区
red-tipped eucosmid	マツツマアカシンムシ	*Rhyacionia duplana* (Hübner)	(チョウ目、ハマキガ科) 日本
red-tipped flower beetle		*Malachius bipustulatus* (Linnaeus)	(コウチュウ目、ジョウカイモドキ科) 旧北区
red-tipped looper moth	ツマアカナミシャク	*Aplocera perelegans* (Warren)	(チョウ目、シャクガ科) 日本
red-tipped swampdamsel		*Leptobasis vacillans* Hagen	(トンボ目、イトトンボ科) 新北区、新熱帯区
red tree ant			red ant (1) を見よ
red tree ants			weaver ants を見よ
red treeticker		*Birrima castanea* (Goding et Froggatt)	(カメムシ目、セミ科) 豪州区
red turnip beetle		*Entomoscelis americana* Brown	(コウチュウ目、ハムシ科) 新北区
red turnip beetle		*Entomoscelis adonidis* Pallas	(コウチュウ目、ハムシ科) 旧北区
red turpentine beetle		*Dendroctonus valens* LeConte	(コウチュウ目、キクイムシ科) 新北区
red twin-spot carpet		*Xanthorhoe spadicearia* (Denis et Schiffermüller)	(チョウ目、シャクガ科) 旧北区
red twin-spot moth			dark twin-spot carpet を見よ
red-undersided drepanid	アカウラカギバ	*Hypsomadius insignis* Butler	(チョウ目、カギバガ科) 日本、東洋区
red underwing		*Catocala nupta* (Linnaeus)	(チョウ目、ヤガ科) 旧北区。日本亜種は *C. n. nozawae* Matsumura エゾベニシタバ
red underwing moth			red underwing を見よ
red underwing moth			rosy underwing を見よ　米国での英名
red underwing skipper	チビチャマダラセセリ	*Spialia sertorius* (Hoffmannsegg)	(チョウ目、セセリチョウ科) 旧北区
red underwings			underwings を見よ
reduntant skipper		*Corticea corticea* (Plötz)	(チョウ目、セセリチョウ科) 新北区、新熱帯区

英名	和名	学名	所属、分布、ほか
red-veind darter	スナアカネ	*Sympetrum fonscolombei* (Selys)	(トンボ目、トンボ科) 日本、旧北区、東洋区、エチオピア区
red-veined dropwing		*Trithemis arteriosa* (Burmeister)	(トンボ目、トンボ科) エチオピア区
red-veined leaf bug	アカミャクカスミカメムシ	*Stenodema rubrinerva* Horvath	(カメムシ目、カスミカメムシ科) 日本、旧北区
red-veined swallowtail		*Graphium cyrnus* (Boisduval)	(チョウ目、アゲハチョウ科) エチオピア区
red velvet-ant		*Dasymutilla magnifica* Mickel	(ハチ目、アリバチ科) 新北区
red velvet ant			common eastern velvet ant を見よ
red-waisted florella moth			orange-spotted flower moth を見よ
red-washed prominent moth		*Oligocentria semirufescens* (Walker)	(チョウ目、シャチホコガ科) 新北区。rusty prominent ともいう
red-washed satyr			Helvina lady slipper を見よ
red wasp		*Vespula rufa* (Linnaeus)	(ハチ目、スズメバチ科) 旧北区
red wax scale	ルビーロウムシ	*Ceroplastes rubens* Maskell	(カメムシ目、カタカイガラムシ科) 日本、汎世界。カンキツ、ナシ、カキなど果樹、ツバキなど樹木の害虫
red-webbed satyr		*Euptychia rubrofasciata* Miller et Miller	(チョウ目、タテハチョウ科) 新熱帯区
red webspinner		*Chelicerca rubra* Ross	(シロアリモドキ目、Anisembiidae) 新北区
red weevil			wheat midge を見よ　米国での英名
red wing hawk		*Phylloxiphia metria* (Jordan)	(チョウ目、スズメガ科) エチオピア区
red-winged grass cicada		*Tibicinoides cupreosparsus* (Uhler)	(カメムシ目、セミ科) 新北区
red-winged grasshopper		*Dissosteira pictipennis* Brunner	(バッタ目、バッタ科) 新北区
red-winged grasshopper		*Oedipoda germanica* (Latreille)	(バッタ目、バッタ科) 旧北区
red-winged grasshopper			bloody-winged grasshopper を見よ
red-winged lycid beetle		*Porrostoma rufipennis* (Fabricius)	(コウチュウ目、ベニボタル科) 豪州区
red-winged migratory locust		*Nomadacris septemfasciata* Serville	(バッタ目、バッタ科) エチオピア区
red-winged wave moth		*Dasyfidonia avuncularia* (Guenée)	(チョウ目、シャクガ科) 新北区
red wood ant			common red ant を見よ
red wood ant			common red ant を見よ
red wood mealybug		*Spirococcus sequoiae* (Coleman)	(カメムシ目、コナカイガラムシ科) 新北区
red wood satyr			red satyr を見よ
redbay scale		*Acutaspis perseae* Ferris	(カメムシ目、マルカイガラムシ科) 全北区
reddish alpine		*Erebia kozhantshikovi* Sheljuzhko	(チョウ目、タテハチョウ科) 新北区

英名	和名	学名	所属、分布、ほか
reddish-back leaf-cut weevil	セアカヒメオトシブミ	*Apoderus geminus* Sharp	(コウチュウ目、オトシブミ科) 日本
reddish brown noctuid	チャイロキリガ	*Orthosia odiosa* (Butler)	(チョウ目、ヤガ科) 日本、旧北区
reddish-brown plum aphid			waterlily aphid を見よ
reddish-brown stag beetle		*Lucanus capreolus* (Linnaeus)	(コウチュウ目、クワガタムシ科) 新北区
reddish buff		*Acosmetia caliginosa* (Hübner)	(チョウ目、ヤガ科) 旧北区
reddish chewing louse of sheep and goats		*Bovicola caprae* Gurlt	(ハジラミ目、ケモノハジラミ科) 旧北区
reddish cochlid			tea cochlid を見よ
reddish-costate noctuid	アカマエヤガ	*Spaelotis valida* (Walker)	(チョウ目、ヤガ科) 日本、旧北区
reddish dotted arctiid			heliotrope moth を見よ
reddish grained moth	アカモクメヨトウ	*Apamea aquila oriens* (Warren)	(チョウ目、ヤガ科) 日本
reddish hairstreak		*Strymon rufofusca* Hewitson	(チョウ目、シジミチョウ科) 新北区、新熱帯区
reddish hawk moth			elephant hawk moth を見よ
reddish heart dart			dull reddish dart moth を見よ
reddish light arches		*Apamea sublustris* (Esper)	(チョウ目、ヤガ科) 旧北区
reddish mapwing		*Hypanartia trimaculata autumna* Willmott, Hall et Lamas	(チョウ目、タテハチョウ科) 新熱帯区。Willmott's admiral を参照
reddish-margined yellow arctiid			clouded buff を見よ
reddish-necked leaf beetle	ムナキルリハムシ	*Smaragdina semiaurantiaca* (Fairmaire)	(コウチュウ目、ハムシ科) 日本、旧北区
reddish oraesia			fruit-piercing moth (9) を見よ
reddish Phaneta moth		*Phaneta raracana* (Kearfott)	(チョウ目、ハマキガ科) 新北区
reddish pine carpet			pine carpet を見よ
reddish potato beetle		*Leptinotarsa rubiginosa* (Rogers)	(コウチュウ目、ハムシ科) 新北区
reddish pubescent dried-fish beetle	ケアカカツオブシムシ	*Dermestes tessellatocollis* Motschulsky	(コウチュウ目、カツオブシムシ科) 日本、旧北区、東洋区
reddish tiger moth	アカヒトリ	*Thanatarctia flammeola* (Moore)	(チョウ目、ヒトリガ科) 日本、旧北区
reddish-tipped prominent			poplar prominent (1)(2) を見よ
reddish yezo bark beetle	アカエゾキクイムシ	*Polygraphus gracilis* Niijima	(コウチュウ目、キクイムシ科) 日本、旧北区
reds	ホソチョウ亜科	Acraeinae	(チョウ目、タテハチョウ科) の昆虫の総称
reduviid bugs			assassin bugs (1) を見よ
reed aphid			bamboo aphid (1) を見よ

英名	和名	学名	所属、分布、ほか
reed beetle	ホソネクイハムシ	*Donacia vulgaris* Zschach	(コウチュウ目、ハムシ科) 日本、旧北区
reed beetle		*Donacia aquatica* (Linnaeus)	(コウチュウ目、ハムシ科) 旧北区
reed beetles			long-horned leaf beetles を見よ
reed-boring crambid moth		*Carectocultus perstrialis* (Hübner)	(チョウ目、メイガ科) 新北区
reed dagger	タテスジケンモン	*Simyra albovenosa* (Goeze)	(チョウ目、ヤガ科) 日本、旧北区
reed damsel bug		*Dolichonabis lineatum* (Linnaeus)	(カメムシ目、マキバサシガメ科) 旧北区
reed lasiocampid			drinker を見よ
reed leopard	ハイイロボクトウ	*Phragmataecia castaneae* (Hübner)	(チョウ目、ボクトウガ科) 日本、旧北区、東洋区、エチオピア区
reed leopard moth			reed leopard を見よ
reed tussock		*Laelia coenosa* (Hübner)	(チョウ目、ドクガ科) 旧北区。日本亜種は *L. c. sangaica* Moore スゲドクガ
reed wainscot		*Archanara algae* (Esper)	(チョウ目、ヤガ科) 旧北区
reedlings		*Indolestes*	(トンボ目、アオイトトンボ科) の昆虫の総称
reedmace bug		*Chilacis typhae* (Perris)	(カメムシ目、ナガカメムシ科) 旧北区
reflective Tortyra moth			Slosson's metalmark moth を見よ
refracted Metarranthis moth		*Metarranthis refractaria* (Guenée)	(チョウ目、シャクガ科) 新北区
refulgent flash		*Rapala refulgens* de Nicéville	(チョウ目、シジミチョウ科) 東洋区
refuse beetle	ゴミムシ	*Anisodactylus signatus* (Panzer)	(コウチュウ目、オサムシ科) 日本、旧北区
regal Apollo	チャールトンウスバ	*Parnassius charltonius* Gray	(チョウ目、アゲハチョウ科) 旧北区、東洋区
regal darner		*Coryphaeschna ingens* (Rambur)	(トンボ目、ヤンマ科) 新北区
regal fritillary	アトグロオオヒョウモン	*Speyeria idalia* (Drury)	(チョウ目、タテハチョウ科) 新北区
regal hairstreak			bank note blue を見よ
regal metalmark		*Metacharis regalis* Butler	(チョウ目、シジミタテハ科) 新熱帯区
regal moth			royal walnut moth を見よ
regal pond cruiser		*Epophthalmia elegans* (Brauer)	(トンボ目、エゾトンボ科) 日本、旧北区、東洋区
regal skimmer		*Orthemis regalis* Ris	(トンボ目、トンボ科) 新熱帯区
regal swallowtail	レックスマダラアゲハ	*Papilio rex* Oberthür	(チョウ目、アゲハチョウ科) エチオピア区
regal walnut moth			royal walnut moth を見よ
regent skipper	ラッフルズセセリ	*Euschemon rafflesia* (Macleay)	(チョウ目、セセリチョウ科) 豪州区
regular grass yellow		*Eurema regularis* (Butler)	(チョウ目、シロチョウ科) エチオピア区
regular sunbeam		*Curetis regula* Evans	(チョウ目、シジミチョウ科) 東洋区
regular woolly legs		*Lachnocnema regularis* Libert	(チョウ目、シジミチョウ科) エチオピア区

英名	和名	学名	所属、分布、ほか
Rehn's shieldback		*Idiostatus rehni* Caudell	（バッタ目、キリギリス科）新北区
Reinhold's creamy glider	キイロタテハ	*Cymothoe reinholdi* (Plötz)	（チョウ、タテハチョウ科）エチオピア区
relict dragonfly	ムカシトンボ	*Epiophlebia superstes* (Seys)	（トンボ目、ムカシトンボ科）日本
relict fritillary			Kriemhild fritillary を見よ
reluctant scaly cricket		*Cycloptilum pigrum* Love et Walker	（バッタ目、コオロギ科）新北区
remarkable felt scale		*Eriococcus insignis* Newstead	（カメムシ目、フクロカイガラムシ科）新北区
Remella skipper		*Remella* sp.	（チョウ目、セセリチョウ科）新熱帯区
Remington's giant-skipper			Coahuila giant skipper を見よ
Remm's rustic		*Mesapamea remmi* Rezbanyai-Reser	（チョウ目、ヤガ科）旧北区
Renata satyr		*Yphthimoides renata* (Stoll)	（チョウ目、タテハチョウ科）新熱帯区
reniform Celaena moth		*Helotropha reniformis* (Grote)	（チョウ目、ヤガ科）新北区
renounced Hydriomena moth		*Hydriomena renunciata* (Walker)	（チョウ目、シャクガ科）新北区
repetitive tachinid fly		*Peleteria iterans* (Walker)	（ハエ目、ヤドリバエ科）新北区、大洋区
replete			plerergate を見よ
residua underwing moth		*Catocala residua* Grote	（チョウ目、ヤガ科）新北区。residua underwing ともいう
resin bees			mason bees (1)(2) を見よ
resin-gall moth			pine resin-gall moth を見よ
resin-gall shoot			pine resin-gall moth を見よ
resin-gall tortrix moth			pine resin-gall moth を見よ
resin gnat		*Retinodiplosis resinicola* (Osten Sacken)	（ハエ目、ユスリカ科）新北区
resin grey		*Eudonia delunella* (Stainton)	（チョウ目、メイガ科）旧北区
resplendent shield bearer		*Coptodisca splendorferella* (Clemens)	（チョウ目、ツヤコガ科）全北区
resplended shield bearer			resplendent shield bearer を見よ
rest harrow			rest-harrow guest を見よ
rest-harrow aphid		*Therioaphis ononidis* (Kaltenbach)	（カメムシ目、アブラムシ科）旧北区
rest-harrow guest		*Aplasta ononaria* (Fuessly)	（チョウ目、シャクガ科）旧北区
rest-harrow piercer		*Collicularia microgrammana* (Guenée)	（チョウ目、ハマキガ科）旧北区
restless beetles			ground beetles を見よ　豪州での英名
restless blue		*Orachrysops lacrimosa* (Bethune-Baker)	（チョウ目、シジミチョウ科）エチオピア区
restless bush cricket		*Hapithus agitator* Uhler	（バッタ目、コオロギ科）新北区

英名	和名	学名	所属、分布、ほか
restless demon		*Indothemis limbata sita* Campion	（トンボ目、トンボ科）東洋区
restricted demon	クロセセリ	*Notocrypta curvifascia* (C. et R. Felder)	（チョウ目、セセリチョウ科）日本、旧北区、東洋区
restricted purple sapphire	ウラフチベニシジミ	*Heliophorus ila malaya* Pendlebury	（チョウ目、シジミチョウ科）東洋区
retarded dagger moth		*Acronicta retardata* Walker	（チョウ目、ヤガ科）新北区
reticulate stag beetle		*Paralissotes reticulatus* (Westwood)	（コウチュウ目、クワガタムシ科）豪州区
reticulate-winged booklouse	ツヤコチャタテ	*Lepinotus reticulatus* Enderlein	（チャタテムシ目、コチャタテ科）日本、汎世界。貯蔵食品害虫
reticulate winged trogiid			reticulate-winged booklouse を見よ
reticulated beetles	ナガヒラタムシ科	Cupedidae	（コウチュウ目）の昆虫の総称
reticulated caddisflies		*Limnephilus*	（トビケラ目、エグリトビケラ科）の昆虫の総称
reticulated Decantha moth		*Decantha boreasella* (Chambers)	（チョウ目、マルハキバガ科）新北区
reticulated fruitworm moth		*Cenopis reticulatana* (Clemens)	（チョウ目、ハマキガ科）新北区
reticulated looper moth			netted carpet を見よ
reticulated mountain satyr		*Lymanopoda labda* Hewitson	（チョウ目、タテハチョウ科）新熱帯区
reticulated netwinged beetle		*Calopteron reticulatum* (Fabricius)	（コウチュウ目、ベニボタル科）新北区
reticulated tortrix			summer fruit tortrix (1)(2) を見よ
reticulated-winged booklouse			reticulate-winged booklouse を見よ
Reverdin's blue		*Lycaeides argyrognomon* (Bergstrasser)	（チョウ目、シジミチョウ科）全北区。日本亜種は *L. a. praelerinsularis* (Verity) ミヤマシジミ
reversed Haploa moth		*Haploa reversa* (Stretch)	（チョウ目、ヒトリガ科）新北区
reversed roadside skipper		*Amblyscirtes reversa* Jones	（チョウ目、セセリチョウ科）新北区
Revillagigedo pipevine swallowtail		*Battus philenor insularis* Vazquez	（チョウ目、アゲハチョウ科）新熱帯区。pipevine swallowtail を参照
rhamnus aphid			sacaline aphid を見よ
rhaphiolepis psylla	サツマキジラミ	*Psylla satsumensis* Kuwayama	（カメムシ目、キジラミ科）日本。シャリンバイ害虫
Rheede's large piercer		*Eucosmomorpha albersana* (Hübner)	（チョウ目、ハマキガ科）旧北区
Rheede's small piercer			fruit mining tortrix を見よ
Rhesus skipper			plains gray skipper を見よ
Rhetenor blue Morpho			blue Morpho (1) を見よ
rhinoceros beetle		*Xylotrupes ulysses* Guérin-Méneville	（コウチュウ目、コガネムシ科）豪州区。common rhinoceros beetle ともいう

英名	和名	学名	所属、分布、ほか
rhinoceros beetle		*Xyloryctes jamaiceasis* (Drury)	（コウチュウ目、コガネムシ科）新北区。サイに似た形に由来
rhinoceros beetle		*Sinodendron cylindricum* (Linnaeus)	（コウチュウ目、クワガタムシ科）旧北区
rhinoceros beetle (1)	タイワンカブトムシ	*Oryctes rhinoceros* (Linnaeus)	（コウチュウ目、コガネムシ科）日本、東洋区。ヤシ害虫
rhinoceros beetle (2)		*Oryctes tarandus* Olivier	（コウチュウ目、コガネムシ科）エチオピア区
rhinoceros beetle (3)		*Oryctes nasicornis illigeri* Minck	（コウチュウ目、コガネムシ科）旧北区
rhinoceros beetle			elephant beetle を見よ
rhinoceros beetles		*Oryctes*	（コウチュウ目、コガネムシ科）の昆虫の総称
rhinoceros beetles		*Dynastes*	（コウチュウ目、コガネムシ科）の昆虫の総称
rhinoceros beetles			Hercules beetles を見よ
rhinocerus beetle			rhinoceros beetle (2)(3) を見よ
rhinocerus beetles			Hercules beetles を見よ
rhinophorid flies			woodlouse flies を見よ
rhinotorid flies		Rhinotoridae	（ハエ目）の昆虫の総称
rhipiphorid beetles			wedge-shaped beetles を見よ
rhizophagid beetles			root-eating beetles を見よ
rhodea thrips	オモトアザミウマ	*Taeniothrips eucharii* (Whetzel)	（アザミウマ目、アザミウマ科）日本、東洋区、新熱帯区
Rhodes grass mealybug			Rhodes grass scale を見よ
Rhodes grass scale	チガヤシロオカイガラムシ	*Antonina graminis* (Maskell)	（カメムシ目、コナカイガラムシ科）。日本、東洋区、新北区、豪州区、エチオピア区、新熱帯区。イネ、シバなどの害虫
rhododendron borer		*Synanthedon rhododendri* (Beutenmüller)	（チョウ目、スカシバガ科）新北区。成虫は rhododendron borer moth
rhododendron bug		*Stephanitis rhododendri* Horvath	（カメムシ目、グンバイムシ科）旧北区
rhododendron gall midge		*Clinodiplosis rhododendri* (Felt)	（ハエ目、タマバエ科）新北区
rhododendron hopper		*Graphocephala fennahi* Young	（カメムシ目、オオヨコバイ科）旧北区
rhododendron lace bug			rhododendron bug を見よ　米国、南アフリカでの英名
rhododendron leafhopper			rhododendron hopper を見よ
rhododendron leafhopper			candy-striped leafhopper を見よ
rhododendron longicorn beetle	ソボリンゴカミキリ	*Oberea sobosana* Ohbayashi	（コウチュウ目、カミキリムシ科）日本。ケヤキ害虫
rhododendron root borer			rhododendron longicorn beetle を見よ
rhododendron slender			azalea leaf miner (2) を見よ

英名	和名	学名	所属、分布、ほか
rhododendron stem borer		*Oberea myops* (Dalman)	（コウチュウ目、カミキリムシ科）新北区。日本亜種は *O. m. japonica* (Thunberg) ニホンチャイロヒメカミキリ
rhododendron whitefly		*Dialeurodes chittendeni* Laing	（カメムシ目、コナジラミ科）日本、全北区
rhododendron whitefly (1)	ツツジコナジラミモドキ	*Odontaleyrodes rhododendri* (Takahashi)	（カメムシ目、コナジラミ科）日本、大洋区。ツツジ害虫
rhombic-marked leafhopper	ヒシモンヨコバイ	*Hishimonus sellatus* Uhler	（カメムシ目、オオヨコバイ科）日本、旧北区、東洋区。カンキツ、クワ害虫
rhombic planthopper	ヒシウンカ	*Pentastridius apicalis* (Uhler)	（カメムシ目、ヒシウンカ科）日本、旧北区。イネ、イネ科牧草害虫
rhomboid tortrix			rhomboid tortrix moth を見よ
rhomboid tortrix moth		*Acleris rhombana* (Denis et Schiffermüller)	（チョウ目、ハマキガ科）全北区
rhopaea canegrub			brown cockchafer を見よ
rhopalosomatid wasps		Rhopalosomatidae	（ハチ目）の昆虫の総称
rhubarb curculio		*Lixus concavus* Say	（コウチュウ目、ゾウムシ科）新北区
rhus aphid	ツルウルシコブアブラムシ	*Juncomyzus rhois* (Takahashi)	（カメムシ目、アブラムシ科）日本
rhus bark beetle	ツタウルシノコキクイムシ	*Cryphalus rhusii* Niijima	（コウチュウ目、キクイムシ科）日本
rhus stink bug	ハサミツノカメムシ	*Acanthosoma labiduroides* Jakovlev	（カメムシ目、ツノカメムシ科）日本、旧北区
rhynchophorans			rhynchophorus beetles を見よ
rhynchophores			rhynchophorus beetles を見よ
rhynchophorus beetles	有吻亜目	Rhynchophora	（コウチュウ目）の昆虫の総称
rhyparida beetles		Rhyparida	（コウチュウ目、ハムシ科）の昆虫の総称
riband wave	エゾキヒメシャク	*Idaea aversata* (Linnaeus)	（チョウ目、シャクガ科）日本、旧北区
ribbed bagworm			ribbed case moth を見よ
ribbed-case bearers			leaf skeletonizer moths (1) を見よ
ribbed case makers			ribbed-cocoon maker moths を見よ
ribbed case moth		*Hyalarcta nigrescens* (Doubleday)	（チョウ目、ミノガ科）豪州区
ribbed cocoon maker			apple bucculatrix を見よ
ribbed cocoon maker moth		*Bucculatrix quadrigemina* Braun	（チョウ目、チビガ科）新北区
ribbed-cocoon maker moths	チビガ科	Bucculatrigidae	（チョウ目）の昆虫の総称
ribbed cocoon makers			leaf skeletonizer moths (1) を見よ
ribbed pine borer		*Rhagium inquisitor* (Linnaeus)	（コウチュウ目、カミキリムシ科）旧北区
ribbed underwing			briseis underwing moth を見よ
Ribbe's glassy Acraea		*Acraea leucographa* Ribbe	（チョウ目、タテハチョウ科）エチオピア区
ribbon-footed corn fly			chloropid gout fly を見よ

英名	和名	学名	所属、分布、ほか
ribbon-winged lacewings			thread-winged lacewings を見よ　中南米での英名
ribwort slender		*Aspilapteryx tringipennella* (Zeller)	（チョウ目、ホソガ科）旧北区
ricaniid planthoppers		*Ricania*	（カメムシ目、ハゴロモ科）の昆虫の総称
ricaniid planthoppers			lantern bugs を見よ
rice aculeated thrips	イネクダアザミウマ	*Haplothrips aculeatus* (Fabricius)	（アザミウマ目、クダアザミウマ科）日本、全北区、東洋区、エチオピア区。イネ、ムギ、ダイズ、サトウキビなど作物害虫
rice armyworm		*Spodoptera mauritia* (Boisduval)	（チョウ目、ヤガ科）東洋区、エチオピア区、豪州区、大洋区。イネ害虫。日本亜種は *S. m. acronyctoides* (Guenée) シロナヨトウ
rice armyworm			armyworm (1)(2) を見よ
rice armyworm			Lorey leafworm を見よ
rice armyworm			common cutworm を見よ
rice bloodworm		*Chironomus tepperi* Skuse	（ハエ目、ユスリカ科）豪州区
rice blue beetle		*Leptispa pygmaea* Baly	（コウチュウ目、ハムシ科）。東洋区。イネ害虫
rice borer			rice stem borer を見よ
rice bug		*Leptocorisa varicornis* Fabricius	（カメムシ目、ホソヘリカメムシ科）東洋区
rice bug (1)	クモヘリカメムシ	*Leptocorisa chinensis* (Dallas)	（カメムシ目、ホソヘリカメムシ科）日本、旧北区。イネ、トウモロコシ、カンキツ、ナシなどの害虫
rice bug (2)	ホソクモヘリカメムシ	*Leptocorisa acuta* (Thunberg)	（カメムシ目、ホソヘリカメムシ科）日本、東洋区、豪州区。イネ、サトウキビ害虫
rice bug (3)	タイワンマルカメムシ	*Megacopta cribraria* (Fabricius)	（カメムシ目、マルカメムシ科）日本、東洋区、豪州区。ダイズ害虫
rice bug			tropical rice bug を見よ
rice bugs		*Leptocorisa*	（カメムシ目、ホソヘリカメムシ科）の昆虫の総称
rice caddisfly (1)	ゴマダラヒゲナガトビケラ	*Oecetis nigropunctata* Ulmer	（トビケラ目、ヒゲナガトビケラ科）日本、旧北区。イネ害虫
rice caddisfly (2)	ギンボシツツトビケラ	*Setodes argentatus* Matsumura	（トビケラ目、ヒゲナガトビケラ科）日本。イネ害虫
rice case bearer			rice case worm (1) を見よ
rice case worm (1)		*Nymphula depunctalis* (Guenée)	（チョウ目、メイガ科）東洋区、エチオピア区。イネ害虫
rice case-worm	シロミズメイガ	*Paraponyx stagnalis* (Zeller)	（チョウ目、メイガ科）日本、東洋区。イネ害虫
rice Chilo			rice stem borer を見よ
rice crane fly	キリウジガガンボ	*Tipula aino* Alexander	（ハエ目、ガガンボ科）日本。イネ、ムギなどの害虫
rice curculio	イネゾウムシ	*Echinocnemus squameus* (Billberg)	（コウチュウ目、ゾウムシ科）日本、東洋区。イネ害虫
rice cutworm			common cutworm を見よ
rice delphacid		*Tagosodes orizicolus* (Muir)	（カメムシ目、ウンカ科）新北区、新熱帯区
rice-ear bug			tropical rice bug を見よ

英名	和名	学名	所属、分布、ほか
rice ear-cutting caterpillar			armyworm (2) を見よ
rice false looper	ウスシロフコヤガ	*Sugia stygia* (Butler)	(チョウ目、ヤガ科) 日本、旧北区
rice false looper (1)	シロマダラコヤガ	*Protodeltote distinguenda* (Staudinger)	(チョウ目、ヤガ科) 日本、旧北区。イネ害虫
rice field fly		*Ephydra macellaria* Egger	(ハエ目、ミギワバエ科) 旧北区
rice gall midge		*Pachydiplosis oryzae* (Wood-Mason)	(ハエ目、タマバエ科) 日本、東洋区、エチオピア区。イネ害虫
rice grasshopper		*Hieroglyphus banian* (Fabricius)	(バッタ目、バッタ科) 東洋区。イネ害虫
rice grasshopper		*Hieroglyphus oryzivorus* Carl	(バッタ目、バッタ科) 東洋区。イネ害虫
rice grasshopper		*Oxya velox* Fabricius	(バッタ目、バッタ科) エチオピア区
rice grasshopper		*Hieroglyphus daganensis* Krauss	(バッタ目、バッタ科) エチオピア区。イネ害虫
rice grasshopper (1)	ハネナガイナゴ	*Oxya japonica* (Thunberg)	(バッタ目、バッタ科) 日本、東洋区、新北区。イネ、サトウキビ、ダイズなどの害虫
rice grasshopper (2)	タイワンハネナガイナゴ	*Patanga chinensis formosana* Shiraki	(バッタ目、バッタ科) 日本、東洋区。イネ害虫
rice grasshopper (3)	チュウゴクハネナガイナゴ	*Oxya chinensis* (Thunberg)	(バッタ目、バッタ科) 旧北区、東洋区。イネ害虫
rice grasshopper (4)	コイナゴ	*Oxya hyla intricata* (Stål)	(バッタ目、バッタ科) 日本、東洋区
rice grasshopper (5)	コバネイナゴ	*Oxya yezoensis* Shiraki	(バッタ目、バッタ科) 日本、東洋区。イネ、トウモロコシなどの害虫
rice grasshoppers			Japanese grasshoppers を見よ　米国での英名
rice green caterpillar			green rice caterpillar を見よ
rice green semilooper		*Naranga diffusa* Walker	(チョウ目、ヤガ科) 東洋区
rice green semilooper			green rice caterpillar を見よ
rice ground beetle		*Gonocephalum simplex* (Fabricius)	(コウチュウ目、ゴミムシダマシ科) 豪州区、エチオピア区。イネ害虫
rice hairy caterpillar		*Psalis pennatula* (Fabricius)	(チョウ目、ドクガ科) 東洋区。イネ害虫
rice hispa	イネトゲハムシ	*Dicladispa armigera* (Olivier)	(コウチュウ目、ハムシ科) 東洋区。イネ害虫
rice hispid			African rice hispa を見よ
rice katydid	ヒメササキリ	*Conocephalus dimidiatus* (Matsumura et Shiraki)	(バッタ目、キリギリス科) 日本
rice leaf beetle	イネクビボソハムシ	*Oulema oryzae* (Kuwayama)	(コウチュウ目、ハムシ科) 日本、旧北区。イネ害虫
rice leaf beetle	クロルリトゲハムシ	*Rhadinosa nigrocyanea* (Motschulsky)	(コウチュウ目、ハムシ科) 日本。ヒエ害虫
rice leaf beetle			rice hispa を見よ
rice leaf bug	アカヒゲボソミドリカスミカメムシ	*Trygonotylus caelestialium* Kirkaldy	(カメムシ目、カスミカメムシ科) 日本。イネ、ムギ、トウモロコシなどの害虫
rice leaf butterfly			evening brown を見よ
rice leaf miner		*Hydrellia scapularis* Loew	(ハエ目、ミギワバエ科) 新北区。イネ害虫
rice leaf roller			Japanese rice leafroller を見よ

英名	和名	学名	所属、分布、ほか
rice leaffolder		*Marasmia patnalis* Bradley	(チョウ目、メイガ科) 東洋区
rice leaffolder		*Lerodea eufala* (Edwards)	(チョウ目、セセリチョウ科) 新北区、新熱帯区
rice leaffolder			rice leafroller を見よ
rice leaffolder			Japanese rice leafroller を見よ
rice leafholder moth			rice leafroller を見よ
rice leafhopper (1)	シロオオヨコバイ	*Cofana spectra* (Distant)	(カメムシ目、オオヨコバイ科) 日本、東洋区、豪州区。イネ、サトウキビ害虫
rice leafhopper (2)		*Nephotettix apicalis* (Motschulsky)	(カメムシ目、オオヨコバイ科) 東洋区、大洋区。イネ害虫
rice leafhopper			green rice leafhopper (1) を見よ　米国での英名
rice leafminer (1)	イネハモグリバエ	*Agromyza oryzae* (Munakata)	(ハエ目、ハモグリバエ科) 日本、東洋区。イネ害虫
rice leafminer (2)	イネミギワバエ	*Hydrellia griseola* Fallén	(ハエ目、ミギワバエ科) 日本、全北区。イネ害虫
rice leafminer		*Hydrellia* sp.	(ハエ目、ミギワバエ科) 豪州区
rice leaf-miner fly			rice leafminer (2) を見よ
rice leafroller	コブノメイガ	*Cnaphalocrocis medinalis* (Guenée)	(チョウ目、メイガ科) 日本、旧北区、東洋区、豪州区。イネ、サトウキビなどの害虫
rice looper			gold spot を見よ
rice maculated leafhopper	イネマダラヨコバイ	*Recilia oryzae* (Matsumura)	(カメムシ目、オオヨコバイ科) 日本、旧北区
rice mealybug (1)		*Brevennia rehi* (Lindinger)	(カメムシ目、コナカイガラムシ科) 東洋区。イネ害虫
rice mealybug (2)	イネノネコナカイガラムシ	*Geococcus oryzae* (Kuwana)	(カメムシ目、コナカイガラムシ科) 日本、旧北区
rice midge	イネユスリカ	*Chironomus oryzae* Matsumura	(ハエ目、ユスリカ科) 日本
rice midge			rice gall midge を見よ
rice moth	ガイマイツヅリガ	*Corcyra cephalonica* (Stainton)	(チョウ目、メイガ科) 日本、汎世界。貯穀害虫
rice moth			rice armyworm を見よ
rice oscinis			rice leafminer (1) を見よ
rice phloeothrips			rice aculeated thrips を見よ
rice plant skipper			rice skipper を見よ
rice plant weevil			rice curculio を見よ
rice root aphid	オカボアカアブラムシ	*Rhopalosiphum rufiabdominalis* (Sasaki)	(カメムシ目、アブラムシ科) 日本、全北区、東洋区、豪州区、エチオピア区。イネ、ムギ、ナシ、モモ、サクラなどの害虫
rice root aphid			Oriental grass root aphid を見よ
rice root maggot	イネミズトゲミギワバエ	*Notiphila sekiyai* Koizumi	(ハエ目、ミギワバエ科) 日本。イネ害虫
rice root maggot	ワタナベトゲミギワバエ	*Notiphila watanabei* Miyagi	(ハエ目、ミギワバエ科) 日本。イネ害虫
rice root mealybug			rice mealybug (2) を見よ
rice root scale			rice mealybug (2) を見よ

英名	和名	学名	所属、分布、ほか
rice root weevil		*Echinocnemus oryzae* Marshall	（コウチュウ目、ゾウムシ科）東洋区
rice root weevil			rice curculio を見よ
rice rootworm	イネネクイハムシ	*Donacia provostii* Fairmaire	（コウチュウ目、ハムシ科）日本、旧北区、東洋区。イネ害虫
rice seed bug		*Oebalus pugnax* (Fabricius)	（カメムシ目、カメムシ科）新北、新熱帯区。イネ害虫
rice seed bug			African rice bug を見よ
rice seed bug			rice bug (2) を見よ
rice seed midge		*Paralauterborniella subcincta* (Townes)	（ハエ目、ユスリカ科）全北区。イネ害虫
rice seed midge (1)		*Cricotopus sylvestris* (Fabricius)	（ハエ目、ユスリカ科）旧北区。イネ害虫
rice seedling flies		*Atherigona*	（ハエ目、イエバエ科）の昆虫の総称
rice seedling fly		*Atherigona exigua* Stein	（ハエ目、イエバエ科）太洋区、東洋区。イネ害虫
rice shield bug		*Diploxys fallax* Stål	（カメムシ目、カメムシ科）エチオピア区
rice shoot fly			rice stem fly (2) を見よ
rice skipper	イチモンジセセリ	*Parnara guttata guttata* (Bremer et Grey)	（チョウ目、セセリチョウ科）日本、東洋区
ricespotting bug		*Eysarcoris trimaculatus* (Distant)	（カメムシ目、カメムシ科）豪州区
rice stalk borer			American rice stem borer を見よ　成虫は rice stalk borer moth
rice stem borer	ニカメイガ	*Chilo suppressalis* (Walker)	（チョウ目、メイガ科）日本、旧北区、東洋区、大洋区。著名イネ害虫
rice stem borer			yellow rice borer を見よ
rice stem case-bearer	ネジロミズメイガ	*Elophila fenguhanalis* (Pryer)	（チョウ目、メイガ科）日本、旧北区
rice stem fly (1)		*Amaurosoma flavipes* (Fallén)	（ハエ目、Cordyluridae）旧北区
rice stem fly (2)	イネクキイエバエ	*Atherigona oryzae* Malloch	（ハエ目、イエバエ科）日本、旧北区、東洋区、豪州区
rice stem fly (3)		*Amaurosoma armillatum* (Zetterstedt)	（ハエ目、Cordyluridae）旧北区
rice stem gall midge		*Orseolia oryzae* (Wood-Mason)	（ハエ目、タマバエ科）エチオピア区、東洋区
rice stem maggot	イネキモグリバエ	*Chlorops oryzae* Matsumura	（ハエ目、キモグリバエ科）日本、旧北区。イネ害虫
rice stink bug	イネカメムシ	*Lagynotomus elongatus* (Dallas)	（カメムシ目、カメムシ科）日本、東洋区。イネ害虫
rice stink bug			rice seed bug を見よ
rice stink bugs			seed bugs (1) を見よ
rice striped borer			rice stem borer を見よ
rice swarming caterpillar			rice armyworm を見よ
rice swift	ユウレイセセリ	*Borbo cinnara* (Wallace)	（チョウ目、セセリチョウ科）日本、東洋区、豪州区

英名	和名	学名	所属、分布、ほか
rice thrips	イネアザミウマ	*Stenchaetothrips biformis* (Bagnall)	(アザミウマ目、アザミウマ科) 日本、旧北区、東洋区、新熱帯区。イネ、ムギ、トウモロコシ、サトウキビ害虫
rice water weevil	イネミズゾウムシ	*Lissorhoptrus oryzophilus* Kuschel	(コウチュウ目、ゾウムシ科) 日本、新北区。イネ害虫
rice webworm			lawn webworm を見よ
rice weevil	ココクゾウムシ	*Sitophilus oryzae* (Linnaeus)	(コウチュウ目、ゾウムシ科) 日本、汎世界。貯穀害虫
rice weevil			maize weevil を見よ
rice whorl maggot (1)	トウヨウイネクキミギワバエ	*Hydrellia philippina* Ferino	(ハエ目、ミギワバエ科) 日本、東洋区。イネ害虫
rice whorl maggot (2)	イネクキミギワバエ	*Hydrellia sasakii* Yuasa et Isitani	(ハエ目、ミギワバエ科) 日本。イネ害虫
rice whorl maggot			rice leafminer (2) を見よ
rice worm moth		*Apamea apamiformis* Guenée	(チョウ目、ヤガ科) 新北区。wild rice worm ともいう
rice-ear bug			tropical rice bug を見よ
Rice's giant-skipper		*Agathymus ricei* Stallings, Turner et Stallings	(チョウ目、セセリチョウ科) 新熱帯区
ricespotting bug		*Eysarcoris trimaculatus* (Distant)	(カメムシ目、カメムシ科) 豪州区
rich sailer		*Neptis anjana* Moore	(チョウ目、タテハチョウ科) 東洋区
richardiid flies		Richardiidae	(ハエ目) の昆虫の総称
Richards' fungus moth		*Metalectra richardsi* Brower	(チョウ目、ヤガ科) 新北区
Richard's Morpho		*Morpho richardus* Fruhstorfer	(チョウ目、タテハチョウ科) 新熱帯区
Richmond birdwing		*Ornithoptera richmondia* (Gray)	(チョウ目、アゲハチョウ科) 豪州区
Ricini longwing			small Heliconius を見よ
ridgewinged fungus beetle		*Thes bergrothi* (Reitter)	(コウチュウ目、ヒメマキムシ科) 新北区
Ridings' fairy moth		*Adela ridingsella* Clemens	(チョウ目、マガリガ科) 新北区
Ridings' forester		*Alypia ridingsi* Grote	(チョウ目、ヤガ科) 新北区。Ridings' forester moth ともいう
Riding's satyr		*Neominois ridingsii* (Edwards)	(チョウ目、タテハチョウ科) 新北区
Ridley's swallowtail			Acraea swordtail を見よ
Riedel's birdwing		*Troides riedeli* Kirsch	(チョウ目、アゲハチョウ科) 東洋区
riffle beetles	ヒメドロムシ科	Elmidae	(コウチュウ目) の昆虫の総称
riffle bugs			ripple bugs を見よ
riffle snaketail		*Ophiogomphus carolus* Needham	(トンボ目、サナエトンボ科) 新北区
rigid sunflower borer moth		*Papaipema rigida* (Grote)	(チョウ目、ヤガ科) 新北区
Riley's clearwing moth		*Synanthedon rileyana* (Edwards)	(チョウ目、スカシバガ科) 新北区

英名	和名	学名	所属、分布、ほか
Riley's constable		*Dichorragia nesseus rileyi* Hall	（チョウ目、タテハチョウ科）旧北区
Riley's copper		*Aloeides rileyi* Tite et Dickson	（チョウ目、シジミチョウ科）エチオピア区
Riley's glory		*Myscelus draudti* Riley	（チョウ目、セセリチョウ科）新熱帯区
Riley's graphium		*Graphium rileyi* Berger	（チョウ目、アゲハチョウ科）エチオピア区
Riley's lappet moth		*Heteropacha rileyana* Harvey	（チョウ目、カレハガ科）新北区
Riley's opal		*Chrysoritis rileyi* (Dickson)	（チョウ目、シジミチョウ科）エチオピア区
Riley's skolly		*Thestor rileyi* Pennington	（チョウ目、シジミチョウ科）エチオピア区
Riley's 13-year cicada		*Magicicada tredecim* (Walsh et Riley)	（カメムシ目、セミ科）新北区。13年の周期セミ
Riley's tree cricket		*Oecanthus rileyi* Baker	（バッタ目、カンタン科）新北区
rind borer			citrus flower moth (1) を見よ
rind-boring orange moth			honeydew moth を見よ
rind thrips		*Elixothrips brevisetis* (Bagnall)	（アザミウマ目、アザミウマ科）日本、東洋区、大洋区、新北区
Rindge's skipper		*Decinea rindgei* Freeman	（チョウ目、セセリチョウ科）新熱帯区
ring borer		*Phassus damor* Moore	（チョウ目、コウモリガ科）東洋区
ring-footed gnat			banded house mosquito を見よ
ring-legged earwig	コヒゲジロハサミムシ	*Euborellia annulipes* (Lucas)	（ハサミムシ目、ハサミムシ科）日本、汎世界
ring-marked noctuid			clay fan-foot (1) を見よ
ring-marked yellow-hindwinged noctuid	ワモンキシタバ	*Catocala fulminea xarippe* (Butler)	（チョウ目、ヤガ科）日本
ring-necked prominent	クビワシャチホコ	*Shaka atrovittatus* (Bremer)	（チョウ目、シャチホコガ科）日本、旧北区
ringant termite		*Neotermes insularis* (Walker)	（シロアリ目、レイビシロアリ科）豪州区
ringbarker phasmatid		*Podacanthus wilkinsoni* Macleay	（ナナフシ目、ナナフシ科）豪州区
ringed Argus	ウラナミベニヒカゲ	*Callerebia annada* (Moore)	（チョウ目、タテハチョウ科）東洋区
ringed beauty			ringed carpet を見よ
ringed border		*Stegania cararia* (Hübner)	（チョウ目、シャクガ科）旧北区
ringed carpet		*Cleora cinctaria* (Denis et Schiffermüller)	（チョウ目、シャクガ科）旧北区。日本亜種は *C. c. superfumata* Inoue キタルリモンエダシャク
ringed cascader		*Zygonyx torridus* (Kirby)	（トンボ目、トンボ科）エチオピア区
ringed china-mark		*Paraponyx stratiotata* Linnaeus	（チョウ目、メイガ科）旧北区
ringed forceptail		*Phyllocycla breviphylla* Belle	（トンボ目、サナエトンボ科）新北区
ringed forceptail		*Aphylla theodorina* (Navas)	（トンボ目、サナエトンボ科）新熱帯区
ringed sawfly		*Pterygophorus cinctus* Klug	（ハチ目、Pergidae）豪州区

英名	和名	学名	所属、分布、ほか
ringed tortoise beetle		*Ischnocodia annulus* (Fabricius)	（コウチュウ目、ハムシ科）新熱帯区
ringed wave		*Sterrha sylvestraria* (Hübner)	（チョウ目、シャクガ科）旧北区
ringed Xenica			eastern ringed Xenica を見よ
ringlet		*Ragadia crito* de Nicéville	（チョウ目、タテハチョウ科）東洋区
ringlet (1)	チョウセンジャノメ	*Aphantopus hyperantus* (Linnaeus)	（チョウ目、タテハチョウ科）旧北区。ringlet butterfly ともいう
ringlet			large heath を見よ　米国での英名
ringlet woodnymph			ringlet (1) を見よ
ringlets	ヒメヒカゲ属	*Coenonympha*	（チョウ目、タテハチョウ科）の昆虫の総称
ringlets (1)	ベニヒカゲ属	*Erebia*	（チョウ目、タテハチョウ科）の昆虫の総称
ringlets			rings を見よ
rings	ウラナミジャノメ属	*Ypthima*	（チョウ目、タテハチョウ科）の昆虫の総称
Rings' cochylid moth		*Cochylis ringsi* Metzler	（チョウ目、ホソハマキガ科）新北区
ringtails		*Erpetogomphus*	（トンボ目、サナエトンボ科）の昆虫の総称
ringworm			sinuate pear tree borer を見よ
Rio Grande thread-legged katydid		*Arethaea phantasma* Rehn et Hebard	（バッタ目、キリギリス科）新北区
Rio Grande virtuoso katydid		*Amblycorypha rivograndis* Walker	（バッタ目、キリギリス科）新北区
Ripart's anomalous blue		*Polyommatus ripartii* (Freyer)	（チョウ目、シジミチョウ科）旧北区
ripped eucosmid	ヤナギサザナミヒメハマキ	*Saliciphaga acharis* (Butler)	（チョウ目、ハマキガ科）日本、旧北区
ripple bugs	カタビロアメンボ科	Veliidae	（カメムシ目）の昆虫の総称
rippled-marked hawk moth	サザナミスズメ	*Dolbina tancrei* Staudinger	（チョウ目、スズメガ科）日本、旧北区
rippled wave moth		*Idaea obfusaria* (Walker)	（チョウ目、シャクガ科）新北区
rippled white looper moth	サザナミシロヒメシャク	*Scopula nupta* (Butler)	（チョウ目、シャクガ科）日本
Rippon's birdwing	サビモンキシタアゲハ	*Troides hypolitus* (Cramer)	（チョウ目、アゲハチョウ科）東洋区
Ris's sanddragon		*Progomphus risi* Williamson	（トンボ目、サナエトンボ科）新熱帯区
Rita blue		*Euphilotes rita* (Barnes et McDunnough)	（チョウ目、シジミチョウ科）新北区
Rita's Remella		*Remella rita* (Evans)	（チョウ目、セセリチョウ科）新熱帯区
river beetles		*Agabus*	（コウチュウ目、ゲンゴロウ科）の昆虫の総称
river bluet		*Enallagma anna* Williamson	（トンボ目、イトトンボ科）新北区
river clubtail			yellow-legged dragonfly を見よ
river cruisers			river skimmers (2) を見よ
river dropwing		*Trithemis pluvialis* Förster	（トンボ目、トンボ科）エチオピア区
river jewelwing		*Calopteryx aequabilis* (Say)	（トンボ目、イトトンボ科）新北区

英名	和名	学名	所属、分布、ほか
river sailer		*Neptis serena* Overlaet	（チョウ目、タテハチョウ科）エチオピア区
river skater		*Aquarius najas* (De Geer)	（カメムシ目、アメンボ科）旧北区
river skimmers (1)	ヤマトンボ科	Macromiidae	（トンボ目）の昆虫の総称
river skimmers (2)		*Macromia*	（トンボ目、ヤマトンボ科）の昆虫の総称
riverhawks		*Onychothemis*	（トンボ目、トンボ科）の昆虫の総称
riverine clubtail		*Stylurus amnicola* (Walsh)	（トンボ目、サナエトンボ科）新北区
riverjack		*Mesocnemis singularis* Karsch	（トンボ目、モノサシトンボ科）エチオピア区
rivulet		*Perizoma affinitata* (Stephens)	（チョウ目、シャクガ科）旧北区
rivulet moth			rivulet を見よ
rivulet tiger		*Gomphidia pearsoni* Fraser	（トンボ目、サナエトンボ科）東洋区
roaches			cockroaches (2) (3) を見よ
roachlike stoneflies	ヒロムネカワゲラ科	Peltoperlidae	（カワゲラ目）の昆虫の総称
roachlike stoneflies		*Peltoperla*	（カワゲラ目、ヒロムネカワゲラ科）の昆虫の総称
roachlike stonefly		*Sierraperla cora* (Needham et Smith)	（カワゲラ目、ヒロムネカワゲラ科）新北区
road beetles			rove beetles を見よ　稀な使用例
road page	タテジマツルギタテハ	*Marpesia chiron* Fabricius	（チョウ目、タテハチョウ科）新北区、新熱帯区
roadside rambler		*Amblyscirtes celia* Skinner	（チョウ目、セセリチョウ科）新北区、新熱帯区
roadside sallow moth		*Metaxaglaea viatica* (Grote)	（チョウ目、ヤガ科）新北区
roadside skipper		*Amblyscirtes vialis* Edwards	（チョウ目、セセリチョウ科）新北区
roadside skippers		*Amblyscirtes*	（チョウ目、セセリチョウ科）の昆虫の総称
robber ant			Amazon ant を見よ
robber dung flies			yellow dung-flies を見よ
robber flies		*Tolmerus*	（ハエ目、ムシヒキアブ科）の昆虫の総称
robber flies (1)	ムシヒキアブ科	Asilidae	（ハエ目）の昆虫の総称　世界に 5,000 種
robber-flies		Asilinae	（ハエ目、ムシヒキアブ科）の昆虫の総称
robber-fly			golden-tabbed robberfly を見よ
robber-fly			kite-tailed robberfly を見よ
robber-fly			downland robberfly を見よ
Robbins' groundstreak		*Lamprospilus* sp.	（チョウ目、シジミチョウ科）新熱帯区
Robbins' hairstreak		*Ocaria* sp.	（チョウ目、シジミチョウ科）新熱帯区
Rober's dartwhite		*Catasticta notha* (Doubleday)	（チョウ目、シロチョウ科）新熱帯区
Rober's skipper		*Elbella miodesmiata* (Rober)	（チョウ目、セセリチョウ科）新熱帯区
Robert's hairstreak		*Iaspis* sp.	（チョウ目、シジミチョウ科）新熱帯区
Robertson's blue		*Lepidochrysops robertsoni* Cottrell	（チョウ目、シジミチョウ科）エチオピア区

英名	和名	学名	所属、分布、ほか
Robertson's brown		*Stygionympha robertsoni* (Riley)	(チョウ目、タテハチョウ科) エチオピア区
robin moth			cecropia moth を見よ
robinia leaf margin gall midge			locust gall midge を見よ
Robin's moth			cecropia moth を見よ
Robin's pincushion			mossy rose gall wasp を見よ
Robinson's Acleris moth		*Acleris robinsoniana* (Forbes)	(チョウ目、ハマキガ科) 新北区
Robinson's Eucosma moth		*Eucosma robinsonana* (Grote)	(チョウ目、ハマキガ科) 新北区
Robinson's underwing moth		*Catocala robinsonii* Grote	(チョウ目、ヤガ科) 新北区
robust apote		*Apote robusta* Caudell	(バッタ目、キリギリス科) 新北区
robust assassin bugs		*Apiomerus*	(カメムシ目、サシガメ科) 新北区
robust baskettail		*Epitheca spinosa* (Hagen)	(トンボ目、エゾトンボ科) 新北区
robust blue-winged grasshopper		*Leprus intermedius* Saussure	(バッタ目、バッタ科) 新北区
robust bot flies			rabbit bots を見よ
robust bush cricket		*Tafalisca lineatipes* Bruner	(バッタ目、コオロギ科) 新北区
robust cicada	ミンミンゼミ	*Hyalessa maculaticollis* (Motschulsky)	(カメムシ目、セミ科) 日本、旧北区
robust ciliate blue		*Cupidesthes robusta* Aurivillius	(チョウ目、シジミチョウ科) エチオピア区
robust click beetles		Cebrionidae	(コウチュウ目) の昆虫の総称
robust conehead		*Neoconocephalus robustus* (Scudder)	(バッタ目、キリギリス科) 新北区
robust ground crickets			ground crickets を見よ
robust hopper		*Platylesches robustus* Neave	(チョウ目、セセリチョウ科) エチオピア区
robust leafhopper		*Stragania robusta* (Uhler)	(カメムシ目、オオヨコバイ科) 新北区
robust noctuid			shaded fanfoot を見よ
robust shieldback		*Atlanticus gibbosus* Scudder	(バッタ目、キリギリス科) 新北区
robust skimmer		*Orthetrum robustum* (Balinsky)	(トンボ目、トンボ科) エチオピア区
robust spreadwing			scarce emerald を見よ
Rocena hairstreak		*Janthecla rocena* (Hewitson)	(チョウ目、シジミチョウ科) 新熱帯区
rock bristletails		Meinertellidae	(イシノミ目) の昆虫の総称
rock bush brown		*Bicyclus pavonis* (Butler)	(チョウ目、タテハチョウ科) エチオピア区
rock crawlers (1)	ガロアムシ目	Grylloblattodea	の昆虫の総称　世界に約20種
rock crawlers (2)	ガロアムシ科	Grylloblattidae	(ガロアムシ目) の昆虫の総称
rock-cress smudge		*Rhigognostis senilella* (Herrich-Schäffer)	(チョウ目、スガ科) 旧北区
rockdwellers		*Bradinopyga*	(トンボ目、トンボ科) の昆虫の総称

英名	和名	学名	所属、分布、ほか
rock grayling	タカネジャノメ	*Hipparchia alcyone* (Denis et Schiffermüller)	（チョウ目、タテハチョウ科）旧北区
rock grayling		*Hipparchia hermione* (Linnaeus)	（チョウ目、タテハチョウ科）旧北区
rock grayling			grayling を見よ
rock hooktail		*Paragomphus cognatus* (Rambur)	（トンボ目、サナエトンボ科）エチオピア区
rockhoppers		*Mesomachilis*	（イシノミ目、イシノミ科）の昆虫の総称
rockhoppers		*Pedetontus*	（イシノミ目、イシノミ科）の昆虫の総称
rock-jumpers			silverfish (2) を見よ
rock lichen moth		*Halone sinuata* (Wallengren)	（チョウ目、ヒトリガ科）豪州区
rock malachite		*Chlorolestes peringueyi* Ris	（トンボ目、Synlestidae）エチオピア区
rock ringlet		*Hypocysta euphemia* Westwood	（チョウ目、タテハチョウ科）豪州区
rockside alpine			Magdalena alpine を見よ
rockslide checkerspot		*Chlosyne whitneyi* (Behr)	（チョウ目、タテハチョウ科）新北区
rock skolly		*Thestor petra* Pennington	（チョウ目、シジミチョウ科）エチオピア区
rock worm			June beetle (1) を見よ　幼虫の英名
rockworms			May beetles (3) を見よ　幼虫の英名
Rocky Mountain arctic blue			Rustic' arctic blue を見よ
Rocky Mountain duskywing		*Erynnis telemachus* Burns	（チョウ目、セセリチョウ科）新北区
Rocky Mountain grasshoppeer		*Melanoplus spretus* (Walsh)	（バッタ目、バッタ科）新北区。絶滅危惧種。Rocky Mountain locust ともいう
Rocky Mountain Parnassian			mountain Parnassian を見よ
Rocky Mountain skipper			Draco skipper を見よ
Rocky Mountain sleepy duskywing		*Erynnis brizo burgessi* (Skinner)	（チョウ目、セセリチョウ科）新熱帯区。sleepy duskywing を参照
rodent beetles			mammal nest beetles を見よ
rodent bot flies		*Cuterebra*	（ハエ目、ヒツジバエ科）の昆虫の総称
rodent bots			rabbit bots を見よ
rodent chewing lice	ナガケモノハジラミ科	Gyropidae	（ハジラミ目）の昆虫の総称
rodent fleas	トゲノミ科	Dolichopsyllidae	（ノミ目）の昆虫の総称
rodent fleas (1)	ケブカノミ科	Hystrichopsyllidae	（ノミ目）の昆虫の総称
rodent fleas			bird fleas を見よ
Rodtenbacher's crow			Malayan crow (1) を見よ
roe biting louse		*Cervicola meyeri* (Taschenberg)	（ハジラミ目、ケモノハジラミ科）旧北区
roe louse		*Solenopotes capreoli* Freund	（シラミ目、ケモノホソジラミ科）旧北区
Roesel's bush-cricket		*Metrioptera roeselii* (Hagenbach)	（バッタ目、キリギリス科）全北区

英名	和名	学名	所属、分布、ほか
Roesel's katydid			Roesel's bush-cricket を見よ
Roesel's signal		*Chrysoesthia roesella* (Linnaeus)	(チョウ目、キバガ科) 旧北区
Roger's Acraea	ゴマダラホソチョウ	*Acraea rogersi* Hewitson	(チョウ目、タテハチョウ科) エチオピア区
Roger's cupid		*Eicochrysops rogersi* Bethune-Baker	(チョウ目、シジミチョウ科) エチオピア区
Roger's gem		*Vansomerenia rogersi* (Riley)	(チョウ目、シジミチョウ科) エチオピア区
Roger's Judy		*Abisara rogersi dollmani* Riley	(チョウ目、シジミタテハ科) エチオピア区
Roger's orange tip		*Colotis rogersi* (Dixey)	(チョウ目、シロチョウ科) エチオピア区
Roger's Pentila		*Pentila rogersi* (Druce)	(チョウ目、シジミチョウ科) エチオピア区
Roger's ranger		*Kedestes rogersi* Druce	(チョウ目、セセリチョウ科) エチオピア区
Roger's sailer		*Neptis rogersi* Eltringham	(チョウ目、タテハチョウ科) エチオピア区
Roibok Acraea		*Acraea oncaea* Hopffer	(チョウ目、タテハチョウ科) エチオピア区
rolled-winged stoneflies	ハラジロオナシカワゲラ科	Leuctridae	(カワゲラ目) の昆虫の総称　多くの種の体長は 10mm 以下
rolled-winged winter stoneflies			rolled-winged stoneflies を見よ
roller moths			leafroller moths を見よ
rolling carrot flat-body		*Agonopterix rotundella* (Douglas)	(チョウ目、マルハキバガ科) 旧北区
roodepoort copper		*Aloeides dentatis* (Swierstra)	(チョウ目、シジミチョウ科) エチオピア区
Rooiberg skolly		*Thestor rooibergensis* Heath	(チョウ目、シジミチョウ科) エチオピア区
root aphids		*Pemphigus*	(カメムシ目、アブラムシ科) の昆虫の総称
root bark channeler			citrus root-bark channeler を見よ
root collar borer moth		*Euzophera ostricolorella* Hulst	(チョウ目、メイガ科) 新北区
root-eating beetles	ネスイムシ科	Rhizophagidae	(コウチュウ目) の昆虫の総称
root-eating flies			anthomyiine flies を見よ
rootfeeding springtails			white blind springtails を見よ
root gnats			dark-winged fungus flies を見よ
root maggot			radish fly を見よ　幼虫の英名
root maggot flies			anthomyiine flies を見よ
root-maggot fly		*Anthomyia ochripes* Thomson	(ハエ目、ハナバエ科) 新北区
root-maggot fly		*Anthomyia oculifera* Bigot	(ハエ目、ハナバエ科) 新北区
root maggots			anthomyiine flies を見よ
root mealybug			citrus mealybug (1) を見よ
root mealybug			ground mealybug (1) を見よ　豪州での英名
root mealybugs		*Rhizoecus*	(カメムシ目、コナカイガラムシ科) の昆虫の総称

英名	和名	学名	所属、分布、ほか
root moth		*Pelochrista medullana* Staudinger	(チョウ目、ハマキガ科) 新北区
root weevil		*Prodecatoma cooki* (Howard)	(ハチ目、カタビロコバチ科) 新北区
root weevil			alfalfa snout beetle を見よ
ropalomerid flies		Ropalomeridae	(ハエ目) の昆虫の総称
roproniid wasps	イシハラクロバチ科	Roproniidae	(ハチ目) の昆虫の総称
roproniids			roproniid wasps を見よ
Roraima skipper		*Cantha roraimae* (Bell)	(チョウ目、セセリチョウ科) 新熱帯区
Rosalia longhorn		*Rosalia alpina* (Linnaeus)	(コウチュウ目、カミキリムシ科) 旧北区。Rosalia longicorn ともいう
Rosa's tree nymph		*Sevenia rosa* (Hewitson)	(チョウ目、タテハチョウ科) エチオピア区
rose aphid		*Chaetosiphon tetrarhodus* (Walker)	(カメムシ目、アブラムシ科) 旧北区
rose aphid	イバラヒゲナガアブラムシ	*Sitobion ibarae* (Matsumura)	(カメムシ目、アブラムシ科) 日本、旧北区、東洋区
rose aphid (1)		*Macrosiphum rosae* (Linnaeus)	(カメムシ目、アブラムシ科) 全北区、豪州区、新熱帯区。温室害虫
rose aphid			rose grain aphid を見よ
rose arge			rose argid sawfly (3) を見よ
rose argid sawfly (1)	アカスジチュウレンジ	*Arge nigrinodosa* (Motschulsky)	(ハチ目、ミフシハバチ科) 日本。バラ害虫
rose argid sawfly (2)	ニホンチュウレンジ	*Arge nipponensis* Rohwer	(ハチ目、ミフシハバチ科) 日本、旧北区。バラ害虫
rose argid sawfly (3)	チュウレンジバチ	*Arge pagana* (Panzer)	(ハチ目、ミフシハバチ科) 日本、旧北区。バラ害虫
rose-banded wave		*Rhodostrophia vibicaria* Clerck	(チョウ目、シャクガ科) 旧北区
rose beauty		*Haematera pyrame* Hübner	(チョウ目、タテハチョウ科) 新熱帯区
rose beetle			rose chafer (1)(2) を見よ
rose bud midge		*Dasineura rhodophaga* (Coquillett)	(ハエ目、タマバエ科) 新北区
rose budworm			bordered sallow を見よ　米国での英名
rose bug			rose chafer (1) を見よ
rose buprestid		*Coraebus rubi* (Linnaeus)	(コウチュウ目、タマムシ科) 旧北区
rose chafer (1)		*Cetonia aurata* (Linnaeus)	(コウチュウ目、コガネムシ科) 旧北区。鮮麗な黄金色の種
rose chafer (2)		*Macrodactylus subspinosus* (Fabricius)	(コウチュウ目、コガネムシ科) 新北区。バラ、ブドウなどの害虫
rose chafer (3)		*Macrodactylus suavis* Bates	(コウチュウ目、コガネムシ科) 新熱帯区。カンキツ害虫
rose chafer			green rose chafer (1) を見よ
rose chafer			western rose chafer を見よ
rose chafers			chafers を見よ
rose clearwing moth	ムナブトヒメスカシバ	*Zenodoxus constricta* (Butler)	(チョウ目、スカシバガ科) 日本、旧北区。バラ害虫

英名	和名	学名	所属、分布、ほか
rose curculio		*Merhynchites bicolor* (Fabricius)	(コウチュウ目、ゾウムシ科) 新北区
rose eucosmid	バラシロハマキ	*Notocelia rosaecolana* (Doubleday)	(チョウ目、ハマキガ科) 日本、全北区。バラ害虫
rose gall wasp	バラトゲタマフシバチ	*Synergus japonicus* Walker	(ハチ目、タマバチ科) 日本
rose grain aphid	ムギウスイロアブラムシ	*Metopolophium dirhodum* (Walker)	(カメムシ目、アブラムシ科) 日本、全北区、エチオピア区、新熱帯区。ムギ、トウモロコシ害虫。rose-grain aphid とも記す
rose hip fly		*Rhagoletis alternata* (Fallén)	(ハエ目、ミバエ科) 旧北区
rose hip fly		*Rhagoletis basiola* (Osten Sacken)	(ハエ目、ミバエ科) 新北区
rose hooktip moth		*Oreta rosea* (Walker)	(チョウ目、カギバガ科) 新北区
rose leaf beetle		*Nodonota puncticollis* (Say)	(コウチュウ目、ハムシ科) 新北区
rose leaf beetle (1)	バラルリツツハムシ	*Cryptocephalus approximatus* Baly	(コウチュウ目、ハムシ科) 日本。リンゴ、ナシ、ウメ、バラなどの害虫
rose leaf midge		*Dasineura rosarum* (Hardy)	(ハエ目、タマバエ科) 旧北区
rose leaf miner			common rose pigmy を見よ
rose leaf sawfly			leaf-rolling rose sawfly を見よ
rose leafcutter	バラハキリバチ	*Megachile nipponica* Cockerell	(ハチ目、ハキリバチ科) 日本、旧北区。バラ害虫
rose leaf-cutting bee			rose leafcutter を見よ
rose leaf-cutting bee			common leaf-cutter bee を見よ
rose leaf-folder			rose tortrix moth を見よ
rose leafhopper	バラヒメヨコバイ	*Edwardsiana rosae* (Linnaeus)	(カメムシ目、オオヨコバイ科) 日本、全北区。バラ害虫
rose leafminer	バラハモグリバエ	*Agromyza potentillae* (Kaltenbach)	(ハエ目、ハモグリバエ科) 日本、全北区。バラ害虫
rose leaf-miner			common rose pigmy を見よ
rose leaf-rolling sawfly			leaf-rolling rose sawfly を見よ
rose looper	ウラベニエダシャク	*Heterolocha aristonaria* (Walker)	(チョウ目、シャクガ科) 日本、旧北区
rose maggots			leafroller moths を見よ　幼虫の英名
rose midge			rose bud midge を見よ　北米での英名
rose midge			rose leaf midge を見よ
rose myrtle lappet moth		*Trabala vishnou* Lefebvre	(チョウ目、カレハガ科) 東洋区
rose of Sharon leaf like moth			fruit-piercing moth (8) を見よ
rose of Sharon pyralid	カクモンノメイガ	*Rehimena surusalis* (Walker)	(チョウ目、メイガ科) 日本、旧北区、東洋区
rose plume	チョウセントリバ	*Platyptilia rhododactyla* (Denis et Schiffermüller)	(チョウ目、トリバガ科) 日本、全北区、エチオピア区。rose plume moth ともいう

英名	和名	学名	所属、分布、ほか
rose root aphid		*Maculolachnus submacula* (Walker)	(カメムシ目、アブラムシ科) 全北区。バラ害虫
rose root gall wasp		*Diplolepis radicum* (Osten Sacken)	(ハチ目、タマバチ科) 全北区。バラ害虫
rose sawfly (1)	クシヒゲハバチ	*Cladius pectinicornis* (Geoffroy)	(ハチ目、ハバチ科) 日本、旧北区。バラ害虫
rose sawfly (2)	オオシロオビクロハバチ	*Allantus meridionalis* Takeuchi	(ハチ目、ハバチ科) 日本。バラ害虫
rose sawfly			large rose sawfly を見よ
rose sawfly			rose slug sawfly を見よ　米国での英名
rose sawfly			rose argid sawfly (1) を見よ
rose scale	バラシロカイガラムシ	*Aulacaspis rosae* (Bouché)	(カメムシ目、マルカイガラムシ科) 日本、新北区、汎熱帯。バラ害虫
rose scale			scurfy scale を見よ
rose shoot sawfly		*Cladardis elongatulus* (Klug)	(ハチ目、ハバチ科) 旧北区
rose slug			rose slug sawfly を見よ
rose slug caterpillar			stinging rose caterpillar を見よ
rose slug sawfly		*Endelomyia aethiops* (Fabricius)	(ハチ目、ハバチ科) 旧北区
rose slug-worm			rose slug sawfly を見よ
rose smooth pea-gall cynipid			smooth pea gall wasp を見よ
rose snout beetle		*Rhynchites bicolor* Fabricius	(コウチュウ目、ゾウムシ科) 新北区
rose spiked pea-gall cynipid		*Diplolepis nervosa* (Curtis)	(ハチ目、タマバチ科) 旧北区。sputnik gall wasp ともいう
rose stem boring sawfly	バラクキハバチ	*Ardis brunniventris* (Hartig)	(ハチ目、ハバチ科) 日本。バラ害虫
rose stem girdler		*Agrilus aurichalceus* Redtenbacher	(コウチュウ目、タマムシ科) 新北区
rose stem sawfly (1)	オオバラクキバチ	*Hartigia agilis* (Smith)	(ハチ目、クキバチ科) 日本。バラ害虫
rose stem sawfly (2)	バラクキバチ	*Syrista similis* Mocsary	(ハチ目、クキバチ科) 日本、旧北区。バラ害虫
rose thrips		*Thrips fuscipennis* Haliday	(アザミウマ目、アザミウマ科) 旧北区
rose tortrix			rose tortrix moth を見よ
rose tortrix moth		*Archips rosana* (Linnaeus)	(チョウ目、ハマキガ科) 全北区。バラ、果樹害虫
rose torymid	バラモンオナガコバチ	*Megastigmus aculeatus* (Swederus)	(ハチ目、オナガコバチ科) 日本、旧北区、エチオピア区
rose twist			rose tortrix moth を見よ
rose web-spinning sawfly		*Pamphilius inanitus* (Villers)	(ハチ目、ヒラタハバチ科) 旧北区
rose windmill	ラトタイユベニモンアゲハ	*Atrophaneura latreillei* (Donovan)	(チョウ目、アゲハチョウ科) 東洋区
roseate skimmer		*Orthemis ferruginea* (Fabricius)	(トンボ目、トンボ科) 新北区、新熱帯区、大洋区
rosemary beetle		*Chrysolina americana* (Linnaeus)	(コウチュウ目、ハムシ科) 旧北区

英名	和名	学名	所属、分布、ほか
Rosenberg's painted Jezebel	ローゼンベルクカザリシロチョウ	*Delias rosenbergi* (Vollenhoven)	（チョウ目、シロチョウ科）東洋区
roseslug			rose slug sawfly を見よ
rosette lac scale		*Tachardina decorella* (Maskell)	（カメムシ目、ラックカイガラムシ科）豪州区
rosewing moth		*Sideridis rosea* (Harvey)	（チョウ目、ヤガ科）新北区
rosinweed moth		*Tebenna silphiella* (Grote)	（チョウ目、ハマキモドキガ科）新北区
Rosita checkerspot			Rosita patch を見よ
Rosita patch		*Chlosyne rosita* Hall	（チョウ目、タテハチョウ科）新北区、新熱帯区
Rossi's walkingstick		*Bacillus rossii* Fabricius	（ナナフシ目、コブナナフシ科）旧北区。Rossi's stickinsect ともいう
Rossouw's blue		*Lepidochrysops rossouwi* Henning et Henning	（チョウ目、シジミチョウ科）エチオピア区
Rossouw's copper		*Aloeides rossouwi* Henning et Henning	（チョウ目、シジミチョウ科）エチオピア区
Rossouw's skolly		*Thestor rossouwi* Dickson	（チョウ目、シジミチョウ科）エチオピア区
Ross's alpine			arctic alpine を見よ
Ross's black scale (1)		*Lindingaspis rossi* (Maskell)	（カメムシ目、マルカイガラムシ科）豪州区。カンキツ害虫
Ross's black scale (2)		*Lindingaspis striata* (Newstead)	（カメムシ目、マルカイガラムシ科）旧北区
rosy Aemilia		*Lophocampa roseata* (Walker)	（チョウ目、ヒトリガ科）新北区
rosy aphid			rosy apple aphid (1) を見よ
rosy aphid			rowan aphid を見よ
rosy apple aphid		*Aphis malifoliae* Fitch	（カメムシ目、アブラムシ科）新北区
rosy apple aphid (1)	オオバコアブラムシ	*Dysaphis plantaginea* (Passerini)	（カメムシ目、アブラムシ科）日本、東洋区、全北区。多食性
rosy banded hawk		*Leucophlebia afra* Karsch	（チョウ目、スズメガ科）エチオピア区
rosy conch		*Cochylis roseana* (Haworth)	（チョウ目、ホソハマキガ科）旧北区
rosy day		*Dasystoma salicella* (Hübner)	（チョウ目、マルハキバガ科）旧北区
rosy ear			rosy rustic を見よ
rosy Euselasia		*Euselasia gelanor* (Stoll)	（チョウ目、シジミタテハ科）新熱帯区
rosy flash		*Rapala rubida* Tytler	（チョウ目、シジミチョウ科）東洋区
rosy-flouced tabby		*Endotricha flammealis* Schiffermüller	（チョウ目、メイガ科）旧北区
rosy footman	ベニヘリコケガ	*Miltochrista miniata* (Forster)	（チョウ目、ヒトリガ科）日本、旧北区
rosy grizzled skipper		*Pyrgus onopordi* (Rambur)	（チョウ目、セセリチョウ科）旧北区
rosy knot-horn		*Oncocera semirubella* (Scopoli)	（チョウ目、メイガ科）旧北区
rosy leaf-curling aphid		*Dysaphis devecta* (Walker)	（カメムシ目、アブラムシ科）旧北区
rosy-legged greenish geometrid	アカアシアオシャク	*Culpinia diffusa* (Walker)	（チョウ目、シャクガ科）日本、旧北区

英名	和名	学名	所属、分布、ほか
rosy maple moth			striped maple worm を見よ　北米での英名
rosy marbled		*Elaphria venustula* (Hübner)	（チョウ目、ヤガ科）旧北区
rosy marsh moth	ノコスジモンヤガ	*Eugraphe subrosea* (Stephens)	（チョウ目、ヤガ科）日本、旧北区
rosy minor		*Mesoligia literosa* (Haworth)	（チョウ目、ヤガ科）旧北区
rosy minor moth			rosy minor を見よ
rosy oakblue		*Panchala alea* (Hewitson)	（チョウ目、シジミチョウ科）東洋区
rosy plume			rose plume を見よ
rosy rustic		*Hydraecia micacea* (Esper)	（チョウ目、ヤガ科）全北区
rosy rustic moth			rosy rustic を見よ
rosy rustic moth			frosted orange (1) を見よ
rosy underwing		*Catocala electa* (Vieweg)	（チョウ目、ヤガ科）旧北区。日本亜種は *C. e. zalmunna* (Butler) ベニシタバ
rosy wave		*Scopula emutaria* (Hübner)	（チョウ目、シャクガ科）旧北区
rosy white			rosy wave を見よ
Rothschild's birdwing	ロスチャイルドトリバネアゲハ	*Ornithoptera rothschildi* (Kenrick)	（チョウ目、アゲハチョウ科）豪州区
Rothschild's saturniid		*Rothschildia jorulla* (Westwood)	（チョウ目、ヤママユガ科）新北区
Rothschild's swordtail		*Protesilaus earis* (Rothschild et Jordan)	（チョウ目、アゲハチョウ科）新熱帯区
rotten-wood termites	シュウカクシロアリ科	Hodotermitidae	（シロアリ目）の昆虫の総称
rotund Idia moth		*Idia rotundalis* (Walker)	（チョウ目、ヤガ科）新北区
Rougeot's sapphire gem		*Iridana rougeoti* Stempffer	（チョウ目、シジミチョウ科）エチオピア区
rough-backed bush cricket		*Uromenus rugosicollis* Serville	（バッタ目、キリギリス科）旧北区
rough blackhardback		*Ligyrus cuniculus* (Fabricius)	（コウチュウ目、コガネムシ科）新熱帯区
rough bollworm		*Earias huegeliana* Gaede	（チョウ目、ヤガ科）豪州区
rough bollworm		*Earias perhuegeli* Holloway	（チョウ目、ヤガ科）豪州区
rough brown weevil		*Baryopadus corrugatus* Pascoe	（コウチュウ目、ゾウムシ科）豪州区
rough harvester ant			harvester ant (3) を見よ
rough-headed beetle of India			sugarcane beetle を見よ
rough-headed corn stalk beetle			sugarcane beetle を見よ
rough-headed cornstalk borer			sugarcane beetle を見よ
rough prominent			tawny prominent (1) を見よ　北米での英名
rough-skinned cutworm			rough-skinned cutworm moth を見よ

英名	和名	学名	所属、分布、ほか
rough-skinned cutworm moth		*Proxenus mindara* Barnes et McDunnough	（チョウ目、ヤガ科）新北区。幼虫はタンポポを食う
rough stink bug		*Brochymena quadripustulata* (Fabricius)	（カメムシ目、カメムシ科）新北区
rough stink bug		*Brochymena affinis* Van Duzee	（カメムシ目、カメムシ科）新北区
rough stink bugs		*Brochymena*	（カメムシ目、カメムシ科）の昆虫の総称
rough strawberry weevil			lesser strawberry weevil を見よ
rough sweetpotato weevil		*Blosyrus asellus* (Olivier)	（コウチュウ目、ゾウムシ科）東洋区
rough-tipped sootywing		*Bolla evippe* (Godman et Salvin)	（チョウ目、セセリチョウ科）新熱帯区
rough-winged conch		*Phtheochroa rugosana* (Hübner)	（チョウ目、ハマキガ科）旧北区
rough-tipped sootywing		*Bolla evippe* (Godman et Salvin)	（チョウ目、セセリチョウ科）新熱帯区
roughened darkling beetle		*Upis ceramboides* (Linnaeus)	（コウチュウ目、ゴミムシダマシ科）旧北区
round bamboo scale			penicillate scale を見よ
round black scale			Ross's black scale (2) を見よ
round darkling beetle		*Coelus ciliatus* Eschscholtz	（コウチュウ目、ゴミムシダマシ科）新北区
round fungus beetles	タマキノコムシ科	Leiodidae	（コウチュウ目）の昆虫の総称
roundhead pine beetle		*Dendroctonus adjunctus* Blandford	（コウチュウ目、キクイムシ科）新北区
roundhead wood borers			long-horned beetles を見よ 幼虫の英名。他に幼虫は sawyer, sawer, wood borer ともいう
roundheaded apple tree borer	リンゴシロスジカミキリ	*Saperda candida* Fabricius	（コウチュウ目、カミキリムシ科）新北区
roundheaded beetles			long-horned beetles を見よ
roundheaded borers			コウチュウ目のカミキリムシ類
round-headed borers			poplar longhorns を見よ
roundheaded cone borer		*Paratimia conicola* Fisher	（コウチュウ目、カミキリムシ科）新北区
round-headed katydid		*Ambrycorypha floridana* Rehn et Hebard	（バッタ目、キリギリス科）新北区
round-headed katydids		*Amblycorypha*	（バッタ目、キリギリス科）の昆虫の総称
round-headed katydids			bush katydids (1) を見よ
roundheaded pasture webworm		*Oncopera brachyphylla* Turner	（チョウ目、コウモリガ科）豪州区
roundheaded pine beetle			roundhead pine beetle を見よ
round-headed squirrel louse		*Enderleinellus nitzschi* Fahrenholz	（シラミ目、ケモノジラミ科）旧北区
round-headed wood borer		*Coptocercus rubripes* Boisduval	（コウチュウ目、カミキリムシ科）豪州区

英名	和名	学名	所属、分布、ほか
roundheaded wood borers			long-horned beetles を見よ
round Japanese cedar scale	スギマルカイガラムシ	*Aspidiotus cryptomeriae* Kuwana	(カメムシ目、マルカイガラムシ科) 日本、旧北区、東洋区
round-necked blister beetle	マルクビツチハンミョウ	*Meloe corvinus* Marseul	(コウチュウ目、ツチハンミョウ科) 日本、旧北区
round-necked flattened longicorn			opaque sawyer を見よ
round-necked longhorns		Cerambycinae	(コウチュウ目、カミキリムシ科) の昆虫の総称
round sand beetles	カワラゴミムシ科	Omophoronidae	(コウチュウ目) の昆虫の総称 オサムシ科に入れられる
round-spotted major		*Oxycera dives* Loew	(ハエ目、ミズアブ科) 旧北区
round-spotted silverdrop		*Epargyreus socus orizaba* Scudder	(チョウ目、セセリチョウ科) 新熱帯区
round-spotted Ticlear			Lycaste tigerwing を見よ
round-tipped bark beetle	アトマルキクイムシ	*Dryocoetes rugicollis* Eggers	(コウチュウ目、キクイムシ科) 日本、旧北区
round-tipped conehead		*Neoconocephalus retusus* (Scudder)	(バッタ目、キリギリス科) 新北区
round-tipped elongate scolytid	マルオナガキクイムシ	*Crossotarsus emancipatus* Murayama	(コウチュウ目、キクイムシ科) 日本、東洋区
round velvety chafer	マルガタビロウドコガネ	*Maladera secreta* (Brenske)	(コウチュウ目、コガネムシ科) 日本
round-winged bluet		*Proischnura rotundipennis* Ris	(トンボ目、イトトンボ科) エチオピア区
round-winged footman			round-winged muslin を見よ
round-winged muslin		*Thumatha senex* (Hübner)	(チョウ目、ヒトリガ科) 旧北区
round winged orange tip	ヒメツマアカシロチョウ	*Colotis evippe* (Linnaeus)	(チョウ目、シロチョウ科) エチオピア区
roundwinged polyptychus		*Pseudandriasa mutata* (Walker)	(チョウ目、スズメガ科) エチオピア区
round-winged skipper		*Thargella caura* (Plötz)	(チョウ目、セセリチョウ科) 新熱帯区
round-winged vagrant		*Nepheronia pharis* (Boisduval)	(チョウ目、シロチョウ科) エチオピア区
rounded Calephelis			lost metalmark を見よ
rounded metalmark		*Calephelis nilus* (C. et R. Felder)	(チョウ目、シジミタテハ科) 新北区
rounded pierrot		*Tarucus extricatus* Butler	(チョウ目、シジミチョウ科) 東洋区
rounded 6-lineblue			six-line blue を見よ
rounded sootywing		*Bolla imbras* (Godman et Salvin)	(チョウ目、セセリチョウ科) 新熱帯区
roundedheaded fir borer		*Tetropium abietis* Fall	(コウチュウ目、カミキリムシ科) 新北区
rove beetle		*Emus hirtus* (Linnaeus)	(コウチュウ目、ハネカクシ科) 旧北区

英名	和名	学名	所属、分布、ほか
rove beetle		*Aleochara bilineata* (Gyllenhal)	（コウチュウ目、ハネカクシ科）旧北区。寄生性
rove beetles	ハネカクシ科	Staphylinidae	（コウチュウ目）の昆虫の総称
Rovena jewel	メガミニシキシジミ	*Hypochrysops polycletus* (Linnaeus)	（チョウ目、シジミチョウ科）豪州区
Rover's skipperling		*Piruna roeveri* (Miller et Miller)	（チョウ目、セセリチョウ科）新熱帯区
rowan aphid		*Dysaphis sorbi* (Kaltenbach)	（カメムシ目、アブラムシ科）旧北区
rowena calligrapha		*Calligrapha rowena* Knab	（コウチュウ目、ハムシ科）新北区
royal Assyrian	ムラサキタテハ	*Terinos terpander* Hewitson	（チョウ目、タテハチョウ科）東洋区。*T. t. robertsia* Butler も同英名
royal blue		*Orachrysops regalis* Henning et Henning	（チョウ目、シジミチョウ科）エチオピア区
royal blue			whitened bluewing を見よ
royal blue pansy	ルリムラサキタテハモドキ（マダガスカルタテハモドキ）	*Precis rhadama* Boisduval	（チョウ目、タテハチョウ科）エチオピア区
royal Cerulean			glistening Cerulean (1) を見よ
royal Elzunia		*Elzunia pavonii* (Doubleday, Hewitson et Westwood)	（チョウ目、タテハチョウ科）新熱帯区
royal hairstreak		*Ostrinotes halciones* (Butler et Druce)	（チョウ目、シジミチョウ科）新熱帯区
royal jewel			Rovena jewel を見よ
royal mantle		*Catarhoe cuculata* (Hufnagel)	（チョウ目、シャクガ科）旧北区
royal moths			giant silkworm moths を見よ
royal palm bug		*Xylastodoris luteolus* Barber	（カメムシ目、Thaumastocoriidae）新北区
royal palm bugs		Thaumastotheriidae	（カメムシ目）の昆虫の総称
royal Perisama		*Perisama calamis* Hewitson	（チョウ目、タテハチョウ科）新熱帯区
royal poinciana caterpillar		*Melipotis acontioides* Guenée	（チョウ目、ヤガ科）新北区、新熱帯区
royal river cruiser		*Macromia taeniolata* Rambur	（トンボ目、ヤマトンボ科）新北区
royal walnut moth		*Citheronia regalis* (Fabricius)	（チョウ目、ヤママユガ科）新北区。royal walnut ともいう。幼虫は hickory horned devil といわれる。開帳 15 cm。クルミ害虫
royals	ヤドリギツバメ属	*Tajuria*	（チョウ目、シジミチョウ科）の昆虫の総称
Roy's mantis		*Sphodromantis royi* La Greca	（カマキリ目、カマキリ科）エチオピア区
rubbed dart moth			sandhill cutworm moth を見よ
rubber bark caterpillar		*Aetherastis circulata* Meyrick	（チョウ目、スガ科）東洋区
rubber flower geometrid		*Hemithea costipunctata* Moore	（チョウ目、シャクガ科）東洋区

英名	和名	学名	所属、分布、ほか
rubber root borer			red-spotted longhorn beetle を見よ
rubber termite		*Coptotermes gestroi* Wasmann	（シロアリ目、ミゾガシラシロアリ科）東洋区
rubber tree caterpillar			Edwards' wasp moth を見よ
Rubeolata Eurybia		*Eurybia rubeolata* Stichel	（チョウ目、シジミタテハ科）新熱帯区
Rubina metalmark			Chinese lantern を見よ
rubus aphid (1)	イチゴアブラムシ	*Aphis ichigo* Shinji	（カメムシ目、アブラムシ科）日本、旧北区
rubus aphid (2)		*Amphorophora rubi* (Kaltenbach)	（カメムシ目、アブラムシ科）全北区
rubus berry beetle	シロオビナガボソタマムシ	*Coroebus quadriundulatus* Motschulsky	（コウチュウ目、タマムシ科）日本
rubus gibbose aphid			southern rubus aphid を見よ
rubus hairy aphid			rubus aphid (1) を見よ
rubus leafhopper		*Macropsis fuscula* (Zetterstedt)	（カメムシ目、オオヨコバイ科）旧北区
rubus leafhopper		*Macropsis scotti* Edwards	（カメムシ目、オオヨコバイ科）旧北区
rubus thrips		*Thrips major* Uzel	（アザミウマ目、アザミウマ科）旧北区
ruby amberwing		*Perithemis rubita* Dunkle	（トンボ目、トンボ科）新熱帯区
ruby ant			red ant (2) を見よ
ruby-collared sapphire		*Gunayan rubricollis* Sepp	（チョウ目、セセリチョウ科）新熱帯区
ruby jewel		*Chlorocypha consueta* (Karsch)	（トンボ目、Chlorocyphidae）エチオピア区
ruby polyptychus		*Afroclanis calcareus* (Rothschild et Jordan)	（チョウ目、スズメガ科）エチオピア区
ruby quaker moth		*Orthosia rubescens* (Walker)	（チョウ目、ヤガ科）新北区。ruby quaker ともいう
ruby scale			red wax scale を見よ
ruby skimmer		*Orthetrum rubens* Barnard	（トンボ目、トンボ科）エチオピア区
rubyspots		*Hetaerina*	（トンボ目、イトトンボ科）の昆虫の総称
ruby-spotted swallowtail	ベニモンマネシアゲハ（ベニモンクロアゲハ）	*Papilio anchisiades* Esper	（チョウ目、アゲハチョウ科）新北区、新熱帯区。*P. a. idaeus* Fabricius も同英名
ruby-tailed wasp		*Chrysis viridula* Linnaeus	（ハチ目、セイボウ科）旧北区
ruby-tailed wasp (1)	リンネセイボウ	*Chrysis ignita* Linnaeus	（ハチ目、セイボウ科）日本、全北区
ruby-tailed wasps			cuckoo wasps (1) を見よ
ruby tiger			flax arctid を見よ
ruby tiger			ruby tiger moth を見よ
ruby tiger moth		*Phragmatobia fuliginosa* (Linnaeus)	（チョウ目、ヒトリガ科）旧北区。日本亜種 *P. f. rubricosa* (Harris) も同英名
ruby wasps			cuckoo wasps (1) を見よ
ruby whiteface		*Leucorrhinia rubicunda* (Linnaeus)	（トンボ目、トンボ科）旧北区
ruddle tiger moth			ruby tiger moth を見よ　北米での英名

英名	和名	学名	所属、分布、ほか
ruddy carpet		*Catarhoe rubidata* (Denis et Schiffermüller)	(チョウ目、シャクガ科) 旧北区
ruddy copper	ヒイロベニシジミ	*Lycaena rubidus* (Behr)	(チョウ目、シジミチョウ科) 新北区
ruddy dagger moth		*Acronicta rubricoma* Guenée	(チョウ目、ヤガ科) 新北区
ruddy daggerwing	ホソオツルギタテハ	*Marpesia petreus* (Cramer)	(チョウ目、タテハチョウ科) 新北区
ruddy darter		*Sympetrum sanguineum* (Müller)	(トンボ目、トンボ科) 旧北区
ruddy hairstreak		*Electrostrymon hugon* (Godart)	(チョウ目、シジミチョウ科) 新北区
ruddy hairstreak		*Electrostrymon sangala* (Hewitson)	(チョウ目、シジミチョウ科) 新北区
ruddy highflier		*Hydriomena ruberata* (Freyer)	(チョウ目、シャクガ科) 全北区。ruddy highflier moth ともいう
ruddy Holomelina		*Holomelina rubicundaria* (Hübner)	(チョウ目、ヒトリガ科) 新北区。ruddy Holomelina moth ともいう
ruddy leafwing		*Fountainea halice* (Godart)	(チョウ目、タテハチョウ科) 新熱帯区
ruddy Metarranthis moth		*Metarranthis duaria* (Guenée)	(チョウ目、シャクガ科) 新北区
ruddy ochreous flat-body		*Agonopterix subpropinquella* (Stainton)	(チョウ目、マルハキバガ科) 旧北区
ruddy sympetrum			ruddy darter を見よ
rufous Antinephele		*Antinephele anomala* (Butler)	(チョウ目、スズメガ科) エチオピア区
rufous-banded crambid moth			barberpole caterpillar moth を見よ
rufous-banded pyralid moth			barberpole caterpillar moth を見よ
rufous brindled flat-body		*Depressaria albipuncta* (Haworth)	(チョウ目、マルハキバガ科) 旧北区
rufous carabid		*Calathus ruficollis* Dejean	(コウチュウ目、オサムシ科) 新北区
rufous geometer moth		*Xanthotype rufaria* Swett	(チョウ目、シャクガ科) 新北区
rufous grasshopper		*Gomphocerus rufus* (Linnaeus)	(バッタ目、バッタ科) 旧北区
rufous leaf sitter		*Gorgyra rubescens* Holland	(チョウ目、セセリチョウ科) エチオピア区
rufous-margined satyr		*Taygetis rufomarginata* Staudinger	(チョウ目、タテハチョウ科) 新熱帯区
rufous minor		*Oligia versicolor* (Borkhausen)	(チョウ目、ヤガ科) 旧北区
rufous moth		*Coenobia rufa* (Haworth)	(チョウ目、ヤガ科) 旧北区
rufous purplewing		*Eunica carias* Hewitson	(チョウ目、タテハチョウ科) 新熱帯区
rufous scale		*Selenaspidus articuclatus* (Morgan)	(カメムシ目、マルカイガラムシ科) 日本、新北区
rufous-shouldered longhorn beetle		*Anaglyptus mysticus* (Linnaeus)	(コウチュウ目、カミキリムシ科) 旧北区

英名	和名	学名	所属、分布、ほか
rufous-tipped Swammerdamia moth			purple-edged ermel を見よ
rufous wainscot			rufous moth を見よ
rufous-winged elfin			rufous-winged flat を見よ
rufous-winged flat		*Eagris nottoana* (Wallengren)	（チョウ目、セセリチョウ科）エチオピア区
rugged darkling beetle		*Noserus plicatus* LeConte	（コウチュウ目、ゴミムシダマシ科）新北区
rugose leaf-roller weevil	サメハダハマキチョッキリ	*Byctiscus rugosus* (Gebler)	（コウチュウ目、オトシブミ科）日本、旧北区
rugose stag beetle		*Sinodendron rugosum* Mannerheim	（コウチュウ目、クワガタムシ科）新北区
rugulose canegrub			southern one-year canegrub を見よ
rumex aphid	ギシギシオマルアブラムシ	*Dysaphis rumecicola* (Hori)	（カメムシ目、アブラムシ科）日本
rumex black cutworm	クロギシギシヤガ	*Naenia contaminata* (Walker)	（チョウ目、ヤガ科）日本、旧北区
rumex cutworm	ギシギシヨトウ	*Atrachea nitens* (Butler)	（チョウ目、ヤガ科）日本、旧北区
rumex leaf beetle	コガタルリハムシ	*Gastrophysa atrocyanea* Motschulsky	（コウチュウ目、ハムシ科）日本、旧北区、東洋区
running bugs			land bugs を見よ
Ruona elfin		*Sarangesa ruona* Evans	（チョウ目、セセリチョウ科）エチオピア区
Ruppell's dotted border		*Mylothris rueppellii* (Koch)	（チョウ目、シロチョウ科）エチオピア区
rural skipper		*Ochlodes agricola* (Boisduval)	（チョウ目、セセリチョウ科）新北区、新熱帯区
Ruricola Argus		*Loxerebia ruricola* Leech	（チョウ目、タテハチョウ科）旧北区
rush-red grain beetle			rusty grain beetle を見よ
rush skeletonweed gall midge			skeleton gall midge を見よ
rush sucker		*Livia juncorum* (Latreille)	（カメムシ目、キジラミ科）旧北区
rush veneer			spruce pyralid を見よ
rush veneer pearl			spruce pyralid を見よ
rush wainscot		*Archanara cannae* (Ochsenheimer)	（チョウ目、ヤガ科）旧北区
rush wainscot			reed wainscot を見よ
russet paradise skipper		*Abantis rubra* Holland	（チョウ目、セセリチョウ科）エチオピア区
russet Protea		*Capys disjunctus* Trimen	（チョウ目、シジミチョウ科）エチオピア区
russet skipper		*Piruna pirus* (Edwards)	（チョウ目、セセリチョウ科）新北区
russet skipperling			russet skipper を見よ
russet-tipped clubtail		*Stylurus plagiatus* (Selys)	（トンボ目、サナエトンボ科）新北区
Russian heath	ロシアヒメヒカゲ	*Coenonympha leander* (Esper)	（チョウ目、タテハチョウ科）旧北区
Russian roach			German cockroach を見よ

英名	和名	学名	所属、分布、ほか
Russian thistle casebearer		*Coleophora klimeschiella* Toll	(チョウ目、ツツミノガ科) 全北区、大洋区。成虫は Russian thistle casebearer moth
Russian thistle stem miner moth		*Coleophora parthenica* Meyrick	(チョウ目、ツツミノガ科) 全北区、東洋区、大洋区
Russian wheat aphid		*Diuraphis noxia* (Mordovilko)	(カメムシ目、アブラムシ科) 東洋区、新北区
Russian-thistle grasshopper		*Aeoloplides turnbulli* (Thomas)	(バッタ目、バッタ科) 新北区
rust flies	ハネオレバエ科	Psilidae	(ハエ目) の昆虫の総称
rust flies			jumping plantlice を見よ　米国での英名
rust fly			carrot rust fly を見よ
rust grain beetle			rusty grain beetle を見よ
rust pine borer	ムネツヤサビカミキリ	*Arhopalus rusticus* (Linnaeus)	(コウチュウ目、カミキリムシ科) 日本、旧北区
rust-red flour beetle			red flour beetle を見よ
rust-red grain beetle			rusty grain beetle を見よ
Rustan hairstreak		*Atlides rustan* (Stoll)	(チョウ目、シジミチョウ科) 新熱帯区
rusted Artena clearwing		*Pteronymia artena praedicta* Maza et Lamas	(チョウ目、タテハチョウ科) 新熱帯区
rusted Salvin's clearwing		*Episcada salvinia portilla* Maza et Lamas	(チョウ目、タテハチョウ科) 新熱帯区。Salvin's clearwing を参照
rusted white-spotted clearwing		*Greta annette moschion* (Godman)	(チョウ目、タテハチョウ科) 新熱帯区。white-spotted clearwing を参照
rusted zea clearwing		*Oleria zea diazi* de la Maza et Lamas	(チョウ目、タテハチョウ科) 新熱帯区。zea clearwing を参照
rustic	タイワンキマダラ	*Cupha erymanthis* (Drury)	(チョウ目、タテハチョウ科) 日本、東洋区。*C. e. lotis* Sulzer も同英名
rustic (1)		*Hoplodrina blanda* (Denis et Schiffermüller)	(チョウ目、ヤガ科) 旧北区
Rustic' arctic blue		*Plebejus glandon rusticus* (W. H. Edwards)	(チョウ目、シジミチョウ科) 新北区。arctic blue を参照
rustic borer		*Xylotrechus colonus* Fabricius	(コウチュウ目、カミキリムシ科) 新北区
rustic grey		*Eudonia truncicolella* (Stainton)	(チョウ目、メイガ科) 旧北区
rustic moth			rustic (1) を見よ
rustic Presba		*Syncordulia serendipator* Dijkstra, Samways et Simaika	(トンボ目、エゾトンボ科) エチオピア区
rustic quaker moth		*Orthodes majuscula* Herrich-Schäffer	(チョウ目、ヤガ科) 新北区、新熱帯区。rustic quaker ともいう
rustic sailor beetle			soldier beetle (1) を見よ
rustic shoulder knot			wheat earworm を見よ
rustic shoulder-knot			ear miner moth を見よ
rustic shoulder knot moth			wheat earworm を見よ
rustic sphinx		*Manduca rustica* (Fabricius)	(チョウ目、スズメガ科) 新北区、新熱帯区。rustic sphinx moth ともいう

英名	和名	学名	所属、分布、ほか
rustics		Caradrinae	（チョウ目、ヤガ科）の昆虫の総称
rusty banded aphid			hawthorn aphid (2) を見よ
rusty broad-mouth weevil	サビクチブトゾウムシ	Canoixus japonicus Roelofs	（コウチュウ目、ゾウムシ科）日本
rusty button		Acleris ferrugana Denis et Schiffermüller	（チョウ目、ハマキガ科）旧北区
rusty clearwing		Greta morgane morgane (Geyer)	（チョウ目、タテハチョウ科）新熱帯区
rusty crescent		Tegosa etia (Hewitson)	（チョウ目、タテハチョウ科）新熱帯区
rusty dot		Pyrausta martialis Guenée	（チョウ目、メイガ科）旧北区
rusty-dot pearl		Udea ferrugalis Hübner	（チョウ目、メイガ科）旧北区
rusty-dot pearl (1)		Udea martialis Scudder	（チョウ目、メイガ科）旧北区
rusty-dotted moth		Mecyna flavidalis Doubleday	（チョウ目、メイガ科）豪州区
rusty forester		Lethe bhairava (Moore)	（チョウ目、タテハチョウ科）東洋区
rusty gourd-shaped weevil	サビヒョウタンゾウムシ	Scepticus griseus (Roelofs)	（コウチュウ目、ゾウムシ科）日本
rusty grain beetle	サビカクムネヒラタムシ	Cryptolestes ferrugineus (Stephens)	（コウチュウ目、ヒラタムシ科）日本、汎世界。貯穀害虫
rusty Holomelina		Holomelina ferruginosa (Walker)	（チョウ目、ヒトリガ科）新北区。rusty Holomelina moth ともいう
rusty-lined leaf-tier		Ichthyura albosigma Fitch	（チョウ目、シャチホコガ科）新北区
rusty longicorn beetle			Far East rusty longicorn beetle を見よ
rusty metalmark		Synargis mycone (Hewitson)	（チョウ目、シジミタテハ科）新熱帯区
rusty mountain satyr		Lymanopoda ferruginosa Butler	（チョウ目、タテハチョウ科）新熱帯区
rusty pierrot		Tarucus alteratus Moore	（チョウ目、シジミチョウ科）東洋区
rusty pine cone moth			webbing coneworm を見よ
rusty plum aphid		Hysteroneura setariae (Thomas)	（カメムシ目、アブラムシ科）新北区。プラム害虫。虫えいを作る
rusty red scale			dictyospermum scale を見よ
rusty sister			Felder's sister (1) を見よ
rusty skipper		Methion melas Godman	（チョウ目、セセリチョウ科）新熱帯区
rusty snaketail		Ophiogomphus rupinsulensis (Walsh)	（トンボ目、サナエトンボ科）新北区
rusty swift		Borbo detecta (Trimen)	（チョウ目、セセリチョウ科）エチオピア区
rusty tigerwing		Olyras theon Bates	（チョウ目、タテハチョウ科）新熱帯区
rusty-tipped page			brown Siproeta を見よ
rusty tussock moth		Orgyia antiqua Linnaeus	（チョウ目、ドクガ科）全北区、新熱帯区。リンゴ、ナシ、バラ他害虫。亜種 O. a. nova (Fitch) も同英名
rusty tussock moth			common vapourer を見よ　北米での英名
rusty wave		Idaea inquinata (Scopoli)	（チョウ目、シャクガ科）旧北区
rutelian beetles			flower beetles (3) を見よ

英名	和名	学名	所属、分布、ほか
Ruth grass scale		*Odonaspis ruthae* Kotinsky	（カメムシ目、マルカイガラムシ科）大洋区
Rutherglen bug		*Nysius vinitor* Bergroth	（カメムシ目、ナガカメムシ科）豪州区。カンキツ害虫
ryegrass mealybug		*Phenacoccus graminicola* Leonardi	（カメムシ目、コナカイガラムシ科）豪州区
Ryphea leafwing			flamingo leafwing を見よ

英名	和名	学名	所属、分布、ほか
S			
s-banded tiger beetle			mudflat tiger beetle を見よ
s-spot darner		*Austroaeschna christine* Theischinger	（トンボ目、ヤンマ科）豪州区
Sabi hooktail		*Paragomphus sabicus* Pinhey	（トンボ目、サナエトンボ科）エチオピア区
Sabi smoky blue		*Euchrysops dolorosa* (Trimen et Bowker)	（チョウ目、シジミチョウ科）エチオピア区
Sabine albatross		*Appias sabina* (C. et R. Felder)	（チョウ目、シロチョウ科）エチオピア区
Sabino dancer		*Argia sabino* Garrison	（トンボ目、イトトンボ科）新北区
sable clubtail		*Gomphus rogersi* Gloyd	（トンボ目、サナエトンボ科）新北区
sabre wasp			sirex parasite (1) を見よ
sabretails		*Megalogomphus*	（トンボ目、サナエトンボ科）の昆虫の総称
sacaline aphid	イタドリオナシアブラムシ	*Macchiatiella itadori* (Shinji)	（カメムシ目、アブラムシ科）日本、旧北区
sacaline sawfly	イタドリクロハバチ	*Ametastegia polygoni* Takeuchi	（ハチ目、ハバチ科）日本
Sacapulas Calephelis		*Calephelis sacapulas* McAlpine	（チョウ目、シジミタテハ科）新熱帯区
sachem			field skipper を見よ
sack-bearer	ミノムシ		bagworm moths を見よ
sack-bearer moths		Mimallonidae	（チョウ目）の昆虫の総称　幼虫は筒状巣を作る
sack-bearers			sack-bearer moths を見よ
Sacken's velvet ant			western velvet ant を見よ
sacred beetle			scarab を見よ
sacred scarab			scarab を見よ
sacred scarabaeus			scarab を見よ
sad goat			quince borer を見よ
sad underwing moth		*Catocala maestosa* Hulst	（チョウ目、ヤガ科）新北区。sad underwing ともいう
saddle back		*Acharia stimulea* (Clemens)	（チョウ目、イラガ科）新北区。saddleback とも記される
saddle casemaker caddisflies	ヤマトビケラ科	Glossosomatidae	（トビケラ目）の昆虫の総称
saddle gall midge		*Haplodiplosis equestris* (Wagner)	（ハエ目、タマバエ科）旧北区
saddle gall midge		*Haplodiplosis marginata* (von Rover)	（ハエ目、タマバエ科）旧北区
saddleback caterpillar			saddle back を見よ
saddleback caterpillar moth			saddle back を見よ
saddleback caterpillars		*Sibine*	（チョウ目、イラガ科）の昆虫の総称

英名	和名	学名	所属、分布、ほか
saddle-back caterpillars			slug caterpillar moths を見よ
saddleback moth			saddle back を見よ
saddle-backed bush cricket		*Ephippiger ephippiger* (Fiebig)	(バッタ目、キリギリス科) 旧北区
saddlebags gliders			gliders (1) を見よ　saddlebags ともいう
saddled leafhopper		*Colladonus elitellarius* (Say)	(カメムシ目、オオヨコバイ科) 新北区
saddled prominent		*Heterocampa guttivitta* (Walker)	(チョウ目、シャチホコガ科) 新北区。saddled prominent moth ともいう
saddled shieldback		*Eremopodes ephippiata* (Scudder)	(バッタ目、キリギリス科) 新北区
Saepiolus blue		*Plebejus saepiolus amica* (Edwards)	(チョウ目、シジミチョウ科) 新北区
safari ants	サスライアリ属	*Dorylus*	(ハチ目、アリ科) の昆虫の総称　大群で移動し、特定の巣を作らない。army ants (1) も参照
safflower skipper		*Pyrgus carthami* (Hübner)	(チョウ目、セセリチョウ科) 旧北区
safflower stem miner		*Melanagromyza splendida* Frick	(ハエ目、ハモグリバエ科) 大洋区
safflower stemminer		*Melanagromyza virens* (Loew)	(ハエ目、ハモグリバエ科) 大洋区
saffron		*Mota massyla* (Hewitson)	(チョウ目、シジミチョウ科) 東洋区
saffron (1)		*Iolaus pallene* (Wallengren)	(チョウ目、シジミチョウ科) エチオピア区
saffron-barred birch pigmy		*Stigmella luteella* (Stainton)	(チョウ目、モグリチビガ科) 旧北区
saffron-headed dwarf		*Elachista luticomella* Zeller	(チョウ目、クサモグリガ科) 旧北区
saffron sapphire			saffron (1) を見よ
saffron skipper		*Poanes aaroni* (Skinner)	(チョウ目、セセリチョウ科) 新北区
Sagana fritillary	メスグロヒョウモン	*Argynnis sagana* (Doubleday)	(チョウ目、タテハチョウ科) 旧北区
Sagba mountain cupid		*Euchrysops sagba* Libert	(チョウ目、シジミチョウ科) エチオピア区
sage plume moth			mint plume moth を見よ
sage skipper		*Syrichtus proto* Esper	(チョウ目、セセリチョウ科) 旧北区
sage sphinx		*Sphinx eremitoides* Strecker	(チョウ目、スズメガ科) 新北区。sage sphinx moth ともいう
sagebrush defoliator		*Aroga websteri* (Clarke)	(チョウ目、キバガ科) 新北区
sagebrush gall midge		*Asphondylia artemisiae* Felt	(ハエ目、タマバエ科) 新北区
sagebrush girdle moth		*Plataea trilinearia* (Packard)	(チョウ目、シャクガ科) 新北区
sagebrush grasshopper		*Melanoplus bowditchi* Scudder	(バッタ目、バッタ科) 新北区
sagebrush grig		*Cyphoderris strepitans* Morris et Gwynne	(バッタ目、Prophalangopsidae) 新北区

英名	和名	学名	所属、分布、ほか
sagebrush sheep moth			Hera moth を見よ
sagebush checkerspot		*Chlosyne acastus* (Edwards)	(チョウ目、タテハチョウ科) 新北区
sagebush sooty hairstreak		*Satyrium semiluna* Klots	(チョウ目、シジミチョウ科) 新北区
sagebush white			Becker's white を見よ
Saghalin black melon beetle			false melon beetle を見よ
Saghalin black-spotted dagger moth	カラフトゴマケンモン	*Panthea coenobita idea* Bryk	(チョウ目、ヤガ科) 日本
Saghalin cimbex			birch sawfly (1) を見よ
Saghalin cutworm			Siberian cutworm を見よ
sago palm scale		*Furchadiaspis zamiae* (Morgan)	(カメムシ目、マルカイガラムシ科) 旧北区
sago palm weevil			Asiatic palm weevil を見よ
Sagra sphinx moth		*Eupyrrhoglossum sagra* (Poey)	(チョウ目、スズメガ科) 新北区、新熱帯区
Sahara bluetail		*Ischnura saharensis* Aguesse	(トンボ目、イトトンボ科) 旧北区
Sahara swallowtail			desert swallowtail (1) を見よ
Saharan leopard butterfly		*Cigaritis nilus* (Hewitson)	(チョウ目、シジミチョウ科) エチオピア区
Saigon cinnamon looper	エグリエダシャク	*Fascellina chromatiaria* Walker	(チョウ目、シャクガ科) 日本、東洋区
sailers	コミスジ属	*Neptis*	(チョウ目、タテハチョウ科) の昆虫の総称 sailors も使われる
sailing bluet			black-tailed bluet を見よ
sailor beetles			soldier beetles (1) を見よ
sainfoin flower midge		*Contarinia onobrychidis* Kieffer	(ハエ目、タマバエ科) 旧北区
sainfoin leaf midge			sainfoin flower midge を見よ
Saint Francis' satyr		*Neonympha mitchellii francisci* Parshall et Kral	(チョウ目、タテハチョウ科) 新北区。Mitchell's marsh satyr を参照
Saint Helena earwig			St. Helena giant earwig を見よ
saints			mantids (2) を見よ
Sakhalin fir bark beetle	ジョウザンコキクイムシ	*Cryphalus piceus* Eggers	(コウチュウ目、キクイムシ科) 日本、旧北区
Sakhalin fir elongate froghopper	トドマツホソアワフキ	*Aphilaenus abieti* (Matsumura)	(カメムシ目、アワフキムシ科) 日本、旧北区
Sakhalin fir engraved weevil	トドマツアナアキゾウムシ	*Dyscerus insularis* Kono	(コウチュウ目、ゾウムシ科) 日本
Sakhalin fir yellow-spotted weevil			yellow-spotted fir weevil を見よ
Sakhalin pine longicorn beetle	カラフトヒゲナガカミキリ	*Monochamus salturarius* (Gebler)	(コウチュウ目、カミキリムシ科) 日本、旧北区。マツ害虫。Sakhalin pine longicorn ともいう
Salambria firewing		*Catonephele salambria* Felder	(チョウ目、タテハチョウ科) 新熱帯区

英名	和名	学名	所属、分布、ほか
Salas skipper		*Typhedanus salas* Freeman	(チョウ目、セセリチョウ科) 新熱帯区
saldids			shore bugs を見よ
Salenus faceted-skipper			Salenus skipper を見よ
Salenus skipper		*Synapte salenus* (Mabille)	(チョウ目、セセリチョウ科) 新北区、新熱帯区
salicet sphinx		*Emerinthus saliceti* Boisduval	(チョウ目、スズメガ科) 新熱帯区
salicet sphinx			unidentified Smerinthus moth を見よ
Salisbury sprite		*Pseudagrion salisburyense* Ris	(トンボ目、イトトンボ科) エチオピア区
salix aphid	ヤナギアブラムシ	*Aphis farinosa yanagicola* Matsumura	(カメムシ目、アブラムシ科) 日本。small willow aphid も参照
salix black hairy aphid	ヤナギフタオアブラムシ	*Cavariella salicicola* (Matsumura)	(カメムシ目、アブラムシ科) 日本、旧北区、東洋区
salix cimbex	オオヒラクチハバチ	*Pseudoclavellaria amerinae* (Linnaeus)	(ハチ目、コンボウハバチ科) 日本、旧北区
salix gall midge	ヤナギコブタマバエ	*Dasineura salicis* (Shrank)	(ハエ目、タマバエ科) 日本、旧北区
salix leaf beetle			poplar leaf beetle を見よ
salix leafhopper	クサビヨコバイ	*Athysanopsis salicis* Matsumura	(カメムシ目、オオヨコバイ科) 日本、旧北区
salix noctuid	キンスジアツバ	*Colobochyla salicalis* (Denis et Schiffermüller)	(チョウ目、ヤガ科) 日本、旧北区
salix wingless aphid	ヤナギケアブラムシ	*Chaitophorus saliapterus* Shinji	(カメムシ目、アブラムシ科) 日本、旧北区
Salle bean weevil		*Mimosestes sallaei* (Sharp)	(コウチュウ目、ハムシ科) 大洋区
sallow	モンキキリガ	*Xanthia icteritia* (Hufnagel)	(チョウ目、ヤガ科) 日本、旧北区
sallow Apotomis moth			sallow marble を見よ
sallow bean-gall sawfly		*Pontania bridgmanii* (Cameron)	(ハチ目、ハバチ科) 旧北区
sallow clearwing		*Synanthedon flaviventris* (Staudinger)	(チョウ目、スカシバガ科) 旧北区
sallow flat body			willow shoot moth を見よ
sallow kitten (1)		*Furcula furcula* (Clerck)	(チョウ目、シャチホコガ科) 旧北区
sallow kitten (2)		*Pterostoma palpina* (Clerck)	(チョウ目、シャチホコガ科) 旧北区
sallow kitten moth			sallow kitten (1) を見よ
sallow leaf aphid		*Chaitophorus capreae* (Mosley)	(カメムシ目、アブラムシ科) 旧北区
sallow leaf beetle			elm tree beetle を見よ
sallow-leaf groundling		*Teleiodes notatella* Pierce et Metcalf	(チョウ目、キバガ科) 旧北区
sallow leaf-mining weevil			willow flea beetle (1) を見よ
sallow leafroller moth			poplar sober を見よ

英名	和名	学名	所属、分布、ほか
sallow little midget		*Phyllonorycter quinqueguttella* (Stainton)	（チョウ目、ホソガ科）旧北区
sallow longhorn		*Adela cuprella* (Denis et Schiffermüller)	（チョウ目、マガリガ科）旧北区
sallow marble	ヤナギツマジロヒメハマキ	*Apotomis capreana* (Hübner)	（チョウ目、ハマキガ科）日本、全北区
sallow moth			sallow を見よ
sallow Nycteoline	ミヤマクロスジキノカワガ	*Nycteola degenerana* (Hübner)	（チョウ目、ヤガ科）日本、旧北区
sallow pea-gall sawfly		*Pontania pedunculi* (Hartig)	（ハチ目、ハバチ科）旧北区
sallow-shoot flat-body			willow shoot moth を見よ
Sally's mottled-skipper		*Codatractus sallyae* Warren	（チョウ目、セセリチョウ科）新熱帯区
salmon-branded bushbrown		*Mycalesis misenus* de Nicéville	（チョウ目、タテハチョウ科）東洋区
salmon Colotis			small salmon Arab を見よ
salmon-flies			stoneflies を見よ
salmon-flies			giant stoneflies (1) を見よ
salmon fly			giant stonefly (1) を見よ
Salome yellow		*Eurema salome* (C. et R. Felder)	（チョウ目、シロチョウ科）新北区。*E. s. jamapa* (Reakirt) も同英名
Salop purple		*Eriocrania salopiella* (Stainton)	（チョウ目、スイコバネガ科）旧北区
Salpensa sailor		*Dynamine tithia salpensa* C. et R. Felder	（チョウ目、タテハチョウ科）新熱帯区
salt-and-pepper looper moth		*Syngrapha rectangula* Kirby	（チョウ目、ヤガ科）新北区。幼虫は salt-and-pepper looper
saltbush blue			chequered blue (1) を見よ
saltbush planthopper		*Privesa pronotalis* Distant	（カメムシ目、ハゴロモ科）豪州区
saltbush sootywing		*Hesperopsis alpheus* (Edwards)	（チョウ目、セセリチョウ科）新北区
salt-grass skipper		*Pseudocopaeodes eunus* Edwards	（チョウ目、セセリチョウ科）新北区、新熱帯区
saltgrass skipper			sandhill skipper を見よ
saltmarsh bell		*Eucosma tripoliana* (Barrett)	（チョウ目、ハマキガ科）旧北区
salt marsh caterpillar			Acraea moth を見よ　成虫は salt marsh caterpillar moth
salt marsh copper		*Lycaena dorcas dospassosi* McDunnough	（チョウ目、シジミチョウ科）新北区。
saltmarsh culex	ヨツボシイエカ	*Culex sitiens* Wiedemann	（ハエ目、カ科）日本、旧北区、東洋区、豪州区、エチオピア区
saltmarsh dwarf		*Biselachista scirpi* (Stainton)	（チョウ目、クサモグリガ科）旧北区
saltmarsh grass veneer		*Pediasia aridella* (Thunberg)	（チョウ目、メイガ科）旧北区

英名	和名	学名	所属、分布、ほか
salt marsh ground cricket		*Allonemobius sparsalus* (Fulton)	（バッタ目、コオロギ科）新北区
saltmarsh horsefly		*Atylotus latistriatus* Brauer	（ハエ目、アブ科）旧北区
saltmarsh meadow katydid		*Conocephalus spartinae* (Fox)	（バッタ目、キリギリス科）新北区
saltmarsh mosquito	ハマベヤブカ	*Aedes vigilax* (Skuse)	（ハエ目、カ科）日本、東洋区、豪州区
salt marsh mosquito		*Aedes (Ochlerotatus) solliicitans* (Walker)	（ハエ目、カ科）新北区．
salt marsh mosquito		*Aedes caspius* (Pallas)	（ハエ目、カ科）旧北区
salt marsh mosquito		*Aedes detritus* Haliday	（ハエ目、カ科）旧北区
salt marsh moth			Acraea moth を見よ
saltmarsh neb		*Monochroa tetragonella* (Stainton)	（チョウ目、キバガ科）旧北区
saltmarsh patch		*Bucculatrix maritima* Stainton	（チョウ目、チビガ科）旧北区
salt marsh ringlet		*Coenonympha tullia nipisiquit* McDunnough	（チョウ目、タテハチョウ科）新北区。large heath を参照
salt marsh ringlet			Nipisquit ringlet を見よ
salt marsh sand fly		*Culicoides furens* (Poey)	（ハエ目、ヌカカ科）新北区、新熱帯区。salt marsh punkie ともいう
saltmarsh scale			cottony saltbush scale を見よ
salt-marsh skipper		*Panoquina panoquin* (Scudder)	（チョウ目、セセリチョウ科）新北区
saltmarsh small case		*Coleophora adjunctella* Hodgkinson	（チョウ目、ツツミノガ科）旧北区
salt marsh water boatman		*Trichocorixa reticulata* (Guérin-Méneville)	（カメムシ目、ミズムシ科）新北区
saltpan blue		*Theclinesthes sulpitius* (Miskin)	（チョウ目、シジミチョウ科）豪州区
saltwater mosquito		*Aedes australis* (Erichson)	（ハエ目、カ科）豪州区
saltern ear	キタショウブヨトウ	*Amphipoea fucosa* (Freyer)	（チョウ目、ヤガ科）日本、旧北区
Salus sister		*Adelpha salus* Hall	（チョウ目、タテハチョウ科）新熱帯区
Salvia blue		*Harpendyreus notoba* (Trimen)	（チョウ目、シジミチョウ科）エチオピア区
salvinia moth		*Samea multiplicalis* Guenée	（チョウ目、メイガ科）新北区、豪州区、新熱帯区、大洋区。salvinia stem-borer moth ともいう
salvinia weevil		*Cyrtobagous salviniae* Calder et Sands	（コウチュウ目、ゾウムシ科）豪州区
Salvin's clearwing		*Episcada salvinia salvinia* (Bates)	（チョウ目、タテハチョウ科）新熱帯区
Salvin's dwarf		*Ipidecla schausi* (Godman et Salvin)	（チョウ目、タテハチョウ科）新熱帯区
Salvin's empress		*Cybdelis boliviana* Salvin	（チョウ目、タテハチョウ科）新熱帯区
Salvin's kite-swallowtail		*Eurytides salvini* (Bates)	（チョウ目、アゲハチョウ科）新熱帯区
Salvin's skipper		*Ouleus salvina* Evans	（チョウ目、セセリチョウ科）新熱帯区
samanea round-headed borer			two-lined albizzia longhorn を見よ

英名	和名	学名	所属、分布、ほか
Samenta skipper		*Ochlodes samenta* Dyar	(チョウ目、セセリチョウ科) 新熱帯区
Sampaguita webworm			jasmine leaf webworm を見よ
samphire blue			saltpan blue を見よ
Samson flasher		*Narcosius samson* (Evans)	(チョウ目、セセリチョウ科) 新熱帯区
San Bruno elfin		*Incisalia mossii bayensis* (Brown)	(チョウ目、シジミチョウ科) 新北区。moss elfin を参照
San Diego meadow katydid		*Conocephalus spinosus* (Morse)	(バッタ目、キリギリス科) 新北区
San Emigdio blue		*Plebulina emigdionis* (Grinnel)	(チョウ目、シジミチョウ科) 新北区
San Francisco forktail		*Ischnura gemina* (Kennedy)	(トンボ目、イトトンボ科) 新北区
San Jacinto shieldback		*Phymonotus jacintotopos* Lightfoot, Weissman et Ueshima	(バッタ目、キリギリス科) 新北区
San Jose scale	ナシマルカイガラムシ (サンホセカイガラムシ)	*Comstockaspis perniciosus* (Comstock)	(カメムシ目、マルカイガラムシ科) 日本、汎世界。米国 San Jose で発見された。ナシ、カシ類害虫
sand ammophila			sand wasp (1) を見よ
sand-cherry weevil		*Coccotorus hirsutus* Brunner	(コウチュウ目、ゾウムシ科) 新北区
sand crickets			cave crickets (2) を見よ
sand crickets			pygmy mole crickets を見よ
sand crickets			camel crickets (1) を見よ
sand dart	ハマヤガ	*Agrotis ripae* (Hübner)	(チョウ目、ヤガ科) 日本、旧北区
sanddragons		*Progomphus*	(トンボ目、サナエトンボ科) の昆虫の総称
sand-dune grasshopper moth		*Areniscythris brachypteris* Powell	(チョウ目、キヌバコガ科) 新北区
sand-dune opal		*Chrysoritis pyroeis* (Trimen)	(チョウ目、シジミチョウ科) エチオピア区
sand dune Phaneta moth		*Phaneta linitipunctana* Blanchard et Knudson	(チョウ目、ハマキガ科) 新北区
sand-dune widow		*Tarsocera cassina* (Butler)	(チョウ目、タテハチョウ科) エチオピア区
sand field cricket		*Gryllus firmus* Scudder	(バッタ目、コオロギ科) 新北区
sand flea			chigoe flea を見よ
sand flies		*Phlebotomus*	(ハエ目、チョウバエ科) の昆虫の総称
sand flies			moth flies を見よ
sandflies			black flies (1) を見よ
sand-flies			biting midges (1) を見よ
sand-flies			punkies を見よ
sandfly			スナバエ、ヌカカ。punkies, sand flies を見よ
sand grasshopper			lesser field grasshopper を見よ
sandgrinder		*Arenopsaltria fullo* (Walker)	(カメムシ目、セミ科) 豪州区
sandgropers		Cylindrachetidae	(バッタ目) の昆虫の総称

英名	和名	学名	所属、分布、ほか
sand-loving Tachytes		*Tachytes distinctus* Smith	(ハチ目、アナバチ科) 新北区
sand-loving wasp			sand-loving Tachytes を見よ
sandloving wasps	ケラトリバチ亜科	Larrinae	(ハチ目、アナバチ科) の昆虫の総称
sand minnow mayflies		Ametropodidae	(カゲロウ目) の昆虫の総称
sand mountain blue		*Euphilotes pallescens arenamontana* Austin	(チョウ目、シジミチョウ科) 新北区。pale blue を参照
sand roaches		*Arenivaga*	(ゴキブリ目、ムカシゴキブリ科) の昆虫の総称
sand roaches		*Eremoblatta*	(ゴキブリ目、ムカシゴキブリ科) の昆虫の総称
sandstone ochre		*Trapezites taori* Atkins	(チョウ目、セセリチョウ科) 豪州区
sand-tailed digger wasp		*Cerceris arenaria* (Linnaeus)	(ハチ目、アナバチ科) 旧北区
sand termites	スナシロアリ亜科	Psammotermitinae	(シロアリ目、ミゾガシラシロアリ科) の昆虫の総称
sand treaders		*Macrobaenetes*	(バッタ目、コロギス科) の昆虫の総称
sand wainscot moth		*Apamea lintneri* Grote	(チョウ目、ヤガ科) 新北区
sand wasp		*Bembix spinolae* Lepeletier	(ハチ目、アナバチ科) 新北区
sand wasp		*Bembix pruinosa* Fox	(ハチ目、アナバチ科) 新北区
sand wasp (1)		*Ammophila sabulosa* (Linnaeus)	(ハチ目、アナバチ科) 日本、旧北区。日本亜種は *A. s. nipponica* Tsuneki サトジガバチ
sand wasps	ハナダカバチ科	Bembicidae	(ハチ目) の昆虫の総称 アナバチ科 に入れられる
sand wasps (1)	アナバチ亜科	Sphecinae	(ハチ目、アナバチ科) の昆虫の総称
sand wasps (2)		*Ammophila*	(ハチ目、アナバチ科) の昆虫の総称
sand wasps		*Bembix*	(ハチ目、アナバチ科) の昆虫の総称
sand wasps			policeman flies を見よ
sand wasps			cicada killers を見よ
sand weevil		*Philopedon plagiatum* (Schaller)	(コウチュウ目、ゾウムシ科) 全北区
sand wireworm		*Horistonotus uhleri* Horn	(コウチュウ目、コメツキムシ科) 新北区
sandal-box hawk moth		*Coenotes eremophilae* (Lucas)	(チョウ目、スズメガ科) 豪州区
sandalid beetles		Sandalidae	(コウチュウ目) の昆虫の総称 通常はクシヒゲムシ科に入れられる
sandalids			sandalid beetles を見よ
sandbar purplewing	クリテイアアイイロタテハ	*Eunica clytia* (Hewitson)	(チョウ目、タテハチョウ科) 新熱帯区
sandhill ant		*Formica bradleyi* Wheeler	(ハチ目、アリ科) 新北区
sandhill bluet		*Enallagma davisi* Westfall	(トンボ目、イトトンボ科) 新北区
sandhill clubtail		*Gomphus cavillaris* Needham	(トンボ目、サナエトンボ科) 新北区
sandhill cutworm moth		*Euxoa detersa* (Walker)	(チョウ目、ヤガ科) 新北区
sandhill rustic		*Luperina nickerlii* (Freyer)	(チョウ目、ヤガ科) 旧北区

英名	和名	学名	所属、分布、ほか
sandhill skipper		*Polites sabuleti* (Boisduval)	（チョウ目、セセリチョウ科）大洋区、新北区。新熱帯区の *P. s. margaretae* Miller et MacNeil も同英名
sandhill virtuoso katydid		*Amblycorypha arenicola* Walker	（バッタ目、キリギリス科）新北区
sandwich caterpillar		*Synchalara rhombota* Meyrick	（チョウ目、マルハキバガ科）東洋区
sandy carpet		*Perizoma flavofasciata* (Thunberg)	（チョウ目、シャクガ科）旧北区
sandy clover weevil			clover leaf weevil (1) を見よ　英国での英名
sandy grizzled skipper		*Pyrgus cinarae* (Rambur)	（チョウ目、セセリチョウ科）新北区
sandy ground bug		*Trapezonotus arenarius* (Linnaeus)	（カメムシ目、ナガカメムシ科）旧北区
sandy skipper		*Zopyrion sandace* Godman et Salvin	（チョウ目、セセリチョウ科）新熱帯区
Sang's purple		*Eriocrania sangii* (Wood)	（チョウ目、スイコバネガ科）旧北区
sanguinary ant			slave-maker ant を見よ
Sanguine Telipna		*Telipna sanguinea* (Plötz)	（チョウ目、シジミチョウ科）エチオピア区
sanguisorba aphid	ワレモコウシリキリアブラムシ	*Aphis sanguisorbicola* Takahashi	（カメムシ目、アブラムシ科）日本、旧北区
Sanoma ground mealybug		*Rhizoecus sonomae* McKenzie	（カメムシ目、コナカイガラムシ科）新北区
Santa Ana tussock moth		*Lophocampa annulosa* (Walker)	（チョウ目、ヒトリガ科）新北区、新熱帯区
Santa Monica shieldback katydid			Malibu shieldback を見よ
Santa Rosa shieldback		*Inyodectes santarosae* (Tinkham)	（バッタ目、キリギリス科）新北区
Santana tufted skipper		*Pellicia santana* Williams et Bell	（チョウ目、セセリチョウ科）新熱帯区
Santineza glasswing		*Oleria santineza* Haensch	（チョウ目、タテハチョウ科）新熱帯区
Sao glasswing		*Pteronymia sao* (Hübner)	（チョウ目、タテハチョウ科）新熱帯区
sap beetles		*Colopterus*	（コウチュウ目、ケシキスイ科）の昆虫の総称
sap beetles		*Conotelus*	（コウチュウ目、ケシキスイ科）の昆虫の総称
sap beetles		*Prometopia*	（コウチュウ目、ケシキスイ科）の昆虫の総称
sap beetles (1)	ケシキスイ科	Nitidulidae	（コウチュウ目）の昆虫の総称
sap chafer			rose chafer (1) を見よ　米国での英名
sap chafers			flower beetles (1)(2) を見よ
sap chafers			chafer beetles を見よ　米国での英名
sap-feeding ant		*Myrmicaria brunnea* Saunders	（ハチ目、アリ科）東洋区
sap-feeding beetle		*Conotelus mexicanus* Murray	（コウチュウ目、ケシキスイ科）新北区
sap-feeding beetles			sap beetles (1) を見よ
sap-flora beetles	エンマムシモドキ科	Synteliidae	（コウチュウ目）の昆虫の総称

英名	和名	学名	所属、分布、ほか
sap flower beetles			sap-flora beetles を見よ
Sapho longwing		*Heliconius sapho* (Drury)	(チョウ目、タテハチョウ科) 新熱帯区。*H. s. leuce* Doubleday も同英名
sapling borer			teak sapling borer を見よ
sapodilla borer moth		*Banisia myrsusalis* (Walker)	(チョウ目、マドガ科) 新北区、新熱帯区、東洋区、豪州区
sapodilla pod borer moth		*Dichrorampha sapodilla* Heppner	(チョウ目、ハマキガ科) 新北区
sapphire		*Iolaus silas* (Westwood)	(チョウ目、シジミチョウ科) エチオピア区
sapphire azure			Cooktown azure を見よ
sapphire beetle			diamond beetle を見よ
sapphire bluet		*Africallagma sapphirinum* (Pinhey)	(トンボ目、イトトンボ科) エチオピア区
sapphire moonbeam		*Philiris sappheira* (Sands)	(チョウ目、シジミチョウ科) 豪州区
sapphire skipper		*Phareas coeleste* Westwood	(チョウ目、セセリチョウ科) 新熱帯区
sapphires		*Heliophorus*	(チョウ目、シジミチョウ科) の昆虫の総称
sapphires	フタオシジミ属	*Iolaus*	(チョウ目、シジミチョウ科) の昆虫の総称
Sappho bent-skipper		*Ebrietas sappho* Steinhauser	(チョウ目、セセリチョウ科) 新熱帯区
sappho underwing moth		*Catocala sappho* Strecker	(チョウ目、ヤガ科) 新北区。sappho underwing ともいう
Sapporo bark beetle	サッポロキクイムシ	*Hypothenemus sapporensis* (Niijima)	(コウチュウ目、キクイムシ科) 日本
sapwood timberworm		*Hylecoetus lugubris* Say	(コウチュウ目、ツツシンクイ科) 新北区
sapygid wasps	ミコバチ科	Sapygidae	(ハチ目) の昆虫の総称
Sara heliconian			Sara longwing を見よ
Sara longwing		*Heliconius sara* (Fabricius)	(チョウ目、タテハチョウ科) 新熱帯区
Sara orange-tip	サラツマキチョウ	*Anthocharis sara* Lucas	(チョウ目、シロチョウ科) 新北区
Sara sailor	サラアカガネタテハ	*Dynamine sara* Bates	(チョウ目、タテハチョウ科) 新熱帯区
Sarah's ranger		*Kedestes sarahae* Henning et Henning	(チョウ目、セセリチョウ科) エチオピア区
Sara's orangetip			Sara orange tip を見よ
Saratoga spittle bug		*Aphrophora saratogensis* (Fitch)	(カメムシ目、アワフキムシ科) 新北区
Sardinian blue		*Pseudophilotes barbagiae* de Prins et Poorten	(チョウ目、シジミチョウ科) 旧北区
Sardinian meadow brown		*Maniola nurag* Ghiliani	(チョウ目、タテハチョウ科) 旧北区
Sarepta glasswing		*Hypoleria sarepta* (Hewitson)	(チョウ目、タテハチョウ科) 新熱帯区
Sarracenia spiketail		*Cordulegaster sarracenia* Abbott et Hibbitts	(トンボ目、オニヤンマ科) 新北区
sart borer			city longicorn beetle を見よ
sart longicorn beetle			city longicorn beetle を見よ

英名	和名	学名	所属、分布、ほか
sasa gall midge	ササウオタマバエ	Hasegawaia sasacola Monzen	(ハエ目、タマバエ科) 日本。ササ害虫
sasa mealy aphid			Japanese horned aphid を見よ
Sasaki cherry aphid	ササキコブアブラムシ	Tuberocephalus sasakii (Matsumura)	(カメムシ目、アブラムシ科) 日本、旧北区
Sasaki pine webworm	マツノイトカケハバチ	Acantholyda sasakii (Yano)	(ハチ目、ヒラタハバチ科) 日本、旧北区
Sasaki siris psylla	ネムノヒゲナガキジラミ	Acizzia sasakii (Miyatake)	(カメムシ目、キジラミ科) 日本。ネムノキ害虫
Sasakia oystershell scale	ササキカキカイガラムシ	Lepidosaphes euryae (Kuwana)	(カメムシ目、マルカイガラムシ科) 日本、東洋区
Sasakia round bark beetle	アカマツザイノキクイムシ	Xyleborus aquilus Blandford	(コウチュウ目、キクイムシ科) 日本、旧北区、東洋区
Sasakia scale	ヒサカキコノハカイガラムシ	Fiorinia euryae Kuwana	(カメムシ目、マルカイガラムシ科) 日本、東洋区
Sasakia xyloterus	キシマキクイムシ	Indocryphalus sordidus (Blandford)	(コウチュウ目、キクイムシ科) 日本
saskatoon borer		Saperda candida bipunctata Hopping	(コウチュウ目、カミキリムシ科) 新北区
sassafras Caloptilia moth		Caloptilia sassafrasella (Chambers)	(チョウ目、ホソガ科) 新北区
satanic skipper		Aethilla eleusinia Hewitson	(チョウ目、セセリチョウ科) 新熱帯区
satellite	エゾミツボシキリガ	Eupsilia transversa (Hufnagel)	(チョウ目、ヤガ科) 日本、旧北区
satellite moth			satellite を見よ
satellite sphinx moth		Eumorpha satellitia (Linnaeus)	(チョウ目、スズメガ科) 新北区、新熱帯区。satellite sphinx ともいう
satin azure			amaryllis azure を見よ
satin beauty	マツオオエダシャク	Deileptenia ribeata (Clerck)	(チョウ目、シャクガ科) 日本、旧北区
satin blue		Nesolycaena albosericea (Miskin)	(チョウ目、シジミチョウ科) 豪州区
satin blue			spotted dusky-blue を見よ
satin carpet			satin lutestring を見よ
satin carpet moth			satin beauty を見よ
satin lift		Heliozela sericiella (Haworth)	(チョウ目、ツヤコガ科) 旧北区
satin lutestring		Tetheella fluctuosa (Hübner)	(チョウ目、トガリバガ科) 旧北区。日本亜種は T. f. isshikii (Matsumura) ヒトテントガリバ
satin moth (1)	ヤナギドクガ	Leucoma salicis (Linnaeus)	(チョウ目、ドクガ科) 日本、全北区。成虫は斑紋のない白色。幼虫はポプラとヤナギ害虫
satin moth (2)	ブチヒゲヤナギドクガ	Leucoma candida (Staudinger)	(チョウ目、ドクガ科) 日本、旧北区。ヤナギ害虫
satin opal			satin blue を見よ
satin silphid			common carrion beetle を見よ
satin wave		Idaea subsericeata (Haworth)	(チョウ目、シャクガ科) 旧北区
satin white Palpita moth		Palpita flegia Cramer	(チョウ目、メイガ科) 新北区

英名	和名	学名	所属、分布、ほか
Satsuma frosted chafer	サツマコフキコガネ	*Melolontha satsumaensis satsumaensis* Niijima et Kinoshita	（コウチュウ目、コガネムシ科）日本
Satsuma pyralid	サツマツトガ	*Calamotropha okanoi* Bleszynski	（チョウ目、メイガ科）日本、旧北区
Satsuma zygaenid	サツマニシキ	*Erasmia pulchella fritzei* Jordan	（チョウ目、マダラガ科）日本
saturn	ルリオビトガリバワモンチョウ	*Zeuxidia amethystus* Butler	（チョウ目、タテハチョウ科）東洋区。saturn butterfly ともいう
saturn butterflies			brush-footed butterflies を見よ
saturniid moths			giant silkworm moths を見よ　saturniids ともいう
saturniid silkmoth			cynthia moth を見よ
Saturnus ringlet		*Euptychoides saturnus* Butler	（チョウ目、タテハチョウ科）新熱帯区
Saturnus skipper		*Callimormus saturnus* (Herrich-Schäffer)	（チョウ目、セセリチョウ科）新熱帯区
satyr			satyr anglewing を見よ
satyr anglewing	サチルアメリカシータテハ	*Polygonia satyrus* (Edwards)	（チョウ目、タテハチョウ科）新北区
satyr butterflies			satyrs を見よ　米国での英名
satyr Charaxes			coast Charaxes を見よ
satyr comma			satyr anglewing を見よ
satyr hairstreak			satyroides hairstreak を見よ
satyr metalmark		*Leucochimona lepida nivalis* (Godman et Salvin)	（チョウ目、シジミタテハ科）新熱帯区
satyr pug		*Eupithecia satyrata* (Hübner)	（チョウ目、シャクガ科）旧北区
satyr skipper		*Timochreon satyrus* (Felder)	（チョウ目、セセリチョウ科）新熱帯区
satyroides hairstreak		*Evenus satyroides* (Hewitson)	（チョウ目、シジミチョウ科）新熱帯区
satyrs	ジャノメチョウ亜科	Satyrinae	（チョウ目、タテハチョウ科）の昆虫の総称
satyrs and woodnymphs			satyrs を見よ
saucer bug		*Aphelocheirus aestivalis* (Fabricius)	（カメムシ目、Aphelocheiridae）旧北区
saucer bug (1)		*Ilyocoris cimicoides* (Linnaeus)	（カメムシ目、コバンムシ科）旧北区
saucer bugs			creeping water bugs を見よ
saucy beauty moth		*Phaloesia saucia* Walker	（チョウ目、ヒトリガ科）新北区
Saunders embiid			orchid embiid を見よ　米国での英名
Saunders' sallow moth		*Sympistis saundersiana* (Grote)	（チョウ目、ヤガ科）新北区。Saunders' Oncocnemis ともいう
Saunder's sister		*Adelpha saundersii* (Hewitson)	（チョウ目、タテハチョウ科）新熱帯区
Saunders's case moth		*Oiketicus elongatus* Saunders	（チョウ目、ミノガ科）豪州区

英名	和名	学名	所属、分布、ほか
Saussure's bush brown		*Bicyclus saussurei* (Dewitz)	(チョウ目、タテハチョウ科) エチオピア区
Saussure's scaly cricket		*Cycloptiloides americanus* (Saussure)	(バッタ目、コオロギ科) 新北区
savage beetles			ground beetles を見よ
savanna brown		*Neita extensa* (Butler)	(チョウ目、タテハチョウ科) エチオピア区
savanna elf		*Eretis lugens* (Rogenhofer)	(チョウ目、セセリチョウ科) エチオピア区
savanna hawkmoth		*Sphingonaepiopsis nana* (Walker)	(チョウ目、スズメガ科) エチオピア区
savannah Charaxes	ルリマダラフタオチョウ	*Charaxes etesipe* (Godart)	(チョウ目、タテハチョウ科) エチオピア区
saw-grass skipper		*Euphyes pilatka* (Edwards)	(チョウ目、セセリチョウ科) 新北区
saw-tailed bush cricket		*Barbitistes serricauda* (Fabricius)	(バッタ目、キリギリス科) 旧北区
saw-tooth grain weevil			saw-toothed grain beetle を見よ
saw-toothed grain beetle	ノコギリヒラタムシ	*Oryzaephilus surinamensis* (Linnaeus)	(コウチュウ目、ホソヒラタムシ科) 日本、汎世界。貯穀害虫
saw-toothed grain beetles		*Oryzaephilus*	(コウチュウ目、ホソヒラタムシ科) の昆虫の総称
saw-toothed grain weevil			saw-toothed grain beetle を見よ
sawtooths	マダラシロチョウ属	*Prioneris*	(チョウ目、シロチョウ科) の昆虫の総称　現在は *Colotis* の亜属とされる
saw-wing		*Euchlaena serrata* Drury	(チョウ目、シャクガ科) 新北区。saw-wing moth ともいう
Sawada oak aphid	サワダブチアブラムシ	*Tuberculatus querciformosanus* (Takahashi)	(カメムシ目、アブラムシ科) 日本、旧北区、東洋区
sawflies	ハバチ上科	Tenthredinoidea	(ハチ目) の昆虫の総称
sawflies (1)	ハバチ科	Tenthredinidae	(ハチ目) の昆虫の総称　雌の産卵管にのこぎり状の歯があることに由来。世界に3,000種以上
sawflies			horntails (1) を見よ
sawflies			horntail wasps を見よ
sawflies			common sawflies を見よ
sawyer			カミキリムシの幼虫
sawyer beetle		*Monochamus galloprovincialis pistor* (Germar)	(コウチュウ目、カミキリムシ科) 旧北区
sawyer beetle		*Monochamus sartor* Fabricius	(コウチュウ目、カミキリムシ科) 旧北区
sawyer beetle			small white-marmorated longicorn を見よ
sawyer beetle			tanner beetle を見よ
sawyer beetles		*Monochamus*	(コウチュウ目、カミキリムシ科) の昆虫の総称
sawyer beetles			long-horned beetles を見よ
Saxena ambrosia beetle	サクセスキクイムシ	*Xyleborus saxeseni* (Ratzeburg)	(コウチュウ目、キクイムシ科) 日本、全北区、東洋区、豪州区。カンキツなど果樹、サクラ害虫

英名	和名	学名	所属、分布、ほか
Saxesen ambrosia beetle			Saxena ambrosia beetle を見よ
Saxon	カラフトシロスジヨトウ	*Hyppa rectilinea* (Esper)	（チョウ目、ヤガ科）日本、旧北区
Saxon moth			Saxon を見よ
Saxon wasp		*Dolichovespula saxonica* (Fabricius)	（ハチ目、スズメバチ科）旧北区。日本には *D. s. nipponica* SK. Yamane ニッポンホオナガスズメバチが分布
Say stink bug			Say's stink bug を見よ
Say's blister beetle		*Lytta sayi* (LeConte)	（コウチュウ目、ツチハンミョウ科）新北区
Say's bush cricket		*Anaxipha exigua* (Say)	（バッタ目、クサヒバリ科）新北区
Say's grasshopper			orange-legged grasshopper を見よ
Say's ground beetle		*Chlaenius pennsylvanicus* Say	（コウチュウ目、オサムシ科）新北区
Say's plant bug			Say's stink bug を見よ
Say's spiketail		*Cordulegaster sayi* Selys	（トンボ目、オニヤンマ科）新北区
Say's stink bug		*Chlorochroa sayi* Stål	（カメムシ目、カメムシ科）新北区。ワタ害虫
Say's trig			Say's bush cricket を見よ
scabious cuckoo bee		*Nomada armata* Herrich-Schäffer	（ハチ目、コシブトハナバチ科）旧北区
scale aphid			fringed orchid aphid を見よ
scale-eating caterpillar		*Catoblemma dubia* (Butler)	（チョウ目、ヤガ科）豪州区
scale-eating ladybird		*Rhyzobius lophanthae* (Blaisdell)	（コウチュウ目、テントウムシ科）豪州区
scale-feeding scavenger moth		*Holcocera coccivorella* (Chambers)	（チョウ目、ネマルハキバガ科）新北区
scale-feeding snout moth			coccid-eating pyralid を見よ
scale insect		*Chrysomphalus rosii* Maskell	（カメムシ目、マルカイガラムシ科）大洋区。カンキツ害虫
scale insect			カイガラムシ類、とくにカタカイガラムシ科の種。単に scale ともいう
scale insects	カイガラムシ上科	Coccoidea	（カメムシ目）の昆虫の総称
scale insects			armored scales を見よ
scale insects			mealybugs (1) を見よ
scale insects			soft scales を見よ
scale parasites			fun flies を見よ
scale parasites			helomyzid flies を見よ
scale parasites			thysanid wasps を見よ　南アフリカでの英名
scale-tooth lacewing		*Phaulernis dentella* (Zeller)	（チョウ目、ササベリガ科）旧北区
scale-winged insects			butterflies and moths を見よ
scaled fleas	ホソノミ科	Leptopsyllidae	（ノミ目）の昆虫の総称
scallop-edged firetip		*Jonaspyge jonas* (Felder et Felder)	（チョウ目、セセリチョウ科）新熱帯区

英名	和名	学名	所属、分布、ほか
scallop moth		*Cepphis armataria* (Herrich-Schäffer)	(チョウ目、シャクガ科) 新北区
scallop-patched Theope		*Theope cratylus* Godman et Salvin	(チョウ目、シジミタテハ科) 新熱帯区
scallop shell	ヤエナミシャク	*Rheumaptera undulata* (Linnaeus)	(チョウ目、シャクガ科) 日本、全北区
scallop-shell moth			scallop shell を見よ
scallop-winged frit			Queen of Spain fritillary を見よ
scalloped Epitola		*Stempfferia marginata* (Kirby)	(チョウ目、シジミチョウ科) エチオピア区
scalloped grass-yellow	アリタキチョウ (エサキキチョウ)	*Eurema alitha* (C. et R. Felder)	(チョウ目、シロチョウ科) 豪州区
scalloped hawk		*Odontosida pusillus* (Felder)	(チョウ目、スズメガ科) エチオピア区
scalloped hazel		*Odontopera bidentata* (Clerck)	(チョウ目、シャクガ科) 旧北区。日本亜種は *O. b. harutai* (Inoue) ウスグロノコバエダシャク
scalloped hazel moth			scalloped hazel を見よ
scalloped hazel thorn			scalloped hazel を見よ
scalloped hooktip		*Drepana lacertinaria* (Linnaeus)	(チョウ目、カギバガ科) 旧北区
scalloped oak	フタモンキバネエダシャク	*Croccallis elinguaria* (Linnaeus)	(チョウ目、シャクガ科) 日本、旧北区
scalloped oak moth			scalloped oak を見よ
scalloped oak thorn			scalloped oak を見よ
scalloped owl-butterfly		*Opsiphanes quiteria quirinus* Godman et Salvin	(チョウ目、タテハチョウ科) 新熱帯区
scalloped owlet		*Opsiphanes quiteria* (Stoll)	(チョウ目、タテハチョウ科) 新熱帯区
scalloped red glider			red glider (1)(2) を見よ
scalloped sack-bearer moth		*Lacosoma chiridota* Grote	(チョウ目、Mimallonidae) 新北区
scalloped sailer	ミスジモドキ	*Neptidopsis ophione* (Cramer)	(チョウ目、タテハチョウ科) エチオピア区
scalloped snout			white-spotted Redectis moth を見よ
scalloped sootywing			southern sootywing を見よ
scalloped spreadwing		*Lestes praemorsus decipiens* Kirby	(トンボ目、アオイトトンボ科) 東洋区
scalloped yellow glider		*Cymothoe fumana* (Westwood)	(チョウ目、タテハチョウ科) エチオピア区
scaly alchemyst		*Catephia squamosa* Wallengren	(チョウ目、ヤガ科) エチオピア区
scaly barklice	ビロウドチャタテ科	Lepidopsocidae	(チャタテムシ目) の昆虫の総称
scaly cricket		*Pseudomogoplistes squamiger* (Fischer)	(バッタ目、コオロギ科) 旧北区
scaly crickets		Mogoplistinae	(バッタ目、コオロギ科) の昆虫の総称　熱帯に多く、鱗でおおわれる
scaly-winged barklice			scaly barklice を見よ

英名	和名	学名	所属、分布、ほか
scaly-winged barklouse		*Echmepteryx hageni* (Packard)	（チャタテムシ目、ビロウドチャタテ科）新北区
scaly-winged psocids		*Echmepteryx*	（チャタテムシ目、ビロウドチャタテ科）の昆虫の総称
Scandinavian green-veined white		*Pieris adalwinda* (Fruhstorfer)	（チョウ目、シロチョウ科）旧北区
scape moths		*Cisseps*	（チョウ目、カノコガ科）の昆虫の総称　北米での英名
scape moths			wasp moths を見よ
scar bank gem		*Ctenoplusia limbirena* (Guenée)	（チョウ目、ヤガ科）旧北区
scar large case		*Coleophora vibicella* (Hübner)	（チョウ目、ツツミノガ科）旧北区
scarab	スカラベ	*Scarabaeus sacer* Linnaeus	（コウチュウ目、コガネムシ科）エチオピア区。神聖甲虫ともいわれる
scarab beetles			chafer beetles を見よ
scarab flies			light flies を見よ
scarabaeid	コガネムシ		コガネムシ科の昆虫。chafer beetles を見よ
scarabaeus			chafer beetles を見よ
scarabee weevil			West Indian sweet potato weevil を見よ
scarabs			chafer beetles を見よ
scarbeids			chafer beetles を見よ
scarce aeschna			migrant hawker を見よ
scarce alder pigmy		*Stigmella glutinosae* (Stainton)	（チョウ目、モグリチビガ科）旧北区
scarce Anastrus		*Anastrus meliboea* (Godman et Salvin)	（チョウ目、セセリチョウ科）新熱帯区
scarce arches		*Luperina zollikoferi* (Freyer)	（チョウ目、ヤガ科）旧北区
scarce awl robberfly		*Neoitamus cothurnatus* (Meigen)	（ハエ目、ムシヒキアブ科）旧北区。日本には *N. c. univittatus* (Loew) が分布
scarce bamboo page			bamboo page を見よ
scarce banded flat		*Celaenorrhinus badia* (Hewitson)	（チョウ目、セセリチョウ科）東洋区
scarce black		*Eupachygaster tarsalis* (Zetterstedt)	（ハエ目、ミズアブ科）旧北区
scarce black arches		*Meganola aerugula* (Hübner)	（チョウ目、コブガ科）旧北区。日本亜種は *M. a. atomosa* (Bremer) カバイロコブガ
scarce black arches		*Celama centonalis* Hübner	（チョウ目、コブガ科）旧北区
scarce blackberry aphid		*Macrosiphum funestum* (Macchiati)	（カメムシ目、アブラムシ科）旧北区
scarce blackneck	ハイイロクビグロクチバ	*Lygephila craccae* (Denis et Schiffermüller)	（チョウ目、ヤガ科）日本、旧北区
scarce blue oakleaf	ホソバルリオビコノハチョウ	*Kallima alompra* Moore	（チョウ目、タテハチョウ科）東洋区
scarce blue tiger	ガウタマコモンマダラ	*Tirumala gautama* (Moore)	（チョウ目、タテハチョウ科）東洋区

英名	和名	学名	所属、分布、ほか
scarce blue-tailed damselfly			small bluetail を見よ
scarce bordered sallow			corn earworm (2) を見よ
scarce bordered straw			corn earworm (2) を見よ
scarce bordered straw moth			corn earworm (2) を見よ
scarce brindle	オオアカヨトウ	*Apamea lateritia* (Hufnagel)	（チョウ目、ヤガ科）日本、旧北区。トウモロコシ害虫
scarce burnished brass	オオキンウワバ	*Diachrysia chryson* (Esper)	（チョウ目、ヤガ科）日本、旧北区
scarce catseye		*Coelites nothis* Westwood	（チョウ目、タテハチョウ科）東洋区
scarce chaser		*Libellula fulva* (Muller)	（トンボ目、トンボ科）旧北区
scarce chocolate-tip			poplar prominent (1) を見よ
scarce clouded yellow			new-clouded yellow を見よ
scarce clubbed sailer		*Neptis nicobule* Holland	（チョウ目、タテハチョウ科）エチオピア区
scarce copper	フチグロベニシジミ	*Lycaena virgaureae* (Linnaeus)	（チョウ目、シジミチョウ科）旧北区
scarce Cornelian		*Deudorix hypargyria* (Elwes)	（チョウ目、シジミチョウ科）東洋区
scarce coronet		*Dianthaecia compta* (Denis et Schiffermüller)	（チョウ目、ヤガ科）旧北区
scarce costus skipper		*Hypoleucis sophia* Evans	（チョウ目、セセリチョウ科）エチオピア区
scarce crescent piercer		*Grapholitha dorsana* Fabricius	（チョウ目、ハマキガ科）旧北区
scarce crimson-and-gold		*Pyrausta sanguinalis* (Linnaeus)	（チョウ目、メイガ科）旧北区
scarce dagger			dagger moth (2) を見よ
scarce dart		*Andronymus hero* Evans	（チョウ目、セセリチョウ科）エチオピア区
scarce duskywing			scarce Anastrus を見よ
scarce emerald	エゾアオイトトンボ	*Lestes dryas* Kirby	（トンボ目、アオイトトンボ科）日本、全北区
scarce emerald damselfly			scarce emerald を見よ
scarce evening brown		*Cyllogenes janetae* de Nicéville	（チョウ目、タテハチョウ科）東洋区
scarce footman		*Eilema complana* (Linnaeus)	（チョウ目、ヒトリガ科）旧北区
scarce footman			dingy footman を見よ
scarce forest horsefly		*Hybomitra solstitialis* (Meigen)	（ハエ目、アブ科）旧北区
scarce forest swift		*Melphina tarace* (Mabille)	（チョウ目、セセリチョウ科）エチオピア区
scarce forester		*Jordanita globulariae* (Hübner)	（チョウ目、マダラガ科）旧北区
scarce fritillary		*Euphydryas maturna* (Linnaeus)	（チョウ目、タテハチョウ科）旧北区
scarce grass yellow	アオジロキチョウ	*Eurema lacteola* (Distant)	（チョウ目、シロチョウ科）東洋区

英名	和名	学名	所属、分布、ほか
scarce grass-veneer		*Chrysocrambus craterellus* (Scopoli)	（チョウ目、メイガ科）旧北区
scarce green cutworm moth		*Melanchra pictula* White	（チョウ目、ヤガ科）豪州区
scarce green lestes			scarce emerald を見よ
scarce green white		*Euchloe falloui* (Allard)	（チョウ目、シロチョウ科）旧北区
scarce green-striped white			scarce green white を見よ
scarce guava blue		*Virachola smilis* (Hewitson)	（チョウ目、シジミチョウ科）東洋区
scarce heath	シロオビヒメヒカゲ	*Coenonympha hero latifasciata* Matsumura	（チョウ目、タテハチョウ科）日本、旧北区。*C. hero* (Linnaeus) の英名
scarce hook-tip	ウスオビカギバ	*Sabra harpagula* (Esper)	（チョウ目、カギバガ科）日本、旧北区
scarce Ischnura			small bluetail を見よ
scarce jester		*Symbrenthia silana* de Nicéville	（チョウ目、タテハチョウ科）東洋区
scarce labyrinth		*Neope pulahina* (Evans)	（チョウ目、タテハチョウ科）東洋区
scarce large blue		*Maculinea teleius* (Bergsträsser)	（チョウ目、シジミチョウ科）旧北区。日本亜種は *M. t. kazamoto* (H. Druce) ゴマシジミ
scarce largest dart		*Paronymus nevea* (Druce)	（チョウ目、セセリチョウ科）エチオピア区
scarce libellula			scarce chaser を見よ
scarce lilacfork		*Lethe dura* (Marshall)	（チョウ目、タテハチョウ科）東洋区
scarce lilacfork	スーラヤマヒカゲ	*Lethe sura* (Doubleday)	（チョウ目、タテハチョウ科）東洋区。lilacfork ともいう
scarce lime bark beetle		*Ernoporus filiae* (Panzer)	（コウチュウ目、キクイムシ科）旧北区
scarce marbled		*Eublemma noctualis* Hübner	（チョウ目、ヤガ科）旧北区
scarce marbled		*Eublemma minutata* (Fabricius)	（チョウ目、ヤガ科）旧北区
scarce Mervel-du-Jour		*Moma orion* Esper	（チョウ目、ヤガ科）旧北区
scarce Merveille du jour			black-spotted dagger moth を見よ
scarce mountain Argus		*Callerebia kalinda* (Moore)	（チョウ目、タテハチョウ科）東洋区
scarce mountain copper		*Trimenia malagrida* (Wallengren)	（チョウ目、シジミチョウ科）エチオピア区
scarce olive		*Baetis buceratus* Eaton	（カゲロウ目、コカゲロウ科）旧北区
scarce olive-tree pearl		*Palpita unionalis* Hübner	（チョウ目、メイガ科）旧北区
scarce orange legionnaire		*Beris clavipes* (Linnaeus)	（ハエ目、ミズアブ科）旧北区
scarce orange tip			bushveld orange tip を見よ
scarce pathfinder skipper		*Pardaleodes sator* (Westwood)	（チョウ目、セセリチョウ科）エチオピア区
scarce polyptychus		*Afrosphinx amabilis* (Jordan)	（チョウ目、スズメガ科）エチオピア区

英名	和名	学名	所属、分布、ほか
scarce prominent		*Odontosia carmelita* (Esper)	(チョウ目、シャチホコガ科) 旧北区
scarce pug	シロマダラカバナミシャク	*Eupithecia extensaria* (Freyer)	(チョウ目、シャクガ科) 日本、旧北区
scarce purple dun		*Paraleptophlebia werneri* Ulmer	(カゲロウ目、トビイロカゲロウ科) 旧北区
scarce ranger		*Kedestes nerva* (Fabricius)	(チョウ目、セセリチョウ科) エチオピア区
scarce red forester		*Lethe distans* Butler	(チョウ目、タテハチョウ科) 東洋区
scarce red-legged robberfly		*Dioctria cothurnata* Meigen	(ハエ目、ムシヒキアブ科) 旧北区
scarce red underwing			rosy underwing を見よ
scarce rose bell		*Eucosma pauperana* (Duponchel)	(チョウ目、ハマキガ科) 旧北区
scarce rush case		*Coleophora taenipennella* Herrich-Schäffer	(チョウ目、ツツミノガ科) 旧北区
scarce sapphire			saffron (1) を見よ
scarce scarlet		*Chrysoritis phosphor* (Trimen)	(チョウ目、シジミチョウ科) エチオピア区
scarce seven-spotted ladybird		*Coccinella magnifica* Redtenbacher	(コウチュウ目、テントウムシ科) 旧北区
scarce shot silverline		*Cigaritis elima* (Moore)	(チョウ目、シジミチョウ科) 東洋区
scarce silver-lines		*Hylophila bicolorana* (Fuessly)	(チョウ目、ヤガ科) 旧北区
scarce silver-lines (1)	アオスジアオリンガ	*Pseudoips fagana* (Fabricius)	(チョウ目、ヤガ科) 日本、旧北区
scarce silver-spotted flambeau	ウラギンドクチョウ	*Dione juno* (Cramer)	(チョウ目、タテハチョウ科) 新北区、新熱帯区
scarce silver Y	ホクトギンウワバ	*Syngrapha interrogationis* (Linnaeus)	(チョウ目、ヤガ科) 日本、旧北区
scarce silverstreak		*Iraota rochana* (Horsfield)	(チョウ目、シジミチョウ科) 東洋区。*I. r. boswelliana* Distant も同英名
scarce sister		*Adelpha nea* (Hewitson)	(チョウ目、タテハチョウ科) 新熱帯区
scarce skipper			European skipper を見よ
scarce skipper			scarce ranger を見よ
scarce slate flash		*Rapala scintilla* de Nicéville	(チョウ目、シジミチョウ科) 東洋区
scarce small recuse		*Leona lota* Evans	(チョウ目、セセリチョウ科) エチオピア区
scarce straw pearl		*Paracorsia repandalis* (Denis et Schiffermüller)	(チョウ目、メイガ科) 旧北区
scarce streaky-skipper			West-Texas streaky-skipper を見よ
scarce summer mayfly		*Siphlonurus armatus* Eaton	(カゲロウ目、フタオカゲロウ科) 旧北区
scarce swallowtail	ヨーロッパタイマイ	*Iphiclides podalirius* (Linnaeus)	(チョウ目、アゲハチョウ科) 全北区
scarce swamp skipper		*Euphyes dukesi* Lindsey	(チョウ目、セセリチョウ科) 新北区
scarce tawny Charaxes		*Charaxes aristogiton* (C. et R. Felder)	(チョウ目、タテハチョウ科) 東洋区

英名	和名	学名	所属、分布、ほか
scarce tissue		*Rheumaptera cervinalis* (Scopoli)	(チョウ目、シャクガ科) 旧北区
scarce tissue		*Rheumaptera certata* (Hübner)	(チョウ目、シャクガ科) 旧北区
scarce tortoiseshell		*Nymphalis xanthomelas* (Esper)	(チョウ目、タテハチョウ科) 東洋区、旧北区。日本亜種 *N. x. japonica* (Stichel) ヒオドシチョウも同英名。
scarce umber		*Agriopis aurantiaria* (Hübner)	(チョウ目、シャクガ科) 旧北区
scarce umber moth			scarce umber を見よ
scarce vapourer		*Orgyia gonostigma* (Scopoli)	(チョウ目、ドクガ科) 旧北区
scarce vapourer			common vapourer を見よ
scarce water-veneer		*Donacaula mucronella* (Denis et Schiffermüller)	(チョウ目、メイガ科) 旧北区
scarce white commodore		*Sumalia zulema* (Doubleday)	(チョウ目、タテハチョウ科) 東洋区
scarce white royal		*Tajuria illurgioides* de Nicéville	(チョウ目、シジミチョウ科) 東洋区
scarce woodbrown	カノクロヒカゲ	*Lethe siderea* Marshall	(チョウ目、タテハチョウ科) 東洋区
scarce wormwood moth		*Cucullia artemisiae* (Hufnagel)	(チョウ目、ヤガ科) 旧北区。日本亜種は *C. a. perspicua* Warnecke トカチセダカモクメ。scarce wormwood ともいう
scarce wormwood shark			scarce wormwood moth を見よ
scarce yellow May dun		*Heptagenia longicauda* (Stephens)	(カゲロウ目、ヒラタカゲロウ科) 旧北区
scarce yellowstreak		*Electrogena affinis* (Eaton)	(カゲロウ目、ヒラタカゲロウ科) 旧北区
scarlet Acraea		*Acraea atolmis* Westwood	(チョウ目、タテハチョウ科) エチオピア区
scarlet admiral			red admiiral (1) を見よ
scarlet-and-green leafhopper			candy-striped leafhopper を見よ
scarlet-backed threadtail		*Protoneura tenuis* Selys	(トンボ目、ミナミイトトンボ科) 新熱帯区
scarlet basker		*Urothemis signata signata* (Rambur)	(トンボ目、トンボ科) 東洋区
scarlet bluet		*Enallagma pictum* Morse	(トンボ目、イトトンボ科) 新北区
scarlet-bodied wasp moth		*Cosmosoma myrodora* Dyar	(チョウ目、ヒトリガ科) 新北区
scarlet butterfly			common scarlet (1) を見よ
scarlet darter		*Crocothemis erythraea* (Brulle)	(トンボ目、トンボ科) 旧北区。common scarlet-darter ともいう
scarlet dragonfly			scarlet darter を見よ
scarlet flash		*Rapala dieneces* (Hewitson)	(チョウ目、シジミチョウ科) 東洋区
scarlet grain			ground pearl (2) を見よ
scarlet Jezebel		*Delias argenthona* (Fabricius)	(チョウ目、シロチョウ科) 豪州区

英名	和名	学名	所属、分布、ほか
scarlet Kauai damselfly		*Megalagrion vagabundum* (Perkins)	(トンボ目、イトトンボ科) 大洋区
scarlet knight	キオビタテハ	*Temenis pulchra* Hewitson	(チョウ目、タテハチョウ科) 新熱帯区
scarlet leafwing		*Siderone galanthis* (Cramer)	(チョウ目、タテハチョウ科) 新熱帯区
scarlet lily beetle		*Lilioceris lilii* (Scopoli)	(コウチュウ目、ハムシ科) 旧北区
scarlet malachite beetle		*Malachius aeneus* (Linnaeus)	(コウチュウ目、ジョウカイモドキ科) 全北区、大洋区
scarlet oak sawfly		*Caliroa quercuscoccineae* (Dyar)	(ハチ目、ハバチ科) 新北区
scarlet peacock		*Anartia amathea* (Linnaeus)	(チョウ目、タテハチョウ科) 新熱帯区
scarlet percher		*Diplacodes haematodes* (Burmeister)	(トンボ目、トンボ科) 豪州区
scarlet skimmer	ショウジョウトンボ	*Crocothemis servilia* (Drury)	(トンボ目、トンボ科) 全北区、東洋区、大洋区
scarlet tiger		*Callimorpha dominula* (Linnaeus)	(チョウ目、ヒトリガ科) 旧北区
scarlet tiger			Jersey tiger (1)(2) を見よ
scarlet tiger moth			scarlet tiger を見よ
scarlet-tip	ダナエツマアカシロチョウ	*Colotis danae* (Fabricius)	(チョウ目、シロチョウ科) 東洋区、エチオピア区
scarlet-tipped loner		*Mysarbiia sejanus* (Hopffer)	(チョウ目、セセリチョウ科) 新熱帯区
scarlet underwing moth		*Catocala coccinata* Grote	(チョウ目、ヤガ科) 新北区。scarlet underwing ともいう
scarlet underwing moth			dark crimson underwing を見よ
scarlet-winged lichen moth			striped footman moth を見よ 北米での英名
scarlets		*Crocothemis*	(トンボ目、トンボ科) の昆虫の総称
scarlets		*Axiocerses*	(チョウ目、シジミチョウ科) の昆虫の総称
scarred goldenrod gall moth		*Eucosma scudderiana* Clemens	(チョウ目、ハマキガ科) 新北区
scarred melolontha			Fuller を見よ
scarred snout beetles	クチブトゾウムシ亜科	Otiorhynchinae	(コウチュウ目、ゾウムシ科) の昆虫の総称
scavenger beetle			*Hydrophilus* 属などガムシ類の昆虫
scavenger beetles			water scavenger beetles (1) を見よ
scavenger flies			flesh flies を見よ 米国での英名
scavenger moths	ネマルハキバガ科	Blastobasidae	(チョウ目) の昆虫の総称
scavenger water beetles			water scavenger beetles (1) を見よ
scavenging beetle			mealybug ladybird を見よ
scelionid wasps	タマゴクロバチ科	Scelionidae	(ハチ目) の昆虫の総称 scelionid flies, scelionids ともいう
scentless plant bug		*Harmostes fraterculus* (Say)	(カメムシ目、ヒメヘリカメムシ科) 新北区、新熱帯区

英名	和名	学名	所属、分布、ほか
scentless plant bugs	ヒメヘリカメムシ科	Rhopalidae	(カメムシ目) の昆虫の総称
Schaffer's neb (1)	ドルリーカザリバ	*Cosmopterix orichalcea* (Stainton)	(チョウ目、カザリバガ科) 日本、旧北区
Schaffer's neb (2)		*Chrysoesthia drurella* (Fabricius)	(チョウ目、キバガ科) 旧北区
Schaffgotsch's swordtail		*Graphium schaffgotschi* (Niepelt)	(チョウ目、アゲハチョウ科) エチオピア区
Schaller's Acleris moth			Logan's button (1) を見よ
Schaus' Emesis		*Emesis vimena* Schaus	(チョウ目、シジミタテハ科) 新熱帯区
Schaus' flasher		*Astraptes janeira* (Schaus)	(チョウ目、セセリチョウ科) 新熱帯区
Schaus' gemmed-satyr		*Cyllopsis schausi* Miller	(チョウ目、タテハチョウ科) 新熱帯区
Schaus' hairstreak			Salvin's dwarf を見よ
Schaus' Ipidecla			Salvin's dwarf を見よ
Schaus' metalmark			red-bordered metalmark を見よ
Schaus' scarlet-eye		*Nascus phintias* Schaus	(チョウ目、セセリチョウ科) 新熱帯区
Schaus' skipperling		*Dalla lethaea* (Schaus)	(チョウ目、セセリチョウ科) 新熱帯区
Schaus' sulphur		*Aphrissa schausi* (Avinoff)	(チョウ目、シロチョウ科) 新熱帯区
Schaus swallowtail		*Papilio aristodemus* Esper	(チョウ目、アゲハチョウ科) 新北区、新熱帯区
Schaus' tussock moth		*Halysidota schausi* Rothschild	(チョウ目、ヒトリガ科) 新北区、新熱帯区
Schlaeger's fruitworm moth		*Antaeotricha schlaegeri* (Zeller)	(チョウ目、マルハキバガ科) 新北区
Schouteden's ciliate blue		*Anthene schoutedeni* (Hulstaert)	(チョウ目、シジミチョウ科) エチオピア区
Schreiber's conch		*Phtheochroa schreibersiana* (Frölich)	(チョウ目、ハマキガ科) 旧北区
Schryver's elfin			moss elfin を見よ
Schuas' Sarota		*Sarota estrada estrada* Schaus	(チョウ目、シジミタテハ科) 新熱帯区
Schultze's Agrias			Godart's Agrias を見よ
Schultze's nymph		*Euriphene schultzei* (Aurivillius)	(チョウ目、タテハチョウ科) エチオピア区
Schwarz's shieldback		*Ateloplus schwarzi* Caudell	(バッタ目、キリギリス科) 新北区
sciarid flies			dark-winged fungus flies を見よ
Scimitra backswimmer		*Buenoa scimitra* Bare	(カメムシ目、マツモムシ科) 新北区
scintillating firefly		*Photinus scintillans* Say	(コウチュウ目、ホタル科) 新北区
Sciras skipper		*Falga sciras* Godman	(チョウ目、セセリチョウ科) 新熱帯区
Sciron ochre			Sciron skipper を見よ
Sciron skipper		*Trapezites sciron* Waterhouse et Lyell	(チョウ目、セセリチョウ科) 豪州区
scissortails		*Microgomphus*	(トンボ目、サナエトンボ科) の昆虫の総称
sclerogibbid wasps	シロアリモドキヤドリバチ科	Sclerogibbidae	(ハチ目) の昆虫の総称

英名	和名	学名	所属、分布、ほか
sclerogibbids			sclerogibbid wasps を見よ
scolia wasp		*Campsomeris marginella modesta* (Smith)	（ハチ目、ツチバチ科）大洋区
scoliid wasps	ツチバチ科	Scoliidae	（ハチ目）の昆虫の総称
scoloid wasps			parasitic wasps を見よ
Scopoli's bell		*Eucosma hohenwartiana* (Denis et Schiffermüller)	（チョウ目、ハマキガ科）旧北区
scopulipede bees	採粉脚類	Podilegina	（ハチ目）の昆虫の総称　脚に花粉を集める構造のあるハチを指す。scopulipeds ともいう
scorched blunt marble		*Endothenia ustulana* Sheldon	（チョウ目、ハマキガ科）旧北区
scorched carpet		*Ligdia adustata* (Denis et Schiffermüller)	（チョウ目、シャクガ科）旧北区
scorched silver			scorched carpet を見よ
scorched wing	ナカキエダシャク	*Plagodis dolabraria* (Linnaeus)	（チョウ目、シャクガ科）日本、旧北区
scorpion flies (1)	シリアゲムシ科	Panorpidae	（シリアゲムシ目）の昆虫の総称
scorpion flies (2)	シリアゲムシ目	Mecoptera	の昆虫の総称　世界に 480 種
scorpionfly		*Panorpa rufescens* Rambur	（シリアゲムシ目、シリアゲムシ科）新北区
scorpionfy		*Panorpa cognata* Rambur	（シリアゲムシ目、シリアゲムシ科）旧北区
scorpionfly			green stigma を見よ
scorpionfly			common scorpion fly を見よ
Scotch annulet		*Gnophos obfuscata* (Denis et Schiffermüller)	（チョウ目、シャクガ科）旧北区。Scottish annulet ともいう
Scotch annulet		*Gnophos myrtillata* Thunberg	（チョウ目、シャクガ科）旧北区
Scotch Argus	ユーラシアベニヒカゲ（スコッチベニヒカゲ）	*Erebia aethiops* (Esper)	（チョウ目、タテハチョウ科）旧北区。Scotch Argus butterfly ともいう
Scotch brindled pearl		*Udea decrepitalis* Herrich-Schäffer	（チョウ目、メイガ科）旧北区
Scotch brown blue			northern brown Argus を見よ
Scotch burnet		*Zygaena exulans* (Hoch)	（チョウ目、マダラガ科）旧北区
Scotch grey		*Aedes alternans* (Westwood)	（ハエ目、カ科）豪州区
Scotch mountain burnet			Scotch burnet を見よ
Scotch ringlet			Scotch Argus を見よ
Scots greys			human lice を見よ
Scots pine adelges			pine adelgid を見よ
Scottish wood ant			northern wood ant を見よ
scraped Pilocrocis moth		*Pilocrocis ramentalis* Lederer	（チョウ目、メイガ科）新北区
scraptiid beetles		Scraptiidae	（コウチュウ目）の昆虫の総称

英名	和名	学名	所属、分布、ほか
scree weta		*Deinacrida connectens* (Ander)	(バッタ目、Stenopelmatidae) 豪州区
screech beetle		*Hygrobia herrmanni* (Fabricius)	(コウチュウ目、Hygrobiidae) 旧北区
screw pine scale			coconut scale (2) を見よ
screwworm		*Cochliomyia hominivorax* (Coquerel)	(ハエ目、クロバエ科) 新北区
screwworm flies		*Cochliomyia*	(ハエ目、クロバエ科) の昆虫の総称　幼虫は screwworms
screw-worm fly			screwworm を見よ
scribbled harvester		*Spalgis lemolea* Druce	(チョウ目、シジミチョウ科) エチオピア区
scribbled sallow moth		*Sympistis perscipta* (Guenée)	(チョウ目、ヤガ科) 新北区
scribbler moth		*Cladara atroliturata* (Walker)	(チョウ目、シャクガ科) 新北区
scribblygum moth		*Ogmograptis scribula* Meyrick	(チョウ目、チビガ科) 豪州区
scrofa hawk moth		*Hippotion scrofa* Boisduval	(チョウ目、スズメガ科) 豪州区
scrollwork pyralid moth		*Neocataclysta magnificalis* (Hübner)	(チョウ目、メイガ科) 新北区
scrub Euchlaena moth		*Euchlaena madusaria* (Walker)	(チョウ目、シャクガ科) 新北区
scrub hoppers		*Aeromachus*	(チョウ目、セセリチョウ科) の昆虫の総称
scruts			horse flies (1) を見よ
Scudder earwig	クギヌキハサミムシ	*Forficula scudderi* Bormans	(ハサミムシ目、クギヌキハサミムシ科) 日本、旧北区
Scudder's bush katydids			fork-tailed bush katydids を見よ
Scudder's decticid			robust shieldback を見よ
Scudder's duskywing		*Erynnis scudderi* (Skinner)	(チョウ目、セセリチョウ科) 新北区
Scudder's mantis		*Oligonicella scudderi* Saussure	(カマキリ目、Thespidae) 新北区
Scudder's scaly cricket		*Cycloptilum squamosum* Scudder	(バッタ目、コオロギ科) 新北区
Scudder's shieldback		*Eremopedes scudderi* Cockerell	(バッタ目、キリギリス科) 新北区
Scudder's willow sulphur	スカダーモンキチョウ	*Colias scudderii* Reakirt	(チョウ目、シロチョウ科) 新北区
sculptured buprestids	タマムシ亜科	Buprestinae	(コウチュウ目、タマムシ科) の昆虫の総称
sculptured moth		*Eumarozia malachitana* (Zeller)	(チョウ目、ハマキガ科) 新北区
sculptured pine beetle		*Chelcophora angulicollis* (LeConte)	(コウチュウ目、タマムシ科) 新北区。sculptured pine borer ともいう
sculptured pine borer			large flathead pine heartwood borer を見よ
scurfy scale		*Chionaspis furfura* (Fitch)	(カメムシ目、マルカイガラムシ科) 新北区
scurfy scale			rose scale を見よ
scurly scale			rose scale を見よ

英名	和名	学名	所属、分布、ほか
scutellerid bugs			shield-backed bugs を見よ
scuttle flies			humpbacked flies を見よ　豪州での英名
scydmaenid beetles			ant-like stone beetles を見よ
Scylla firetip		*Elbella scylla* (Ménétriès)	（チョウ目、セセリチョウ科）新熱帯区
scythrid moths			flower moths を見よ
seabeach flies			helcomyzid flies を見よ
seabord case		*Coleophora salinella* Stainton	（チョウ目、ツツミノガ科）旧北区
sea buckthorn hawkmoth		*Deilephila hippophaes* (Esper)	（チョウ目、スズメガ科）旧北区
sea flat-body		*Agonopterix cnicella* (Treitschke)	（チョウ目、マルハキバガ科）旧北区
seagrape borer moth		*Hexeris enhydris* (Grote)	（チョウ目、マドガ科）新北区
sea-grape scale		*Ephedraspis ephedrarum* (Lindinger)	（カメムシ目、マルカイガラムシ科）旧北区
seagrape spanworm moth		*Ametris nitocris* (Cramer)	（チョウ目、シャクガ科）新北区、新熱帯区
sea green swallowtail		*Papilio lorquinianus* C. et R. Felder	（チョウ目、アゲハチョウ科）東洋区
seashore earwig		*Anisolabis littorea* (White)	（ハサミムシ目、オオハサミムシ科）豪州区
sea-shore earwig			maritime earwig を見よ
seashore flies			beach flies (1) を見よ
seashore meadow katydid		*Conocephalus aigialus* Rehn et Hebard	（バッタ目、キリギリス科）新北区
seashore springtail		*Anurida maritima* (Guérin)	（トビムシ目、ヒメトビムシ科）新北区
seashore springtails		Anuridae	（トビムシ目）の昆虫の総称
seaside dragonlet		*Erythrodiplax berenice* (Drury)	（トンボ目、トンボ科）新北区、新熱帯区
seaside earwig			maritime earwig を見よ　米国での英名
seaside goldenrod borer moth		*Papaipema duovata* (Bird)	（チョウ目、ヤガ科）新北区
seaside grasshopper		*Trimerotropis maritima* (Harris)	（バッタ目、バッタ科）新北区
seaside meadow katydid		*Orchelimum fidicinium* Rehn et Hebard	（バッタ目、キリギリス科）新北区
sea-side plume		*Agdistis bennettii* (Curtis)	（チョウ目、トリバガ科）旧北区
seaside Thestor		*Thestor brachycerus* (Trimen)	（チョウ目、シジミチョウ科）エチオピア区
sea skaters		*Halobates*	（カメムシ目、アメンボ科）の昆虫の総称
seaweed flies	ハマベバエ科	Coelopidae	（ハエ目）の昆虫の総称
seaweed fly		*Coelopa frigida* (Fabricius)	（ハエ目、ハマベバエ科）旧北区
seaweed fly		*Coelopa pilipes* Haliday	（ハエ目、ハマベバエ科）旧北区
sea-wormwood case	ヨモギケブカツツミノガ	*Coleophora artemisiella* Scott	（チョウ目、ツツミノガ科）旧北区
seal lice			spiny sucking lice を見よ

英名	和名	学名	所属、分布、ほか
seal louse		*Echinophthirius horridus* (von Olfers)	(シラミ目、カイジュウジラミ科) 全北区
seam squirrel			body louse を見よ
searchers		Carabini	(コウチュウ目、オサムシ科) の昆虫の総称
seasonal leafwing		*Zaretis ellops* (Ménétriès)	(チョウ目、タテハチョウ科) 新熱帯区
seathorn hawk		*Celerio hippophaes* Esper	(チョウ目、スズメガ科) 旧北区。seathorn hawk-moth ともいう
secondary screwworm		*Cochliomyia macellaris* (Fabricius)	(ハエ目、クロバエ科) 新北区
secondary screw worms			blow flies を見よ
seconds			alienicola を見よ
secret lotus midge			lotus lily midge を見よ
sector-spotted piercer		*Pammene germmana* (Hübner)	(チョウ目、ハマキガ科) 旧北区
secucuni shadefly		*Coenyra rufiplaga* Trimen	(チョウ目、タテハチョウ科) エチオピア区
sedge flies			caddisflies (1) を見よ
sedge moths	ホソハマキモドキガ科	Glyphipterygidae	(チョウ目) の昆虫の総称
sedge root worm			variable reed beetle を見よ
sedge skipper		*Euphyes dion* (Edwards)	(チョウ目、セセリチョウ科) 新北区、新熱帯区
sedge sprite		*Nehalennia irene* (Hagen)	(トンボ目、イトトンボ科) 新北区
sedge tussock moth			reed tussock を見よ
sedge witch			Dun skipper (1) を見よ
sedgeling			pygmy damselfly を見よ　sedgling ともいう
seed bean maggot			bean seed fly を見よ　南アフリカでの幼虫の英名
seed beetle			groundnut bruchid を見よ
seed beetles			pea weevils を見よ
seed bugs		*Nysius*	(カメムシ目、ナガカメムシ科) の昆虫の総称
seed bugs (1)		*Oebalus*	(カメムシ目、カメムシ科) の昆虫の総称
seed bugs			chinch bugs (1) を見よ
seed chalcid fly			clover seed chalcid　(1) を見よ
seed chalcids (1)	オナガコバチ科	Torymidae	(ハチ目) の昆虫の総称
seed chalcids		*Megastigmus*	(ハチ目、オナガコバチ科) の昆虫の総称
seed chalcids			jointworms (2) を見よ
seed chalcids			beautiful parasites を見よ
seed-corn beetle		*Stenolophus lecontei* (Chaudoir)	(コウチュウ目、オサムシ) 新北区。播種したトウモロコシの種子を食べる
seed-corn beetle		*Clivina impressifrons* LeConte	(コウチュウ目、オサムシ科) 新北区。トウモロコシの種子害虫
seed-corn fly			bean seed fly を見よ
seedcorn maggot		*Hylemya cilicrura* (Rondani)	(ハエ目、ハナバエ科) 新北区。イネ、トウモロコシ害虫
seed-corn maggot (1)		*Delia floralis* (Fallén)	(ハエ目、ハナバエ科) 全北区

英名	和名	学名	所属、分布、ほか
seed corn maggot			bean seed fly を見よ　北米での幼虫の英名。seedcorn maggot とも記す。
seed-eating ground beetles		Amarini	(コウチュウ目、オサムシ科) の昆虫の総称
seed flies			fruit flies を見よ
seed harvester ant		*Pogonomyrmex californicus* (Buckley)	(ハチ目、アリ科) 新北区
seed midge			clover seed midge を見よ
seed moth			large common tubic を見よ
seed potato maggot		*Phorbia florilega* (Zetterstedt)	(ハエ目、ハナバエ科) 全北区。米国での幼虫の英名
seed wasps			jointworms (2) を見よ　米国での英名
seed wasps			seed chalcids (1) を見よ
seed weevils (1)		Lariidae	(コウチュウ目、ハムシ科) の昆虫の総称　マメゾウムシ科の異名
seed weevils (2)	ホソクチゾウムシ亜科	Apioninae	(コウチュウ目、ホソクチゾウムシ科) の昆虫の総称
seed weevils			leaf beetles　(1) を見よ
seed weevils			pea weevils を見よ
seed weevils			fruit weevils を見よ
seedharvesting ant		*Pheidole ampla* Forel	(ハチ目、アリ科) 豪州区
seedharvesting ant		*Pheidole anthracina* Forel	(ハチ目、アリ科) 豪州区
seedling bean midge		*Smittia aterima* (Meigen)	(ハエ目、ユスリカ科) 豪州区
seedling-gum moth		*Meganola metallopa* Meyrick	(チョウ目、コブガ科) 豪州区
seedling springtail			garden springtail を見よ
seedpod weevils	コバンゾウムシ亜科	Gymnaetrinae	(コウチュウ目、ゾウムシ科) の昆虫の総称
seepage dancer		*Argia bipunctulata* Hagen	(トンボ目、イトトンボ科) 新北区
Seever's short-winged katydid		*Dichopetala seeversi* Strohecker	(バッタ目、キリギリス科) 新北区
seine-making caddisflies			net-spinning caddisflies を見よ
Seiryori ambrosia beetle	セイリョウリキクイムシ	*Xyleborus seiryorensis* Murayama	(コウチュウ目、キクイムシ科) 日本、旧北区。クリ害虫。韓国の清涼里に由来
Sela skipper		*Mimardaris sela* Hewitson	(チョウ目、セセリチョウ科) 新熱帯区
Selene crescent		*Tegosa selene* (Rober)	(チョウ目、タテハチョウ科) 新熱帯区
self-heal pigmy		*Fedalmia headleyella* (Stainton)	(チョウ目、ホソガ科) 旧北区
Selicanis moth		*Selicanis cinereola* Smith	(チョウ目、ヤガ科) 新北区
Selina tiger-stripe		*Theritas selina* Hewitson	(チョウ目、シジミチョウ科) 新熱帯区
sellate thread-legged katydid		*Arethaea sellata* Rehn	(バッタ目、キリギリス科) 新北区
selva firetail		*Telebasis racenisi* Bick et Bick	(トンボ目、イトトンボ科) 新熱帯区
sematurid moths			American swallowtail moths を見よ

英名	和名	学名	所属、分布、ほか
semi-alate thread-legged katydid		*Arethaea semialata* Rehn et Hebard	(バッタ目、キリギリス科) 新北区
semi-barred slender		*Caloptilia semifascia* (Haworth)	(チョウ目、ホソガ科) 旧北区
semibrown Epitola		*Cerautola semibrunnea* (Bethune-Baker)	(チョウ目、シジミチョウ科) エチオピア区
semi-loopers			owlet moths を見よ　英国では Plusiinae に使用する
Seminole crescent		*Phyciodes texana seminole* (Edwards)	(チョウ目、タテハチョウ科) 新北区。Texan crescentspot を参照
semirelict underwing moth		*Catocala semirelicta* Grote	(チョウ目、ヤガ科) 新北区。semirelict underwing ともいう
semirufous tiger		*Premolis semirufa* Walker	(チョウ目、ヒトリガ科) 新熱帯区
semisocial sweat bees		*Lasioglossum*	(ハチ目、コハナバチ科) の昆虫の総称
semisocial sweat bees			sweat bees (2) を見よ
semi-tropical army worm		*Planotortrix excessana* Walker	(チョウ目、ハマキガ科) 豪州区
Semones hairstreak		*Semonina semones* (Godman et Salvin)	(チョウ目、シジミチョウ科) 新熱帯区
sendatsu bark beetle	センダツキクイムシ	*Xyleborus praevius* Blandford	(コウチュウ目、キクイムシ科) 日本、旧北区
senecio moth		*Nyctemera amica* (White)	(チョウ目、ヒトリガ科) 豪州区
Senegal golden dartlet			common bluetail (1) を見よ
sensitive fern borer moth		*Papaipema inquaesita* (Grote et Robinson)	(チョウ目、ヤガ科) 新北区
Senta skipper		*Neohesperilla senta* (Miskin)	(チョウ目、セセリチョウ科) 豪州区
Sentia sister		*Adelpha nea sentia* Godman et Salvin	(チョウ目、タテハチョウ科) 新熱帯区
sentinel arctic		*Oeneis excubitor* Troubridge, Philip, Scott et Shepart	(チョウ目、タテハチョウ科) 新北区
sentinel arctic (1)		*Oeneis alpina* Kurentzov	(チョウ目、タテハチョウ科) 全北区
Separata spritestreak			zebra teaser を見よ
separated sphinx		*Sphinx separata* Neumoegen	(チョウ目、スズメガ科) 新北区
sepia baskettail		*Epitheca sepia* (Gloyd)	(トンボ目、エゾトンボ科) 新北区
sepia dun		*Leptophlebia marginata* (Linnaeus)	(カゲロウ目、トビイロカゲロウ科) 旧北区
sepia tip hawk		*Likoma apicalis* Rothschild et Jordan	(チョウ目、スズメガ科) エチオピア区
Sepp's gold		*Micropteryx aruncella* (Scopoli)	(チョウ目、コバネガ科) 旧北区
Sepp's speedster		*Pseudonascus paulliniae* (Sepp)	(チョウ目、セセリチョウ科) 新熱帯区
sepsid flies			black scavenger flies を見よ
September thorn			cotton leaf caterpillar (2) を見よ

英名	和名	学名	所属、分布、ほか
Septima's clubtail		*Gomphus septima* Westfall	（トンボ目、サナエトンボ科）新北区
sequoia bark beetle		*Phloeosinus sequoiae* Hopkins	（コウチュウ目、キクイムシ科）新北区
sequoia pitch moth		*Synanthedon sequoiae* (H. Edwards)	（チョウ目、スカシバガ科）新北区
sequoia sphinx		*Sphinx sequoiae* Boisduval	（チョウ目、スズメガ科）新北区。sequoia sphinx moth ともいう
seram crow	セラムシロモンルリマダラ	*Euploea dentiplaga* Rothschild	（チョウ目、タテハチョウ科）東洋区
seram small tree-nymph		*Ideopsis klassika* Martin	（チョウ目、タテハチョウ科）東洋区
seraph moth		*Olceclostera seraphica* (Dyar)	（チョウ目、カイコガ科）新北区
seraphim		*Gonerilia seraphim* (Oberthür)	（チョウ目、シジミチョウ科）旧北区
seraphim (1)		*Lobophora halterata* (Hufnagel)	（チョウ目、シャクガ科）旧北区。日本亜種は *L. h. ijimai* Inoue シロシタヒメナミシャク
seraphim moth			seraphim (1) を見よ
Serene underwing moth		*Catocala serena* Edwards	（チョウ目、ヤガ科）新北区。Serene underwing ともいう
sergeant ants			bulldog ants を見よ
sergeant emperor		*Mimathyma chevana* (Moore)	（チョウ目、タテハチョウ科）旧北区、東洋区
sergeant-major	オスアカミスジ	*Abrota ganga* Moore	（チョウ目、タテハチョウ科）東洋区
Sergestus ruby-eye		*Talides sergestus* (Cramer)	（チョウ目、セセリチョウ科）新熱帯区
Sergia Euselasia		*Euselasia sergia sergia* (Godman et Salvin)	（チョウ目、シジミタテハ科）新熱帯区
Sergius firetip		*Pyrrhopyge sergius* Hopffer	（チョウ目、セセリチョウ科）新熱帯区
seriated Aethes moth		*Aethes seriatana* (Zeller)	（チョウ目、ホソハマキガ科）新北区、新熱帯区
seriate-punctate tiger moth	スジモンヒトリ	*Spilarctia seriatopunctata* Motschulsky	（チョウ目、ヒトリガ科）日本
Sericas		*Serica*	（コウチュウ目、コガネムシ科）の昆虫の総称
sericostomatids			bushtailed caddisflies を見よ
seroot flies			horse flies (1) を見よ
serpent ringtail		*Erpetogomphus lampropeltis* Kennedy	（トンボ目、サナエトンボ科）新北区
serpentflies			snake-flies (3) を見よ
serpent-flies			snake-flies (1) を見よ　serpentflies とも記される
serpentine leaf miner		*Liriomyza pusilla* (Meigen)	（ハエ目、ハモグリバエ科）旧北区
serpentine leaf miner			crucifer leafminer を見よ
serpentine leafminer			legume leafminer を見よ
serpentine leafminer moths	コハモグリガ科	Phyllocnistidae	（チョウ目）の昆虫の総称
serpentine miners			midget moths を見よ

英名	和名	学名	所属、分布、ほか
serpentine webworm moth		*Herpetogramma aeglealis* (Walker)	(チョウ目、メイガ科) 新北区
serphid wasps	シリボソクロバチ科	Serphidae	(ハチ目) の昆虫の総称
Serpia crescent		*Tegosa serpia* Higgins	(チョウ目、タテハチョウ科) 新熱帯区
serrate citrus stink bug		*Rhynchocoris serratus* Donovan	(カメムシ目、カメムシ科) 東洋区。カンキツ害虫
serrate-hindwinged geometrid	トビカギバエダシャク	*Luxiaria amasa* (Butler)	(チョウ目、シャクガ科) 日本、東洋区
serrate-horned sawfly			rose sawfly (1) を見よ
serrate longicorn beetle	ノコギリカミキリ	*Prinosus insularis* Motschulsky	(コウチュウ目、カミキリムシ科) 日本。マツ、モミなどの害虫。serrate longicorn ともいう
serrate margined stink bug	ノコギリカメムシ	*Megymenum gracilicornis* Dallas	(カメムシ目、キンカメムシ科) 日本、旧北区、東洋区
serrate-margined looper moth	ツマトビキエダシャク	*Bizia aexaria* Walker	(チョウ目、シャクガ科) 日本、旧北区、東洋区
serrate-winged hawk moth	ノコギリスズメ	*Laothoe amurensis* (Staudinger)	(チョウ目、スズメガ科) 日本、旧北区
serrated Eulithis moth		*Eulithis serrataria* (Barnes et McDunnough)	(チョウ目、シャクガ科) 新北区
serropalpid beetles		Serropalpidae	(コウチュウ目) の昆虫の総称　ナガクチキムシ科の異名
serviceberry leafroller moth		*Olethreutes appendiceum* (Zeller)	(チョウ目、ハマキガ科) 新北区
Serville swordtail	セルビレオナガタイマイ	*Eurytides serville* (Godart)	(チョウ目、アゲハチョウ科) 新熱帯区
Serville's lantern fly		*Laterinaria servillei* Spinosa	(カメムシ目、ビワハゴロモ科) 新熱帯区
Serville's piercer		*Cydia servillana* (Duponchel)	(チョウ目、ハマキガ科) 旧北区
sesame gall midge		*Asphondylia sesami* Felt	(ハエ目、タマバエ科) 東洋区、エチオピア区
sesame leafroller moth			sesame webworm を見よ
sesame-spotted longicor beetle	ゴマフカミキリ	*Mesosa myops* (Dalman)	(コウチュウ目、カミキリムシ科) 日本、旧北区。カンキツ、クリ、サクラなどの害虫。sesame-spotted longicorn ともいう
sesame webworm			gingelly borer moth を見よ
sesostris cattle-heart		*Parides sesostris* (Cramer)	(チョウ目、アゲハチョウ科) 新熱帯区
setaceous hebrew character			spotted cutworm moth (1) を見よ　英国での英名
seven-spot ladybird	ナナホシテントウ	*Coccinella septempunctata* Linnaeus	(コウチュウ目、テントウムシ科) 日本、全北区、東洋区。seven-spot ladybird beetle, seven-spotted beetle, seven-spotted lady beetle, seven-spotted ladybird, seven-spotted ladybug beetle (米国での英名) ともいう
seven-spurred mole crickets		*Gryllotalpa*	(バッタ目、ケラ科) の昆虫の総称
seventeen year cicada			seventeen-year locust (1) を見よ

英名	和名	学名	所属、分布、ほか
seventeen-year locust (1)		*Magicicada septendecim* (Linnaeus)	(カメムシ目、セミ科) 新北区。周期セミといわれる6種のうち、17年性3種のうちの1種
seventeen-year locust			little 17-year cicada を見よ
seventeen-year locust			Cassin's 17-year cicada を見よ
seventeen-year locust			little 17-year cicada を見よ
sewage springtail		*Hypogastrura viatica* (Tullberg)	(トビムシ目、ヒメトビムシ科) 豪州区
sexes			sexuales を見よ
sexton beetle		*Necrophorus vespilloides* Herbst	(コウチュウ目、シデムシ科) 旧北区
sexton beetle		*Necrophorus investigator* Zetterstedt	(コウチュウ目、シデムシ科) 新北区
sexton beetles			burying beetles (2) を見よ
sexuales	有性型		アブラムシ類などで雄虫と卵生無翅雌虫を指す
sexuparae	産性虫		単性雌虫で無翅幹母の卵から生ずる
Seychelles fluted scale			yellow cottony-cussion scale を見よ Seychelles scale ともいう
shad flies			mayflies を見よ　shad-flies とも記す
shade tree bagworm			bagworm (1) を見よ
shade-tree longhorn		*Stenodontes dasytomus* (Say)	(コウチュウ目、カミキリムシ科) 新北区
shaded broad-bar		*Scotopteryx chenopodiata* (Linnaeus)	(チョウ目、シャクガ科) 旧北区
shaded fanfoot	トビスジアツバ	*Herminia tarsicrinalis* (Knoch)	(チョウ目、ヤガ科) 日本、旧北区
shaded pug		*Eupithecia subumbrata* (Denis et Schiffermüller)	(チョウ目、シャクガ科) 旧北区
shaded Theope		*Theope publius incompositus* Hall	(チョウ目、シジミタテハ科) 新熱帯区
shadow damsels		*Drepanosticta*	(トンボ目、Platystictidae) の昆虫の総称
shadow darner		*Aeschna umbrosa* (Walker)	(トンボ目、ヤンマ科) 新北区
shadowy arches moth		*Drasteria adumbrata* (Behr)	(チョウ目、ヤガ科) 新北区。shadowy arches ともいう
shaft louse			chicken louse を見よ
shagbark hickory leafroller moth		*Pseudexentera cressoniana* (Clemens)	(チョウ目、ハマキガ科) 新北区
shagreened cutworm			pale-sided cutworm moth を見よ
shagreened slug moth		*Apoda biguttata* (Packard)	(チョウ目、イラガ科) 新北区
Shaka's ranger		*Kedestes chaca* (Trimen)	(チョウ目、セセリチョウ科) エチオピア区
Shaka's ranger			mountain sylph を見よ
Shaka's skipper			Shaka's ranger を見よ
Shaler's Fabiola moth		*Fabiola shaleriella* (Chambers)	(チョウ目、マルハキバガ科) 新北区
shallot aphid	ワケギコブアブラムシ	*Myzus ascalonicus* Doncaster	(カメムシ目、アブラムシ科) 日本、全北区、豪州区

英名	和名	学名	所属、分布、ほか
shallot fly			bean seed fly を見よ　北米での英名
Shandur fritillary		*Melitaea shandura* Evans	（チョウ目、タテハチョウ科）東洋区
Shandur rockbrown		*Chazara heydenreichi* (Lederer)	（チョウ目、タテハチョウ科）東洋区
Shannon-moth		*Asmicridea grisea* (Mosely)	（トビケラ目、シマトビケラ科）豪州区
shard beetles			dor beetles を見よ
shardborn beetle			dor beetle (1) を見よ
shark moth		*Cucullia umbratica* (Linnaeus)	（チョウ目、ヤガ科）旧北区。shark ともいう
sharks	セダカモクメ亜科	Cucullinae	（チョウ目、ヤガ科）の昆虫の総称
sharp angle shades moth		*Conservula anodonta* (Guenée)	（チョウ目、ヤガ科）新北区。sharp angle shades ともいう
sharp-angled carpet		*Euphyia unangulata* (Haworth)	（チョウ目、シャクガ科）旧北区。日本亜種は *E. u. gracilaria* (Bang Haas) フタテンツマジロナミシャク
sharp-angled carpet moth		*Euphyia intermediata* (Guenée)	（チョウ目、シャクガ科）新北区。sharp-angled carpet ともいう
sharp-angled Euselasia		*Euselasia angulata* (Bates)	（チョウ目、シジミタテハ科）新熱帯区
sharp-angled peacock		*Macaria alternata* (Denis et Schiffermüller)	（チョウ目、シャクガ科）旧北区
sharp-angled roller		*Ancylis subarcuana* Douglas	（チョウ目、ハマキガ科）旧北区
sharp-angled roller			double-arched roller を見よ
sharp banded skipper			spike-banded skipper を見よ
sharp-blotched Nola moth		*Nola pustulata* (Walker)	（チョウ目、コブガ科）新北区
sharp-dentated bark beetle			pine ips を見よ
sharp diving beetle	シャープゲンゴロウモドキ	*Dytiscus sharpi* Wehncke	（コウチュウ目、ゲンゴロウ科）日本
sharp green Hydriomena moth		*Hydriomena pluviata* (Guenée)	（チョウ目、シャクガ科）新北区
sharp-headed grain leafhopper			watercress sharpshooter (1) を見よ
sharp-lined powder moth		*Eufidonia discospilata* (Walker)	（チョウ目、シャクガ科）新北区
sharp-lined tussock moth			hardwood tussock moth を見よ
sharp-lined yellow moth		*Sicya macularia* (Harris)	（チョウ目、シャクガ科）新北区
sharp peacock angle			sharp-angled peacock を見よ
sharpshooter parasitoid		*Gonatocerus ashmeadi* Girault	（ハチ目、ホソハネコバチ科）新北区
sharp shooters			leafhoppers (1) を見よ
sharpshooters			leafhoppers (2) を見よ　米国での英名

英名	和名	学名	所属、分布、ほか
sharp-stigma looper moth		*Ctenoplusia oxygramma* (Geyer)	(チョウ目、ヤガ科) 新北区、新熱帯区。sharp stigma looper ともいう
sharp-tailed grasshopper		*Euchorthippus declivus* (Brisout de Barneville)	(バッタ目、バッタ科) 旧北区
sharp-toothed bark beetle			pine ips を見よ
sharp-veined white			common white を見よ　*P. rapae* Linnaeus の英名
sharp-veined white			mustard white を見よ
sharp-winged drill		*Dichrorampha acuminatana* (Liebig et Zeller)	(チョウ目、ハマキガ科) 旧北区
sharp-winged hawk moth			vine hawk-moth (1) を見よ
sharpen banded skipper		*Autochton zarex* (Hübner)	(チョウ目、セセリチョウ科) 新熱帯区
Shasta blue		*Aricia shasta* (W. H. Edwards)	(チョウ目、シジミチョウ科) 新北区。*A. s. pitkinensis* Ferris も同英名
Shasta daisy midge		*Contarinia chrysanthemi* (Kieffer)	(ハエ目、タマバエ科) 旧北区
Shasta Pyrausta moth		*Pyrausta perrubralis* (Packard)	(チョウ目、メイガ科) 新北区
shattered Hydriomena moth		*Hydriomena perfracta* Swett	(チョウ目、シャクガ科) 新北区
Shaw's dwarf		*Baltia shawi* (Bates)	(チョウ目、シロチョウ科) 東洋区
shears		*Hada plebeja* Linnaeus	(チョウ目、ヤガ科) 旧北区
shears (1)		*Hada nana* Hufnagel	(チョウ目、ヤガ科) 旧北区
shears moth			shears (1) を見よ
sheathed quaker moth		*Ulolonche culea* (Guenée)	(チョウ目、ヤガ科) 新北区
shedder bug		*Creontiades pallidus* (Rambur)	(カメムシ目、カスミカメムシ科) エチオピア区。ワタ、ピーナッツ害虫
sheep biting louse	ヒツジハジラミ	*Damalinia ovis* (Schrank)	(ハジラミ目、ケモノハジラミ科) 日本、汎世界。ヒツジに寄生
sheep blow fly			green blow fly (1) を見よ
sheep body louse			sheep biting louse を見よ
sheep bot fly	ヒツジバエ	*Oestrus ovis* (Linnaeus)	(ハエ目、ヒツジバエ科) 日本、汎世界。ヒツジに寄生。sheep bot ともいう
sheep chewing louse			sheep biting louse を見よ
sheep face louse		*Linognathus ovilus* (Newman)	(シラミ目、ケモノホソジラミ科) 汎世界
sheep foot louse		*Linognathus pedalis* (Osborn)	(シラミ目、ケモノホソジラミ科) 汎世界
sheep gadfly			sheep bot fly を見よ
sheep ked	ヒツジシラミバエ	*Melophagus ovinus* (Linnaeus)	(ハエ目、シラミバエ科) 日本、全北区。sheep ked fly ともいう
sheep louse			sheep biting louse を見よ
sheep louse			sheep ked を見よ

英名	和名	学名	所属、分布、ほか
sheep maggot			sheep bot fly を見よ 幼虫の英名
sheep maggot fly			green blow fly (1) を見よ
sheep maggot fly			sheep bot fly を見よ
sheep moth			common sheep moth を見よ
sheep nasal bot fly			sheep bot fly を見よ
sheep nasalfly			sheep bot fly を見よ
sheep nose bot			sheep bot fly を見よ 幼虫の英名
sheep nose fly			sheep bot fly を見よ
sheep nostril fly			sheep bot fly を見よ
sheep skipper		*Atrytonopsis edwardsii* Barnes et McDunnough	（チョウ目、セセリチョウ科）新熱帯区
sheep skipper (1)		*Atrytonopsis ovinia* (Hewitson)	（チョウ目、セセリチョウ科）新北区、新熱帯区
sheep spider-fly			sheep ked を見よ
sheep sucking body louse			sheep face louse を見よ
sheep sucking louse			sheep face louse を見よ
sheep tick			sheep ked を見よ
Sheldon's sallow tort			sallow Nycteoline を見よ
shell cricket		*Gryllus testaceous* Walker	（バッタ目、コオロギ科）旧北区
shell scallop			scallop shell を見よ
Shepherd's button	コアミメチャハマキ	*Acleris shepherdana* (Stephens)	（チョウ目、ハマキガ科）日本、旧北区
Shepherd's footman		*Termessa shepherdi* Newmann	（チョウ目、ヒトリガ科）豪州区
Shepherd's fritillary	ヒメヒョウモン	*Boloria pales* (Schiffermüller)	（チョウ目、タテハチョウ科）旧北区
Sheppard's plume moth		*Geina sheppardi* Landry	（チョウ目、トリバガ科）新北区
Sheridan's hairstreak			white-lined green hairstreak を見よ
sherry spinner			blue winged olive を見よ
shichito mat grass pyralid			mat grass pyralid を見よ 七島藺に由来
shichito mat grass sawfly	オオオスグロハバチ	*Dorelus harukawai* Waterston	（ハチ目、ハバチ科）日本
shield-back bugs		*Stiretrus*	（カメムシ目、カメムシ科）の昆虫の総称
shieldback mayflies		Prosopistomatoidea	（カゲロウ目）の昆虫の総称
shield-backed bug		*Stethaulax marmorata* (Say)	（カメムシ目、キンカメムシ科）新北区
shield-backed bugs	キンカメムシ科	Scutelleridae	（カメムシ目）の昆虫の総称 shield-back bugs ともいう
shield-backed grasshoppers			crickets (3) を見よ

英名	和名	学名	所属、分布、ほか
shield-backed grasshoppers			shield-backed katydids を見よ
shield-backed grasshoppers			long-horned grasshoppers (1) を見よ
shield-backed katydids		Decticinae	（バッタ目、キリギリス科）の昆虫の総称　キリギリスよりコオロギに似る。捕まえるとかむことがあるので、欧州では wart biters という
shield-backed katydids			crickets (3) を見よ
shieldbacked pine seedbug		*Tetyra bipunctata* (Herrich-Schäffer)	（カメムシ目、カメムシ科）新北区
shield bearer moths	ツヤコガ科	Heliozelidae	（チョウ目）の昆虫の総称
shield bearers		*Coptodisca*	（チョウ目、ホソガ科）の昆虫の総称
shield bearers			shield bearer moths を見よ
shield beetles			leaf beetles (1) を見よ
shield bugs			stink bugs を見よ
shield bugs			true bugs を見よ
shield bugs			shield-backed bugs を見よ
shield bugs			ground bugs を見よ
shieldbugs	ツノカメムシ科	Acanthosomatidae	（カメムシ目）の昆虫の総称
shieldbugs			stink bugs を見よ　英国での英名
shiia bark beetle			castanopsis ambrosia beetle (1) を見よ
shiia longicorn beetle	ヒシカミキリ	*Microlera ptinoides* Bates	（コウチュウ目、カミキリムシ科）日本
shiia oystershell scale	ハムグリカキカイガラムシ	*Andaspis crawii* (Cockerell)	（カメムシ目、マルカイガラムシ科）日本、東洋区
shiitake crane fly	シイタケガガンボ	*Ula shiitakea* Nobuchi	（ハエ目、ガガンボ科）日本。シイタケ害虫
shiitake fungus gnat	シイタケトンボキノコバエ	*Exechia shiitakevora* Okada	（ハエ目、キノコバエ科）日本。シイタケ害虫
shiitake fungus moth	シイタケオオヒロズコガ	*Morophagoides moriutii* Robinson	（チョウ目、ヒロズコガ科）日本。シイタケ害虫
shimmering Adoxophyes moth		*Adoxophyes negundana* (McDunnough)	（チョウ目、ハマキガ科）新北区
shimmering forester		*Bebearia micans* (Aurivillius)	（チョウ目、タテハチョウ科）エチオピア区
shiner			German cockroach を見よ
shiner steamfly			German cockroach を見よ
shineworm			glow-worm (2) を見よ
shineworms			fireflies (1) を見よ
shining amazon			shining slave maker を見よ
shining blue forester		*Bebearia carshena* (Hewitson)	（チョウ目、タテハチョウ科）エチオピア区
shining blue nymph	メスキオビタテハ	*Cynandra opis* Drury	（チョウ目、タテハチョウ科）エチオピア区
shining brown dwarf		*Perittia obscurepunctella* (Stainton)	（チョウ目、クサモグリガ科）旧北区
shining-brown smoke		*Bacotia sepium* (Speyer)	（チョウ目、ミノガ科）旧北区

英名	和名	学名	所属、分布、ほか
shining Cerulean		*Jamides amarauge* Druce	（チョウ目、シジミチョウ科）豪州区
shining clubtail		*Stylurus ivae* Williamson	（トンボ目、サナエトンボ科）新北区
shining Dichomeris moth		*Dichomeris ochripalpella* (Zeller)	（チョウ目、キバガ科）新北区
shining flower beetles	ヒメハナムシ科	Phalacridae	（コウチュウ目）の昆虫の総称
shining fungus beetles	デオキノコムシ科	Scaphidiidae	（コウチュウ目）の昆虫の総称
shining fungus beetles			shining flower beetles を見よ
shining fungus beetles			rove beetles を見よ　稀な使用
shining Gossamerwing		*Euphaea splendens* Hagen	（トンボ目、Euphaeidae）東洋区
shining groundstreak		*Calycopis demonassa* (Hewitson)	（チョウ目、シジミチョウ科）新熱帯区
shining jet black ant			field ant を見よ　英国での英名
shining leaf beetles			asparagus beetles を見よ
shining leaf chafers			flower beetles (3) を見よ
shining Macromia		*Macromia splendens* (Pictet)	（トンボ目、ヤマトンボ科）旧北区。shining Macromia dragonfly ともいう
shining oak-blue			common oak blue を見よ
shining pencil-blue		*Candalides helenita* (Semper)	（チョウ目、シジミチョウ科）豪州区
shining pitch bell		*Epinotia caprana* (Fabricius)	（チョウ目、ハマキガ科）旧北区
shining plushblue		*Flos diardi* (Hewitson)	（チョウ目、シジミチョウ科）東洋区。*F. d. capeta* Hewitson も同英名
shining plushblue (1)		*Flos fulgida* (Hewitson)	（チョウ目、シジミチョウ科）東洋区
shining purplewing	ホソバアイイロタテハ	*Eunica alpais excelsa* Godman et Salvin	（チョウ目、タテハチョウ科）新熱帯区
shining red Charaxes		*Charaxes zingha* (Stoll)	（チョウ目、タテハチョウ科）エチオピア区
shining slave maker		*Polyerges lucidus* (Mayr)	（ハチ目、アリ科）新北区
shining sweep		*Psyche casta* (Pallas)	（チョウ目、ミノガ科）全北区
shining wasps		Masaridae	（ハチ目）の昆虫の総称
shining whin midget		*Phyllonorycter ulicicolella* (Stainton)	（チョウ目、ホソガ科）旧北区
shiny bluebottle fly			blow fly を見よ
shiny cereal weevils		*Nematocerus*	（コウチュウ目、ゾウムシ科）の昆虫の総称
shiny chafer	ツヤコガネ	*Anomala lucens* Ballion	（コウチュウ目、コガネムシ科）日本。ムギ、トウモロコシ、リンゴ、スギなどの害虫
shiny gray carpet moth		*Stamnodes gibbicostata* (Walker)	（チョウ目、シャクガ科）新北区
shiny head-standing moths	メムシガ科	Argyresthiidae	（チョウ目）の昆虫の総称
shiny pasture scarab		*Scitala sericans* Erichson	（コウチュウ目、コガネムシ科）豪州区
shiny spider beetle		*Mezium affine* Boieldieu	（コウチュウ目、シバンムシ科）豪州区

英名	和名	学名	所属、分布、ほか
shiny-spotted bob		*Isoteinon lamprospilus formosanus* Fruhstorfer	(チョウ目、セセリチョウ科) 東洋区
shiny wings		Stilbopterygidae	(アミメカゲロウ目) の昆虫の総称
ship cockroach			American cockroach を見よ
ship-timber beetle			timber beetle を見よ
ship-timber beetles	ツツシンクイ科	Lymexylonidae	(コウチュウ目) の昆虫の総称
shipworm			timber beetle を見よ　幼虫の英名
Shirai Chinese sumac gall aphid	ヌルデノハナフシアブラムシ	*Nurudea shiraii* (Matsumura)	(カメムシ目、アブラムシ科) 日本
Shirane bamboo scale	ハラビロナガカイガラムシ	*Nikkoaspis shiranensis* Kuwana	(カメムシ目、マルカイガラムシ科) 日本
Shiva's sunbeam	インドウラギンシジミ	*Curetis siva* Evans	(チョウ目、シジミチョウ科) 東洋区
shogun ambrosia beetle	ショウグンキクイムシ	*Scolytoplatypus shogun* Blandford	(コウチュウ目、キクイムシ科) 日本、東洋区。木材害虫。将軍に由来
shogun bark beetle			shogun ambrosia beetle を見よ
shootfly			sorghum shoot fly を見よ
shoot-tip-fly		*Lonchaea chalybea* Wiedemann	(ハエ目、クロツヤバエ科) 新熱帯区。キャッサバ害虫
shore bug			common shore bug を見よ
shore bugs	ミズギワカメムシ科	Saldidae	(カメムシ目) の昆虫の総称
shore doubtful marble			thrift tortrix moth を見よ
shore flies	ミギワバエ科	Ephydridae	(ハエ目) の昆虫の総称
shore fly			brine fly (1) を見よ
shore wainscot		*Mythimna litoralis* (Curtis)	(チョウ目、ヤガ科) 旧北区
short-banded sailer		*Phaedyma columella* (Cramer)	(チョウ目、タテハチョウ科) 東洋区。*P. c. singa* Fruhstorfer、*P. c. parvimacula* Pendlebury も同英名
short-barred grey marble		*Apotomis semifasciana* (Haworth)	(チョウ目、ハマキガ科) 旧北区
short-barred sapphire		*Iolaus aemulus* Trimen	(チョウ目、シジミチョウ科) エチオピア区
shortbeaked clover aphid			clover aphid を見よ
short-circuit beetle			lead cable borer を見よ
short-clasped treble-bar		*Anaitis efformata* Guenée	(チョウ目、シャクガ科) 旧北区
short-cloaked black arches			short-cloaked moth を見よ
short-cloaked moth		*Nola cucullatella* (Linnaeus)	(チョウ目、コブガ科) 旧北区
short-faced scorpionflies			scorpion flies (1) を見よ
short flash		*Rapala buxaria* de Nicéville	(チョウ目、シジミチョウ科) 東洋区
short-haired bat bug		*Cimex dissimilis* (Horvath)	(カメムシ目、トコジラミ科) 旧北区

英名	和名	学名	所属、分布、ほか
short-haired humble-bee		*Bombus latreillellus* Kirby	（ハチ目、ミツバチ科）旧北区、豪州区
short heel flies			small dung flies を見よ
shorthorn grasshoppers			grasshoppers (3) を見よ
short-horned black legionnaire	ヒゲブトルリミズアブ	*Beris fuscipes* Meigen	（ハエ目、ミズアブ科）日本、旧北区
short-horned bugs		*Cryptocerata*	（カメムシ目、ミズギワカメムシ科）の昆虫の総称　water bugs ともいう
short-horned flies	短角亜目	*Brachycera*	（ハエ目）の昆虫の総称
short-horned grasshopper		*Metalepta brevicornis* (Johnson)	（バッタ目、バッタ科）新北区
short-horned grasshoppers			grasshoppers (1) を見よ
short-horned leaf beetles			case-bearing leaf beetles (1) を見よ
short-horned locusts			grasshoppers (1) を見よ
short-horned springtails	ミジントビムシ科	*Neelidae*	（トビムシ目）の昆虫の総称　朽木や洞に生息。日本から2種
short-horns			short-horned flies を見よ
short leafcutter bee		*Megachile brevis* Say	（ハチ目、ハキリバチ科）新北区
shortleaf pine cone borer		*Eucosma cocana* Kearfott	（チョウ目、ハマキガ科）新北区。shortleaf pinecone borer moth ともいう
short-lined chocolate moth		*Argyrostrotis anilis* (Drury)	（チョウ目、ヤガ科）新北区。short-lined chocolate ともいう
short-lined kite-swallowtail	アゲシラウスオナガタイマイ	*Eurytides agesilaus fortis* (Rothschild et Jordan)	（チョウ目、アゲハチョウ科）新熱帯区。*E. a. neosilaus* (Hopffer)　も同英名
short nosed cattle louse	ウシジラミ	*Haematopinus eurysternus* (Nitzsch)	（シラミ目、ケモノジラミ科）日本、汎世界。ウシに寄生
short-nosed ox louse			short nosed cattle louse を見よ　米国での英名
short-nosed weevil		*Brachyderes incanus* (Linnaeus)	（コウチュウ目、ゾウムシ科）旧北区
short-nosed weevils			scarred snout beetles を見よ
short-palped craneflies		*Limnophiinae*	（ハエ目、ガガンボ科）の昆虫の総称
short-tail skipper		*Polythrix caunus* (Herrich-Schäffer)	（チョウ目、セセリチョウ科）新熱帯区
short-tail thread-legged katydid		*Arethaea brevicauda* (Scudder)	（バッタ目、キリギリス科）新北区
short-tailed admiral	アカタテハモドキ	*Antanartia hippomene hippomene* (Hübner)	（チョウ目、タテハチョウ科）エチオピア区
short-tailed Arizona skipper		*Zestusa dorus* (Edwards)	（チョウ目、セセリチョウ科）新北区、エチオピア区
short-tailed black swallowtail			Indra swallowtail を見よ
short-tailed blue		*Everes argiades* (Pallas)	（チョウ目、シジミチョウ科）旧北区、東洋区。日本亜種は *E. a. hellotia* (Ménétriès) ツバメシジミ。short-tailed blue butterfly ともいう

英名	和名	学名	所属、分布、ほか
short-tailed bumblebee			short-haired bumblebee を見よ
short-tailed cricket		*Anurogryllus muticus* (De Geer)	(バッタ目、コオロギ科) 新北区
short-tailed crickets		Brachytrupinae	(バッタ目、コオロギ科) の昆虫の総称
short-tailed crickets		*Anurogryllus*	(バッタ目、コオロギ科) の昆虫の総称
short-tailed crickets			crickets (1) を見よ
short-tailed damselflies			dancers (1) を見よ
short-tailed field cricket		*Gryllus brevicaudus* Weissman, Rentz, Alexander et Loher	(バッタ目、コオロギ科) 新北区
short-tailed flasher		*Astraptes brevicauda* (Plötz)	(チョウ目、セセリチョウ科) 新熱帯区
short-tailed ichneumons			common ichneumonids を見よ
short-tailed line-blue			Felder's line blue (2) を見よ
short-tailed shieldback		*Decticita brevicauda* (Caudell)	(バッタ目、キリギリス科) 新北区
short-tailed sister		*Adelpha diocles* Godman et Salvin	(チョウ目、タテハチョウ科) 新熱帯区。tailed sister ともいう
short-tailed skipper			short-tailed Arizona skipper を見よ
short-tailed stonefly	ミジカオカワゲラ	*Strophopteryx nohirae* (Okamoto)	(カワゲラ目、ミジカオカワゲラ科) 日本
short-tailed swallowtail		*Papilio brevicauda* Saunders	(チョウ目、アゲハチョウ科) 新北区
short tongued bees			andrenid bees を見よ
short tongued bees			yellow-faced bees (1) を見よ
short tongued bees			sweat bees (1) を見よ
short-tongued burrowing bees			andrenid bees を見よ
short-winged beetles			staphylinoid beetles を見よ
short-winged blister beetle		*Meloe angusticollis* Say	(コウチュウ目、ツチハンミョウ科) 新北区
short-winged bush cricket		*Hapithus brevipennis* (Saussure)	(バッタ目、コオロギ科) 新北区
short-winged cone-head		*Conocephalus dorsalis* (Latreille)	(バッタ目、キリギリス科) 旧北区
short-winged coneheads		*Belocephalus*	(バッタ目、キリギリス科) の昆虫の総称
short-winged crickets			scaly crickets を見よ
short-winged earwig (1)		*Apterygida albipennis* Megerle	(ハサミムシ目、クギヌキハサミムシ科) 旧北区
short-winged earwig (2)		*Apterygida media* (Hagenbach)	(ハサミムシ目、クギヌキハサミムシ科) 旧北区
short-winged firefly			lesser glow-worm を見よ

英名	和名	学名	所属、分布、ほか
short-winged gourd-shaped bug	コバネヒョウタンナガカメムシ	*Togo hemipterus* (Scott)	（カメムシ目、ナガカメムシ科）日本、旧北区
short-winged katydid	コバネヒメギス	*Chizuella bonneti* (Bolivar)	（バッタ目、キリギリス科）日本、旧北区
short-winged katydids		*Dichopetala*	（バッタ目、キリギリス科）の昆虫の総称
short-winged longicorn beetle	コバネカミキリ	*Psephactus remiger remiger* Harold	（コウチュウ目、カミキリムシ科）日本、東洋区
short-winged meadow katydid			meadow grasshopper (1) を見よ
short-winged mold beetles			antloving beetles (1) を見よ
short-winged mole cricket		*Scapteriscus abbreviatus* Scudder	（バッタ目、ケラ科）新北区
short-winged rice katydid	コバネササキリ	*Conocephalus japonicus* (Redtenbacher)	（バッタ目、キリギリス科）日本、東洋区
short-winged shieldback			short-tailed shieldback を見よ
short-winged weevils		Pterocolinae	（コウチュウ目、ゾウムシ科）の昆虫の総称
shot-borer			shot-hole borer (1)(3) を見よ
shot borers			pinhole beetles を見よ
shothole borer (1)	リンゴカワノキクイムシ	*Scolytus rugulosus* (Ratzeburg)	（コウチュウ目、キクイムシ科）全北区
shot-hole borer (2)		*Xyleborus affinis* Eichhoff	（コウチュウ目、キクイムシ科）旧北区
shot-hole borer (3)		*Xyleborus dispar* (Fabricius)	（コウチュウ目、キクイムシ科）旧北区
shot hole borers			ambrosia beetles を見よ
shot-hole borers			false powderpost beetles を見よ
shot-hole borers			bark beetles (1)(2) を見よ
shot-hole borers			pinhole beetles を見よ
shot-hole gall midge		*Helicomyia pierrei* (Kieffer)	（ハエ目、タマバエ科）旧北区
shot hole midge		*Rhabdophaga purpureaperda* Barnes	（ハエ目、タマバエ科）旧北区
shoulder-streaked firetip		*Pyrrhopyge papius* Hopffer	（チョウ目、セセリチョウ科）新熱帯区
shoulder-stripe moth		*Anticlea badiata* (Denis et Schiffermüller)	（チョウ目、シャクガ科）旧北区。shoulder stripe ともいう
shoulder-striped carpet			shoulder-stripe moth を見よ
shoulder-striped clover			flax budworm を見よ
shoulder-striped clover moth			flax budworm を見よ
shoulder-striped wainscot		*Mythimna comma* (Linnaeus)	（チョウ目、ヤガ科）旧北区
shouldered brown		*Heteronympha penelope* Waterhouse	（チョウ目、タテハチョウ科）豪州区。shouldered brown butterfly ともいう

英名	和名	学名	所属、分布、ほか
shower of gold		*Argyrogrammana sticheli* (Talbot)	(チョウ目、シジミタテハ科) 新熱帯区
showy emerald moth		*Dichorda iridaria* (Guenée)	(チョウ目、シャクガ科) 新北区
showy Holomelina		*Holomelina ostenta* (H. Edwards)	(チョウ目、ヒトリガ科) 新北区
showy wave moth		*Idaea ostentaria* (Walker)	(チョウ目、シャクガ科) 新北区。showy wave ともいう
shrew louse		*Polyplax reclinata* (Nitzsch)	(シラミ目、フトゲシラミ科) 旧北区
shrill carder-bee		*Bombus sylvarum* Linnaeus	(ハチ目、ミツバチ科) 旧北区
shrill house cricket		*Brachytrypes megacephalus* Lefebvre	(バッタ目、コオロギ科) 旧北区
shrub katydid	ヤブキリ	*Tettigonia orientalis orientalis* Uvarov	(バッタ目、キリギリス科) 日本
shuttle-shaped dart		*Agrotis puta* (Hübner)	(チョウ目、ヤガ科) 旧北区
shy bug			hop capsid を見よ
shy cosmet	ガマトガリホソガ	*Limnaecia phragmitella* Stainton	(チョウ目、カザリバガ科) 日本、全北区。shy cosmet moth ともいう
shy emerald damselfly			southern emerald damselfly を見よ
shy Saliana		*Saliana longirostris* (Sepp)	(チョウ目、セセリチョウ科) 新熱帯区
shy yellow		*Eurema messalina* (Fabricius)	(チョウ目、シロチョウ科) 新北区、新熱帯区
sialids			alderflies (1) を見よ
Siam babul blue		*Azanus urios* Riley et Godfrey	(チョウ目、シジミチョウ科) 東洋区
Siam tree nymph	オオゴマダラ	*Idea leuconoe* Erichson	(チョウ目、タテハチョウ科) 東洋区
Siamese grain beetle	ホソチビコクヌスト	*Lophocateres pusillus* (Klug)	(コウチュウ目、コクヌスト科) 日本、全北区、東洋区、エチオピア区。貯穀害虫
Siberian cutworm	ウスグロヤガ	*Euxoa sibirica* Boisduval	(チョウ目、ヤガ科) 日本、旧北区。トウモロコシ、ダイズ、アブラナ科野菜などの害虫
Siberian fir bark-beetle			larch ips を見よ
Siberian hawker	オオルリボシヤンマ	*Aeschna crenata* Hagen	(トンボ目、ヤンマ科) 日本、旧北区
Siberian onion moth		*Acrolepia alliella* Kuznetsov	(チョウ目、スガ科) 旧北区
Siberian silkworm moth		*Dendrolimus sibiricus* Chetverikov	(チョウ目、カレハガ科) 旧北区
Siberian white admiral		*Limenitis sydyi* Lederer	(チョウ目、タテハチョウ科) 旧北区
Siberian winter	オツネントンボ	*Sympecma paedisca* (Brauer)	(トンボ目、アオイトトンボ科) 日本、旧北区。Siberian winter damsel ともいう
sibling Mellana		*Quasimellana siblinga* Burns	(チョウ目、セセリチョウ科) 新熱帯区
Sichuan black-veined white		*Aporia martineti* (Oberthür)	(チョウ目、シロチョウ科) 旧北区

英名	和名	学名	所属、分布、ほか
Sicilian marbled white		*Melanargia pherusa* (Boisduval)	（チョウ目、タテハチョウ科）旧北区
Sicilian mealybug parasite		*Leptomastidea abnormis* (Girault)	（ハチ目、トビコバチ科）新北区、大洋区
sickle			outlet sword を見よ
sickle-bearing bush cricket			black-horned katydid を見よ
sickle spreadwing		*Lestes uncifer* Karsch	（トンボ目、アオイトトンボ科）エチオピア区
sickle-winged Morpho			blue Morpho (1) を見よ
sickle-winged skipper		*Achlyodes thraso* (Hübner)	（チョウ目、セセリチョウ科）新北区
sickle-winged skipper			northern sicklewing を見よ
sierolomorphid wasps		Sierolomorphidae	（ハチ目）の昆虫の総称
Sierra blue		*Agriades glandon podarce* (C. et R. Felder)	（チョウ目、シジミチョウ科）新北区。
Sierra green sulphur			Behr's sulphur を見よ
Sierra Madre dancer		*Argia lacrimans* (Hagen)	（トンボ目、イトトンボ科）新北区
Sierra Nevada blue		*Plebejus podarce* (C. et R. Felder)	（チョウ目、シジミチョウ科）新北区
Sierra Nevada blue			Nevada blue を見よ
Sierra Nevada Parnassian		*Parnassius behrii* W. H. Edwards	（チョウ目、アゲハチョウ科）新北区
Sierra shieldback		*Neduba sierranus* (Rehn et Hebard)	（バッタ目、キリギリス科）新北区
Sierra skipper		*Hesperia miriamae* MacNeill	（チョウ目、セセリチョウ科）新北区
Sierra worm-lion		*Vermileo comstocki* Wheeler	（ハエ目、シギアブ科）新北区。山地に生息
Sierran blue-winged grasshopper		*Circotettix thalassinus* Saussure	（バッタ目、バッタ科）新北区
Sierran dampwood termite			Nevada dampwood termite を見よ
Sierran pericopid		*Gnophaela latipennis* (Boisduval)	（チョウ目、ヒトリガ科）新北区。Pericopinae に由来
Sifax skipper		*Phlebodes campo sifax* Evans	（チョウ目、セセリチョウ科）新熱帯区
sigma darner		*Austroaeschna sigma* Theischinger	（トンボ目、ヤンマ科）豪州区
sigmoid fungus beetle		*Cryptophagus varus* Woodroffe et Coombs	（コウチュウ目、キスイムシ科）新北区
sigmoid prominent			rusty-lined leaf-tier を見よ　sigmoid prominent moth ともいう。
signals			sun moths (2) を見よ
signate lady beetle		*Hyperaspis signata* (Olivier)	（コウチュウ目、テントウムシ科）新北区、大洋区

英名	和名	学名	所属、分布、ほか
signate looper moth		*Rindgea s-signata* (Packard)	(チョウ目、シャクガ科) 新北区
signate Melanolophia moth		*Melanolophia signataria* (Walker)	(チョウ目、シャクガ科) 新北区
signate quaker moth		*Tricholita signata* (Walker)	(チョウ目、ヤガ科) 新北区
signiphorid parasites		Signiphoridae	(ハチ目) の昆虫の総称
Sikkim ace		*Halpe sikkima* Moore	(チョウ目、セセリチョウ科) 東洋区
Sikkim dart		*Potanthus mara* Evans	(チョウ目、セセリチョウ科) 東洋区
Sikkim freak		*Calinaga gautama* Moore	(チョウ目、タテハチョウ科) 東洋区
Sikkim white flat		*Seseria sambara* (Moore)	(チョウ目、セセリチョウ科) 東洋区
silent leaf-runner			dwarf cricket を見よ
silent scaly cricket		*Oligacanthopus prograptus* Rehn et Hebard	(バッタ目、コオロギ科) 新北区
silk button spangle gall			silk-botton spangle gall wasp がカシ類の葉の下面につくる虫えいで絹毛に覆われる。旧北区
silk-button spangle gall wasp			spangle gall wasp を見よ
silk lacewings	キヌバカゲロウ科	Psychopsidae	(カゲロウ目) の昆虫の総称
silken fungus beetles	キスイムシ科	Cryptophagidae	(コウチュウ目) の昆虫の総称
silken-tube spinners			finger-net caddisflies を見よ
silkmoth			silkworm を見よ
silkmoths		*Argema*	(チョウ目、ヤママユガ科) の昆虫の総称
silkmoths			giant silkworm moths を見よ
silkworm	カイコ	*Bombyx mori* (Linnaeus)	(チョウ目、カイコガ科) 日本、汎世界
silkworm maggot			silkworm tachina fly (2) を見よ
silkworm moth	カイコガ		silkworm を見よ
silkworm moths	カイコガ科	Bombycidae	(チョウ目) の昆虫の総称　世界に約 100 種
silkworm seed			蚕卵。silkworm を見よ
silkworm tachina fly (1)	カイコノクロウジバエ	*Pales pavida* (Meigen)	(ハエ目、ヤドリバエ科) 日本、旧北区。カイコに寄生
silkworm tachina fly (2)	カイコノウジバエ	*Blepharipa sericareae* (Rondani)	(ハエ目、ヤドリバエ科) 日本、旧北区、東洋区。カイコなどに寄生
silkworm uji-fly			silkworm tachina fly (2) を見よ
silky ammophila		*Sphex holoserica* Fabricius	(ハチ目、アナバチ科) 新北区
silky ant		*Formica fusca* Linnaeus	(ハチ目、アリ科) 新北区。体表に微毛をもつことから
silky azure		*Ogyris oroetes* Hewitson	(チョウ目、シジミチョウ科) 豪州区
silky cane weevil		*Metamasius hemipterus sericeus* (Olivier)	(コウチュウ目、ゾウムシ科) 新北区。*M. hemipterus* (Linnaeus) は West Indian cane weevil といわれる
silky cane weevil			West Indian sugarcane weevil を見よ
silky ciliate blue		*Anthene lachares lachares* (Hewitson)	(チョウ目、シジミチョウ科) エチオピア区
silky Emesis		*Emesis brimo progne* (Godman)	(チョウ目、シジミタテハ科) 新熱帯区。

英名	和名	学名	所属、分布、ほか
silky hairstreak		*Pseudalmenus chlorinda* (Blanchard)	(チョウ目、シジミチョウ科) 豪州区
silky jewel			Diggle's blue を見よ
silky lacewings			silk lacewings を見よ
silky oak leafminer		*Peraglyphis atimana* (Meyrick)	(チョウ目、ハマキガ科) 豪州区
silky oakblue		*Amblypodia alax* Evans	(チョウ目、シジミチョウ科) 東洋区。
silky owl		*Taenaris catops* Westwood	(チョウ目、タテハチョウ科) 豪州区
silky ringlet	キヌツヤベニヒカゲ	*Erebia gorge* (Esper)	(チョウ目、タテハチョウ科) 旧北区
silky sallow moth		*Chaetaglaea sericea* (Morrison)	(チョウ目、ヤガ科) 新北区。silky sallow ともいう
silky skipper		*Semalea pulvina* (Plötz)	(チョウ目、セセリチョウ科) エチオピア区
silky skipper			dark skipper を見よ
silky wainscot		*Chilodes maritimus* (Tauscher)	(チョウ目、ヤガ科) 旧北区
silky wanderer	キイロトガリバシロチョウ	*Leptophobia eleone* (Doubleday)	(チョウ目、シロチョウ科) 新熱帯区
silky wave		*Idaea dilutaria* (Hübner)	(チョウ目、シャクガ科) 旧北区
silphius borer moth		*Papaipema silphii* Bird	(チョウ目、ヤガ科) 新北区
silurian		*Eriopygodes imbecilla* (Fabricius)	(チョウ目、ヤガ科) 旧北区。silurian moth ともいう
silver			fir budworm を見よ
silver arrowhead copper			foxtrot copper を見よ
silver-banded hairstreak		*Chlorostrymon simaethis* (Drury)	(チョウ目、シジミチョウ科) 新北区。新熱帯区。*C. s. sartia* (Skinner) も同英名
silver-banded sister			Ethelda sister を見よ
silver-banded snipefly		*Chrysopilus erythrophthalmus* Loew	(ハエ目、シギアブ科) 旧北区
silver-barred Charaxes		*Charaxes druceanus* Butler	(チョウ目、タテハチョウ科) エチオピア区
silver barred Nephele		*Nephele argentifera* (Walker)	(チョウ目、スズメガ科) エチオピア区
silver-barred sable		*Pyrausta cingulata* (Linnaeus)	(チョウ目、メイガ科) 旧北区
silver bars	フタスジコヤガ	*Deltote bankiana* (Fabricius)	(チョウ目、ヤガ科) 日本、旧北区。silver barred ともいう
silver-belly dagger moth	シロハラケンモン	*Plataplecta pulverosa* (Hampson)	(チョウ目、ヤガ科) 日本、旧北区
silver-berry capitophorus			artichoke aphid を見よ
silver-berry curculio	シロモンシギゾウムシ	*Curculio alboscutellatus* (Roelofs)	(コウチュウ目、ゾウムシ科) 日本、東洋区
silver-berry elongate scale			false silver berry scale を見よ
silver-berry sucker	グミキジラミ	*Psylla eleagni* Kuwayama	(カメムシ目、キジラミ科) 日本、旧北区
silver berry scale			shiia oystershell scale を見よ

英名	和名	学名	所属、分布、ほか
silver birch aphid		*Euceraphis betulae* (Koch)	（カメムシ目、アブラムシ科）旧北区
silverbirch branchcutter		*Strongylurus cretifer* Hope	（コウチュウ目、カミキリムシ科）豪州区
silver blotch-backed drill		*Dichrorampha sequana* Curtis	（チョウ目、ハマキガ科）旧北区
silver-bordered fritillary	ナカギンヒョウモン	*Clossiana selene atrocostalis* (Huard)	（チョウ目、タテハチョウ科）新北区。small pearl-bordered fritillary を参照
silver-bottom brown		*Pseudonympha magus* (Fabricius)	（チョウ目、タテハチョウ科）エチオピア区
silver butterfly	チリギンジャノメ	*Argyrophorus argenteus* Blanchard	（チョウ目、タテハチョウ科）新熱帯区
silver cloud		*Egira conspicillaris* (Linnaeus)	（チョウ目、ヤガ科）旧北区。silver cloud moth ともいう
silver colonel		*Odontomyia argentata* (Fabricius)	（ハエ目、ミズアブ科）旧北区
silver-dotted hawk moth	ギンボシスズメ	*Parum colligata* (Walker)	（チョウ目、スズメガ科）日本、東洋区
silver-drop skipper			broken silverspot を見よ
silver-edged knot-horn		*Pima boisduvaliella* Guenée	（チョウ目、メイガ科）旧北区
silver eight		*Polychrysia moneta* (Fabricius)	（チョウ目、ヤガ科）旧北区
silver emperor			Laure を見よ
silver fern moth		*Selidosema aristarcha* Meyrick	（チョウ目、シャクガ科）豪州区
silver fir adelges			balsam woolly adelgid を見よ
silver fir adelgid		*Adelges nuesslini* (Börner)	（カメムシ目、カサアブラムシ科）旧北区
silver fir adelgid			balsam woolly adelgid を見よ
silver fir bark beetle		*Ips curvidens* Germar	（コウチュウ目、キクイムシ科）旧北区
silver fir migratory adelges			silver fir adelgid を見よ
silver fir weevil		*Pissodes piceae* Gyllenhal	（コウチュウ目、ゾウムシ科）旧北区
silver fish moth			silverfish (3) を見よ
silver fish moths			silverfish (2)(4) を見よ
silver forget-me-not		*Catochrysops lithargyria* (Moore)	（チョウ目、シジミチョウ科）東洋区
silver forget-me-not (1)		*Catochrysops panormus* (C. Felder)	（チョウ目、シジミチョウ科）東洋区。*C. p. exiguus* (Distant) ウスアオオナガウラナミシジミ は silver forget-me-not blue という
silver fritillary		*Clossiana selene myrina* (Cramer)	（チョウ目、タテハチョウ科）新北区。small pearl-bordered fritillary を参照
silver-green leaf weevil		*Phyllobius argentatus* (Linnaeus)	（コウチュウ目、ゾウムシ科）旧北区
silver grey silverline		*Cigaritis nipalicus* (Moore)	（チョウ目、シジミチョウ科）東洋区
silver ground carpet		*Xanthorhoe montanata* (Denis et Schiffermüller)	（チョウ目、シャクガ科）旧北区
silver hairstreak		*Chrysozephyrus syla* (Küllar)	（チョウ目、シジミチョウ科）東洋区

英名	和名	学名	所属、分布、ほか
silver hook	スジコヤガ	*Deltote uncula* (Clerck)	(チョウ目、ヤガ科) 日本、旧北区
silver leafhopper	ギンイロヒメヨコバイ	*Aguriahana stellulata* (Burmeister)	(カメムシ目、オオヨコバイ科) 日本、全北区
silverleaf whitefly	シルバーリーフコナジラミ	*Bemisia argentifolii* Bellows et Perring	(カメムシ目、コナジラミ科) 日本。多くの作物、花卉の害虫
silverleaf whitefly			tobacco whitefly を見よ
silverlines			leopard butterflies を見よ
silver-marked bell moth		*Pyrgotis plinthoglypta* Meyrick	(チョウ目、ハマキガ科) 豪州区
silver-marked yellow moth		*Diptychophora auriscriptella* Walker	(チョウ目、メイガ科) 豪州区
silver meadow fritillary			small pearl-bordered fritillary を見よ
silverpin			dragonflies (1) を見よ
silver-plated skipper		*Vettius coryna conka* Evans	(チョウ目、セセリチョウ科) 新熱帯区。Coryna skipper を参照
silver prominent	ギンシャチホコ	*Harpyia umbrosa* (Staudinger)	(チョウ目、シャチホコガ科) 日本、旧北区
silver-rayed skipper		*Adopaeoides bistriata* (Godman)	(チョウ目、セセリチョウ科) 新熱帯区
silver ringlet		*Ypthima albida* Butler	(チョウ目、タテハチョウ科) エチオピア区
silver royal	ウラギンルリフタオシジミ	*Ancema blanka* (de Nicéville)	(チョウ目、シジミチョウ科) 東洋区
silver-spot frit			dark green fritillary を見よ
silver spots		*Aphnaeus*	(チョウ目、シジミチョウ科) の昆虫の総称
silver-spotted dwarf		*Elachista albifrontella* (Hübner)	(チョウ目、クサモグリガ科) 旧北区
silver-spotted fern moth		*Callopistria cordata* (Ljungh)	(チョウ目、ヤガ科) 新北区
silver-spotted ghost moth		*Sthenopis argenteomaculatus* (Harris)	(チョウ目、コウモリガ科) 新北区
silver-spotted grey		*Crudaria leroma* (Wallengren)	(チョウ目、シジミチョウ科) エチオピア区
silver-spotted Halisidota		*Lophocampa argentata* (Packard)	(チョウ目、ヒトリガ科) 新北区。*Halisidota* 属に入っていたことに由来
silver spotted lancer		*Plastingia naga* (de Nicéville)	(チョウ目、セセリチョウ科) 東洋区
silver-spotted midget		*Phyllonorycter mespilella* (Hübner)	(チョウ目、ホソガ科) 旧北区
silver-spotted ochre		*Trapezites argenteoornatus* (Hewitson)	(チョウ目、セセリチョウ科) 豪州区
silver-spotted prominent	ギンモンシャチホコ	*Spatalia dives* Oberthür	(チョウ目、シャチホコガ科) 日本、旧北区
silver-spotted skipper (1)		*Epargyreus clarus clarus* (Cramer)	(チョウ目、セセリチョウ科) 新北区、大洋区
silver-spotted skipper (2)	アムールアカセセリ	*Hesperia comma* (Linnaeus)	(チョウ目、セセリチョウ科) 旧北区。silver-spotted skipper butterfly ともいう
silver-spotted skipper			silver-spotted ochre を見よ

英名	和名	学名	所属、分布、ほか
silver-spotted tiger moth			silver-spotted Halisidota を見よ
silverstreak blue		*Iraota timoleon* (Stoll)	(チョウ目、シジミチョウ科) 旧北区、東洋区
silverstreak blues		*Iraota*	(チョウ目、シジミチョウ科) の昆虫の総称
silver-streaked acacia blue		*Zinaspa todara* (Moore)	(チョウ目、シジミチョウ科) 東洋区
silver-streaked case		*Coleophora ochrea* (Haworth)	(チョウ目、ツツミノガ科) 旧北区
silver-streaked forester		*Lethe argentata* Leech	(チョウ目、タテハチョウ科) 旧北区
silver-streaked tubic		*Schiffermuelleria grandis* (Desvignes)	(チョウ目、マルハキバガ科) 旧北区
silver striped caddisflies		*Hesperophylax*	(トビケラ目、エグリトビケラ科) の昆虫の総称
silver-striped Charaxes		*Charaxes lasti* Grose-Smith	(チョウ目、タテハチョウ科) エチオピア区
silver-striped hawk			vine hawk-moth (1) を見よ　silver-striped hawkmoth ともいう
silver-striped hawk-moth			elephant hawk moth を見よ
silver-striped hawk-moth			vine hawk-moth (1) を見よ
silver-striped vine moth			vine hawk-moth (1) を見よ　豪州での英名
silver striped sedges			silver striped caddisflies を見よ
silver striped webworm			common grass-veneer moth を見よ
silver striped webworm			lesser vagabond sod webworm を見よ
silver-strips black		*Neopachygaster meromelas* (Dufour)	(ハエ目、ミズアブ科) 旧北区
silver-studded blue		*Plebejus argus* (Linnaeus)	(チョウ目、シジミチョウ科) 旧北区。日本のヒメシジミ *P. a. micrargus* (Butler) も同英名。silver-studded blue butterfly ともいう
silver-studded leafwing		*Hypna clytemnestra mexicana* Hall	(チョウ目、タテハチョウ科) 新熱帯区。marbled leafwing を参照
silver-studded ochre			iacchoides skipper を見よ
silver twig weevil		*Ophryastes argentatus* LeConte	(コウチュウ目、ゾウムシ科) 新北区
silver-washed fritillary	ミドリヒョウモン	*Argynnis paphia* (Linnaeus)	(チョウ目、タテハチョウ科) 旧北区
silver water beetle			great black water beetle を見よ
silver water beetles			water scavenger beetles (1) を見よ
silver Y			cabbage looper (1) を見よ
silver Y moth (1)	ガンマキンウワバ	*Autographa gamma* (Linnaeus)	(チョウ目、ヤガ科) 日本、旧北区。翅にγ型の紋がある。ガンマ蛾ともいう。トウモロコシ、ダイズ、アブラナ科野菜や花卉類の害虫。silver Y ともいう
silver Y moth (2)	イチジクキンウワバ	*Chrysodeixis eriosoma* (Doubleday)	(チョウ目、ヤガ科) 日本、東洋区、大洋区、豪州区。ダイズ害虫

英名	和名	学名	所属、分布、ほか
silver Y moth (3)		*Plusia chalcites* Esper	(チョウ目、ヤガ科) 旧北区
silvered carrot conch		*Aethes williana* (Brahm)	(チョウ目、ホソハマキガ科) 旧北区
silvered case		*Coleophora argentula* (Stephens)	(チョウ目、ツツミノガ科) 旧北区
silvered case		*Coleophora discordella* Zeller	(チョウ目、ツツミノガ科) 旧北区
silvered Haimbachia moth		*Haimbachia albescens* Capps	(チョウ目、メイガ科) 新北区
silvered marble		*Olethreutes olivana* Rebel	(チョウ目、ハマキガ科) 旧北区
silvered prominent moth		*Didugua argentilinea* Druce	(チョウ目、シャチホコガ科) 新北区
silvered ruby-eye		*Lycas argentea* (Hewitson)	(チョウ目、セセリチョウ科) 新熱帯区
silvered skipper		*Hesperilla crypsargyra* Meyrick	(チョウ目、セセリチョウ科) 豪州区
silverfish		*Acrotelsa collaris* (Fabricius)	(シミ目、シミ科) 豪州区
silverfish (1)	シミ目	Thysanura	の昆虫の総称 世界に約 320 種
silverfish (2)	セイヨウシミ	*Lepisma saccharina* Linnaeus	(シミ目、シミ科) 日本、汎世界。紙類を食べる汎世界種
silverfish (3)		*Ctenolepisma lineata* (Fabricius)	(シミ目、シミ科) 豪州区。亜種 *C. l. pilifera* Lucas は four-lined silverfish という
silverfish (4)	シミ科	Lepismatidae	(シミ目) の昆虫の総称
silverfish			gray silverfish を見よ
silverfish			firebrat (1) を見よ
silverfish			long-tailed silverfish を見よ
silverfish moths			silverfish (1) を見よ
silverfishes			silverfish (1) (4) を見よ
silvery arches		*Polia hepatica* (Clerck)	(チョウ目、ヤガ科) 旧北区
silvery arches		*Polia trimaculosa* (Esper)	(チョウ目、ヤガ科) 旧北区
silvery Argus		*Aricia nicias* (Meigen)	(チョウ目、シジミチョウ科) 旧北区
silvery bar			silvery barred-blue を見よ
silvery barred-blue		*Cigaritis phanes* (Trimen)	(チョウ目、シジミチョウ科) エチオピア区
silvery blue		*Lepidochrysops glauca* (Trimen)	(チョウ目、シジミチョウ科) エチオピア区
silvery blue (1)	アメリカカバイロシジミ	*Glaucopsyche lygdamus* (Doubleday)	(チョウ目、シジミチョウ科) 新北区
silvery broad-barred conch		*Commophila aeneana* Villers	(チョウ目、ハマキガ科) 旧北区。*Cydia pomonella* (Linnaeus) codling moth の異名
silvery checkerspot			silvery crescentspot を見よ
silvery crescentspot	ニクテウスアメリカヒョウモンモドキ	*Chlosyne nycteis* (Doubleday)	(チョウ目、タテハチョウ科) 新北区、新熱帯区
silvery demon Charaxes		*Charaxes catachrous* van Someren et Jackson	(チョウ目、タテハチョウ科) エチオピア区
silvery Epitola		*Cerautola ceraunia* (Hewitson)	(チョウ目、シジミチョウ科) エチオピア区
silvery Epitola		*Geritola subargentea* (Jackson)	(チョウ目、シジミチョウ科) エチオピア区

英名	和名	学名	所属、分布、ほか
silvery Euselasia		*Euselasia eusepus* (Hewitson)	(チョウ目、シジミタテハ科) 新熱帯区
silvery hedge blue		*Celastrina argiolus ladonides* (d'Orza)	(チョウ目、シジミチョウ科) 旧北区。holly blue を参照
silvery metalmark		*Chimastrum argentea argentea* (H. Bates)	(チョウ目、シジミタテハ科) 新熱帯区
silvery moth			silver Y moth (1) を見よ
silvery springtail		*Lepidocyrtus cyaneus* Tullberg	(トビムシ目、アヤトビムシ科) 旧北区
silvery tiger		*Halysidota argentata* Packard	(チョウ目、ヒトリガ科) 新北区
Simaethis green hairstreak			silver-banded hairstreak を見よ
similar awlking		*Choaspes xanthopogon* (Kollar)	(チョウ目、セセリチョウ科) 東洋区
similar shieldback		*Idionotus similis* Caudell	(バッタ目、キリギリス科) 新北区。灯火に飛来する
similar skipper		*Corticea similea* (Bell)	(チョウ目、セセリチョウ科) 新熱帯区
similar underwing moth		*Catocala similis* Edwards	(チョウ目、ヤガ科) 新北区。similar underwing ともいう
Simius roadside skipper			orange roadside skipper を見よ
Simius skipper		*Notamblyscirtes simius* (Edwards)	(チョウ目、セセリチョウ科) 新熱帯区
simple checkerspot		*Chlosyne hippodrome hippodrome* (Geyer)	(チョウ目、タテハチョウ科) 新熱帯区
simple clearwing		*Pteronymia simplex* (Salvin)	(チョウ目、タテハチョウ科) 新熱帯区。*P. s. fenochioi* Lamas、*P. s. timagenes* Godman et Salvin も同英名
simple-dot slender	ヤナギコハモグリ	*Phyllocnistis saligna* (Zeller)	(チョウ目、コハモグリガ科) 日本、東洋区、旧北区
simple orange forester		*Euphaedra simplex* Hecq	(チョウ目、タテハチョウ科) エチオピア区
simple patch			simple checkerspot を見よ
simple satyr		*Magneuptychia alcinoe* (C. et R. Felder)	(チョウ目、タテハチョウ科) 新熱帯区
simple skipper		*Tigasis simplex* (Bell)	(チョウ目、セセリチョウ科) 新熱帯区
simple wave moth		*Scopula junctaria* (Walker)	(チョウ目、シャクガ科) 新北区。simple wave ともいう
Sinai Baton blue		*Pseudophilotes sinaicus* Nakamura	(チョウ目、シジミチョウ科) 旧北区
Sinaloan Calephelis		*Calephelis sinaloensis sinaloensis* McAlpine	(チョウ目、シジミタテハ科) 新熱帯区。*C. s. nuevoleon* McAlpine も同英名
Sinaloan skipperling		*Piruna maculata* Freeman	(チョウ目、セセリチョウ科) 新熱帯区
Singapore ant			destroyer ant を見よ
Singapore oak blue		*Arhopala aedias* (Hewitson)	(チョウ目、シジミチョウ科) 東洋区
Singapore oakblue		*Amblypodia yendava* Grose-Smith	(チョウ目、シジミチョウ科) 東洋区。

英名	和名	学名	所属、分布、ほか
Singhalese hedge blue		*Polyommatus singalensis* Moore	（チョウ目、シジミチョウ科）東洋区
singing bush-crickets			crickets (3) を見よ
single-banded backswimmer		*Notonecta unifasciata* Guérin-Méneville	（カメムシ目、マツモムシ科）新北区
single-banded groundstreak		*Lamprospilus arza* (Hewitson)	（チョウ目、シジミチョウ科）新熱帯区
single-dotted wave		*Idaea dimidiata* (Hufnagel)	（チョウ目、シャクガ科）旧北区。成虫は single-dotted wave moth
single-lined emerald moth		*Nemoria unitaria* (Packard)	（チョウ目、シャクガ科）新北区
single ring bushbrown	コヒトツジャノメ	*Mycalesis sangaica mara* Fruhstorfer	（チョウ目、タテハチョウ科）東洋区
single silverstripe		*Lethe ramadeva* (de Nicéville)	（チョウ目、タテハチョウ科）東洋区
singleton			Una を見よ
singular lady beetles		Coccidulini	（コウチュウ目、テントウムシ科）の昆虫の総称 singular ladybird beetles ともいう
singular owl-butterfly		*Mielkella singularis* (Weymer)	（チョウ目、タテハチョウ科）新熱帯区
Sinhalese shadowdamsel		*Drepanosticta sinhalensis* Lieftinck	（トンボ目、Platystictidae）東洋区
sinister moth		*Pholodes sinistraria* Guenée	（チョウ目、シャクガ科）豪州区
sinuate clubtail		*Burmagomphus pyramidalis sinuatus* Fraser	（トンボ目、サナエトンボ科）東洋区
sinuate lady beetle		*Hyppodamia sinuata* Mulsant	（コウチュウ目、テントウムシ科）新北区
sinuate pear borer			sinuate pear tree borer を見よ
sinuate pear tree borer		*Agrilus sinuatus* (Olivier)	（コウチュウ目、タマムシ科）全北区。日本亜種は *A. s. yokoyamai* Y. Kurosawa アカバナガタタマムシ。ナシ害虫
sinuate-striped flea beetle		*Phyllotreta zimmermanni* (Crotch)	（コウチュウ目、ハムシ科）新北区。野菜害虫
sinuate-striped flea beetle			striped flea beetle (1) を見よ
sinuous Lytrosis moth		*Lytrosis sinuosa* Rindge	（チョウ目、シャクガ科）新北区
sinuous snaketail		*Ophiogomphus occidentis* (Hagen)	（トンボ目、サナエトンボ科）新北区
Sioux snaketail		*Ophiogomphus smithi* Tennessen et Vogt	（トンボ目、サナエトンボ科）新北区
siphunculina		*Siphunculina fimicola* Fly	（ハエ目、キモグリバエ科）東洋区
Siren	ゴマダラチョウ	*Hestina japonica* (C. et R. Felder)	（チョウ目、タテハチョウ科）日本。Siren butterfly ともいう
Siren		*Hestina persimilis* (Westwood)	（チョウ目、タテハチョウ科）東洋区
Sirens	ゴマダラチョウ属	*Hestina*	（チョウ目、タテハチョウ科）の昆虫の総称

英名	和名	学名	所属、分布、ほか
sirex parasite (1)		*Rhyssa persuasoria* (Linnaeus)	(ハチ目、ヒメバチ科) 豪州区
sirex parasite		*Ibalia leucospoides* Hochenwarth	(ハチ目、ヒラタタマバチ科) 豪州区
sirex parasite		*Schlettererius cinctipes* (Cresson)	(ハチ目、ツノヤセバチ科) 豪州区
sirex parasite			western giant ichneumon を見よ
sirex sawfly			horntail (2) を見よ
sirex wasp			horntail (2) を見よ
sirex wood wasp			horntail (2) を見よ
siricid woodwasp			horntail (1) を見よ
siris bean weevil	ネムノキマメゾウムシ	*Bruchidius terrenus* (Sharp)	(コウチュウ目、ハムシ科) 日本。ダイズ害虫
siris leaf-like moth			snout (2) を見よ
siris psylla	ヤマトキジラミ	*Acizzia jamatonica* (Kuwayama)	(カメムシ目、キジラミ科) 日本。ネムノキ害虫
sisal weevil		*Scyphophorus interstitialis* Gyllenhal	(コウチュウ目、ゾウムシ科) エチオピア区、東洋区、豪州区
Sisamnus dartwhite	ヘリブトタカネマダラシロチョウ	*Catasticta sisamnus* (Fabricius)	(チョウ目、シロチョウ科) 新熱帯区
Siskiyou shieldback		*Idionotus siskiyou* Hebard	(バッタ目、キリギリス科) 新北区
sisters		*Adelpha*	(チョウ目、タテハチョウ科) の昆虫の総称
Sitala ace			Tamil ace を見よ
sitka spruce gall aphid			Douglas fir adelges を見よ
sitka spruce weevil			white pine weevil を見よ
sitona egg parasite		*Anaphes diana* (Girault)	(ハチ目、ホソハネコバチ科) 豪州区
sitona weevil		*Sitona discoideus* Gyllenhal	(コウチュウ目、ゾウムシ科) 豪州区
sitona weevil parasite		*Microctonus aethiopoides* Loan	(ハチ目、コマユバチ科) 豪州区
Siva juniper hairstreak		*Callophrys gryneus siva* (W. H. Edwards)	(チョウ目、シジミチョウ科) 新北区、新熱帯区。Siva hairstreak ともいう。juniper hairstreak (1) を参照
sixbar swordtail	アサクラアゲハ	*Graphium eurous* (Leech)	(チョウ目、アゲハチョウ科) 東洋区
six-belted clearwing		*Bembecia ichneumoniformis* (Denis et Schiffermüller)	(チョウ目、スカシバガ科) 旧北区
six-belted clearwing		*Bembecia scopigera* Scopoli	(チョウ目、スカシバガ科) 旧北区
six-continent butterfly			danaid eggfly を見よ
six-dentated bark beetle			six-toothed bark beetle (1) を見よ
six-line blue		*Nacaduba berenice* (Herrich-Schäffer)	(チョウ目、シジミチョウ科) 豪州区。*N. b. icena* Fruhstorfer も同英名
six-plume moth		*Alucita montana* Barnes et Lindsey	(チョウ目、ニジュウシトリバガ科) 新北区

英名	和名	学名	所属、分布、ほか
six-punctate leafhopper	ムツボシヒメヨコバイ	*Limnavouriana sexmaculata* (Hardy)	（カメムシ目、オオヨコバイ科）日本、旧北区
six-spined engraver beetle			six-toothed bark beetle (2) を見よ
sixspined ips		*Ips calligraphus* (Germar)	（コウチュウ目、キクイムシ科）新北区
six-spot burnet		*Zygaena filipendulae* (Linnaeus)	（チョウ目、マダラガ科）旧北区
six-spot burnet moth			six-spot burnet を見よ　six-spotted burnet moth ともいう
six spot skipper		*Hesperilla sexguttata* (Herrich-Schäffer)	（チョウ目、セセリチョウ科）豪州区。six-spotted skipper ともいう
six-spotted angle moth		*Macaria sexmaculata* Packard	（チョウ目、シャクガ科）新北区。six-spotted angle ともいう
six-spotted Aroga moth		*Aroga compositella* Walker	（チョウ目、キバガ科）新北区
six-spotted buprestid	ムツボシタマムシ	*Chrysobothris succedanea* E. Saunders	（コウチュウ目、タマムシ科）日本
six-spotted Digrammia moth		*Digrammia sexpunctata* (Bates)	（チョウ目、シャクガ科）新北区
six-spotted eighty-eight			Aegina numberwing を見よ
six-spotted gray moth		*Spargaloma sexpunctata* Grote	（チョウ目、ヤガ科）新北区
six-spotted green tiger beetle			six-spotted tiger beetle を見よ
six-spotted ladybird		*Chilomenes sexmaculata* Fabricius	（コウチュウ目、テントウムシ科）東洋区
six-spotted leaf-cut weevil	ムツモンオトシブミ	*Apoderus praecellens* Sharp	（コウチュウ目、オトシブミ科）日本
six-spotted leafhopper			aster leafhopper (1) を見よ
six-spotted longhorn beetle		*Anoplodera sexguttata* (Fabricius)	（コウチュウ目、カミキリムシ科）旧北区
six-spotted pear sucker	ムツボシキジラミ	*Cyamophila hexastigma* (Horvath)	（カメムシ目、キジラミ科）日本、旧北区
six-spotted pot beetle	ムツボシツツハムシ	*Cryptocephalus sexpunctatus* (Linnaeus)	（コウチュウ目、ハムシ科）日本、旧北区
six-spotted sphinx			rustic sphinx を見よ　北米での英名
six-spotted thrips		*Scolothrips sexmaculatus* (Pergande)	（アザミウマ目、アザミウマ科）全北区
six-spotted tiger beetle		*Cicindela (Cicindela) sexguttata* (Fabricius)	（コウチュウ目、ハンミョウ科）新北区
six-spotted treehopper	ムツテンヨコバイ	*Macrosteles sexnotatus* (Fallén)	（カメムシ目、オオヨコバイ科）日本、全北区
six-striped rustic		*Xestia sexstrigata* (Haworth)	（チョウ目、ヤガ科）旧北区
six-tailed Helicopis		*Helicopis endymion elegans* Kaye	（チョウ目、シジミタテハ科）新熱帯区
six-toothed bark beetle (1)	ホシガタキクイムシ	*Pityogenes chalcographus* (Linnaeus)	（コウチュウ目、キクイムシ科）旧北区

英名	和名	学名	所属、分布、ほか
six-toothed bark beetle (2)		*Ips sexdentatus* Boern	（コウチュウ目、キクイムシ科）旧北区
six-toothed engraver beetle			six-toothed bark beetle (2) を見よ
six-toothed spruce bark beetle			six-toothed bark beetle (1) を見よ
six-toothed spruce engraver beetle			six-toothed bark beetle (1) を見よ
sixteen-spot ladybird		*Tytthaspis sedecimpunctata* (Linnaeus)	（コウチュウ目、テントウムシ科）旧北区
Sjostedt's sprite		*Pseudagrion sjoestedti* Förster	（トンボ目、イトトンボ科）エチオピア区
skater			common water strider を見よ
skater hawk			dragonflies (1) を見よ
skeleton gall midge		*Cystiphora schmidti* (Rubsaamen)	（ハエ目、タマバエ科）豪州区
skeletonising leaf beetles			cucumber beetles を見よ
skeletonweed gall midge			skeleten gall midge を見よ
skiff beetles	デオミズムシ科	Hydroscaphidae	（コウチュウ目）の昆虫の総称　微小な昆虫
skiff moth		*Prolimacodes badia* (Hübner)	（チョウ目、イラガ科）新北区
skiff moth			crowned slug moth を見よ
skillet clubtail		*Gomphus ventricosus* Walsh	（トンボ目、サナエトンボ科）新北区
skimmer dragonflies			common skimmers を見よ
skimmers (1)	トンボ上科	Libelluloidea	（トンボ目、トンボ科）の昆虫の総称
skimmers		*Orthetrum*	（トンボ目、トンボ科）の昆虫の総称
skimmers			common skimmers を見よ
skimming bluet		*Enallagma geminatum* Kellicott	（トンボ目、イトトンボ科）新北区
skin beetle	マルコブスジコガネ	*Trox mitis* Barthasar	（コウチュウ目、コガネムシ科）日本、旧北区
skin beetles		Troginae	（コウチュウ目、コガネムシ科）の昆虫の総称
skin beetles		*Trox*	（コウチュウ目、コガネムシ科）の昆虫の総称
skin beetles			larder beetles を見よ
skin beetles			hide beetles (1) を見よ
skin miners	ホソガ科	Gracilariidae	（チョウ目）の昆虫の総称
skin miners		*Marmara*	（チョウ目、ホソガ科）の昆虫の総称
skin miners			leafblotch miners (1) を見よ
skin moth		*Monopis laevigella* (Denis et Schiffermüller)	（チョウ目、ヒロズコガ科）旧北区
skin moth			dark brindled clothes moth を見よ
Skinner's cloudywing		*Achallarus albociliatus* (Mabille)	（チョウ目、セセリチョウ科）新北区、新熱帯区

英名	和名	学名	所属、分布、ほか
Skinner's hairstreak		*Mitoura loki* Skinner	（チョウ目、シジミチョウ科）新北区
skipjack			common click beetle (1) を見よ　幼虫の英名
skipjack			obscure click beetle を見よ　幼虫の英名
skipjack			click beetle (1)(2) を見よ
skipjack beetles			click beetles (1) を見よ
skipjacks			click beetles (1) を見よ
skipper			common pond skater を見よ
skipper butterflies			skippers (1) を見よ
skipper flies	チーズバエ科	Piophilidae	（ハエ目）の昆虫の総称
skippers		Hesperioidea	（チョウ目）の昆虫の総称
skippers (1)	セセリチョウ科	Hesperiidae	（チョウ目）の昆虫の総称　世界に3,000種以上
skippers			water striders (1) を見よ
skippers			skipper flies を見よ
skullcap skeletonizer moth		*Prochoreutis inflatella* (Clemens)	（チョウ目、ハマキモドキガ科）新北区
skunk moth		*Polix coloradella* (Walsingham)	（チョウ目、マルハキバガ科）新北区
skunkback Monopis moth		*Monopis dorsistrigella* (Clemens)	（チョウ目、ヒロズコガ科）新北区
sky beetle			Asian long-horned beetle を見よ
sky blue		*Jamides caeruleus caeruleus* Druce	（チョウ目、シジミチョウ科）東洋区
sky-blue groundstreak		*Ziegleria syllis* (Godman et Salvin)	（チョウ目、シジミチョウ科）新熱帯区
sky-blue hairstreak		*Pseudolycaena damo* (Druce)	（チョウ目、シジミチョウ科）新北区、新熱帯区
slant-faced grasshoppers		*Syrbula montezuma* (Saussure)	（バッタ目、バッタ科）新北区
slant-faced grasshoppers			true grasshoppers を見よ
slant-faced pasture grasshopper		*Orphulella speciosa* (Scudder)	（バッタ目、バッタ科）新北区
slant-lined owlet moth		*Macrochilo absorptalis* (Walker)	（チョウ目、ヤガ科）新北区。slant-lined fan-foot ともいう
slash pine flower thrips		*Gnophothrips fuscus* (Morgan)	（アザミウマ目、クダアザミウマ科）新北区
slash pine sawfly		*Neodiprion merkeli* Ross	（ハチ目、マツハバチ科）新北区
slash pine seedworm		*Cydia anaranjada* (Miller)	（チョウ目、ハマキガ科）新北区。slash pine seedworm moth ともいう
slate awl	オナシビロウドセセリ	*Hasora anura* de Nicéville	（チョウ目、セセリチョウ科）東洋区
slate flash	マネアシジミ	*Rapala manea* (Hewitson)	（チョウ目、シジミチョウ科）東洋区。*R. m. chozeba* (Hewitson) も同英名
slate flash		*Rapala schistacea* (Moore)	（チョウ目、シジミチョウ科）東洋区
slate royal		*Maneca bhotea* (Moore)	（チョウ目、シジミチョウ科）東洋区

英名	和名	学名	所属、分布、ほか
slate sober		*Syncopacma albipalpella* (Herrich-Schäffer)	(チョウ目、キバガ科) 旧北区
slaty roadside skipper		*Amblyscirtes nereus* Edwards	(チョウ目、セセリチョウ科) 新北区、新熱帯区
slaty skipper		*Chiomara mithrax* (Möschler)	(チョウ目、セセリチョウ科) 新北区、新熱帯区
slave dart moth		*Euxoa servitus* (Smith)	(チョウ目、ヤガ科) 新北区
slave-maker ant	アカヤマアリ	*Formica sanguinea* Latreille	(ハチ目、アリ科) 日本、旧北区。他種のアリの巣に集団で入り、蛹を盗んできて育て、奴隷にする。サムライアリと Amazon ant も同習性
slave-making ant	サムライアリ	*Polyergus samurai* Yano	(ハチ目、アリ科) 日本、旧北区
slave-making ant		*Harpagoxenus americanus* (Emery)	(ハチ目、アリ科) 新北区
slave-making ant			slave-maker ant を見よ
slave-making ant			European Amazon ant を見よ
sleeping Baileya moth		*Baileya dormitans* (Guenée)	(チョウ目、ヤガ科) 新北区
sleepy duskywing	ネムリミヤマセセリ	*Erynnis brizo* (Boisduval et LeConte)	(チョウ目、セセリチョウ科) 新北区、新熱帯区
sleepy orange		*Eurema nicippe* (Cramer)	(チョウ目、シロチョウ科) 新北区、新熱帯区
slender banded hornet		*Polistes sagittarius* Saussure	(ハチ目、スズメバチ科) 東洋区
slender baskettail		*Epitheca costalis* (Selys)	(トンボ目、エゾトンボ科) 新北区
slender black and yellow bamboo longicorn			bamboo longhorn beetle を見よ
slender blue-winged grasshopper			blue-winged locust を見よ
slender bluet		*Enallagma traviatum* Selys	(トンボ目、イトトンボ科) 新北区
slender bluet		*Africallagma elongatum* (Martin)	(トンボ目、イトトンボ科) エチオピア区
slender-bodied digger wasp		*Crabro cribrarius* (Linnaeus)	(ハチ目、アナバチ科) 旧北区
slender bottletail		*Olpogastra lugubris* (Karsch)	(トンボ目、トンボ科) エチオピア区
slender brindle	セスジヨトウ	*Apamea scolopacina* (Esper)	(チョウ目、ヤガ科) 日本、旧北区
slender burnished brass	ネッタイキクキンウワバ	*Thysanoplusia orichalcea* (Fabricius)	(チョウ目、ヤガ科) 日本、東洋区、豪州区、エチオピア区。ラッカセイ、レタス、ニンジンの害虫
slender checkered beetles		*Cymatodera*	(コウチュウ目、カッコウムシ科) の昆虫の総称
slender clearwing moth		*Hemaris gracilis* (Grote et Robinson)	(チョウ目、スズメガ科) 新北区。slender clearwing ともいう
slender conehead		*Neoconocephalus lyristes* (Rehn et Hebard)	(バッタ目、キリギリス科) 新北区。灯火に飛来する
slender dipluran		*Holjapyx diversiunguis* (Silvestri)	(コムシ目、ハサミコムシ科) 新北区
slender duck louse	アヒルナガハジラミ	*Anaticola crassicornis* (Scopoli)	(ハジラミ目、チョウカクハジラミ科) 日本、汎世界。カモ類に寄生

英名	和名	学名	所属、分布、ほか
slender entotrophs	ナガコムシ科	Campodeidae	(コムシ目) の昆虫の総称
slender-footed robberfly		*Leptarthus brevirostris* (Meigen)	(ハエ目、ムシヒキアブ科) 旧北区
slender goose louse		*Anaticola anseris* (Linnaeus)	(ハジラミ目、チョウカクハジラミ科) 全北区、豪州区。北米での英名
slender grasshopper		*Leptysma marginicollis* (Serville)	(バッタ目、バッタ科) 新北区
slender grasshopper			granulated grouse locust (1) を見よ
slender green longicorn beetle	ミドリカミキリ	*Chloridolum viride* (Thomson)	(コウチュウ目、カミキリムシ科) 日本、旧北区。イスノキ害虫。slender green longicorn ともいう
slender groundhopper			granulated grouse locust (1) を見よ
slender Guinea louse		*Lipeurus numidas* (Denny)	(ハジラミ目、チョウカクハジラミ科) 全北区
slender Guineapig louse			Guinea pig louse (1) を見よ
slender gumleaf grasshopper		*Goniaea vocans* (Fabricius)	(バッタ目、バッタ科) 豪州区
slender hairy aphid	シイニセケクダアブラムシ	*Mollitrichosiphum tenuicorpus* (Okajima)	(カメムシ目、アブラムシ科) 日本
slender-horned flour beetle	コツノコクヌストモドキ	*Gnathocerus maxillosus* (Fabricius)	(コウチュウ目、ゴミムシダマシ科) 日本、汎世界
slender-horned horsefly	キボシアブ	*Hybomitra montana* (Meigen)	(ハエ目、アブ科) 日本、全北区
slender Kauai damselfly		*Megalagrion oresitrophum* (Perkins)	(トンボ目、イトトンボ科) 大洋区
slender lacewings		Nymphidae	(アミメカゲロウ目) の昆虫の総称
slender longwing		*Capnobotes attenuatus* Rentz et Birchim	(バッタ目、キリギリス科) 新北区
slender meadow katydid		*Conocephalus fasciatus* (De Geer)	(バッタ目、キリギリス科) 新北区
slender moths			leafblotch miners (1) を見よ
slender mudnest builders		*Sceliphron*	(ハチ目、アナバチ科) の昆虫の総称
slender owlet moth		*Rhapsa scotosialis* Walker	(チョウ目、ヤガ科) 豪州区
slender pigeon louse		*Lipeurus baculus* Nitzsch	(ハジラミ目、チョウカクハジラミ科) 大洋区
slender pigeon louse	ハトナガハジラミ	*Columbicola columbae* (Linnaeus)	(ハジラミ目、チョウカクハジラミ科) 日本、汎世界。ハトに寄生
slender pigeon louse (1)		*Colomerus gardeniella* (Kieffer)	(ハジラミ目、チョウカクハジラミ科) 新北区
slender Pseudomyrmex		*Pseudomyrmex gracilis* (Fabricius)	(ハチ目、アリ科) 新熱帯区
slender pug		*Eupithecia tenuiata* (Hübner)	(チョウ目、シャクガ科) 旧北区
slender rice bug	ヒメハリカメムシ	*Cletus trigonus* (Thunberg)	(カメムシ目、ヘリカメムシ科) 日本。イネ、ダイズ害虫
slender scarlet			Divisa scarlet を見よ　slender scarlet darter ともいう

英名	和名	学名	所属、分布、ほか
slender Scotch burnet		*Zygaena loti scotica* (Rowland-Brown)	(チョウ目、マダラガ科) 旧北区
slender seed-corn beetle		*Clivina impressifrons* LeConte	(コウチュウ目、オサムシ科) 新北区。トウモロコシの芽出しを食べる
slender skimmer			green skimmer を見よ
slender soft scale			narrow brown scale を見よ
slender sphagnum springtail			orchid springtail を見よ
slender spreadwing		*Lestes rectangularis* Say	(トンボ目、アオイトトンボ科) 新北区
slender springtails	アヤトビムシ科	Entomobryidae	(トビムシ目) の昆虫の総称　日本から約 50 種
slender-striped rufous		*Coenocalpe lapidata* (Hübner)	(チョウ目、シャクガ科) 旧北区
slender Texas longhorn		*Psyrassa texana* Schaeffer	(コウチュウ目、カミキリムシ科) 新北区
slender treble-bar		*Analtis plagiata* Linnaeus	(チョウ目、シャクガ科) 旧北区
slender turkey louse		*Goniodes stylifer* Nitzsch	(ハジラミ目、チョウカクハジラミ科) 旧北区
slender turkey louse (1)		*Oxylipeurus polytrapezius* (Burmeister)	(ハジラミ目、チョウカクハジラミ科) 新北区
slender turkey louse			turkey louse を見よ
slender velvety longicorn beetle	ニセビロウドカミキリ	*Acalolepta sejuncta* (Fabricius)	(コウチュウ目、カミキリムシ科) 日本。イチョウ、イチイなどの害虫。slender velvety longicorn ともいう
slenders			leafblotch miners (1) を見よ
slicker			silverfish (3) を見よ
slickers			silverfish (2) (4) を見よ
slickers bristletail			silverfish (3) を見よ
slight-barred ermel		*Paraswammerdamia albicapitella* (Scharfenberg)	(チョウ目、スガ科) 旧北区
slight-barred ermel	ニセウスグロヒメスガ	*Paraswammerdamia caesiella* (Hübner)	(チョウ目、スガ科) 旧北区
slightly musical conehead		*Neoconocephalus exiliscanorus* (Davis)	(バッタ目、キリギリス科) 新北区
slim scarlet			little scarlet を見よ　slim scarlet-darter ともいう
slims		*Aciagrion*	(トンボ目、イトトンボ科) の昆虫の総称
slippery dart moth		*Anicla lubricans* (Guenée)	(チョウ目、ヤガ科) 新北区
slitworm			potato tuber moth (1) を見よ　幼虫の英名
Sloane's urania	ニシキアオツバメガ	*Uranus sloanus* Cramer	(チョウ目、ツバメガ科) 新熱帯区
sloe bug	ブチヒゲカメムシ	*Dorycoris baccarum* (Linnaeus)	(カメムシ目、カメムシ科) 日本、旧北区。多くの作物、野菜、花卉の害虫
sloe carpet			sloe pug (1) を見よ
sloe hairstreak		*Satyrium acaciae* (Fabricius)	(チョウ目、シジミチョウ科) 旧北区
sloe midget		*Phyllonorycter spinicolella* Zeller	(チョウ目、ホソガ科) 旧北区

英名	和名	学名	所属、分布、ほか
sloe pug (1)		*Aleucis distinctata* (Herrich-Schäffer)	（チョウ目、シャクガ科）旧北区
sloe pug		*Pasiphila chloerata* (Mabille)	（チョウ目、シャクガ科）旧北区
Slosson's metalmark moth		*Tortyra slossonia* (Fernald)	（チョウ目、ハマキモドキガ科）新北区
Slosson's scaly cricket		*Cycloptilum slossoni* (Scudder)	（バッタ目、コオロギ科）新北区
slough amberwing		*Perithemis domitia* (Drury)	（トンボ目、トンボ科）新北区、新熱帯区
slow scaly cricket		*Cycloptilum tardum* Love et Walker	（バッタ目、コオロギ科）新北区
slow-tinkling trig		*Anaxipha* sp.	（バッタ目、クサヒバリ科）新北区
slowpoke moth		*Athetis tarda* (Guenée)	（チョウ目、ヤガ科）新北区
slug caterpillar		*Thosea sinensis* (Walker)	（チョウ目、イラガ科）旧北区、東洋区
slug caterpillar moths	イラガ科	Limacodidae	（チョウ目）の昆虫の総称　世界に400種
slug caterpillars			slug caterpillar moths を見よ
slug moths		Eucleidae	（チョウ目）の昆虫の総称
slug moths			slug caterpillar moths を見よ
sluggish plume	ヨモギトリバ	*Leioptilus lienigianus* (Zeller)	（チョウ目、トリバガ科）日本、旧北区、東洋区、豪州区、エチオピア区
sluggish weevil		*Cleonus piger* (Scopoli)	（コウチュウ目、ゾウムシ科）新北区
slugworm			pear sawfly (1)(2) を見よ
small angle shades	アカガネヨトウ	*Euplexia lucipara* (Linnaeus)	（チョウ目、ヤガ科）日本、旧北区。ダイズ、アブラナ科野菜の害虫
small angle-shades moth			small angle shades を見よ
small angle-wing moth		*Gymnobathra hyetodes* Meyrick	（チョウ目、マルハキバガ科）豪州区
small ant-blue		*Acrodipsas myrmecophila* (Waterhouse et Lyell)	（チョウ目、シジミチョウ科）豪州区
small Apollo			large parnassian を見よ
small argent-and-sable		*Epirrhoe tristata* (Linnaeus)	（チョウ目、シャクガ科）旧北区
small asparagus fly			asparagus miner を見よ
small aspen leaftier moth		*Acleris fuscana* (Barnes et Busck)	（チョウ目、ハマキガ科）新北区
small aspen pigmy		*Stigmella assimilella* (Zeller)	（チョウ目、モグリチビガ科）旧北区
small autumnal moth		*Epirrita filigrammaria* (Herrich-Schäffer)	（チョウ目、シャクガ科）旧北区
small backswimmers		*Buenoa*	（カメムシ目、マツモムシ科）の昆虫の総称
small Baileya moth		*Baileya australis* (Grote)	（チョウ目、ヤガ科）新北区
small bamboo borers		*Dinoderus*	（コウチュウ目、ナガシンクイムシ科）の昆虫の総称
small banded flat		*Celaenorrhinus nigricans* (de Nicéville)	（チョウ目、セセリチョウ科）東洋区

英名	和名	学名	所属、分布、ほか
small banded grasshopper		*Arcyptera microptera* (Fischer von Waldheim)	(バッタ目、バッタ科) 旧北区
small banded pine weevil		*Pissodes castaneus* (De Geer)	(コウチュウ目、ゾウムシ科) 旧北区
small banded-wing grasshopper		*Oedaleus abruptus* (Thunberg)	(バッタ目、バッタ科) 大洋区
small bath white	ヨーロッパシロチョウ	*Pontia chloridice* (Hübner)	(チョウ目、シロチョウ科) 旧北区、東洋区
small bean bug	メダカナガカメムシ	*Chauliopus fallax* Scott	(カメムシ目、メダカナガカメムシ科) 日本、東洋区。イネ、ダイズなどの害虫
small birch leafminer moth		*Ectoedemia lindguisti* (Freeman)	(チョウ目、モグリチビガ科) 新北区
small birch pigmy		*Stigmella sakhalinella* (Puplesis)	(チョウ目、モグリチビガ科) 旧北区
small bird-dropping moth		*Ponometia erastrioides* (Guenée)	(チョウ目、ヤガ科) 新北区。small bird lime moth ともいう
small black arches		*Meganola strigula* (Denis et Schiffermüller)	(チョウ目、コブガ科) 旧北区
small black blister beetle		*Lytta nigripilis* (Fall)	(コウチュウ目、ツチハンミョウ科) 新北区
small black evening moth	ヒメクロホウジャク	*Macroglossum bombylans* Boisduval	(チョウ目、スズメガ科) 日本、東洋区
small black longhorn beetle		*Stenurella nigra* (Linnaeus)	(コウチュウ目、カミキリムシ科) 旧北区
small black-specked groundling		*Exoteleia dodecella* (Linnaeus)	(チョウ目、キバガ科) 旧北区
small black-spotted conch		*Hysterophora maculosana* (Haworth)	(チョウ目、ハマキガ科) 旧北区
small blackish cochlid			persimmon cochlid を見よ
small blister beetle	ヒメツチハンミョウ	*Meloe coarctatus* Motschulsky	(コウチュウ目、ツチハンミョウ科) 日本
small blood-vein		*Scopula imitaria* (Hübner)	(チョウ目、シャクガ科) 旧北区
small blood-veined wave			small blood-vein を見よ
small blue	チビシジミ	*Philotiella speciosa* (H. Edwards)	(チョウ目、シジミチョウ科) 新北区
small blue butterfly	コマシジミ	*Cupido minimus* (Fuessly)	(チョウ目、シジミチョウ科) 旧北区。small blue ともいう
small blue cattle louse	ケナガウシジラミ	*Solenopotes capillatus* Enderlein	(シラミ目、ケモノホソジラミ科) 日本、旧北区、豪州区
small blue Grecian			Sara longwing を見よ
small blue horntail	コルリキバチ	*Sirex juvencus juvencus* (Linnaeus)	(ハチ目、キバチ科) 日本、旧北区
small blue Lema			wheat leaf beetle を見よ
small blue longicorn	ヒゲナガヒメルリカミキリ	*Praolia citrinipes* Bates	(コウチュウ目、カミキリムシ科) 日本
small bluetail		*Ischnura pumilio* (Charpentier)	(トンボ目、イトトンボ科) 旧北区。small bluetip ともいう。英国の絶滅危惧種
small body louse			chicken louse を見よ　米国での英名

英名	和名	学名	所属、分布、ほか
small-bordered fritillary			small pearl-bordered fritillary を見よ
small bottle cicada		*Chlorocysta vitripennis* (Westwood)	(カメムシ目、セミ科) 豪州区
small branded swift		*Pelopidas mathias* (Fabricius)	(チョウ目、セセリチョウ科) 東洋区、エチオピア区。日本亜種 *P. m. oberthureri* Evans チャバネセセリも同英名
small branded swift			Chinese swift を見よ
small brindled beauty		*Apocheima hispidaria* Denis et Schiffermüller	(チョウ目、シャクガ科) 旧北区
small brocade moth		*Oligia minuscula* (Morrison)	(チョウ目、ヤガ科) 新北区
small bronze azure			small brown azure を見よ
small brown azure		*Ogyris otanes* C. et R. Felder	(チョウ目、シジミチョウ科) 豪州区
small brown-backed planthopper			small brown planthopper を見よ
small brown-barred conch			grape cochylid を見よ
small brown beetle		*Serica castanea* Arrow	(コウチュウ目、コガネムシ科) 全北区。カンキツ害虫
small brown crow			dwarf crow を見よ
small brown crow			Darwin brown crow を見よ
small brown mantids		*Bolbe*	(カマキリ目、カマキリ科) の昆虫の総称
small brown planthopper	ヒメトビウンカ	*Laodelphax striatellus* (Fallén)	(カメムシ目、ウンカ科) 日本、旧北区、東洋区。イネ、ムギなどの大害虫。smaller brown planthopper が適切との意見あり
small brown quaker moth		*Pseudorthodes vecors* (Guenée)	(チョウ目、ヤガ科) 新北区。smaller brown quaker ともいう
small brown rice leaf roller			rice leafroller を見よ
small brown rice planthopper			small brown planthopper を見よ
small cabbage bug	ヒメナガメ	*Eurydema dominulus* (Scopoli)	(カメムシ目、カメムシ科) 日本、東洋区。アブラナ科野菜害虫
small cabbage moth			diamondback moth を見よ
small cabbage white			common white を見よ *P. rapae* Linnaeus の英名
small camphor bark beetle	ツヅミキクイムシ	*Xyleborus amputatus* Blandford	(コウチュウ目、キクイムシ科) 日本、東洋区
small carpenter bees		Ceratinini	(ハチ目、ミツバチ科) の昆虫の総称
small carpenter bees (1)		*Ceratina*	(ハチ目、ミツバチ科) の昆虫の総称
small carrion beetles		Leptodiridae	(コウチュウ目) の昆虫の総称
small celery aphid			bean aphid を見よ
small chafer			summer chafer を見よ
small checkered skipper		*Pyrgus scriptura* (Boisduval)	(チョウ目、セセリチョウ科) 新北区
small chestnut weevil		*Curculio sayi* (Gyllenhal)	(コウチュウ目、ゾウムシ科) 新北区

英名	和名	学名	所属、分布、ほか
small china mark		*Cataclysta lemnata* Linnaeus	(チョウ目、メイガ科) 旧北区
small chocolate-tip		*Clostera pigra* (Hufnagel)	(チョウ目、シャチホコガ科) 旧北区
small chrysanthemum aphid			pale chrysanthemum aphid を見よ
small citrus butterfly			dainty swallowtail を見よ
small citrus butterfly			chequered swallowtail を見よ
small citrus looper	ソトカバナミシャク	*Eupithis signigera* Butler	(チョウ目、シャクガ科) 日本
small claw			wanstead grey を見よ
small clouded brindle		*Apamea unanimis* (Hübner)	(チョウ目、ヤガ科) 全北区。small clouded brindle moth ともいう
small clouded knot-horn		*Homoeosoma nimbella* Duponchel	(チョウ目、メイガ科) 旧北区
small cluster fly	ネアブラキモグリバエ	*Thaumatomyia notata* (Meigen)	(ハエ目、キモグリバエ科) 日本、旧北区
small coconut leaf moth			coconut leafminer (1) を見よ
small coconut leaf moth			coconut leaf moth (1)(2) を見よ
small common carpet moth		*Chloroclystis semialbata* Walker	(チョウ目、シャクガ科) 豪州区
small copper (1)		*Lycaena phlaeas* (Linnaeus)	(チョウ目、シジミチョウ科) 旧北区。日本亜種は *L. p. daimio* (Matsumura) ベニシジミ。small copper butterfly ともいう
small copper (2)		*Lucia limbaria* (Swainson)	(チョウ目、シジミチョウ科) 豪州区
small cotton bug	ヒメアカホシカメムシ	*Dysdercus poecilus* (Herrich-Schäffer)	(カメムシ目、ホシカメムシ科) 日本、東洋区。ワタ害虫
small cupid		*Chilades contracta* (Butler)	(チョウ目、シジミチョウ科) 旧北区、東洋区
small currant and lettuce aphid			currant-lettuce aphid を見よ
small cypress jewel beetle		*Diadoxus erythrurus* (White)	(コウチュウ目、タマムシ科) 豪州区
small dark-banded looper moth		*Xanthorhoe lucidata* Walker	(チョウ目、シャクガ科) 豪州区
small dark brown plume moth		*Platyptilia aeolodes* Meyrick	(チョウ目、トリバガ科) 豪州区
small dark olive		*Baetis scambus* Eaton	(カゲロウ目、コカゲロウ科) 旧北区
small dark yellow underwing			dark yellow underwing を見よ
small darkling ground beetle		*Metoponium abnorme* LeConte	(コウチュウ目、ゴミムシダマシ科) 新北区
small darter		*Telicota brachydesma* Lower	(チョウ目、セセリチョウ科) 豪州区
small death's head hawkmoth		*Acherontia styx* (Westwood)	(チョウ目、スズメガ科) 旧北区、東洋区。death's head を参照
small dingy skipper		*Hesperilla crypsigramma* (Meyrick et Lower)	(チョウ目、セセリチョウ科) 豪州区

英名	和名	学名	所属、分布、ほか
small Diopetes		*Pilodeudorix deritas* (Hewitson)	(チョウ目、シジミチョウ科) エチオピア区
small domestic buff			small dotted buff を見よ
small dotted buff		*Photedes minima* (Haworth)	(チョウ目、ヤガ科) 旧北区
small dotted footman		*Pelosia obtusa* (Herrich-Schäffer)	(チョウ目、ヒトリガ科) 旧北区。日本亜種は *P. o. sutchana* (Staudinger) ヒメクロスジホソガ
small dull bell moth		*Cnephasia imbriferana* Meyrick	(チョウ目、ハマキガ科) 豪州区
small dung beetles			Aphodian dung beetles を見よ
small dung beetles			dung beetles (2) を見よ
small dung flies	ハヤトビバエ科	Sphaeroceridae	(ハエ目) の昆虫の総称
small duskhawker		*Gynacantha bayadera* Selys	(トンボ目、ヤンマ科) 東洋区
small dusky blue		*Candalides erinus* (Fabricius)	(チョウ目、シジミチョウ科) 豪州区。small dusky-blue butterfly ともいう
small dusty wave		*Idaea seriata* (Schrank)	(チョウ目、シャクガ科) 旧北区
small earth humble bee			white-tailed bumble bee を見よ
small earwig			lesser earwig を見よ　米国での英名
small eastern caddisflies		Beraeidae	(トビケラ目) の昆虫の総称
small eggar			birch lasiocampid を見よ　birch eggar moth ともいう
small elephant hawk		*Deilephila porcellus* (Linnaeus)	(チョウ目、スズメガ科) 旧北区。small elephant hawkmoth ともいう
small elephant hawk-moth		*Chaerocampa elpenor* (Linnaeus)	(チョウ目、スズメガ科) 旧北区
small elfin			orange flat を見よ
small elm bark beetle			elm bark beetle (2) を見よ
small emerald		*Hemistola chrysoprasaria* (Esper)	(チョウ目、シャクガ科) 旧北区
small emerald damselfly		*Lestes virens* (Charpentier)	(トンボ目、アオイトトンボ科) 旧北区
small engrailed	フトフタオビエダシャク	*Ectropis crepuscularia* (Denis et Schiffermüller)	(チョウ目、シャクガ科) 日本、全北区。small engrailed moth ともいう
small engrailed geometer			small engrailed を見よ
small ermine			orchard ermine を見よ
small ermine knot-horn		*Phycitodes binaevella* (Hübner)	(チョウ目、メイガ科) 旧北区
small ermine moth			orchard ermine を見よ
small ermine moth			apple ermine moth を見よ
small ermine moths			ermine moths (1)(2) を見よ　米国での英名
small European raspberry aphid			raspberry aphid (1) を見よ

英名	和名	学名	所属、分布、ほか
small evergreen oak scolytid	カシノコナガキクイムシ	*Crossotarsus simplex* Murayama	(コウチュウ目、ナガキクイムシ科) 日本、旧北区
small eyed click beetle			eyed elater を見よ　small-eyed click beetle とも記す
small-eyed flour beetle	ヒメコクヌストモドキ	*Palorus ratzeburgii* (Wissmann)	(コウチュウ目、ゴミムシダマシ科) 日本、汎世界。貯穀害虫
small eyed moth		*Austramathes purpurea* Butler	(チョウ目、ヤガ科) 豪州区
small-eyed sailor			artemisia sailor を見よ
small-eyed sphinx		*Paonias myops* (J. E. Smith)	(チョウ目、スズメガ科) 新北区、新熱帯区、大洋区。small-eyed sphinx moth ともいう
small faggot-worm			faggot-worm (1) を見よ
small false wireworm			false wireworm (1)(2)(3) を見よ
small fanfoot		*Herminia grisealis* (Denis et Schiffermüller)	(チョウ目、ヤガ科) 旧北区
small fanfoot (1)	クロスジアツバ	*Herminia nemoralis* (Fabricius)	(チョウ目、ヤガ科) 日本、旧北区、東洋区
small fan-footed wave	ウスキヒメシャク	*Idaea biselata* (Hufnagel)	(チョウ目、シャクガ科) 日本、旧北区
small field cricket		*Modicogryllus conspersus* (Schaum)	(バッタ目、コオロギ科) 大洋区
small field crickets		*Bobilla*	(バッタ目、コオロギ科) の昆虫の総称
small fiery Acraea	ツマグロベニホソチョウ	*Acraea chaeribula* Oberthür	(チョウ目、タテハチョウ科) エチオピア区
small flambeau	ベニスジドクチョウ	*Eueides heliconioides* C. et R. Felder	(チョウ目、タテハチョウ科) 新熱帯区
small flame bordered Charaxes	カバヘリヒメフタオチョウ	*Charaxes anticlea* (Drury)	(チョウ目、タテハチョウ科) エチオピア区
small flat		*Celaenorrhinus cynapes cynapes* (Hewitson)	(チョウ目、セセリチョウ科) 新熱帯区
small flattened bark beetles		Monotomidae	(コウチュウ目) の昆虫の総称　ネスイムシ科の異名
small flattened bark beetles	デオネスイムシ亜科	Monotominae	(コウチュウ目、ネスイムシ科) の昆虫の総称
small flax flea beetle		*Longitarsus parvulus* (Paykull)	(コウチュウ目、ハムシ科) 旧北区
small fleck-winged snipefly		*Rhagio lineola* Fabricius	(ハエ目、シギアブ科) 旧北区
small flitter		*Zographetus rama* (Mabille)	(チョウ目、セセリチョウ科) 東洋区
small flower click beetle	コガタクシコメツキ	*Melanotus erythropygus* Candèze	(コウチュウ目、コメツキムシ科) 日本、旧北区
small forester		*Aglaope infausta* Linnaeus	(チョウ目、マダラガ科) 旧北区
small four-eyed geometrid	コヨツメヒメシャク	*Ophithalmitis irroraria* (Bremer et Grey)	(チョウ目、シャクガ科) 日本、旧北区
small 4-lineblue	ウラマダラシジミ	*Nacaduba pavana* (Horsfield)	(チョウ目、シジミチョウ科) 東洋区
small frosted wave moth		*Scopula lautaria* (Hübner)	(チョウ目、シャクガ科) 新北区

英名	和名	学名	所属、分布、ほか
small frosted weevil	チビコフキゾウムシ	*Sitona japonicus* Roelofs	（コウチュウ目、ゾウムシ科）日本
small fruit bark beetle			shothole borer (1) を見よ
small fruit flies			vinegar flies (1)(2) を見よ
small fruit-tree borer		*Cryptophasa albacosta* Lewin	（チョウ目、マルハキバガ科）豪州区
small garden humble-bee			large garden bumble bee を見よ
small gatekeeper			small heath を見よ
small gold		*Micropteryx calthella* Linnaeus	（チョウ目、コバネガ科）旧北区
small gold grasshopper		*Chrysochraon brachypterus* (Ocskay)	（バッタ目、バッタ科）旧北区
small goldenfork		*Lethe atkinsonia* (Hewitson)	（チョウ目、タテハチョウ科）東洋区
small gooseberry sawfly		*Nematus appendiculatus* Hartig	（ハチ目、ハバチ科）旧北区
small gooseberry sawfly		*Pristiphora pallipes* Lepeletier	（ハチ目、ハバチ科）旧北区
small gooseberry sawfly (1)		*Pristiphora carinata* (Hartig)	（ハチ目、ハバチ科）旧北区
small grained noctuid			purple cloud を見よ
small grape plume moth			grape plume moth (1) を見よ
small grass blue			black-spotted grass-blue を見よ
small grass emerald		*Chlorissa viridata* (Linnaeus)	（チョウ目、シャクガ科）旧北区
small grass yellow	ミナミキチョウ	*Eurema smilax* (Donovan)	（チョウ目、シロチョウ科）豪州区
small grass yellow (1)	ホシボシキチョウ	*Eurema brigitta* (Stoll)	（チョウ目、シロチョウ科）日本、豪州区、新熱帯区。東洋区の *E. b. senna* (C. et R. Felder) も同英名
small grasshopper			field grasshopper を見よ
small grass-root moth		*Borkhausenia chlorodelpha* Meyrick	（チョウ目、マルハキバガ科）豪州区
small greasy			glasswing を見よ
small green awlet		*Burara amara* (Moore)	（チョウ目、セセリチョウ科）東洋区
small green chrysanthemum aphid			lesser rose aphid を見よ　米国での英名
small green leafhopper	ミドリヒメヨコバイ	*Edwardsiana flavescens* (Fabricius)	（カメムシ目、オオヨコバイ科）日本、旧北区。smaller green leafhopper ともいう
small green sapphire		*Iolaus catori* Bethune-Baker	（チョウ目、シジミチョウ科）エチオピア区
small green tortoise beetle			beet tortoise beetle (2) を見よ
small green undeerwing		*Lycaena metallica* Felder et Felder	（チョウ目、シジミチョウ科）東洋区

英名	和名	学名	所属、分布、ほか
small green-banded blue		*Danis hymetus* Felder	(チョウ目、シジミチョウ科) 豪州区
small green-banded blue		*Psychonotis caelius* (C. et R. Felder)	(チョウ目、シジミチョウ科) 豪州区
small green-streaked playboy		*Pilodeudorix virgata* (Druce)	(チョウ目、シジミチョウ科) エチオピア区
small grey		*Eudonia mercurella* (Linnaeus)	(チョウ目、メイガ科) 旧北区
small grey looper moth		*Xanthorhoe cinerearia* Doubleday	(チョウ目、シャクガ科) 豪州区
small ground crickets		*Neonemobius*	(バッタ目、コオロギ科) の昆虫の総称
small hair-grass dwarf		*Cosmiotes stahilella* (Stainton)	(チョウ目、クサモグリガ科) 旧北区
small hairy maggot blowfly		*Chrysomya varipes* (Macquart)	(ハエ目、クロバエ科) 豪州区
small harvester		*Megalopalpus zymna* (Westwood)	(チョウ目、シジミチョウ科) エチオピア区
small hawk moth	コスズメ	*Theretra japonica* (Boisduval)	(チョウ目、スズメガ科) 日本、旧北区
small-headed flies	コガシラアブ科	Acroceridae	(ハエ目) の昆虫の総称
small-headed froghopper	コガシラアワフキ	*Eoscartopis assimilis* (Uhler)	(カメムシ目、コガシラアワフキ科) 日本、旧北区、東洋区
small heath	チャイロヒメヒカゲ	*Coenonympha pamphilus* (Linnaeus)	(チョウ目、タテハチョウ科) 旧北区。small heath butterfly ともいう
small heath bumblebee			heath bumblebee を見よ
small Heliconius		*Heliconius ricini* (Linnaeus)	(チョウ目、タテハチョウ科) 新熱帯区
small Heterocampa moth		*Heterocampa subrotata* Harvey	(チョウ目、シャチホコガ科) 新北区
small hillside brown			spotted-eye brown を見よ
small hive beetle		*Aethina tumida* Murray	(コウチュウ目、ケシキスイ科) 全北区、豪州区、大洋区、エチオピア区
small hook-tip moth		*Asaphodes megaspilata* Walker	(チョウ目、シャクガ科) 豪州区
small hopper		*Platylesches tina* Evans	(チョウ目、セセリチョウ科) エチオピア区
small horse-fly		*Tabanus bromius* Linnaeus	(ハエ目、アブ科) 旧北区
small hydrophilid	コガムシ	*Hydrochara affinis* (Sharp)	(コウチュウ目、ガムシ科) 日本、旧北区
small Indian bee			Indian honey bee を見よ
small Japanese cedar longicorn beetle	ヒメスギカミキリ	*Callidiellum rufipenne* (Motschulsky)	(コウチュウ目、カミキリムシ科) 日本、旧北区。イブキ、スギ、ヒノキなどの害虫
small jewel blue		*Plebejus christophe* (Staudinger)	(チョウ目、シジミチョウ科) 東洋区
small lacewing		*Actinote pellenea* Hübner	(チョウ目、タテハチョウ科) 新熱帯区
small lappet		*Phyllodesma ilicifolia* (Linnaeus)	(チョウ目、カレハガ科) 旧北区
small larch sawfly		*Pristiphora laricis* (Hartig)	(ハチ目、ハバチ科) 旧北区

英名	和名	学名	所属、分布、ほか
small leaf-cut weevil	ヒメオトシブミ	*Attelabus montanus* Roelofs	（コウチュウ目、オトシブミ科）日本
small-leaved oak cottony aphid	コナラハアブラムシ	*Diphyllaphis konarae* (Shinji)	（カメムシ目、アブラムシ科）日本
small leaved oak curculio			chestnut curculio を見よ
small-leaved oak gall wasp	コナラフシバチ	*Callirhytis glauduliferae* Monzen	（ハチ目、タマバチ科）日本
small-leaved oak leaf roller	オオクシヒゲシマメイガ	*Datanoides fasciatus* Butler	（チョウ目、メイガ科）日本、旧北区
small-leaved oak leaf-cut weevil	コナライクビチョッキリ	*Deporaus unicolor* (Roelofs)	（コウチュウ目、オトシブミ科）日本、旧北区、東洋区
small leaved oak scale	サイタマシロカイガラムシ	*Chionaspis saitamensis* Kuwana	（カメムシ目、マルカイガラムシ科）日本
small leopard	ヒメウラベニヒョウモン	*Phalanta alcippe* (Stoll)	（チョウ目、タテハチョウ科）東洋区。*P. a. alcesta* Corbet も同英名
small lighting beetle		*Phausis splendidula* (Linnaeus)	（コウチュウ目、ホタル科）旧北区
small long-brand bushbrown		*Mycalesis igilia* Fruhstorfer	（チョウ目、タテハチョウ科）東洋区
small lucerne weevil		*Atrichonotus taeniatulus* (Berg)	（コウチュウ目、ゾウムシ科）豪州区
small magpie			small magpie moth を見よ
small magpie moth	イラクサノメイガ	*Eurrhypara hortulata* (Linnaeus)	（チョウ目、メイガ科）日本、旧北区
small-mammal sucking lice	フトケジラミ科	Hoplopleuridae	（シラミ目）の昆虫の総称　世界に約500種。ネズミ、モグラなどに寄生
small mango tipborer			mango shoot caterpillar (1) を見よ
small marbled		*Eublemma parva* (Hübner)	（チョウ目、ヤガ科）旧北区
small marbled bush brown		*Bicyclus auricruda* (Butler)	（チョウ目、タテハチョウ科）エチオピア区
small marbled elfin		*Eretis umbra* (Trimen)	（チョウ目、セセリチョウ科）エチオピア区。small marbled elf ともいう
small mayflies (1)	コカゲロウ科	Baetidae	（カゲロウ目）の昆虫の総称
small mayflies		*Baetis*	（カゲロウ目、コカゲロウ科）の昆虫の総称
small meadow white		*Pontia distorta* (Butler)	（チョウ目、シロチョウ科）エチオピア区
small milkweed bug		*Lygaeus kalmii* Stål	（カメムシ目、ナガカメムシ科）新北区
small minnow mayflies			small mayflies (1) を見よ
small Mocis moth			South American semilooper を見よ
small monarch butterfly			African monarch を見よ　ニュージーランドでの英名
small mossy Lithacodia moth		*Lithacodia musta* (Grote et Robinson)	（チョウ目、ヤガ科）新北区。small mossy Lithacodia ともいう
small moth borer			sugarcane borer (1) を見よ
small moth borer			yellow-headed borer を見よ
small mottled bell moth			oblique tortrix を見よ

英名	和名	学名	所属、分布、ほか
small mottled drab moth		*Barea confusella* Walker	(チョウ目、マルハキバガ科) 豪州区
small mottled willow			beet armyworm を見よ
small mottled willow caterpillar			beet armyworm を見よ
small mottled willow moth			beet armyworm を見よ
small mountain ringlet			mountain ringlet を見よ　small mountain ringlet butterfly ともいう
small mulberry longicorn beetle	マルモンサビカミキリ	*Pterolophia angusta* (Bates)	(コウチュウ目、カミキリムシ科) 日本、旧北区。クワ、フジ害虫。small mulberry longicorn ともいう
small narcissus bulb fly			onion bulb fly を見よ
small narcissus fly			onion bulb fly を見よ
small necklace moth		*Hypsoropha hormos* Hübner	(チョウ目、ヤガ科) 新北区
small network-marked geometrid			latticed heath を見よ
small nutmeg			nutmeg を見よ
small oak blue		*Arhopala wildei* Miskin	(チョウ目、シジミチョウ科) 豪州区
small ochreous pearl		*Ebulea crocealis* (Hübner)	(チョウ目、メイガ科) 旧北区
small orange Acraea (1)		*Acraea eponina* (Cramer)	(チョウ目、タテハチョウ科) エチオピア区
small orange Acraea (2)	アフリカホソチョウ	*Acraea serena* (Fabricius)	(チョウ目、タテハチョウ科) エチオピア区
small orange ochre			orange white-spot skipper を見よ
small orange-tip		*Colotis evagore antigone* (Boisduval)	(チョウ目、シロチョウ科) エチオピア区
small orange tip			little orange tip を見よ
small orange tips			orange tips を見よ
small Oriental cotton stainer			small cotton bug を見よ
small owl-butterfly		*Narope minor* Casagrande	(チョウ目、タテハチョウ科) 新熱帯区
small paintbrush swift		*Baoris chapmani* Evans	(チョウ目、セセリチョウ科) 東洋区
small paintbrush swift			black paintbrush swift を見よ
small pale geometrid	ヒメウスアオシャク	*Jodis putata orientalis* Wehrli	(チョウ目、シャクガ科) 日本、旧北区
small palm bob		*Suastus minutus* (Moore)	(チョウ目、セセリチョウ科) 東洋区
small palm nightfighter		*Zophopetes ganda* Evans	(チョウ目、セセリチョウ科) エチオピア区
small pasture scarab		*Sericesthis nigra* (Lea)	(コウチュウ目、コガネムシ科) 豪州区
small-patched metalmark		*Caria domitianus vejento* Clench	(チョウ目、シジミタテハ科) 新熱帯区
small pearl-bordered fritillary	ナカギンコヒョウモン	*Clossiana selene* (Denis et Schiffermüller)	(チョウ目、タテハチョウ科) 全北区

英名	和名	学名	所属、分布、ほか
small pearl-white		*Elodina walkeri* Butler	(チョウ目、シロチョウ科) 豪州区
small Phigalia moth		*Phigalia strigataria* (Minot)	(チョウ目、シャクガ科) 新北区
small phoenix		*Ecliptopera silaceata* (Denis et Schiffermüller)	(チョウ目、シャクガ科) 旧北区。日本亜種は *E. s. leuca* (Diakonov) ヒメハガタナミシャク。small phoenix moth ともいう
small pied blue		*Megisba strongyle* (C. Felder)	(チョウ目、シジミチョウ科) 豪州区
small pigeon louse (1)	ハトマルハジラミ	*Campanulotes bidentatus compar* (Burmeister)	(ハジラミ目、チョウカクハジラミ科) 日本、汎世界
small pigeon louse (2)		*Goniodes minor* Piaget	(ハジラミ目、チョウカクハジラミ科) 新北区
small pincertail		*Onychogomphus forcipatus* (Linnaeus)	(トンボ目、サナエトンボ科) 旧北区
small pine looper moth		*Eupithecia palpata* Packard	(チョウ目、シャクガ科) 新北区
small plague grasshopper		*Austroicetes cruciata* (Saussure)	(バッタ目、バッタ科) 豪州区
small plain stiletto		*Thereva fulva* (Meigen)	(ハエ目、ツルギアブ科) 旧北区
small pond dragonfly			graphic flutterer を見よ
small poplar beetle			small poplar longhorn を見よ
small poplar leaf beetle		*Phyllodecta laticollis* Suffrian	(コウチュウ目、ハムシ科) 旧北区
small poplar longhorn		*Saperda populnea* (Linnaeus)	(コウチュウ目、カミキリムシ科) 全北区。small poplar borer ともいう
small postman			common longwing を見よ
small powderpost beetle		*Lyctus discedens* Blackburn	(コウチュウ目、ナガシンクイムシ科) 豪州区
small-pox bug			subterranean chinch bug を見よ
small prominent	ヒナシャチホコ	*Micromelalopha troglodyla* (Graeser)	(チョウ目、シャチホコガ科) 日本、旧北区
small purple barred		*Phytometra viridaria* (Clerck)	(チョウ目、ヤガ科) 旧北区
small purple bars			small purple barred を見よ
small purple line blue		*Prosotas dubiosa* (Semper)	(チョウ目、シジミチョウ科) 豪州区。東洋区の *P. d. lumpura* Corbet も同英名
small purple-lined borer		*Chilo agamemnon* Bleszynski	(チョウ目、メイガ科) 旧北区、エチオピア区。イネ害虫
small purplish gray moth		*Iridopsis humaria* (Guenée)	(チョウ目、シャクガ科) 新北区
small quaker		*Orthosia cruda* (Denis et Schiffermüller)	(チョウ目、ヤガ科) 旧北区
small ranunculus moth		*Hecatera dysodea* (Denis et Schiffermüller)	(チョウ目、ヤガ科) 全北区。small ranuncule, small ranunculus ともいう
small raspberry aphid		*Amphorophora rubicola* Oestlund	(カメムシ目、アブラムシ科) 新北区
small raspberry aphid			raspberry aphid (1) を見よ

英名	和名	学名	所属、分布、ほか
small raspberry sawfly (1)		*Priophorus brullei* Dahlbom	(ハチ目、ハバチ科) 旧北区
small raspberry sawfly (2)		*Priophorus morio* (Lepeletier)	(ハチ目、ハバチ科) 旧北区
small red bob		*Idmon obliquans* (Mabille)	(チョウ目、セセリチョウ科) 東洋区
small red damselfly		*Ceriagrion tenellum* (de Villers)	(トンボ目、イトトンボ科) 旧北区
small red playboy		*Pilodeudorix aruma aruma* (Hewitson)	(チョウ目、シジミチョウ科) エチオピア区
small red-belted clearwing			apple clearwing moth を見よ
small reddish-hindwinged arctiid	コベニシタヒトリ	*Rhyparioides metelkana flavidus* (Bremer)	(チョウ目、ヒトリガ科) 日本、旧北区
small redeye		*Gangara sanguinocculus* (Martin)	(チョウ目、セセリチョウ科) 東洋区
small red-eyed damselfaly		*Erythromma viridulum* (Charpentier)	(トンボ目、イトトンボ科) 旧北区
small red-eyed damselfaly			small bluetail を見よ
small rice case worm			small rice casebearer を見よ
small rice casebearer	イネコミズメイガ	*Paraponyx vittalis* (Bremer)	(チョウ目、メイガ科) 日本、旧北区。イネ害虫
small rice froghopper	ヒメフタテンナガアワフキ	*Clovia punctata* (Walker)	(カメムシ目、アワフキムシ科) 日本、東洋区。イネ害虫
small rice grasshopper			rice grasshopper (2)(3)(4) を見よ
small rice grasshoppers			Japanese grasshoppers を見よ
small rice stink bug			South American rice stink bug を見よ
small rice weevil			rice weevil を見よ
small ringlet			large heath を見よ
small rivulet		*Perizoma alchemillata* (Linnaeus)	(チョウ目、シャクガ科) 全北区。small rivulet moth ともいう
small rufous			rufous moth を見よ
small runic smudge		*Ypsolopha sequella* Denis et Schiffermüller	(チョウ目、スガ科) 旧北区
small rustic smudge		*Ypsolophus sequellus* (Clerck)	(チョウ目、クチブサガ科) 旧北区
small salmon Arab	カライスツマアカシロチョウ	*Colotis calais* (Cramer)	(チョウ目、シロチョウ科) エチオピア区
small salmon Arab			Topax Arab を見よ
small scallop		*Idaea emarginata* (Linnaeus)	(チョウ目、シャクガ科) 旧北区
small scallop wave			small scallop を見よ
small sedge neb		*Monochroa arundinetella* (Stainton)	(チョウ目、キバガ科) 旧北区
small seraphim		*Pterapherapteryx sexalata* (Retzius)	(チョウ目、シャクガ科) 旧北区

英名	和名	学名	所属、分布、ほか
small shiia bark beetle			castanopsis ambrosia beetle (2) を見よ
small shot-hole borer			shot-hole borer (3) を見よ
small silkmoth		*Ocinara ficicola* (Westwood)	(チョウ目、カイコガ科) 旧北区
small silky skipper		*Semalea atrio* (Mabille)	(チョウ目、セセリチョウ科) エチオピア区
small silverfork		*Lethe jalaurida* (de Nicéville)	(チョウ目、タテハチョウ科) 東洋区
small skipper	キマダラセセリ	*Potanthus flavus flavus* (Murray)	(チョウ目、セセリチョウ科) 日本、旧北区
small skipper (1)	ヘリグロチャバネセセリ	*Thymelicus sylvestris* Poda	(チョウ目、セセリチョウ科) 旧北区
small skipper butterfly			small skipper (1) を見よ
small southern pine engraver		*Ips avulsus* (Eichhoff)	(コウチュウ目、キクイムシ科) 新北区
small-spot palm bob		*Suastus minutus flemingi* Eliot	(チョウ目、セセリチョウ科) 東洋区。small palm bob を参照
small spotted flasher		*Astraptes egregius* Butler	(チョウ目、セセリチョウ科) 新北区、新熱帯区
small-spotted silverdrop		*Epargyreus clavicornis gaumeri* Godman et Salvin	(チョウ目、セセリチョウ科) 新熱帯区
small spotted skipperling		*Piruna microsticta* (Godman)	(チョウ目、セセリチョウ科) 新北区
small-spotted skipperling			hour-glass skipperling を見よ
small spreadwing			small emerald damselfly を見よ
small sprite streak			pallid hairstreak を見よ
small spurwing		*Centroptilum luteolum* (Müller)	(カゲロウ目、コカゲロウ科) 旧北区
small spurwing		*Antigonus corrosus* Mabille	(チョウ目、セセリチョウ科) 新熱帯区
small squaregill mayflies	ヒメカゲロウ科	Caenidae	(カゲロウ目) の昆虫の総称
small square-spot		*Diarsia rubi* (Vieweg)	(チョウ目、ヤガ科) 旧北区
small stable tabby			small tabby を見よ
small staff sergeant		*Athyma zeroca* Moore	(チョウ目、タテハチョウ科) 東洋区
small stink bug	ヒメカメムシ	*Rubiconia intermedia* Wolff	(カメムシ目、カメムシ科) 日本、旧北区
small streaked sailor			streaked sailor を見よ
small striped flea beetle			striped flea beetle (2) を見よ
small striped grey moth		*Diptychophora elaina* Meyrick	(チョウ目、メイガ科) 豪州区
small striped hawk		*Hippotion roseipennis* (Butler)	(チョウ目、スズメガ科) エチオピア区
small-striped swordtail			common swordtail (1) を見よ
small sugarcane borer			sugarcane borer (1) を見よ

英名	和名	学名	所属、分布、ほか
small swift		*Borbo perobscura* (Druce)	(チョウ目、セセリチョウ科) エチオピア区
small tabby		*Aglossa caprealis* (Hübner)	(チョウ目、メイガ科) 汎世界
small tan German cockroach			German cockroach を見よ
small tawny wall	モーレイキンジャノメ	*Rhaphicera moorei* (Butler)	(チョウ目、タテハチョウ科) 東洋区
small Telemiades		*Telemiades fides* Bell	(チョウ目、セセリチョウ科) 新熱帯区
small Thais			southern festoon を見よ
small three-ring		*Ypthima norma posticalis* Matsumura	(チョウ目、タテハチョウ科) 東洋区
small Thyridia		*Thyridia themisto* Hübner	(チョウ目、タテハチョウ科) 新熱帯区
small tiger longicorn	コトラカミキリ	*Plagionotus pulcher* Blessig	(コウチュウ目、カミキリムシ科) 日本、旧北区
small tiger moth	コトラガ	*Mimeusemia persimilis* Butler	(チョウ目、トラガ科) 日本、旧北区
small tineid moth			apple fruit moth を見よ
small Tolype moth		*Tolype notialis* Franclemont	(チョウ目、カレハガ科) 新北区
small tortoiseshell		*Aglais urticae* (Linnaeus)	(チョウ目、タテハチョウ科) 全北区。日本亜種は *A. u. connexa* (Butler) コヒオドシ。small tortoiseshell butterfly ともいう
small tussock caterpillar	コシロモンドクガ	*Orgyia postica* Walker	(チョウ目、ドクガ科) 日本、旧北区、東洋区、豪州区。ダイズ、リンゴ、チャ他害虫
small twin-spot carpet		*Calostigia didymata* Linnaeus	(チョウ目、シャクガ科) 旧北区
small two-banded prominent	コフタオビシャチホコ	*Gluphisia crenata japonica* Wileman	(チョウ目、シャチホコガ科) 日本
small Verdant hawk		*Basiothia medea* (Fabricius)	(チョウ目、スズメガ科) エチオピア区
small wainscot		*Photedes pygmina* (Haworth)	(チョウ目、ヤガ科) 旧北区
small water scorpion			water scorpion (3) を見よ
small water strider		*Velia currens* (Fabricius)	(カメムシ目、カタビロアメンボ科) 旧北区
small waved umber		*Horisme vitalbata* Denis et Schiffermüller	(チョウ目、シャクガ科) 旧北区。日本亜種は *H. v. staudingeri* Prout ボタンズルナミシャク
small white			common white を見よ。*P. rapae* Linnaeus の英国での英名。small white butterfly ともいう
small white			southern cabbage worm を見よ
small white-banded froghopper	ヒメシロオビアワフキ	*Aphrophora obliqua* Uhler	(カメムシ目、アワフキムシ科) 日本、旧北区
small white grass-veneer moth		*Crambus albellus* Clemens	(チョウ目、メイガ科) 新北区。small white grass-veneer ともいう
small white grass yellow		*Eurema lirina* (Bates)	(チョウ目、シロチョウ科) 新熱帯区
small white-hindwinged catocala	ヒメシロシタバ	*Catocala nagioides* Wileman	(チョウ目、ヤガ科) 日本、旧北区
small white-lady swordtail			white lady swordtail を見よ

英名	和名	学名	所属、分布、ほか
small white-marmorated longicorn	ヒメシラフヒゲナガカミキリ	*Monochamus sutor* (Linnaeus)	（コウチュウ目、カミキリムシ科）日本、旧北区
small white wave		*Asthena candidata* (Denis et Schiffermüller)	（チョウ目、シャクガ科）旧北区
small white wave (1)		*Asthena albulata* (Hufnagel)	（チョウ目、シャクガ科）旧北区
small whiteface			white-faced darter (1) を見よ
small willow aphid		*Aphis farinosa* Gmelin	（カメムシ目、アブラムシ科）旧北区
small willow aphid			willow aphid を見よ
small willow-clearwing			red-tipped clearwing を見よ
small winter moth			winter moth を見よ
small winter stoneflies			smoky stoneflies を見よ
small wood nymph			dark wood nymph を見よ
small woodbrown		*Lethe nicetella* de Nicéville	（チョウ目、タテハチョウ科）東洋区
small woodland mosquito		*Aedes cinerous* Meigen	（ハエ目、カ科）新北区
small yellow grass dart			dingy grass dart を見よ
small yellow-hindwinged catocala	コガタキシタバ	*Catocala praegnax esther* Butler	（チョウ目、ヤガ科）日本。smaller yellow-hindwinged catocala ともいう
small yellow jacket	コガタスズメバチ	*Vespa analis insularis* Dalla Torre	（ハチ目、スズメバチ科）日本
small yellow-legged robberfly		*Dioctria linearis* (Fabricius)	（ハエ目、ムシヒキアブ科）旧北区
small yellow sailer	ネッタイキイロミスジ	*Neptis miah* Moore	（チョウ目、タテハチョウ科）東洋区
small yellow-spotted leaf beetle	キボシツツハムシ	*Cryptocephalus perelegans perelegans* Baly	（コウチュウ目、ハムシ科）日本、東洋区
small yellow underwing		*Panemeria tenebrata* (Scopoli)	（チョウ目、ヤガ科）旧北区
small yellow wave	キオビナミシャク	*Hydrelia flammeolaria* (Hufnagel)	（チョウ目、シャクガ科）日本、旧北区
smaller akebia leaf-like moth			fruit-piercing moth (1) を見よ
smaller alnus bark beetle			alnus ambrosia beetle を見よ
smaller angular-winged katydid			angular-winged katydid (2) を見よ
smaller auricles leafhopper	コミミズク	*Ledropsis discolor* (Uhler)	（カメムシ目、オオヨコバイ科）日本、旧北区
smaller bamboo shot-hole borer			bamboo powderpost beetle を見よ　米国での英名
smaller bean narrow-mouth weevil			bean narrow-mouth weevil を見よ
smaller biting knotweed leaf beetle	ジュンサイハムシ	*Galerucella nipponensis* (Laboissier)	（コウチュウ目、ハムシ科）日本、旧北区

英名	和名	学名	所属、分布、ほか
smaller black leaf-cut weevil			yellow-legged black leafcut weevil を見よ
smaller black rice bug	ヒメクロカメムシ	Scotinophara scotti Horvath	(カメムシ目、カメムシ科) 日本、旧北区
smaller blotch-marked bell		Epiblema cirsiana Zeller	(チョウ目、ハマキガ科) 旧北区
smaller brownish chafer	ハイイロビロウドコガネ	Paraserica grisea (Motschulsky)	(コウチュウ目、コガネムシ科) 日本
smaller cherry chafer	ヒメサクラコガネ	Anomala geniculata (Motschulsky)	(コウチュウ目、コガネムシ科) 日本
smaller chrysanthemum aphid	ヨモギコブアブラムシ	Micraphis artemisiae (Takahashi)	(カメムシ目、アブラムシ科) 日本、旧北区、東洋区
smaller citrus cottony scale	ミカンヒモワタカイガラムシ	Saisetia citricola (Kuwana)	(カメムシ目、カタカイガラムシ科) 日本、旧北区
smaller citrus dog			citrus swallowtail (1) を見よ
smaller citrus leafhopper	ミカンヒメヨコバイ	Zyginella citri (Matsumura)	(カメムシ目、オオヨコバイ科) 日本。カンキツ害虫
smaller cucurbit stem borer	カノコサビカミキリ	Apomecyna naevia Bates	(コウチュウ目、カミキリムシ科) 日本。ウリ類害虫
smaller diving beetle	コガタノゲンゴロウ	Cybister tripunctatus orientalis Gschwendtner	(コウチュウ目、ゲンゴロウ科) 日本、旧北区、東洋区
smaller elm bark beetle			elm bark beetle (2) を見よ
smaller European elm bark beetle			elm bark beetle (2) を見よ
smaller evening moth	ホシヒメホウジャク	Aspledon himachala sangaica (Butler)	(チョウ目、スズメガ科) 日本、東洋区
smaller false chinch bug	ヒメナガカメムシ	Nysius plebejus Distant	(カメムシ目、ナガカメムシ科) 日本、大洋区
smaller flea beetle	コカミナリハムシ	Altica viridicyanea (Baly)	(コウチュウ目、ハムシ科) 日本、旧北区
smaller flower chafer	ヒメアシナガコガネ	Ectinohoplia obducta (Motschulsky)	(コウチュウ目、コガネムシ科) 日本
smaller four-spotted leafhopper	ヨツモンヒメヨコバイ	Empoascanara limbata (Matsumura)	(カメムシ目、オオヨコバイ科) 日本、旧北区、東洋区
smaller globular stink bug	ヒメマルカメムシ	Coptosoma biguttulum Motschulsky	(カメムシ目、マルカメムシ科) 日本、旧北区
smaller grape hawk moth	ヒメスズメ	Deilephila askoldensis (Oberthür)	(チョウ目、スズメガ科) 日本、旧北区
smaller green flower chafer			citrus flower chafer を見よ
smaller Hawaiian cutworm		Agrotis dislocata (Walker)	(チョウ目、ヤガ科) 新北区
smaller hiba bark beetle	ヒバノコキクイムシ	Phloeosinus lewisi Chapuis	(コウチュウ目、キクイムシ科) 日本、東洋区
smaller Japanese cedar chafer	ヒメスジコガネ	Mimela flavifabris (Waterhouse)	(コウチュウ目、コガネムシ科) 日本
smaller Japanese cedar mealybug	スギヒメコナカイガラムシ	Spirococcus flavidus (Kanda)	(カメムシ目、コナカイガラムシ科) 日本

英名	和名	学名	所属、分布、ほか
smaller Japanese ceder longicorn	ヒメスギカミキリ	Palaeocallidium rufipenne (Motschulsky)	(コウチュウ目、カミキリムシ科) 日本、旧北区。イブキ、スギ、ヒノキ害虫
smaller Japanese larch torymid			larch torymid を見よ
smaller Japanese silk moth	ヒメヤママユ	Caligula jonasi Butler	(チョウ目、ヤママユガ科) 日本
smaller lantana butterfly	ランタナコツバメ	Strymon bazochii (Godart)	(チョウ目、シジミチョウ科) 新北区、新熱帯区。大洋区の亜種 S. b. gundlachiamus (Bates) も同英名。ランタナ除除のためにメキシコからハワイに導入された
smaller lasiocampid	ヒメカレハ	Phyllodesma japonica (Leech)	(チョウ目、カレハガ科) 日本
smaller long-headed locust	ショウリョウバッタモドキ	Gonista bicolor (de Haan)	(バッタ目、バッタ科) 日本、東洋区
smaller maize borer			peach moth を見よ
smaller marmorated grasshopper	マダラバッタ	Aiolopus tamulus (Fabricius)	(バッタ目、バッタ科) 日本、東洋区、豪州区、大洋区
smaller marmorated leafhopper	マダラヒメヨコバイ	Platytettix pulchra (Matsumura)	(カメムシ目、オオヨコバイ科) 日本、東洋区
smaller meadow katydids			long-horned grasshoppers (2) を見よ
smaller mulberry leaf roller	コクワヒメハマキ	Olethreutes morivora (Matsumura)	(チョウ目、ハマキガ科) 日本
smaller network-marked leafhopper	コチャイロヨコバイ	Matsumurella kogatensis (Matsumura)	(カメムシ目、オオヨコバイ科) 日本、旧北区
smaller oraesia			fruit-piercing moth (10) を見よ
smaller Parasa moth			green slug moth を見よ
smaller pear aphid	ナシオナガアブラムシ	Melanaphis siphonella (Essig et Kuwana)	(カメムシ目、アブラムシ科) 日本
smaller refuse beetle			constricted refuse beetle を見よ
smaller rice crane fly	ヒメキリウジガガンボ	Tipula latemarginata Alexander	(ハエ目、ガガンボ科) 日本、旧北区
smaller rice leafminer			rice leafminer (2) を見よ
smaller-ripped hawk moth	ヒメサザナミスズメ	Dolbina exacta Staudinger	(チョウ目、スズメガ科) 日本、旧北区
smaller rusty grain beetle	コカクムネヒラタムシ	Cryptolestes turcicus Grouvelle	(コウチュウ目、ヒラタムシ科) 日本、汎世界
smaller sand cricket		Ellipes minuta (Scudder)	(バッタ目、ノミバッタ科) 新北区。体長 5 mm
smaller sand scarab		Pericoptus punctatus (White)	(コウチュウ目、コガネムシ科) 豪州区
smaller sugarcane chafer	ヒメカンショコガネ	Apogonia amida Lewis	(コウチュウ目、コガネムシ科) 日本
smaller sugarcane tortoise beetle			tortoise beetle (1) を見よ
smaller tea tortrix	チャノコカクモンハマキ	Adoxophyes honmai Yasuda	(チョウ目、ハマキガ科) 日本。果樹、チャ、バラなどの害虫
smaller tortoise beetle	ヒメジンガサハムシ	Cassida fuscorufa Motschulsky	(コウチュウ目、ハムシ科) 日本、旧北区、東洋区

英名	和名	学名	所属、分布、ほか
smaller turnip maggot			radish root maggot を見よ
smaller velvety chafer	ヒメビロウドコガネ	*Maladera orientalis* (Motschulsky)	(コウチュウ目、コガネムシ科) 日本
smaller water strider bugs			ripple bugs を見よ
smaller water striders			ripple bugs を見よ
smaller wax moth			lesser wax moth を見よ
smaller white tussock moth	ヒメシロドクガ	*Arctornis chichibense* (Matsumura)	(チョウ目、ドクガ科) 日本
smaller willow curculio	ヤドリチビシギゾウ	*Curculio parsiticus* Morimoto	(コウチュウ目、ゾウムシ科) 日本
smaller wood nymph	ヒメゴマダラ	*Ideopsis gaura perakana* Fruhstorfer	(チョウ目、タテハチョウ科) 東洋区
smaller yellow ant		*Acanthomyops claviger* (Roger)	(ハチ目、アリ科) 新北区
smaller yellow jacket	ヒメスズメバチ	*Vespa tropica pulchra* Buysson	(ハチ目、スズメバチ科) 日本
smaller yellow-legged grass cricket	クロヒバリモドキ	*Trigonidium cicindeloides* Rambur	(バッタ目、クサヒバリ科) 日本、旧北区、東洋区、エチオピア区
smaller yellow-striped prominent	クロツマキシャチホコ	*Phalera minor* Nagano	(チョウ目、シャチホコガ科) 日本
smaller Yezo cicada	コエゾゼミ	*Lyristes bihamatus* (Motschulsky)	(カメムシ目、セミ科) 日本、旧北区
smallest hydrohili	ヒメガムシ	*Sternolophus rufipes* (Fabricius)	(コウチュウ目、ガムシ科) 日本、旧北区、東洋区
smart-banded hunchback		*Ogcodes gibbosus* (Linnaeus)	(ハエ目、コガシラアブ科) 旧北区
smartweed borer		*Ostrinia obumbratalis* (Lederer)	(チョウ目、メイガ科) 新北区。成虫は smartweed borer moth
smartweed caterpillar			smeared dagger moth を見よ
smartweed copper			Gorgon copper を見よ
smartweed flea beetle		*Systena hudsonias* Förster	(コウチュウ目、ハムシ科) 新北区。トウモロコシ害虫
smeared dagger moth		*Acronicta oblinita* (Smith)	(チョウ目、ヤガ科) 新北区。幼虫はタデ類を加害
smear-spotted skipper		*Xeniades orchamus orchamus* (Cramer)	(チョウ目、セセリチョウ科) 新熱帯区
Smeathman's Aethes moth			Smeathman's conch を見よ
Smeathman's conch		*Aethes smeathmanniana* (Fabricius)	(チョウ目、ホソハマキガ科) 旧北区
smilax aphid	サルトリトックリアブラムシ	*Rhopalomyzus smilacis* (Matsumura)	(カメムシ目、アブラムシ科) 日本
smilax long-horned aphid	ホウセンカヒゲナガアブラムシ	*Impatientinum impatiens* (Shinji)	(カメムシ目、アブラムシ科) 日本、旧北区、東洋区
smilax scale	サルトリイバラノカイガラムシ	*Aulacaspis spinosa* Maskell	(カメムシ目、マルカイガラムシ科) 日本、旧北区、東洋区
smilax thrips			banana thrips を見よ

英名	和名	学名	所属、分布、ほか
sminthurid springtails			globular springtails を見よ
sminthurids			globular springtails を見よ
Smith bamboo scale	アリノタカラカイガラムシ	*Eumyrmococcus smithii* Silvestri	(カメムシ目、コナカイガラムシ科) 日本、東洋区
Smith sawfly	シマヒラタハバチ	*Pamphilius volatilis* (Smith)	(ハチ目、ヒラタハバチ科) 日本、旧北区。ナシ、モモ、サクラなどの害虫
Smith's darkwing moth		*Gondysia smithii* (Guenée)	(チョウ目、ヤガ科) 新北区。Smith's darkwing ともいう
Smith's giant-skipper		*Stallingsia smithi* (Druce)	(チョウ目、セセリチョウ科) 新熱帯区
Smith's Pixie		*Melanis smithiae* (Westwood)	(チョウ目、シジミタテハ科) 新熱帯区
Smith's strepsicrates moth			bayberry leaftier moth を見よ
smoke beetle		*Melanophila consputa* LeConte	(コウチュウ目、タマムシ科) 新北区
smoke flies			flat-footed flies を見よ
smoke fly		*Microsania australis* (Walker)	(ハエ目、ヒラタアシバエ科) 豪州区
smoked leafroller moth		*Archips infumatana* (Zeller)	(チョウ目、ハマキガ科) 新北区
smoked sallow moth		*Enargia infumata* (Grote)	(チョウ目、ヤガ科) 新北区
smokes		Talaeporiinae	(チョウ目、ミノガ科) の昆虫の総称
smoky alderfly		*Sialis infumata* Newman	(アミメカゲロウ目、センブリ科) 新北区。ハンノキに多い
smoky arches moth		*Drasteria fumosa* (Strecker)	(チョウ目、ヤガ科) 新北区。smoky arches ともいう
smoky bean cupid			smoky blue を見よ
smoky blue		*Euchrysops malathana* (Boisduval)	(チョウ目、シジミチョウ科) エチオピア区
smoky-brown cockroach	クロゴキブリ	*Periplaneta fuliginosa* (Serville)	(ゴキブリ目、ゴキブリ科) 日本、汎世界。日本では暖地の屋内普通種
smoky buckeye			West Indian buckeye を見よ
smoky buzzer		*Cicadetta waterhousei* (Distant)	(カメムシ目、セミ科) 豪州区
smoky chalk blue		*Thermoniphas fumosa* Stempffer	(チョウ目、シジミチョウ科) エチオピア区
smoky crane fly		*Tipula cunctans* Say	(ハエ目、ガガンボ科) 新北区
smoky dusk-darter		*Zyxomma atlanticum* Selys	(トンボ目、トンボ科) エチオピア区
smoky eyed brown		*Satyrodes eurydice fumosa* (Leussler)	(チョウ目、タテハチョウ科) 新北区。eyed brown を参照
smoky Idia moth		*Idia scobialis* (Grote)	(チョウ目、ヤガ科) 新北区。smoky Idia ともいう
smoky marbled piercer		*Cydia fagiglandana* (Zeller)	(チョウ目、ハマキガ科) 旧北区
smoky moth		*Lexis bicolore* (Grote)	(チョウ目、ヒトリガ科) 新北区
smoky moths		Ctenuchinae	(チョウ目、カノコガ科) の昆虫の総称
smoky moths (1)		Pyromorphidae	(チョウ目) の昆虫の総称　北米での英名

英名	和名	学名	所属、分布、ほか
smoky moths			leaf skeletonizer moths (2) を見よ
smoky orange tip	ヒメツマアカシロチョウ	*Colotis euippe omphale* (Godart)	(チョウ目、シロチョウ科) エチオピア区。round-winged orange tip を参照
smoky rubyspot		*Hetaerina titia* (Drury)	(トンボ目、イトトンボ科) 新北区
smoky sage		*Psaltoda fumipennis* Ashton	(カメムシ目、セミ科) 豪州区
smoky spreadwing		*Lestes virgatus* (Burmeister)	(トンボ目、アオイトトンボ科) エチオピア区
smoky stoneflies	クロカワゲラ科	Capniidae	(カワゲラ目) の昆虫の総称
smoky Temnora		*Temnora fumosa* (Walker)	(チョウ目、スズメガ科) エチオピア区
smoky Tetanolita moth		*Tetanolita mynesalis* Walker	(チョウ目、ヤガ科) 新北区。smoky Tetanolita ともいう
smoky wainscot	ヨシノキヨトウ	*Mythimna impura* (Hübner)	(チョウ目、ヤガ科) 日本、旧北区
smoky wave		*Scopula ternata* (Schrank)	(チョウ目、シャクガ科) 旧北区
smoky-winged argia		*Argia fumipennis* (Burmeister)	(トンボ目、イトトンボ科) 新北区
smoky-winged darner		*Epiaeschna heros* (Fabricius)	(トンボ目、ヤンマ科) 新北区。開帳 12 cm
smoky-winged threadtail		*Elattoneura leucostigma* (Fraser)	(トンボ目、モノサシトンボ科) 東洋区
smooth Amphipyra moth		*Amphipyra glabella* Morrison	(チョウ目、ヤガ科) 新北区
smooth-banded sister			five continent を見よ
smooth broad-nosed weevil			strawberry fruit weevil を見よ
smooth cockroach		*Symploce pallens* (Stephens)	(ゴキブリ目、チャバネゴキブリ科) 東洋区
smooth emerald		*Omphax plantaria* Guenée	(チョウ目、シャクガ科) エチオピア区
smooth-legged tree crickets		*Neoxabea*	(バッタ目、コオロギ科) の昆虫の総称
smooth pea gall wasp		*Diplolepis eglanteriae* (Hartig)	(ハチ目、タマバチ科) 旧北区。本種がバラにつくる虫えいは smooth pea gall といい、1 枚の葉に数個生じる
smooth rustic			rustic (1) を見よ
smooth shieldback		*Atlanticus glaber* Rehn et Hebard	(バッタ目、キリギリス科) 新北区
smooth spider beetle			museum beetle (1) を見よ
smooth springtails	ツチトビムシ科	Isotomidae	(トビムシ目) の昆虫の総称　日本から約 70 種
smooth stick-insect			common stick insect を見よ
smooth sucking lice	ケモノホソジラミ科	Linognathidae	(シラミ目) の昆虫の総称　家畜と一部の野生動物に寄生。世界に約 70 種
smooth water boatmen		*Trichocorixa*	(カメムシ目、ミズムシ科) の昆虫の総称
smudged crescent			mimic crescent (1) を見よ
smudged hairstreak		*Rekoa stagira* (Hewitson)	(チョウ目、シジミチョウ科) 新熱帯区

英名	和名	学名	所属、分布、ほか
smudged Temnora		*Temnora subapicalis* Rothschild et Jordan	（チョウ目、スズメガ科）エチオピア区
smudges			ermine moths (1) を見よ　スガ科の異名 Plutellidae の英名
smut beetle		*Phalacrus politus* Melsheimer	（コウチュウ目、ヒメハナムシ科）新北区
snailcase bagworm moth		*Apterona helix* (Siebold)	（チョウ目、ミノガ科）新北区
snail-case caddisflies	カタツムリトビケラ科	Helicopsychidae	（トビケラ目）の昆虫の総称
snail casemaker caddisflies			snail-case caddisflies を見よ
snaileaters		*Scaphinotus*	（コウチュウ目、オサムシ科）の昆虫の総称
snail-eating carabid		*Brennus ventricosus* (Dejean)	（コウチュウ目、オサムシ科）新北区
snail eating ground beetle		*Scaphinotus angulatus* Harris	（コウチュウ目、オサムシ科）新北区。貝を捕食
snail-eating ground beetles		Cychrini	（コウチュウ目、オサムシ科）の昆虫の総称
snail-killing flies			marsh flies を見よ　幼虫が貝を捕食することに由来
snail killing fly		*Hydromya dorsalis* (Fabricius)	（ハエ目、ヤチバエ科）日本、旧北区
snail parasitic blowfly		*Amenia imperialis* Robineau-Desvoidy	（ハエ目、クロバエ科）豪州区
snailshell caddisflies			snail-case caddisflies を見よ
snake fly		*Raphidia notata* Fabricius	（アミメカゲロウ目、キスジラクダムシ科）旧北区
snake worm			army worm (3) を見よ　幼虫の英名
snakedoctors			dragonflies (1) を見よ
snake-feeders			dragonflies (1) を見よ
snake-flies (1)	ラクダムシ亜目	Raphidiodea	（アミメカゲロウ目）の昆虫の総称
snake-flies (2)	ラクダムシ科	Inocellidae	（アミメカゲロウ目）の昆虫の総称
snake-flies (3)	キスジラクダムシ科	Raphididae	（アミメカゲロウ目）の昆虫の総称
snakeflies		*Raphidia*	（アミメカゲロウ目、キスジラクダムシ科）の昆虫の総称
snakeflies		*Inocellia*	（アミメカゲロウ目、ラクダムシ科）の昆虫の総称
snakeflies			common snakeflies を見よ
snakeflies			lacewings (1) を見よ
snakeflies		Raphidioptera	snake-flies (1) を見よ　ラクダムシ亜目 Raphidiodea の異名
snakefly		*Raphidia ophiopsis* Linnaeus	（アミメカゲロウ目、キスジラクダムシ科）旧北区
snakefly		*Raphidia maculicollis* (Stephens)	（アミメカゲロウ目、キスジラクダムシ科）旧北区
snake-gourd stem borer		*Melittia bombyliformis* Cramer	（チョウ目、スカシバガ科）東洋区

英名	和名	学名	所属、分布、ほか
snake-stang			dragonflies (1) を見よ
snakeweed borer moth		*Eucosma ridingsana* (Robinson)	(チョウ目、ハマキガ科) 新北区
snakeweed grasshopper		*Hesperotettix viridis* (Thomas)	(バッタ目、バッタ科) 新北区
snap beetles			click beetles (1) を見よ
snapdragon plume moth		*Stenoptilodes antirrhina* (Lange)	(チョウ目、トリバガ科) 新北区
snapping beetles			click beetles (1) を見よ
sneg			pear sawfly (1) を見よ　幼虫の英名
Snellens Cerulean		*Jamides snelleni* (Rober)	(チョウ目、シジミチョウ科) 東洋区
snipe flies	シギアブ科	Rhagionidae	(ハエ目) の昆虫の総称
snipe fly (1)		*Rhagio scolopacea* (Linnaeus)	(ハエ目、シギアブ科) 全北区
snipe fly (2)		*Rhagio mystacea* (Macquart)	(ハエ目、シギアブ科) 新北区
snout (1)	フタオビアツバ	*Hypena proboscidalis* (Linnaeus)	(チョウ目、ヤガ科) 日本、旧北区
snout (2)	カキバトモエ	*Hypopyra vespertilio* (Fabricius)	(チョウ目、ヤガ科) 日本、東洋区。突出した下唇鬚に由来
snout			nettle-tree butterfly を見よ
snout beetle		*Brachyrhinus cribricollis* (Gyllenhal)	(コウチュウ目、ゾウムシ科) 新北区。カンキツ害虫
snout beetle			black weevil (1) を見よ
snout beetles			weevils (1) を見よ
snout beetles			rhynchophorous beetles を見よ
snout butterflies			snouts を見よ
snout butterfly		*Libythea labdaca laius* Trimen	(チョウ目、タテハチョウ科) エチオピア区。African beak を参照
snout butterfly (1)	アメリカテングチョウ	*Libytheana bachmanni bachmanni* Kirtland	(チョウ目、タテハチョウ科) 東洋区
snout moth			snout (1) を見よ
snout moths			tent caterpillar moths (1) を見よ　豪州での英名
snout moths			grass moths (1)(2) を見よ
snout moths			snouts (1) を見よ
snout prominent	オオエグリシャチホコ	*Pterostoma sinicum* Moore	(チョウ目、シャチホコガ科) 日本、旧北区
snout weevils			weevils (1) を見よ
snouted grasshopper		*Acrida hungarica mediterranea* Dirsh	(バッタ目、バッタ科) 旧北区
snouted skipper		*Anisochoria pedaliodina* (Butler)	(チョウ目、セセリチョウ科) 新熱帯区
snouts	テングチョウ亜科	Libytheinae	(チョウ目、タテハチョウ科) の昆虫の総称
snouts (1)	アツバ亜科	Hypeninae	(チョウ目、ヤガ科) の昆虫の総称
snouts			beaks を見よ

英名	和名	学名	所属、分布、ほか
snow Apollos			Apollo butterflies を見よ
snow ball aphid			viburnum aphid (1) を見よ
snow flats		*Tagiades*	（チョウ目、セセリチョウ科）の昆虫の総称
snow flea		*Entomobrya nivalis* (Linnaeus)	（トビムシ目、アヤトビムシ科）旧北区
snow flea		*Boreus mucronata* Byers	（シリアゲムシ目、ユキシリアゲムシ科）旧北区
snowflea (1)		*Hypogastrura nivicola* (Fitch)	（トビムシ目、ヒメトビムシ科）新北区。雪上でかたまってみつかる
snow flea			snow scorpionfly を見よ
snowfleas	ミズトビムシ科	Poduridae	（トビムシ目）の昆虫の総称　水面を跳ぶ
snow fleas		*Chionea*	（ハエ目、ガガンボ科）の昆虫の総称
snowfleas			springtails を見よ
snowfleas			blind springtails を見よ
snow fleas			snow flies を見よ
snow flies		*Boreus*	（シリアゲムシ目、ユキシリアゲムシ科）の昆虫の総称
snow flies			snow scorpion flies を見よ
snow-flies			scorpion flies (2) を見よ
snow-flies			whiteflies を見よ
snowfly			snowflea (1) を見よ
snow fly			snow gnat を見よ
snow gnat		*Chionea nivicola* Doane	（ハエ目、ガガンボ科）新北区
snow gnat flies			snow scorpion flies を見よ
snow gnats			snow scorpion flies を見よ
snow horned skipper		*Chondrolepis niveicornis* (Plötz)	（チョウ目、セセリチョウ科）エチオピア区
snow mosquito	トカチヤブカ	*Aedes communis* (De Geer)	（ハエ目、カ科）日本、全北区
snow pool mosquitoes		*Aedes*	（ハエ目、カ科）のカ類が北半球で雪解け水に発生する時に使用される。field mosquitoes を見よ
snow scale			citrus snow scale を見よ
snow scorpion flies	ユキシリアゲムシ科	Boreidae	（シリアゲムシ目）の昆虫の総称
snow scorpionfly		*Boreus hyemalis* (Linnaeus)	（シリアゲムシ目、ユキシリアゲムシ科）旧北区
Snow skipper		*Ochlodes snowi* Edwards	（チョウ目、セセリチョウ科）新北区、新熱帯区
snowball aphid		*Ceruraphis viburnicola* (Gillette)	（カメムシ目、アブラムシ科）新北区
snowball-spotted skipper		*Paratrytone aphractoia* Dyar	（チョウ目、セセリチョウ科）新熱帯区
snowberry checkerspot		*Euphydryas chalcedona colon* (W. H. Edwards)	（チョウ目、タテハチョウ科）新北区。Chalcedon を参照
snowberry clearwing			honeysuckle clearwing を見よ
snowberry clearwing moth			bumble bee moth を見よ　snowberry clearwing とも記す

英名	和名	学名	所属、分布、ほか
snowberry clearwing sphinx			bumble bee moth を見よ
snowberry fly		*Rhagoletis zephyria* Snow	（ハエ目、ミバエ科）新北区
snowbush caterpillar		*Melanchroia chephise* Cramer	（チョウ目、シャクガ科）新熱帯区
snowflake		*Leucidia brephos* (Hübner)	（チョウ目、シロチョウ科）新熱帯区
snowflake caddisflies		*Smicridea*	（トビケラ目、シマトビケラ科）豪州区
snow-fringed skipper		*Poanes niveolimbus* (Mabille)	（チョウ目、セセリチョウ科）新熱帯区
Snow's copper		*Lycaena cupreus snowi* (W. H. Edwards)	（チョウ目、シジミチョウ科）新北区。lustrous copper を参照
Snow's owlet moth		*Isogona snowi* (Smith)	（チョウ目、ヤガ科）新北区。Snow's owlet ともいう
snowthistle aphid			currant-lettuce aphid を見よ
snow-white linden moth			elm spanworm moth を見よ
snowy carpet		*Lithostege farinata* (Hufnagel)	（チョウ目、シャクガ科）旧北区
snowy Eupseudosoma		*Eupseudosoma involutum* Druce	（チョウ目、ヒトリガ科）新北区
snowy Eupseudosoma moth			snowy Eupseudosoma を見よ
snowy fly			cabbage whitefly (1) を見よ
snowy geometer moth		*Eugonobapta nivosaria* (Guenée)	（チョウ目、シャクガ科）新北区。snowy geometer ともいう
snowy-shouldered Acleris moth		*Acleris nivisellana* (Walsingham)	（チョウ目、ハマキガ科）新北区
snowy tree cricket		*Oecanthus fultoni* Walker	（バッタ目、カンタン科）新北区。アブラムシを捕食。気温により鳴声が変わるので temperature cricket ともいわれる
snowy tree cricket (1)		*Oecanthus niveus* (De Geer)	（バッタ目、カンタン科）新北区。果樹害虫
snowy-veined Apamea moth		*Apamea niveivenosa* Grote	（チョウ目、ヤガ科）新北区。snowy-veined Apamea ともいう
Snyder's angel insect		*Zoratypus snyderi* Caudell	（ジュズヒゲムシ目、ジュズヒゲムシ科）新北区
soapberry bug			red-shouldered bug を見よ
soapberry hairstreak		*Phaeostrymon alcestis* (Edwards)	（チョウ目、シジミチョウ科）新北区
soapberry noctuid	ネジロフトクチバ	*Serrodes campana* Guenée	（チョウ目、ヤガ科）東洋区
sober Renia moth		*Renia sobrialis* (Walker)	（チョウ目、ヤガ科）新北区。sober Renia ともいう
sociable glider			ferruginous glider を見よ
sociable Pyrausta moth		*Pyrausta socialis* (Grote)	（チョウ目、メイガ科）新北区
sociable Renia moth		*Renia factiosalis* (Walker)	（チョウ目、ヤガ科）新北区
social bees			bees (1) を見よ
social bees			honey bees を見よ

英名	和名	学名	所属、分布、ほか
social parasitic ant		*Anergates atratulus* Schrank	(ハチ目、アリ科) 旧北区、新北区。米国での英名
social pear sawfly		*Neurotoma flaviventris* (Retzius)	(ハチ目、ヒラタハバチ科) 旧北区
social pear sawfly (1)		*Neurotoma saltuum* (Linnaeus)	(ハチ目、ヒラタハバチ科) 旧北区
social skipper		*Hyalothyrus neleus* (Linnaeus)	(チョウ目、セセリチョウ科) 新熱帯区
social wasps			hornets (1)(2) を見よ 英国、豪州での英名
social wasps			papernest wasps (1)(2)(3) を見よ
social white		*Eucheira socialis socialis* Westwood	(チョウ目、シロチョウ科) 新熱帯区。*E. s. westwoodi* Beutelspacher も同英名
sod webworm			elegant grass-veneer moth を見よ
sod webworm			grass webworm (1) (2) を見よ
sod webworm moth		*Pediasia trisecta* (Walker)	(チョウ目、メイガ科) 新北区
sod webworm moths		*Crambus*	(チョウ目、メイガ科) の昆虫の総称 北米での英名
sod webworms			grass moths (1) を見よ 米国での英名
sod webworms			sod webworm moths を見よ 北米での英名
sod webworms			grass moths (2) を見よ カナダでの英名
soft bamboo scale			bamboo fringed scale を見よ
soft-bodied plant beetles	ナガハナノミ科	Ptilodactylidae	(コウチュウ目) の昆虫の総称
soft brown scale			brown soft scale を見よ 豪州での英名
soft green scale		*Coccus alpinus* De Lotto	(カメムシ目、カタカイガラムシ科) エチオピア区
soft green scale			green scale を見よ
soft insects			scale insects を見よ
soft-lined wave moth		*Scopula inductata* (Guenée)	(チョウ目、シャクガ科) 新北区。soft-lined wave ともいう
soft scale			brown soft scale を見よ
soft scales	カタカイガラムシ科	Coccidae	(カメムシ目) の昆虫の総称 世界に 1,000 種。果樹、園芸植物の害虫が多い
soft-winged flower beetles (1)	ジョウカイモドキ科	Melyridae	(コウチュウ目) の昆虫の総称
soft-winged flower beetles	ジョウカイモドキ亜科	Malachiinae	(コウチュウ目、ジョウカイモドキ科) の昆虫の総称
soil proturan		*Acerella*	(カマアシムシ目、クシカマアシ科) の昆虫の総称
solanaceous fruit fly			solanum fruit fly (2) を見よ
solanaceous treehopper		*Antianthe expansa* (Germar)	(カメムシ目、ツノゼミ科) 新北区、大洋区
Solander's bell	セウスモンヒメハマキ	*Epinotia solandriana* (Linnaeus)	(チョウ目、ハマキガ科) 日本、全北区
Solander's brown		*Heteronympha solandri* Waterhouse	(チョウ目、タテハチョウ科) 豪州区
Solandra long-horned beetle			artocarpus long-horned beetle を見よ

英名	和名	学名	所属、分布、ほか
solanum aphid			potato aphid (2) を見よ
solanum flea beetle	ナスノミハムシ	*Psylliodes angusticollis* Baly	(コウチュウ目、ハムシ科) 日本、旧北区、東洋区。ジャガイモ、ナス、トマト害虫
solanum fruit fly (1)		*Dacus cacuminatus* (Hering)	(ハエ目、ミバエ科) 豪州区
solanum fruit fly (2)		*Bactrocera latifrons* (Hendel)	(ハエ目、ミバエ科) 日本、東洋区、大洋区
solanum mealybug	ナスコナカイガラムシ	*Phenacoccus solani* Ferris	(カメムシ目、コナカイガラムシ科) 日本、新北区。ナス、トウガラシ害虫
solanum root aphid			bean root aphid を見よ
soldier			アリ、シロアリの兵アリで大腮が発達
soldier (1)	ミナミカバマダラ	*Danaus eresimus* (Cramer)	(チョウ目、タテハチョウ科) 新北区、新熱帯区。*D. e. montezuma* (Talbot)、*D. e. xanthippus* (Cramer) も同英名
soldier beetle		*Tegrodera erosa* Casey	(コウチュウ目、ツチハンミョウ科) 新北区
soldier beetle		*Cantharis fusca* Linnaeus	(コウチュウ目、ジョウカイボン科) 旧北区
soldier beetle (1)		*Cantharis rustica* Fallén	(コウチュウ目、ジョウカイボン科) 旧北区
soldier beetle			blood sucker を見よ
soldier beetles		*Chauliognathus*	(コウチュウ目、ジョウカイボン科) の昆虫の総称
soldier beetles (1)	ジョウカイボン科	Cantharidae	(コウチュウ目) の昆虫の総称 世界に500種。成虫は花上に多い
soldier blister beetles		*Tegrodera*	(コウチュウ目、ツチハンミョウ科) の昆虫の総称
soldier caerulean		*Jamides malaccanus malaccanus* (Rober)	(チョウ目、シジミチョウ科) 東洋区
soldier commodore	カバタテハモドキ	*Junonia terea* (Drury)	(チョウ目、タテハチョウ科) エチオピア区
soldier flies	ミズアブ科	Stratiomyidae	(ハエ目) の昆虫の総称 世界に1,300種
soldier-flies		*Stratiomys*	(ハエ目、ミズアブ科) の昆虫の総称
soldier flies		*Hedriodiscus*	(ハエ目、ミズアブ科) の昆虫の総称
soldier fly		*Hedriodiscus trivittatus* (Say)	(ハエ目、ミズアブ科) 新北区
soldier fly (1)		*Stratiomys chameleon* (Linnaeus)	(ハエ目、ミズアブ科) 全北区
soldier-fly			broad centurion を見よ
soldierless termites		*Anoplotermes*	(シロアリ目、シロアリ科) の昆虫の総称
soldierless termites (1)	シロアリ科	Termitidae	(シロアリ目) の昆虫の総称 世界に1,500種以上
soldiers and sailors			soldier beetles (1) を見よ
solenophagous louse	ハツカネズミジラミ	*Polyplax serrata* (Burmeister)	(シラミ目、ホソゲジラミ科) 日本、汎世界
solidago Eucosma moth		*Eucosma cataclystiana* (Walker)	(チョウ目、ハマキガ科) 新北区
solidago gall moth			goldenrod gall moth (1) を見よ
solidago root moth		*Grapholita eclipsana* Zeller	(チョウ目、ハマキガ科) 新北区
solitaire clubtail		*Anisogomphus solitaris* Lieftinck	(トンボ目、サナエトンボ科) 東洋区

英名	和名	学名	所属、分布、ほか
solitary ants			velvet ants (1) を見よ
solitary bees			yellow-faced bees (1) を見よ
solitary bees			andrenas を見よ
solitary blue			Reakirt's blue を見よ
solitary midges	ユスリカバエ科	Thaumaleidae	(ハエ目) の昆虫の総称
solitary oak leaf miner		*Cameraria hamadryadella* (Clemens)	(チョウ目、ホソガ科) 新北区。solitary oak leafminer moth ともいう
solitary sootywing		*Bolla solitaria* Steinhauser	(チョウ目、セセリチョウ科) 新熱帯区
solitary wasps			potter wasps を見よ
Solomon's seal sawfly		*Phymatocera aterrima* (Klug)	(ハチ目、ハバチ科) 旧北区
Somali gem		*Chloroselas esmeralda* Butler	(チョウ目、シジミチョウ科) エチオピア区
Somali marbled sapphire		*Stugeta somalina* Stempffer	(チョウ目、シジミチョウ科) エチオピア区
Somali silverline		*Cigaritis somalina* (Butler)	(チョウ目、シジミチョウ科) エチオピア区
somber carpet moth		*Disclisioprocta stellata* (Guenée)	(チョウ目、シャクガ科) 新北区。somber carpet ともいう
somber skipper		*Papias phainis* Godman	(チョウ目、セセリチョウ科) 新熱帯区
sombre Emesis		*Emesis temesa* (Hewitson)	(チョウ目、シジミタテハ科) 新熱帯区
sombre goldenring		*Cordulegaster bidentata* Selys	(トンボ目、オニヤンマ科) 旧北区
sombre lieutenant		*Brachydiplax sobrina* (Rambur)	(トンボ目、トンボ科) 東洋区
sonchus fly		*Ensina sonchi* (Linnaeus)	(ハエ目、ミバエ科) 新北区
Sonora skipper		*Polites sonora* (Scudder)	(チョウ目、セセリチョウ科) 新北区、新熱帯区。Sonoran skipper ともいう
Sonoran banded skipper			false golden-banded skipper を見よ
Sonoran blue	ヒメマルヒメシジミ	*Philotes sonorensis* (C. et R. Felder)	(チョウ目、シジミチョウ科) 新北区。Sonora blue ともいう
Sonoran bumble bee		*Bombus sonorus* Say	(ハチ目、ミツバチ科) 新北区。メキシコの地名に由来
Sonoran carpenter bee		*Xylocopa sonorina* F. Smith	(ハチ目、ミツバチ科) 大洋区
Sonoran hairstreak		*Hypostrymon critola* (Hewitson)	(チョウ目、シジミチョウ科) 新北区
Sonoran marble		*Euchloe guaymasensis* Opler	(チョウ目、シロチョウ科) 新熱帯区
Sonoran metalmark		*Apodemia mejicanus* (Behr)	(チョウ目、シジミタテハ科) 新熱帯区
Sonoran satyr		*Cyllopsis henshawi* (Edwards)	(チョウ目、タテハチョウ科) 新北区
Sonoran sphinx		*Ceratomia sonorensis* Hodges	(チョウ目、スズメガ科) 新北区。Sonora sphinx moth ともいう
Sonoran tent caterpillar		*Malacosoma tigris* (Dyar)	(チョウ目、カレハガ科) 新北区。Sonoran tent caterpillar moth ともいう
Soothsayers			mantids (2) を見よ

英名	和名	学名	所属、分布、ほか
sooty azure		*Celastrina ebenina* Clench	(チョウ目、シジミチョウ科) 新北区
sooty azure			dusky azure を見よ
sooty blue			African grass blue を見よ
sooty Bomolocha moth			flowing-line Bomolocha moth を見よ
sooty copper	クロベニシジミ	*Heodes tityrus* Poda	(チョウ目、シジミチョウ科) 旧北区
sooty dancer		*Argia lugens* (Hagen)	(トンボ目、イトトンボ科) 新北区
sooty hairstreak		*Satyrium fuliginosum* (W. H. Edwards)	(チョウ目、シジミチョウ科) 新北区
sooty Lipocosmodes moth		*Lipocosmodes fuliginosalis* (Fernald)	(チョウ目、メイガ科) 新北区
sooty longwing		*Capnobotes fuliginosus* (Thomas)	(バッタ目、キリギリス科) 新北区
sooty orange-tip		*Zegris eupheme* (Esper)	(チョウ目、シロチョウ科) 旧北区
sooty ringlet	ススグロベニヒカゲ	*Erebia pluto* (de Prunner)	(チョウ目、タテハチョウ科) 旧北区
sooty saddlebag		*Tramea binotata* (Rambur)	(トンボ目、トンボ科) 新北区、新熱帯区
sooty threadtail		*Elattoneura frenulata* (Hagen)	(トンボ目、ミナミイトトンボ科) エチオピア区
sootywings		*Pholisora*	(チョウ目、セセリチョウ科) の昆虫の総称
Sophora long-horned beetle			Hawaiian mamane long-horned beetle を見よ
Sophora worm			genista broom moth を見よ
Sophus forester		*Bebearia sophus* (Fabricius)	(チョウ目、タテハチョウ科) エチオピア区
Sorana eighty-eight		*Callicore sorana* (Godart)	(チョウ目、タテハチョウ科) 新熱帯区
sordid Bomolocha moth		*Hypena sordidula* Grote	(チョウ目、ヤガ科) 新北区
sordid emperor		*Chitoria sordida* (Moore)	(チョウ目、タテハチョウ科) 東洋区
sordid underwing moth		*Catocala sordida* Grote	(チョウ目、ヤガ科) 新北区。sordid underwing ともいう
sorghum aphid			cane aphid を見よ
sorghum caterpillar		*Stenachroia elongella* Chambers	(チョウ目、メイガ科) 東洋区
sorghum grain midge			sorghum midge を見よ
sorghum head moth		*Cryptoblabes adoceta* Turner	(チョウ目、メイガ科) 豪州区。幼虫は sorghum head caterpillar
sorghum midge	ソルガムタマバエ	*Stenodiplosis sorghicola* (Coquillett)	(ハエ目、タマバエ科) 日本、全北区、東洋区、豪州区、エチオピア区、新熱帯区。モロコシ害虫
sorghum plant bug	アカスジカスミカメムシ	*Stenotus rubrovittatus* (Matsumura)	(カメムシ目、カスミカメムシ科) 日本、旧北区。イネ、ムギ、モロコシ害虫
sorghum shoot fly		*Atherigona soccata* Rondani	(ハエ目、イエバエ科) 東洋区、エチオピア区
sorghum stem borer			pink borer (2) を見よ
sorghum webworm moth		*Celama sorghiella* (Riley)	(チョウ目、コブガ科) 新北区、新熱帯区。モロコシ害虫。北米での英名

英名	和名	学名	所属、分布、ほか
sorghum webworm moth		*Nola cereella* (Bosc)	(チョウ目、コブガ科) 新北区、新熱帯区。幼虫は sorghum webworm。
sororia canegrub		*Lepidiota sororia* Moser	(コウチュウ目、コガネムシ科) 豪州区
sorred cutworm			dusky knot-grass dagger を見よ
sorrel copper		*Lycaena orus* (Stoll)	(チョウ目、シジミチョウ科) エチオピア区
sorrel cutworm	ナシケンモン	*Xanthodes rumicis* (Linnaeus)	(チョウ目、ヤガ科) 日本。ナシなど果樹、花卉、サクラ、ツツジなどの害虫
sorrel cutworm			knotgrass を見よ
sorrel leaf beetle			pot-bellied emerald beetle を見よ
sorrel looper	キヒメシャク	*Idaea nudaria infuscaria* (Leech)	(チョウ目、シャクガ科) 日本、旧北区
sorrel pigmy		*Nepticula acetosae* Stainton	(チョウ目、モグリチビガ科) 旧北区
sorrel sapphire		*Heliophorus sena* (Kollar)	(チョウ目、シジミチョウ科) 東洋区
sorrel sawfly	ハグロハバチ	*Allantus luctifer* (Smith)	(ハチ目、ハバチ科) 日本
Sosis satyr		*Taygetis sosis* Hopffer	(チョウ目、タテハチョウ科) 新熱帯区
Sotik Acraea		*Acraea sotikensis* Sharpe	(チョウ目、タテハチョウ科) エチオピア区
souls			diurnal Lepidoptera を見よ
sour bugs			sap beetles (1) を見よ　南アフリカでの英名
sourbush seed fly		*Acinia picturata* (Snow)	(ハエ目、ミバエ科) 新北区
souring beetle parasite		*Zeteticontus utilis* Noyes	(ハチ目、トビコバチ科) 大洋区
South African carnation tortrix			African carnation tortrix を見よ
South African citrus thrips		*Scirtothrips aurantii* (Faure)	(アザミウマ目、アザミウマ科) エチオピア区。カンキツ害虫
South African desert locust			desert locust を見よ　南アフリカでの英名
South African emex weevil		*Apion antiquum* Gyllenhal	(コウチュウ目、ホソクチゾウムシ科) 新北区、豪州区
South African heel-walker			Biedouw Mantophasma を見よ
South African migratory locust			brown locust を見よ
South African praying mantis		*Miomantis caffra* Saussure	(カマキリ目、カマキリ科) 豪州区、エチオピア区
South American bollworm moth		*Helicoverpa gelotopoeon* (Dyar)	(チョウ目、ヤガ科) 新熱帯区
South American cactus moth			cactus moth を見よ
South American cockroach		*Blaptica dubia* Serville	(ゴキブリ目、ブラベルスゴキブリ科) 新熱帯区
South American cucurbit fly			Bumelia fruit fly など南米のミバエ数種について使用
South American dead leaf mantis		*Acanthops falcataria* Goeze	(カマキリ目、Acanthopidae) 新熱帯区
South American dead leaf mantis (1)		*Acanthops falcata* Stål	(カマキリ目、Acanthopidae) 新熱帯区

英名	和名	学名	所属、分布、ほか
South American eighty-eight butterfly			eighty eight butterfly (2) を見よ
South American fruit fly			mango fruit fly (1) を見よ
South American fruit fly			Bumelia fruit fly を見よ
South American grasshopper		*Schistocera cancellata* (Serville)	（バッタ目、バッタ科）新熱帯区
South American grasshopper			South American migratory locust を見よ
South American green mantis		*Oxyopsis gracilis* Giglio-Tos	（カマキリ目、カマキリ科）新熱帯区
South American lantern fly		*Laternaria phosphorea* Linnaeus	（カメムシ目、ビワハゴロモ科）新熱帯区
South American leafminer	アシグロハモグリバエ	*Liriomyza huidobrensis* (Blanchard)	（ハエ目、ハモグリバエ科）日本、全北区、新熱帯区。ジャガイモ、ダイズなど作物、花卉の害虫
South American locust			American grasshopper (1) を見よ
South American locust			South American migratory locust を見よ
South American locust			South American grasshopper を見よ
South American longhorn beetle			titan beetle を見よ
South American malaria vector		*Anopheles darlingi* Root	（ハエ目、カ科）新熱帯区
South American migratory locust		*Schistocera paranensis* Burmeister	（バッタ目、バッタ科）新熱帯区
South American moth		*Cyanotricha necyria* (Felder et Rogenhofer)	（チョウ目、シャチホコガ科）新熱帯区
South American palm weevil		*Rhynchophorus palmarum* (Linnaeus)	（コウチュウ目、オサゾウムシ科）新熱帯区
South American red bollworm		*Sacadodes pyralis* Dyar	（チョウ目、ヤガ科）新北区、新熱帯区。ワタ害虫
South American rice stink bug		*Oebalus poecilus* (Dallas)	（カメムシ目、カメムシ科）新北区、新熱帯区
South American semilooper		*Mocis latipes* (Guenée)	（チョウ目、ヤガ科）新北区。幼虫はタデ類につく
South American tomato pinworm		*Scrobipalpula absoluta* Meyrick	（チョウ目、キバガ科）新熱帯区
South American white borer		*Rupela albinella* (Cramer)	（チョウ目、メイガ科）新北区、新熱帯区。イネ害虫。South American white rice borer ともいう
South China bushbrown		*Mycalesis zonata* Matsumura	（チョウ目、タテハチョウ科）東洋区
south coastal coneworm		*Dioryctria ebeli* Mutuura et Munroe	（チョウ目、メイガ科）新北区。成虫は south coastal coneworm moth
South Indian blue oakleaf			blue oakleaf (2) を見よ
South Indian palmfly		*Elymnias caudata* Butler	（チョウ目、タテハチョウ科）東洋区

英名	和名	学名	所属、分布、ほか
South Island zebra moth		*Declana egregia* Felder	(チョウ目、シャクガ科) 豪州区
south sea fly			island fruit fly (1) を見よ
southeastern blueberry bee		*Habropoda laboriosa* (Fabricius)	(ハチ目、コシブトハナバチ科) 新北区
southeastern bush katydid		*Scudderia cuneata* Morse	(バッタ目、キリギリス科) 新北区
southeastern drywood termite		*Kalotermes snyderi* Light	(シロアリ目、レイビシロアリ科) 新北区
southeastern field cricket		*Gryllus rubens* (Scudder)	(バッタ目、コオロギ科) 新北区
southeastern lubber grasshopper			eastern lubber grasshopper を見よ
south-eastern petaltail		*Petalura gigantea* Leach	(トンボ目、ムカシヤンマ科) 豪州区
southeastern spinyleg		*Dromogomphus armatus* (Selys)	(トンボ目、サナエトンボ科) 新北区
southern aeschna			southern hawker を見よ
southern angle moth		*Speranza varadaria* (Walker)	(チョウ目、シャクガ科) 新北区
southern armyworm (1)		*Spodoptera eridania* (Cramer)	(チョウ目、ヤガ科) 新北区。トウモロコシ害虫。southern armyworm moth ともいう
southern armyworm (2)		*Persectania ewingii* (Westwood)	(チョウ目、ヤガ科) 豪州区
southern armyworm			black army cutworm moth を見よ
southern bamboo aphid	ササマルアブラムシ	*Glyphinaphis bambusae* Van der Goot	(カメムシ目、アブラムシ科) 日本、東洋区
southern bamboo planthopper			bamboo planthopper (1) を見よ
southern beet webworm		*Herpetogramma bipunctalis* (Fabricius)	(チョウ目、メイガ科) 東洋区、エチオピア区、新北区、新熱帯区。幼虫はビート他多くの植物につく。成虫は southern beet webworm moth
southern birdwing		*Troides minos* Cramer	(チョウ目、アゲハチョウ科) 東洋区
southern blue		*Lepidochrysops australis* Tite	(チョウ目、シジミチョウ科) エチオピア区
southern broken-dash			broken dash を見よ
southern brown Argus		*Aricia cramera* (Eschscholtz)	(チョウ目、シジミチョウ科) 旧北区
southern buffalo gnat			buffalo gnat (1) を見よ
southern bush cricket		*Meconema meridionale* (Costa)	(バッタ目、キリギリス科) 旧北区
southern cabbage worm		*Pieris prodice* (Boisduval et LeConte)	(チョウ目、シロチョウ科) 新北区。southern cabbage-worm butterfly ともいう
southern California mealybug		*Spirococcus atriplicis* (Cockerell)	(カメムシ目、コナカイガラムシ科) 新北区
southern carpenter bee		*Xylocopa (Schoenherria) micans* Lepeletier	(ハチ目、ミツバチ科) 新北区

英名	和名	学名	所属、分布、ほか
southern checkered skipper		*Pyrgus xanthus* Edwards	(チョウ目、セセリチョウ科) 新北区
southern chestnut		*Agrochola haematidea* Duponchel	(チョウ目、ヤガ科) 旧北区
southern chicken flea			sticktight flea を見よ　米国での英名
southern chinch bug			lawn chinch bug を見よ
southern Chinese peacock		*Papilio dialis tatsuta* Murayama	(チョウ目、アゲハチョウ科) 東洋区
southern chocolate angle moth		*Macaria distribuaria* (Hübner)	(チョウ目、シャクガ科) 新北区
southern cloudywing		*Thorybes bathyllus* (Smith)	(チョウ目、セセリチョウ科) 新北区
southern cloudywing		*Thorybes dunus* Cramer	(チョウ目、セセリチョウ科) 新北区
southern coenagrion			southern damselfly を見よ
southern comma			eastern comma を見よ
southern corn billbug		*Sphenophorus callosus* (Olivier)	(コウチュウ目、ゾウムシ科) 新北区。トウモロコシ害虫
southern corn gall midge		*Cecidomyia bisetosa* Gagné	(ハエ目、タマバエ科) 新北区
southern corn rootworm			spotted cucumber beetle を見よ
southern corn stalk borer		*Diatraea crambidoides* (Grote)	(チョウ目、メイガ科) 新北区。トウモロコシ害虫。成虫は　southern cornstalk borer moth
southern corn stalk borer (1)		*Diatraea lineolata* Walker	(チョウ目、メイガ科) 新北区、新熱帯区
southern cowpea weevil			cowpea weevil を見よ
southern cowpea weevil			adzuki weevil を見よ　北米での英名
southern cyan tiger moth		*Macrocneme chrysitis* (Guérin-Méneville)	(チョウ目、ヒトリガ科) 新北区、新熱帯区
southern damselfly		*Coenagirion mercuriale* (Charpentier)	(トンボ目、イトトンボ科) 旧北区
southern dart		*Ocybadistes walkeri hypochlorus* Lower	(チョウ目、セセリチョウ科) 豪州区。yellow-banded dart を参照
southern darter		*Sympetrum meridionale* (Selys)	(トンボ目、トンボ科) 旧北区
southern dogface	イヌモンキチョウ	*Colias caesonia* (Stoll)	(チョウ目、シロチョウ科) 新北区、新熱帯区
southern dragon		*Antipodogomphus acolythus* (Martin)	(トンボ目、サナエトンボ科) 豪州区
southern dry-wood termite		*Kalotermes hubbardi* Banks	(シロアリ目、レイビシロアリ科) 新北区
southern duffer		*Discophora lepida* (Moore)	(チョウ目、タテハチョウ科) 東洋区
southern duskywing		*Erynnis meridianus* Bell	(チョウ目、セセリチョウ科) 新北区
southern dusted skipper		*Atrytonopsis loammi* (Whitney)	(チョウ目、セセリチョウ科) 新北区
southern emerald damselfly		*Lestes barbarus* (Fabricius)	(トンボ目、アオイトトンボ科) 旧北区、東洋区

英名	和名	学名	所属、分布、ほか
southern emerald moth		*Synchlora frondaria* (Guenée)	（チョウ目、シャクガ科）新北区
southern false wireworm		*Gonocephalum macleayi* (Blackburn)	（コウチュウ目、ゴミムシダマシ科）豪州区
southern festoon	タイスアゲハ	*Zerynthia polyxena* (Denis et Schiffermüller)	（チョウ目、アゲハチョウ科）旧北区
southern field cricket			two-spotted cricket を見よ
southern fire ant		*Solenopsis xyloni* (McCook)	（ハチ目、アリ科）新北区
southern flannel moth			puss caterpillar moth を見よ
southern garden leafhopper		*Empoasca solana* DeLong	（カメムシ目、オオヨコバイ科）新北区
southern garden leafhopper			Stephen's leafhopper を見よ
southern garden leafhopper			green leafhopper (1) を見よ　米国での英名
southern gatekeeper	ヘリブトチャイロジャノメ	*Pyronia cecilia* (Vallantin)	（チョウ目、タテハチョウ科）旧北区
southern gold-dot bentwing		*Leucoptera lathyrifoliella* (Stainton)	（チョウ目、ハモグリガ科）旧北区
southern grass-dart			southern dart を見よ
southern grass worm			black armyworm を見よ
southern grayling		*Hipparchia aristaeus* (Bonelli)	（チョウ目、タテハチョウ科）旧北区
southern green looper moth		*Xanthorhoe beata* Butler	（チョウ目、シャクガ科）豪州区
southern green stink bug	ミナミアオカメムシ	*Nezara viridula* (Linnaeus)	（カメムシ目、カメムシ科）日本、全北区、東洋区、新熱帯区、豪州区。イネ、ダイズ、カンキツなど多くの作物、果樹害虫
southern grizzled skipper		*Pyrgus malvoides* (Elwes et Edwards)	（チョウ目、セセリチョウ科）旧北区
southern ground cricket		*Allonemobius socius* (Scudder)	（バッタ目、コオロギ科）新北区
southern hairstreak	フロリダフタオカラスシジミ	*Satyrium favonius* (Smith)	（チョウ目、シジミチョウ科）新北区。*S. f. autolycus* (W. H. Edwards) 新熱帯区も同英名
southern harvester ant			Florida harvester ant を見よ
southern hawker		*Aeschna cyanea* (Müller)	（トンボ目、ヤンマ科）旧北区
southern hermit		*Chazara prieuri* (Pierret)	（チョウ目、タテハチョウ科）旧北区
southern house mosquito	ネッタイイエカ	*Culex pipiens quinquefasciatus* Say	（ハエ目、カ科）日本、東洋区、豪州区、エチオピア区、新北区、新熱帯区。バンクロフト糸状虫を媒介する
southern iron blue		*Nigrobaetis niger* (Linnaeus)	（カゲロウ目、コカゲロウ科）旧北区
southern ladybird		*Cleobora mellyi* Mulsant	（コウチュウ目、テントウムシ科）豪州区
southern lappet moth		*Phyllodesma occidentis* (Walker)	（チョウ目、カレハガ科）新北区
southern longhorn moth		*Adela caeruleella* Walker	（チョウ目、マガリガ科）新北区

英名	和名	学名	所属、分布、ほか
southern lyctus beetle		*Lyctus planicollis* LeConte	（コウチュウ目、ナガシンクイムシ科）全北区
southern marbled skipper		*Carcharodus baeticus* (Rambur)	（チョウ目、セセリチョウ科）旧北区
southern marbled white	フチグロシロジャノメ	*Melanargia lachesis* (Hübner)	（チョウ目、タテハチョウ科）旧北区
southern masked chafer		*Cyclocephala immaculata* (Olivier)	（コウチュウ目、コガネムシ科）新北区
southern migrant hawker		*Aeschna affinis* Van der Linden	（トンボ目、ヤンマ科）旧北区
southern milkweed butterfly		*Danaus chrysippus aegyptius* (Schreiber)	（チョウ目、タテハチョウ科）エチオピア区。African monarch を参照
southern mole cricket		*Scapteriscus acletus* Rehn et Hebard	（バッタ目、ケラ科）新北区
southern mole cricket		*Scapteriscus borellii* Giglio-Tos	（バッタ目、ケラ科）新北区
southern monarch		*Danaus erippus* (Cramer)	（チョウ目、タテハチョウ科）新熱帯区
southern oak bush cricket			southern bush cricket を見よ
southern oak dagger moth		*Acronicta increta* (Morrison)	（チョウ目、ヤガ科）新北区
southern oak looper moth		*Lambdina pultaria* (Guenée)	（チョウ目、シャクガ科）新北区
southern old lady moth		*Dasypodia selenophora* Guenée	（チョウ目、ヤガ科）豪州区。southern old lady ともいう
southern one-year canegrub		*Antitrogus mussoni* (Blackburn)	（コウチュウ目、コガネムシ科）豪州区
southern orange dart		*Suniana sunias nola* (Waterhouse)	（チョウ目、セセリチョウ科）豪州区
southern pear scale		*Quadraspidiotus marani* Zahradnik	（カメムシ目、マルカイガラムシ科）旧北区
southern pearl-white			common pearl white を見よ
southern pearly eye			pearly eye (2) を見よ
southern pied woolly legs		*Lachnocnema laches* (Fabricius)	（チョウ目、シジミチョウ科）エチオピア区
southern pine beetle		*Elacatis fasciatus* Bland	（コウチュウ目、Othniidae）新北区
southern pine beetle (1)		*Dendroctonus frontalis* Zimmermann	（コウチュウ目、キクイムシ科）新北区
southern pine catkinworm moth		*Satronia tantilla* Heinrich	（チョウ目、ハマキガ科）新北区
southern pine coneworm		*Dioryctria amatella* (Hulst)	（チョウ目、メイガ科）新北区。成虫は southern pine coneworm moth
southern pine looper moth		*Caripeta aretaria* (Walker)	（チョウ目、シャクガ科）新北区
southern pine root weevil		*Hylobius aliradicus* Warner	（コウチュウ目、ゾウムシ科）新北区
southern pine sawyer		*Monochamus titilator* Fabricius	（コウチュウ目、カミキリムシ科）新北区

英名	和名	学名	所属、分布、ほか
southern pine sphinx			Abbott's pine sphinx を見よ　southern pine sphinx moth ともいう
southern pine-tip moth		*Rhyacionia neomexicana* (Dyar)	(チョウ目、ハマキガ科) 新北区
southern pitch-blister moth		*Retinia taedana* (Miller)	(チョウ目、ハマキガ科) 新北区
southern potato wireworm		*Conoderus falli* Lane	(コウチュウ目、コメツキムシ科) 新北区
southern powder post beetle			southern lyctus beetle を見よ　米国での英名
southern Ptichodis moth		*Ptichodis bistrigata* Hübner	(チョウ目、ヤガ科) 新北区
southern purple		*Aslauga australis* Cottrell	(チョウ目、シジミチョウ科) エチオピア区
southern purple azure			Genoveva azure を見よ
southern purple mint moth		*Pyrausta laticlavia* (Grote et Robinson)	(チョウ目、メイガ科) 新北区
southern pygmy clubtail		*Lanthus vernalis* Carle	(トンボ目、サナエトンボ科) 新北区
southern rain beetle		*Pleocoma australis* Fall	(コウチュウ目、コガネムシ科) 新北区
southern red ant			common red ant を見よ
southern rice katydid	クサキリ	*Homorocoryphus lineosus* (Walker)	(バッタ目、キリギリス科) 日本
southern rubus aphid	イチゴハトゲアブラムシ	*Matsumuraja rubifoliae* Takahashi	(カメムシ目、アブラムシ科) 日本、旧北区、東洋区
southern rustic		*Rhyacia lucipeta* (Denis et Schiffermüller)	(チョウ目、ヤガ科) 全北区
southern saltmarsh mosquito		*Aedes camptorhynchus* (Thomson)	(ハエ目、カ科) 豪州区
southern salt marsh mosquito			black salt marsh mosquito を見よ
southern sandy dung beetle		*Euoniticellus pallipes* (Fabricius)	(コウチュウ目、コガネムシ科) 豪州区
southern saw-tailed bush cricket		*Barbitistes obtusus* Targioni-Tozzetti	(バッタ目、キリギリス科) 旧北区
southern scalloped sootywing			Mazans sootywing を見よ
southern scallopwing			Mazans sootywing を見よ
southern scarce swallowtail		*Iphiclides feisthamelii* (Duponchel)	(チョウ目、アゲハチョウ科) 旧北区
southern short-tailed admiral		*Vanessa hippomene* (Hübner)	(チョウ目、タテハチョウ科) エチオピア区
southern sicklewing		*Eantis thraso* (Hübner)	(チョウ目、セセリチョウ科) 新熱帯区
southern silver ochre		*Trapezites praxedus* Plötz	(チョウ目、セセリチョウ科) 豪州区
southern silver-stiletto		*Cliorismia rustica* (Panzer)	(ハエ目、ツルギアブ科) 旧北区
southern skimmer		*Orthetrum brunneum brunneum* (Fonscolombe)	(トンボ目、トンボ科) 東洋区

英名	和名	学名	所属、分布、ほか
southern skipperling	チビギンイチモンジセセリ	*Copaeodes minimus* (Edwards)	(チョウ目、セセリチョウ科) 新北区、新熱帯区
southern small white		*Pieris mannii* (Mayer)	(チョウ目、シロチョウ科) 旧北区
southern snaketail		*Ophiogomphus australis* Carle	(トンボ目、サナエトンボ科) 新北区
southern snout		*Libytheana carinenta* (Cramer)	(チョウ目、タテハチョウ科) 新北区、新熱帯区。southern snout butterfly ともいう
southern sootywing	ミナミスケセセリ	*Staphylus hayhurstii* (Edwards)	(チョウ目、セセリチョウ科) 新北区
southern spotted ace		*Thoressa astigmata* (Swinhoe)	(チョウ目、セセリチョウ科) 東洋区
southern Spragueia moth		*Spragueia dama* (Guenée)	(チョウ目、ヤガ科) 新北区
southern spreadwing		*Lestes australis* Walker	(トンボ目、アオイトトンボ科) 新北区
southern spring porina moth		*Wiseana jacosa* Meyrick	(チョウ目、コウモリガ科) 豪州区
southern sprite		*Nehalennia integricollis* Calvert	(トンボ目、イトトンボ科) 新北区
southern sullied sailer			clear sailer (1) を見よ
southern swallowtail	トラフキアゲハ	*Papilio alexanor* Esper	(チョウ目、アゲハチョウ科) 旧北区
southern tiger moth			Phyllira moth を見よ
southern tobacco worm			tobacco worm を見よ
southern Tolype moth		*Tolype minta* Dyar	(チョウ目、カレハガ科) 新北区
southern tussock moth		*Dasychira meridionalis* (Barnes et McDunnough)	(チョウ目、ドクガ科) 新北区
southern ugly-nest caterpillar moth		*Archips rileyana* (Grote)	(チョウ目、ハマキガ科) 新北区
southern variable dart moth		*Xestia elimata* Guenée	(チョウ目、ヤガ科) 新北区。southern variable dart ともいう
southern vicetail		*Hemigomphus gouldii* (Selys)	(トンボ目、サナエトンボ科) 豪州区
southern wainscot		*Mythimna straminea* (Treitschke)	(チョウ目、ヤガ科) 旧北区
southern whistling moth		*Hecatesia thyridion* Feisthamel	(チョウ目、ヤガ科) 豪州区
southern white admiral	アオイチモンジ	*Limenitis reducta* Staudinger	(チョウ目、タテハチョウ科) 旧北区
southern white page		*Graphium telesilans* Felder	(チョウ目、アゲハチョウ科) 新熱帯区
southern wood cricket		*Gryllus fultoni* (Alexander)	(バッタ目、コオロギ科) 新北区
southern wood nymph			large wood nymph を見よ
southern yellowjack		*Notogomphus praetorius* (Selys)	(トンボ目、サナエトンボ科) エチオピア区
southern Zestusa		*Zestusa staudingeri* (Mabille)	(チョウ目、セセリチョウ科) 新熱帯区

英名	和名	学名	所属、分布、ほか
Southey's blue		*Lepidochrysops southeyae* Pennington	（チョウ目、シジミチョウ科）エチオピア区
Southey's brown		*Pseudonympha southeyi* (Pennington)	（チョウ目、タテハチョウ科）エチオピア区
Southey's widow		*Tarsocera southeyae* Dickson	（チョウ目、タテハチョウ科）エチオピア区
southwest-Mexican faceted-skipper		*Synapte silna* Evans	（チョウ目、セセリチョウ科）新熱帯区
southwest-Mexican skipperling			small spotted skipperling を見よ
southwestern angle-wing		*Microcentrum latifrons* Spooner	（バッタ目、キリギリス科）新北区
southwestern azure		*Celastrina echo cinerea* (W. H. Edwards)	（チョウ目、シジミチョウ科）新熱帯区。Pacific azure を参照
southwestern big red			red skimmer を見よ
southwestern corn borer		*Zeadiatraea grandiosella* (Dyar)	（チョウ目、メイガ科）新北区。トウモロコシ害虫
southwestern corn borer moth			western corn borer を見よ
southwestern eyed click beetle		*Alaus zunianus* Casey	（コウチュウ目、コメツキムシ科）新北区
southwestern fritillary	ウスグロギンボシヒョウモン	*Speyeria atlantis nausicaa* (W. H. Edwards)	（チョウ目、タテハチョウ科）新北区。Atlantis fritillary を参照
southwestern Hercules beetle		*Dynastes granti* Horn	（コウチュウ目、コガネムシ科）新北区
southwestern orangetip		*Anthocharis thoosa inghami* (Gunder)	（チョウ目、シロチョウ科）新熱帯区
southwestern pine tip moth			southern pine-tip moth を見よ
southwestern sack-bearer moth		*Lacosoma arizonicum* Dyar	（チョウ目、Mimallonidae）新北区
southwestern squash vine borer		*Melittia calabaza* Duckworth et Eichlin	（チョウ目、スカシバガ科）新北区
southwestern tent caterpillar		*Malacosoma incurvum* (Edwards)	（チョウ目、カレハガ科）新北区。southwestern tent caterpillar moth ともいう
sow thistle aphid			currant-lettuce aphid を見よ
sow thistle aphid			currant aphid を見よ
soybean aphid	ダイズアブラムシ	*Aphis glycines* Matsumura	（カメムシ目、アブラムシ科）日本、旧北区、東洋区
soybean beetle	ヒメコガネ	*Anomala rufocuprea* Motschulsky	（コウチュウ目、コガネムシ科）日本。イネ、ダイズ、リンゴなど多くの作物、果樹、樹木の害虫
soybean felt scale			soybean scale を見よ
soybean flea beetle	キアシノミハムシ	*Luperomorpha tenebrosa* (Jacoby)	（コウチュウ目、ハムシ科）日本。ダイズ害虫
soybean fly			soybean stem miner を見よ　豪州での英名
soybean girdler		*Oberopsis brevis* Swederus	（コウチュウ目、ヒゲナガカミキリムシ科）東洋区
soybean green semilooper			Tunbridge wells gem を見よ

英名	和名	学名	所属、分布、ほか
soybean hairy caterpillar		*Spilarctia casigneta* Kollar	(チョウ目、ヒトリガ科) 東洋区
soybean leaf beetle		*Plagiodera inclusa* Stål	(コウチュウ目、ハムシ科) 東洋区。ダイズ害虫
soybean leaf folder		bean-leaf webworm moth を見よ	
soybean leafminer	ダイズクロハモグリバエ	*Japanagromyza tristella* (Thomson)	(ハエ目、ハモグリバエ科) 日本、東洋区。ダイズ害虫
soybean leafroller	マメホソガ	*Caloptilia soyella* (Deventer)	(チョウ目、ホソガ科) 日本。ダイズ、ナシ害虫
soybean leafstalk midge	ダイズクキタマバエ	*Resseliella soya* (Monzen)	(ハエ目、タマバエ科) 日本、旧北区。ダイズ害虫
soybean looper		*Pseudoplusia includens* (Walker)	(チョウ目、ヤガ科) 新北区、新熱帯区。成虫は soybean looper moth
soybean looper		slender burnished brass を見よ	
soybean moth		*Stomopteryx simplexella* (Walker)	(チョウ目、キバガ科) 豪州区
soybean moth		kidney-vetch sober を見よ	
soybean moth		soybean pod borer を見よ	
soybean nodule fly		*Rivellia quadrifasciata* (Macquart)	(ハエ目、ヒロクチバエ科) 新北区
soybean pod borer	マメシンクイガ	*Leguminivora glycinivorella* (Matsumura)	(チョウ目、ハマキガ科) 日本、旧北区。ダイズ害虫
soybean pod gall midge	ダイズサヤタマバエ	*Asphondylia yushimai* Yukawa et Uechi	(ハエ目、タマバエ科) 日本。ダイズ害虫
soybean podworm	ダイズサヤムシガ	*Matsumuraeses falcana* Walsingham	(チョウ目、ハマキガ科) 日本、東洋区。ダイズ害虫
soybean podworm	マメヒメサヤムシガ	*Matsumuraeses phaseoli* (Matsumura)	(チョウ目、ハマキガ科) 日本、旧北区。ダイズ、エンドウ、マメ科牧草の害虫
soybean root miner	ダイズネモグリバエ	*Ophiomyia shibatsujii* (Kato)	(ハエ目、ハモグリバエ科) 日本。ダイズ害虫
soybean root-gall fly	ダイズコンリュウバエ	*Rivellia apicalis* Hendel	(ハエ目、ヒロクチバエ科) 日本、旧北区。ダイズ害虫
soybean sawfly	ダイズハバチ	*Takeuchiella pentagona* Malaise	(ハチ目、ハバチ科) 日本。ダイズ害虫
soybean scale	ダイズフクロカイガラムシ	*Eriococcus sojae* Kuwana	(カメムシ目、フクロカイガラムシ科) 日本。ダイズ害虫
soybean stem fly		soybean stem miner を見よ	
soybean stem gall midge		soybean leafstalk midge を見よ	
soybean stem miner	ダイズクキモグリバエ	*Melanagromyza sojae* (Zehntner)	(ハエ目、ハモグリバエ科) 日本、旧北区、東洋区、豪州区。ダイズ害虫
soybean thrips		*Sericothrips variabilis* (Beach)	(アザミウマ目、アザミウマ科) 新北区
soybean thrips	ダイズアザミウマ	*Micterothrips glycines* (Okamoto)	(アザミウマ目、アザミウマ科) 日本。サツマイモ、ダイズ、ウリ類害虫
soybean thrips		*Neohydatothrips variabilis* (Beach)	(アザミウマ目、アザミウマ科) 新北区
soybean top miner	ダイズメモグリバエ	*Melanagromyza koizumii* Kato	(ハエ目、ハモグリバエ科) 日本。ダイズ害虫
soybean webworm		bean-leaf webworm moth を見よ	

英名	和名	学名	所属、分布、ほか
soybean webworm moth		*Brachyacma palpigera* (Walsingham)	（チョウ目、キバガ科）新北区
Spalding's blue		*Euphilotes spaldingi* (Barnes et McDunnough)	（チョウ目、シジミチョウ科）新北区
Spalding's dart		*Agrotis herzogi* Rebel	（チョウ目、ヤガ科）旧北区
spang beetle			June beetle (1) を見よ
spangle		*Papilio protenor* Cramer	（チョウ目、アゲハチョウ科）東洋区、エチオピア区。*P. p. demetries* Stoll クロアゲハ 日本も同英名
spangle gall wasp		*Neuroterus numismalis* (Geoffroy)	（ハチ目、タマバチ科）旧北区
spangled cupid			gold drop Helicopis を見よ
spangled flower beetle		*Euphoria sepulcralis* (Fabricius)	（コウチュウ目、コガネムシ科）新北区
spangled golden Emesis		*Emesis cerea* (Linnaeus)	（チョウ目、シジミタテハ科）新熱帯区。
spangled green moth		*Cosmodes elegans* Donovan	（チョウ目、ヤガ科）豪州区
spangled plushblue		*Nilasera asoka* de Nicéville	（チョウ目、シジミチョウ科）東洋区
Spanish Argus		*Aricia morronensis* (Ribbe)	（チョウ目、シジミチョウ科）旧北区
Spanish bean beetle		*Bruchus affinis* Frölich	（コウチュウ目、ハムシ科）旧北区
Spanish brassy ringlet	スペインベニヒカゲ	*Erebia hispania* Butler	（チョウ目、タテハチョウ科）旧北区
Spanish carpet		*Scotopteryx peribolata* (Hübner)	（チョウ目、シャクガ科）旧北区
Spanish chalk-hill blue		*Polyommatus albicans* Gerhard	（チョウ目、シジミチョウ科）旧北区
Spanish copris		*Copris hispanus* (Linnaeus)	（コウチュウ目、コガネムシ科）旧北区
Spanish cricket		*Gryllus hispanicus* Rambur	（バッタ目、コオロギ科）旧北区
Spanish festoon	マドタイスアゲハ	*Zerynthia rumina* (Linnaeus)	（チョウ目、アゲハチョウ科）旧北区
Spanish fly			blister beetle を見よ
Spanish fritillary		*Euphydryas desfontainii* (Godart)	（チョウ目、タテハチョウ科）旧北区
Spanish gatekeeper		*Pyronia bathseba* (Fabricius)	（チョウ目、タテハチョウ科）旧北区
Spanish greenish black-tip		*Euchloe bazae* Fabiano	（チョウ目、シロチョウ科）旧北区
Spanish heath	スペインヒメヒカゲ	*Coenonympha iphioides* Staudinger	（チョウ目、タテハチョウ科）旧北区
Spanish marbled white		*Melanargia ines* (Hoffmannsegg)	（チョウ目、タテハチョウ科）旧北区
Spanish moon moth		*Actias isabellaea* (Graells)	（チョウ目、ヤママユガ科）旧北区
Spanish moth		*Xanthopastis timais* (Cramer)	（チョウ目、ヤガ科）全北区
Spanish purple hairstreak	ウチムラサキウラキンシジミ	*Laeosopis roboris* (Esper)	（チョウ目、シジミチョウ科）旧北区

英名	和名	学名	所属、分布、ほか
Spanish red scale			dictyospermum scale を見よ　南アフリカ、豪州での英名
Spanish swallowtail			southern scarce swallowtail を見よ
Spanish zephyr blue		*Plebejus hespericus* (Rambur)	（チョウ目、シジミチョウ科）旧北区
spanworms			geometer moths を見よ　幼虫の英名でいわゆるシャクトリムシ
Sparganothis fruitworm moth		*Sparganothis sulfureana* (Clemens)	（チョウ目、ハマキガ科）新北区、新熱帯区
sparkler			green tiger beetle を見よ
sparkling archaic sun moths			primitive moths (1) を見よ
sparkling Aterpia moth		*Aterpia approximana* (Heinrich)	（チョウ目、ハマキガ科）新北区
sparkling jewelwing		*Calopteryx dimidiata* Burmeister	（トンボ目、イトトンボ科）新北区
sparks		*Sinthusa*	（チョウ目、シジミチョウ科）の昆虫の総称
sparrow-nest clothes moth		*Tinea columbariella* Wooke	（チョウ目、ヒロズコガ科）旧北区
spatterdock darner		*Rhionaeschna mutata* (Hagen)	（トンボ目、ヤンマ科）新北区
spatulate sootywing		*Bolla eusebius* (Plötz)	（チョウ目、セセリチョウ科）新熱帯区
spear-marked black moth			argent and sable を見よ　北米での英名
spear-thistle lace bug		*Tingis cardui* (Linnaeus)	（カメムシ目、グンバイムシ科）旧北区
spear-winged flies	ヤリバエ科	Lonchopteridae	（ハエ目）の昆虫の総称
spearhead bluet		*Coenagrion hastulatum* Charpentier	（トンボ目、イトトンボ科）旧北区
speared dagger moth		*Acronicta hasta* Guenée	（チョウ目、ヤガ科）新北区
speckle-winged grasshopper			orange-winged grasshopper を見よ
speckled Acleris moth		*Acleris negundana* (Busck)	（チョウ目、ハマキガ科）新北区
speckled Argyresthia moth		*Argyresthia subreticulata* Walsingham	（チョウ目、メムシガ科）新北区
speckled beauty		*Fagivorina arenaria* (Hufnagel)	（チョウ目、シャクガ科）旧北区
speckled black Pyla moth			brown muslin sweep を見よ
speckled buff moth		*Dipaustica epiastra* Meyrick	（チョウ目、ヤガ科）豪州区
speckled bush-cricket		*Leptophyes punctatissimus* (Bosc)	（バッタ目、キリギリス科）旧北区
speckled casebearer moth		*Coleophora elaegnisella* Kearfott	（チョウ目、ツツミノガ科）新北区
speckled cockroach			cinereous cockroach を見よ
speckled dun		*Callibaetis pictus* Eaton	（カゲロウ目、コカゲロウ科）新北区

英名	和名	学名	所属、分布、ほか
speckled Eucosma moth		*Eucosma guttulana* Blanchard	(チョウ目、ハマキガ科) 新北区
speckled footman			black-speckled flunkey を見よ
speckled grasshopper		*Bryodema tuberculata* (Fabricius)	(バッタ目、バッタ科) 旧北区
speckled green fruitworm		*Orthosia hibisci* Guenée	(チョウ目、ヤガ科) 新北区
speckled hawk		*Praedora plagiata* Rothschild et Jordan	(チョウ目、スズメガ科) エチオピア区
speckled lilac nymph		*Credidomimas concordia* Hoppfer	(チョウ目、タテハチョウ科) エチオピア区
speckled line blue		*Catopryops florinda* (Butler)	(チョウ目、シジミチョウ科) 豪州区
speckled longhorn beetle		*Pachytodes cerambyciformis* (Schrank)	(コウチュウ目、カミキリムシ科) 旧北区
speckled longicorn		*Paradisperna plumifera* (Pascoe)	(コウチュウ目、カミキリムシ科) 豪州区
speckled nettle moth		*Abrostola urentis* Guenée	(チョウ目、ヤガ科) 新北区
speckled ochre		*Trapezites atkinsi* (Williams, Williams et Hay)	(チョウ目、セセリチョウ科) 豪州区
speckled oil beetle		*Meloe variegatus* Donovan	(コウチュウ目、ツチハンミョウ科) 旧北区
speckled Phaneta moth		*Phaneta verna* Miller	(チョウ目、ハマキガ科) 新北区
speckled red Acraea		*Acraea violarum* Boisduval	(チョウ目、タテハチョウ科) エチオピア区
speckled Renia moth		*Renia adspergillus* (Bosc)	(チョウ目、ヤガ科) 新北区
speckled rufous nymph		*Bebearia plistonax* (Hewitson)	(チョウ目、タテハチョウ科) エチオピア区
speckled Sereda moth		*Sereda tautana* (Clemens)	(チョウ目、ハマキガ科) 新北区
speckled snout beetle		*Eremnus cerealis* Marshall	(コウチュウ目、ゾウムシ科) エチオピア区
speckled sootywing		*Staphylus minor* Schaus	(チョウ目、セセリチョウ科) 新熱帯区
speckled sulphur-tip	ウスズミツマアカシロチョウ	*Colotis agoye* (Wallengren)	(チョウ目、シロチョウ科) エチオピア区
speckled Talponia moth		*Talponia plummeriana* (Busck)	(チョウ目、ハマキガ科) 新北区
speckled wood	キマダラジャノメ	*Pararge aegeria* (Linnaeus)	(チョウ目、タテハチョウ科) 旧北区。speckled wood butterfly ともいう
speckled yellow		*Pseudopanthera macularia* (Linnaeus)	(チョウ目、シャクガ科) 旧北区
spectacle		*Abrostola tripartita* Hufnagel	(チョウ目、ヤガ科) 旧北区
spectacle (1)	イラクサマダラウワバ	*Abrostola trigemina* (Werneburg)	(チョウ目、ヤガ科) 日本、旧北区
spectacle (2)	ハガタキリガ	*Scolyopterix libatrix* (Linnaeus)	(チョウ目、ヤガ科) 日本、全北区。リンゴ、ナシ、モモ、ブドウ害虫
spectacle swordtail		*Graphium glycerion* (Gray)	(チョウ目、アゲハチョウ科) 東洋区
spectacle swordtail			mandarine swallowtail を見よ

英名	和名	学名	所属、分布、ほか
spectacled nettle moth			speckled nettle moth を見よ
spectacled skimmer		*Orthetrum icteromelas* Ris	(トンボ目、トンボ科) エチオピア区
spectacular grasshopper			Leichardt's grasshopper を見よ
spectacular scaly cricket		*Cycloptilum spectabile* Strohecker	(バッタ目、コオロギ科) 新北区
Sperry's lawn moth		*Crambus sperryellus* Klots	(チョウ目、メイガ科) 新北区
sphaeriid beetles			minute bog beetles を見よ
sphaerocerids			small dung flies を見よ
sphagnum bug		*Hebrus ruficeps* Thomson	(カメムシ目、ケシミズカメムシ科) 旧北区
sphagnum bugs			velvet water bugs を見よ
sphagnum ground cricket		*Neonemobius palustris* (Blatchley)	(バッタ目、コオロギ科) 新北区
sphagnum sprite		*Nehalennia gracilis* (Morse)	(トンボ目、イトトンボ科) 新北区
sphecid wasps			cicada killers を見よ
sphecoid wasps	アナバチ上科	Sphecoidea	(ハチ目) の昆虫の総称
spherical mealybug		*Nipaecoccus viridis* (Newstead)	(カメムシ目、コナカイガラムシ科) 新北区、東洋区、豪州区、エチオピア区
spherical mealybug			hibiscus mealybug (1) を見よ
sphindid beetles			dry-fungus beetles を見よ
sphindids			dry-fungus beetles を見よ
sphingids			hawk moths を見よ
sphinx horn worms			hawk moths を見よ
sphinx moth			rippled-marked hawk moth を見よ
sphinx moths			hawk moths を見よ
spicebush silk moth		*Callosamia promethea* (Drury)	(チョウ目、ヤママユガ科) 新北区
spicebush silkworm			spicebush silk moth を見よ
spicebush swallowtail	クスノキアゲハ	*Papilio troilus* Linnaeus	(チョウ目、アゲハチョウ科) 新北区
spider beetle (1)	カバイロヒョウホンムシ	*Pseudeurostus hilleri* (Reitter)	(コウチュウ目、ヒョウホンムシ科) 日本、全北区
spider beetle		*Sphaericus gibboides* Boieldieu	(コウチュウ目、ヒョウホンムシ科) 汎世界
spider beetle			brown spider beetle を見よ
spider beetle			museum beetle (1)(2) を見よ　豪州での英名
spider beetle			white-marked spider beetle を見よ
spider beetle			golden spider beetle を見よ
spider beetles	ヒョウホンムシ科	Ptinidae	(コウチュウ目) の昆虫の総称
spider flies			small-headed flies を見よ
spider flies			louse flies を見よ
spider fly			forest fly を見よ
spider hunters			spider wasps を見よ

英名	和名	学名	所属、分布、ほか
spider-hunting wasps			spider wasps を見よ
spider-like spine-waisted ant		*Aphaenogaster araneoides* Emery	(ハチ目、アリ科) 新熱帯区
spider longicorn			citrus longhorn を見よ
spider mite destroyer		*Stethorus picipes* Casey	(コウチュウ目、テントウムシ科) 新北区
spider-parasite flies			small-headed flies を見よ
spider-tail thread-legged katydid		*Arethaea arachnopyga* Rehn et Hebard	(バッタ目、キリギリス科) 新北区
spider wasp		*Dipogon calipterus* (Say)	(ハチ目、ベッコウバチ科) 新北区
spider wasps	ベッコウバチ科	Pompilidae	(ハチ目) の昆虫の総称
spider wasps			ground nesting wasps を見よ
spider-wing cupid		*Helicopis gnidus* (Fabricius)	(チョウ目、シジミタテハ科) 新熱帯区
spiders			crane flies (1) を見よ
spike-banded skipper		*Autochton longipennis* (Plötz)	(チョウ目、セセリチョウ科) 新熱帯区
spike-marked bell		*Dichrorampha montanana* (Duponchel)	(チョウ目、ハマキガ科) 旧北区
spiked magician			matriarcha katydid を見よ
spiked pea gall		*Diplolepis nervosa* (Curtis)	(ハチ目、タマバチ科) がバラの葉の下面につくる虫えいでトゲがある。旧北区
Spiller's Canary yellow		*Dixeia spilleri* (Spiller)	(チョウ目、シロチョウ科) エチオピア区
Spiller's sulphur yellow			Spiller's Canary yellow を見よ
Spiller's yellow			Spiller's Canary yellow を見よ
spinach aphid			green peach aphid を見よ
spinach aphis of America			green peach aphid を見よ
spinach carrion beetle		*Silpha bituberosa* LeConte	(コウチュウ目、シデムシ科) 新北区
spinach caterpillar			beautiful brocade を見よ
spinach flea beetle		*Disonycha xanthomelas* (Dalman)	(コウチュウ目、ハムシ科) 新北区。野菜害虫
spinach leafminer	アカザモグリハナバエ	*Pegomya exilis* (Meigen)	(ハエ目、ハナバエ科) 日本。ホウレンソウ害虫
spinach leafminer			beet leaf miner (1) を見よ　spinach leaf miner とも記す
spinach maggot			beet leaf miner (1) を見よ
spinach moth		*Eulithis mellinata* (Fabricius)	(チョウ目、シャクガ科) 旧北区。spinach ともいう
spinach stem fly		*Hylemya echinata* (Séguy)	(ハエ目、ハナバエ科) 旧北区
spindle			dragonflies (1) を見よ
spindle ermine		*Yponomeuta cagnagella* (Hübner)	(チョウ目、スガ科) 全北区。成虫は spindle ermine moth
spindle tree aphid			euonymus aphid を見よ
spindle worm			elder shoot borer moth を見よ

英名	和名	学名	所属、分布、ほか
spine-breasted grasshoppers		Catantopidae	（バッタ目）の昆虫の総称
spine-crowned clubtail		*Gomphus abbreviatus* Hagen	（トンボ目、サナエトンボ科）新北区
spinehead dragon		*Antipodogomphus proselythus* (Martin)	（トンボ目、サナエトンボ科）豪州区
spine-horned bark borer		*Elaphidion spinicorne* (Drury)	（コウチュウ目、カミキリムシ科）新熱帯区
spine-horned tiger longicorn	トゲヒゲトラカミキリ	*Demonax transilis* Bates	（コウチュウ目、カミキリムシ科）日本。スギ害虫。spine-horned tiger longicorn beetle ともいう
spinelegged citrus weevil		*Maleuterpes spinipes* Blackburn	（コウチュウ目、ゾウムシ科）豪州区。カンキツ害虫
spine-legged redbolt		*Rhodothemis rufa* (Rambur)	（トンボ目、トンボ科）東洋区
spine-tailed earwig		*Doru aculeatum* (Scudder)	（ハサミムシ目、クギヌキハサミムシ科）新北区。北米の普通種
spinetailed weevil		*Desiantha caudata* Pascoe	（コウチュウ目、ゾウムシ科）豪州区
spine-tipped dancer		*Argia extranea* (Hagen)	（トンボ目、イトトンボ科）新北区、新熱帯区
spine-tufted skimmer		*Orthetrum chrysis* (Selys)	（トンボ目、トンボ科）東洋区
spine-winged tiger longicorn beetle	エグリトラカミキリ	*Chlorophorus japonicus* (Chevrolat)	（コウチュウ目、カミキリムシ科）日本、旧北区、東洋区。カンキツ、カキ、フジなどの害虫。spine-winged tiger longicorn ともいう
spined assassin bug		*Sinea diadema* (Fabricius)	（カメムシ目、サシガメ科）新北区。体長 12～14 mm
spined citrus bug		*Biprorulus bibax* Breddin	（カメムシ目、カメムシ科）豪州区。カンキツ害虫
spined elongate scolytid	トゲナガキクイムシ	*Diapus aculeatus* Blandford	（コウチュウ目、ナガキクイムシ科）日本、東洋区
spined fairytail		*Lestinogomphus angustus* Martin	（トンボ目、サナエトンボ科）エチオピア区
spined grouse locust	トゲヒシバッタ	*Criotettix japonicus* (de Haan)	（バッタ目、ヒシバッタ科）日本、東洋区
spined rat louse	イエネズミジラミ	*Polyplax spinulosa* (Burmeister)	（シラミ目、ホソケジラミ科）日本、汎世界。ドブネズミ、クマネズミに寄生
spined silverdrop		*Epargyreus spina spina* Evans	（チョウ目、セセリチョウ科）新熱帯区
spined silverfish		*Allacrotelsa spinulata* (Packard)	（シミ目、シミ科）新北区
spined soldier bug		*Podisus maculiventris* (Say)	（カメムシ目、カメムシ科）新北区。北米の普通種
spined stilt bug		*Jalysus wickkami* Van Duzee	（カメムシ目、イトカメムシ科）新北区。北米の普通種
spined stilt bug		*Jalysus spinosus* (Say)	（カメムシ目、イトカメムシ科）新北区。北米の普通種
spined stink bug			spined soldier bug を見よ
spined woodborer			pine sawyer beetle を見よ
spineless silverdrop		*Epargyreus aspina* Evans	（チョウ目、セセリチョウ科）新熱帯区
spinifex sand-skipper			Polysema skipper を見よ
spinifex termite			cathedral termite を見よ

英名	和名	学名	所属、分布、ほか
spinner			dragonflies (1) を見よ
spinneroo			dragonflies (1) を見よ
spinners			pronggill mayflies を見よ
spinners			mayflies を見よ
spinnin' jennie			dragonflies (1) を見よ
spinning jennies			crane flies (1) を見よ
spinose ground mealybug		*Rhizoecus spinosus* McKenzie	（カメムシ目、コナカイガラムシ科）新北区
spinose scale	ヤシトビイロマルカイガラムシ	*Oceanaspidiotus spinosus* (Comstock)	（カメムシ目、マルカイガラムシ科）日本
spinose skipper		*Muschampia cribrellum* (Eversmann)	（チョウ目、セセリチョウ科）旧北区。spinous skipper とも記す
spiny assassin bug	トゲサシガメ	*Polididus armatissimus* Stål	（カメムシ目、サシガメ科）日本、東洋区、新北区
spiny baskettail		*Epitheca spinigera* (Selys)	（トンボ目、エゾトンボ科）新北区
spiny bollworm			Egyptian bollworm を見よ
spiny bollworm moth		*Earias biplaga* Walker	（チョウ目、ヤガ科）旧北区。spiny bollworm ともいう
spiny bollworms		*Earias*	（チョウ目、ヤガ科）の昆虫の総称
spiny brown bugs		*Acanthomia*	（カメムシ目、ヘリカメムシ科）の昆虫の総称
spiny crawler mayflies	マダラカゲロウ科	Ephemerellidae	（カゲロウ目）の昆虫の総称
spiny digger wasps		Oxybelini	（ハチ目、アナバチ科）の昆虫の総称
spiny digger wasps		Oxybelinae	（ハチ目、アナバチ科）の昆虫の総称
spiny flower mantis		*Pseudocreobotra wahlbergi* Stål	（カマキリ目、Hymenopodidae）エチオピア区
spinyheaded burrower mayflies		Palingeniidae	（カゲロウ目）の昆虫の総称
spiny leaf beetle		*Hispa atra* (Linnaeus)	（コウチュウ目、ハムシ科）旧北区
spiny leaf insect			giant prickly stick insect を見よ　spiny-leaf insect とも記される
spiny leaf phasmid			giant prickly stick insect を見よ
spiny-legged bugs			spiny shore bugs を見よ
spiny-legged flies			black scavenger flies を見よ
spiny-legged rove beetles	セスジハネカクシ亜科	Oxytelinae	（コウチュウ目、ハネカクシ科）の昆虫の総称
spiny mason wasp			mason wasp (1) を見よ
spiny oak caterpillar			orange-tipped oakworm moth を見よ
spiny oak-slug moth		*Euclea delphinii* (Boisduval)	（チョウ目、イラガ科）新北区。spiny oak slug ともいう
spiny oakworm		*Anisota stigma* (Fabricius)	（チョウ目、ヤママユガ科）新北区。成虫は spiny oakworm moth
spiny rainforest katydid		*Phricta aberrans* Brunner von Wattenwyl	（バッタ目、キリギリス科）豪州区
spiny rat lice	ホソケジラミ科	Polyplacidae	（シラミ目）の昆虫の総称　ネズミ、ウサギ、サルなどに寄生。世界に 175 種

英名	和名	学名	所属、分布、ほか
spiny rat louse			spined rat louse を見よ
spiny shore bug		*Patapius spinosus* (Rossi)	(カメムシ目、Leptopodidae) 新北区
spiny shore bugs		Leptopodidae	(カメムシ目) の昆虫の総称 米国に1種
spiny stick insect	サカダチコノハムシ	*Heteropteryx dilatata* (Parkinson)	(ナナフシ目、ナナフシ科) 東洋区。体長25cmの大型種
spiny sucking lice	カイジュウジラミ科	Echinophthiridae	(シラミ目) の昆虫の総称 アザラシ他に寄生
spiny tachina fly		*Paradejeania rutilioides* Jaennicke	(ハエ目、ヤドリバエ科) 新北区、新熱帯区
spiny twig pruner		*Elaphidionoides mucronatus* (Fabricius)	(コウチュウ目、カミキリムシ科) 新北区
spiny whitefly			citrus spiny whitefly を見よ
spiny-winged moths		Eriocottidae	(チョウ目) の昆虫の総称
Spio grizzled skipper			mountain sandman を見よ
spiraea aphid			spirea aphid を見よ
spiraea sawfly		*Nematus spiraeae* Zaddach	(ハチ目、ハバチ科) 旧北区
spiral faggot-worm		*Cryptothelea holmesi* Watt	(チョウ目、ミノガ科) 東洋区
spiraling whitefly		*Aleurodicus dispersus* Russell	(カメムシ目、コナジラミ科) 東洋区、豪州区、大洋区
spirea aphid	ユキヤナギアブラムシ	*Aphis citricola* Van der Goot	(カメムシ目、アブラムシ科) 日本、全北区。ユキヤナギ、サクラ、バラ、カンキツなどの害虫
spirea leaftier moth		*Evora hemidesma* (Zeller)	(チョウ目、ハマキガ科) 新北区
spitfire grubs		*Perga*	(ハチ目、Pergidae) の昆虫の総称
spittle bugs (1)	コガシラアワフキ科	Cercopidae	(カメムシ目) の昆虫の総称
spittle bugs (2)	アワフキムシ科	Aphrophoridae	(カメムシ目) の昆虫の総称 泡吹き虫で spit, spittle, spittle insect といい、ときに froghopper も使われる
spittle insects	アワフキムシ上科	Cercopoidea	(カメムシ目) の昆虫の総称
spittle insects			spittle bugs (2) を見よ
spittlebug			cuckoo spit insect を見よ
splashed leafhopper	カスリヨコバイ	*Balclutha punctata* Matsumura	(カメムシ目、オオヨコバイ科) 日本、汎世界
splayed deerfly		*Chrysops caecutiens* (Linnaeus)	(ハエ目、アブ科) 旧北区
splended clubtail		*Gomphus lineatifrons* Calvert	(トンボ目、サナエトンボ科) 新北区
splended cruiser			shining Macromia を見よ
splended knot-horn		*Dioryctria splendidella* Herrich-Schäffer	(チョウ目、メイガ科) 旧北区
splended mapwing			Godman's mapwing を見よ
splended Palpita moth		*Palpita magniferalis* (Walker)	(チョウ目、メイガ科) 新北区。splended Palpita snout moth, ash pyralid, ash leafroller ともいう
splended piercer			chestnut tortrix を見よ
splended shield-back katydid		*Neduba ovata* Scudder	(バッタ目、キリギリス科) 新北区
splended tiger beetle		*Cicindela splendida* Hentz	(コウチュウ目、ハンミョウ科) 新北区

英名	和名	学名	所属、分布、ほか
splendid dagger moth		*Acronicta superans* Guenée	（チョウ目、ヤガ科）新北区
splendid dung beetle		*Phanaeus vindex* Macleay	（コウチュウ目、コガネムシ科）新北区。成虫は深い穴を掘り、産卵した糞球を入れる
splendid emerald wasp	オオセイボウ	*Stilbum cyanurum pacificum* Linsenmeier	（ハチ目、セイボウ科）日本
splendid emerald wasp		*Stilbum splendidum* Fabricius	（ハチ目、セイボウ科）豪州区
splendid ghost moths		*Aenetus*	（チョウ目、コウモリガ科）の昆虫の総称
splendid jewel		*Hypochrysops cleon* (Grose-Smith)	（チョウ目、シジミチョウ科）豪州区
splendid ochre			Symmomus skipper を見よ
splendid shieldback		*Ateloplus splendidus* Hebard	（バッタ目、キリギリス科）新北区
splendid tenebrionid		*Cyphaleus mastersi* Pascoe	（コウチュウ目、ゴミムシダマシ科）豪州区
splendid tiger beetle		*Cicindela (Cicindela) splendida* Hentz	（コウチュウ目、ハンミョウ科）新北区。体長 12〜15 mm
splendour beetles			flat-headed beetles を見よ
splitback mayflies	ヒラタカゲロウ亜目	Schistonata	（カゲロウ目）の昆虫の総称
split-banded owl-butterfly		*Opsiphanes cassina* C. et R. Felder	（チョウ目、タテハチョウ科）新北区。新熱帯区の *O. c. fabricii* (Boisduval) も同英名。split-banded owlet ともいう
split-lined angle moth		*Speranza bitactata* (Walker)	（チョウ目、シャクガ科）新北区
sponged treehopper		*Sphongophorus mexicana* (Guérin-Méneville)	（カメムシ目、ツノゼミ科）新北区
sponged treehoppers			antlered treehoppers を見よ
spongefly			common spongilla fly を見よ
spongilla-flies	ミズカゲロウ科	Sisyridae	（アミメカゲロウ目）の昆虫の総称　sponge-flies ともいう
spoon-headed leafhopper	サジヨコバイ	*Hecalus prasinus* (Matsumura)	（カメムシ目、オオヨコバイ科）日本、旧北区、東洋区
spoon-tailed short-winged katydid		*Dichopetala catinata* Rehn et Hebard	（バッタ目、キリギリス科）新北区
spoonwinged lacewing		*Chasmoptera hutii* Westwood	（アミメカゲロウ目、Nemopteridae）豪州区
spoon-winged lacewings			thread-winged lacewings を見よ
spot and cream-stripe moth		*Monopis ethelella* Newmann	（チョウ目、ヒロズコガ科）豪州区
spot-banded longtail		*Urbanus pronta* Evans	（チョウ目、セセリチョウ科）新熱帯区
spot-banded small noctuid	モンオビヒメヨトウ	*Dysmilichia gemella* (Leech)	（チョウ目、ヤガ科）日本、旧北区
spot-celled sister		*Adelpha basiloides* (Bates)	（チョウ目、タテハチョウ科）新北区
spot-edged Dyspyralis moth		*Dyspyralis puncticosta* Smith	（チョウ目、ヤガ科）新北区
spot Judy		*Abisara chela* de Nicéville	（チョウ目、シジミタテハ科）東洋区
spot puffin		*Appias lalage* (Doubleday)	（チョウ目、シロチョウ科）東洋区

英名	和名	学名	所属、分布、ほか
spot swordtail		*Graphium nomius* (Esper)	(チョウ目、アゲハチョウ科) 東洋区
spot-tailed darter		*Micrathyria aequalis* (Hagen)	(トンボ目、トンボ科) 新熱帯区
spot-winged glider		*Pantala hymenaea* (Say)	(トンボ目、トンボ科) 新北区
spotless anglewing		*Polygonia haroldii* (Dewitz)	(チョウ目、タテハチョウ科) 新熱帯区
spotless fall webworm		*Hyphantria textor* Harris	(チョウ目、ヒトリガ科) 新北区
spotless grass yellow	ツマグロキチョウ	*Eurema laeta* Boisduval	(チョウ目、シロチョウ科) 日本、旧北区、東洋区、豪州区。line grass yellow を参照
spotless hog			small skipper (1) を見よ
spotless lady beetle		*Cycloneda sanguinea* (Linnaeus)	(コウチュウ目、テントウムシ科) 新北区、新熱帯区
spotless oakblue		*Arhopala fulla* (Hewitson)	(チョウ目、シジミチョウ科) 東洋区
spotless policeman		*Coeliades libeon* (Druce)	(チョウ目、セセリチョウ科) エチオピア区
spotted Adonis blue		*Polyommatus punctifera* (Oberthür)	(チョウ目、シジミチョウ科) 旧北区
spotted alfalfa aphid		*Therioaphis maculata* (Buckton)	(カメムシ目、アブラムシ科) 新北区
spotted alfalfa aphid			yellow clover aphid を見よ　米国での英名
spotted alfalfa aphid parasite		*Trioxys complanatus* Quilis	(ハチ目、コマユバチ科) 豪州区
spotted alfalfa-like aphid	アルファルファアブラムシ	*Therioaphis trifolii* (Monell)	(カメムシ目、アブラムシ科) 日本、全北区、エチオピア区、豪州区。マメ科牧草の害虫。外国の本種は yellow clover aphid という
spotted angle		*Caprona agama* (Moore)	(チョウ目、セセリチョウ科) 東洋区
spotted Anopheles		*Anopheles maculatus* Theobald	(ハエ目、カ科) 太洋区
spotted Apatelodes moth		*Apatelodes torrefacta* (Smith)	(チョウ目、カイコガ科) 新北区。spotted Apatelodes ともいう
spotted apple budworm			marbled orchard tortrix をみよ
spotted asparagus beetle		*Crioceris duodecimpunctata* (Linnaeus)	(コウチュウ目、ハムシ科) 新北区。アスパラガス害虫
spotted aspen leafroller			poplar leafroller moth を見よ
spotted assassin		*Rhynocoris ventralis* (Say)	(カメムシ目、サシガメ科) 新北区
spotted bean weevil			spotted pea weevil を見よ
spotted beet webworm		*Hymenia perspectalis* (Hübner)	(チョウ目、メイガ科) 新北区。成虫は spotted beet webworm moth
spotted black crow	クラメルルリマダラ	*Euploea crameri* Lucas	(チョウ目、タテハチョウ科) 東洋区。*E. c. bremeri* C. et R. Felder も同英名
spotted black plume		*Capperia britanniodactyla* Gregs	(チョウ目、トリバガ科) 旧北区
spotted blister beetle		*Epicauta maculata* (Say)	(コウチュウ目、ツチハンミョウ科) 新北区。ジャガイモ害虫
spotted bollworm (1)	クサオビリンガ	*Earias vittella* (Fabricius)	(チョウ目、ヤガ科) 日本、東洋区、大洋区、豪州区

1045

英名	和名	学名	所属、分布、ほか
spotted bollworm (2)	ワタリンガ	*Earias cupreoviridis* Walker	(チョウ目、ヤガ科) 日本、東洋区、エチオピア区。ワタ害虫
spotted bronze bright		*Lampronia capitella* (Clerck)	(チョウ目、マガリガ科) 旧北区
spotted brown		*Heteronympha paradelpha* (Lower)	(チョウ目、タテハチョウ科) 豪州区
spotted buff		*Pentila tropicalis* (Boisduval)	(チョウ目、シジミチョウ科) エチオピア区
spotted camel cricket		*Ceuthophilus maculatus* (Harris)	(バッタ目、カマドウマ科) 新北区
spotted chafer			Asiatic beetle を見よ
spotted ciliate blue		*Anthene larydas* (Cramer)	(チョウ目、シジミチョウ科) エチオピア区
spotted clover	ヨモギガ	*Heliothis scutosa* (Denis et Schiffermüller)	(チョウ目、ヤガ科) 日本、旧北区
spotted cowpea bruchid			cowpea weevil を見よ
spotted crane fly		*Nephrotoma appendiculata* (Pierre)	(ハエ目、ガガンボ科) 旧北区
spotted cranefly (1)		*Pales maculatus* Meigen	(ハエ目、ガガンボ科) 旧北区
spotted crow eggfly	ヤエヤママラサキ	*Hypolimnas antilope* (Cramer)	(チョウ目、タテハチョウ科) 豪州区
spotted cuckoo bees		*Thyreus*	(ハチ目、ミツバチ科) の昆虫の総称　豪州での英名
spotted cucumber beetle			twelve-spotted cucumber beetle を見よ
spotted cutworm			spotted cutworm moth (1)(2) を見よ　北米での英名
spotted cutworm moth (1)	シロモンヤガ	*Xestia c-nigrum* (Linnaeus)	(チョウ目、ヤガ科) 日本、全北区。トウモロコシ害虫
spotted cutworm moth (2)		*Xestia dolosa* Franclemont	(チョウ目、ヤガ科) 新北区。タバコ害虫
spotted cutworms		*Xestia*	(チョウ目、ヤガ科) の昆虫の総称
spotted cymatophorid	モントガリバ	*Thyatira batis japonica* Werny	(チョウ目、トガリバガ科) 日本
spotted dark bright		*Lampronia luzella* (Hübner)	(チョウ目、マガリガ科) 旧北区
spotted darter	タイリクアキアカネ	*Sympetrum depressiusculum* (Selys)	(トンボ目、トンボ科) 日本、旧北区
spotted Datana moth		*Datana perspicua* Grote et Robinson	(チョウ目、シャチホコガ科) 新北区
spotted demon		*Notocrypta feisthamelii* (Boisduval)	(チョウ目、セセリチョウ科) 東洋区。*N. f. alysos* (Moore) も同英名
spotted Dichomeris moth		*Dichomeris punctidiscella* Clemens	(チョウ目、キバガ科) 新北区
spotted dung beetle	エンマハバビロガムシ	*Sphaeridium scarabaeoides* (Linnaeus)	(コウチュウ目、ガムシ科) 新北区
spotted dusky-blue		*Candalides delospila* (Waterhouse)	(チョウ目、シジミチョウ科) 豪州区
spotted earwig			ring-legged earwig を見よ　米国での英名

英名	和名	学名	所属、分布、ほか
spotted-eye brown		*Pseudonympha narycia* Wallengren	(チョウ目、タテハチョウ科) エチオピア区
spotted fireworm moth		*Choristoneura parallela* (Robinson)	(チョウ目、ハマキガ科) 新北区
spotted flats		*Celaenorrhinus*	(チョウ目、セセリチョウ科) の昆虫の総称
spotted flower buprestids		*Acmaeodera*	(コウチュウ目、タマムシ科) の昆虫の総称
spotted flower chafer			punctate flower chafer を見よ
spotted fritillary	ヒメフチグロヒョウモンモドキ	*Melitaea didyma* (Esper)	(チョウ目、タテハチョウ科) 旧北区
spotted froghopper	ホシアワフキ	*Aphrophora stictica* Matsumura	(カメムシ目、アワフキムシ科) 日本、旧北区
spotted geometrid moth			speckled yellow を見よ　米国での英名
spotted giant hornet			hornet を見よ
spotted grape beetle		*Pelidnota punctata* Linnaeus	(コウチュウ目、コガネムシ科) 新北区
spotted grass-blue			dark grass blue (1) を見よ
spotted grass moth		*Rivula propinqualis* Guenée	(チョウ目、ヤガ科) 新北区
spotted ground cricket		*Allonemobius maculatus* (Blatchley)	(バッタ目、コオロギ科) 新北区
spotted gum lerp psyllid		*Eucalyptolyma maideni* Froggatt	(カメムシ目、キジラミ科) 新北区、豪州区
spotted hairy fungus beetle		*Mycetophagus quadriguttatus* Müller	(コウチュウ目、コキノコムシ科) 新北区
spotted Halisidota			yellow-spotted tiger moth を見よ
spotted Jay	ヒメタイマイ	*Graphium arycles* (Boisduval)	(チョウ目、アゲハチョウ科) 東洋区
spotted jester			Himalayan jester (1) を見よ
spotted Jezebel			wood white (1) を見よ
spotted joker			joker (1) を見よ
spotted June beetle			spotted grape beetle を見よ
spotted katydid		*Epipitytha trigintiduoguttata* (Serville)	(バッタ目、キリギリス科) 豪州区
spotted lady beetle		*Coleomegilla fuscilabris* (Mulsant)	(コウチュウ目、テントウムシ科) 新北区。spotted ladybird beetle ともいう
spotted large lacewing		*Polystoechotes punctatus* (Fabricius)	(アミメカゲロウ目、Polystoechotidae) 新北区
spotted leafhopper		*Austroagallia torrida* Evans	(カメムシ目、オオヨコバイ科) 豪州区
spotted leafhopper		*Rhopalogonia scita* (Walker)	(カメムシ目、オオヨコバイ科) 旧北区
spotted-legged cutworm			old man dart を見よ
spotted-legged meadow katydid		*Conocephalus stictomerus* Rehn et Hebard	(バッタ目、キリギリス科) 新北区

英名	和名	学名	所属、分布、ほか
spotted lilac tree nymph	アカムラサキタテハ	*Sevenia pechueli* (Dewitz)	(チョウ目、タテハチョウ科) エチオピア区
spotted loblolly pine sawfly		*Neodiprion taedae taedae* Ross	(ハチ目、マツハバチ科) 新北区。亜種 *N. t. linearis* Ross は loblolly pine sawfly という
spotted longhorn		*Rutpela maculata* (Poda)	(コウチュウ目、カミキリムシ科) 旧北区
spotted maize beetle		*Astylus atromaculatus* (Blanchard)	(コウチュウ目、ジョウカイモドキ科) エチオピア区、新熱帯区
spotted manuka moth		*Declana leptomera* Walker	(チョウ目、シャクガ科) 豪州区
spotted mediterranean cockroach			tawny cockroach (2) を見よ
spotted milkweed bug			large milkweed bug を見よ
spotted mystic		*Lethe tristigmata* Elwes	(チョウ目、タテハチョウ科) 東洋区
spotted oleander caterpillar		*Empyreuma affinis* Rothschild	(チョウ目、ヒトリガ科) 新熱帯区
spotted oleander caterpillar moth		*Empyreuma pugione* (Linnaeus)	(チョウ目、ヒトリガ科) 新北区
spotted opal		*Nesolycaena urumelia* (Tindale)	(チョウ目、シジミチョウ科) 豪州区
spotted palmfly	ツマムラサキヒカゲ	*Elymnias malelas* (Hewitson)	(チョウ目、タテハチョウ科) 東洋区
spotted pea-blue			gram blue を見よ
spotted pea weevil		*Sitona crinita* Dejean	(コウチュウ目、ゾウムシ科) 旧北区
spotted Pelidnota			spotted grape beetle を見よ
spotted peppergrass moth		*Eustixia pupula* Hübner	(チョウ目、メイガ科) 新北区
spotted Phosphila moth		*Phosphila miselioides* (Guenée)	(チョウ目、ヤガ科) 新北区。spotted Phosphila ともいう
spotted pierrot		*Tarucus callinara* Butler	(チョウ目、シジミチョウ科) 東洋区
spotted pierrot		*Tarucus nigra* Bethune-Baker	(チョウ目、シジミチョウ科) 東洋区
spotted pine sawyer		*Monochamus mutator* LeConte	(コウチュウ目、カミキリムシ科) 新北区
spotted pine sawyer (1)		*Monochamus maculosus* Haldeman	(コウチュウ目、カミキリムシ科) 新北区
spotted pink bollworm			pink-spotted bollworm を見よ
spotted redeye		*Pudicitia pholus* (de Nicéville)	(チョウ目、セセリチョウ科) 東洋区
spotted royal		*Tajuria maculata* (Hewitson)	(チョウ目、シジミチョウ科) 東洋区
spotted rustic			leopard を見よ
spotted sable pigmy		*Ectoedemia subbimaculella* (Haworth)	(チョウ目、モグリチビガ科) 旧北区
spotted saffron			brimstone を見よ
spotted sailer		*Neptis magadha* (Stoll)	(チョウ目、タテハチョウ科) 東洋区。*N. m. khasiana* Moore も同英名

英名	和名	学名	所属、分布、ほか
spotted sailer		*Neptis saclava* Boisduval	(チョウ目、タテハチョウ科) エチオピア区。*N. s. marpessa* Hopffer は small-spotted sailer という
spotted sawtooth	マダラシロチョウ	*Prioneris thestylis* (Doubleday)	(チョウ目、シロチョウ科) 東洋区
spotted sawyer			spotted pine sawyer (1) を見よ
spotted scarlet-eye		*Bungalotis erythus* (Cramer)	(チョウ目、セセリチョウ科) 新熱帯区
spotted sergeant		*Athyma sulpitia* (Cramer)	(チョウ目、タテハチョウ科) 東洋区
spotted shoot	ニセマツアカヒメハマキ	*Rhyacionia pinivorana* (Lienig et Zeller)	(チョウ目、ハマキガ科) 日本、旧北区。spotted shoot moth ともいう
spotted-sided cutworm			pale-banded dart moth を見よ
spotted-sided cutworm moth		*Xestia badinosis* (Grote)	(チョウ目、ヤガ科) 新北区。北米での英名。spotted-sided cutworm ともいう
spotted skimmers		*Celithemis*	(トンボ目、トンボ科) の昆虫の総称
spotted skipper		*Hesperilla ornata* (Leach)	(チョウ目、セセリチョウ科) 豪州区
spotted skipper			chequered skipper を見よ
spotted skipperling		*Piruna polingii* (Barnes)	(チョウ目、セセリチョウ科) 新北区、新熱帯区
spotted small flat		*Sarangesa purendra* (Moore)	(チョウ目、セセリチョウ科) 東洋区
spotted snow flat		*Tagiades menaka* (Moore)	(チョウ目、セセリチョウ科) 東洋区
spotted soldier fly		*Stratiomys maculosa* (Loew)	(ハエ目、ミズアブ科) 新北区
spotted Spragueia moth		*Spragueia guttata* Grote	(チョウ目、ヤガ科) 新北区
spotted spreadwing		*Lestes tridens* McLachlan	(トンボ目、アオイトトンボ科) エチオピア区
spotted spreadwing			dark Lestes を見よ
spotted springtail		*Papirius maculosus* Schøtt	(トビムシ目、アヤトビムシ科) 新北区
spotted stalk borer			pink borer (2) を見よ
spotted stem borer		*Proceras sacchariphagus* (Bojer)	(チョウ目、メイガ科) エチオピア区。サトウキビ害虫
spotted stem borer			pink borer (2) を見よ
spotted straw moth		*Heliocheilus turbata* (Walker)	(チョウ目、ヤガ科) 新北区
spotted sugar ant		*Camponotus maculatus* (Fabricius)	(ハチ目、アリ科) 新熱帯区
spotted sulphur		*Emmelia sulphuralis* (Linnaeus)	(チョウ目、ヤガ科) 旧北区。spotted sulphur moth ともいう
spotted-sulphur			yellow-spotted small noctuid を見よ
spotted sylph		*Astictopterus stellata* (Mabille)	(チョウ目、セセリチョウ科) エチオピア区
spotted tan skipper		*Calleagris jamesoni jamesoni* Sharpe	(チョウ目、セセリチョウ科) エチオピア区
spotted tentiform leaf miner			apple-leaf midget (1) を見よ
spotted tentiform leafminer			apple leaf miner (1) を見よ

英名	和名	学名	所属、分布、ほか
spotted Thyris moth		*Thyris maculata* Harris	(チョウ目、マドガ科) 新北区
spotted tiger moth		*Arctia purpurea* (Linnaeus)	(チョウ目、ヒトリガ科) 旧北区
spotted Toxonprucha moth		*Toxonprucha pardalis* (Smith)	(チョウ目、ヤガ科) 新北区
spotted tree borer beetle		*Synaphaeta guexi* (LeConte)	(コウチュウ目、カミキリムシ科) 新北区
spotted tussock moth			yellow-spotted tiger moth を見よ　spotted tussock ともいう
spotted vegetable weevil		*Desiantha diversipes* (Pascoe)	(コウチュウ目、ゾウムシ科) 豪州区
spotted velvet skipper		*Abantis tettensis* Hopffer	(チョウ目、セセリチョウ科) エチオピア区
spotted violet bright		*Incurvaria praelatella* (Denis et Schiffermüller)	(チョウ目、マガリガ科) 旧北区
spotted water beetle			marbled diving beetle を見よ
spotted white plume		*Pterophorus galactodactyla* Denis et Schiffermüller	(チョウ目、トリバガ科) 旧北区
spotted willow leaf-beetle		*Melasoma lapponicum* (Linnaeus)	(コウチュウ目、ハムシ科) 旧北区
spotted willow-twig aphid			black willow aphid を見よ
spotted-winged antlion		*Dendroleon obsoletus* (Say)	(アミメカゲロウ目、ウスバカゲロウ科) 新北区
spotted-winged grasshopper		*Cordillacris occipitalis* (Thomas)	(バッタ目、バッタ科) 新北区
spotted-winged raspberry aphid		*Illinoia rubicola* (Oestlund)	(カメムシ目、アブラムシ科) 新北区
spotted yellow clothes		*Tenaga nigripunctella* (Haworth)	(チョウ目、ヒロズコガ科) 旧北区
spotted yellow lancer		*Salanoemia noemi* (de Nicéville)	(チョウ目、セセリチョウ科) 東洋区
spotted yellow tussock moth	マガリキドクガ	*Euproctis curvata* Wileman	(チョウ目、ドクガ科) 日本
spotted zebra		*Graphium megarus* (Westwood)	(チョウ目、アゲハチョウ科) 東洋区
spotted Zulu		*Alaena amazoula* Boisduval	(チョウ目、シジミチョウ科) エチオピア区
Sprague's Pygarctia		*Pygarctia spraguei* (Grote)	(チョウ目、ヒトリガ科) 新北区。Sprague's Pygarctia moth ともいう
sprawler		*Brachionycha sphinx* (Hufnagel)	(チョウ目、ヤガ科) 旧北区
sprawler moth			sprawler を見よ
spread-winged damselflies			emerald damselflies を見よ
spread-winged skippers			open-winged skippers を見よ
spreadwings		*Lestes*	(トンボ目、アオイトトンボ科) の昆虫の総称
spreadwings			emerald damselflies を見よ

英名	和名	学名	所属、分布、ほか
sprigged green button		*Acleris literana* (Linnaeus)	（チョウ目、ハマキガ科）旧北区
spring azure		*Celastrina ladon* (Cramer)	（チョウ目、シジミチョウ科）新北区。亜種 *C. l. lucia* (Kirby) も同英名
spring azure			holly blue を見よ　米国での英名
spring beetle		*Colymbomorpha vittata* Britton	（コウチュウ目、コガネムシ科）豪州区
spring bollworm			spiny bollworm moth を見よ
spring bollworm			spotted bollworm (1) を見よ
spring bollworm			Egyptian bollworm を見よ
spring cankerworm		*Paleacrita vernata* (Peck)	（チョウ目、シャクガ科）新北区。spring cankerworm moth ともいう
spring cicada	ハルゼミ	*Tarpnosia vacua* (Olivier)	（カメムシ目、セミ科）日本、旧北区
spring cranefly		*Tipula vernalis* Meigen	（ハエ目、ガガンボ科）旧北区
spring dead-leaf roller moth		*Cenopis diluticostana* (Walsingham)	（チョウ目、ハマキガ科）新北区
spring field cricket		*Gryllus veletis* Alexander et Bigelow	（バッタ目、コオロギ科）新北区
spring grain aphid			greenbug を見よ
spring green carpet		*Calostigia pectinataria* Knoch	（チョウ目、シャクガ科）旧北区
spring grey smoke		*Dahlica inconspicuella* (Stainton)	（チョウ目、ミノガ科）旧北区
spring hawker			brown hawker を見よ
spring heath robberfly		*Lasiopogon cinctus* (Fabricius)	（ハエ目、ムシヒキアブ科）旧北区
spring mountain fritillary	チャマダラモンギンボシヒョウモン	*Speyeria zerene carolae* (dos Passos et Grey)	（チョウ目、タテハチョウ科）新北区。Zerene fritillary を参照
spring oak leafroller moth		*Chionodes continuella* (Zeller)	（チョウ目、キバガ科）全北区
spring oak leafroller moth		*Chionodes formosella* (Murtfeldt)	（チョウ目、キバガ科）新北区
spring porina moth		*Wiseana cervinata* Meyrick	（チョウ目、コウモリガ科）豪州区
spring rat louse			spined rat louse を見よ
spring redtail			large red damselfly を見よ
spring ringlet	ユキワリベニヒカゲ	*Erebia epistygne* (Hübner)	（チョウ目、タテハチョウ科）旧北区
spring sooty			dusky azure を見よ
spring sruce needle moth		*Archips packardiana* (Fernald)	（チョウ目、ハマキガ科）新北区
spring stoneflies		*Nemoura*	（カワゲラ目、オナシカワゲラ科）の昆虫の総称
spring stoneflies			thread-tailed stoneflies を見よ
spring trig		*Anaxipha* sp.	（バッタ目、クサヒバリ科）新北区
spring usher	シロフフユエダシャク	*Agriopis leucophaearia* (Denis et Schiffermüller)	（チョウ目、シャクガ科）日本、旧北区

英名	和名	学名	所属、分布、ほか
spring webworm		*Ocnogyna loewii* Zeller	(チョウ目、ヒトリガ科) 旧北区
spring white	カリフォルニアシロチョウ	*Pontia sisymbrii* (Boisduval)	(チョウ目、シロチョウ科) 新北区
spring widow		*Tarsocera cassus* (Linnaeus)	(チョウ目、タテハチョウ科) エチオピア区
springflies			predatory stoneflies を見よ
springing beetles			click beetles (1) を見よ
springtails	トビムシ目	Collembola	の昆虫の総称 0.25〜10 mm の小型昆虫
springtails			snowfleas を見よ
springtails			globular springtails を見よ
springtime darner		*Basiaeschna janata* (Say)	(トンボ目、ヤンマ科) 新北区
springtime dung beetle		*Trypocopris vernalis* (Linnaeus)	(コウチュウ目、センチコガネ科) 旧北区
springwater dancer		*Argia plana* (Calvert)	(トンボ目、イトトンボ科) 新北区
springwater sprite		*Pseudagrion caffrum* Burmeister	(トンボ目、イトトンボ科) エチオピア区
sprinkled locust		*Chloealtis conspersa* Harris	(バッタ目、バッタ科) 新北区
sprinkled scots-fir argent		*Cedestis gysselinella* Zeller	(チョウ目、スガ科) 旧北区
sprinkler sewage filter fly	ホシチョウバエ	*Tinearia alternata* (Say)	(ハエ目、チョウバエ科) 日本、汎世界
sprinkling sewage filter fly			trickling filter fly を見よ
sprites		*Archibasis*	(トンボ目、イトトンボ科) の昆虫の総称
sprites		*Pseudagrion*	(トンボ目、イトトンボ科) の昆虫の総称
spruce adelgids			conifer aphids を見よ
spruce aphid		*Elatobium abietinum* (Walker)	(カメムシ目、アブラムシ科) 全北区
spruce aphid moth			Ratzeburg's bell を見よ
spruce aphid scale			spruce bud scale (1) を見よ
spruce aphids			conifer aphids を見よ
spruce bark beetle	ウスイロキクイムシ	*Hylurgops palliatus* (Gyllenhal)	(コウチュウ目、キクイムシ科) 日本、旧北区
spruce bark beetle			eight-spined ips を見よ 原名亜種の英名
spruce beetle		*Dendroctonus rufipennis* (Kirby)	(コウチュウ目、ゾウムシ科) 新北区
spruce beetle		*Hylastes cumicularius* Erichson	(コウチュウ目、キクイムシ科) 旧北区
spruce beetles			weevils (1) を見よ 稀な使用
spruce bell moth		*Epinotia tedella* (Clerck)	(チョウ目、ハマキガ科) 旧北区
spruce bog alpine	キシタシベリアベニヒカゲ	*Erebia disa* (Thunberg)	(チョウ目、タテハチョウ科) 新北区
spruce bud gall midge	エゾマツノシントメタマバエ	*Dasineura ezomatsue* (Uchida et Inouye)	(ハエ目、タマバエ科) 日本。エゾマツ害虫
spruce bud midge		*Dasineura swainei* (Felt)	(ハエ目、タマバエ科) 新北区

英名	和名	学名	所属、分布、ほか
spruce bud moth		*Zeiraphera canadensis* Mutuura et Freeman	(チョウ目、ハマキガ科) 新北区
spruce bud moth (1)	トウヒノシントメハマキ	*Choristoneura fumiferana* (Clemens)	(チョウ目、ハマキガ科) 新北区。モミ、トウヒなど樹木害虫。spruce budowrm, spruce budworm moth ともいう
spruce bud moth			Ratzeburg's bell を見よ
spruce bud scale	トウヒタマカイガラモドキ	*Physokermes jezoensis* Siraiwa	(カメムシ目、カタカイガラムシ科) 日本。トウヒ、エゾマツ害虫
spruce bud scale (1)		*Physokermes piceae* (Schrank)	(カメムシ目、カタカイガラムシ科) 全北区
spruce budworm			Ratzeburg's bell を見よ
spruce bug			spruce cone bug を見よ
spruce carpet (1)		*Thera britanica* Turner	(チョウ目、シャクガ科) 旧北区
spruce carpet	キオビハガタナミシャク	*Thera variata bellisi* Viidalepp	(チョウ目、シャクガ科) 日本。原名亜種の英名
spruce cone bug		*Gastrodes abietum* (Bergroth)	(カメムシ目、ナガカメムシ科) 旧北区
spruce cone looper moth		*Eupithecia mutata* Pearsall	(チョウ目、シャクガ科) 新北区。幼虫は spruce cone looper
spruce cone moth		*Laspeyresia strobilella* Linnaeus	(チョウ目、ハマキガ科) 旧北区
spruce cone moth			pine knot-horn を見よ
spruce cone pyralid			pine knot-horn を見よ
spruce coneworm		*Dioryctria reniculelloides* Mutuura et Munroe	(チョウ目、メイガ科) 新北区。成虫は spruce coneworm moth
spruce cone worm			pine knot-horn を見よ
spruce engraver beetles			pine engraver beetles を見よ
spruce false looper moth		*Syngrapha viridisigma* (Grote)	(チョウ目、ヤガ科) 新北区。spruce false looper ともいう
spruce gall adelges		*Adelges viridis* (Ratzeburg)	(カメムシ目、カサアブラムシ科) 旧北区
spruce gall adelgid			spruce pineapple gall adelges を見よ
spruce gall aphid			spruce pineapple gall adelges を見よ
spruce gall louse			spruce pineapple gall adelges を見よ
spruce gall midge		*Dasineura piceae* (Felt)	(ハエ目、タマバエ科) 新北区
spruce gall mite			spruce pineapple gall adelges を見よ
spruce hawk			spurge hawkmoth を見よ
spruce leafminer		*Coleophora piceaella* Kearfott	(チョウ目、ツツミノガ科) 全北区
spruce leaf miner			Minnion's groundling を見よ　カナダでの英名
spruce leafroller	トウヒハマキ	*Cymolomia hartigiana* (Saxesen)	(チョウ目、ハマキガ科) 日本、旧北区。トドマツ、トウヒ、エゾマツ害虫
spruce longicorn beetle			black spruce beetle を見よ　トドマツ害虫。spruce longhorn beetle, spruce longicorn ともいう
spruce mealybug		*Puto sandini* Washburn	(カメムシ目、コナカイガラムシ科) 新北区

英名	和名	学名	所属、分布、ほか
spruce needle miner			spruce tortrix を見よ
spruce needle miner			spruce bell moth を見よ
spruce needle tortrix			spruce bell moth を見よ
spruce needleminer moth		*Taniva albolineana* (Kearfott)	(チョウ目、ハマキガ科) 新北区
spruce needleworm moth		*Dolichomia thymetusalis* (Walker)	(チョウ目、メイガ科) 新北区
spruce pineapple gall adelges	トウヒアブラムシ	*Adelges abietis* (Linnaeus)	(カメムシ目、カサアブラムシ科) 日本、全北区。spruce pineapple gall aphid ともいう
spruce pineapple gall adelges			spruce gall adelges を見よ
spruce psylla	ハイマツキジラミ	*Psylla haimatsucola* Miyatake	(カメムシ目、キジラミ科) 日本。ハイマツ害虫
spruce pyralid	ワモンノメイガ	*Nomophila noctuella* (Denis et Schiffermüller)	(チョウ目、メイガ科) 日本、全北区、大洋区。アブラナ科野菜、ホウレンソウ、エゾマツなどの害虫
spruce root aphid		*Pachypappa tremulae* (Linnaeus)	(カメムシ目、アブラムシ科) 旧北区
spruce root bark beetle	トウヒノネキクイムシ	*Dryocoetes autographus* (Ratzeburg)	(コウチュウ目、キクイムシ科) 日本、全北区、東洋区。トウヒ、マツ類の害虫
spruce sawfly	トウヒノクロハバチ	*Gilpinia tohi* Takeuchi	(ハチ目、マツハバチ科) 日本。トウヒ害虫
spruce sawfly (1)		*Gilpinia hercyniae* (Hartig)	(ハチ目、マツハバチ科) 旧北区、新北区
spruce seed chalcid		*Megastigmus suspectus* Borries	(ハチ目、オナガコバチ科) 旧北区
spruce seed moth		*Cydia youngana* Kearfott	(チョウ目、ハマキガ科) 新北区
spruce seed moth (1)	エゾマツカサヒメハマキ	*Cydia strobiella* (Linnaeus)	(チョウ目、ハマキガ科) 日本、全北区
spruce shoot aphid	エゾアメイロオオアブラムシ	*Cinara pilicornis* (Hartig)	(カメムシ目、アブラムシ科) 日本、全北区
spruce-shoot argent		*Blastotere glabratella* (Zeller)	(チョウ目、メムシガ科) 旧北区
spruce shortwing beetle		*Molorchus minor* (Linnaeus)	(コウチュウ目、カミキリムシ科) 旧北区。日本には *M. m. fuscus* Hayashi シラホシヒゲナガコバネカミキリ、*M. m. ikedai* Takakuwa シナノシラホシヒゲナガコバネカミキリが分布する。*M. m. fuscus* は white-spotted short-wing longicorn という
spruce tortrix		*Endothenia albolineana* Kearfott	(チョウ目、ハマキガ科) 新北区
spruce web-spinning sawfly		*Cephalcia fallenii* Dalm	(ハチ目、ヒラタハバチ科) 旧北区
spruce webworm	ヒロバビロウドハマキ	*Eurydaxa advena* Filipjev	(チョウ目、ハマキガ科) 日本、旧北区。トドマツ、エゾマツ害虫
spruce weevil		*Pissodes harcyniae* (Herbst)	(コウチュウ目、ゾウムシ科) 旧北区
spruce wood engraver			six-toothed bark beetle (1) を見よ
spun glass slug moth		*Isochaetes beutenmuelleri* (H. Edwards)	(チョウ目、イラガ科) 新北区

英名	和名	学名	所属、分布、ほか
spur-throated grasshopper		*Melanoplus ponderosus* (Scudder)	（バッタ目、バッタ科）新北区、新熱帯区、大洋区
spur-throated grasshoppers		Cyrtacanthacridinae	（バッタ目、バッタ科）の昆虫の総称
spur-throated grasshoppers (1)		Melanoplinae	（バッタ目、バッタ科）の昆虫の総称
spur-throated locust		*Nomadacris guttulosa* (Walker)	（バッタ目、バッタ科）豪州区
spurge bugs	ナガヘリカメムシ科	Stenocephalidae	（カメムシ目）の昆虫の総称
spurge hawkmoth		*Hyles euphorbiae* (Linnaeus)	（チョウ目、スズメガ科）全北区。米国での英名。spurge hawk ともいう
spurge spanworm moth		*Oxydia vesulia* (Cramer)	（チョウ目、シャクガ科）新北区
spurgebug		*Stenocephalus agilis* Scopoli	（カメムシ目、ヘリカメムシ科）旧北区
Spurina hairstreak		*Thereus spurina* (Hewitson)	（チョウ目、シジミチョウ科）新熱帯区
spurious sphinx moth		*Cautethia spuria* (Boisduval)	（チョウ目、スズメガ科）新北区、新熱帯区
spurlegged phasmatid		*Didymuria violescens* (Leach)	（ナナフシ目、ナナフシ科）豪州区。spurlegged phasmid ともいう
spurred wave moth		*Drepanulatrix unicalcararia* (Guenée)	（チョウ目、シャクガ科）新北区
spying eyed-metalmark		*Mesosemia ceropia* Druce	（チョウ目、シジミタテハ科）新熱帯区
spying hairstreak		*Atlides inachus* (Cramer)	（チョウ目、シジミチョウ科）新熱帯区
squalid duck louse			slender duck louse を見よ　米国での英名
squamulata canegrub		*Lepidiota squamulata* Waterhouse	（コウチュウ目、コガネムシ科）豪州区
square-barred bell		*Epinotia tetraquetrana* (Haworth)	（チョウ目、ハマキガ科）旧北区
square-blotched bell		*Epiblema tetragonana* (Stephens)	（チョウ目、ハマキガ科）旧北区
squared bent-skipper		*Helias cama* Evans	（チョウ目、セセリチョウ科）新熱帯区
square-headed snakeflies			snake-flies (2)(3) を見よ
square-headed wasps			crabronid wasps を見よ
square-necked grain beetle	タバコヒラタムシ	*Cathartus quadricollis* Guérin-Méneville	（コウチュウ目、ホソヒラタムシ科）日本、全北区。貯穀害虫
squarenosed fungus beetle	クロヒメマキムシ	*Lathridius minutus* (Linnaeus)	（コウチュウ目、ヒメマキムシ科）日本、汎世界
square-patched carpet moth		*Perizoma basaliata* (Walker)	（チョウ目、シャクガ科）新北区
square-spot	シナトビスジエダシャク	*Paradarisa consonaria* (Hübner)	（チョウ目、シャクガ科）日本、旧北区
square-spot beauty			square-spot を見よ
square-spot dart		*Euxoa obelisca* (Denis et Schiffermüller)	（チョウ目、ヤガ科）旧北区

英名	和名	学名	所属、分布、ほか
square-spot deerfly		*Chrysops viduatus* (Fabricius)	（ハエ目、アブ科）旧北区
square-spot rustic		*Xestia xanthographa* (Denis et Schiffermüller)	（チョウ目、ヤガ科）全北区。square-spot rustic moth ともいう
square-spotted blue		*Euphilotes battoides* (Behr)	（チョウ目、シジミチョウ科）新熱帯区、新熱帯区。square spotted blue とも記す
square-spotted clay		*Xestia rhomboidea* (Esper)	（チョウ目、ヤガ科）旧北区
square-spotted clay		*Photedes stigmatica* Eversmann	（チョウ目、ヤガ科）旧北区
square-spotted ministreak		*Ministrymon zilda* (Hewitson)	（チョウ目、シジミチョウ科）新熱帯区
square-tipped crescent			longwing crescent を見よ
square-winged red Charaxes		*Charaxes pleione* (Godart)	（チョウ目、タテハチョウ科）エチオピア区
squash aphid			cotton aphid を見よ
squash bee		*Peponapis pruinosa* (Say)	（ハチ目、ミツバチ科）新北区
squash beetle	ウリテントウ	*Epilachna borealis* (Fabricius)	（コウチュウ目、テントウムシ科）新北区
squash borer		*Melittia cucurbitae* (Harris)	（チョウ目、スカシバガ科）新北区
squash bug		*Anasa tristis* (De Geer)	（カメムシ目、ヘリカメムシ科）新北区
squash bug		*Phthia picta* (Drury)	（カメムシ目、ヘリカメムシ科）新北区、新熱帯区
squash bug			leaf-footed plant bug (2) を見よ
squash bugs			coreid bugs を見よ　豪州での英名
squash root maggot		*Muscina assimilis* (Fallén)	（ハエ目、イエバエ科）全北区、大洋区
squash root maggot			false stable fly を見よ
squash vine borer		*Melittia satyriniformis* (Hübner)	（チョウ目、スカシバガ科）東洋区、新北区、新熱帯区
squash vine borer			squash borer を見よ
squawcarpet mealybug		*Spirococcus ceanothi* McKenzie	（カメムシ目、コナカイガラムシ科）新北区
squeak beetle		*Agabus undulatus* (Schrank)	（コウチュウ目、ゲンゴロウ科）旧北区
squeak beetle			screech beetle を見よ
squeaker beetles			hygrobiid water beetles を見よ
squinting bush brown		*Bicyclus anynana* (Butler)	（チョウ目、タテハチョウ科）エチオピア区
squirrel flea		*Monopsyllus sciurorum* Schrank	（ノミ目、ナガノミ科）旧北区
squirrel lice		Enderleinellidae	（シラミ目）の昆虫の総称
squirrel louse		*Neuhaematopinus sciuri* Iarucke	（シラミ目、ケモノジラミ科）旧北区
squrred Panurgus		*Panurgus calcaratus* (Scopoli)	（ハチ目、ヒメハナバチ科）旧北区
Sri Lanka albatross		*Appias galene* (C. et R. Felder)	（チョウ目、シロチョウ科）東洋区

英名	和名	学名	所属、分布、ほか
Sri Lanka cascader		*Zygonyx iris ceylonicus* (Kirby)	（トンボ目、トンボ科）東洋区
Sri Lanka cruiser		*Macromia zeylanica* Fraser	（トンボ目、ヤマトンボ科）東洋区
Sri Lanka forktail		*Macrogomphus lankanensis* Fraser	（トンボ目、サナエトンボ科）東洋区
Sri Lanka grappletail		*Heliogomphus ceylonicus* (Selys)	（トンボ目、サナエトンボ科）東洋区
Sri Lanka midget		*Mortonagrion ceylonicum* Lieftinck	（トンボ目、イトトンボ科）東洋区
Sri Lanka orange-faced sprite		*Pseudagrion rubriceps ceylonicum* (Kirby)	（トンボ目、イトトンボ科）東洋区
Sri Lanka sabretail		*Megalogomphus ceylonicus* (Laidlaw)	（トンボ目、サナエトンボ科）東洋区
Sri Lankan relict ant		*Aneuretus simoni* Emery	（ハチ目、アリ科）東洋区
St. Andrews cotton bug			cotton stainer (1) を見よ　St. Andrew's cotton bug とも記す
St. Croix snaketail		*Ophiogomphus susbehcha* Vogt et Smith	（トンボ目、サナエトンボ科）新北区
St. Helena giant earwig		*Labidura herculeana* (Fabricius)	（ハサミムシ目、オオハサミムシ科）新北区、エチオピア区。セントヘレナ島の体長8cmの絶滅危惧種
St. Johnswort beetle		*Chrysolina varians* (Schaller)	（コウチュウ目、ハムシ科）旧北区
St. Johnswort beetle			Klamath weed beetle (2) を見よ
St. John's wort leaf beetle			Klamath weed beetle (1)(2) を見よ　Klamath weed の別名 St. Johnswort に由来
St. Johnswort midge		*Zeuxidiplosis giardi* (Kieffer)	（ハエ目、タマバエ科）新北区、豪州区。St. John's wort midge とも記される
St. John's wort rootborer		*Agrilus hyperici* (Creutzer)	（コウチュウ目、タマムシ科）豪州区
St. Lawrence tiger moth			giant tiger moth を見よ
St. Lucia basker		*Urothemis luciana* Balinsky	（トンボ目、トンボ科）エチオピア区
St. Lucia widow			portia widow を見よ
St. Mark's flies			march flies を見よ
St. Mark's fly		*Bibio marci* (Linnaeus)	（ハエ目、ケバエ科）旧北区
stable flies	サシバエ科	Stomoxydidae	（ハエ目）の昆虫の総称　厩舎に多いことから
stable flies			house flies (1) を見よ
stable fly	サシバエ	*Stomoxys calcitrans* (Linnaeus)	（ハエ目、サシバエ科）日本、汎世界。ウシ、ウマなど動物から吸血
stack bug			debris bug を見よ
staff sergeant		*Athyma selenophora* (Kollar)	（チョウ目、タテハチョウ科）東洋区。日本亜種は *A. s. ishiana* Fruhstorfer ヤエヤマイチモンジ。*A. s. amharina* (Moore) も同英名
stag beetle	ヨーロッパミヤマクワガタ	*Lucanus cervus* (Linnaeus)	（コウチュウ目、クワガタムシ科）旧北区
stag beetle borer			antelope beetle を見よ
stag beetles (1)	クワガタムシ科	Lucanidae	（コウチュウ目）の昆虫の総称

英名	和名	学名	所属、分布、ほか
stag beetles		*Pseudolucanus*	(コウチュウ目、クワガタムシ科) の昆虫の総称
staghorn Cholla moth		*Euscirrhopterus cosyra* Druce	(チョウ目、ヤガ科) 新北区
staghorn fern beetle		*Halticorcus platycerii* Lea	(コウチュウ目、ハムシ科) 豪州区
stagira hairstreak			smudged hairstreak を見よ
stained-back leafroller moth		*Acleris maculidorsana* (Clemens)	(チョウ目、ハマキガ科) 新北区
stained greenstreak		*Cyanophrys agricolor* (Butler et Druce)	(チョウ目、シジミチョウ科) 新熱帯区
stained Lophosis moth		*Lophosis labeculata* (Hulst)	(チョウ目、シャクガ科) 新北区
stainers			red bugs (1)(2) を見よ
Stainton's catchfly case		*Coleophora silenella* Herrich-Schäffer	(チョウ目、ツツミノガ科) 旧北区
Stainton's golden-rod case		*Coleophora virgaureella* Doubleday	(チョウ目、ツツミノガ科) 旧北区
Stainton's smoke		*Bankesia douglasii* Stainton	(チョウ目、ミノガ科) 旧北区
stalk borer			common stalk borer を見よ
stalk borer moth			common stalk borer を見よ
stalk-eyed borer		*Diopsis macrophthalma* Dalman	(ハエ目、トビメバエ科) エチオピア区。イネ害虫
stalk-eyed borer			stalk-eyed fly (1) を見よ
stalk-eyed flies	トビメバエ科	Diopsidae	(ハエ目) の昆虫の総称
stalk-eyed fly		*Sphyracephala brevicornis* (Say)	(ハエ目、トビメバエ科) 新北区
stalk-eyed fly (1)		*Diopsis thoracica* Westwood	(ハエ目、トビメバエ科) エチオピア区
stalk-eyed rice borer		*Diopsis tenuipes* Westwood	(ハエ目、トビメバエ科) エチオピア区。イネ害虫
stalk-eyed rice borer			stalk-eyed fly (1) を見よ
stalk-marsh caterpillar			Acraea moth を見よ
stalked-winged damselflies			emerald damselflies を見よ
Stallings' Calephelis		*Calephelis stallingsi* McAlpine	(チョウ目、シジミタテハ科) 新熱帯区
Stallings' flat		*Celaenorrhinus stallingsi* Freeman	(チョウ目、セセリチョウ科) 新北区、新熱帯区
stangin ether			dragonflies (1) を見よ
staphylinoid beetles	ハネカクシ上科	Staphylinoidea	(コウチュウ目) の昆虫の総称
star blue		*Lepidochrysops asteris* (Godart)	(チョウ目、シジミチョウ科) エチオピア区
star jasmine thrips		*Thrips orientalis* (Bagnall)	(アザミウマ目、アザミウマ科) 新北区、大洋区
star sandman		*Spialia asterodia* (Trimen)	(チョウ目、セセリチョウ科) エチオピア区
star scale		*Asterolecanium coffeae* Newstead	(カメムシ目、フサカイガラムシ科) エチオピア区

英名	和名	学名	所属、分布、ほか
star scales			pit scales を見よ
staring-eye spreadwing			Myris skipper を見よ
Starmuehlner's shadowdamsel		*Drepanosticta starmuehlneri* St. Quentin	(トンボ目、Platystictidae) 東洋区
starred Oxeo			yellow-patched satyr を見よ
starred skipper		*Arteurotia tractipennis* Butler et Druce	(チョウ目、セセリチョウ科) 豪州区
starry bob		*Iambrix stellifer* (Butler)	(チョウ目、セセリチョウ科) 東洋区
starry brindled pearl		*Cynaeda dentalis* (Denis et Schiffermüller)	(チョウ目、メイガ科) 旧北区
starry night cracker	マルバネカスリタテハ	*Hamadryas laodamia* (Cramer)	(チョウ目、タテハチョウ科) 新熱帯区。*H. l. saurites* (Fruhstorfer) は starry cracker という
starry night metalmark		*Echydna punctata* (C. et R. Felder)	(チョウ目、シジミタテハ科) 新熱帯区。
starry sky			Asian long-horned beetle を見よ
starry tortoise beetle		*Chelymorpha constellata* (Klug)	(コウチュウ目、ハムシ科) 新熱帯区
startled bush brown		*Bicyclus angulosa* (Butler)	(チョウ目、タテハチョウ科) エチオピア区
starwort		*Cucullia asteris* (Denis et Schiffermüller)	(チョウ目、ヤガ科) 旧北区。成虫は starwort moth ともいう
starwort shark			starwort を見よ
stately nawab	ウスイロフタオチョウ	*Polyura dolon* (Westwood)	(チョウ目、タテハチョウ科) 旧北区、東洋区
stately Rajah			stately nawab を見よ
stathmopodid moths	ニセマイコガ科	Stathmopodidae	(チョウ目) の昆虫の総称 Stathmopodids ともいう
statice thrips		*Haplothrips leucanthemi* (Schrank)	(アザミウマ目、クダアザミウマ科) 全北区
Statira			pale sulphur を見よ
Statira sulphur			pale sulphur を見よ
Staudinger's longtail		*Erateina staudingeri* Snellen	(チョウ目、シャクガ科) 旧北区
Staudinger's marbled white		*Melanargia epimede* Staudinger	(チョウ目、タテハチョウ科) 旧北区
Staudinger's nawab		*Polyura alphius* (Staudinger)	(チョウ目、タテハチョウ科) 東洋区
Staudinger's owl-butterfly		*Mimoblepia staudingeri mexicana* (Maza et Maza)	(チョウ目、タテハチョウ科) 新熱帯区
Staudinger's owlet		*Narope syllabus* Staudinger	(チョウ目、タテハチョウ科) 新熱帯区
Staudinger's Raven	ベニオビシロチョウ	*Pereute callinira* Staudinger	(チョウ目、シロチョウ科) 新熱帯区
steam-bug			German cockroach を見よ
steamfly			German cockroach を見よ
steel beetles			hister beetles を見よ

英名	和名	学名	所属、分布、ほか
steelblue blowfly		*Chrysomya saffranea* (Bigot)	（ハエ目、クロバエ科）豪州区
steel-blue lady beetle		*Halmus chalybeus* (Boisduval)	（コウチュウ目、テントウムシ科）新北区、豪州区。steel-blue ladybird ともいう
steel-blue ladybird		*Orcus janthinus* Mulsant	（コウチュウ目、テントウムシ科）東洋区
steel-blue mud dauber		*Sphex luctuosa* Smith	（ハチ目、アナバチ科）新北区
steelblue sawfly		*Perga dorsalis* Leach	（ハチ目、Pergidae）豪州区
steelblue sawfly		*Perga affinis affinis* Kirby	（ハチ目、Pergidae）豪州区
steel-blue sawfly			horntail (2) を見よ
steel-blue wood wasp			blue horntail を見よ
steel-blue wood wasp			horntail (2) を見よ
steel-blue wood wasp			polished horntail を見よ
steel-blue wood wasps		*Sirex*	（ハチ目、キバチ科）の昆虫の総称
steel-brown longicorn beetle	テツイロヒメカミキリ	*Cerseium sinicum* White	（コウチュウ目、カミキリムシ科）日本、東洋区
steely blue beetle		*Korynetes caeruleus* (De Geer)	（コウチュウ目、カッコウムシ科）旧北区
steely flasher		*Narcosius parisi helen* (Evans)	（チョウ目、セセリチョウ科）新熱帯区
steely underwing			little lined underwing moth を見よ
Steindachner's shieldback		*Neduba steindachneri* (Herman)	（バッタ目、キリギリス科）新北区
Steinhauser's blue-skipper		*Pythonides rosa* Steinhauser	（チョウ目、セセリチョウ科）新熱帯区
Steinhauser's flasher		*Astraptes* sp.	（チョウ目、セセリチョウ科）新熱帯区
Steinhauser's gemmed-satyr		*Cyllopsis steinhauserorum* Miller	（チョウ目、タテハチョウ科）新熱帯区
Steinhauser's skipperling		*Dalla steinhauseri* Freeman	（チョウ目、セセリチョウ科）新熱帯区
Steinkellner's flat-body		*Semioscopis steinkellneriana* (Denis et Schiffermüller)	（チョウ目、マルハキバガ科）旧北区
Stella orangetip		*Anthocharis stella* (Edwards)	（チョウ目、シロチョウ科）新北区
stem borer			pink borer (2) を見よ
stem borers			fruit flies を見よ
stem flies			frit flies を見よ
stem grasshopper			common Adreppus を見よ
stem mother	幹母		アブラムシ類の越冬卵から孵化した無翅胎生で単性生殖する雌をさす。fundatrigenia, fundatrix, fundatricis ともいう。幹雌ともいう
stem moths			cereal stem moths を見よ
stem sawflies	クキバチ上科	Cephoidea	（ハチ目）の昆虫の総称
stem sawflies	クキバチ科	Cephidae	（ハチ目）の昆虫の総称

英名	和名	学名	所属、分布、ほか
Stempffer's on-off		*Tetrahanis stempfferi* (Berger)	（チョウ目、シジミチョウ科）エチオピア区
stencilled hairstreak			Ictinus blue を見よ
stephanid wasps	ツノヤセバチ科	Stephanidae	（ハチ目）の昆虫の総称　stephanids ともいう
Stephen's gem			bilobed looper moth を見よ
Stephens Island weta			Cook Strait giant weta を見よ
Stephen's leafhopper		*Empoasca stephensi* Young	（カメムシ目、オオヨコバイ科）新北区、大洋区。Stephens leafhopper とも記される
Stephen's skolly		*Thestor stepheni* Swanepoel	（チョウ目、シジミチョウ科）エチオピア区
steppe bush cricket		*Platycleis montana* (Kollar)	（バッタ目、キリギリス科）旧北区
steppe grasshopper		*Chorthippus dorsatus* (Zetterstedt)	（バッタ目、バッタ科）旧北区
sterile	不妊性型		シロアリ類の職蟻と兵蟻で、無翅で生殖器官が退化している
sternoxid beetles		Sternoxia	（コウチュウ目）の昆虫の総称
Stevenson's copper		*Aloeides stevensoni* Tite et Dickson	（チョウ目、シジミチョウ科）エチオピア区
Stevenson's shieldback		*Pediodectes stevensonii* (Thomas)	（バッタ目、キリギリス科）新北区
Stevenson's Temnora			yellow winged Temnora を見よ
stiched sister			Erotia sister を見よ
Stichel's blue Emesis		*Emesis orichalceus* Stichel	（チョウ目、シジミタテハ科）新熱帯区
Stichel's doctor		*Ancyluris melior* Stichel	（チョウ目、シジミタテハ科）新熱帯区
Stichel's Euselasia		*Euselasia eustola* Stichel	（チョウ目、シジミタテハ科）新熱帯区
Stichel's eyemark		*Mesosemia pacifica* Stichel	（チョウ目、シジミタテハ科）新熱帯区
Stichel's hermit		*Zelotaea nivosa* (Stichel)	（チョウ目、シジミタテハ科）新熱帯区
Stichel's jewelmark		*Anteros cruentatus* Stichel	（チョウ目、シジミタテハ科）新熱帯区
Stichel's Nymphidium		*Nymphidium plinthobaphis* Stichel	（チョウ目、シジミタテハ科）新熱帯区。
Stichel's tailed owlet		*Opoptera aorsa* Latreille et Godart	（チョウ目、タテハチョウ科）新熱帯区
stick-and-leaf insects			walking sticks (2) を見よ
stick caterpillars			geometer moths を見よ　幼虫の英名
stick insect		*Clonopsis gallica* (Charpentier)	（ナナフシ目、コノハムシ科）旧北区
stick insect		*Ctenomorpha chronus* Gray	（ナナフシ目、ナナフシ科）豪州区
stick insect			Indian stick-insect を見よ
stick insects	コブナナフシ科	Bacillidae	（ナナフシ目）の昆虫の総称
stick insects			walking sticks (1)(2) を見よ
stick katydids		Saginae	（バッタ目、キリギリス科）の昆虫の総称
stick locusts			grasshoppers (2) を見よ
stick mantis		*Archimantis latistyla* (Serville)	（カマキリ目、カマキリ科）豪州区

英名	和名	学名	所属、分布、ほか
sticktight flea	ニワトリフトノミ	*Echidnophaga gallinacea* (Westwood)	（ノミ目、ヒトノミ科）日本、汎世界。ニワトリに寄生。sticktight ともいう
sticktight fleas			chigoe fleas を見よ
sticktights			common fleas を見よ
stiff-winged crickets	ケマスズムシ科	Scleropteridae	（バッタ目）の昆虫の総称
stiletto flies	ツルギアブ科	Therevidae	（ハエ目）の昆虫の総称　成虫の形が小剣に似ることから
stiletto fly		*Ozodiceromyia nigrimana* (Kröber)	（ハエ目、ツルギアブ科）新北区
stillwater clubtail		*Arigomphus lentulus* (Needham)	（トンボ目、サナエトンボ科）新北区
stilt bugs	イトカメムシ科	Berytidae	（カメムシ目）の昆虫の総称
stilt-legged flies	マルズヤセバエ科	Micropezidae	（ハエ目）の昆虫の総称
stilt-legged flies		Calobatidae	（ハエ目）の昆虫の総称
stilt-walker katydid		*Arethaea grallator* (Scudder)	（バッタ目、キリギリス科）新北区
stinging caterpillar		*Parasa virida* (Walker)	（チョウ目、イラガ科）エチオピア区
stinging rose caterpillar		*Parasa indetermina* (Boisduval)	（チョウ目、イラガ科）新北区。stinging rose caterpillar moth ともいう
stingless bees (1)	ハリナシバチ亜科	Meliponinae	（ハチ目、ミツバチ科）の昆虫の総称
stingless bees (2)		*Trigona*	（ハチ目、ミツバチ科）の昆虫の総称
stingless bees		*Austroplebeia*	（ハチ目、ミツバチ科）の昆虫の総称
stink ants			tapinoma ants を見よ
stink beetle		*Nomius pygmaeus* (Dejean)	（コウチュウ目、オサムシ科）新北区。捕まえると悪臭を出す
stink beetles		Psydrini	（コウチュウ目、オサムシ科）の昆虫の総称
stink beetles			olive beetles を見よ
stink bug		*Rhynchocoris longirostris* Stål	（カメムシ目、カメムシ科）東洋区。カンキツ害虫
stink bug		*Scutellera perplexa* Westwood	（カメムシ目、キンカメムシ科）東洋区。カンキツ害虫
stink bug egg parasite		*Trissolcus basalis* (Wollaston)	（ハチ目、タマゴクロバチ科）大洋区
stink bugs	カメムシ科	Pentatomidae	（カメムシ目）の昆虫の総称　中大型のカメムシで世界に 5,000 種
stink bugs			ground bugs を見よ　豪州での英名
stink bugs			plataspid bugs を見よ
stink flies			green lacewings (1) を見よ　捕まえると異臭をだすための俗称
stink fly			golden-eye lacewing (2) を見よ
stinking rice bug			rice bug (2) を見よ
stinky leafwing			Orion を見よ
Stockholm button		*Acleris holmiana* (Linnaeus)	（チョウ目、ハマキガ科）旧北区
Stoffberg widow		*Dingana fraterna* Henning et Henning	（チョウ目、タテハチョウ科）エチオピア区

英名	和名	学名	所属、分布、ほか
Stola flat		*Celaenorrhinus stola* Evans	(チョウ目、セセリチョウ科) 新熱帯区
Stoll's Sarota			tailed jewelmark を見よ
stomach botflies			horse bot flies を見よ
stomoxid flies			stable flies を見よ
stone beetles			ant like stone beetles を見よ
stone crickets			cave crickets (2) を見よ
stone fly		*Perla maxima* Scopoli	(カワゲラ目、カワゲラ科) 旧北区
stonefly		*Perla bipunctata* Pictet	(カワゲラ目、カワゲラ科) 旧北区
stone humble bee			red-tailed bumble bee (1) を見よ
stone leek aphid			onion aphid を見よ
stone leek leaf beetle	ネギハムシ	*Galeruca extensa* Motschulsky	(コウチュウ目、ハムシ科) 日本、旧北区。ネギ害虫
stone leek leafminer	ネギハモグリバエ	*Liriomyza chinensis* (Kato)	(ハエ目、ハモグリバエ科) 日本、旧北区。ネギ害虫
stone leek leafminer	ニッポンネギハモグリバエ	*Liriomyza nipponallia* Sasakawa	(ハエ目、ハモグリバエ科) 日本。ネギ害虫
stone leek miner			allium leafminer を見よ
stone leek plume moth	オダマキトリバ	*Platyptilia jezoensis* Matsumura	(チョウ目、トリバガ科) 日本、全北区
stonecrop blue			Sonoran blue を見よ
stonecrop elfin			moss elfin を見よ
stoneflies	カワゲラ目	Plecoptera	の昆虫の総称　世界に約1500種
stoneflies			giant stoneflies (1) を見よ
stone-winged owlet moth			morbid owlet moth を見よ
Storax skipper		*Parphorus storax* (Mabille)	(チョウ目、セセリチョウ科) 新熱帯区
stored grain fungus beetle		*Litargus balteatus* LeConte	(コウチュウ目、コキノコムシ科) 豪州区
stored grain moth			small tabby を見よ
stored nut moth	ツヅリガ	*Paralipsa gularis* (Zeller)	(チョウ目、メイガ科) 日本、全北区、東洋区、大洋区。貯穀害虫
stored tobacco moth			tobacco moth (1) を見よ
storehouse beetle			museum beetle (1) を見よ　米国、南アフリカでの英名
stormy arches moth		*Polia nimbosa* (Guenée)	(チョウ目、ヤガ科) 新北区
stout barklice	マドチャタテ科	Peripsocidae	(チャタテムシ目) の昆虫の総称　ケチャタテムシ科 Caeciliidae の異名
stout-bodied grasshopper			Italian locust を見よ
stout dart		*Spaelotis ravida* (Denis et Schiffermüller)	(チョウ目、ヤガ科) 旧北区。日本亜種は *S. r. nipona* (Felder et Rogenhofer) ヒメアカマエヤガ。成虫は stout dart moth
stout snout		*Hypena obesali* Treitschke	(チョウ目、ヤガ科) 旧北区
stout spanworm moth		*Lycia ursaria* (Walker)	(チョウ目、シャクガ科) 新北区。幼虫は spruce cone looper

英名	和名	学名	所属、分布、ほか
stout vicetail		*Hemigomphus heteroclytus* Selys	（トンボ目、サナエトンボ科）豪州区
stouts			horse flies (1) を見よ
stouts			clegs を見よ
Stout's bostrichid			Stout's branch borer を見よ
Stout's branch borer		*Polycaon stouti* (LeConte)	（コウチュウ目、ナガシンクイムシ科）新北区
Straatman's oakblue		*Arhopala straatmani* Nieuwenhuis	（チョウ目、シジミチョウ科）東洋区
straight-banded tree brown	シロオビクロヒカゲ	*Lethe verma* (Kollar)	（チョウ目、タテハチョウ科）東洋区
straight-barred grass yellow		*Eurema elathea* (Cramer)	（チョウ目、シロチョウ科）新熱帯区
straight-barred marble		*Celypha striana* (Denis et Schiffermüller)	（チョウ目、ハマキガ科）旧北区
straight-edged skipper		*Gindanes brontinus brontinus* Godman et Salvin	（チョウ目、セセリチョウ科）新熱帯区
straight-lanced meadow katydid		*Conocephalus strictus* (Scudder)	（バッタ目、キリギリス科）新北区
straight line map	ツマグロイシガケチョウ	*Cyrestis nivea* (Zinken-Sommer)	（チョウ目、タテハチョウ科）東洋区
straight line mapwing	スジブトイシガケチョウ	*Cyrestis maenalis martini* Hartert	（チョウ目、タテハチョウ科）東洋区
straightline royal	カレンコウシジミ	*Tajuria diaeus* (Hewitson)	（チョウ目、シジミチョウ科）東洋区
straight-line sulphur		*Phoebis trite* (Linnaeus)	（チョウ目、シロチョウ科）新熱帯区
straight-lined Argyria moth		*Argyria critica* Forbes	（チョウ目、メイガ科）新北区
straight-lined Cydosia moth		*Cydosia aurivitta* Grote et Robinson	（チョウ目、ヤガ科）新北区
straight-lined looper moth		*Pseudeva purpurigera* (Walker)	（チョウ目、ヤガ科）新北区。western straight-lined looper ともいう
straight lined mallow moth		*Bagisara rectifascia* (Grote)	（チョウ目、ヤガ科）新北区
straight-lined Plagodis moth		*Plagodis phlogosaria* (Guenée)	（チョウ目、シャクガ科）新北区
straight-lined seed moth		*Eublemma recta* (Guenée)	（チョウ目、ヤガ科）新北区、新熱帯区。ワタ害虫
straight-lined tiger moth			Oithon tiger moth を見よ
straight-lined wave moth		*Lobocleta plemyraria* (Guenée)	（チョウ目、シャクガ科）新北区。straight-lined wave ともいう
straight pierrot	ロクスシジミ	*Caleta roxus* (Godart)	（チョウ目、シジミチョウ科）東洋区。*C. r. pothus* Fruhstorfer も同英名
straight pierrot		*Pycnophallium roxus* (Godart)	（チョウ目、シジミチョウ科）東洋区
straight plum Judy	シロオビヒョットコシジミタテハ	*Abisara kausambi* C. et R. Felder	（チョウ目、シジミタテハ科）東洋区
straight snouted weevils	ミツギリゾウムシ科	Brentidae	（コウチュウ目）の昆虫の総称

英名	和名	学名	所属、分布、ほか
straight swift		*Parnara bada sida* (Waterhouse)	(チョウ目、セセリチョウ科) 豪州区
straight swift (1)		*Parnara naso* (Fabricius)	(チョウ目、セセリチョウ科) 豪州区
straight swift			rice skipper を見よ
straight-tipped ringtail		*Erpetogomphus elaps* Selys	(トンボ目、サナエトンボ科) 新熱帯区
straight treebrown		*Lethe verma robinsoni* Pendlebury	(チョウ目、タテハチョウ科) 東洋区。straight-banded tree brown を参照
straightwing blue	ウラジロアカスジミツオシジミ	*Orthomiella pontis* (Elwes)	(チョウ目、シジミチョウ科) 東洋区
strange moth			stranger を見よ
strange wave		*Sterrha laevigata* Scopoli	(チョウ目、シャクガ科) 旧北区
stranger		*Lacanobia blenna* (Hübner)	(チョウ目、ヤガ科) 旧北区
Strauch's outer barklouse		*Ectopsocus strauchi* Enderlein	(チャタテムシ目、Ectopsocidae) 全北区、新熱帯区
straw-barred sward pearl		*Pyrausta cespitalis* Denis et Schiffermüller	(チョウ目、メイガ科) 旧北区
straw belle		*Aspilates gilvaria* (Denis et Schiffermüller)	(チョウ目、シャクガ科) 旧北区
straw Besma moth		*Besma endropiaria* (Grote et Robinson)	(チョウ目、シャクガ科) 新北区
straw-coloured grass-veneer		*Agriphila culmella* (Linnaeus)	(チョウ目、メイガ科) 旧北区
straw-coloured tortrix moth			cyclamen tortrix を見よ
straw dot			leguminosae rivula を見よ
straw fly			chloropid gout fly を見よ
straw May			heath fritillary を見よ
straw oblique-barred twist			cyclamen tortrix を見よ
straw point			leguminosae rivula を見よ
straw short-barred conch		*Cochylimorpha straminea* (Haworth)	(チョウ目、ハマキガ科) 旧北区
straw tortrix			maple twist を見よ
straw underwing		*Thalpophila matura* (Hufnagel)	(チョウ目、ヤガ科) 旧北区
straw wave moth		*Idaea eremiata* (Hulst)	(チョウ目、シャクガ科) 新北区
straw worms			jointworms (1)(2) を見よ
strawberry aphid	イチゴケナガアブラムシ	*Chaetosiphon fragaefolii* (Cockerell)	(カメムシ目、アブラムシ科) 日本、汎世界。イチゴ害虫
strawberry beetles			common carabids を見よ
strawberry black leaf-cut weevil	クロケシツブチョッキリ	*Auletobius uniformis* (Roelofs)	(コウチュウ目、オトシブミ科) 日本
strawberry blossom weevil (1)		*Anthonomus bisignifer* Schenkling	(コウチュウ目、ゾウムシ科) 日本、旧北区
strawberry blossom weevil (2)		*Anthonomus rubi* (Herbst)	(コウチュウ目、ゾウムシ科) 旧北区

英名	和名	学名	所属、分布、ほか
strawberry bud weevil			strawberry weevil (1) を見よ
strawberry bug		*Euander lacertosus* (Erichson)	(カメムシ目、ナガカメムシ科) 豪州区
strawberry button			strawberry tortricid を見よ
strawberry capitophorus	イチゴクビケアブラムシ	*Chaetosiphon minor* (Forbes)	(カメムシ目、アブラムシ科) 日本、新北区
strawberry capsid			potato capsid を見よ
strawberry crown borer		*Tyloderma fragariae* (Riley)	(コウチュウ目、ゾウムシ科) 新北区。イチゴ害虫
strawberry crown girdler			strawberry root weevil (1) を見よ
strawberry crown miner		*Anacampsis fragariella* Busck	(チョウ目、キバガ科) 新北区。成虫は strawberry crown miner moth
strawberry crown moth		*Synanthedon bibionipennis* (Boisduval)	(チョウ目、スカシバガ科) 新北区、大洋区
strawberry curculio		*Coenorrhinus aeneovirens* Marsham	(コウチュウ目、オトシブミ科) 旧北区
strawberry cutworm	イチゴキリガ	*Orbona fragariae* (Vieweg)	(チョウ目、ヤガ科) 日本
strawberry cutworm moth			interoceanic ear moth を見よ　幼虫は strawberry cutworm
strawberry cymatophorid			buff arches を見よ　原名亜種の英名
strawberry false-worm		*Emphytus maculatus* Norton	(ハチ目、ハバチ科) 旧北区
strawberry flea beetle		*Altica ignita* Illiger	(コウチュウ目、ハムシ科) 新北区
strawberry fruit weevil		*Barypeithes araneiformis* (Schrank)	(コウチュウ目、ゾウムシ科) 旧北区
strawberry garden tortrix			garden tortrix を見よ
strawberry ground beetle		*Pterostichus melanarius* (Illiger)	(コウチュウ目、オサムシ科) 旧北区
strawberry ground beetle (1)		*Pterostichus madidus* (Fabricius)	(コウチュウ目、オサムシ科) 旧北区
strawberry leaf beetle	イチゴハムシ	*Galerucella grisescens* (Joannis)	(コウチュウ目、ハムシ科) 日本、旧北区、東洋区。イチゴ害虫
strawberry leaf beetle		*Galerucella tenella* (Linnaeus)	(コウチュウ目、ハムシ科) 旧北区。strawberry leaf-beetle とも記す
strawberry leaf folder		*Ancylis comptana* (Frölich)	(チョウ目、ハマキガ科) 旧北区
strawberry leaf roller	オオクリモンヒメハマキ	*Olethreutes transversana* (Christoph)	(チョウ目、ハマキガ科) 日本、旧北区
strawberry leaf roller		*Ancylis comptana fragariae* (Walsh et Riley)	(チョウ目、ハマキガ科) 新北区。イチゴ害虫。米国での英名。strawberry leafroller moth ともいう
strawberry leaf roller			strawberry crown miner を見よ
strawberry leafhopper		*Aphrodes bicinctus* (Schrank)	(カメムシ目、オオヨコバイ科) 旧北区
strawberry looper			beautiful carpet を見よ
strawberry moth			strawberry tortricid を見よ

英名	和名	学名	所属、分布、ほか
strawberry prominent	ヒメシャチホコ	*Neostauropus basalis* (Moore)	(チョウ目、シャチホコガ科) 日本、旧北区
strawberry rhynchites		*Coenorrhinus germanicus* (Herbst)	(コウチュウ目、オトシブミ科) 旧北区
strawberry root aphid	イチゴネアブラムシ	*Aphis forbesi* Weed	(カメムシ目、アブラムシ科) 日本、新北区。イチゴ害虫
strawberry root louse			strawberry root aphid を見よ
strawberry root weevil		*Sciaphilus asperatus* (Bonsdorff)	(コウチュウ目、ゾウムシ科) 旧北区
strawberry root weevil		*Otiorhynchus rugifrons* (Gyllenhal)	(コウチュウ目、ゾウムシ科) 旧北区
strawberry root weevil (1)	イチゴクチブトゾウムシ	*Brochyrhinus ovatus* (Linnaeus)	(コウチュウ目、ゾウムシ科) 全北区。イチゴ害虫
strawberry rootworm		*Paria fragariae* Wilcox	(コウチュウ目、ハムシ科) 新北区
strawberry rootworm		*Paria canella* (Fabricius)	(コウチュウ目、ハムシ科) 新北区
strawberry rootworm		*Graphops pubescens* Melsheimer	(コウチュウ目、ハムシ科) 新北区。イチゴ害虫
strawberry rootworm			grape colaspis (2) を見よ
strawberry sap beetle		*Stelidota geminata* (Say)	(コウチュウ目、ケシキスイ科) 新北区
strawberry sawfly	イチゴハバチ	*Allantus albicinctus* (Matsumura)	(ハチ目、ハバチ科) 日本、旧北区。イチゴ害虫
strawberry sawfly		*Monophadnoides confusus* Förster	(ハチ目、ハバチ科) 旧北区
strawberry seed beetle		*Harpalus rufipes* (De Geer)	(コウチュウ目、オサムシ科) 旧北区
strawberry thrips			yellow tea thrips (1) を見よ　豪州での英名
strawberry tortricid	バラハマキ	*Acleris comariana* (Zeller)	(チョウ目、ハマキガ科) 日本、全北区。バラ、イチゴ、リンゴ害虫。バラモンハマキともいう
strawberry tortrix			strawberry tortricid を見よ
strawberry tortrix moth			strawberry tortricid を見よ
strawberry weevil (1)		*Anthonomus signatus* Say	(コウチュウ目、ゾウムシ科) 新北区
strawberry weevil		*Rhinaria perdix* (Pascoe)	(コウチュウ目、ゾウムシ科) 豪州区
strawberry weevil			strawberry root weevil (1) を見よ
strawberry weevil			strawberry blossom weevil (1) を見よ
strawberry weevil			black vine weevil を見よ
strawberry whitefly	イチゴコナジラミ	*Trialeurodes packardi* (Morrill)	(カメムシ目、コナジラミ科) 日本、大洋区、新北区。イチゴ害虫
strawberry whitefly		*Aleyrodes fragariae* Walker	(カメムシ目、コナジラミ科) 旧北区
strawberry white-margined bug	シロヘリナガカメムシ	*Panaorus japonicus* (Stål)	(カメムシ目、ナガカメムシ科) 日本、旧北区
streak (1)		*Chesias spartiata* Fuessley	(チョウ目、シャクガ科) 旧北区
streak (2)		*Chesias legatella* (Denis et Schiffermüller)	(チョウ目、シャクガ科) 旧北区

英名	和名	学名	所属、分布、ほか
streak moth			streak (1) (2) を見よ
streak-winged red skimmer			cardinal meadowhawk を見よ
streaked Aethes moth		*Aethes sonorae* (Walsingham)	（チョウ目、ホソハマキガ科）新北区、新熱帯区
streaked armyworm moth		*Persectania aversa* Walker	（チョウ目、ヤガ科）豪州区
streaked baron		*Euthalia alpheda* (Godart)	（チョウ目、タテハチョウ科）東洋区
streaked Calidota		*Calidota strigosa* (Walker)	（チョウ目、ヒトリガ科）新北区
streaked Calidota moth		*Calidota laqueata* (Edwards)	（チョウ目、ヒトリガ科）新北区
streaked carpet			streak (2) を見よ
streaked Coleophora moth		*Coleophora cratipennella* Clemens	（チョウ目、ツツミノガ科）新北区
streaked copper		*Aloeides pierus* (Cramer)	（チョウ目、シジミチョウ科）エチオピア区
streaked dagger moth		*Acronicta lithospila* Grote	（チョウ目、ヤガ科）新北区
streaked Ethmia moth		*Ethmia longimaculella* (Chambers)	（チョウ目、スエヒロキバガ科）新北区
streaked hooded owlet moth		*Cucullia strigata* (Smith)	（チョウ目、ヤガ科）新北区
streaked orange moth		*Nascia acutella* (Walker)	（チョウ目、メイガ科）新北区
streaked pale mountain satyr		*Lymanopoda huilana* Maassen et Weymer	（チョウ目、タテハチョウ科）新熱帯区
streaked paradise skipper		*Abantis leucogaster* (Mabille)	（チョウ目、セセリチョウ科）エチオピア区
streaked pine bell			spruce bell moth を見よ
streaked Plusia		*Trichoplusia vittata* Wallengren	（チョウ目、ヤガ科）旧北区
streaked sailer		*Neptis melicerta* (Drury)	（チョウ目、タテハチョウ科）エチオピア区
streaked sailor		*Neptis goochi* Trimen	（チョウ目、タテハチョウ科）エチオピア区
streaked silver case		*Coleophora pennella* (Denis et Schiffermüller)	（チョウ目、ツツミノガ科）旧北区
streaked sphinx		*Protambulyx strigilis* (Linnaeus)	（チョウ目、スズメガ科）新北区、新熱帯区。streaked sphinx moth ともいう
streaked tussock moth		*Dasychira obliquata* (Grote et Robinson)	（チョウ目、ドクガ科）新北区
streaked wave	コヒメシャク	*Scopula virgulata* (Denis et Schiffermüller)	（チョウ目、シャクガ科）旧北区
streaky leafwing		*Memphis philumena* Doubleday	（チョウ目、タテハチョウ科）新熱帯区
streaky skippers		*Celotes*	（チョウ目、セセリチョウ科）の昆虫の総称
stream bluet		*Enallagma exsulans* (Hagen)	（トンボ目、イトトンボ科）新北区
stream cruiser		*Didymops transversa* (Say)	（トンボ目、ヤマトンボ科）新北区
stream-damsels			threadtails (1) を見よ

英名	和名	学名	所属、分布、ほか
stream hawker		*Pinheyschna subpupillata* McLachlan	（トンボ目、ヤンマ科）エチオピア区
stream mayflies	ヒラタカゲロウ科	Heptageniidae	（カゲロウ目）の昆虫の総称
stream midges			solitary midges を見よ
streamer		*Anticlea derivata* (Denis et Schiffermüller)	（チョウ目、シャクガ科）旧北区
streamer carpet			streamer を見よ
streamer moth			streamer を見よ
Strecker's giant-skipper		*Megathymus streckeri* (Skinner)	（チョウ目、セセリチョウ科）新北区。*M. s. texana* Barns et McDunnough も同英名
strelitzia nightfighter			banana nightfighter を見よ
strepsipterans			stylops を見よ
Stretch's underwing			joined underwing moth を見よ
striate false wireworm		*Pterohelaeus alternatus* Pascoe	（コウチュウ目、ゴミムシダマシ科）豪州区
striate-punctate leafhopper	マダラヨコバイ	*Psammotettix striatus* (Linnaeus)	（カメムシ目、オオヨコバイ科）日本、東洋区
striated cantharid			leather-winged beetle を見よ
striated chafer			lineate chafer を見よ
striated Emesis			Lacrines metalmark を見よ
striated leafroller			striated Tortrix moth を見よ
striated lichen moth		*Cisthene striata* Ottolengui	（チョウ目、ヒトリガ科）新北区
striated pearl-white			chalk white を見よ
striated pellucid hawk moth			honeysuckle clearwing を見よ
striated Phaneta moth		*Phaneta striatana* (Clemens)	（チョウ目、ハマキガ科）新北区
striated planthopper	シマウンカ	*Nisia nervosa* (Motschulsky)	（カメムシ目、シマウンカ科）日本、旧北区、豪州区
striated satyr		*Aulocera saraswati* (Kollar)	（チョウ目、タテハチョウ科）東洋区
striated sister		*Adelpha radiata* Fruhstorfer	（チョウ目、タテハチョウ科）新熱帯区
striated Tortrix moth		*Archips strianus* Fernald	（チョウ目、ハマキガ科）新北区
striated white	スジグロシロチョウ	*Pieris melete melete* Ménétriès	（チョウ目、シロチョウ科）日本、旧北区
Strigate sailer		*Neptis strigata* Aurivillius	（チョウ目、タテハチョウ科）エチオピア区
Strigosa metalmark		*Pheles strigosa strigosa* (Staudinger)	（チョウ目、シジミタテハ科）新熱帯区
string cottony scale	ヒモワタカイガラムシ	*Takahashia japonica* (Cockerell)	（カメムシ目、カタカイガラムシ科）日本、旧北区。カンキツ、リンゴ、クワ、カエデなど果樹、樹木害虫
string maggot			hop string midge を見よ
stripe-backed moth		*Arogalea cristifasciella* (Chambers)	（チョウ目、キバガ科）新北区
stripe-backed moth		*Battaristis vittella* (Busck)	（チョウ目、キバガ科）新北区

英名	和名	学名	所属、分布、ほか
stripe-faced meadow katydid		*Orchelimum concinnum* Scudder	（バッタ目、キリギリス科）新北区
stripe-headed threadtail		*Prodasineura sita* (Kirby)	（トンボ目、モノサシトンボ科）東洋区
stripe-legged robberfly		*Dioctria baumbaueri* Meigen	（ハエ目、ムシヒキアブ科）全北区
stripe swordtail		*Graphium aristeus hermocrates* (C. et R. Felder)	（チョウ目、アゲハチョウ科）東洋区。five-bar swordtail (1) を参照
stripe-winged baskettail			slender baskettail を見よ
striped albatross	ヤエヤマシロチョウ	*Appias libythea* (Fabricius)	（チョウ目、シロチョウ科）日本、東洋区
striped albatross			eastern striped albatross を見よ
striped alder sawfly		*Hemichroa crocea* (Geoffroy)	（ハチ目、ハバチ科）全北区
striped ambrosia beetle	シラベザイノキクイムシ	*Trypodendron lineatum* (Olivier)	（コウチュウ目、キクイムシ科）日本、全北区
striped artichoke caterpillar			common pinkband moth を見よ
striped-back pyralid	キスジノメイガ	*Sinibotys evenoralis* (Walker)	（チョウ目、メイガ科）日本、旧北区
striped bean weevil			pea leaf weevil を見よ
striped birch pyralid moth		*Ortholepis pasadamia* (Dyar)	（チョウ目、メイガ科）新北区
striped black crow	ダブルデールリマダラ	*Euploea doubledayi* C. et R. Felder	（チョウ目、タテハチョウ科）東洋区
striped black crow		*Euploea eyndhovii gardineri* (Fruhstorfer)	（チョウ目、タテハチョウ科）東洋区
striped black crow (1)	クロムラサキマダラ	*Euploea alcathoe* (Godart)	（チョウ目、タテハチョウ科）東洋区、豪州区
striped black moth			copper underwing (2) を見よ
striped blister beetle		*Epicauta albovittata* (Gestro)	（コウチュウ目、ツチハンミョウ科）エチオピア区
striped blister beetle (1)		*Epicauta vittata* (Fabricius)	（コウチュウ目、ツチハンミョウ科）新北区
striped blue			marine blue を見よ
striped blue crow		*Euploea mulciber* (Cramer)	（チョウ目、タテハチョウ科）東洋区。日本亜種は *E. m. barsine* Fruhstorfer ツマムラサキマダラ
striped blue skipper		*Quadrus contubernalis* (Mabille)	（チョウ目、セセリチョウ科）新熱帯区
striped buff moth		*Hydriomena subochraria* Doubleday	（チョウ目、シャクガ科）豪州区
striped bush cricket		*Leptophyes albovittata* (Kollar)	（バッタ目、キリギリス科）旧北区
striped bush hopper		*Ampittia virgata miyakei* Matsumura	（チョウ目、セセリチョウ科）東洋区
striped cabbage flea beetle			striped flea beetle (1) を見よ　米国での英名

英名	和名	学名	所属、分布、ほか
striped chocolate-tip moth		*Clostera strigosa* (Grote)	（チョウ目、シャチホコガ科）新北区
striped click beetle			common click beetle (1) を見よ
striped comb			vine leaf-roller moth を見よ
striped cucumber beetle		*Acalymma vittatum* (Fabricius)	（コウチュウ目、ハムシ科）新北区
striped cutworm			striped cutworm moth を見よ
striped cutworm moth		*Euxoa tessellata* (Harris)	（チョウ目、ヤガ科）新北区
striped dagger moth	シマケンモン	*Craniophora fasciata* (Moore)	（チョウ目、ヤガ科）日本、東洋区
striped dawnfly		*Capila jayadeva* (Moore)	（チョウ目、セセリチョウ科）東洋区
striped earwig		*Labidura bidens* (Olivier)	（ハサミムシ目、オオハサミムシ科）新北区
striped earwig	オオハサミムシ	*Labidium riparia japonica* (de Haan)	（ハサミムシ目、オオハサミムシ科）日本、汎世界
striped earwigs			long-horned earwigs を見よ
striped elaterid beetle			common click beetle (1) を見よ
striped Eudonia moth		*Eudonia strigalis* (Dyar)	（チョウ目、メイガ科）新北区
striped falcon		*Corades ulema* Hewitson	（チョウ目、タテハチョウ科）新熱帯区
striped firetail		*Telebasis filiola* (Perty)	（トンボ目、イトトンボ科）新北区
striped flea beetle	スジカミナリハムシ	*Altica latericosta* (Jacoby)	（コウチュウ目、ハムシ科）日本、旧北区
striped flea beetle (1)	キスジノミハムシ	*Phyllotreta striolata* (Fabricius)	（コウチュウ目、ハムシ科）日本、全北区、東洋区。野菜害虫
striped flea beetle (2)		*Phyllotreta undulata* (Kutschera)	（コウチュウ目、ハムシ科）豪州区
striped flea beetle			large striped flea beetle を見よ　豪州での英名
striped footman moth		*Hypoprepia miniata* (Kirby)	（チョウ目、ヒトリガ科）新北区
striped fruit fly	ミスジミバエ	*Zeugodacus scutellatus* (Hendel)	（ハエ目、ミバエ科）日本、東洋区
striped garden caterpillar			striped garden cutworm moth を見よ　成虫は striped garden caterpillar moth
striped garden cutworm moth		*Lacanobia legitima* (Grote)	（チョウ目、ヤガ科）新北区。幼虫は striped garden cutworm
striped glider	シロランウスキタテハ	*Cymothoe oemilius* (Doumet)	（チョウ、タテハチョウ科）エチオピア区
striped gold midget		*Phyllonorycter emberizaepenella* (Bouché)	（チョウ目、ホソガ科）旧北区
striped gourd-shaped weevil	クワヒョウタンゾウムシ	*Scepticus insularis* (Roelofs)	（コウチュウ目、ゾウムシ科）日本
striped grass looper		*Mocis repanda* Fabricius	（チョウ目、ヤガ科）新熱帯区、エチオピア区
striped grasshopper		*Ramburiella hispanica* (Rambur)	（バッタ目、バッタ科）旧北区
striped grasshopper		*Amphitornus coloradus* (Thomas)	（バッタ目、バッタ科）新北区

英名	和名	学名	所属、分布、ほか
striped grayling	クモガタジャノメ	*Hipparchia fidia* (Linnaeus)	（チョウ目、タテハチョウ科）旧北区
striped green longicorn			two-lined albizzia longhorn を見よ
striped grey		*Eudonia lineola* (Curtis)	（チョウ目、メイガ科）旧北区
striped ground cricket		*Allonemobius fasciatus* (De Geer)	（バッタ目、コオロギ科）新北区。アブラムシを捕食。気温により鳴声が変わるので temperature cricket ともいわれる
striped hairstreak	ウラナミカラスシジミ	*Satyrium liparops* (LeConte)	（チョウ目、シジミチョウ科）新北区
striped hawk		*Hyles lineata* Fabricius	（チョウ目、スズメガ科）全北区。日本亜種は *H. l. livornica* (Esper) アカオビスズメ。
striped hawkmoth		*Hyles livornica* (Esper)	（チョウ目、スズメガ科）旧北区
striped hawkmoth			striped hawk を見よ
striped horse fly		*Tabanus lineola* Fabricius	（ハエ目、アブ科）新北区
striped horsefly		*Hybomitra expollicata* (Pandelle)	（ハエ目、アブ科）旧北区
striped Jay		*Graphium chironides malayanum* Eliot	（チョウ目、アゲハチョウ科）東洋区
striped ladybird		*Micraspis frenata* (Erichson)	（コウチュウ目、テントウムシ科）豪州区
striped ladybird		*Myzia oblongoguttata* (Linnaeus)	（コウチュウ目、テントウムシ科）全北区
striped lerp psylla	クロオビカイガラキジラミ	*Cacopsylla usubai* (Miyatake)	（カメムシ目、キジラミ科）日本。エノキ害虫
striped long-headed leafhopper	ナカグロホソサジヨコバイ	*Nirvana suturalis* Matsumura	（カメムシ目、オオヨコバイ科）日本、東洋区
striped longhorn		*Navomorpha lineata* (Fabricius)	（コウチュウ目、カミキリムシ科）豪州区
striped longicorn beetle	タテジマカミキリ	*Aulaconotus pachypezoides* Thomson	（コウチュウ目、カミキリムシ科）日本、東洋区
striped love beetle		*Eudicella gralli* (Buquet)	（コウチュウ目、コガネムシ科）エチオピア区
striped lychnis shark		*Cucullia lychnitis* Rambur	（チョウ目、ヤガ科）旧北区。striped lychnis ともいう
striped maple worm		*Dryocampa rubicunda* Fabricius	（チョウ目、ヤママユガ科）新北区
striped mayfly		*Ephemera lineata* Eaton	（カゲロウ目、モンカゲロウ科）旧北区
striped mealybug	フタスジコナカイガラムシ	*Ferrisia virgata* (Cockerell)	（カメムシ目、コナカイガラムシ科）日本、汎熱帯。カンキツ、マンゴー，ハイビスカスなどの害虫
striped morning sphinx			striped hawk を見よ　米国での英名
striped mosquito		*Aedes notoscriptus* (Skuse)	（ハエ目、カ科）豪州区
striped oak webworm moth		*Pococera expandens* (Walker)	（チョウ目、メイガ科）新北区
striped orange moth		*Lythria perornata* Walker	（チョウ目、シャクガ科）豪州区
striped ox warble fly			cattle warble fly (2) を見よ
striped pea weevil			pea leaf weevil を見よ
striped Phaneta moth		*Phaneta clavana* (Fernald)	（チョウ目、ハマキガ科）新北区

英名	和名	学名	所属、分布、ほか
striped pierrot		*Tarucus nara* (Kollar)	(チョウ目、シジミチョウ科) 東洋区
striped pine scale		*Toumeyella pini* (King)	(カメムシ目、カタカイガラムシ科) 新北区
striped policeman		*Coeliades forestan* (Stoll)	(チョウ目、セセリチョウ科) エチオピア区
striped porina moth		*Wiseana umbraculata* Guenée	(チョウ目、コウモリガ科) 豪州区
striped psocid		*Psocus longicornis* Fabricius	(チャタテムシ目、チャタテ科) 旧北区
striped punch	アドニラシジミタテハ	*Dodona adonira* Hewitson	(チョウ目、シジミタテハ科) 東洋区
striped rice borer			rice stem borer を見よ
striped rice katydid	ササキリ	*Conocephalus melas* (de Haan)	(バッタ目、キリギリス科) 日本、東洋区
striped ringlet		*Ragadia crisilda* Hewitson	(チョウ目、タテハチョウ科) 東洋区
striped ringlet			Malayan ringlet を見よ
striped saddlebags		*Tramea calverti* Muttowski	(トンボ目、トンボ科) 新熱帯区
striped sand grasshopper		*Melanoplus foedus* Scudder	(バッタ目、バッタ科) 新北区
striped shield bug		*Tetroda histeroides* (Fabricius)	(カメムシ目、カメムシ科) 東洋区。イネ害虫
striped slender robberfly		*Leptogaster cylindrica* (De Geer)	(ハエ目、ムシヒキアブ科) 旧北区
striped sod webworm			changeable grass veneer moth を見よ
striped sodworm			striped webworm を見よ
striped stalk-borer			common stalk borer を見よ
striped stem borer			rice stem borer を見よ
striped swarming leaf beetle		*Rhyparida didyma* (Fabricius)	(コウチュウ目、ハムシ科) 豪州区
striped sweet potato weevil		*Alcidodes dentipes* (Olivier)	(コウチュウ目、ゾウムシ科) エチオピア区
striped thrips	シマアザミウマ	*Aeolothrips fasciatus* (Linnaeus)	(アザミウマ目、シマアザミウマ科) 日本、全北区
striped tiger			common tiger (3) を見よ
striped tortoise beetle		*Cassida bivittata* Say	(コウチュウ目、ハムシ科) 新北区。サツマイモ害虫
striped tortoise beetle	クロスジカメノコハムシ	*Cassida lineola* Creutzer	(コウチュウ目、ハムシ科) 日本、旧北区、東洋区
striped tortoise beetle			gold-striped tortoise beetle を見よ
striped tortrix	タテスジハマキ	*Archips pulcher* (Butler)	(チョウ目、ハマキガ科) 日本、旧北区
striped turnip flea beetle			large striped flea beetle を見よ
striped twin-spot carpet		*Nebula salicata* (Hübner)	(チョウ目、シャクガ科) 旧北区
striped veronica moth		*Epirrhanthis veronicae* Prout	(チョウ目、シャクガ科) 豪州区
striped vine-hawk moth			vine hawk-moth (1) を見よ

英名	和名	学名	所属、分布、ほか
striped wainscot		*Mythimna pudorina* (Denis et Schiffermüller)	(チョウ目、ヤガ科) 旧北区。日本亜種は *M. p. subrosea* (Matsumura) ウスベニキヨトウ
striped walking sticks		Pseudophasmatidae	(ナナフシ目) の昆虫の総称　現在は Bacunculidae とされる
striped webworm		*Crambus mutabilis* Clemens	(チョウ目、メイガ科) 新北区。トウモロコシ他害虫
striped-winged grasshopper			lined grasshopper を見よ　stripe-winged grasshopper ともいう
strong-barred conch	ニセフトオビホソハマキ	*Phalonidia curvistrigana* (Stainton)	(チョウ目、ホソハマキガ科) 日本、旧北区
strong blues			hairstreaks を見よ
strong skimmer		*Orthetrum stemmale* (Burmeister)	(トンボ目、トンボ科) エチオピア区
strongylophthalmyiid flies		Strongylophthalmyiidae	(ハエ目) の昆虫の総称
Strophius hairstreak		*Allosmaitia strophius* (Godart)	(チョウ目、シジミチョウ科) 新北区
Strutt's skolly		*Thestor strutti* van Son	(チョウ目、シジミチョウ科) エチオピア区
stub moth		*Ephestia vapidella* Mannerheim	(チョウ目、メイガ科) 旧北区。カンキツ害虫
stub-tailed gemmed-satyr		*Cyllopsis hedemanni hedemanni* R. Felder	(チョウ目、タテハチョウ科) 新熱帯区。*C. h. tamaulipensis* Miller も同英名
stub-tailed Morpho		*Morpho theseus justitiae* (Salvin et Godman)	(チョウ目、タテハチョウ科) 新熱帯区。*M. t. oaxacensis* (Le Moult et Real), *M. t. schweizeri* Maza も同英名
stub-tailed skipper		*Chrysoplectrum epicincea* (Butler et Druce)	(チョウ目、セセリチョウ科) 新熱帯区
studded sergeant	ナカグロミスジ	*Athyma asura* Moore	(チョウ目、タテハチョウ科) 東洋区。*A. a. idita* Moore も同英名
stump borer		*Centrodera spurca* LeConte	(コウチュウ目、カミキリムシ科) 新北区
stumpstabber			western giant ichneumon を見よ
Sturnula metalmark		*Calydna sturnula* (Geyer)	(チョウ目、シジミタテハ科) 新熱帯区
Stygian ringlet	ステイックスベニヒカゲ	*Erebia styx* (Freyer)	(チョウ目、タテハチョウ科) 旧北区
stylopids	ハチネジレバネ科	Stylopidae	(ネジレバネ目) の昆虫の総称
stylopids			stylops を見よ
stylops	ネジレバネ目	Strepsiptera	の昆虫の総称　世界に約 450 種
styphnolobium leaf-cut weevil	オオケブカチョッキリゾウムシ	*Involvulus amabilis* (Roelofs)	(コウチュウ目、オトシブミ科) 日本、旧北区
styrax gall aphid	エゴアブラムシ	*Astegopteryx styracophila* Karsh	(カメムシ目、アブラムシ科) 日本、東洋区
Styrian ringlet		*Erebia stirius* (Godart)	(チョウ目、タテハチョウ科) 旧北区
Suave citril		*Ceriagrion suave* Ris	(トンボ目、イトトンボ科) エチオピア区
sub-angled-wave		*Scopula nigropunctata* (Hufnagel)	(チョウ目、シャクガ科) 旧北区。日本亜種は *S. n. imbella* (Warren) マエキヒメシャク
subarctic darner			bog hawker を見よ
subflava sedge borer moth		*Capsula subflava* Grote	(チョウ目、ヤガ科) 新北区。subflava sedge borer ともいう

英名	和名	学名	所属、分布、ほか
subgothic dart moth			dingy cutworm moth (1) を見よ
subhyaline darter		*Ochlodes subhyalina* (Bremer et Grey)	(チョウ目、セセリチョウ科) 東洋区
subject lichen moth		*Cisthene subjecta* Walker	(チョウ目、ヒトリガ科) 新北区
submetallic flea beetle		*Nisotra submetallica* Blackburn	(コウチュウ目、ハムシ科) 豪州区
subreticulate skipper		*Polites subreticulata* (Plötz)	(チョウ目、セセリチョウ科) 新熱帯区
subrufescens skipper		*Vertica subrufescens* (Schaus)	(チョウ目、セセリチョウ科) 新熱帯区
subsocial burrower bug	シロヘリツチカメムシ	*Canthaphorus niveimarginatus* (Scott)	(カメムシ目、ツチカメムシ科) 日本、旧北区
sub-tailed satyr			Virgilia wood nymph を見よ
subterranean chinch bug		*Cyrtomenus bergi* Froeschner	(カメムシ目、ツチカメムシ科) 新熱帯区
subterranean clover weevil		*Listroderes delaiguei* Germain	(コウチュウ目、ゾウムシ科) 豪州区
subterranean coccid		*Margarodes capensis* Giard	(カメムシ目、ワタフキカイガラムシ科) エチオピア区
subterranean dart			granulate cutworm を見よ　subterranean dart moth ともいう
subterranean silverfishes	メナシシミ科	Nicoletiidae	(シミ目) の昆虫の総称　旧世界に多い
subterranean termite		*Heterotermes ferax* (Froggatt)	(シロアリ目、ミゾガシラシロアリ科) 豪州区
subterranean termite		*Coptotermes acinaciformis* (Froggatt)	(シロアリ目、ミゾガシラシロアリ科) 豪州区
subterranean termite		*Coptotermes frenchi* Hill	(シロアリ目、ミゾガシラシロアリ科) 豪州区。地下を 70 m も進む種
subterranean termite			eastern subterranean termite を見よ
subterranean termites		*Coptotermes, Heterotermes, Schedorhinotermes, Reticulitermes*	(シロアリ目、ミゾガシラシロアリ科) の昆虫の総称
subterranean termites (1)	ミゾガシラシロアリ科	Rhinotermitidae	(シロアリ目) の昆虫の総称
subtriple-spot button		*Acleris aspersana* Denis et Schiffermüller	(チョウ目、ハマキガ科) 旧北区
subtropical pine-tip moth			pine tip moth (4) を見よ
suckers			leaf insects を見よ
suckers			jumping plantlice を見よ
suckfly		*Cyrtopeltis notatus* (Distant)	(カメムシ目、カスミカメムシ科) 新北区
sucking dog louse			dog sucking louse を見よ　米国での英名
sucking goat louse	ヤギジラミ	*Linognathus stenopsis* (Burmeister)	(シラミ目、ケモノホソジラミ科) 日本、汎世界。米国での英名
sucking horse louse			horse sucking louse を見よ　米国での英名
sucking lice	シラミ目	Anoplura	の昆虫の総称　世界に約 500 種

英名	和名	学名	所属、分布、ほか
sucking louse of horse			horse sucking louse を見よ　米国での英名
sucking louse of the dog			dog sucking louse を見よ
sucking rabbit louse			rabbit louse を見よ　米国での英名
sucking sheep louse			sheep face louse を見よ　米国での英名
Sucova skipper		*Sucova sucova* (Schaus)	（チョウ目、セセリチョウ科）新熱帯区
Sudan bollworm			red bollworm (1) を見よ
Sudan cotton worm		*Xanthodes graellsii* Feisthamel	（チョウ目、ヤガ科）東洋区、エチオピア区
Sudan sprite		*Pseudagrion sudanicum* Le Roi	（トンボ目、イトトンボ科）エチオピア区
Sudan's caper white		*Belenois sudanensis* (Talbot)	（チョウ目、シロチョウ科）エチオピア区
Sudetan ringlet	ズデーテンベニヒカゲ	*Erebia sudetica* Staudinger	（チョウ目、タテハチョウ科）旧北区。Sudeten ringlet ともいう
suffused Acraea		*Acraea stenobea* Wallengren	（チョウ目、タテハチョウ科）エチオピア区
suffused flash		*Rapala suffusa* (Moore)	（チョウ目、シジミチョウ科）東洋区。*R. s. barthema* Distant も同英名
suffused Saliana		*Saliana fusta* Evans	（チョウ目、セセリチョウ科）新熱帯区
suffused silverdrop		*Epargyreus spinosa* Evans	（チョウ目、セセリチョウ科）新熱帯区
suffused snow flat			immaculate suffused snow flat を見よ
sugar ant		*Camponotus nigripes* Smith	（ハチ目、アリ科）豪州区
sugar ant (1)		*Camponotus consobrinus* (Erichson)	（ハチ目、アリ科）豪州区
sugar ant			fire ant (1) を見よ
sugar-apple mealybug			coconut mealybug を見よ
sugarbag bees			stingless bees (2) を見よ　豪州での英名
sugarbeet Aphomia	フタテンツヅリガ	*Aphomia sapozhnikovi* (Krulikowski)	（チョウ目、メイガ科）日本、東洋区。ダイズ、テンサイ害虫
sugarbeet armyworm			beet armyworm を見よ
sugar beet beetle			beet tortoise beetle (2) を見よ
sugarbeet caterpillar			white dot を見よ
sugarbeet crane fly	アカホソガガンボ	*Nephrotoma minuticornis* Alexander	（ハエ目、ガガンボ科）日本、旧北区
sugarbeet crown borer		*Hulstia undulatella* (Clemens)	（チョウ目、メイガ科）新北区。成虫は sugarbeet crown borer moth
sugarbeet cutworm			small angle shades を見よ
sugar beet flea beetle			beet flea beetle を見よ
sugarbeet leaf beetle			brassy flea beetle (1) を見よ
sugarbeet leaf bug	テンサイカスミカメムシ	*Orthotylus flavosparsus* (Sahlberg)	（カメムシ目、カスミカメムシ科）日本、全北区。テンサイ害虫
sugar-beet leafhopper			beet leafhopper を見よ

英名	和名	学名	所属、分布、ほか
sugarbeet moth		*Scrobipalpula ocellatella* (Boyd)	(チョウ目、キバガ科) 旧北区
sugarbeet root aphid		*Pemphigus populivenae* Fitch	(カメムシ目、アブラムシ科) 新北区
sugarbeet root aphid		*Pemphigus betae* Doane	(カメムシ目、アブラムシ科) 新北区
sugarbeet root maggot		*Tetanops myopaeformis* (Roder)	(ハエ目、ハネフリバエ科) 新北区
sugarbeet thrips			banded glasshouse thrips を見よ
sugarbeet webworm			beet webworm (2) を見よ　幼虫の英名
sugarbeet weevil		*Bothynoderes punctiventris* Germain	(コウチュウ目、ゾウムシ科) 旧北区
sugar-beet wireworm		*Limonius californicus* Mannerheim	(コウチュウ目、コメツキムシ科) 新北区
sugarcane and maize stemborer			cane-moth borer を見よ
sugarcane aphid			cane aphid を見よ
sugarcane armyworm		*Leucania stenographa* Lower	(チョウ目、ヤガ科) 豪州区
sugarcane beetle		*Euetheola rugiceps* (LeConte)	(コウチュウ目、コガネムシ科) 新北区、新熱帯区。サトウキビ、トウモロコシの害虫
sugarcane beetles			black maize beetles を見よ
sugarcane borer		*Sesamia uniformis* Dudgeon	(チョウ目、ヤガ科) 東洋区
sugarcane borer		*Lithophane unimoda* Lintner	(チョウ目、ヤガ科) 新北区
sugarcane borer (1)		*Diatraea saccharalis* (Fabricius)	(チョウ目、メイガ科) 新北区、新熱帯区。サトウキビ、イネ害虫。成虫は　sugarcane borer moth。sugarcane moth stalk borer ともいう
sugar cane borer			banana moth (1) を見よ
sugarcane bud moth		*Opogona glycyphaga* Meyrick	(チョウ目、ヒロズコガ科) 豪州区
sugarcane bud moth		*Decadarchis flavistriata* (Walsingham)	(チョウ目、ヒロズコガ科) 東洋区、大洋区、新北区
sugarcane budworm		*Mythimna humidicola* Guenée	(チョウ目、ヤガ科) 新熱帯区。成虫は sugarcane budworm moth
sugarcane butt weevil		*Leptopius maleficus* Lea	(コウチュウ目、ゾウムシ科) 豪州区
sugarcane butterfly		*Calisto archebates* Ménétriès	(チョウ目、メイガ科) 新熱帯区
sugarcane cicada	イワサキクサゼミ	*Mogannia minuta* Matsumura	(カメムシ目、セミ科) 日本、東洋区
sugarcane click beetle		*Melanotus okinawaensis* Ohira	(コウチュウ目、コメツキムシ科) 日本
sugarcane cottony aphid	サトウキビコナフキツノアブラムシ	*Ceratovacuna lanigera* Zehntner	(カメムシ目、アブラムシ科) 日本、東洋区。サトウキビ害虫
sugarcane delphacid			sugarcane planthopper を見よ
sugarcane delphacid egg sucker			sugarcane planthopper mirid を見よ
sugarcane elongate weevil	カムチャホソゾウムシ	*Gasteroclisus auriculatus* (Sahlberg)	(コウチュウ目、ゾウムシ科) 日本、東洋区

1077

英名	和名	学名	所属、分布、ほか
sugarcane froghopper		*Euryaulax carnifes* (Fabricius)	（カメムシ目、コガシラアワフキ科）日本、豪州区
sugarcane froghopper		*Tomaspis saccharina* Distant	（カメムシ目、コガシラアワフキ科）新熱帯区
sugarcane internodal borer			gold-fringed rice borer を見よ
sugarcane lace bug		*Leptodictya tabida* (Herrich-Schäffer)	（カメムシ目、グンバイムシ科）大洋区、新北区
sugarcane leaf beetle		*Rhyparida dimidiata* Baly	（コウチュウ目、ハムシ科）豪州区
sugarcane leaf miner		*Cosmopterix dulcivora* Meyrick	（チョウ目、カザリバガ科）東洋区、豪州区、大洋区
sugarcane leafhopper			sugarcane planthopper を見よ
sugarcane leafminer moth		*Elachista saccharella* (Busck)	（チョウ目、クサモグリガ科）新北区
sugarcane leafmining beetle		*Aphanisticus cochinchinae* Obenberger	（コウチュウ目、ハムシ科）大洋区
sugarcane leafroller		*Hedylepta accepta* (Butler)	（チョウ目、メイガ科）新北区、大洋区
sugarcane leaf-roller		*Pseudopyrausta acutangulalis* Snellen	（チョウ目、メイガ科）大洋区、新熱帯区
sugarcane leafroller worm			maize webworm を見よ
sugarcane leafroller worm			ground-pearl (1) を見よ
sugarcane long-winged planthopper	サトウマダラハネナガウンカ	*Kamendaka saccharivora* Matsumura	（カメムシ目、ハネナガウンカ科）日本、東洋区。サトウキビ害虫
sugarcane looper		*Mocis frugalis* (Fabricius)	（チョウ目、ヤガ科）東洋区、豪州区
sugarcane looper			South American semilooper を見よ
sugarcane mealybug			pink sugarcane mealybug を見よ
sugar-cane mealybug			citrophilus mealybug (2) を見よ
sugarcane midget moth		*Elaphria nucicolora* (Guenée)	（チョウ目、ヤガ科）新北区、新熱帯区、大洋区。sugarcane midget ともいう
sugarcane moth borer of India			rice stem borer を見よ
sugarcane planthopper	クロフツノウンカ	*Perkinsiella saccharicida* Kirkaldy	（カメムシ目、ウンカ科）日本、東洋区、大洋区、エチオピア区。サトウキビ、トウモロコシ害虫
sugarcane planthopper mirid		*Tytthus mundulus* (Breddin)	（カメムシ目、カスミカメムシ科）大洋区
sugarcane rhinoceros beetle		*Strategus simson* (Linnaeus)	（コウチュウ目、コガネムシ科）東洋区
sugarcane root aphid	サトウキビネワタムシ	*Geoica lucifuga* (Zehntner)	（カメムシ目、アブラムシ科）日本、東洋区。サトウキビ害虫
sugarcane root borer		*Diaprepes abbreviatus* (Linnaeus)	（コウチュウ目、ゾウムシ科）新熱帯区。カンキツ害虫
sugarcane root borers		*Diaprepes*	（コウチュウ目、ゾウムシ科）の昆虫の総称
sugar-cane root bug		*Stibaropus molginus* Schiodte	（カメムシ目、カメムシ科）東洋区
sugarcane sandgrub		*Dipelicus optatus* (Sharp)	（コウチュウ目、コガネムシ科）豪州区

英名	和名	学名	所属、分布、ほか
sugarcane scale		*Aspidiella sacchari* (Cockerell)	（カメムシ目、マルカイガラムシ科）新北区
sugarcane scale		*Aulacaspis madiunensis* (Zehntner)	（カメムシ目、マルカイガラムシ科）豪州区
sugarcane scale (1)		*Aulacaspis tegalensis* (Zehntner)	（カメムシ目、マルカイガラムシ科）エチオピア区
sugarcane shoot borer	カンショシンクイハマキ	*Tetramoera schistaceana* (Snellen)	（チョウ目、ハマキガ科）日本、東洋区、大洋区、エチオピア区。サトウキビ害虫
sugarcane shoot-borer			early shoot borer を見よ
sugarcane shot-hole borer			island pinhole borer を見よ
sugarcane soldier fly		*Inopus rubriceps* (Macquart)	（ハエ目、ミズアブ科）豪州区
sugarcane spittlebugs		*Tomaspis*	（カメムシ目、コガシラアワフキ科）新熱帯区
sugarcane spittlebugs		*Aeneolamia*	（カメムシ目、コガシラアワフキ科）新熱帯区
sugarcane springtail		*Salina wolcotti* Folsom	（トビムシ目、ヒゲナガトビムシ科）新熱帯区
sugarcane stalk borer		*Eldana saccharina* Walker	（チョウ目、メイガ科）エチオピア区
sugarcane stalk borer		*Chilo sacchariphagus* Bojer	（チョウ目、メイガ科）東洋区、エチオピア区
sugarcane stalk-borer			gold-fringed rice borer を見よ
sugarcane stem maggot	サトウキビクキイエバエ	*Atherigona boninensis* Snyder	（ハエ目、イエバエ科）日本、東洋区、豪州区。サトウキビ害虫
sugarcane stool moth		*Acrolophus sacchari* Busck	（チョウ目、メイガ科）新熱帯区
sugarcane termite		*Pseudacanthotermes militaris* (Hagen)	（シロアリ目、シロアリ科）エチオピア区
sugarcane thrips		*Stenchaetothrips minutus* (van Deventer)	（アザミウマ目、アザミウマ科）日本、新北区、大洋区
sugarcane thrips	サトウキビチビアザミウマ	*Fulmekiola serrata* (Kobus)	（アザミウマ目、アザミウマ科）日本、東洋区
sugar cane top borer		*Tryporyza nivella* (Fabricius)	（チョウ目、メイガ科）東洋区
sugarcane weevil borer	カンショオサゾウムシ	*Rhabdoscelus obscurus* (Boisduval)	（コウチュウ目、オサゾウムシ科）日本、東洋区、大洋区、新北区、豪州区。サトウキビ害虫
sugarcane whitefly	カンショコナジラミ	*Neomaskellia bergii* (Signoret)	（カメムシ目、コナジラミ科）日本、東洋区、大洋区、エチオピア区。サトウキビ害虫
sugarcane whitefly		*Aleurolobus barodensis* (Maskell)	（カメムシ目、コナジラミ科）豪州区
sugarcane whitegrub		*Cochliotis melolonthoides* (Gerstaecker)	（コウチュウ目、コガネムシ科）エチオピア区
sugarcane wireworm		*Agrypnus variabilis* (Candèze)	（コウチュウ目、コメツキムシ科）豪州区
sugarcane wooly aphid			sugarcane cottony aphid を見よ
sugar-iced bug			Jacarand bug を見よ　米国、南アフリカでの英名
sugar lerp insect		*Spondyliaspis eucalypti* Dobson	（カメムシ目、キジラミ科）豪州区
sugar maple borer		*Glycobius speciosus* (Say)	（コウチュウ目、カミキリムシ科）新北区。sugar-maple borer とも記す

英名	和名	学名	所属、分布、ほか
sugar maple scale		*Cryptococcus williamsi* Kosztarab et Hale	（カメムシ目、Cryptococcidae）新北区
sugar pine cone beetle		*Conophthorus lambertianae* Hopkins	（コウチュウ目、キクイムシ科）新北区
sugi bark borer			cryptomeria bark borer を見よ
sugi leaf beetle	ウスイロサルハムシ	*Basilepta pallidula* (Baly)	（コウチュウ目、ハムシ科）日本、旧北区。モモ、マツ、スギ害虫
sugi tussock moth			cedar tussock moth を見よ
Sugiyama's admiral		*Limenitis misuji* Sugiyama	（チョウ目、タテハチョウ科）旧北区
Suk playboy		*Deudorix suk* Stempffer	（チョウ目、シジミチョウ科）エチオピア区
Sulanus owl-butterfly		*Caligo brasiliensis sulanus* Fruhstorfer	（チョウ目、タテハチョウ科）新熱帯区
Sulawesi banded swallowtail	オオオビモンアゲハ	*Papilio gigon* C. et R. Felder	（チョウ目、アゲハチョウ科）東洋区
Sulawesi blue Mormon	セレベスアゲハ	*Papilio ascalaphus* Boisduval	（チョウ目、アゲハチョウ科）東洋区
Sulawesi blue nawab		*Polyura cognatus* Vollenhoven	（チョウ目、タテハチョウ科）東洋区
Sulawesi blue tiger	セレベスコモンマダラ	*Tirumala choaspes* (Butler)	（チョウ目、タテハチョウ科）東洋区
Sulawesi blue triangle		*Graphium monticolus* (Fruhstorfer)	（チョウ目、アゲハチョウ科）東洋区
Sulawesi clubtail		*Atrophaneura palu* (Martin)	（チョウ目、アゲハチョウ科）東洋区
Sulawesi Gandy baron	シタベニイナズマ	*Euthalia amanda* (Hewitson)	（チョウ目、タテハチョウ科）東洋区
Sulawesi jungle brown	ヨパスニセコジャノメ	*Orsotriaena jopas* (Hewitson)	（チョウ目、タテハチョウ科）東洋区
Sulawesi Marquis		*Bassarona labotas* (Hewitson)	（チョウ目、タテハチョウ科）東洋区
Sulawesi pied crow			Vanoort's crow を見よ
Sulawesi quaker		*Pithecops phoenix* (Rober)	（チョウ目、シジミチョウ科）東洋区
Sulawesi rose	セレベスベニモンアゲハ	*Atrophaneura polyphontes* (Boisduval)	（チョウ目、アゲハチョウ科）東洋区
Sulawesi sergeant	オオアカミスジ	*Athyma eulimene* (Godart)	（チョウ目、タテハチョウ科）東洋区
Sulawesi sorcerer		*Hestina divona* (Hewitson)	（チョウ目、タテハチョウ科）東洋区
Sulawesi striped blue crow	コンフィギュラータルリマダラ	*Euploea configurata* C. et R. Felder	（チョウ目、タテハチョウ科）東洋区
Sulawesi tabby		*Pseudergolis avesta* C. et R. Felder	（チョウ目、タテハチョウ科）東洋区
Sulawesi tawny Rajah		*Charaxes musashi* Tsukada	（チョウ目、タテハチョウ科）東洋区
Sulawesi white emperor		*Helcyra celebensis* Martin	（チョウ目、タテハチョウ科）東洋区
Sulawesi zebra		*Graphium encelades* (Boisduval)	（チョウ目、アゲハチョウ科）東洋区
sulfur butterflies			cabbage whites を見よ　米国での英名
sulfur whites			cabbage whites を見よ　米国での英名

英名	和名	学名	所属、分布、ほか
Sulkowsky's Morpho		*Morpho sulkowskyi* (Kollar)	(チョウ目、タテハチョウ科) 新熱帯区
sullied Mellana		*Quasimellana balsa* (Bell)	(チョウ目、セセリチョウ科) 新熱帯区
sullied sailer	スズキミスジ	*Neptis soma* Moore	(チョウ目、タテハチョウ科) 東洋区
sullied Saliana			Cramer's proboscis skipper を見よ
sulphur			brimstone を見よ　sulfur と記されることあり
sulphur angle moth		*Speranza sulphurea* (Packard)	(チョウ目、シャクガ科) 新北区
sulphur butterflies			cabbage whites を見よ　米国での英名
sulphur butterfly			Oriental clouded yellow を見よ
sulphur butterfly			brimstone を見よ
sulphur Esperia moth		*Esperia sulphurella* (Fabricius)	(チョウ目、マルハキバガ科) 全北区
sulphur metalmark		*Baeotis sulphurea sulphurea* (R. Felder)	(チョウ目、シジミタテハ科) 新熱帯区。*B. s. macularia* (Boisduval) も同英名
sulphur moth		*Hesperumia sulphuraria* Packard	(チョウ目、シャクガ科) 新北区
sulphur orange-tip		*Colotis aurora* (Cramer)	(チョウ目、シロチョウ科) エチオピア区。亜種 *C. a. dissociatus* (Butler) も同英名
sulphur orange tip			plain orange tip を見よ
sulphur pigmy		*Baeotis macularia* (Boisduval)	(チョウ目、シジミタテハ科) 新熱帯区
sulphur thorn			brimstone moth を見よ
sulphur-pearl	ウラグロシロノメイガ	*Sitochroa palealis* (Denis et Schiffermüller)	(チョウ目、メイガ科) 日本、旧北区、東洋区
sulphur-tipped clubtail		*Gomphus militaris* Hagen	(トンボ目、サナエトンボ科) 新北区
sulphur-winged locust		*Arphia sculphurea* (Fabricius)	(バッタ目、バッタ科) 新北区。sulphur-winged grasshopper ともいう
sulphurs	トガリキチョウ属	*Dercas*	(チョウ目、シロチョウ科) の昆虫の総称
sulphury threadtail		*Protoneura sulfurata* Donnelly	(トンボ目、ミナミイトトンボ科) 新熱帯区
Sulzer's lady slipper		*Pierella lamia* Sulzer	(チョウ目、タテハチョウ科) 新熱帯区
sumac caterpillar			spotted Datana moth を見よ　成虫は sumac caterpillar moth
sumac caterpillar moth			spotted Datana moth を見よ
sumac leafblotch miner moth		*Caloptilia rhoifoliella* (Chambers)	(チョウ目、ホソガ科) 新北区。幼虫は sumac leafblotch miner
sumac leaftier moth			grape berry moth (2) を見よ
sumac stem borer		*Oberea ocellata* Haldeman	(コウチュウ目、カミキリムシ科) 新北区
Sumatran bob		*Arnetta verones* (Hewitson)	(チョウ目、セセリチョウ科) 東洋区
Sumatran chocolate tiger		*Parantica tityoides* (Hagen)	(チョウ目、タテハチョウ科) 東洋区
Sumatran crow		*Euploea martinii* de Nicéville	(チョウ目、タテハチョウ科) 東洋区

英名	和名	学名	所属、分布、ほか
Sumatran gem		*Poritia sumatrae sumatrae* (C. et R. Felder)	(チョウ目、シジミチョウ科) 東洋区
Sumatran sunbeam		*Curetis saronis sumatrana* Corbet	(チョウ目、シジミチョウ科) 東洋区。Burmese sunbeam を参照
Sumbawa tiger		*Parantica philo* (Grose-Smith)	(チョウ目、タテハチョウ科) 東洋区
Sumichrast toothpick grasshopper		*Achurum sumichrasti* (Saussure)	(バッタ目、バッタ科) 新北区。Sumichrast's toothpick grasshopper とも記す
summer azure		*Celastrina ladon neglecta* (Edwards)	(チョウ目、シジミチョウ科) 新北区。spring azure を参照
summer cabbage fly			seed-corn maggot (1) を見よ
summer chafer		*Amphimallon solstitialis* (Linnaeus)	(コウチュウ目、コガネムシ科) 全北区
summer chafer beetle			summer chafer を見よ
summer cricket			Bordeaux cricket を見よ
summer fruit tortricid			summer fruit tortrix (1) を見よ　米国での英名
summer fruit tortrix (1)		*Capua reticulana* (Hübner)	(チョウ目、ハマキガ科) 旧北区
summer fruit tortrix (2)		*Adoxophyes orana* (Fischer von Rösslerstamm)	(チョウ目、ハマキガ科) 旧北区。日本亜種は *A. o. fasciata* Walsingham リンゴコカクモンハマキで、リンゴ、ナシなど果樹、クワ、チャ、バラなど多くの植物害虫
summer fruit tortrix moth			summer fruit tortrix (2) を見よ
summer fruit twist			summer fruit tortrix (1) を見よ
summer grey smoke		*Luffia lapidella* (Goeze)	(チョウ目、ミノガ科) 旧北区
summer mayfly		*Siphlonurus lacustris* Eaton	(カゲロウ目、フタオカゲロウ科) 旧北区
summer porina moth		*Wiseana despecta* Walker	(チョウ目、コウモリガ科) 豪州区
sun beetle	イグチマルガタゴミムシ	*Amara communis* (Panzer)	(コウチュウ目、オサムシ科) 日本、旧北区
sun beetle		*Amara plebeja* Gyllenhal	(コウチュウ目、オサムシ科) 旧北区
sun beetle		*Pachnoda marginata peregrina* Kolbe	(コウチュウ目、コガネムシ科) エチオピア区
sun flies			helomyzid flies を見よ
sun-fly		*Helophilus intentus* Curran et Fluke	(ハエ目、ハナアブ科) 新北区
sun moths (1)	カストニアガ科	Castniidae	(チョウ目) の昆虫の総称
sun moths (2)	マイコガ科	Heliodinidae	(チョウ目) の昆虫の総称
sunbeam butterflies			sunbeams を見よ
sunbeams	ウラギンシジミ属	*Curetis*	(チョウ目、シジミチョウ科) の昆虫の総称
sunburst glory		*Myscelus phoronis* (Hewitson)	(チョウ目、セセリチョウ科) 新熱帯区
sunburst satyr		*Pedaliodes hopfferi* Staudinger	(チョウ目、タテハチョウ科) 新熱帯区
sundew plume moth		*Buckleria parvulus* (Barnes et Lindsey)	(チョウ目、トリバガ科) 新北区
sunflower bee		*Svastra obliqua* (Say)	(ハチ目、ミツバチ科) 新北区

英名	和名	学名	所属、分布、ほか
sunflower beetle		*Zygogramma exclamationis* (Fabricius)	(コウチュウ目、ハムシ科) 新北区
sunflower borer moth		*Papaipema necopina* (Grote)	(チョウ目、ヤガ科) 新北区
sunflower bud moth		*Suleima helianthana* (Riley)	(チョウ目、ハマキガ科) 新北区
sunflower headclipping weevil		*Haplorhynchites aeneus* (Boheman)	(コウチュウ目、ゾウムシ科) 新北区
sunflower maggot		*Stranzia longipennis* (Wiedemann)	(ハエ目、ミバエ科) 新北区
sunflower midge		*Contarinia schulzi* Gagné	(ハエ目、タマバエ科) 新北区
sunflower moth		*Homoeosoma electellum* (Hulst)	(チョウ目、メイガ科) 新北区
sunflower moth (1)		*Homoeosoma nebulellum* Denis et Schiffermüller	(チョウ目、メイガ科) 旧北区
sunflower peacock fly		*Strauzia langipennis* (Wiedemann)	(ハエ目、ミバエ科) 新北区
sunflower root weevil		*Baris strenua* (LeConte)	(コウチュウ目、ゾウムシ科) 新北区
sunflower seed maggot fly		*Neotephritis finalis* (Loew)	(ハエ目、ミバエ科) 新北区
sunflower seed midge		*Neolasioptera murtfeldtiana* (Felt)	(ハエ目、タマバエ科) 新北区
sunflower spittle bug		*Clastoptera xanthocephala* Germar	(カメムシ目、コガシラアワフキ科) 新北区
sunflower stem weevil		*Cylindrocopturus adspersus* (LeConte)	(コウチュウ目、ゾウムシ科) 新北区
sunn pest	ムギチャイロカメムシ	*Eurygaster integriceps* Puton	(カメムシ目、キンカメムシ科) 旧北区
sunrise skipper		*Adopaeoides prittwitzi* (Plötz)	(チョウ目、セセリチョウ科) 新北区、新熱帯区
sunset daggerwing		*Marpesia furcula* (Fabricius)	(チョウ目、タテハチョウ科) 新熱帯区
sunset Morpho		*Morpho hecuba* (Linnaeus)	(チョウ目、タテハチョウ科) 新熱帯区
sunset moth		*Chrysiridia madagascarensis* (Lesson)	(チョウ目、ツバメガ科) エチオピア区
superb banner	シロオビムラサキタテハ	*Epiphile eriopis* Hewitson	(チョウ目、タテハチョウ科) 新熱帯区
superb cycadian			great cycadian を見よ
superb jewelwing		*Calopteryx amata* (Hagen)	(トンボ目、イトトンボ科) 新北区
superb leafwing		*Fountainea nessus* Latreille	(チョウ目、タテハチョウ科) 新熱帯区
superb meadow katydid		*Orchelimum superbum* Rehn et Hebard	(バッタ目、キリギリス科) 新北区
superb numberwing		*Callicore excelsior* (Hewitson)	(チョウ目、タテハチョウ科) 新熱帯区
superb plant bug		*Adelphocoris superbus* (Uhler)	(カメムシ目、カスミカメムシ科) 新北区
superb white Charaxes		*Charaxes superbus* Schultze	(チョウ目、タテハチョウ科) エチオピア区

英名	和名	学名	所属、分布、ほか
Suraka silk moth		*Antherina suraka* (Boisduval)	（チョウ目、ヤママユガ科）エチオピア区
surface beetles			darkling beetles (1) を見よ
surface caterpillars		*Scotia*	（チョウ目、ヤガ科）の昆虫の総称　幼虫の英名
surface caterpillars			owlet moths を見よ　幼虫の英名
surface swimmers			whirligig beetles を見よ
Surinam cockroach	オガサワラゴキブリ	*Pycnoscelis surinamensis* (Linnaeus)	（ゴキブリ目、オガサワラゴキブリ科）日本、新北区、汎熱帯。Surinam roach ともいう
Surinam long-horned grasshopper			Caribbean meadow katydid を見よ
Suroifui hairstreak		*Teratozephyrus doni* (Tytler)	（チョウ目、シジミチョウ科）東洋区
Surrey scabious midget		*Phyllonorycter scabiosella* (Douglas)	（チョウ目、ホソガ科）旧北区
Susan's copper		*Aloeides susanae* Tite et Dickson	（チョウ目、シジミチョウ科）エチオピア区
suspected	ハイイロヨトウ	*Parastichtis suspecta* (Hübner)	（チョウ目、ヤガ科）日本、新北区。suspected moth ともいう
Sussex case		*Coleophora salicorniae* Heinemann et Wocke	（チョウ目、ツツミノガ科）旧北区
Sussex emerald		*Thalera fimbrialis* (Scopoli)	（チョウ目、シャクガ科）旧北区
Sussex spruce argent		*Blastotere illuminatella* (Zeller)	（チョウ目、メムシガ科）旧北区
Sussex wainscot			white-mottled wainscot を見よ
Sutherland long-cloak			dark long-cloaked marble を見よ
Sutrina moth		*Hada sutrina* (Grote)	（チョウ目、ヤガ科）新北区
Suva nightfighter		*Porphyrogenes suva* Evans	（チョウ目、セセリチョウ科）新熱帯区
Suwako cottony-cussion scale	スワコワタカイガラモドキ	*Coccura suwakoensis* (Kuwana et Toyoda)	（カメムシ目、コナカイガラムシ科）日本、旧北区。リンゴ、ナシ、カキ、サクラなどの害虫
Suzuki leafhopper	スズキヒメヨコバイ	*Arboridia suzukii* (Matsumura)	（カメムシ目、オオヨコバイ科）日本、旧北区
Suzuki's Promalactis moth	シロスジカバマルハキバガ	`*Promalactis suzukiella* (Matsumura)	（チョウ目、マルハキバガ科）日本、全北区
Svensson's copper underwing		*Amphipyra berbera svenssoni* Fletcher	（チョウ目、ヤガ科）旧北区
swag-lined wave moth		*Scopula umbilicata* (Fabricius)	（チョウ目、シャクガ科）新北区、新熱帯区
swaine Jack pine sawfly		*Neodiprion swainei* Middleton	（ハチ目、マツハバチ科）新北区
Swale fly		*Dictya umbroides* Curran	（ハエ目、ヤチバエ科）新北区
swallow bug		*Oeciacus vicarius* Horvath	（カメムシ目、トコジラミ科）新北区。ニワトリに寄生
swallow bug (1)		*Oeciacus hirundinis* Jenyns	（カメムシ目、トコジラミ科）旧北区
swallow bugs			bed bugs を見よ

英名	和名	学名	所属、分布、ほか
swallow flea		*Ceratophyllus farreni* Rothschild	（ノミ目、ナガノミ科）旧北区
swallow flea		*Ceratophyllus hirundinis* (Curtis)	（ノミ目、ナガノミ科）旧北区
swallow kitten		*Harpyia furcula* (Clerck)	（チョウ目、シャチホコガ科）旧北区
swallow louse-fly		*Ornithomyia biloba* Dufour	（ハエ目、シラミバエ科）旧北区
swallow louse-fly		*Stenopteryx hirundinis* Linnaeus	（ハエ目、シラミバエ科）旧北区
swallow prominent		*Pheosia tremula* (Clerck)	（チョウ目、シャチホコガ科）旧北区
swallowtail			common yellow swallowtail を見よ　swallowtail butterfly ともいう
swallowtail butterflies			swallowtails を見よ
swallowtail moths	ツバメガ科	Uraniidae	（チョウ目）の昆虫の総称
swallow-tailed elder			swallow-tailed moth を見よ
swallow-tailed moth		*Ourapteryx sambucaria* (Linnaeus)	（チョウ目、シャクガ科）旧北区。swallowtail geometrid moth, swallow-tail moth ともいう
swallow-tailed moth			tailed geometrid (1) を見よ
swallowtails	アゲハチョウ科	Papilionidae	（チョウ目）の昆虫の総称　世界に500種以上
swallowtails			swordtails を見よ
Swammerdamm's longhorn		*Nematopogon swammerdammella* (Linnaeus)	（チョウ目、マガリガ科）旧北区
swamp ash bark beetle	ヤチダモノキクイムシ	*Hylesinus laticollis* Blandford	（コウチュウ目、キクイムシ科）日本、旧北区
swamp ash longicorn			slender velvety longicorn beetle を見よ
swamp Belle			yellow-lined owlet moth を見よ
swamp bluet		*Africallagma glaucum* (Burmeister)	（トンボ目、イトトンボ科）エチオピア区
swamp cicada		*Tibicen chloromera* (Walker)	（カメムシ目、セミ科）新北区
swamp damsels		*Leptobasis*	（トンボ目、イトトンボ科）の昆虫の総称
swamp darner		*Austroaeschna parvistigma* Selys	（トンボ目、ヤンマ科）豪州区
swamp darner			smoky-winged darner を見よ
swamp darter			affinis skipper を見よ
swamp metalmark		*Calephelis muticum* McAlpine	（チョウ目、シジミタテハ科）新北区
swamp palm forester		*Bebearia paludicola* Holmes	（チョウ目、タテハチョウ科）エチオピア区
swamp patroller		*Heteropsis perspicua* (Trimen)	（チョウ目、タテハチョウ科）エチオピア区
swamp ringlet		*Ypthimomorpha itonia* (Hewitson)	（チョウ目、タテハチョウ科）エチオピア区
swamp spreadwing		*Lestes vigilax* Selys	（トンボ目、アオイトトン科）新北区
swamp tiger			Malay tiger を見よ

英名	和名	学名	所属、分布、ほか
swan-feather dwarf		*Elachista argentella* (Clerck)	(チョウ目、クサモグリガ科) 旧北区
swan moth			browntail moth (2) を見よ
Swanepoel's blue		*Lepidochrysops swanepoeli* (Pennington)	(チョウ目、シジミチョウ科) エチオピア区
Swanepoel's brown		*Pseudonympha swanepoeli* van Son	(チョウ目、タテハチョウ科) エチオピア区
Swanepoel's copper		*Aloeides swanepoeli* Tite et Dickson	(チョウ目、シジミチョウ科) エチオピア区
Swanepoel's opal		*Chrysoritis swanepoeli* (Dickson)	(チョウ目、シジミチョウ科) エチオピア区
Swanepoel's widow		*Dina swanepoeli* (van Son)	(チョウ目、タテハチョウ科) エチオピア区
swarming leaf beetles			rhyparida beetles を見よ
Swartberg blue		*Lepidochrysops swartbergensis* Swanepoel	(チョウ目、シジミチョウ科) エチオピア区
Swartberg copper			Pyramus opal を見よ
swarthy skipper	ナストラセセリ	*Nastra lherminier* (Latreille)	(チョウ目、セセリチョウ科) 新北区
Sweadner's hairstreak		*Callophrys gryneus sweadneri* (Chermock)	(チョウ目、シジミチョウ科) 新北区。juniper hairstreak (1) を参照
Sweadner's juniper hairstreak			Sweadner's hairstreak を見よ
sweat bee		*Agopostemon texanus* Cresson	(ハチ目、コハナバチ科) 新北区
sweat bee		*Halictus farinosus* Smith	(ハチ目、コハナバチ科) 新北区
sweat bee		*Halictus tripartitus* Cockerell	(ハチ目、コハナバチ科) 新北区
sweat bees (1)	コハナバチ科	Halictidae	(ハチ目) の昆虫の総称
sweat bees (2)		*Halictus*	(ハチ目、コハナバチ科) の昆虫の総称
sweat flies		*Hydrotaea*	(ハエ目、イエバエ科) の昆虫の総称 人の汗を吸いにくる
sweat flies			flower flies を見よ
sweat fly		*Hydrotaea irritans* (Fallén)	(ハエ目、イエバエ科) 旧北区
swede midge		*Contarinia nasturtii* (Kieffer)	(ハエ目、タマバエ科) 全北区
swede turnip midge			swede midge を見よ
swede turnip midge			cabbage midge を見よ
sweeps			bagworm moths を見よ
sweet bees			sweat bees (1) を見よ
sweet clover weevil	ナガチビコフキゾウムシ	*Sitona cylindricollis* Fahraeus	(コウチュウ目、ゾウムシ科) 日本、全北区。マメ科牧草害虫
sweet gale moth		*Acronicta euphorbiae* (Denis et Schiffermüller)	(チョウ目、ヤガ科) 旧北区。亜種 *A. e. myricae* (Guenée) も同英名
sweet pepperbush Nola moth		*Nola clethrae* Dyar	(チョウ目、コブガ科) 新北区
sweet underwing moth		*Catocala dulciola* Grote	(チョウ目、ヤガ科) 新北区。sweet underwing ともいう

英名	和名	学名	所属、分布、ほか
sweetbay silkmoth		*Callosamia securifera* (Maassen)	(チョウ目、ヤママユガ科) 新北区、新熱帯区
sweetclover aphid		*Therioaphis riehmi* (Börner)	(カメムシ目、アブラムシ科) 全北区。米国での英名
sweetclover root borer		*Walshia miscecolorella* (Chambers)	(チョウ目、カザリキバガ科) 新北区。sweetclover root borer moth ともいう
sweetclover root borer moth		*Triclonella pergandeella* Busck	(チョウ目、カザリバガ科) 新北区
sweetfern geometer moth		*Cyclophora pendulinaria* (Guenée)	(チョウ目、シャクガ科) 新北区
sweetfern leaf casebearer		*Acrobasis comptoniella* Hulst	(チョウ目、メイガ科) 新北区。成虫は sweetfern leaf casebearer moth
sweetfern underwing moth		*Catocala antinympha* Hübner	(チョウ目、ヤガ科) 新北区。sweetfern underwing ともいう
sweetflag spreadwing		*Lestes forcipatus* Rambur	(トンボ目、アオイトトンボ科) 新北区
sweet-gale dagger		*Apatele euphorbiae* Denis et Schiffermüller	(チョウ目、ヤガ科) 旧北区
sweetgum defoliator			large Paectes moth を見よ
sweetgum leafroller moth		*Sciota uvinella* (Ragonot)	(チョウ目、メイガ科) 新北区
sweetheart underwing		*Catocala amatrix* (Hübner)	(チョウ目、ヤガ科) 新北区。sweetheart underwing moth ともいう
sweet-oil tiger			common mechanitis を見よ
sweet-oil tiger			Lysimnia tigerwing を見よ
sweet potato Acraea		*Acraea acerata* Hewitson	(チョウ目、タテハチョウ科) エチオピア区。sweet potato butterfly ともいう
sweet-potato and yam weevil			sweetpotato weevil (1) を見よ
sweet potato army worm			nutmeg を見よ
sweetpotato beetles			サツマイモ害虫で、ハムシ科 のジンガサハムシ類。北米での英名
sweet potato beetles			tortoise beetles (1) を見よ
sweet potato borer			West Indian sweet potato weevil を見よ
sweet potato clearwing		*Synanthedon dasysceles* Bradley	(チョウ目、スカシバガ科) エチオピア区
sweetpotato delphacid			sweetpotato leafhopper を見よ
sweetpotato flea beetle	サツマイモトビハムシ	*Chaetocnema confinis* Crotch	(コウチュウ目、ハムシ科) 日本、新北区。サツマイモ害虫
sweetpotato hawk-moth	エビガラスズメ	*Agrius convolvuli* (Linnaeus)	(チョウ目、スズメガ科) 日本、東洋区、豪州区、大洋区、全北区、エチオピア区。サツマイモ、アサガオなどの害虫。sweet potato moth, sweetpotato horn worm ともいう
sweet potato hornworm			pink spotted hawk-moth を見よ
sweetpotato leaf beetle		*Typophorus nigritus viridicyaneus* (Crotch)	(コウチュウ目、ハムシ科) 新北区
sweetpotato leaf beetle	イモサルハムシ	*Colasposoma dauricum* Mannerheim	(コウチュウ目、ハムシ科) 日本、旧北区。サツマイモ害虫

英名	和名	学名	所属、分布、ほか
sweetpotato leaf beetle		*Colasposoma sellatum* Baly	（コウチュウ目、ハムシ科）豪州区
sweet-potato leaf beetle		*Typophorus nigritus* Fabricius	（コウチュウ目、ハムシ科）新北区、新熱帯区
sweet potato leaf-eating beetle		*Cerotoma ruficornis* (Olivier)	（コウチュウ目、ハムシ科）新熱帯区
sweet potato leaf-folder		*Microthyris helcitalis* Walker	（チョウ目、メイガ科）新熱帯区。sweet potato webworm ともいう
sweetpotato leaf folder	イモキバガ	*Brachmia macroscopa* Meyrick	（チョウ目、キバガ科）日本、東洋区。サツマイモ害虫
sweetpotato leaf worm	ナカジロシタバ	*Aedia leucomelas* (Linnaeus)	（チョウ目、ヤガ科）日本、旧北区、東洋区、大洋区、豪州区、エチオピア区。サツマイモ害虫
sweetpotato leafhopper		*Aloha ipomoeae* Kirkaldy	（カメムシ目、ウンカ科）大洋区
sweetpotato leafminer		*Bedellia orchilella* Walsingham	（チョウ目、ハムグリガ科）新北区、大洋区
sweetpotato leafminer			morning-glory leaf miner を見よ
sweetpotato leafroller		*Pilocrocis tripunctata* (Fabricius)	（チョウ目、メイガ科）新北区。サツマイモ害虫。成虫は sweetpotato leafroller moth
sweetpotato leafroller			sweetpotato webworm moth を見よ
sweetpotato moth			sweetpotato webworm moth を見よ
sweetpotato plume moth	サツマイモトリバ	*Ochyrotica concursa* (Walsingham)	（チョウ目、トリバガ科）日本、東洋区、豪州区。サツマイモ害虫
sweet-potato plume moth			common plume を見よ　米国での英名
sweet potato root borer			sweetpotato weevil (1) を見よ
sweet potato stem borer			sweetpotato vine borer を見よ
sweetpotato tortoise beetle	タテスジヒメジンガサハムシ	*Cassida circumdata* Herbst	（コウチュウ目、ハムシ科）日本、東洋区
sweetpotato tortoise beetles		*Aspidomorpha*	（コウチュウ目、ハムシ科）の昆虫の総称
sweetpotato vine borer	サツマイモノメイガ	*Omphisa anastomosalis* (Guenée)	（チョウ目、メイガ科）日本、東洋区、豪州区、大洋区、新北区。サツマイモ害虫。成虫は sweetpotato vine borer moth
sweetpotato webworm moth		*Helcystogramma convolvuli* (Walsingham)	（チョウ目、キバガ科）全北区、大洋区、新熱帯区
sweetpotato weevil		*Cylas degantulus* (Summers)	（コウチュウ目、ゾウムシ科）新北区。サツマイモ害虫
sweetpotato weevil (1)	アリモドキゾウムシ	*Cylas fromicarius* (Fabricius)	（コウチュウ目、ミツギリゾウムシ科）日本、東洋区、豪州区、エチオピアピア区、新北区、新熱帯区。亜種 *C. f. elegantulus* (Summers) も同英名。サツマイモ害虫
sweetpotato weevils	アリモドキゾウムシ亜科	Cyladinae	（コウチュウ目、ゾウムシ科）の昆虫の総称
sweetpotato whitefly		*Aleurotrachelus trachoides* (Back)	（カメムシ目、コナジラミ科）新北区
sweetpotato whitefly		*Bemisia tuberculata* Bondar	（カメムシ目、コナジラミ科）新熱帯区

英名	和名	学名	所属、分布、ほか
sweetpotato whitefly			tobacco whitefly を見よ
sweetpotato wireworm	マルクビクシコメツキ	*Melanotus caudex* Lewis	(コウチュウ目、コメツキムシ科) 日本。サツマイモ、トウモロコシ、ナス、カンキツ、タバコなどの害虫
swift		*Baoris penicillata* (Moore)	(チョウ目、セセリチョウ科) 東洋区
swift			ghost moth を見よ
swift conehead		*Neoconocephalus velox* Rehn et Hebard	(バッタ目、キリギリス科) 新北区
swift flea		*Ceratophyllus styx* Rothschild	(ノミ目、ナガノミ科) 旧北区
swift forktail		*Ischnura erratica* Calvert	(トンボ目、イトトンボ科) 新北区
swift ked		*Crataerina melbae* (Rondani)	(ハエ目、シラミバエ科) 旧北区
swift louse-fly		*Crataerina pallida* (Olivier)	(ハエ目、シラミバエ科) 旧北区
swift moth	コウモリガ	*Endoclita excrescens* (Butler)	(チョウ目、コウモリガ科) 日本、旧北区。多くの作物、果樹、樹木の害虫。アマツバメに似た飛翔に由来
swift moth			wood swift を見よ
swift moths			ghost moths を見よ
swift river cruiser		*Macromia illinoiensis* (Walsh)	(トンボ目、ヤマトンボ科) 新北区
swift scaly cricket		*Cycloptilum velox* Love et Walker	(バッタ目、コオロギ科) 新北区
swifts		*Baoris*	(チョウ目、セセリチョウ科) の昆虫の総称
swifts		*Borbo*	(チョウ目、セセリチョウ科) の昆虫の総称
swifts		*Caltoris*	(チョウ目、セセリチョウ科) の昆虫の総称
swifts		*Parnara*	(チョウ目、セセリチョウ科) の昆虫の総称
swifts		*Polytremis*	(チョウ目、セセリチョウ科) の昆虫の総称
swifts			ghost moths を見よ
swifts			leafwings を見よ
swimming bugs			water bugs を見よ
Swinhoe's chocolate tiger		*Parantica swinhoei* (Moore)	(チョウ目、タテハチョウ科) 東洋区
Swinhoe's hedge blue		*Monodontides musina* (Snellen)	(チョウ目、シジミチョウ科) 東洋区
Swiss brassy ringlet	スイスベニヒカゲ	*Erebia tyndarus* (Esper)	(チョウ目、タテハチョウ科) 旧北区
swollen Epitola		*Stempfferia tumentia* (Druce)	(チョウ目、シジミチョウ科) エチオピア区
swollen silver-stiletto		*Dialineura anilis* (Linnaeus)	(ハエ目、ツルギアブ科) 旧北区
sword-bearing crickets			bush crickets (3) を見よ
sword-bearing katydid		*Neoconocephalus ensiger* (Harris)	(バッタ目、キリギリス科) 新北区。北米のトウモロコシ畑で普通。sword-bearing conehead ともいう
sword-grass		*Xylena exsoleta* (Linnaeus)	(チョウ目、ヤガ科) 旧北区。成虫は sword-grass moth

英名	和名	学名	所属、分布、ほか
sword grass brown		*Tisiphone abeona* (Donovan)	（チョウ目、タテハチョウ科）豪州区。sword-grass brown butterfly ともいう
sword-tail crickets			bush crickets (3) を見よ
sword-tailed crickets			bush crickets (4) を見よ
sword-tailed doctor			long-tailed metalmark を見よ
sword-tailed flash			Australian plane を見よ
swordtails			kite swallowtails を見よ
Swynnerton's Temnora		*Temnora swynnertoni* Stevenson	（チョウ目、スズメガ科）エチオピア区
sycamore			sycamore moth を見よ
sycamore aphid		*Drepanosiphum platanoidis* (Schrank)	（カメムシ目、アブラムシ科）旧北区
sycamore aphid			black hairy aphid を見よ
sycamore borer moth			California sycamore borer を見よ
sycamore dagger	クロハナコヤガ	*Aventiola pusilla* (Butler)	（チョウ目、ヤガ科）日本、旧北区。sycamore moth ともいう
sycamore lace bug		*Corythucha ciliata* (Say)	（カメムシ目、グンバイムシ科）全北区、豪州区。スズカケにつく
sycamore lace bug		*Corythucha confraterna* Gibson	（カメムシ目、グンバイムシ科）新北区
sycamore leaf skeletonizer		*Gelechia desiliens* Meyrick	（チョウ目、キバガ科）新北区
sycamore leafhopper		*Edwardsiana nigriloba* (Edwards)	（カメムシ目、オオヨコバイ科）旧北区
sycamore leaf-roll gall midge		*Contarinia acerplicans* (Kieffer)	（ハエ目、タマバエ科）旧北区
sycamore maple aphid		*Periphyllus acericola* (Walker)	（カメムシ目、アブラムシ科）旧北区
sycamore maple aphid			sycamore aphid を見よ
sycamore moth		*Acronicta aceris* (Linnaeus)	（チョウ目、ヤガ科）旧北区
sycamore scale		*Stomacoccus platani* Ferris	（カメムシ目、ワタフキカイガラムシ科）新北区
sycamore tussock moth		*Halysidota harrisii* Walsh	（チョウ目、ヒトリガ科）新北区
sycamore webworm moth		*Pococera militella* (Zeller)	（チョウ目、メイガ科）新北区
Sydney azure		*Ogyris ianthis* Waterhouse	（チョウ目、シジミチョウ科）豪州区
Sydney kelp fly		*Chaetocoelopa sydneyensis* (Schiner)	（ハエ目、ハマベバエ科）豪州区
Sydney Mountain darner		*Austroaeschna obscura* Theischinger	（トンボ目、ヤンマ科）豪州区
Syedra hairstreak		*Strephonota syedra* (Hewitson)	（チョウ目、シジミチョウ科）新熱帯区
Sykes' Acraea		*Acraea sykesi* Sharpe	（チョウ目、タテハチョウ科）エチオピア区
Sylhet oakblue		*Arhopala silhetensis* (Hewitson)	（チョウ目、シジミチョウ科）東洋区。*A. s. adorea* de Nicéville も同英名

英名	和名	学名	所属、分布、ほか
Sylphina angel	オオスカシツバメシジミタテハ	*Chorinea sylphina* (H. Bates)	(チョウ目、シジミタテハ科) 新熱帯区
sylvan anglewing		*Polygonia sylvius* (Edwards)	(チョウ目、タテハチョウ科) 新北区
sylvan anglewing			Oreas anglewing を見よ
sylvan hairstreak		*Satyrium sylvinus desertorum* (F. Grinnell)	(チョウ目、シジミチョウ科) 新熱帯区
sylvan waved carpet		*Hydrelia testaceata* (Donovan)	(チョウ目、シャクガ科) 旧北区
sylvestris aphid		*Cinara pini* (Linnaeus)	(カメムシ目、アブラムシ科) 日本、全北区
Sylvia wood nymph		*Taygetis sylvia* Bates	(チョウ目、タテハチョウ科) 新熱帯区
sylvian hairstreak	シルバンカラスシジミ	*Satyrium sylvinus* (Boisduval)	(チョウ目、シジミチョウ科) 新北区
Syma sister		*Adelpha syma* (Godart)	(チョウ目、タテハチョウ科) 新熱帯区
Symmomus skipper		*Trapezites symmomus* Hübner	(チョウ目、セセリチョウ科) 豪州区
sympetrums			darter dragonflies を見よ
sympherobiids			lesser brown lacewings を見よ
symphytans			horntails (1) を見よ
syncopated scaly cricket		*Cycloptilum comprehendens* Hebard	(バッタ目、コオロギ科) 新北区
synneurid gnats		Synneuridae	(ハエ目) の昆虫の総称
syntexid wasps			cedar wood wasps を見よ　Anaxyelidae の異名 Syntexidae にちなむ。syntexids ともいう
syntomid tiger moths		Syntominae	(チョウ目、カノコガ科) の昆虫の総称　北米での英名
Syrian silkworm		*Pachypasa otus* Drury	(チョウ目、カレハガ科) 旧北区
syringa aphid	ハシドイヒゲナガアブラムシ	*Acyrthosiphon syringae* (Matsumura)	(カメムシ目、アブラムシ科) 日本
syrphian	ヒラタアブ		flower flies を見よ　syrphids、syrphid flies ともいう
systates weevil		*Systates pollinosus* Gerstaecker	(コウチュウ目、ゾウムシ科) エチオピア区
syzygium leaf psyllid			Eugenia psyllid を見よ
Szechuan Argus		*Loxerebia sylvicola* (Oberthür)	(チョウ目、タテハチョウ科) 旧北区
Szechwan whitefly		*Aleurolobus szechwanensis* B. Young	(カメムシ目、コナジラミ科) 東洋区。カンキツ害虫

英名	和名	学名	所属、分布、ほか

T

英名	和名	学名	所属、分布、ほか
tabanids			horse flies (1) を見よ
tabano negro			coliguacho を見よ
Tabasco skipper		*Turesis tabascoensis* Freeman	（チョウ目、セセリチョウ科）新熱帯区
tabbies			grass moths (2) を見よ
tabby		*Pseudergolis wedah* (Kollar)	（チョウ目、タテハチョウ科）旧北区、東洋区
tabby knot-horn		*Euzophera pinguis* Haworth	（チョウ目、メイガ科）旧北区
tabby moth			small tabby を見よ
tabby moth			large tabby を見よ
Tabebuia long-horned beetle			artocarpus long-horned beetle を見よ
Tabitha's swordtail	ドルクスオオオナガタイマイ	*Graphium dorcus* (de Haan)	（チョウ目、アゲハチョウ科）東洋区
Table Mountain beauty		*Aeropetes tulbaghia* (Linnaeus)	（チョウ目、タテハメチョウ科）エチオピア区
tableland pasture scarab		*Antitrogus morbillosus* (Blackburn)	（コウチュウ目、コガネムシ科）豪州区
Tabora swallowtail		*Graphium taboranus* (Oberthür)	（チョウ目、アゲハチョウ科）エチオピア区
tachina flies	ヤドリバエ科	Tachinidae	（ハエ目）の昆虫の総称　tachinids ともいう。世界に 6,000 種以上
tachina fly		*Gonia porca* Williston	（ハエ目、ヤドリバエ科）新北区
tachinid flies			tachina flies を見よ
tachinid-fly		*Tachina fera* Linnaeus	（ハエ目、ヤドリバエ科）旧北区
tacitum wood cricket		*Gryllus ovisopis* Walker	（バッタ目、コオロギ科）新北区
taeniostigmatids		Taeniostigmatidae	（ハチ目）の昆虫の総称
Tagyra hairstreak		*Evenus tagyra* (Hewitson)	（チョウ目、シジミチョウ科）新熱帯区
Tahiti coconut weevil		*Diocalandra tahitense* (Guérin-Méneville)	（コウチュウ目、オサゾウムシ科）東洋区、新北区、エチオピア区。Tahitian coconut weevil ともいう
Tahoe wingless stonefly		*Capnia lacustra* Jewett	（カワゲラ目、クロカワゲラ科）新北区。カリフォルニアの Tahoe 湖に由来
taiga alpine		*Erebia mancinus* Doubleday	（チョウ目、タテハチョウ科）新北区
taiga bluet		*Coenagrion resolutum* (Hagen)	（トンボ目、イトトンボ科）新北区
tail louse			cattle-tail louse を見よ
tail switch louse			cattle-tail louse を見よ　豪州での英名
tailed Aguna		*Aguna metophis* (Latreille)	（チョウ目、セセリチョウ科）新北区、新熱帯区
tailed bark beetle	ナガオキクイムシ	*Xyleborus calamoides* Murayama	（コウチュウ目、キクイムシ科）日本
tailed black-eye		*Leptomyrina hirundo* (Wallengren)	（チョウ目、シジミチョウ科）エチオピア区
tailed blue			pea blue を見よ

英名	和名	学名	所属、分布、ほか
tailed bush brown		*Bicyclus sambulos* (Hewitson)	(チョウ目、タテハチョウ科) エチオピア区
tailed caterpillar		*Epicampoptera andersoni* Tams	(チョウ目、カギバガ科) エチオピア区
tailed caterpillar		*Epicampoptera marantica* Tams	(チョウ目、カギバガ科) エチオピア区
tailed caterpillars		*Epicampoptera*	(チョウ目、カギバガ科) の昆虫の総称
tailed cecropian			cadmus を見よ
tailed coffee mealybug			striped mealybug を見よ
tailed copper	オナガベニシジミ	*Lycaena arota* (Boisduval)	(チョウ目、シジミチョウ科) 新北区、新熱帯区
tailed cupid		*Elkalyce argiades* (Pallas)	(チョウ目、シジミチョウ科) 東洋区
tailed cupid (1)		*Everes lacturnus* (Godart)	(チョウ目、シジミチョウ科) 東洋区、豪州区。*E. l. australis* Couchman も同英名
tailed cupid			Indian cupid (1) を見よ　豪州での英名
tailed cupid			short-tailed blue を見よ
tailed disc oakblue		*Arhopala atosia* (Hewitson)	(チョウ目、シジミチョウ科) 東洋区。*A. a. malayana* Bethune-Baker も同英名
tailed emperor	ミナミフタオチョウ	*Polyura pyrrhus* (Linnaeus)	(チョウ目、タテハチョウ科) 豪州区
tailed emperor butterfly		*Polyura sempronius* (Fabricius)	(チョウ目、タテハチョウ科) 豪州区。tailed emperor ともいう
tailed geometric (1)	シロツバメエダシャク	*Ourapteryx maculicaudaria* (Motschulsky)	(チョウ目、シャクガ科) 日本、旧北区
tailed geometric (2)	ウスキツバメエダシャク	*Ourapteryx nivea* Butler	(チョウ目、シャクガ科) 日本
tailed giant sulphur		*Phoebis neocypris* (Hübner)	(チョウ目、シロチョウ科) 新北区、新熱帯区
tailed green-barded blue		*Danis cyanea* (Cramer)	(チョウ目、シジミチョウ科) 豪州区
tailed green-barded blue			green-banded line-blue を見よ
tailed green-barded swallowtail	カロプスルリアゲハ	*Papilio charopus* Westwood	(チョウ目、アゲハチョウ科) エチオピア区
tailed jay	コモンタイマイ	*Graphium agamemnon* (Linnaeus)	(チョウ目、アゲハチョウ科) 東洋区、豪州区
tailed jewelmark		*Sarota chrysus* (Stoll)	(チョウ目、シジミタテハ科) 新熱帯区
tailed Judy	シロオビオナガシジミタテハ	*Abisara neophron* (Hewitson)	(チョウ目、シジミタテハ科) 東洋区
tailed labyrinth		*Neope bhadra* (Moore)	(チョウ目、タテハチョウ科) 東洋区
tailed meadow blue		*Cupidopsis jobates* (Hopffer)	(チョウ目、シジミチョウ科) エチオピア区
tailed moth	ウスバツバメガ	*Elcysma westwoodii* (Vollenhoven)	(チョウ目、マダラガ科) 日本、旧北区。モモ、ウメ、サクラなどの害虫
tailed net-winged beetle		*Lycus trabeatus* Guérin-Méneville	(コウチュウ目、ベニボタル科) エチオピア区
tailed orange		*Eurema proterpia* (Fabricius)	(チョウ目、シロチョウ科) 新北区

英名	和名	学名	所属、分布、ほか
tailed punch	シジミタテハ	*Dodona eugenes* Bates	(チョウ目、シジミタテハ科) 東洋区。punch ともいう
tailed redbreast		*Papilio bootes* Westwood	(チョウ目、アゲハチョウ科) 東洋区
tailed redbreast		*Papilio janaka* Moore	(チョウ目、アゲハチョウ科) 東洋区
tailed rustic			Australian vagrant を見よ
tailed sulphur	トガリキチョウ	*Dercas verhuelli* (van der Hoeven)	(チョウ目、シロチョウ科) 東洋区
tailed sulphur			tailed giant sulphur を見よ
tailed switch louse			cattle-tail louse を見よ
tailed wasps			petiolate hymenoptera を見よ
tailless bushblue		*Panchala ganesa* (Moore)	(チョウ目、シジミチョウ科) 東洋区。日本亜種は *P. g. loomisi* (H. Pryer) ルーミスシジミ
tailless line blue			small purple line blue を見よ
tailless metallic green hairstreak	カシミドリシジミ	*Chrysozephyrus khasia* (de Nicéville)	(チョウ目、シジミチョウ科) 東洋区。tailless metallic hairstreak ともいう
tailless plushblue		*Flos areste* (Hewitson)	(チョウ目、シジミチョウ科) 旧北区、東洋区
tailless scrub-hairstreak			Cestri hairstreak を見よ
tailless swallowtail			Polydamas swallowtail を見よ
tailor ants			weaver ants を見よ
Taiwan mulberry psyllid	タイワンキジラミ	*Paurocephala psylloptera* Crowford	(カメムシ目、キジラミ科) 日本、東洋区
Taiwanese bamboo carpenter bee		*Xylocopa (Biluna) tranquebarorum* (Swederus)	(ハチ目、ミツバチ科) 東洋区
Taiwanese stingless bee	タイワンハリナシバチ	*Trigona ventralis hoozana* Strand	(ハチ目、ミツバチ科) 東洋区
Takahashi false cottony scale	タカハシワタカイガラモドキ	*Heliococcus takahashii* Kanda	(カメムシ目、コナカイガラムシ科) 日本
Takahashi weevil	タカハシトゲゾウムシ	*Dinorhopala takahashii* (Kono)	(コウチュウ目、ゾウムシ科) 日本
Takamuku lasiocampid	タカムクカレハ	*Cosmotriche lunigera takamukuana* (Matsumura)	(チョウ目、カレハガ科) 日本
Talaud black birdwing		*Troides dohertyi* Rippon	(チョウ目、アゲハチョウ科) 東洋区
Talbot's ciliate blue		*Anthene talboti* Stempffer	(チョウ目、シジミチョウ科) エチオピア区
tallow beetles			larder beetles を見よ
tamarac sawfly			larch sawfly (2) を見よ　米国での英名
tamarac tree cricket		*Oecanthus laricis* Walker	(バッタ目、カンタン科) 新北区
tamarind seed weevil			groundnut bruchid を見よ
tamarind weevil		*Bruchus gonager* Fabricius	(コウチュウ目、ハムシ科) 東洋区
tamarind weevil		*Sitophilus linearis* (Herbst)	(コウチュウ目、オサゾウムシ科) 新北区
tamarind weevil			groundnut bruchid を見よ
tamarindi owlet		*Opsiphanes tamarindi* Felder	(チョウ目、タテハチョウ科) 新北区、新熱帯区

英名	和名	学名	所属、分布、ほか
tamarisk manna scale		*Trabutina mannipara* (Ehrenberg)	（カメムシ目、タマカイガラムシ科）旧北区。マンナをとる
tamarisk plume		*Agdistis tamaricis* (Zeller)	（チョウ目、トリバガ科）旧北区
tamarisk pug		*Eupithecia tamarisciata* Freyer	（チョウ目、シャクガ科）旧北区
tamarisk scale		*Rugaspidiotus tamaricicola* (Malenotti)	（カメムシ目、マルカイガラムシ科）旧北区
tamarix leafhopper		*Opsius stactogalus* Fieber	（カメムシ目、オオヨコバイ科）新北区
Tamaulipas clubtail		*Gomphus gonzalezi* Dunkle	（トンボ目、サナエトンボ科）新北区
Tamaulipas skipper		*Parphorus* sp.	（チョウ目、セセリチョウ科）新熱帯区
tambo cricket	タンボコオロギ	*Velarifictorus parvus* (Chopard)	（バッタ目、コオロギ科）日本、東洋区
Tamil ace		*Thoressa sitala* (de Nicéville)	（チョウ目、セセリチョウ科）東洋区
Tamil bushbrown		*Mycalesis visala subdita* Moore	（チョウ目、タテハチョウ科）東洋区
Tamil catseye		*Zipaetis saitis* Hewitson	（チョウ目、タテハチョウ科）東洋区
Tamil dartlet		*Oriens concinna* (Elwes et Edwards)	（チョウ目、セセリチョウ科）東洋区
Tamil grass dart		*Taractrocera ceramas* (Hewitson)	（チョウ目、セセリチョウ科）東洋区
Tamil lacewing		*Cethosia nietneri* C. et R. Felder	（チョウ目、タテハチョウ科）東洋区
Tamil oakblue		*Arhopala bazaloides* (Hewitson)	（チョウ目、シジミチョウ科）東洋区
Tamil peacock		*Papilio paris tamilana* Moore	（チョウ目、アゲハチョウ科）東洋区。Paris peacock を参照
Tamil spotted flat		*Celaenorrhinus ruficornis* (Mabille)	（チョウ目、セセリチョウ科）東洋区
Tamil treebrown		*Lethe drypetis* (Hewitson)	（チョウ目、タテハチョウ科）東洋区
Tamil Yeomen	インドネッタイヒョウモン	*Cirrochroa thais* (Fabricius)	（チョウ目、タテハチョウ科）東洋区
Tamos hairstreak		*Calycopis tamos* (Godman et Salvin)	（チョウ目、シジミチョウ科）新熱帯区
tampa moth		*Tampa dimediatella* Ragonot	（チョウ目、メイガ科）新北区
tanacetum root moth		*Dichrorampha vancouverana* McDunnough	（チョウ目、ハマキガ科）全北区
Tanais hairstreak		*Symbiopsis tanais* (Godman et Salvin)	（チョウ目、シジミチョウ科）新熱帯区
tanaostigmatids	マメトビコバチ科	Tanaostigmatidae	（ハチ目）の昆虫の総称　Eupelmidae ナガコバチ科の異名
tanbark beetle	チャイロホソヒラタカミキリ	*Phymatodes testaceus* (Linnaeus)	（コウチュウ目、カミキリムシ科）日本、全北区
tanbark borer			blue-winged dry-wooden longicorn を見よ
tangle-veined flies	ツリアブモドキ科	Nemestrinidae	（ハエ目）の昆虫の総称　tanglevein flies ともいう

英名	和名	学名	所属、分布、ほか
tangle-wing fly		*Neorhynchocephalus sackenii* (Williston)	(ハエ目、ツリアブモドキ科) 新北区、新熱帯区
tanguru chafer		*Stethaspis suturalis* Fabricius	(コウチュウ目、コガネムシ科) 豪州区
Tanna longtail		*Urbanus tanna* Evans	(チョウ目、セセリチョウ科) 新北区、新熱帯区。Tanna ともいう
tanned blue-skipper		*Quadrus lugubris* (Felder)	(チョウ目、セセリチョウ科) 新北区、新熱帯区
tanned hoary-skipper		*Carrhenes fuscescens fuscescens* (Mabille)	(チョウ目、セセリチョウ科) 新熱帯区
tanner beetle		*Prionus coriarius* (Linnaeus)	(コウチュウ目、カミキリムシ科) 旧北区
tansy aphid		*Dactynotus tanaceti* (Linnaeus)	(カメムシ目、アブラムシ科) 旧北区
tansy beetle		*Chrysolina graminis* (Linnaeus)	(コウチュウ目、ハムシ科) 旧北区
tansy leaf beetle		*Galeruca tanaceti* (Linnaeus)	(コウチュウ目、ハムシ科) 旧北区
tantale sphinx		*Aellopos tantalus* (Linnaeus)	(チョウ目、スズメガ科) 新北区。Tantalus sphinx moth, Tantalus sphinx ともいう
tanypezid flies		Tanypezidae	(ハエ目) の昆虫の総称
Tanzanian fiery Acraea		*Acraea utengulensis* Thurau	(チョウ目、タテハチョウ科) エチオピア区
tapestry beetle			black carpet beetle (1) を見よ
tapestry clothes moth			carpet moth を見よ
tapestry moth			carpet moth を見よ
tapinoma ants	カタアリ亜科	Dolichoderinae	(ハチ目、アリ科) の昆虫の総称
tarantula hawk	オオベッコウバチ	*Pepsis formosa* (Say)	(ハチ目、ベッコウバチ科) 新北区
tarantula hawks		*Pepsis*	(ハチ目、ベッコウバチ科) の昆虫の総称
tarata looper moth		*Epirrhanthis ustaria* Walker	(チョウ目、シャクガ科) 豪州区
tardy Renia			chocolate Renia moth を見よ
tarnished plant bug (1)	サビイロカスミカメムシ	*Lygus lineolaris* (Palisot et Beauvois)	(カメムシ目、カスミカメムシ科) 新北区
tarnished plant bug (2)	ミドリカスミカメムシ	*Lygus pratensis* Linnaeus	(カメムシ目、カスミカメムシ科) 日本。イネ、ダイズ、ウリ類、キクなどの害虫
tarnished plant bug			pasture leaf bug を見よ
tarnished plant bug			common green capsid (1) を見よ
tarnished plant bugs			lygus bugs を見よ
taro delphacid		*Tarophagus colocasiae* (Matsumura)	(カメムシ目、ウンカ科) 大洋区
taro delphacid egg parasite			taro leafhopper egg parasite を見よ
taro leafhopper			taro planthopper を見よ
taro leafhopper egg parasite		*Ootetrastichus megameli* Fullaway	(ハチ目、ヒメコバチ科) 大洋区
taro planthopper	タロイモウンカ	*Tarophagus proserpina* (Kirkaldy)	(カメムシ目、ウンカ科) 日本、東洋区、大洋区、豪州区

英名	和名	学名	所属、分布、ほか
taro root aphid		*Patchiella reamuri* (Kaltenbach)	(カメムシ目、アブラムシ科) 大洋区
Tarricina tiger		*Tithorea tarricina* Hewitson	(チョウ目、タテハチョウ科) 新熱帯区
Tasar silk insect		*Antheraea mylitta* Drury	(チョウ目、ヤママユガ科) 東洋区
Tasmania skipper		*Pasma tasmanica* (Miskin)	(チョウ目、セセリチョウ科) 豪州区
Tasmanian brown		*Argynnina hobartia* (Westwood)	(チョウ目、タテハチョウ科) 豪州区
Tasmanian cushion plant moth		*Nemotyla oribates* Nielsen, McQuillan et Common	(チョウ目、マルハキバガ科) 豪州区
Tasmanian cutworm		*Dasygaster padockina* (Le Guillou)	(チョウ目、ヤガ科) 豪州区
Tasmanian darner		*Austroaeschna tasmanica* Tillyard	(トンボ目、ヤンマ科) 豪州区
Tasmanian devil			devil's coach-horse (1) を見よ
Tasmanian eucalyptus leaf beetle		*Chrysophtharta bimaculata* (Olivier)	(コウチュウ目、ハムシ科) 豪州区
Tasmanian grass grub beetle		*Acrossidius tasmaniae* (Hope)	(コウチュウ目、コガネムシ科) 豪州区
Tasmanian hairy cicada		*Tettigarcta tomentosa* White	(カメムシ目、Tettigarctidae) 豪州区
Tasmanian inchman		*Myrmecia esuriens* Smith	(ハチ目、アリ科) 豪州区
Tasmanian redspot		*Archipetalia auriculata* Tillyard	(トンボ目、Archipetaliidae) 豪州区
Tasmanian swamp tigertail		*Synthemis tasmanica* Tillyard	(トンボ目、Synthemistidae) 豪州区
Tasmanian Xenica			Leprea brown を見よ
tassah silkworm			tassar silkworm を見よ
tassar silkworm		*Antheraea paphia* (Linnaeus)	(チョウ目、ヤママユガ科) 東洋区
tau emerald		*Hemicordulia tau* Selys	(トンボ目、エゾトンボ科) 豪州区
tau emperor		*Aglia tau* (Linnaeus)	(チョウ目、ヤママユガ科) 旧北区。斑紋がギリシャ語のタウに似ることから
Tauscher's alpine ringlet			Theano alpine を見よ
Tavela silverline		*Cigaritis tavetensis* (Lathy)	(チョウ目、シジミチョウ科) エチオピア区
tawny angle		*Ctenoptilum vasava* (Moore)	(チョウ目、セセリチョウ科) 東洋区
tawny-barred angle		*Chiasmia liturata* (Clerck)	(チョウ目、シャクガ科) 旧北区。日本亜種は *C. l. pressaria* (Christoph) チャオビエダシャク
tawny burrowing bee			tawny mining bee を見よ
tawny cockroach		*Ectobius pallidus* (Olivier)	(ゴキブリ目、チャバネゴキブリ科) 旧北区
tawny cockroach (1)		*Ectobius lapponicus* (Linnaeus)	(ゴキブリ目、チャバネゴキブリ科) 全北区。欧州から米国に入る
tawny cockroach (2)		*Ectobius lucidus* (Hagenbach)	(ゴキブリ目、チャバネゴキブリ科) 旧北区
tawny-cornered satyr		*Euptychia hilara* (C. et R. Felder)	(チョウ目、タテハチョウ科) 新熱帯区

英名	和名	学名	所属、分布、ほか
tawny cosmet		*Mompha epilobiella* (Roemer)	（チョウ目、カザリバガ科）旧北区
tawny coster		*Acraea terpsicore* (Linnaeus)	（チョウ目、タテハチョウ科）東洋区
tawny coster		*Acraea violae* (Fabricius)	（チョウ目、タテハチョウ科）東洋区
tawny crescent		*Phyciodes batesii* (Reakirt)	（チョウ目、タテハチョウ科）新北区
tawny-dotted marble		*Olethreutes mygindiana* (Denis et Schiffermüller)	（チョウ目、ハマキガ科）旧北区
tawny earwig			common brown earwig (1) を見よ
tawny-edged skipper		*Polites themistocles* (Latreille)	（チョウ目、セセリチョウ科）新北区
tawny emperor		*Chitoria ulupi* (Doherty)	（チョウ目、タテハチョウ科）旧北区、東洋区
tawny emperor (1)	キマダラエノキタテハ	*Asterocampa clyton* (Boisduval et LeConte)	（チョウ目、タテハチョウ科）新北区。新熱帯区の *A. c. texana* (Skinner) も同英名
tawny Eupithecia moth		*Eupithecia ravocostaliata* Packard	（チョウ目、シャクガ科）新北区。tawny Eupithecia ともいう
tawny Holomelina		*Holomelina opella* (Grote)	（チョウ目、ヒトリガ科）新北区。tawny Holomelina moth ともいう
tawny longhorn beetle		*Paracorymbia fulva* (De Geer)	（コウチュウ目、カミキリムシ科）旧北区
tawny marbled minor			tawny minor を見よ
tawny meadowbrown		*Hyponephele pulchra* (C. et R. Felder)	（チョウ目、タテハチョウ科）東洋区
tawny metalmark		*Notheme erota diadema* Stichel	（チョウ目、シジミタテハ科）新熱帯区。Erota metalmark を参照
tawny mime	カバシタアゲハ	*Chilasa agestor* (Gray)	（チョウ目、アゲハチョウ科）東洋区
tawny mining bee		*Andrena armata* (Gmelin)	（ハチ目、ヒメハナバチ科）旧北区
tawny minor		*Oligia latruncula* (Denis et Schiffermüller)	（チョウ目、ヤガ科）旧北区
tawny mole cricket			changa を見よ
tawny palmfly		*Elymnias panthera* (Fabricius)	（チョウ目、タテハチョウ科）東洋区
tawny pennant		*Brachymesia herbida* (Gundlach)	（トンボ目、トンボ科）新北区、新熱帯区
tawny pinion		*Lithophane semibrunnea* (Haworth)	（チョウ目、ヤガ科）旧北区
tawny prominent		*Harpyia milhauseri* (Fabricius)	（チョウ目、シャチホコガ科）旧北区
tawny prominent (1)		*Nadata gibbosa* (J. E. Smith)	（チョウ目、シャチホコガ科）新北区
tawny Rajah	チャイロフタオチョウ	*Charaxes bernardus* (Fabricius)	（チョウ目、タテハチョウ科）旧北区、東洋区
tawny rockbrown			dark grayling を見よ
tawny sanddragon		*Progomphus alachuensis* Byers	（トンボ目、サナエトンボ科）新北区
tawny shears			pod lover を見よ

英名	和名	学名	所属、分布、ほか
tawny shoulder			granulate cutworm を見よ
tawny silverline		*Cigaritis acamas* (Klug)	(チョウ目、シジミチョウ科) 旧北区、東洋区、エチオピア区
tawny-speckled pug		*Eupithecia icterata* (Villers)	(チョウ目、シャクガ科) 旧北区
tawny spreadwing		*Lestes ictericus* Gerstäcker	(トンボ目、アオイトトンボ科) エチオピア区
tawny-striped sister		*Adelpha leucerioides leucerioides* Beutelspacher	(チョウ目、タテハチョウ科) 新熱帯区
tawny treble-barred midget		*Phyllonorycter trifasciella* (Haworth)	(チョウ目、ホソガ科) 旧北区
tawny wave		*Scopula rubiginata* (Hufnagel)	(チョウ目、シャクガ科) 旧北区
Taxiles skipper			golden skipper をみよ
taxus mealybug			pear mealybug を見よ
Taygetos blue			Eugene's blue を見よ
tea amatid	キハダカノコ	*Amata germana nigricauda* (Miyake)	(チョウ目、カノコガ科) 日本、旧北区、東洋区
tea aphid			black citrus aphid を見よ　アジアでの英名
tea bagworm	チャミノガ	*Eumeta minuscula* Butler	(チョウ目、ミノガ科) 日本、旧北区。チャはじめ多くの果樹、植物の害虫
tea black scale			tea scale (1) を見よ
tea bunch caterpillar		*Andraca bipunctata* Walker	(チョウ目、カイコガ科) 東洋区。チャ害虫
tea cochlid	アカイラガ	*Phrixolepia sericea* Butler	(チョウ目、イラガ科) 日本、旧北区。リンゴ、ウメ、サクラ、カエデなどの害虫
tea flush worm			leaf longitudinal roller を見よ
tea flushworm			tea tortrix を見よ
tea geometrid	チャエダシャク	*Megabiston plumosaria* (Leech)	(チョウ目、シャクガ科) 日本。チャ、カンキツ、サクラなどの害虫
tea geometrid (1)	チャノウンモンエダシャク	*Jankowskia fuscaria* (Leech)	(チョウ目、シャクガ科) 日本。チャ、ツバキの害虫
tea green fly			green leafhopper (1) を見よ
tea green leafhopper	チャノミドリヒメヨコバイ	*Empoasca onukii* Matsuda	(カメムシ目、オオヨコバイ科) 日本。チャ、カンキツ害虫
tea green leafhopper			green leafhopper (1) を見よ
tea ground mealybug	チャノネコナカイガラムシ	*Rhizoecus theae* Kawai et Takagi	(カメムシ目、コナカイガラムシ科) 日本。チャ害虫
tea leafminer	チャノハモグリバエ	*Tropicomyia theae* (Cotes)	(ハエ目、ハモグリバエ科) 日本、旧北区、エチオピア区。チャ、ツバキ害虫
tea leaf miner			tea leafroller を見よ
tea leafroller	チャノホソガ	*Caloptilia theivora* (Walsingham)	(チョウ目、ホソガ科) 日本、東洋区。チャ、ツバキ害虫
tea mosquito bug		*Holopeltis theivora* Waterhouse	(カメムシ目、カスミカメムシ科) 東洋区。tea mosquito ともいう
tea moth		*Parametriotes theae* Kuznetsov	(チョウ目、Momphidae) 旧北区
tea nettle grub			tea cochlid を見よ
tea parlatoria scale			tea scale (1) を見よ

英名	和名	学名	所属、分布、ほか
tea pyralid	ギンモンシマメイガ	*Pyralis regalis* (Denis et Schiffermüller)	（チョウ目、メイガ科）日本
tea root borer		*Xyleborus germanus* Blandford	（コウチュウ目、キクイムシ科）日本、全北区、東洋区
tea root weevil		*Aperitmetus brunneus* (Hustache)	（コウチュウ目、ゾウムシ科）エチオピア区
tea scale	チャコノハカイガラムシ	*Fiorinia theae* Green	（カメムシ目、マルカイガラムシ科）日本、東洋区。チャ害虫
tea scale (1)	チャクロホシカイガラムシ	*Parlatoria theae* Cockerell	（カメムシ目、マルカイガラムシ科）日本、汎世界。チャ、ツバキ、ナシ、バラ、カエデなどの害虫
tea seed bug		*Poecilocoris nepalensis* Herrich-Schäffer	（カメムシ目、カメムシ科）東洋区
tea seed fly		*Adrama determinata* Walker	（ハエ目、ミバエ科）東洋区
tea shot-hole borer		*Xyleborus fornicatus* (Eichhoff)	（コウチュウ目、キクイムシ科）東洋区、エチオピア区
tea stem borer			castanopsis ambrosia beetle (2) を見よ
tea tabby			black rice worm を見よ　英国での英名
tea thrips		*Scirtothrips bispinosus* (Bagnall)	（アザミウマ目、アザミウマ科）東洋区
tea thrips		*Lefroyothrips lefroyi* (Bagnall)	（アザミウマ目、アザミウマ科）東洋区
tea thyridid	アミメマドガ	*Striglina suzukii* Matsumura	（チョウ目、マドガ科）日本。チャ、カキ害虫
tea tortrix		*Homona coffearia* Nietner	（チョウ目、ハマキガ科）東洋区
tea tussock moth	チャドクガ	*Euproctis pseudoconspersa* (Strand)	（チョウ目、ドクガ科）日本、東洋区。チャ、ツバキ、カキ害虫
tea tussock moth		*Dasychira mendosa* Hübner	（チョウ目、ドクガ科）東洋区、豪州区
tea unmon geometrid			tea geometrid (1) を見よ
teak beehole borer			beehole borer を見よ
teak defoliator moth	キオビセセリモドキ	*Hyblaea puera* (Cramer)	（チョウ目、セセリモドキガ科）日本、東洋区、大洋区、新北区、エチオピア区。teak defoliator, teak moth ともいう
teak-leaf gall-maker			foreign grain beetle を見よ　南アフリカでの英名
teak leaf skeletonizer		*Eutectona machaeralis* Walker	（チョウ目、メイガ科）東洋区
teak moths	セセリモドキガ科	Hyblaeidae	（チョウ目）の昆虫の総称
teak sapling borer		*Sahyadrassus malabaricus* Moore	（チョウ目、コウモリガ科）東洋区
teak termite		*Calotermes tectonae* Dammerman	（シロアリ目、レイビシロアリ科）東洋区
tear-drop skipper		*Eracon paulinus* (Stoll)	（チョウ目、セセリチョウ科）新熱帯区
tearful underwing moth		*Catocala lacrymosa* Guenée	（チョウ目、ヤガ科）新北区。tearful underwing ともいう
teatree moth		*Catamola marmorea* (Warren)	（チョウ目、メイガ科）豪州区

英名	和名	学名	所属、分布、ほか
teatree web moth		*Catamola thyrisalis* Walker	(チョウ目、メイガ科) 豪州区
teazle long-cloaked marble	ツマキヒメハマキ	*Endothenia gentianaeana* (Hübner)	(チョウ目、ハマキガ科) 日本、旧北区
tecate cypress hairstreak		*Mitoura thornei* Brown	(チョウ目、シジミチョウ科) 新北区
teddy bear bee		*Amegilla bombiformis* Smith	(ハチ目、ミツバチ科) 豪州区
tee-banded Euselasia		*Euselasia eurypus* (Hewitson)	(チョウ目、シジミタテハ科) 新熱帯区
tegula Emesis			dark falcate Emesis を見よ
Tehachapi shieldback		*Idionotus tehachapi* Hebard	(バッタ目、キリギリス科) 新北区
Tehuacan cloudywing		*Achalarus tehuacana* (Draudt)	(チョウ目、セセリチョウ科) 新熱帯区
Teita glider		*Cymothoe teita* van Someren	(チョウ目、タテハチョウ科) エチオピア区
Telassa firetip	アカモンスソアカセセリ	*Pyrrhopyge telassa* Hewitson	(チョウ目、セセリチョウ科) 新熱帯区
Telata skipper		*Monca telata* (Herrich-Schäffer)	(チョウ目、セセリチョウ科) 新熱帯区
Telea hairstreak		*Chlorostrymon telea* (Hewitson)	(チョウ目、シジミチョウ科) 新北区、新熱帯区
telegeusid beetles		Telegeusidae	(コウチュウ目) の昆虫の総称
Telemus hairstreak		*Paiwarria telemus* (Cramer)	(チョウ目、シジミチョウ科) 新熱帯区
telephone pile beetles	チビナガヒラタムシ科	Micromalthidae	(コウチュウ目) の昆虫の総称
Telesiphe longwing		*Heliconius telesiphe* Doubleday	(チョウ目、タテハチョウ科) 新熱帯区
Teleus longtail		*Urbana teleus* (Hübner)	(チョウ目、セセリチョウ科) 新北区、新熱帯区
telini flies			blister beetles (3) を見よ
telini fly		*Mylabris cichorii* (Linnaeus)	(コウチュウ目、ツチハンミョウ科) 旧北区
telini fly		*Mylabris floralis* Pallas	(コウチュウ目、ツチハンミョウ科) 旧北区
Tellane white		*Leodonta tellane* (Hewitson)	(チョウ目、シロチョウ科) 新熱帯区
telson-tails			proturans を見よ
Temesa hairstreak		*Iaspis temesa* (Hewitson)	(チョウ目、シジミチョウ科) 新熱帯区
temperature cricket			snowy tree cricket を見よ
ten-dotted long-corn beetle	トホシカミキリ	*Saperda alberti* Plavistchikov	(コウチュウ目、カミキリムシ科) 日本、旧北区
ten-lined giant chafer			ten-lined June beetle を見よ
ten-lined June beetle		*Polyphylla decemlineata* (Say)	(コウチュウ目、コガネムシ科) 新北区
ten-lined potato beetle			Colorado potato beetle を見よ
ten-spot		*Libellula pulchella* Drury	(トンボ目、トンボ科) 新北区

英名	和名	学名	所属、分布、ほか
ten-spotted honeysuckle moth			dotted grey groundling を見よ
ten-spotted lady beetle		*Coelophora pupillata* (Swartz)	(コウチュウ目、テントウムシ科) 新北区。ten-spotted ladybug ともいう
ten-spotted lady beetle			cucumber-feeder ladybeetle を見よ
ten-spotted ladybird		*Adalia decempunctata* (Linnaeus)	(コウチュウ目、テントウムシ科) 旧北区。ten-spot ladybird ともいう
ten-spotted leaf beetle	クロモンハムシ	*Gonioctena springlovae* (Bechyne)	(コウチュウ目、ハムシ科) 日本
ten-spotted leaf-cutting bee			wool carder bee を見よ
ten-spotted lema	トホシクビボソハムシ	*Lema decempunctata* Gebler	(コウチュウ目、ハムシ科) 日本、旧北区
ten-spotted pot beetle		*Cryptocephalus decemmaculatus* (Linnaeus)	(コウチュウ目、ハムシ科) 旧北区
ten-spotted psocid		*Peripsocus californicus* (Banks)	(チャタテムシ目、マドチャタテ科) 新北区
ten-spotted stink bug	トホシカメムシ	*Lelia decempunctata* (Motschulsky)	(カメムシ目、カメムシ科) 日本、旧北区
ten-striped spearman			Colorado potato beetle を見よ
Tennessee clubtail		*Gomphus sandrius* Tennessen	(トンボ目、サナエトンボ科) 新北区
tenspot skimmer			ten-spot を見よ
tendu defoliator	インドキシタクチバ	*Hypocala rostrata* (Fabricius)	(チョウ目、ヤガ科) 日本、東洋区、エチオピア区。カキノキ科の植物の地方名に由来
tenebrionid beetles			darkling beetles (1) を見よ
tent caterpillar	オビカレハ	*Malacosoma neustria testacea* (Motschulsky)	(チョウ目、カレハガ科) 日本。ウメ、リンゴなど果樹、バラ、サクラなどの害虫
tent caterpillar		*Taragama repanda* (Hübner)	(チョウ目、カレハガ科) 旧北区。カンキツ害虫
tent caterpillar			forest tent caterpillar を見よ
tent caterpillar			lackey を見よ
tent caterpillar moths (1)	カレハガ科	Lasiocampidae	(チョウ目) の昆虫の総称　世界に 1,500 種
tent caterpillar moths		*Malacosoma*	(チョウ目、カレハガ科) の昆虫の総称　幼虫はテンマクケムシ tent caterpillar
tent caterpillars			tent caterpillar moths (1) を見よ
tentacled prominent		*Cerura scitiscripta multiscripta* Riley	(チョウ目、シャチホコガ科) 新北区
tentacled prominents			American puss moths を見よ　北米での英名
tenthredinid sawflies			sawflies (1) を見よ
Tephraeus hairstreak		*Siderus tephraeus* (Geyer)	(チョウ目、シジミチョウ科) 新北区、新熱帯区
Tepper's skipper		*Motasingha trimaculata* (Tepper)	(チョウ目、セセリチョウ科) 豪州区
Tequila giant-skipper		*Aegiale hesperiaris* (Walker)	(チョウ目、セセリチョウ科) 新北区、新熱帯区

英名	和名	学名	所属、分布、ほか
terebrant thrips	アザミウマ亜目	Terebrantia	(アザミウマ目) の昆虫の総称
terebrantian thrips			western flower thrips (2) を見よ
termatophylid bugs		Termatophylidae	(カメムシ目) の昆虫の総称
termite			cathedral termite を見よ
termites	シロアリ目	Isoptera	の昆虫の総称　世界に約 2,200 種
ternstroemia sucker	モッコクトガリキジラミ	*Trioza ternstroemiae* Matsumoto	(カメムシ目、トガリキジラミ科) 日本。モッコク害虫
terrapin scale		*Mesolecanium nigrofasciatum* (Pergande)	(カメムシ目、カタカイガラムシ科) 新北区
terrenella bee moth		*Aphomia terrenella* Zeller	(チョウ目、メイガ科) 新北区
terrestrial bugs			land bugs を見よ
terrestrial satyr			Butler's ringlet を見よ
terrestrial turtle bugs		Podopinae	(カメムシ目、カメムシ科) の昆虫の総称
territorial ant		*Azteca trigona* Emery	(ハチ目、アリ科) 新熱帯区
tersa sphinx		*Xylophanes tersa* (Drury)	(チョウ目、スズメガ科) 新北区。北米での英名。tersa sphinx moth ともいう
tesselated skipper		*Muschampia tessellum* (Hübner)	(チョウ目、セセリチョウ科) 旧北区
tessellate bush katydid		*Insara tessellata* Hebard	(バッタ目、キリギリス科) 新北区
tessellate dart moth			striped cutworm moth を見よ
tessellated conch		*Aethes tesserana* (Denis et Schiffermüller)	(チョウ目、ホソハマキガ科) 旧北区
tessellated phasmatid		*Ctenomorphodes tessulatus* (Gray)	(ナナフシ目、ナナフシ科) 豪州区
tessellated scale	カメノコロウカタカイガラムシ	*Eucalymnatus tessellatus* (Signoret)	(カメムシ目、カタカイガラムシ科) 日本 (小笠原)、新北区、汎熱帯。樹木害虫
tessellated shieldback		*Tessellana tessellata* (Charpentier)	(バッタ目、キリギリス科) 新北区
Tessmann's forester		*Bebearia tessmanni* (Grunberg)	(チョウ目、タテハチョウ科) エチオピア区
tethinid flies		Tethinidae	(ハエ目) の昆虫の総称
tetramerous beetles		Tetramera	(コウチュウ目) の昆虫の総称　tetramerans ともいう
tetrastichid parasites		Tetrastichidae	(ハチ目) の昆虫の総称
tetrastigma skipper		*Zera tetrastigma* (Sepp)	(チョウ目、セセリチョウ科) 新熱帯区
Tetrio sphinx moth			giant grey sphinx を見よ　Tetrio sphinx ともいう
tettigonoid grasshoppers			long-horned grasshoppers (3) を見よ
Teucer owl butterfly		*Caligo teucer* (Linnaeus)	(チョウ目、タテハチョウ科) 新熱帯区。Teucer giant owl ともいう
Texa eighty-eight			Texa numberwing を見よ
Texa numberwing		*Callicore texa* (Hewitson)	(チョウ目、タテハチョウ科) 新熱帯区
Texan crescent			Texan crescentspot を見よ
Texan crescentspot		*Anthanassa texana* (Edwards)	(チョウ目、タテハチョウ科) 新北区

英名	和名	学名	所属、分布、ほか
Texana clearwing moth		*Carmenta texana* (Edwards)	（チョウ目、スカシバガ科）新北区
Texas angle-wing		*Microcentrum minus* Strohecker	（バッタ目、キリギリス科）新北区
Texas beetle		*Brachypsectra fulva* LeConte	（コウチュウ目、Brachypsectridae）新北区
Texas beetles		Brachypsectridae	（コウチュウ目）の昆虫の総称
Texas blue		*Plebejus lupini texanus* (Goodpasture)	（チョウ目、シジミチョウ科）新熱帯区
Texas bush katydid		*Scudderia texensis* Saussure et Pictet	（バッタ目、キリギリス科）新北区
Texas carpenter ant		*Camponotus festinatus* (Buckley)	（ハチ目、アリ科）新北区
Texas emperor		*Asterocampa texana* (Skinner)	（チョウ目、タテハチョウ科）新北区
Texas false katydid		*Amblycorypha huasteca* (Saussure)	（バッタ目、キリギリス科）新北区
Texas field cricket		*Gryllus texensis* Cade et Otte	（バッタ目、コオロギ科）新北区
Texas fly			horn fly (1) を見よ
Texas fly			cattle biting fly を見よ　米国での英名
Texas grass tubeworm moth		*Acrolophus texanella* (Chambers)	（チョウ目、Acrolophidae）新北区
Texas gray moth		*Glenoides texanaria* (Hulst)	（チョウ目、シャクガ科）新北区
Texas heel fly			cattle warble fly (2) を見よ　米国での英名
Texas juniper hairstreak	オリーブスギカラスシジミ	*Callophrys gryneus castalis* (W. H. Edwards)	（チョウ目、シジミチョウ科）新熱帯区。juniper hairstreak (1) を参照
Texas leaf cutting ant	テキサスハキリアリ	*Atta texana* (Buckley)	（ハチ目、アリ科）新北区、新熱帯区
Texas meadow katydid		*Orchelimum bullatum* Rehn et Hebard	（バッタ目、キリギリス科）新北区
Texas mealybug			pear mealybug を見よ
Texas Mocis moth		*Mocis texana* (Morrison)	（チョウ目、ヤガ科）新北区
Texas powdered skipper		*Systasea pulverulenta* (Felder)	（チョウ目、セセリチョウ科）新北区、新熱帯区
Texas roadside skipper		*Amblyscirtes texanae* Bell	（チョウ目、セセリチョウ科）新北区、新熱帯区
Texas spotted range grasshopper		*Psoloessa texana* Scudder	（バッタ目、バッタ科）新北区
Texas tree cricket		*Oecanthus texensis* Symes et Collins	（バッタ目、カンタン科）新北区
Texas unicorn mantis		*Phyllovates chlorophaena* Blanchard	（カマキリ目、カマキリ科）新熱帯区
Texas wasp moth		*Horama panthalon* (Fabricius)	（チョウ目、ヒトリガ科）新北区、新熱帯区
Tezpi dancer		*Argia tezpi* Calvert	（トンボ目、イトトンボ科）新北区
Thales blackstreak		*Ocaria thales* Fabricius	（チョウ目、シジミチョウ科）新熱帯区

英名	和名	学名	所属、分布、ほか
thalictrum looper			marsh carpet を見よ
Thamyra satyr			Andromeda wood nymph を見よ
Thara hairstreak		*Enos thara* (Hewitson)	(チョウ目、シジミチョウ科) 新熱帯区
Thasus metalmark		*Cremna thasus subrutila* Stichel	(チョウ目、シジミタテハ科) 新熱帯区
Thasus metalmark		*Hyphilaria thasus* (Stoll)	(チョウ目、シジミタテハ科) 新熱帯区
thatch slender		*Acrocercops brongniardella* (Fabricius)	(チョウ目、ホソガ科) 旧北区
thatched bagworm		*Acanthopsyche tristis* Janse	(チョウ目、ミノガ科) エチオピア区
thatching ant (1)		*Formica obscripes* Forel	(ハチ目、アリ科) 新北区
thatching ant		*Formica haemorrhoidalis* Emery	(ハチ目、アリ科) 新北区
thaumetopoeid silkmoth		*Anaphe reticulata* Walker	(チョウ目、シャチホコガ科) エチオピア区
Theano alpine		*Erebia theano* (Tauscher)	(チョウ目、タテハチョウ科) 旧北区
Thecla banner		*Ectima thecla* (Fabricius)	(チョウ目、タテハチョウ科) 新熱帯区
Thecla della quercia			purple hairstreak を見よ
thelaxine plantlice		Thelaxinae	(カメムシ目、アブラムシ科) の昆虫の総称
Thelma's Agonopterix moth		*Agonopterix thelmae* Clarke	(チョウ目、マルハキバガ科) 新北区
Theodora metalmark		*Chalodeta theodora* (C. et R. Felder)	(チョウ目、シジミタテハ科) 新熱帯区
Theogenis skipper		*Cymaenes theogenis* (Capronnier)	(チョウ目、セセリチョウ科) 新熱帯区
Theona checkerspot		*Chlosyne theona* (Ménétriès)	(チョウ目、タテハチョウ科) 新北区、新熱帯区
Thericles firetip		*Pyrrhopyge thericles* Mabille	(チョウ目、セセリチョウ科) 新熱帯区
Thermus skipper		*Phocides thermus thermus* (Mabille)	(チョウ目、セセリチョウ科) 新熱帯区
Theseus Morpho		*Morpho theseus* Deyrolle	(チョウ目、タテハチョウ科) 新熱帯区
Thesprotia sister		*Adelpha thesprotia* (C. et R. Felder)	(チョウ目、タテハチョウ科) 新熱帯区
Thessalia sister		*Adelpha thessalia* (C. et R. Felder)	(チョウ目、タテハチョウ科) 新熱帯区
Theste skipper		*Turesis theste* Godman	(チョウ目、セセリチョウ科) 新熱帯区
thick barklice	フトチャタテ科	Pachytroctidae	(チャタテムシ目) の昆虫の総称
thick-bordered kite-swallowtail		*Eurytides dioxippus lacandones* (H. Bates)	(チョウ目、アゲハチョウ科) 新熱帯区
thick-bordered swordtail		*Neographium dioxippus* Hewitson	(チョウ目、アゲハチョウ科) 新熱帯区
thick-edged longwing		*Eueides lineata* Salvin et Godman	(チョウ目、タテハチョウ科) 新熱帯区
thick-headed flies	メバエ科	Conopidae	(ハエ目) の昆虫の総称
thick-headed fly			common thick-headed fly を見よ
thick-legged hoverfly		*Syritta pipiens* (Linnaeus)	(ハエ目、ハナアブ科) 全北区、東洋区、新熱帯区

英名	和名	学名	所属、分布、ほか
thick-legged moth	アシブトクチバ	*Parallelia stuposa* (Fabricius)	（チョウ目、ヤガ科）日本、旧北区、東洋区
thick-rimmed sailor		*Dynamine chryseis* Bates	（チョウ目、タテハチョウ科）新熱帯区
thick-tailed hairstreak		*Paiwarria umbratus* (Geyer)	（チョウ目、シジミチョウ科）新熱帯区
thick-tipped Greta			darkened rusty clearwing を見よ
thicket hairstreak	ヤブカラスシジミ	*Callophrys spinetorum* (Hewitson)	（チョウ目、シジミチョウ科）新北区
thicket twist	アミメトビハマキ	*Pandemis dumetana* (Treitscke)	（チョウ目、ハマキガ科）日本、旧北区。ダイズ、キク、バラなどの害虫
thickety knob-tipped shadowdamsel		*Drepanosticta montana* (Hagen)	（トンボ目、Platystictidae）東洋区
thickset scolytid borer		*Xyleborus soldidus* Eichhoff	（コウチュウ目、キクイムシ科）豪州区
thief ant		*Solenopsis fugax* (Latreille)	（ハチ目、アリ科）旧北区
thief ant			fire ant (1) を見よ
thief ants			fire ants を見よ　他種のアリの巣の近くに巣を作り、食料を盗むことに由来
Thieme's satyr		*Pedaliodes asconia* Thieme	（チョウ目、タテハチョウ科）新熱帯区
thin-banded lichen moth		*Cisthene tenuifascia* Harvey	（チョウ目、ヒトリガ科）新北区、新熱帯区
thin-footed thread-legged katydid		*Arethaea gracilipes* (Thomas)	（バッタ目、キリギリス科）新北区
thin-lined Chlorochlamys moth		*Chlorochlamys phyllinaria* (Zeller)	（チョウ目、シャクガ科）新北区
thin-lined Erastria moth		*Erastria cruentaria* (Hübner)	（チョウ目、シャクガ科）新北区
thin-lined owlet moth		*Isogona tenuis* (Grote)	（チョウ目、ヤガ科）新北区。thin-lined owlet ともいう
thin-lined tree cricket		*Oecanthus leptogrammus* Walker	（バッタ目、カンタン科）新北区
thin strawberry weevil		*Rhadinosomus lacordairei* Pascoe	（コウチュウ目、ゾウムシ科）豪州区
thin-winged longicorn beetle	ウスバカミキリ	*Megopis sinica sinica* White	（コウチュウ目、カミキリムシ科）日本、旧北区。リンゴ、サクラ、ポプラなどの害虫。thin-winged longicorn ともいう
thin-winged owlet moth		*Nigetia formosalis* Walker	（チョウ目、ヤガ科）新北区
thinker moth		*Lacinipolia meditata* (Grote)	（チョウ目、ヤガ科）新北区
thirteen-spotted lady beetle		*Hyppodamia tredecimpunctata* (Linnaeus)	（コウチュウ目、テントウムシ科）全北区。thirteen-spot ladybird ともいう。日本に *H. t. timberlakei* Capra ジュウサンホシテントウが分布
thirteen-year locust			Riley's 13-year cicada を見よ
thirteen-year locust			Cassin's 13-year cicada を見よ
thirteen-year locust			little 13-year cicada を見よ
thistle aphid	アザミオマルアブラムシ	*Brachycaudus cardui* (Linnaeus)	（カメムシ目、アブラムシ科）日本、汎世界

英名	和名	学名	所属、分布、ほか
thistle aphid			artichoke aphid を見よ
thistle butterfly			painted lady を見よ
thistle crescent	ミカズキタテハ	Phyciodes mylitta (W. H. Edwards)	（チョウ目、タテハチョウ科）新北区、新熱帯区
thistle-down mutilid			glorious velvet ant を見よ　thistle down velvet ant ともいう
thistle ermin		Myelois circumvolula (Fourceroy)	（チョウ目、メイガ科）旧北区
thistle ermin			burdock pyralid を見よ
thistle fall-fly			thistle stem gall を見よ
thistle-feeding tortoise beetle	セスジカメノコハムシ	Cassida vibex Linnaeus	（コウチュウ目、ハムシ科）日本、旧北区
thistle flat-body		Agonopterix carduella (Hübner)	（チョウ目、マルハキバガ科）旧北区
thistle lace bug	アザミグンバイ	Tingis ampliata (Herrich-Schäffer)	（カメムシ目、グンバイムシ科）日本、旧北区、東洋区
thistle lace bug			spear-thistle lace bug を見よ
thistle lady beetle	エゾアザミテントウ	Epilachna pustulosa Kono	（コウチュウ目、テントウムシ科）日本
thistle leaf beetle	アザミオオハムシ	Galeruca vicina Solsky	（コウチュウ目、ハムシ科）日本、旧北区
thistle mantis			devil's flower mantis（1）を見よ
thistle root aphid		Dysaphis lappae (Koch)	（カメムシ目、アブラムシ科）豪州区
thistle stem gall		Urophora cardui (Linnaeus)	（ハエ目、ミバエ科）がツルアザミの葉につくる虫えいで 10 mm。緑から褐色になる。旧北区
thistle tortoise beetle	アオカメノコハムシ	Cassida rubiginosa Müller	（コウチュウ目、ハムシ科）日本、旧北区
Thoas swallowtail			King page を見よ
Thoasa sister		Adelpha thoasa (Hewitson)	（チョウ目、タテハチョウ科）新熱帯区
Thomas's trig		Anaxipha sp.	（バッタ目、クサヒバリ科）新北区
Thomson's meadow brown		Maniola halicarnassus Thomson	（チョウ目、タテハチョウ科）旧北区
thorax fly		Argyramoela trifasciata Meigen	（ハエ目、ツリアブ科）旧北区。ハナアブを捕食
Thordesa hairstreak		Michaelus thordesa (Hewitson)	（チョウ目、シジミチョウ科）新熱帯区
Thoria hairstreak		Gargina thoria (Hewitson)	（チョウ目、シジミチョウ科）新熱帯区
thorn bug		Umbonia spinosa Fabricius	（カメムシ目、ツノゼミ科）新熱帯区
thornbug			horned treehopper（1）を見よ
thorn-scrub Emesis		Emesis poeas Godman	（チョウ目、シジミタテハ科）新熱帯区
thorn-scrub leafwing		Fountainea halice martinezi (Maza et Diaz)	（チョウ目、タテハチョウ科）新熱帯区。ruddy leafwing を参照
thorn-tree bush-blue			black-bordered Babul blue を見よ
Thorne's hairstreak		Callophrys gryneus thornei (Brown)	（チョウ目、シジミチョウ科）新北区。juniper hairstreak（1）を参照
thorns	エダシャク亜科	Ennominae	（チョウ目、シャクガ科）の昆虫の総称

英名	和名	学名	所属、分布、ほか
thorny devil stick insect			New Guinea spiny stick insect を見よ
Thor's fritillary		*Clossiana thore* (Hübner)	(チョウ目、タテハチョウ科) 旧北区。日本亜種は *C. t. jezoensis* (Matsumura) ホソバヒョウモン
thoughtful Apamea moth		*Apamea cogitata* Smith	(チョウ目、ヤガ科) 新北区。thoughtful Apamea ともいう
Thrasea skipper		*Thracides thrasea* (Hewitson)	(チョウ目、セセリチョウ科) 新熱帯区
thread bug	マダラカモドキサシガメ	*Empicoris rubromaculatus* (Blackburn)	(カメムシ目、サシガメ科) 日本。新北区
thread scale			black thread scale を見よ
thread-horned flies	糸角亜目	Nematocera	(ハエ目) の昆虫の総称
thread-legged assassin bug		*Empicoris vagabundus* (Linnaeus)	(カメムシ目、サシガメ科) 旧北区
thread-legged bug		*Emesaya brevipennis* (Say)	(カメムシ目、サシガメ科) 新北区
thread-legged bug		*Ploiariola culiciformis* De Geer	(カメムシ目、Ploiariidae) 旧北区
thread-legged bug			thread bug を見よ
thread-legged bugs		Ploiariidae	(カメムシ目) の昆虫の総称　サシガメ科と近縁
thread-legged bugs		Emesinae	(カメムシ目、サシガメ科) の昆虫の総称
thread-legged bugs			assassin bugs (1) を見よ
thread-legged katydid		*Arethaea gracilipes papago* (Hebard)	(バッタ目、キリギリス科) 新北区。灯火に飛来する
thread-legged katydids		*Arethaea*	(バッタ目、キリギリス科) の昆虫の総称
thread-tailed stoneflies	オナシカワゲラ科	Nemouridae	(カワゲラ目) の昆虫の総称　多くの種は体長 10mm 以下
threadtailed stonefly		*Soyedina vallicularia* (Wu)	(カワゲラ目、オナシカワゲラ科) 新北区
threadtails (1)	ミナミイトトンボ科	Protoneuridae	(トンボ目) の昆虫の総称
threadtails		*Elattoneura*	(トンボ目、ミナミイトトンボ科) の昆虫の総称
thread-waisted conopid		*Physocephala texana* (Williston)	(ハエ目、メバエ科) 新北区
thread-waisted digger wasps			sand wasps (2) を見よ
thread-waisted wasp		*Sphex procerus* (Dahlbom)	(ハチ目、アナバチ科) 新北区
thread-waisted wasps			sand wasps (1) を見よ
thread-waisted wasps			cicada killers を見よ
thread-winged lacewings		Nemopteridae	(アミメカゲロウ目) の昆虫の総称
three-banded crescent			variable mimic (1) を見よ
three-banded grasshopper		*Hadrotettix trifasciatus* (Say)	(バッタ目、バッタ科) 新北区
three-banded ladybird			maculate ladybird を見よ
three-banded ladybird beetle		*Coccinella trifasciata* Linnaeus	(コウチュウ目、テントウムシ科) 新北区

英名	和名	学名	所属、分布、ほか
three-banded leafhopper		*Erythroneura tricincta* Fitch	（カメムシ目、オオヨコバイ科）新北区。ブドウ害虫
three-banded lichen moth			thin-banded lichen moth を見よ
three-banded longhorn beetle	シララカハナカミキリ	*Judolia sexmaculata* (Linnaeus)	（コウチュウ目、カミキリムシ科）日本、旧北区
three-black-spotted leaf beetle	アトボシハムシ	*Paridea angulicollis* (Motschulsky)	（コウチュウ目、ハムシ科）日本、旧北区、東洋区
three-cornered alfalfa hopper		*Spissistilus festinus* (Say)	（カメムシ目、ツノゼミ科）新北区。three-cornered alfalfa treehopper ともいう
three-dentate bark beetle	ミツハリキクイムシ	*Scolytus trispinosus* Strohmeyer	（コウチュウ目、キクイムシ科）日本、旧北区
three-dotted rose bell		*Pardia cynosbatella* Linnaeus	（チョウ目、ハマキガ科）旧北区
three-horned bush cricket		*Orocharis tricomis* Walker	（バッタ目、コオロギ科）新北区
three-humped prominent		*Notodonta tritophus* (Denis et Schiffermüller)	（チョウ目、シャチホコガ科）旧北区
three-lined angle moth		*Digrammia eremiata* (Guenée)	（チョウ目、シャクガ科）新北区
three-lined fig-tree borer		*Ptychodes vittatus* Fabricius	（コウチュウ目、カミキリムシ科）新北区。イチジク害虫
three-lined Grapholita moth		*Grapholita tristrigana* (Clemens)	（チョウ目、ハマキガ科）新北区
three-lined hover fly		*Helophilus trilineata* Linnaeus	（ハエ目、ハナアブ科）豪州区
threelined leafroller		*Pandemis limitata* (Robinson)	（チョウ目、ハマキガ科）新北区。three-lined leafroller moth ともいう
three lined potato beetle		*Lema trilinea* White	（コウチュウ目、ハムシ科）新北区、大洋区。potato beetle, potato bug ともいう
three-lined potato beetle		*Lema trilineata* (Olivier)	（コウチュウ目、ハムシ科）新北区、新熱帯区
three-lined potato beetle		*Lema daturaphia* Kogan et Goeden	（コウチュウ目、ハムシ科）新北区
three lined potato beetle (1)		*Lema trivittata* Say	（コウチュウ目、ハムシ科）豪州区
three-lined shieldback		*Steiroxys trilineatus* (Thomas)	（バッタ目、キリギリス科）新北区
three-lined soldier		*Oxycera trilineata* (Linnaeus)	（ハエ目、ミズアブ科）旧北区
three-lined truncate-tipped geometrid			larch geometrid moth を見よ
three-part sister			Naxia sister を見よ
three-patched bigwing moth		*Heterophleps refusaria* (Walker)	（チョウ目、シャクガ科）新北区
three-pronged bristletails			silverfish (2) を見よ
three-row dwarf		*Elachista dispunctella* (Duponchel)	（チョウ目、クサモグリガ科）旧北区

英名	和名	学名	所属、分布、ほか
three-serrate-striped geometrid			great oak beauty を見よ
three-spined darners		*Triacanthagyna*	（トンボ目、ヤンマ科）の昆虫の総称
three spot crimson tip		*Colotis eunoma* (Hopffer)	（チョウ目、シロチョウ科）エチオピア区
three-spot grass yellow		*Eurema blanda* Boisduval	（チョウ目、シロチョウ科）東洋区。日本亜種は *E. b. arsakia* (Fruhstorfer) タイワンキチョウ
three-spot sylph		*Metisella trisignatus* (Neave)	（チョウ目、セセリチョウ科）エチオピア区
three-spotted backswimmer	マツモムシ	*Notonecta triguttata* Motschulsky	（カメムシ目、マツモムシ科）日本、旧北区
three-spotted bell		*Epiblema costipunctana* (Haworth)	（チョウ目、ハマキガ科）旧北区
three-spotted Crambus moth		*Catoptria latiradiellus* (Walker)	（チョウ目、メイガ科）新北区
three-spotted Fillip moth		*Heterophleps triguttaria* (Herrich-Schäffer)	（チョウ目、シャクガ科）新北区。three-spotted Fillip ともいう
threespotted flea beetle		*Disonycha triangularis* (Say)	（コウチュウ目、ハムシ科）新北区
three-spotted Nephele		*Nephele peneus* (Cramer)	（チョウ目、スズメガ科）エチオピア区
three-spotted Nola moth		*Nola triquetrana* (Fitch)	（チョウ目、コブガ科）新北区。three-spotted Nola ともいう
three-spotted phytometra			three-spotted plusia を見よ
three-spotted plusia	ミツモンキンウワバ	*Ctenoplusia agnata* (Staudinger)	（チョウ目、ヤガ科）日本、旧北区、東洋区。サツマイモ、ダイズ、キクなど野菜、花の害虫
three-spotted prominent			brown-spotted prominent を見よ
three-spotted sallow moth		*Eupsilia tristigmata* (Grote)	（チョウ目、ヤガ科）新北区
three-spotted skipper		*Cymaenes tripunctus* (Herrich-Schäffer)	（チョウ目、セセリチョウ科）新北区
three-spotted specter		*Euerythra trimaculata* Smith	（チョウ目、ヒトリガ科）新北区。three-spotted specter moth ともいう
three-staff underwing moth		*Catocala amestris* Strecker	（チョウ目、ヤガ科）新北区。three-staff underwing ともいう
three-streaked Sparganothis moth		*Sparganothis tristriata* Kearfott	（チョウ目、ハマキガ科）新北区
threestriped blister beetle		*Epicauta lemniscata* (Fabricius)	（コウチュウ目、ツチハンミョウ科）新北区
three striped blue dart		*Pseudagrion decorum* (Rambur)	（トンボ目、イトトンボ科）東洋区
three-striped dasher		*Micrathyria didyma* (Selys)	（トンボ目、トンボ科）新北区、新熱帯区
threestriped lady beetle		*Brumoides suturalis* (Fabricius)	（コウチュウ目、テントウムシ科）新北区
three-striped longhorn		*Adela trigrapha* Zeller	（チョウ目、マガリガ科）新北区

英名	和名	学名	所属、分布、ほか
three-striped maple leafhopper			hyaline maple leafhopper を見よ
three-striped powdered geometrid			common wave を見よ
three-striped pyralid	ホソミスジノメイガ	*Pleuroptya chlorophanta* (Butler)	(チョウ目、メイガ科) 日本、旧北区、東洋区
three-tailed hairstreak		*Oxylides faunus albata* (Aurivillius)	(チョウ目、シジミチョウ科) エチオピア区
three-tailed sapphire		*Iolaus silanus* Grose-Smith	(チョウ目、シジミチョウ科) エチオピア区
three-tailed swallowtail	ミツオトラフアゲハ	*Papilio pilumnus* Boisduval	(チョウ目、アゲハチョウ科) 新北区、新熱帯区
three-tailed tiger swallowtail			moss peacock を見よ
three-tailed tiger swallowtail			three-tailed swallowtail を見よ
three-toned Prepona		*Archaeoprepona meander phoebus* (Boisduval)	(チョウ目、タテハチョウ科) 新熱帯区
thrice-striped cosmet		*Dystebenna stephensi* (Stainton)	(チョウ目、Agonoxenidae) 旧北区
thrift clearwing		*Bembecia muscaeformis* (Esper)	(チョウ目、スカシバガ科) 旧北区
thrift neb		*Aristotelia brizella* (Treitschke)	(チョウ目、キバガ科) 旧北区
thrift tortrix moth		*Lobesia littoralis* (Humphreys et Westwood)	(チョウ目、ハマキガ科) 旧北区
thripids			thrips (2) を見よ
thrips		*Frankliniella insularis* Franklin	(アザミウマ目、アザミウマ科) 新熱帯区。カンキツ害虫
thrips		*Frankliniella rodeos* Moulton	(アザミウマ目、アザミウマ科) 新熱帯区。カンキツ害虫
thrips		*Frankliniella varipes* Moulton	(アザミウマ目、アザミウマ科) 新熱帯区。カンキツ害虫
thrips		*Heterothrips moreirai* (Moulton)	(アザミウマ目、Heterothripidae) 新北区。カンキツ害虫
thrips		*Taeniothrips frisi* Uzel	(アザミウマ目、アザミウマ科) 旧北区。カンキツ害虫
thrips		Physapoda	(アザミウマ目) の昆虫の総称　Thysanoptera の異名
thrips (1)	アザミウマ科	Thripidae	(アザミウマ目) の昆虫の総称　世界に 1,500 種
thrips (2)	アザミウマ目	Thysanoptera	の昆虫の総称
throat bot			throat bot fly を見よ　幼虫の英名
throat bot fly	ムネアカウマバエ	*Gasterophilus nasalis* (Linnaeus)	(ハエ目、ウマバエ科) 日本、汎世界
throat fly			throat bot fly を見よ
throscid beetles	ヒゲブトコメツキ科	Throscidae	(コウチュウ目) の昆虫の総称
thuja bark beetle	ヒバノキクイムシ	*Phloeosinus perlatus* Chapuis	(コウチュウ目、キクイムシ科) 日本、旧北区、東洋区

英名	和名	学名	所属、分布、ほか
thuja torymid	アスナロモンオナガコバチ	*Megastigmus thuyopsis* Yano	（ハチ目、オナガコバチ科）日本
Thula sulphur		*Colias thula* Hovanitz	（チョウ目、シロチョウ科）新北区
thumb-bearing short-winged katydid		*Dichopetala pollicifera* Rehn et Hebard	（バッタ目、キリギリス科）新北区
Thunberg's flat-body		*Enicostoma lobella* Denis et Schiffermüller	（チョウ目、マルハキバガ科）旧北区
thunder fly			grain thrips を見よ
thunderbugs			thrips (2) を見よ
thunderflies			thrips (2) を見よ
thurberia bollworm		*Thurberiphaga diffusa* Barnes	（チョウ目、ヤガ科）新北区
thurberia weevil		*Anthonomus grandis thurbeiriae* Pierce	（コウチュウ目、ゾウムシ科）新北区
thuya aphid	ネズミサシオオアブラムシ	*Cinara juniperi* (De Geer)	（カメムシ目、アブラムシ科）日本、旧北区、豪州区。イブキ害虫
thyatirid moths			false owlet moths を見よ
thylacine darner		*Acanthaeschna victoria* Martin	（トンボ目、ヤンマ科）豪州区
thyme lacebug		*Lasiacantha capucina* Germar	（カメムシ目、グンバイムシ科）旧北区
thyme leafhopper		*Dikraneura mollicula* (Boheman)	（カメムシ目、オオヨコバイ科）旧北区
thyme moth			knot-horn を見よ
thyme pug		*Eupithecia distinctaria constrictata* Guenée	（チョウ目、シャクガ科）旧北区
thyreophorid flies		Thyreophoridae	（ハエ目）の昆虫の総称
Thyridia glasswing			Melantho tigerwing を見よ
thyridid moths			window-winged moths を見よ
Thyrrhus skipper		*Toxidia thyrrhus* Mabille	（チョウ目、セセリチョウ科）豪州区
thysanid parasites			thysanid wasps を見よ　米国での英名
thysanid wasps		Thysanidae	（ハチ目）の昆虫の総称　thysanids ともいう
thysanopterans			thrips (2) を見よ
thysanopters			thrips (2) を見よ
thysanurans			silverfish (2) を見よ
thysanures			silverfish (2) を見よ
ti scale			coconut scale (2) を見よ
ti thrips			banded-winged palm thrips を見よ
Tibet blackvein		*Aporia peloria* Hewitson	（チョウ目、シロチョウ科）東洋区
Tibet marbled satyr		*Tatinga thibetanus* Oberthür	（チョウ目、タテハチョウ科）旧北区
Tibetan black-veined white		*Aporia procris* Leech	（チョウ目、シロチョウ科）東洋区
Tibetan cupid		*Tongeia zuthus* Leech	（チョウ目、シジミチョウ科）旧北区
Tibetan glory		*Bhutanitis thaidina* (Blanchard)	（チョウ目、アゲハチョウ科）東洋区

英名	和名	学名	所属、分布、ほか
Tibetan heath		*Coenonympha sinica* Alphéraky	（チョウ目、タテハチョウ科）東洋区
Tibetan labyrinth		*Neope simulans* Leech	（チョウ目、タテハチョウ科）旧北区
Tibetan white			striated white を見よ
Ticida glasswing		*Pteronymia ticida* (Hewitson)	（チョウ目、タテハチョウ科）新熱帯区
tick flies			pupiparous flies を見よ
tickseed moth			goldenrod stowaway moth を見よ
tick-tock		*Cicadetta quadricincta* (Walker)	（カメムシ目、セミ科）豪州区
ticker		*Pauropsalta mneme* (Walker)	（カメムシ目、セミ科）豪州区
tidewater meadow katydid		*Conocephalus nigropleuroides* (Fox)	（バッタ目、キリギリス科）新北区
tidewater trig		*Anaxipha* sp.	（バッタ目、クサヒバリ科）新北区
Tien Shan blue		*Agriades pheretiades* Eversmann	（チョウ目、シジミチョウ科）旧北区。
Tiessa ringlet		*Magneuptychia tiessa* Hewitson	（チョウ目、タテハチョウ科）新熱帯区
tiger	キボシトンボマダラ	*Tithorea harmonia megara* Godart	（チョウ目、タテハチョウ科）新熱帯区。tiger butterfly を参照
tiger and footman moths			tiger moths (1) を見よ
tiger bamboo longhorn			bamboo longhorn beetle をみよ
tiger beauty			zebra sapseeker を見よ
tiger bee fly		*Anthrax tigrinus* De Geer	（ハエ目、ツリアブ科）新北区、新熱帯区、大洋区
tiger beetle			green tiger beetle を見よ
tiger beetles (1)	ハンミョウ科	Cicindelidae	（コウチュウ目）の昆虫の総称
tiger beetles		*Cicindela*	（コウチュウ目、ハンミョウ科）の昆虫の総称
tiger beetles			ground beetles を見よ
tiger blue butterfly			blue tiger (1) を見よ
tiger brown		*Orinoma damaris* Gray	（チョウ目、タテハチョウ科）旧北区、東洋区
tiger bug mimic			lycid-mimic moth を見よ
tiger butterfly	キボシカバタテハ	*Tithorea harmonia* (Cramer)	（チョウ目、タテハチョウ科）新熱帯区
tiger crescent		*Eresia eunice* Hübner	（チョウ目、タテハチョウ科）新熱帯区
tiger-eye hairstreak		*Rekoa meton* (Cramer)	（チョウ目、シジミチョウ科）新北区、新熱帯区
tiger fly		*Coenosia tigrina* (Fabricius)	（ハエ目、イエバエ科）全北区
tiger heliconian		*Heliconius ismenius* Latreille	（チョウ目、タテハチョウ科）新熱帯区。longwing や ismenius tiger ともいう。*H. i. telchinia* Doubleday は tiger-striped longwing という
tiger heliconian			longwing crescent を見よ
tiger hopper		*Ochus subvittatus* (Moore)	（チョウ目、セセリチョウ科）東洋区

英名	和名	学名	所属、分布、ほか
tiger leafwing		*Consul fabius* (Cramer)	（チョウ目、タテハチョウ科）新熱帯区
tiger longicorn		*Aridaeus thoracicus* (Donovan)	（コウチュウ目、カミキリムシ科）豪州区。tiger longhorn ともいう
tiger longicorn beetle	トラフカミキリ	*Xylotrechus chinensis* Chevrolat	（コウチュウ目、カミキリムシ科）日本、旧北区、東洋区。クワ害虫。tiger longicorn ともいう
tiger longwing		*Podotricha judith* (Guérin-Méneville)	（チョウ目、タテハチョウ科）新熱帯区
tiger longwing			polymorphic longwing を見よ
tiger mimic longwing			Numata longwing を見よ
tiger mimic-queen (1)		*Lycorea cleobaea* (Godart)	（チョウ目、タテハチョウ科）新北区、新熱帯区
tiger mimic-queen (2)		*Lycorea halia* (Hübner)	（チョウ目、タテハチョウ科）新北区、新熱帯区。*L. h. atergatis* Doubleday も同英名
tiger mimic-white	ツマキコバネシロチョウ	*Dismorphia amphione isolda* Llorente	（チョウ目、シロチョウ科）新熱帯区。*D. a. lupita* Lamas も同英名。tiger Pierid を参照
tiger mosquito	ヒトスジシマカ	*Aedes albopictus* (Skuse)	（ハエ目、カ科）日本、東洋区、大洋区、豪州区。人家、墓地に普通
tiger moth		*Spilosoma metarhoda* Walker	（チョウ目、ヒトリガ科）東洋区。カンキツ害虫
tiger moth		*Arctia fasciata* (Esper)	（チョウ目、ヒトリガ科）旧北区
tiger moth		*Arctia opulenta* (Edwards)	（チョウ目、ヒトリガ科）新北区
tiger moth		*Amata trigonophora* (Turner)	（チョウ目、カノコガ科）豪州区
tiger moth		*Arachnis aulaea* (Geyer)	（チョウ目、ヒトリガ科）新北区、新熱帯区
tiger moth (1)		*Apantesis ornata* (Packard)	（チョウ目、ヒトリガ科）新北区
tiger moth			red-costate tiger moth を見よ
tiger moth			black woolly-bear を見よ
tiger moths (1)	ヒトリガ科	Arctiidae	（チョウ目）の昆虫の総称　世界に10,000種以上
tiger moths	ヒトリガ亜科	Arctiinae	（チョウ目、ヒトリガ科）の昆虫の総称
tiger palmfly	スジマネシヒカゲ	*Elymnias nesaea* (Linnaeus)	（チョウ目、タテハチョウ科）東洋区
tiger palmfly	アオスジルリモンジャノメ	*Elymnias nesaea lioneli* Fruhstorfer	（チョウ目、タテハチョウ科）東洋区
tiger pear cochineal		*Dactylopius austrinus* De Lotto	（カメムシ目、コチニールカイガラムシ科）豪州区
tiger Pierid	キマダラコバネシロチョウ	*Dismorphia amphione* (Cramer)	（チョウ目、シロチョウ科）新熱帯区
tiger prince		*Macrotristria godingi* Distant	（カメムシ目、セミ科）豪州区
tiger spiketail		*Cordulegaster erronea* Selys	（トンボ目、オニヤンマ科）新北区
tiger-spotted flower lepturine		*Stenostrophia tribalteata* LeConte	（コウチュウ目、カミキリムシ科）新北区
tiger-striped Euselasia		*Euselasia toppini* Sharpe	（チョウ目、シジミタテハ科）新熱帯区
tiger-striped leafwing		*Consul fabius cecrops* (Doubleday)	（チョウ目、タテハチョウ科）新熱帯区。tiger leafwing を参照

英名	和名	学名	所属、分布、ほか
tiger swallowtail	トラフアゲハ	*Papilio glaucus* Linnaeus	(チョウ目、アゲハチョウ科) 新北区。tiger swallowtail butterfly ともいう
tigertail			common tigertail を見よ
tiger tiphiids		*Myzinum*	(ハチ目、コツチバチ科) の昆虫の総称
tiger wasp moth		*Amata germana* (Felder)	(チョウ目、カノコガ科) 東洋区
tiger white	カバイロシロチョウ	*Charonias eurytele nigrescens* (Salvin et Godman)	(チョウ目、シロチョウ科) 新熱帯区
tiger with tails			tiger leafwing を見よ
tigers		*Gomphidia*	(トンボ目、サナエトンボ科) の昆虫の総称
tigers (1)	カバマダラ属	*Danaus*	(チョウ目、タテハチョウ科) の昆虫の総称
tigers and crows			monarchs を見よ
Tikal Calephelis		*Calephelis tikal* Austin	(チョウ目、シジミタテハ科) 新熱帯区
til leafroller			gingelly borer moth を見よ
tile-horned Prionus		*Prionus (Neopolyarthron) imbricornis* (Linnaeus)	(コウチュウ目、カミキリムシ科) 新北区
tilia aphid	マエグロハネマダラアブラムシ	*Therioaphis tilicola* Shinji	(カメムシ目、アブラムシ科) 日本
tilia hawk moth	ヒサゴスズメ	*Mimas christophi* (Staudinger)	(チョウ目、スズメガ科) 日本、旧北区
Tillyard's skipper		*Anisynta tillyardi* Waterhouse	(チョウ目、セセリチョウ科) 豪州区
Tilodi copper			Waterberg copper を見よ
timber beetle		*Lymexylon navale* (Linnaeus)	(コウチュウ目、ツツシンクイ科) 旧北区
timber beetles			long-horned beetles を見よ
timber beetles			ship-timber beetles を見よ
timber beetles			bark beetles (1) を見よ
timber borer			wharf borer を見よ
timber borer rhipiphorid	キクイオオハナノミ	*Micropelecotomoides japonicus* (Pic)	(コウチュウ目、オオハナノミ科) 日本、東洋区
timber borers			long-horned beetles を見よ
timberman		*Acanthocinus aedilis* (Linnaeus)	(コウチュウ目、カミキリムシ科) 旧北区。timberman beetle ともいう
timemas		Timematidae	(ナナフシ目) の昆虫の総称
timemas		*Timema*	(ナナフシ目、Timematidae) の昆虫の総称
timothy aphid		*Brachycolus muehlei* (Börner)	(カメムシ目、アブラムシ科) 旧北区
timothy billbug		*Calendra zeae* (Walsh)	(コウチュウ目、ゾウムシ科) 新北区。トウモロコシ害虫
timothy flies		*Amaurosoma*	(ハエ目、フンバエ科) の昆虫の総称
timothy plant bug	フタスジカスミカメムシ	*Stenotus binolatus* (Fabricius)	(カメムシ目、カスミカメムシ科) 日本、旧北区。イネ科牧草の害虫
timothy thrips	ヒゲブトアザミウマ	*Chirothrips manicatus* (Haliday)	(アザミウマ目、アザミウマ科) 日本、全北区、東洋区、豪州区、新熱帯区。ムギ、トウモロコシなどの害虫

英名	和名	学名	所属、分布、ほか
timothy tortrix		*Aphelia paleana* (Hübner)	（チョウ目、ハマキガ科）旧北区。timothy tortrix moth ともいう
Tindale's grassgrub		*Oncopera tindalei* Common	（チョウ目、コウモリガ科）豪州区
tineid moths			clothes moths を見よ　tineids ともいう
Tingo spreadwing		*Potamanaxas paralus* (Godman et Salvin)	（チョウ目、セセリチョウ科）新熱帯区
Tinkham's shieldback		*Pediodectes tinkhami* Hebard	（バッタ目、キリギリス科）新北区
tinkling ground cricket		*Allonemobius tinnulus* (Fulton)	（バッタ目、コオロギ科）新北区
tinktinkie blue		*Brephidium metophis* (Wallengren)	（チョウ目、シジミチョウ科）エチオピア区
Tinsels		*Catapaecilma*	（チョウ目、シジミチョウ科）の昆虫の総称
tiny Acraea	ウブイカザリヒメホソチョウ	*Acraea uvui* Grose-Smith	（チョウ目、タテハチョウ科）エチオピア区
tiny blue			cyna blue を見よ
tiny checkerspot			Dymas checkerspot を見よ
tiny ciliate blue		*Anthene contrastata* Ungemach	（チョウ目、シジミチョウ科）エチオピア区
tiny crescent			Dymas checkerspot を見よ
tiny flat		*Sarangesa sati* de Nicéville	（チョウ目、セセリチョウ科）東洋区
tiny gem		*Chloroselas minima* Jackson	（チョウ目、シジミチョウ科）エチオピア区
tiny grass blue (1)	ホリイコシジミ	*Zizula hylax* (Fabricius)	（チョウ目、シジミチョウ科）日本（小笠原）東洋区、豪州区、エチオピア区
tiny grass blue (2)		*Zizula gaika* (Trimen)	（チョウ目、シジミチョウ科）東洋区、新熱帯区
tiny hedge blue		*Lycaenopsis minima* (Evans)	（チョウ目、シジミチョウ科）東洋区
tiny hunter		*Austrogomphus pusillus* Sjöstedt	（トンボ目、サナエトンボ科）豪州区
tiny metalmark		*Adelotypa eudocia* (Godman et Salvin)	（チョウ目、シジミタテハ科）新熱帯区
tiny mountain Acraea			tiny Acraea を見よ
tiny nymph underwing moth		*Catocala micronympha* Guenée	（チョウ目、ヤガ科）新北区
tiny orange tip			round winged orange tip を見よ
tiny orange tip			desert orange tip (2) を見よ
tiny sailer		*Neptis nina* Staudinger	（チョウ目、タテハチョウ科）エチオピア区
tiny yellow house ant			ghost ant を見よ
tip moth			pine tip moth (3) を見よ
tiphia wasp		*Tiphia segregata* Crawford	（ハチ目、コツチバチ科）大洋区
tiphiid wasp		*Methocha ichneumonides* Latreille	（ハチ目、コツチバチ科）旧北区
tiphiid wasps	コツチバチ科	Tiphiidae	（ハチ目）の昆虫の総称
tipulid gnats		Tipulariae	（ハエ目）の昆虫の総称

英名	和名	学名	所属、分布、ほか
tipulids			crane flies (1) を見よ
tischerid moths			trumpet leafminer moths を見よ
Tisias bent-skipper		*Cycloglypha tisias* (Godman et Salvin)	(チョウ目、セセリチョウ科) 新熱帯区
Tissoides crescent		*Tegosa tissoides* (Hall)	(チョウ目、タテハチョウ科) 新熱帯区
tissue		*Triphosa dubitata* (Linnaeus)	(チョウ目、シャクガ科) 旧北区。tissue moth ともいう。日本亜種は *T. d. amblychiles* Prout ウスグロオオナミシャク で dark wavy-striped geometrid という
tissue moth		*Triphosa haesitata* (Guenée)	(チョウ目、シャクガ科) 新北区
titan beetle	タイタンオオウスバカミキリ	*Titanus giganteus* (Linnaeus)	(コウチュウ目、カミキリムシ科) 東洋区、新熱帯区。体長 17 cm に達する世界最大のカミキリムシ
titan sphinx moth		*Aellopos titan* (Cramer)	(チョウ目、スズメガ科) 新北区、新熱帯区。titan sphinx ともいう
titan stick insect		*Acrophylla titan* Macleay	(ナナフシ目、ナナフシ科) 豪州区。オーストラリア最長のナナフシ
titan underwing			Alabama underwing moth を見よ
Titania's fritillary	アルタイヒメヒョウモン	*Clossiana titania* (Esper)	(チョウ目、タテハチョウ科) 新北区。Titania fritillary とも記す
Tite's blue		*Lepidochrysops titei* Dickson	(チョウ目、シジミチョウ科) エチオピア区
Tite's copper		*Aloeides titei* Henning	(チョウ目、シジミチョウ科) エチオピア区
Tite's zebra blue		*Leptotes brevidentatus* (Tite)	(チョウ目、シジミチョウ科) エチオピア区
Tithonus birdwing	チトヌストリバネアゲハ	*Ornithoptera tithonus* De Haan	(チョウ目、アゲハチョウ科) 東洋区
tits		*Chliaria*	(チョウ目、シジミチョウ科) の昆虫の総称
tits (1)	フタオツバメ属	*Hypolycaena*	(チョウ目、シジミチョウ科) の昆虫の総称
tizi			saddle-backed bush cricket を見よ　tizzi (大あわて) 由来か
Tlascala moth		*Tlascala reductella* (Walker)	(チョウ目、メイガ科) 新北区
toad bug		*Gelastocoris oculatus* (Fabricius)	(カメムシ目、アシブトメミズムシ科) 新北区
toad bug		*Nertha grandis* (Montandon)	(カメムシ目、アシブトメミズムシ科) 豪州区
toad bugs	アシブトメミズムシ科	Gelastocoridae	(カメムシ目) の昆虫の総称
toad fly		*Lucilia bufonivora* Moniez	(ハエ目、クロバエ科) 旧北区
toad weevils		Tachygoninae	(コウチュウ目、ゾウムシ科) の昆虫の総称
toadflax brocade		*Calophasia lunula* (Hufnagel)	(チョウ目、ヤガ科) 旧北区
toadflax pug		*Eupithecia linariata* (Denis et Schiffermüller)	(チョウ目、シャクガ科) 旧北区
Toba bamboo scale	タケノコギリカイガラムシ	*Serrolecanium tobai* (Kuwana)	(カメムシ目、コナカイガラムシ科) 日本
tobacco ant			fire ant (2) を見よ

英名	和名	学名	所属、分布、ほか
tobacco aphid		*Myzus nicotianae* Blackman	（カメムシ目、アブラムシ科）新北区
tobacco aphid			green peach aphid を見よ
tobacco beetle		*Catorama tabaci* Guérin	（コウチュウ目、シバンムシ科）旧北区
tobacco beetle			cigarette beetle を見よ　豪州での英名
tobacco budworm	ニセアメリカタバコガ	*Heliothis virescens* (Fabricius)	（チョウ目、ヤガ科）新北区、新熱帯区、大洋区。tobacco budworm moth ともいう
tobacco budworm			corn earworm (2) を見よ　幼虫の英名
tobacco capsid			tobacco leaf bug を見よ
tobacco caterpillar			common cutworm を見よ
tobacco coloured longhorn	ホクチチビハナカミキリ	*Alosterna tabacicolor* (De Geer)	（コウチュウ目、カミキリムシ科）日本、旧北区
tobacco cricket		*Brachytrypes membranaceus* (Drury)	（バッタ目、コオロギ科）エチオピア区
tobacco cutworm			common cutworm を見よ
tobacco cutworm			cutworm (2) を見よ
tobacco flea beetle		*Epitrix hirtipennis* (Melsheimer)	（コウチュウ目、ハムシ科）新北区
tobacco flea beetle		*Epitrix fasciata* Blatchley	（コウチュウ目、ハムシ科）大洋区
tobacco flies			米国でタバコにつくスズメガ類の英名
tobacco fly			tobacco worm を見よ　北米での英名
tobacco ground beetle	コガシラスナゴミムシダマシ	*Mesomorhus villiger* (Blanchard)	（コウチュウ目、ゴミムシダマシ科）日本、旧北区、東洋区、大洋区、豪州区、エチオピア区
tobacco hornworm			death's head hawkmoth (1) を見よ
tobacco hornworm			tobacco worm を見よ　成虫は北米で tobacco hornworm moth という
tobacco hornworms		*Protoparce*	（チョウ目、スズメガ科）の昆虫の総称
tobacco leaf bug	タバコカスミカメムシ	*Nesisiocoris tenuis* (Reuter)	（カメムシ目、カスミカメムシ科）日本、東洋区、大洋区。タバコ、トマト、ウリ類の害虫
tobacco leaf caterpillar			common cutworm を見よ　幼虫の英名
tobacco leaf miner			tobacco stem borer (1) を見よ
tobacco leaf worm	ホソバハイイロハマキ	*Cnephasia cinereipalpana* Razowski	（チョウ目、ハマキガ科）日本、旧北区。タバコ、イチゴ、リンゴなどの害虫
tobacco leaf-folder		*Psara periusalis* Walker	（チョウ目、メイガ科）新北区
tobacco leafhopper	トバヨコバイ	*Alobaldia tobae* (Matsumura)	（カメムシ目、オオヨコバイ科）日本、旧北区
tobacco leafminer			potato tuber moth (1) を見よ　豪州での英名
tobacco looper		*Chrysodeixis argentifera* (Guenée)	（チョウ目、ヤガ科）豪州区
tobacco moth		*Setomorpha tineoides* Walsingham	（チョウ目、Setomorphidae）東洋区
tobacco moth (1)	チャマダラメイガ	*Ephestia elutella* (Hübner)	（チョウ目、メイガ科）日本、汎世界。タバコ、貯穀害虫
tobacco moth			almond moth を見よ
tobacco seed beetle			wardrobe beetle を見よ

英名	和名	学名	所属、分布、ほか
tobacco slitworm			potato tuber moth (1) を見よ
tobacco split worm			tobacco stem borer (1) を見よ
tobacco splitworm			potato tuber moth (1) を見よ　米国での英名
tobacco stalk borer		*Trichobaris mucorea* (LeConte)	（コウチュウ目、ゾウムシ科）新北区
tobacco stem borer (1)		*Scrobipalpa heliopa* (Lower)	（チョウ目、キバガ科）旧北区、東洋区、豪州区、エチオピア区
tobacco stemborer		*Scrobipalpa aptotella* (Lower)	（チョウ目、キバガ科）豪州区
tobacco striped caterpillar			bordered sallow を見よ
tobacco thrips		*Frankliniella fusca* (Hinds)	（アザミウマ目、アザミウマ科）日本、新北区
tobacco thrips		*Anaphothrips cecili* Girault	（アザミウマ目、アザミウマ科）豪州区
tobacco thrips			onion thrips を見よ
tobacco webworm		*Crambus caliginosellus* Clemens	（チョウ目、メイガ科）新北区
tobacco whitefly	タバココナジラミ	*Bemisia tabaci* (Gennadius)	（カメムシ目、コナジラミ科）日本、汎世界。タバコ、ナス、ワタなど多くの作物、花害虫
tobacco wireworm		*Conoderus vespertinus* (Fabricius)	（コウチュウ目、コメツキムシ科）新北区
tobacco worm		*Manduca sexta* (Linnaeus)	（チョウ目、スズメガ科）新北区、新熱帯区。tobacco hornworm, goliath worm ともいう
tobira psylla	トベラキジラミ	*Psylla tobirae* Miyatake	（カメムシ目、キジラミ科）日本。トベラ害虫
Todd's skipper		*Ridens toddi* Steinhauser	（チョウ目、セセリチョウ科）新熱帯区
toe biter		*Abedus indentatus* (Haldeman)	（カメムシ目、コオイムシ科）新北区
toe-biter			water scorpion (1) を見よ
toe-biters			giant water bugs (1) を見よ
toe-biters			creeping water bugs を見よ
toed-winged beetles			soft-bodied plant beetles を見よ
Togarna hairstreak		*Arawacus togarna* (Hewitson)	（チョウ目、シジミチョウ科）新熱帯区。Togarna stripestreak ともいう
tokoriro			Auckland tree weta を見よ
toktokkieo		*Psammodes reichei* Solier	（コウチュウ目、ゴミムシダマシ科）エチオピア区。無翅。腹部で地面をたたき求愛音を出す
Tokyo scale	トウキョウカキカイガラムシ	*Andaspis tokyoensis* Takagi et Kawai	（カメムシ目、マルカイガラムシ科）日本
Tolima numberwing		*Callicore tolima* (Hewitson)	（チョウ目、タテハチョウ科）新熱帯区
Tolosa tigerwing		*Napeogenes tolosa* (Hewitson)	（チョウ目、タテハチョウ科）新熱帯区
Toltec Emesis		*Emesis toltec* Reakirt	（チョウ目、シジミタテハ科）新熱帯区
Toltec roadside-skipper		*Amblyscirtes tolteca* Scudder	（チョウ目、セセリチョウ科）新北区
Toltec scoliid		*Campsomeris tolteca* (Saussure)	（ハチ目、ツチバチ科）新北区

英名	和名	学名	所属、分布、ほか
toltecan grouse locust		*Paratettix toltecus* (Saussure)	（バッタ目、ヒシバッタ科）新北区
tom breeze			dragonflies (1) を見よ
Tom-spinners			crane flies (1) を見よ
Tom's mountain satyr		*Eretris depresissima* Pyrcz	（チョウ目、タテハチョウ科）新熱帯区
tomato		*Temenis laothoe* Cramer	（チョウ目、タテハチョウ科）新熱帯区
tomato and bean bug			southern green stink bug を見よ
tomato aphid			potato aphid (1)(2) を見よ
tomato armyworm			Egyptian cotton leafworm を見よ
tomato budworm			tobacco budworm を見よ
tomato bug		*Cyrtopeltis modestus* (Distant)	（カメムシ目、カスミカメムシ科）新北区。トマト害虫
tomato caterpillar			common cutworm を見よ　幼虫の英名
tomato fruit borer		*Neoleucinodes elegantalis* Guenée	（チョウ目、メイガ科）新熱帯区。成虫は tomato fruit borer moth
tomato fruitworm			corn earworm (1)(2) を見よ
tomato fruitworm moth			corn earworm (2) を見よ
tomato grub			corn earworm (2) を見よ
tomato hornworm		*Protoparce quinquemaculata* (Haworth)	（チョウ目、スズメガ科）新北区。トマト他の害虫。tomato worm ともいう
tomato hornworm			tobacco hornworm を見よ
tomato leaf-folder			eggplant leaf miner (1) を見よ
tomato leafhopper			vegetable leafhopper を見よ
tomato leaf miner		*Liriomyza solani* Hering	（ハエ目、ハモグリバエ科）旧北区
tomato leafminer			bryony leafminer を見よ
tomato leafminer			vegetable leafminer を見よ
tomato leafminer			South American tomato pinworm を見よ
tomato mirid		*Engytatus nicotianae* (Koningsberger)	（カメムシ目、カスミカメムシ科）豪州区
tomato mirid			tobacco leaf bug を見よ
tomato moth			bright-line brown-eye を見よ
tomato moth			corn earworm (2) を見よ　豪州での英名
tomato pinworm		*Keiferia lycopersicella* (Walsingham)	（チョウ目、キバガ科）全北区、大洋区、エチオピア区、新熱帯区、大洋区。成虫は tomato pinworm moth
tomato psylla			tomato psyllid を見よ
tomato psyllid		*Paratrioza cockerelli* (Suke)	（カメムシ目、キジラミ科）新北区
tomato sphinx			tobacco worm を見よ
tomato sphinx moth			tobacco hornworm を見よ
tomato stalk-borer			common stalk borer を見よ
tomato stemborer		*Symmetrischema tangolias* (Geyen)	（チョウ目、キバガ科）豪州区

英名	和名	学名	所属、分布、ほか
tomato thrips			yellow flower thrips を見よ
tomato thrips			grain thrips を見よ
tomato worm			Egyptian cotton leafworm を見よ
tomato worm			corn earworm (2) を見よ
Tommy-longlegs			crane flies (1) を見よ
tomocerid springtails	トゲトビムシ科	Tomoceridae	（トビムシ目）の昆虫の総称
Tongida hairstreak		*Micandra tongida* Clench	（チョウ目、シジミチョウ科）新熱帯区
tonto dancer		*Argia tonto* Calvert	（トンボ目、イトトンボ科）新北区
toog long-winged planthopper	アカメガシワハネビロウンカ	*Vekunta malloti* Matsumura	（カメムシ目、ハネナガウンカ科）日本、東洋区
toog pyrrhocorid	オオホシカメムシ	*Physopella gutta* (Burmeister)	（カメムシ目、ホシカメムシ科）日本、東洋区、豪州区
toon borer			mahogany shoot borer (1) を見よ
toon shoot borer			mahogany shoot borer (1) を見よ
toothlegged flea beetle			brassy flea beetle (1) を見よ
tooth-legged grasshoppers		Gomphocerinae	（バッタ目、バッタ科）の昆虫の総称
tooth-necked fungus beetles	マキムシモドキ科	Derodontidae	（コウチュウ目）の昆虫の総称
tooth-necked long-horns			prionine longhorns を見よ
tooth-nosed snout beetles			leaf-and-bud weevils を見よ
tooth-nosed weevils			weevils (1) を見よ
tooth-streaked hooked smudge		*Harpipteryx xylostella* (Linnaeus)	（チョウ目、スガ科）旧北区、新北区
toothed Apharetra moth		*Sympistis dentata* (Grote)	（チョウ目、ヤガ科）新北区
toothed brown carpet moth		*Xanthorhoe lacustrata* (Guenée)	（チョウ目、シャクガ科）新北区
toothed citrus swallowtail		*Papilio erithonioides* Grose-Smith	（チョウ目、アゲハチョウ科）エチオピア区
toothed cream stripe		*Centroctena imitaus* (Butler)	（チョウ目、スズメガ科）エチオピア区
toothed earwig		*Spongovostox apicedentatus* (Caudell)	（ハサミムシ目、チビハサミムシ科）新北区
toothed flea beetle		*Chaetocnema denticulata* (Illiger)	（コウチュウ目、ハムシ科）新北区。トウモロコシの病害を伝搬
toothed hunter		*Austrogomphus arbustorum* Tillyard	（トンボ目、サナエトンボ科）豪州区
toothed-legged turnip beetle			brassy flea beetle (1) を見よ
toothed Phigalia moth		*Phigalia denticulata* Hulst	（チョウ目、シャクガ科）新北区
toothed pondskater		*Gerris odontogaster* (Zetterstedt)	（カメムシ目、アメンボ科）旧北区
toothed snout-moth			dimorphic Bomolocha moth を見よ

英名	和名	学名	所属、分布、ほか
toothed somberwing moth		*Euclidia cuspidea* (Hübner)	（チョウ目、ヤガ科）新北区。toothed somberwing ともいう
toothed sunbeam		*Curetis dentata* Moore	（チョウ目、シジミチョウ科）東洋区
top borer	シロオオメイガ	*Scirpophaga excerptalis* (Walker)	（チョウ目、メイガ科）日本、東洋区。サトウキビ害虫
top borer (1)	ツマキオオメイガ	*Scirpophaga nivella* (Fabricius)	（チョウ目、メイガ科）日本、旧北区、東洋区。サトウキビ害虫
top end dragon		*Antipodogomphus dentosus* Watson	（トンボ目、サナエトンボ科）豪州区
top-horned hunchback		*Paracrocera orbiculus* (Fabricius)	（ハエ目、コガシラアブ科）旧北区
Topax Arab	アマタツマアカシロチョウ	*Colotis amata* (Fabricius)	（チョウ目、シロチョウ科）東洋区、エチオピア区。亜種 *C. a. calais* (Cramer) も同英名
topaz Arab			small salmon Arab を見よ
topaz blue		*Azanus jesous* (Guérin-Méneville)	（チョウ目、シジミチョウ科）エチオピア区
topaz-spotted bush-blue			Natal Babul blue を見よ
topers			common skimmers を見よ
topiary grass-veneer moth			cranberry girdler を見よ
toringo crab bark beetle	オオザイノキクイムシ	*Xyleborus montanus* Niijima	（コウチュウ目、キクイムシ科）日本
tornado wasp		*Episyron biguttatus biguttatus* (Fabricius)	（ハチ目、ベッコウバチ科）新北区
torpedo bug		*Siphonta acuta* (Walker)	（カメムシ目、アオバハゴロモ科）大洋区
torpedo bug			green leafhopper (3) を見よ
torpedo bug planthopper			green leafhopper (3) を見よ
Torquatus swallowtail	マルバネキオビアゲハ	*Papilio torquatus* (Cramer)	（チョウ目、アゲハチョウ科）新熱帯区。*P. t. mazai* Beutelspacher, *P. t. tolus* (Godman et Salvin) も同英名
Torquoise eyed-metalmark		*Mesosemia gemina* Maza et Maza	（チョウ目、シジミタテハ科）新熱帯区
torsalo			human bot fly を見よ
tortoise beetle	ジンガサハムシ	*Aspidomorpha indica* Boheman	（コウチュウ目、ハムシ科）日本、旧北区、東洋区
tortoise beetle (1)	ヒメカメノコハムシ	*Cassida piperata* Hope	（コウチュウ目、ハムシ科）日本、旧北区、東洋区
tortoise beetle			mottled tortoise beetle を見よ
tortoise beetle			gold bug (1) を見よ
tortoise beetle			artichoke tortoise beetle を見よ
tortoise beetle			green tortoise beetle (1) を見よ
tortoise beetle			beet tortoise beetle (2) を見よ
tortoise beetles (1)	カメノコハムシ亜科	Cassidinae	（コウチュウ目、ハムシ科）の昆虫の総称
tortoise beetles		*Cassida*	（コウチュウ目、ハムシ科）の昆虫の総称
tortoise beetles			leaf beetles (1) を見よ

英名	和名	学名	所属、分布、ほか
tortoise beetles			sweetpotato tortoise beetles を見よ
tortoise scale			brown soft scale を見よ
tortoise scales			soft scales を見よ
tortoise-shaped scale			tesserated scale を見よ
tortoise-shell butterfly			タテハチョウ科のチョウ。とくにヒメヒオドシの類 (*Aglais* 属)
tortoise-shelled ladybird		*Harmonia testudinaria* (Mulsant)	（コウチュウ目、テントウムシ科）東洋区、豪州区
tortoiseshells			anglewings を見よ
tortricid moths			leafroller moths を見よ　tortricids ともいう
tortrix beetles			sweetpotato beetles を見よ
tortrix moth			summer fruit tortrix (1)(2) を見よ
tortrix moths			leafroller moths を見よ
torts		Nycteolinae	（チョウ目、ヤガ科）の昆虫の総称
torymid wasps			seed chalcids (1) を見よ　torymids ともいう
touch bell			common earwig を見よ
tow bug			cigarette beetle を見よ
town longhorn beetle			city longicorn beetle を見よ
Towne's clubtail		*Stylurus townesi* Gloyd	（トンボ目、サナエトンボ科）新北区
Toxomerus hover flis		*Toxomerus*	（ハエ目、ハナアブ科）の昆虫の総称
Toxopeus tiger	トクソペイアサギマダラ	*Parantica toxopei* (Nieuwenthuis)	（チョウ目、タテハチョウ科）東洋区
Toxopeus' yellow tiger			Toxopeus tiger を見よ
toy beetle of the Philippines		*Leucopholis irrorata* (Chevrolat)	（コウチュウ目、コガネムシ科）東洋区。サトウキビなどの害虫
toyon leaf miner		*Stigmella heteromelis* Newton et Wilkinson	（チョウ目、モグリチビガ科）新北区
Tracta sister		*Adelpha tracta* (Butler)	（チョウ目、タテハチョウ科）新北区、新熱帯区
trail gnat		*Amiota picta* (Coquillett)	（ハエ目、ショウジョウバエ科）新北区、新熱帯区
trailside skipper			dimorphic grass skipper (1) を見よ
Trajan's forest queen		*Charaxes trajanus* (Ward)	（チョウ目、タテハチョウ科）エチオピア区
tramp ant			ghost ant を見よ
transfigured Hydriomena moth		*Hydriomena transfigurata* Swett	（チョウ目、シャクガ科）新北区
translucent straw pearl		*Microstega hyalinalis* (Hübner)	（チョウ目、メイガ科）旧北区
transparent burnet		*Zygaena purpuralis* (Brunnich)	（チョウ目、マダラガ科）旧北区
transparent scale			coconut scale (1) を見よ　豪州での英名
transparent sixline blue	アマミウラナミシジミ	*Nacaduba kurava* (Moore)	（チョウ目、シジミチョウ科）東洋区。日本亜種は *N. k. septentrionalis* Shirozu アマミウラナミシジミ
transparent 6-lineblue			white-line blue (1) を見よ

英名	和名	学名	所属、分布、ほか
transparent sweep		*Epichnopteryx plumella* (Denis et Schiffermüller)	（チョウ目、ミノガ科）旧北区
transparent sweep		*Epichnopteryx pulla* Esper	（チョウ目、ミノガ科）旧北区
transparent tussock	スキバドクガ	*Perina nuda* (Fabricius)	（チョウ目、ドクガ科）日本
transparent-winged plant bug		*Hyalopelpus pellucidus* (Stål)	（カメムシ目、カスミカメムシ科）新北区、大洋区
Transvaal copper		*Aloeides dryas* Tite et Dickson	（チョウ目、シジミチョウ科）エチオピア区
transverse flower fly		*Eristalis transversa* (Wiedemann)	（ハエ目、ハナアブ科）新北区
transverse ladybird		*Coccinella transversalis* Fabricius	（コウチュウ目、テントウムシ科）豪州区
transvestite clubtail		*Cyclogomphus gynostylus* Fraser	（トンボ目、サナエトンボ科）東洋区
transvolcanic Zestusa		*Zestusa* sp.	（チョウ目、セセリチョウ科）新熱帯区
trapdoor spider wasp		*Psammochares plantus* (Fox)	（ハチ目、Psammocharidae）新北区
trapeze moth			maize webworm を見よ
trapezites			skippers (1) を見よ　豪州での英名
Traun's black & white piercer		*Pammene regiana* (Zeller)	（チョウ目、ハマキガ科）旧北区
Travancore evening brown		*Parantirrhoea marshalli* Wood-Mason	（チョウ目、タテハチョウ科）東洋区
traveller			bent-line carpet moth を見よ
treble-atomed dwarf		*Elachista triatomea* Meyrick	（チョウ目、クサモグリガ科）旧北区
treble-bar		*Aplocera plagiata* (Linnaeus)	（チョウ目、シャクガ科）全北区。treble-bar moth ともいう
treble brown-spot		*Idaea trigeminata* (Haworth)	（チョウ目、シャクガ科）旧北区
treble lines		*Charanyca trigrammica* (Hufnagel)	（チョウ目、ヤガ科）旧北区。treble-line ともいう
treble silverstripe		*Lethe baladeva* (Moore)	（チョウ目、タテハチョウ科）東洋区
treble spot button		*Acleris tripunctana* Hübner	（チョウ目、ハマキガ科）旧北区
treble-spot button		*Acleris notana* (Donovan)	（チョウ目、ハマキガ科）旧北区
treble-spot wave			treble brown spot を見よ
treble-striped midget		*Phyllonorycter tristrigella* (Haworth)	（チョウ目、ホソガ科）旧北区
Trebula groundstreak		*Calycopis trebula* (Hewitson)	（チョウ目、シジミチョウ科）新熱帯区
tree-beetle			June beetle (1) を見よ
tree-browns			browns (1) を見よ
tree bumblebee		*Bombus hypnorum* Linnaeus	（ハチ目、ミツバチ科）旧北区。日本には *B. h. koropokkrus* Sakagami et Ishikawa アカマルハナバチが分布
tree-chafer			June beetle (1) を見よ

英名	和名	学名	所属、分布、ほか
tree cricket		*Paragryllacris combusta* (Gerstaecker)	（バッタ目、コロギス科）豪州区
tree cricket		*Oecanthus turanicus* Uvarov	（バッタ目、カンタン科）旧北区
tree cricket (1)		*Oecanthus pellucens* Scopoli	（バッタ目、カンタン科）旧北区
tree cricket (2)	タイワンカンタン	*Oecanthus rufescens* Serville	（バッタ目、カンタン科）日本、東洋区、豪州区
tree cricket (3)	カンタン	*Oecanthus longicaudus* Matsumura	（バッタ目、カンタン科）日本、旧北区
tree crickets		Oecanthinae	（バッタ目、カンタン科）の昆虫の総称　単眼を欠き、鳴声は大きい
tree crickets (1)		*Oecanthus*	（バッタ目、カンタン科）の昆虫の総称
tree crickets			cave crickets (1) を見よ
tree crickets			white tree crickets を見よ
tree crickets			long-horned grasshoppers (3) を見よ
tree damsel bug	ハラビロマキバサシガメ	*Himacerus apterus* (Fabricius)	（カメムシ目、マキバサシガメ科）日本、旧北区
tree flitter	ニセクロセセリ	*Hyarotis adrastus* (Stoll)	（チョウ目、セセリチョウ科）東洋区。*H. a. praba* Moore も同英名
tree grayling	クロイチモジヒカゲ	*Hipparchia statilinus* (Hufnagel)	（チョウ目、タテハチョウ科）旧北区
tree-hole mosquito		*Aedes triseriatus* (Say)	（ハエ目、カ科）新北区
treehopper			horned treehopper (1) を見よ
tree hopper		*Hoplophorion (Metcalfiella) pertusa* Germar	（カメムシ目、ツノゼミ科）旧北区。カンキツ害虫
treehoppers		Eurymelidae	（カメムシ目）の昆虫の総称
treehoppers (1)	ツノゼミ科	Membracidae	（カメムシ目）の昆虫の総称　世界に 2,400 種
tree lice			aphids (2) を見よ
tree-lichen beauty		*Cryphia algae* (Fabricius)	（チョウ目、ヤガ科）旧北区
tree-lichens		Cryphinae	（チョウ目、ヤガ科）の昆虫の総称
treeline emerald		*Somatochlora sahlbergi* Trybom	（トンボ目、エゾトンボ科）旧北区
tree lobster			Lord Howe Island stick insect を見よ
tree locust			Egyptian grasshopper を見よ
tree lucerne moth		*Uresiphita ornithopteralis* (Guenée)	（チョウ目、メイガ科）豪州区
tree nymph	ウスイロオオゴマダラ	*Idea lynceus* (Drury)	（チョウ目、タテハチョウ科）東洋区
tree nymphs		*Sallya*	（チョウ目、タテハチョウ科）の昆虫の総称
tree nymphs (1)	オオゴマダラ属	*Idea*	（チョウ目、タテハチョウ科）の昆虫の総称
tree of heaven Eligma	シンジュキノカワガ	*Eligma narcissus* (Cramer)	（チョウ目、ヤガ科）日本
tree pear beetle			opuntia biocontrol beetle を見よ
tree snipefly		*Chrysopilus laetus* Zetterstedt	（ハエ目、シギアブ科）旧北区
tree termite			nest building termite を見よ

英名	和名	学名	所属、分布、ほか
tree thrips		Heterothripidae	（アザミウマ目）の昆虫の総称　南北アメリカに80種
tree top Acraea		*Hyalites cerasa* Hewitson	（チョウ目、タテハチョウ科）エチオピア区。tree-top Acraea とも記す
treetop bush katydid		*Scudderia fasciata* Beutenmüller	（バッタ目、キリギリス科）新北区
tree wasp		*Dolichovespula sylvestris* (Scopoli)	（ハチ目、スズメバチ科）旧北区
tree weta		*Hemideina*	（バッタ目、Stenopelmatidae）の昆虫の総称
tree yellow	ムモンキチョウ	*Gandaca harina* (Horsfield)	（チョウ目、シロチョウ科）東洋区。*G. h. distanti* Fruhstorfer も同英名
tree yellows	ムモンキチョウ属	*Gandaca*	（チョウ目、シロチョウ科）の昆虫の総称
trefoil flower midge		*Contarinia loti* (De Geer)	（ハエ目、タマバエ科）旧北区
trefoil seed chalcid			clover seed chalcid (2) を見よ
trefoil seed chalcid		*Bruchophagus kolovae* Fedoseeva	（ハチ目、カタビロコバチ科）旧北区
Trembath's gem		*Chloroselas trembathi* Collins et Larsen	（チョウ目、シジミチョウ科）エチオピア区
Tremona glasswing		*Oleria tremona* Haensch	（チョウ目、タテハチョウ科）新熱帯区
trent double stripe		*Clytie illunaris* (Hübner)	（チョウ目、ヤガ科）旧北区
Tres cruces skipper		*Serdis viridicans* (Felder et Felder)	（チョウ目、セセリチョウ科）新熱帯区
Tres Marias pipevine swallowtail		*Battus philenor orsua* (Godman et Salvin)	（チョウ目、アゲハチョウ科）新熱帯区。pipevine swallowtail を参照
triangle		*Heterogenea asellana* Hübner	（チョウ目、イラガ科）旧北区
triangle (1)	カギバイラガ	*Heterogenea asella* (Denis et Schiffermüller)	（チョウ目、イラガ科）日本、旧北区
triangle-backed Eucosma moth		*Eucosma dorsisignatana* (Clemens)	（チョウ目、ハマキガ科）新北区
triangle-marked piercer		*Pammene spiniana* (Duponchel)	（チョウ目、ハマキガ科）旧北区
triangle-marked plume			triangle plume を見よ
triangle-marked slender	ヤナギハマキホソガ	*Caloptilia stigmatella* (Fabricius)	（チョウ目、ホソガ科）日本、旧北区、東洋区、エチオピア区
triangle moth			triangle (1) を見よ
triangle plume		*Platyptilia gonodactyla* Denis et Schiffermüller	（チョウ目、トリバガ科）旧北区
triangle skimmer		*Orthetrum triangulare triangulare* (Selys)	（トンボ目、トンボ科）東洋区
triangular Eumorpha		*Eumorpha triangulum* (Rothschild et Jordan)	（チョウ目、スズメガ科）新熱帯区
triangular Saliana		*Saliana triangularis* (Kaye)	（チョウ目、セセリチョウ科）新熱帯区
trichiine beetles	トラハナムグリ亜科	Trichiinae	（コウチュウ目、コガネムシ科）の昆虫の総称
trichodectid lice			mammal chewing lice を見よ
trichogrammatid egg parasites			minute egg parasites を見よ

英名	和名	学名	所属、分布、ほか
trichogrammatid wasps			minute egg parasites を見よ
trichogrammatids			minute egg parasites を見よ
trichopterans			caddisflies (1) を見よ
trickling filter fly		*Psychoda alternata* (Say)	（ハエ目、チョウバエ科）汎世界
tricolor buck moth		*Hemileuca tricolor* (Packard)	（チョウ目、ヤママユガ科）新北区、新熱帯区
tricolored Acrobasis moth		*Acrobasis tricolorella* Grote	（チョウ目、メイガ科）新北区
tri-colored bumble bee			red-tailed bumble bee (2) を見よ
tricolored Cosipara moth		*Cosipara tricoloralis* (Dyar)	（チョウ目、メイガ科）新北区
tricolored metalmark		*Symmachia tricolor* Hewitson	（チョウ目、シジミタテハ科）新熱帯区
tricolour pied flat		*Coladenia indrani* (Moore)	（チョウ目、セセリチョウ科）東洋区
tricose dart			confused dart moth を見よ
trident pencil-blue		*Candalides margarita* (Semper)	（チョウ目、シジミチョウ科）豪州区
trifid duskhawker		*Agyrtacantha dirupta* Karsch	（トンボ目、ヤンマ科）豪州区
trigonalid wasps			trigonalids を見よ
trigonalids	カギバラバチ科	Trigonalidae	（ハチ目）の昆虫の総称
trigonidiid crickets			bush crickets (3) を見よ
trigs			bush crickets (3) を見よ
trilobe scale	コバンマルカイガラムシ	*Pseudaonidia trilobitiformis* (Green)	（カメムシ目、マルカイガラムシ科）日本、汎熱帯。カンキツ害虫
trilobite cockroach		*Laxta* sp.	（ゴキブリ目、ブラベルスゴキブリ科）豪州区。生活史の殆どを落葉、朽木下で過ごす豪州固有種
trilobite scale			trilobe scale を見よ
Trimen's Acraea		*Acraea trimeni* Aurivillius	（チョウ目、タテハチョウ科）エチオピア区
Trimen's blue		*Lepidochrysops trimeni* (Bethune-Baker)	（チョウ目、シジミチョウ科）エチオピア区
Trimen's brown		*Pseudonympha trimenii* Butler	（チョウ目、タテハチョウ科）エチオピア区
Trimen's ciliate blue		*Anthene otacilia* (Trimen)	（チョウ目、シジミチョウ科）エチオピア区
Trimen's copper		*Aloeides trimeni* Tite et Dickson	（チョウ目、シジミチョウ科）エチオピア区
Trimen's dotted border		*Mylothris trimenia* (Butler)	（チョウ目、シロチョウ科）エチオピア区
Trimen's false Acraea			Boisduval's false Acraea を見よ
Trimen's knob		*Acantholipes trimeni* Felder et Rogenhofer	（チョウ目、ヤガ科）エチオピア区
Trimen's opal		*Chrysoritis trimeni* (Riley)	（チョウ目、シジミチョウ科）エチオピア区
Trimen's sapphire		*Iolaus trimeni* Wallengren	（チョウ目、シジミチョウ科）エチオピア区
trimerans			trimerous beetles を見よ

英名	和名	学名	所属、分布、ほか
trimerous beetles		Trimera	チャタテムシ目の異節亜目 Heterotecnomera の異名
trimmed skipper		*Cogia aventinus* (Godman et Salvin)	(チョウ目、セセリチョウ科) 新熱帯区
Trinidad and Tobago silk-moth		*Rothschildia hesperus* Linnaeus	(チョウ目、ヤママユガ科) 新熱帯区
Trinidad bollworm			South American red bollworm を見よ
Trinidad stink bug tachinid		*Trichopoda pennipes pilipes* (Fabricius)	(ハエ目、ヤドリバエ科) 大洋区
triple-barred		*Mocis trifasciata* (Stephens)	(チョウ目、ヤガ科) 旧北区
triple-blotched bell		*Epiblema trimaculana* (Haworth)	(チョウ目、ハマキガ科) 旧北区
triple-dot Iliana		*Iliana purpurascens* (Mabille et Boullet)	(チョウ目、セセリチョウ科) 新熱帯区
triple-spotted clay		*Xestia ditrapezium* (Denis et Schiffermüller)	(チョウ目、ヤガ科) 旧北区。日本亜種は *X. d. orientalis* (Strand) タンポヤガ
triple-spotted dwarf		*Elachista cerusella* (Hübner)	(チョウ目、クサモグリガ科) 旧北区
triple-spotted pug		*Eupithecia trisignaria* Herrich-Schäffer	(チョウ目、シャクガ科) 旧北区
triple-striped piercer		*Cydia compositella* Fabricius	(チョウ目、ハマキガ科) 旧北区
triple swallowtail			three-tailed swallowtail を見よ
triplex cutworm moth		*Micrathetis triplex* (Walker)	(チョウ目、ヤガ科) 新北区
Triton dagger moth		*Acronicta tritona* (Hübner)	(チョウ目、ヤガ科) 新北区
triungulids			blister beetles (1) を見よ　幼虫の英名
trixoscelid flies		Trixoscelidae	(ハエ目) の昆虫の総称　Trichoscelidae の異名
trixoscelidid flies			trixoscelid flies を見よ
trixoscelids			trixoscelid flies を見よ
trogiid booklice			granary booklice を見よ
trogiids			booklice (2) を見よ
trogiids			granary booklice を見よ
trogositid beetles			gnawing beetles (1) を見よ
tropic queen			soldier (1) を見よ
tropical American silkworm moths		Oxytenidae	(チョウ目) の昆虫の総称
tropical and subtropical green bottle-flies		*Chrysomya*	(ハエ目、クロバエ科) の昆虫の総称
tropical Anomis moth			cotton semi-looper を見よ　tropical Anomis ともいう
tropical barklice	ウロコチャタテ科	Amphientomidae	(チャタテムシ目) の昆虫の総称　主に熱帯に分布
tropical bed bug			bed bug (2) を見よ
tropical blue wave			whitened bluewing を見よ

英名	和名	学名	所属、分布、ほか
tropical buckeye		*Junonia genoveva zonalis* (C. et R. Felder)	（チョウ目、タテハチョウ科）新北区。black buckeye を参照
tropical buckeye			West Indian buckeye を見よ
tropical burnet moths		Lacturidae	（チョウ目）の昆虫の総称
tropical bush cricket		*Phlugiolopsis henryi* Zeuner	（バッタ目、キリギリス科）旧北区
tropical carpenter bee		*Xylocopa latipes* (Drury)	（ハチ目、ミツバチ科）東洋区
tropical carpenterworm moths		Metarbelidae	（チョウ目）の昆虫の総称
tropical case-bearing clothes moth			casemaking clothes moth (2) を見よ
tropical checkered skipper		*Pyrgus oileus* (Linnaeus)	（チョウ目、セセリチョウ科）新北区、新熱帯区
tropical citrus aphid			citrus brown aphid を見よ
tropical cockroach			brown-banded cockroach (1) を見よ
tropical cockroach			discoid cockroach を見よ
tropical cut-worm		*Spodoptera albula* Walker	（チョウ目、ヤガ科）新熱帯区
tropical damselflies			threadtails (1) を見よ
tropical dotted border		*Mylothris rhodope* (Fabricius)	（チョウ目、シロチョウ科）エチオピア区
tropical ermine moths		Attevidae	（チョウ目）の昆虫の総称
tropical fire ant			fire ant (2) を見よ
tropical fruit piercer			fruit-piercing moth (4) を見よ
tropical fruitworm moths		Copromorphidae	（チョウ目）の昆虫の総称
tropical gray chaff			chaff scale (2) を見よ　tropical grey chaff の使用もあり
tropical greenstreak		*Cyanophrys herodotus* (Fabricius)	（チョウ目、シジミチョウ科）新北区
tropical grey chaff scale			chaff scale (2) を見よ
tropical house cricket		*Gryllodes supplicans* (Walker)	（バッタ目、コオロギ科）旧北区
tropical house-cricket			Indian house cricket を見よ
tropical king skimmers		Orthemis	（トンボ目、トンボ科）の昆虫の総称
tropical lattice moths			lattice moths を見よ
tropical leafwing	ベニイロキノハタテハ	*Anaea aidea* Guérin-Méneville	（チョウ目、タテハチョウ科）新北区
tropical least skipperling		*Ancyloxypha arene* Edwards	（チョウ目、セセリチョウ科）新北区、新熱帯区
tropical legume leaf beetle			bean leaf beetle (1) を見よ

英名	和名	学名	所属、分布、ほか
tropical log beetles	ミジンムシダマシ科	Discolomidae	（コウチュウ目）の昆虫の総称
tropical longhorned moths			longhorned moths を見よ
tropical meal moth		Pyralis manihotalis Guenée	（チョウ目、メイガ科）全北区、東洋区、豪州区、大洋区、エチオピア区、新熱帯区
tropical milkweed butterfly			tiger mimic-queen（1）(2) を見よ
tropical mound-building termite			mound building termite (1) を見よ
tropical nasutiform termite		Nasutitermes corniger (Motschulsky)	（シロアリ目、シロアリ科）新北区
tropical nut borer			apple twig borer (1) を見よ
tropical palm scale	ジャワマルカイガラムシ	Dynaspidiotus palmae (Cockerell)	（カメムシ目、マルカイガラムシ科）日本、汎熱帯。バナナ、パイナップル害虫
tropical paper wasp		Polistes stigma (Fabricius)	（ハチ目、スズメバチ科）豪州区
tropical Pentila			spotted buff を見よ
tropical plume moths		Oxychirotidae	（チョウ目）の昆虫の総称
tropical queen			soldier を見よ
tropical rat flea			Oriental rat flea を見よ
tropical rat louse		Hoplopleura pacifica Ewing	（シラミ目、フトゲシラミ科）新北区
tropical rat louse		Hoplopleura oenomydis Ferris	（シラミ目、フトゲシラミ科）旧北区
tropical rice bug	タイワンクモヘリカメムシ	Leptocorisa oratorius (Fabricius)	（カメムシ目、ホソヘリカメムシ科）日本、東洋区、豪州区。イネ害虫
tropical rice bug			rice bug (3) を見よ
tropical rough-headed drywood termite			West Indian drywood termite を見よ
tropical shield mantis		Choevadodis stalii Wood-Mason	（カマキリ目、カマキリ科）新熱帯区
tropical shutwing		Cordulephya bidens Sjöstedt	（トンボ目、エゾトンボ科）豪州区
tropical slug caterpillars		Dalceridae	（チョウ目）の昆虫の総称
tropical smooth-headed drywood termite		Cryptotermes cavifrons Banks	（シロアリ目、レイビシロアリ科）新北区
tropical sod webworm	ケナシクロオビクロノメイガ	Herpetogramma phaeopteralis Guenée	（チョウ目、メイガ科）日本、東洋区、豪州区、新北区、新熱帯区
tropical soda apple leaf beetle		Gratiana boliviana Spaeth	（コウチュウ目、ハムシ科）新熱帯区
tropical striped blue			cassius blue を見よ
tropical swallowtail moth	オオツバメガ	Lyssa zampa (Butler)	（チョウ目、ツバメガ科）日本、東洋区
tropical tobacco moth		Setomorpha rutella Zeller	（チョウ目、ヒロズコガ科）全北区、東洋区、大洋区、豪州区、新熱帯区、エチオピア区

英名	和名	学名	所属、分布、ほか
tropical unicorn darner		*Austroaeschna speciosa* Sjöstedt	(トンボ目、ヤンマ科) 豪州区
tropical warble-fly			human bot fly を見よ
tropical warehouse moth			almond moth を見よ　豪州での英名
tropical white		*Appias drusilla* (Cramer)	(チョウ目、シロチョウ科) 新北区
tropical yellow			Xanthochlora grass yellow を見よ
tropiduchid planthoppers	グンバイウンカ科	Tropiduchidae	(カメムシ目) の昆虫の総称　熱帯に多い
troscid beetles	ヒゲブトコメツキ科	Troscidae	(コウチュウ目) の昆虫の総称
trout-stream beetles		Amphizoidae	(コウチュウ目) の昆虫の総称　成虫幼虫とも半水生
true armyworm			armyworm (1) を見よ
true bark beetles		Scolytinae	(コウチュウ目、キクイムシ科) の昆虫の総称
true bed bug			bed bug (1) を見よ
true bees			bees (1) を見よ
true bees			honey bees を見よ
true brushfoots			brushfoots を見よ
true bugs	異翅亜目	Heteroptera	(カメムシ目) の昆虫の総称　世界に約 50,000 種
true cattleheart		*Parides eurimedes mylotes* (H. W. Edwards)	(チョウ目、アゲハチョウ科) 新熱帯区
true cave crickets		Dolichopodini	(バッタ目、コロギス科) の昆虫の総称
true crickets			crickets (1) を見よ
true crickets			field crickets (1) を見よ
true death's head cockroach			death's head cockroach を見よ
true flies			flies を見よ
true forester		*Euphaedra cyparissa* (Cramer)	(チョウ目、タテハチョウ科) エチオピア区
true fritillaries			fritillaries (1) を見よ
true grasshoppers		Acridinae	(バッタ目、バッタ科) の昆虫の総称
true ichneumon flies		Ichneumoninae	(ハチ目、ヒメバチ科) の昆虫の総称
true katydid		*Pterophylla camellifolia* (Fabricius)	(バッタ目、キリギリス科) 新北区。鳴声が katy-did, katy-didn't として知られる
true katydids	ヒラタツユムシ亜科	Pseudophyllinae	(バッタ目、キリギリス科) の昆虫の総称
true leaf hoppers		Jassinae	(カメムシ目、オオヨコバイ科) の昆虫の総称
true lice			sucking lice を見よ
true lover's knot		*Lycophotia porphyrea* (Denis et Schiffermüller)	(チョウ目、ヤガ科) 全北区、大洋区
true lover's knot		*Lycophotia varia* Villers	(チョウ目、ヤガ科) 旧北区
true meal moth			meal moth を見よ
true mosquitoes			malaria mosquitoes を見よ
true mosquitos	カ亜科	Culicinae	(ハエ目、カ科) の昆虫の総称

英名	和名	学名	所属、分布、ほか
true powder-post beetle			oak lyctid を見よ
true powder post beetles			lyctids を見よ
true sawflies			sawflies (1) を見よ
true scorpionflies			scorpion flies (1) を見よ
true snout beetles			weevils (1) を見よ　南アフリカでの英名
true stoneflies			common stoneflies (1) を見よ
true wasps			hornets (2) を見よ
true wasps			wasps (1) を見よ
true water beetles			diving beetles (1) を見よ
true water bugs			short-horned bugs を見よ
true weevils			weevils (1) を見よ
true wireworm			sugarcane wireworm を見よ
trumpet fly			sheep bot fly を見よ
trumpet leaf miner		*Tischeria complanella* (Hübner)	（チョウ目、ムモンハモグリガ科）全北区、豪州区
trumpet leaf miner			red-feather carl を見よ
trumpet leafminer moths	ムモンハモグリガ科	Tischeriidae	（チョウ目）の昆虫の総称
trumpet-net caddisflies (1)	クダトビケラ科	Psychomyidae	（トビケラ目）の昆虫の総称
trumpet-net caddisflies (2)	イワトビケラ科	Polycentropodidae	（トビケラ目）の昆虫の総称
trumpet vine moth		*Clydonopteron sacculana* (Bosc)	（チョウ目、メイガ科）新北区
truncate imperial		*Cheritrella truncipennis* de Nicéville	（チョウ目、シジミチョウ科）東洋区
truncate-tipped geometrid	ツマキリエダシャク	*Endropiodes abjectus* (Butler)	（チョウ目、シャクガ科）日本、旧北区
truncate-tipped leaf-like moth			spectacle (2) を見よ
truncated true katydid		*Paracyrtophyllus robustus* Caudell	（バッタ目、キリギリス科）新北区
trunk borer		*Diploschema rotundicolle* Serville	（コウチュウ目、カミキリムシ科）新熱帯区。カンキツ害虫
trunk borer			horse-bean longhorn を見よ
tsetse	ツエツエバエ	*Glossina morsitans* Westwood	（ハエ目、イエバエ科）エチオピア区。trypanosome の中間宿主となり、家畜にナガナ病を、人間に眠り病を伝染させる。tsetse はアフリカ南部の現地語
tse-tse flies		*Glossina*	（ハエ目、イエバエ科）の昆虫の総称
tsetse-flies		Glossininae	（ハエ目、イエバエ科）の昆虫の総称
tsetse-fly		*Glossina palpalis* (Robineau-Desvoidy)	（ハエ目、イエバエ科）エチオピア区
tsetse-fly			tsetse を見よ

英名	和名	学名	所属、分布、ほか
Tsomo blue		*Harpendyreus tsomo* (Trimen)	(チョウ目、シジミチョウ科) エチオピア区
Tsomo River copper		*Chrysoritis lyncurium* (Trimen)	(チョウ目、シジミチョウ科) エチオピア区
Tsomo River opal			Tsomo River copper を見よ
Tsukada's hedge blue		*Uranobothria tsukadai* Eliot et Kawazoe	(チョウ目、シジミチョウ科) 東洋区
tuart bud weevil		*Cryptoplus tibialis* (Lea)	(コウチュウ目、ゾウムシ科) 豪州区
tuart longicorn		*Phoracantha impavida* Newman	(コウチュウ目、カミキリムシ科) 豪州区
tube beetle		*Agonum maculicolle* Dejean	(コウチュウ目、オサムシ科) 新北区。ホタルイ付近に生息
tube beetles		*Agonum*	(コウチュウ目、オサムシ科) の昆虫の総称
tube billbug		*Sphenophorus discolor* Mannerheim	(コウチュウ目、ゾウムシ科) 新北区
tube-forming desert termite		*Gnathamitermes tubiformans* (Buckley)	(シロアリ目、シロアリ科) 新北区
tube-making caddisflies			trumpet-net caddisflies (1)(2) を見よ
tube moths			burrowing webworm moths を見よ
tube-net caddisflies			trumpet-net caddisflies (1)(2) を見よ
tube thrips			tube-tailed thrips を見よ
tuber flea beetle		*Epitrix tuberis* Gentner	(コウチュウ目、ハムシ科) 新北区。野菜害虫
tuber mealybug		*Pseudococcus affinis* (Maskell)	(カメムシ目、コナカイガラムシ科) 全北区、豪州区
tube-tailed thrips	クダアザミウマ科	Phlaeothripidae	(アザミウマ目) の昆虫の総称
tubercle-bearing louse			small blue cattle louse を見よ 豪州での英名
tubular black thrips		*Haplothrips victoriensis* Bagnall	(アザミウマ目、クダアザミウマ科) 豪州区
tubuliferous thrips		Tubulifera	(アザミウマ目) の昆虫の総称 tubuliferans ともいう
Turker's awlet		*Bibasis tuckeri* Elwes et Edwards	(チョウ目、セセリチョウ科) 東洋区
Tucuti flasher		*Astraptes tucuti* (Williams)	(チョウ目、セセリチョウ科) 新熱帯区
tufted apple bud moth		*Platynota idaeusalis* (Walker)	(チョウ目、ハマキガ科) 新北区
tufted bird-dropping moth		*Cerma cerintha* (Treitschke)	(チョウ目、ヤガ科) 新北区
tufted bush brown		*Bicyclus trilophus* (Rebel)	(チョウ目、タテハチョウ科) エチオピア区
tufted forest sylph		*Ceratrichia semilutea* Mabille	(チョウ目、セセリチョウ科) エチオピア区
tufted green-streaked playboy		*Pilodeudorix camerona* (Plötz)	(チョウ目、シジミチョウ科) エチオピア区
tufted jungle king		*Thauria aliris pseudaliris* Butler	(チョウ目、タテハチョウ科) 東洋区。tufted jungle queen を参照
tufted jungle queen	シロオビワモンチョウ	*Thauria aliris* (Westwood)	(チョウ目、タテハチョウ科) 東洋区

英名	和名	学名	所属、分布、ほか
tufted marble skipper		*Carcharodes flocciferus* (Zeller)	(チョウ目、セセリチョウ科) 旧北区。tufted marbled skipper ともいう
tufted royals	クロボシルリシジミ属	*Pratapa*	(チョウ目、シジミチョウ科) の昆虫の総称
tufted skipper		*Carcharodus floccifera* (Zeller)	(チョウ目、セセリチョウ科) 旧北区
tufted snout			pale Phalaenostola moth を見よ
tufted spruce caterpillar			black zigzag moth を見よ
tufted swift		*Caltoris plebeia* (de Nicéville)	(チョウ目、セセリチョウ科) 東洋区
tufted white pine caterpillar			eastern Panthea moth を見よ
Tulbagh sylph		*Tsitana tulbagha* Evans	(チョウ目、セセリチョウ科) エチオピア区
Tulcis crescent		*Phyciodes frisia tulcis* (Bates)	(チョウ目、タテハチョウ科) 新北区。Cuban crescent を参照
tule beetle			tube beetle を見よ
tule bluet		*Enallagma carunculatum* Morse	(トンボ目、イトトンボ科) 新北区
tulip aphid	チューリップネアブラムシ	*Dysaphis tulipae* (Boyer et Fonscolombe)	(カメムシ目、アブラムシ科) 日本、汎世界。チューリップ、アイリス害虫
tulip bulb aphid			tulip aphid を見よ
tuliptree aphid		*Macrosiphum liriodendri* (Monell)	(カメムシ目、アブラムシ科) 新北区
tulip-tree beauty		*Epimecis hortaria* (Fabricius)	(チョウ目、シャクガ科) 新北区。tulip-tree beauty moth ともいう
tuliptree borer			root collar borer moth を見よ
tulip tree leaf spot gall midge		*Resseliella liriodendri* (Osten Sacken)	(ハエ目、タマバエ科) 新北区
tulip-tree leaftier moth		*Paralobesia liriodendrana* (Kearfott)	(チョウ目、ハマキガ科) 新北区
tuliptree scale		*Toumeyella liriodendri* (Gmelin)	(カメムシ目、カタカイガラムシ科) 新北区
tulip-tree silkmoth		*Callosamia angulifera* (Walker)	(チョウ目、ヤママユガ科) 新北区
tumblebug		*Canthon septemmaculatus* Latreille	(コウチュウ目、コガネムシ科) 新熱帯区
tumblebug (1)		*Copris lunaris* (Linnaeus)	(コウチュウ目、コガネムシ科) 旧北区
tumblebugs		*Copris*	(コウチュウ目、コガネムシ科) の昆虫の総称
tumblebugs		*Deltochilum*	(コウチュウ目、コガネムシ科) の昆虫の総称
tumble-bugs (1)		*Canthon*	(コウチュウ目、コガネムシ科) の昆虫の総称
tumble-bugs			dung beetles (3) を見よ
tumblebugs			tumble-bugs (1) を見よ
tumbling flower beetle		*Mordella atrata* Melsheimer	(コウチュウ目、ハナノミ科) 新北区
tumbling flower beetles (1)	ハナノミ科	Mordellidae	(コウチュウ目) の昆虫の総称

英名	和名	学名	所属、分布、ほか
tumbling flower beetles		Mordella	(コウチュウ目、ハナノミ科) の昆虫の総称
Tumbu-fly		Cordylobia anthropophaga (Blanchard)	(ハエ目、イエバエ科) エチオピア区
Tunbridge wells gem	ホソバネキンウワバ	Chrysodeixis acuta (Walker)	(チョウ目、ヤガ科) 日本、東洋区、エチオピア区。ダイズ害虫
Tuneta Arcas			frosted greentail を見よ
tung oil tree measuring worm		Buzura suppressaria Guenée	(チョウ目、シャクガ科) 東洋区
Tuolumne shieldback		Idionotus tuolumne Hebard	(バッタ目、キリギリス科) 新北区
tupelo clearwing moth		Synanthedon rubrofascia (Edwards)	(チョウ目、スカシバガ科) 新北区
tupelo leaf miner		Antispila nysaefoliella Clemens	(チョウ目、ツヤコガ科) 新北区。tupelo ヌマミズキに由来。tupelo leafminer moth ともいう
tupelo leaffolder moth		Actrix nyssaecolella (Dyar)	(チョウ目、メイガ科) 新北区
tur plume			tur pod caterpillar を見よ
tur plume moth	チャイロトリバ	Marasmarcha pumilio Zeller	(チョウ目、トリバガ科) 日本、東洋区
tur pod caterpillar		Exelastis atomosa Walsingham	(チョウ目、トリバガ科) 東洋区、エチオピア区
turbulent Phosphila moth		Phosphila turbulenta Hübner	(チョウ目、ヤガ科) 新北区
turf ant			pavement ant を見よ
turf ant			yellow meadow ant を見よ
turf planthopper		Toya dryope (Kirkaldy)	(カメムシ目、ウンカ科) 豪州区
Turkana pierrot		Tarucus kulala Evans	(チョウ目、シジミチョウ科) エチオピア区
Turkestan cockroach		Blatta (Sherfoldella) lateralis (Walker)	(ゴキブリ目、ゴキブリ科) 新北区
Turkey brown		Paraleptophlebia	(カゲロウ目、トビイロカゲロウ科) の昆虫の総称
Turkey brown mayfly		Paraleptophlebia submarginata (Stephens)	(カゲロウ目、トビイロカゲロウ科) 旧北区
turkey gnat		Simulium meridionale Riley	(ハエ目、ブユ科) 新北区
turkey gnats			black flies (2) を見よ
turkey louse		Lipeurus gallipavonis Geoffroy	(ハジラミ目、チョウカクハジラミ科) 新北区
turkey louse			large turkey louse を見よ
turkey wing louse			slender turkey louse (1) を見よ
Turkish clubtail		Gomphus schneiderii Selys	(トンボ目、サナエトンボ科) 旧北区
Turkish goldenring		Corudulegaster picta Selys	(トンボ目、オニヤンマ科) 旧北区
Turkish meadow brown		Maniola telmessia (Zeller)	(チョウ目、タテハチョウ科) 旧北区
Turk's-cap white skipper		Heliopetes macaira Reakirt	(チョウ目、セセリチョウ科) 新北区、新熱帯区
Turk's groundhopper		Tetrix tuerki (Krauss)	(バッタ目、ヒシバッタ科) 旧北区

英名	和名	学名	所属、分布、ほか
Turlough spreadwing			scarce emerald を見よ
Turner's giant-skipper		*Turnerina mejicanus* (Bell)	（チョウ目、セセリチョウ科）新熱帯区
Turner's opal		*Chrysoritis turneri* (Riley)	（チョウ目、シジミチョウ科）エチオピア区
turnip and cabbage gall weevil			turnip gall weevil を見よ
turnip and cabbage seed midge			cabbage gall midge を見よ
turnip aphid		*Rhopalosiphum pseudobrassicae* (Davis)	（カメムシ目、アブラムシ科）新北区。キャベツ害虫
turnip aphid (1)	ニセダイコンアブラムシ	*Lipaphis erysimi* (Kaltenbach)	（カメムシ目、アブラムシ科）日本、汎世界。アブラナ科野菜害虫
turnip beetle			large striped flea beetle を見よ
turnip ceutorhynchus			rape stem weevil を見よ
turnip dart			cutworm (2) を見よ
turnip flea			large striped flea beetle を見よ
turnip flea beetle (1)		*Phyllotreta consobrina* (Curtis)	（コウチュウ目、ハムシ科）旧北区
turnip flea beetle (2)		*Phyllotreta cruciferae* (Goeze)	（コウチュウ目、ハムシ科）旧北区
turnip flea beetle			large striped flea beetle を見よ
turnip flower beetle			pollen beetle を見よ
turnip fly			seed-corn maggot (1) を見よ
turnip fly			turnip flea beetle (1) を見よ
turnip fly			large striped flea beetle を見よ
turnip gall weevil		*Ceutorhynchus pleurostigma* (Marsham)	（コウチュウ目、ゾウムシ科）旧北区
turnip jack			large striped flea beetle を見よ
turnip maggot			radish root maggot を見よ
turnip maggot			cabbage root fly (1) を見よ　幼虫の英名
turnip moth			cutworm (2) を見よ
turnip moth			cross-striped cabbage worm を見よ
turnip mud beetle		*Helophorus rufipes* Bosc	（コウチュウ目、ガムシ科）旧北区
turnip mud beetle		*Megempleurus porculus* (Marsham)	（コウチュウ目、ガムシ科）旧北区
turnip mud beetle		*Helophorus rugosus* (Olivier)	（コウチュウ目、ガムシ科）旧北区
turnip-root fly			cabbage root fly (1) を見よ
turnip root maggot			cabbage root fly (1) を見よ
turnip sawflies		*Athalia*	（ハチ目、ハバチ科）の昆虫の総称
turnip sawfly			cabbage sawfly (1)(2)(3) を見よ
turnip sawfly			nigger-worm を見よ
turnip seed weevil			cabbage seed weevil を見よ
turnip weevil			cabbage seed weevil を見よ

英名	和名	学名	所属、分布、ほか
turpentine borer		*Buprestis apricans* Herbst	(コウチュウ目、タマムシ科) 新北区
turquoise-banded shoemaker		*Archaeoprepona amphimachus* (Fabricius)	(チョウ目、タテハチョウ科) 新熱帯区
turquoise blue	ドリーラスヒメシジミ (ドルスシジミ)	*Plebejus dorylas* Denis et Schiffermüller	(チョウ目、シジミチョウ科) 旧北区
turquoise bluet		*Enallagma divagans* Selys	(トンボ目、イトトンボ科) 新北区
turquoise emperor		*Doxocopa laurentia* (Godart)	(チョウ目、タテハチョウ科) 新熱帯区
turquoise emperor			Cherubina emperor を見よ
turquoise eyed-metalmark		*Mesosemia gemina* Maza et Maza	(チョウ目、シジミタテハ科) 新熱帯区
turquoise hairstreak		*Jalmenus clementi* (Druce)	(チョウ目、シジミチョウ科) 豪州区
turquoise jewel		*Argyrogrammana caelestina* Hall et Willmott	(チョウ目、シジミタテハ科) 新熱帯区
turquoise jewel			western jewel を見よ
turquoise longtail		*Urbanus evona* Evans	(チョウ目、セセリチョウ科) 新熱帯区
turquoise Perisama		*Perisama comnena* Hewitson	(チョウ目、タテハチョウ科) 新熱帯区
turquoise sapphire		*Heliophorus viridipunctata* de Nicéville	(チョウ目、シジミチョウ科) 旧北区
turquoise-tipped darner		*Rhionaeschna psilus* (Calvert)	(トンボ目、ヤンマ科) 新北区、新熱帯区
turret wasps			mason wasps を見よ
turtle beetles	ダエンマルトゲムシ科	Chelonariidae	(コウチュウ目) の昆虫の総称
turtle head borer moth		*Papaipema nepheleptena* (Dyar)	(チョウ目、ヤガ科) 新北区。turtle head borer ともいう
tusked mayflies			burrowing mayflies を見よ
tusked weta			elephant weta を見よ
tuskless burrower mayflies			hacklegill mayflies (1) を見よ
tussah			ヤママユガ科の *Antheraea paphia* 或いは *A. mylitta*、サクサン (*A. pernyi*)、その繭からとった絹糸、その糸で織った絹布
tusser silkworm			tassar silkworm を見よ
tussock moth		*Hemerocampa plagiata* (Walker)	(チョウ目、ドクガ科) 新北区
tussock moth			definite tussock moth を見よ
tussock moths		*Orgyia*	(チョウ目、ドクガ科) の昆虫の総称
tussock moths (1)	ドクガ科	Lymantriidae	(チョウ目) の昆虫の総称 世界に2,500種以上
tussock tiger moth		*Metacrias strategica* Hudson	(チョウ目、ヒトリガ科) 豪州区
tussore silkworm			Chinese oak silkmoth を見よ
Tutia clearwing		*Ceratinia tutia* (Hewitson)	(チョウ目、タテハチョウ科) 新熱帯区
twelve-spot skimmer			ten-spot を見よ

英名	和名	学名	所属、分布、ほか
twelve-spotted asparagus beetle			spotted asparagus beetle を見よ
twelve-spotted cucumber beetle		*Diabrotica undecimpunctata* (Fabricius)	（コウチュウ目、ハムシ科）新北区。亜種 *D. u. howardi* Barber は spotted cucumber beetle という
twelve spotted cucumber beetle			spotted cucumber beetle を見よ
twelve spotted ladybird beetle	シロホシテントウ	*Vibidia duodecimguttata* (Poda)	（コウチュウ目、テントウムシ科）日本、旧北区
twenty-eight-spotted ladybird	ニジュウヤホシテントウ	*Epilachna vigintioctopunctata* (Fabricius)	（コウチュウ目、テントウムシ科）日本、東洋区、豪州区
twenty-eight-spotted lady beetle			twenty-eight-spotted ladybird を見よ
twentyeight spotted potato ladybird			twenty-eight-spotted ladybird を見よ　豪州での英名
twentyfour-spot ladybird			Jacque's beetle を見よ
twenty-plume moth		*Alucita hexadactyla* Linnaeus	（チョウ目、ニジュウシトリバガ科）旧北区。twenty-plume ともいう
twenty-plumes			many plume moths を見よ
twentysix-spotted potato ladybird	ジュウイチホシウリハムシ	*Epilachna vigintioctopunctata pardalis* (Boisduval)	（コウチュウ目、テントウムシ科）豪州区
twenty-spot ermel		*Yponomeuta sedella* Treitschke	（チョウ目、スガ科）旧北区
twenty-spot ermel			twenty spot ermine を見よ
twenty spot ermine	ベンケイソウスガ	*Yponomeuta vigintipunctata* (Retzius)	（チョウ目、スガ科）日本、旧北区
twenty-spot moth			twenty spot ermine を見よ
twenty-spotted asparagus beetle			spotted asparagus beetle を見よ
twenty-spotted lady beetle		*Psyllobora vigintimacutata* (Say)	（コウチュウ目、テントウムシ科）新北区
twenty-two-spot ladybird		*Psyllobora vigintiduopunctata* (Linnaeus)	（コウチュウ目、テントウムシ科）旧北区
twice-constricted myrmicine ant		*Pheidole biconstricta* Mayr	（ハチ目、アリ科）新熱帯区
twice-stabbed lady beetle		*Chilocorus stigma* (Say)	（コウチュウ目、テントウムシ科）新北区
twice-stabbed ladybird beetle			twice-stabbed lady beetle を見よ
twig-and-stem weevils		Zygopinae	（コウチュウ目、ゾウムシ科）の昆虫の総称
twig cutter		*Coenorrhinus interpunctatus* (Stephens)	（コウチュウ目、オトシブミ科）旧北区
twig cutter			apple twig cutter を見よ
twig-cutting weevil			apple twig cutter を見よ
twig girdler		*Oncideres cingulata* (Say)	（コウチュウ目、カミキリムシ科）新北区

英名	和名	学名	所属、分布、ほか
twig girdlers		Agrilus	(コウチュウ目、タマムシ科) の昆虫の総称
twig looper		Ectropis excursaria (Guenée)	(チョウ目、シャクガ科) 豪州区
twig pruner		Elaphidionoides villosus (Fabricius)	(コウチュウ目、カミキリムシ科) 新北区
twig pruners		Elaphidionoides	(コウチュウ目、カミキリムシ科) の昆虫の総称
twilight brown			evening brown を見よ 南アでの英名
twilight darner		Gynacantha nervosa Rambur	(トンボ目、ヤンマ科) 新北区、新熱帯区
twilight moth		Lycia rachelae (Hulst)	(チョウ目、シャクガ科) 新北区
twin-banded golden flat		Celaenorrhinus suthina (Hewitson)	(チョウ目、セセリチョウ科) 新熱帯区
twin-barred knot-horn		Homoeosoma sinuella (Fabricius)	(チョウ目、メイガ科) 旧北区
twin-dotted border		Mylothris rueppellii haemus (Trimen)	(チョウ目、シロチョウ科) エチオピア区
twin-dotted Macrochilo moth		Macrochilo hypocritalis Ferguson	(チョウ目、ヤガ科) 新北区
twin dusky-blue		Candalides geminus Edwards et Kerr	(チョウ目、シジミチョウ科) 豪州区
twin-eyed mountain satyr		Lymanopoda panacea (Hewitson)	(チョウ目、タテハチョウ科) 新熱帯区
twin gold spot			two-spotted looper moth を見よ
twin-line lappet		Pachypasa bilinea Walker	(チョウ目、シャクガ科) 旧北区
twin-lobed deerfly		Chrysops relictus Meigen	(ハエ目、アブ科) 旧北区
twin-spot banded skipper		Autochton bipunctatus (Gmelin)	(チョウ目、セセリチョウ科) 新熱帯区
twin-spot blue		Lepidochrysops plebeia plebeia (Butler)	(チョウ目、シジミチョウ科) エチオピア区
twin-spot carpet		Perizoma didymata (Linnaeus)	(チョウ目、シャクガ科) 旧北区
twin-spot centurion		Sargus bipunctatus (Scopoli)	(ハエ目、ミズアブ科) 旧北区
twin-spot duke		Siseme alectryo Westwood	(チョウ目、シジミタテハ科) 新熱帯区
twin-spot fritillary	スジグロコヒョウモン	Brenthis hecate (Schiffermüller)	(チョウ目、タテハチョウ科) 旧北区。twin-spotted fritillary ともいう
twinspot hunter		Austroepigomphus melaleucae (Tillyard)	(トンボ目、サナエトンボ科) 豪州区
twin spot longhorn beetle		Oberea oculata (Linnaeus)	(コウチュウ目、カミキリムシ科) 旧北区
twin-spot skipper		Oligoria maculata (Edwards)	(チョウ目、セセリチョウ科) 新北区
twin-spot sphinx		Smerinthus jamaicensis (Drury)	(チョウ目、スズメガ科) 新北区。twin-spot sphinx moth ともいう
twin-spot stiletto		Thereva bipunctata Meigen	(ハエ目、ツルギアブ科) 旧北区
twin-spot wainscot		Archanara geminipuncta (Haworth)	(チョウ目、ヤガ科) 旧北区。twin-spotted wainscot ともいう

英名	和名	学名	所属、分布、ほか
twinspotted budworm		*Hedya chionosema* (Zeller)	（チョウ目、ハマキガ科）新北区
twin-spotted major		*Oxycera leonina* (Panzer)	（ハエ目、ミズアブ科）旧北区
twin-spotted quaker	スモモキリガ	*Orthosia munda* (Denis et Schiffermüller)	（チョウ目、ヤガ科）日本、東洋区、旧北区。リンゴ、スモモ、カキ、サクラなどの害虫。twin-spot quaker ともいう
twin-spotted quaker moth			twin-spotted quaker を見よ
twin-spotted sphinx			twin-spot sphinx を見よ
twin-spotted spiketail		*Cordulegaster maculata* Selys	（トンボ目、オニヤンマ科）新北区
twin-stigma dwarf		*Elachista biatomella* (Stainton)	（チョウ目、クサモグリガ科）旧北区
twin-striped clubtail		*Gomphus geminatus* Carle	（トンボ目、サナエトンボ科）新北区
twin-striped tabby			dried leaf moth を見よ
twin swift		*Borbo gemella* (Mabille)	（チョウ目、セセリチョウ科）エチオピア区
twirler moth		*Symmetrischema striatella* (Murtfeldt)	（チョウ目、キバガ科）新北区
twirler moths	キバガ科	Gelechiidae	（チョウ目）の昆虫の総称　世界に 4000 種以上
twisted-wing parasites			stylopids を見よ
twisted-winged flies			stylops を見よ
twisted-winged insects			stylopids を見よ
twisted-winged insects			stylops を見よ
twisted-winged parasites			stylopids を見よ
twister		*Tholymis tillarga* (Fabricius)	（トンボ目、トンボ科）東洋区、大洋区、エチオピア区
twisters		*Tholymis*	（トンボ目、トンボ科）の昆虫の総称
twitch-ballock			common earwig を見よ
two-banded Catoptria			three-spotted Crambus moth を見よ
two-banded checkered skipper		*Pyrgus ruralis* (Boisduval)	（チョウ目、セセリチョウ科）新北区
two-banded cruiser		*Phyllomacromia contumax* Selys	（トンボ目、ヤマトンボ科）エチオピア区
two-banded Ectypia		*Ectypia bivittata* Clemens	（チョウ目、ヒトリガ科）新北区
two-banded fruit weevil			strawberry blossom weevil (1) を見よ
two-banded fungus beetle			waste grain beetle を見よ
two-banded hairstreak		*Michaelus phoenissa* (Hewitson)	（チョウ目、シジミチョウ科）新熱帯区
twobanded Japanese weevil		*Callirhopalus bifasciatus* (Roelofs)	（コウチュウ目、ゾウムシ科）新北区

英名	和名	学名	所属、分布、ほか
two-banded longhorn		*Rhagium bifasciatum* Fabricius	（コウチュウ目、カミキリムシ科）旧北区。two-banded longhorn beetle ともいう
two-banded Petrophila moth		*Petrophila bifascialis* (Robinson)	（チョウ目、メイガ科）新北区
two-banded quercus leafhopper	フタオビハトムネヨコバイ	*Macropsis matsumurana* China	（カメムシ目、オオヨコバイ科）日本、旧北区
two-banded satyr			white satyr を見よ
two-banded tortrix	フタスジクリイロハマキ	*Acleris platynotana* (Walsingham)	（チョウ目、ハマキガ科）日本、旧北区
two-barred flasher			flashing Astraptes を見よ
two-black-banded tiger moth	フタスジヒトリ	*Spilarctia bifasciata* (Butler)	（チョウ目、ヒトリガ科）日本
two-black-dotted cymatophorid			common lutestring を見よ
two-brand crow	ルリマダラ	*Euploea sylvester* (Fabricius)	（チョウ目、マダラチョウ科）豪州区。東洋区の *E. s. swinhoei* Wallace も同英名。two-brand crow butterfly ともいう
two brand crow			double-branded blue crow を見よ　豪州での使用
two-brand grass skipper			Dominula skipper を見よ
two-clawed mole crickets		*Scapteriscus*	（バッタ目、ケラ科）の昆虫の総称
two-coloured bush cricket	ショウリョウバッタモドキ	*Metrioptera bicolor* (de Haan)	（バッタ目、キリギリス科）日本
two-coloured coconut leaf beetle		*Plesispa reichei* Chapuis	（コウチュウ目、ハムシ科）東洋区
two-coloured grasshopper			field grasshopper を見よ
two circuli mealybug		*Phenacoccus dearnessi* King	（カメムシ目、コナカイガラムシ科）新北区。two-circuit mealybug ともいう
two-dot alpine			arctic alpine を見よ
two-dotted flea beetle	フタホシオオノミハムシ	*Pseudodera xanthospila* Baly	（コウチュウ目、ハムシ科）日本、東洋区
two-dotted large geometrid	フタテンオエダシャク	*Chasmia defixaria* (Walker)	（チョウ目、シャクガ科）日本、旧北区
two-eyed eighty-eight		*Callicore pitheas* (Latreille)	（チョウ目、タテハチョウ科）新熱帯区
two-eyed small longicorn	フタツメケシカミキリ	*Phloeopsis bioculata* (Matsumura et Matsushita)	（コウチュウ目、カミキリムシ科）日本
two-gemmed sawfly		*Tenthredo bigemina* (Dyar)	（ハチ目、ハバチ科）新北区
two-horned stem-boring beetle	カキノフタツノナガシンクイ	*Sinoxylon japonicus* Lesne	（コウチュウ目、ナガシンクイムシ科）日本
two-lined albizzia longhorn	アオスジカミキリ	*Xystrocera globosa* (Olivier)	（コウチュウ目、カミキリムシ科）日本、全北区、東洋区、大洋区、エチオピア区。ネムノキ害虫
two-lined chestnut borer		*Agrilus bilineatus* (Weber)	（コウチュウ目、タマムシ科）新北区
two-lined collops		*Collops vittatus* Say	（コウチュウ目、ジョウカイモドキ科）新北区

英名	和名	学名	所属、分布、ほか
two-lined hooktip moth		*Drepana bilineata* (Packard)	（チョウ目、カギバガ科）新北区
two-lined shieldback		*Eremopedes bilineatus* (Thomas)	（バッタ目、キリギリス科）新北区
two-lined spittle bug		*Prosapia bicincta* Say	（カメムシ目、コガシラアワフキ科）新北区、新熱帯区
two-marked treehopper		*Enchenopa binotata* (Say)	（カメムシ目、ツノゼミ科）新北区
two pip policeman	シロオビアオバセセリ	*Coeliades pisistratus* (Fabricius)	（チョウ目、セセリチョウ科）エチオピア区
two-pronged bristletails			entotrophs を見よ
two-pupil satyr		*Cissia themis* (Butler)	（チョウ目、タテハチョウ科）新熱帯区
two seta pine scale		*Matsucoccus bisetosus* Morrison	（カメムシ目、ワタフキカイガラムシ科）新北区
two-spined auger beetle		*Xylobosca bispinosa* (Macleay)	（コウチュウ目、ナガシンクイムシ科）豪州区
two spot Charaxes	コモンカスリフタオチョウ	*Charaxes bipunctatus* Rothschild	（チョウ目、タテハチョウ科）エチオピア区
two-spot dart moth		*Eueretagrotis perattentus* (Grote)	（チョウ目、ヤガ科）新北区。two-spot dart ともいう
two-spot ladybird	フタホシテントウ	*Hyperaspis japonica* (Crotch)	（コウチュウ目、テントウムシ科）日本、旧北区
two-spot oak splendour beetle		*Agrilus biguttatus* (Fabricius)	（コウチュウ目、タマムシ科）旧北区
two-spotted bean weevil	サムライマメゾウムシ	*Bruchidius japonicus* (Harold)	（コウチュウ目、ハムシ科）日本、旧北区
two-spotted Carolella moth		*Carolella bimaculana* (Robinson)	（チョウ目、ハマキガ科）新北区。two-spotted Carolella ともいう
two-spotted cricket	フタホシコオロギ	*Gryllus bimaculatus* (De Geer)	（バッタ目、コオロギ科）日本、旧北区、東洋区、エチオピア区
two-spotted dragonfly		*Epitheca bimaculata* (Charpentier)	（トンボ目、エゾトンボ科）旧北区
two-spotted dragonfly			Eurasian baskettail を見よ
two-spotted eucosmid	フタホシヒメハマキ	*Ancylis selenana* (Guenée)	（チョウ目、ハマキガ科）日本、旧北区
two-spotted forester moth		*Alypiodes bimaculata* (Herrich-Schäffer)	（チョウ目、ヤガ科）新北区
two-spotted grass-skipper			Tasmania skipper を見よ
two-spotted groundhopper		*Tetrix bipunctata* (Linnaeus)	（バッタ目、ヒシバッタ科）旧北区
two-spotted ladybird		*Adalia bipunctata* (Linnaeus)	（コウチュウ目、テントウムシ科）全北区。two-spot ladybird, two-spot ladybird beetle、two-spotted lady beetle, two-spotted ladybird beetle ともいう
two-spotted leaf bug	フタモンカスミカメムシ	*Adelphocoris variabilis* (Uhler)	（カメムシ目、カスミカメムシ科）日本、旧北区
two spotted leafhopper		*Sophonia rufofascia* (Kuoh et Kuoh)	（カメムシ目、オオヨコバイ科）東洋区、大洋区

英名	和名	学名	所属、分布、ほか
two spotted leafhopper		*Sophonia orientalis* (Matsumura)	（カメムシ目、オオヨコバイ科）東洋区、大洋区
two-spotted leafhopper			aster leafhopper (1) を見よ
two-spotted line-blue			double-spotted lineblue を見よ
two-spotted lithosid			four-spotted footman を見よ
two-spotted looper moth		*Autographa bimaculata* Stephens	（チョウ目、ヤガ科）新北区
two-spotted Mea moth		*Mea bipunctella* (Dietz)	（チョウ目、ヒロズコガ科）新北区。two-spotted Mea ともいう
two-spotted Prepona		*Archaeoprepona demophoon gulina* (Fruhstorfer)	（チョウ目、タテハチョウ科）新熱帯区。*A. d. mexicana* Llorente, Descimon et Johnson も同英名
two-spotted pumpkinfly			pumpkin fly を見よ
two-spotted rice bug	ハリカメムシ	*Cletus rusticus* Stål	（カメムシ目、ヘリカメムシ科）日本、旧北区
two-spotted scarlet-eye		*Dyscophellus nicephorus* (Hewitson)	（チョウ目、セセリチョウ科）新熱帯区
two-spotted skipper		*Euphyes bimacula* (Grote et Robinson)	（チョウ目、セセリチョウ科）新北区
two-spotted stink bug		*Perillus bioculatus* (Fabricius)	（カメムシ目、カメムシ科）新北区。コロラドハムシを捕食
two-spotted threadtail		*Elattoneura bigemmata* Lieftinck	（トンボ目、モノサシトンボ科）東洋区
two-spotted tree cricket		*Neoxabea bipunctata* (De Geer)	（バッタ目、コオロギ科）新北区
two-spotted white looper moth	フタホシシロエダシャク	*Lomographa bimaculata subnotata* (Warren)	（チョウ目、シャクガ科）日本
two-stabbed ladybird		*Chilocorus orbus* Casey	（コウチュウ目、テントウムシ科）新北区。two-stabbed ladybeetle ともいう
two-striped cricket			two-spotted cricket を見よ
two-striped emerald			red-fringed emerald moth を見よ
two-striped forceptail		*Aphylla williamsoni* (Gloyd)	（トンボ目、サナエトンボ科）新北区、新熱帯区
two-striped grasshopper		*Melanoplus bivittatus* (Say)	（バッタ目、バッタ科）新北区。トウモロコシ害虫
two-striped green buprestid			Japanese jewel beetle を見よ
two-striped leaf beetle	フタスジヒメハムシ	*Medythia nigrobilineata* (Motschulsky)	（コウチュウ目、ハムシ科）日本、旧北区
two-striped owlet moth		*Macrochilo bivittata* Grote	（チョウ目、ヤガ科）新北区。two-striped snout-moth ともいう
two-striped skimmer		*Orthetrum caffrum* (Burmeister)	（トンボ目、トンボ科）エチオピア区
two-striped slantfaced grasshopper		*Mermiria bivittata* (Serville)	（バッタ目、バッタ科）新北区
two-striped walkingstick		*Anisomorpha buprestoides* (Stoll)	（ナナフシ目、Pseudophasmatidae）新北区。southern two-striped walkingstick ともいう

英名	和名	学名	所属、分布、ほか
two-tailed buprestid	フタオタマムシ	*Dicerca furcato aino* Lewis	(コウチュウ目、タマムシ科) 日本、旧北区
two-tailed Pasha	ヨーロッパフタオチョウ	*Charaxes jasius* (Linnaeus)	(チョウ目、タテハチョウ科) 旧北区。エチオピア区
two-tailed swallowtail		*Papilio multicaudatus* Kirby	(チョウ目、アゲハチョウ科) 新北区、新熱帯区
two-tailed tiger swallowtail			two-tailed swallowtail を見よ
two-tone skipper		*Remella remus* (Fabricius)	(チョウ目、セセリチョウ科) 新熱帯区
two-toned Ancylis moth		*Ancylis divisana* (Walker)	(チョウ目、ハマキガ科) 新北区
two-toned caterpillar parasite		*Heleropelma scaposum* (Morley)	(ハチ目、ヒメバチ科) 豪州区
two-toned gemmed-satyr		*Cyllopsis hilaria* (Godman)	(チョウ目、タテハチョウ科) 新熱帯区
two-toned groundstreak		*Lamprospilus collucia* (Hewitson)	(チョウ目、シジミチョウ科) 新熱帯区
two-toned skipper		*Vettius lafrenaye pica* (Herrich-Schäffer)	(チョウ目、セセリチョウ科) 新熱帯区
two-tooth longicorn		*Ambeodontus tristis* (Fabricius)	(コウチュウ目、カミキリムシ科) 豪州区
two-toothed pine beetle		*Pityogenes bidentatus* (Herbst)	(コウチュウ目、キクイムシ科) 旧北区
two-toothed scaly cricket		*Cycloptilum bidens* Hebard	(バッタ目、コオロギ科) 新北区
two-wavy-lined geometrid	フタナミトビヒメシャク	*Pylargosceles steganioides* (Butler)	(チョウ目、シャクガ科) 日本、旧北区
two-winged flies			flies を見よ
two-year-cycle budworm moth			western spruce budworm (1) を見よ
two-yellow-striped weevil	フタキボシゾウムシ	*Lepyrus japonicus* Roelofs	(コウチュウ目、ゾウムシ科) 日本、旧北区
tycon bark beetle	タイコンキクイムシ	*Scolytoplatypus tycon* Blandford	(コウチュウ目、キクイムシ科) 日本、旧北区、東洋区
typewriter		*Pauropsalta extrema* (Distant)	(カメムシ目、セミ科) 豪州区
Typhaon skipper		*Azonax typhaon* (Hewitson)	(チョウ目、セセリチョウ科) 新熱帯区
typhiid wasp			tiphiid wasp を見よ
typhoid fly			house fly を見よ
Typhon skipper		*Methionopsis typhon* Godman	(チョウ目、セセリチョウ科) 新熱帯区
Typhon sphinx moth		*Eumorpha typhon* (Klug)	(チョウ目、スズメガ科) 新北区、新熱帯区。Typhon sphinx ともいう
typical ants	ヤマアリ亜科	Formicinae	(ハチ目、アリ科) の昆虫の総称
typical darkling beetles	ゴミムシダマシ亜科	Tenebrioninae	(コウチュウ目、ゴミムシダマシ科) の昆虫の総称
typical sawflies			sawflies (1) を見よ
typical snout beetles			acorn weevils (2) を見よ

英名	和名	学名	所属、分布、ほか
typograph			eight-spined ips を見よ
typographer beetle			eight-spined ips を見よ
Tyrolean earwig		*Chelidurella mutica* (Krauss)	（ハサミムシ目、クギヌキハサミムシ科）旧北区
Tytler's bushbrown		*Mycalesis evansii* Tytler	（チョウ目、タテハチョウ科）東洋区
Tytler's dull oakblue		*Arhopala ace* de Nicéville	（チョウ目、シジミチョウ科）東洋区
Tytler's emperor		*Eulaceura manipuriensis* Tytler	（チョウ目、タテハチョウ科）東洋区
Tytler's hairstreak	ホソオビミドリシジミ	*Chrysozephyrus vittatus* (Tytler)	（チョウ目、シジミチョウ科）東洋区
Tytler's jester		*Symbrenthia doni* Tytler	（チョウ目、タテハチョウ科）東洋区
Tytler's lascar	ウスイロキンミスジ	*Pantoporia bieti* (Oberthür)	（チョウ目、タテハチョウ科）東洋区
Tytler's rosy oakblue		*Arhopala allata* (Staudinger)	（チョウ目、シジミチョウ科）東洋区
Tytler's rosy oakblue		*Amblypodia suffusa* Tytler	（チョウ目、シジミチョウ科）東洋区。
Tytler's royal		*Tajuria sebonga* Tytler	（チョウ目、シジミチョウ科）東洋区
Tytler's treebrown	アリサンチャイロヒカゲ	*Lethe gemina* Leech	（チョウ目、タテハチョウ科）東洋区
Tytler's windmill			Pemberton's windmill を見よ

英名	和名	学名	所属、分布、ほか
U			
ubiquitous bluetail			common bluetail (1) を見よ
ubiquitous footman		*Ovenna vicaria* (Walker)	(チョウ目、ヒトリガ科) エチオピア区
Uddmann's bell			bramble shoot moth を見よ
udo aphid	ハゼアブラムシ	*Toxoptera odinae* (Van der Goot)	(カメムシ目、アブラムシ科) 日本、東洋区、新北区
udo leaf roller			aralia leafroller を見よ
udo longicorn beetle	センノカミキリ	*Acalolepta luxuriosa* (Bates)	(コウチュウ目、カミキリムシ科) 日本、旧北区。ウド、タラノキ、ヤツデの害虫。udo longicorn ともいう
udo weevil	ヒメシロコブゾウムシ	*Dermatoxenus caesicollis* (Gyllenhal)	(コウチュウ目、ゾウムシ科) 日本、東洋区
Uffen's claw			peach twig borer を見よ
Uganda sapphire		*Iridana hypocala* Eltringham	(チョウ目、シジミチョウ科) エチオピア区
uglynest caterpillar			cherry leaf roller を見よ
uglynest caterpillar moth			cherry leaf roller を見よ
Uhler's arctic	コロラドタカネヒカゲ	*Oeneis uhleri* (Reakirt)	(チョウ目、タテハチョウ科) 新北区
Uhler's camel cricket		*Ceuthophilus uhleri* Scudder	(バッタ目、カマドウマ科) 新北区。朽木にすむ
Uhler's virtuoso katydid		*Amblycorypha uhleri* Stål	(バッタ目、キリギリス科) 新北区
Uitenhage sylph		*Tsitana uitenhaga* Evans	(チョウ目、セセリチョウ科) エチオピア区
ulalume underwing moth		*Catocala ulalume* Strecker	(チョウ目、ヤガ科) 新北区。ulalume underwing ともいう
ulmus Sapporo aphid	アカダモブチアブラムシ	*Sappocallis ulmicola* Matsumura	(カメムシ目、アブラムシ科) 日本、旧北区、東洋区
ulmus woolly aphid	アキニレスジワタムシ	*Tetraneura akinire* Sasaki	(カメムシ目、アブラムシ科) 日本
Ulphila skipper		*Poanes ulphila* (Plötz)	(チョウ目、セセリチョウ科) 新熱帯区
ultima gem		*Libellago finalis* (Hagen)	(トンボ目、Chlorocyphidae) 東洋区
ultraviolet sulphur			Queen Alexandra's sulphur を見よ
ultronia underwing moth			plum tree underwing を見よ
Ulysses butterfly	オオルリアゲハ	*Papilio ulysses* (Linnaeus)	(チョウ目、アゲハチョウ科) 豪州区。*P. u. joesa* (Butler) も同英名
Ulysses swallowtail			Ulysses butterfly を見よ
umbellifer aphid			celery aphid (1) を見よ
umbellifer borer moth		*Papaipema birdi* (Dyar)	(チョウ目、ヤガ科) 新北区。umbellifer borer ともいう
umbellifer longhorn beetle		*Phytoecia cylindrica* (Linnaeus)	(コウチュウ目、カミキリムシ科) 旧北区
umber moth		*Hypomecis umbrosaria* (Hübner)	(チョウ目、シャクガ科) 新北区
umber skipper		*Poanes melane* (Edwards)	(チョウ目、セセリチョウ科) 新北区、新熱帯区

英名	和名	学名	所属、分布、ほか
umber underwing moth		*Catocala umbrosa* Brou	(チョウ目、ヤガ科) 新北区
umber waved carpet			small waved umber を見よ
umbrella ants			leafcutter ants を見よ
umbrella wasps			papernest wasps (1) を見よ
ume aphid	ウメコブアブラムシ	*Myzus mumecola* (Matsumura)	(カメムシ目、アブラムシ科) 日本、旧北区
ume bark beetle	ウメノキクイムシ	*Scolytus aratus* Blandford	(コウチュウ目、キクイムシ科)日本、旧北区。リンゴ、ウメ、オウトウの害虫
ume bud moth			prunus bud moth を見よ
ume cankerworm			plum cankerworm を見よ
ume globose scale	タマカタカイガラムシ	*Lecanium kunoensis* Kuwana	(カメムシ目、カタカイガラムシ科) 日本、全北区
ume leaf roller			holly tortrix を見よ
ume oystershell scale			mulberry oystershell scale を見よ
Una		*Una usta* (Distant)	(チョウ目、シジミチョウ科) 東洋区
unadorned carpet moth		*Hydrelia inornata* (Hulst)	(チョウ目、シャクガ科) 新北区
unarmed shieldback		*Idiostatus inermis* (Scudder)	(バッタ目、キリギリス科) 新北区
unarmed stick-insect		*Acanthoxyla inermis* Salmon	(ナナフシ目、ナナフシ科) 旧北区
unbranded ace			southern spotted ace を見よ
unbranded recluse		*Caenides xychus* (Mabille)	(チョウ目、セセリチョウ科) エチオピア区
unbroken sergeant	プラバラミスジ	*Athyma pravara* Moore	(チョウ目、タテハチョウ科) 東洋区
Uncas skipper		*Hesperia uncas* Edwards	(チョウ目、セセリチョウ科) 新北区、新熱帯区
uncertain		*Hoplodrina alsines* Brahm	(チョウ目、ヤガ科) 旧北区。uncertain moth ともいう
uncertain blue tiger	イスモイデスコモンマダラ (ニセミナミコモンマダラ)	*Tirumala ishmoides* Moore	(チョウ目、タテハチョウ科) 東洋区
uncertain owlet			confused owlet を見よ
uncertain royal		*Tajuria ister* (Hewitson)	(チョウ目、シジミチョウ科) 東洋区
uncertain satyr			yellow barred を見よ
uncle sam moth			faithful beauty moth を見よ
unclear dagger moth		*Acronicta inclara* Smith	(チョウ目、ヤガ科) 新北区
uncompaghre fritillary		*Clossiana acrocnema* (Gall et Sperling)	(チョウ目、タテハチョウ科) 新北区
underground ant			yellow meadow ant を見よ
underground grass caterpillar		*Oncopera fasciculata* (Walker)	(チョウ目、コウモリガ科) 豪州区
underground grassgrub		*Oncopera rufobrunnea* Tindale	(チョウ目、コウモリガ科) 豪州区
underground grassgrub			underground grass caterpillar を見よ
underwing moths	シタバガ亜科	Catocalinae	(チョウ目、ヤガ科) の昆虫の総称

英名	和名	学名	所属、分布、ほか
underwings		*Catocala*	（チョウ目、ヤガ科）の昆虫の総称
underwings			owlet moths を見よ
underwings			underwing moths を見よ
unequal bark-beetle			shot-hole borer (3) を見よ
unequal smudge		*Ypsolophus lucellus* (Fabricius)	（チョウ目、クチブサガ科）旧北区
unexpected Cycnia		*Cycnia inopinatus* (H. Edwards)	（チョウ目、ヒトリガ科）新北区。unexpected Cycnia moth ともいう
unexpected lotus pigmy		*Trifurcula eurema* (Tutt)	（チョウ目、モグリチビガ科）旧北区
unexpected tiger blue		*Hewitsonia inexpectata* Bouyer	（チョウ目、シジミチョウ科）エチオピア区
unexplained sober		*Sophronia humerella* (Denis et Schiffermüller)	（チョウ目、キバガ科）旧北区
unfolded Zestusa		*Zestusa* sp.	（チョウ目、セセリチョウ科）新熱帯区
Ungemach's pierrot		*Tarucus ungemachi* Stempffer	（チョウ目、シジミチョウ科）エチオピア区
ungulate lice			wrinkled sucking lice を見よ
unibanded leaf beetle	イチモンジハムシ	*Morphosphaera japonica* (Hornstedt)	（コウチュウ目、ハムシ科）日本、旧北区
unibanded leafhopper	イチモンジヨコバイ	*Limotettix striola* (Fallén)	（カメムシ目、オオヨコバイ科）日本、旧北区
unibanded stink bug	イチモンジカメムシ	*Piezodorus hybneri* (Gmelin)	（カメムシ目、カメムシ科）日本、東洋区、豪州区、エチオピア区
unibanded tortoise beetle	イチモンジカメノコハムシ	*Thlaspida cribrosa* (Boheman)	（コウチュウ目、ハムシ科）日本、旧北区、東洋区
unicorn beetle		*Dynastes tityus* (Linnaeus)	（コウチュウ目、コガネムシ科）新北区。一角獣に似ることから
unicorn beetles			Hercules beetles を見よ
unicorn caterpillar moth		*Schizura unicornis* (Smith)	（チョウ目、シャチホコガ科）新北区。幼虫は unicorn caterpillar。
unicorn clubtail		*Arigomphus villosipes* (Selys)	（トンボ目、サナエトンボ科）新北区
unicorn cruiser		*Phyllomacromia monoceros* (Förster)	（トンボ目、ヤマトンボ科）エチオピア区
unicorn darner		*Austroaeschna unicornis* (Martin)	（トンボ目、ヤンマ科）豪州区
unicorn hunter		*Austrogomphus cornutus* Watson	（トンボ目、サナエトンボ科）豪州区
unidentified Smerinthus moth		*Smerinthus saliceti* Boisduval	（チョウ目、スズメガ科）新北区、新熱帯区
uniform lichen moth		*Crambidia uniformis* Dyar	（チョウ目、ヒトリガ科）新北区
uniform shieldback		*Idiostatus aequalis* (Scudder)	（バッタ目、キリギリス科）新北区
Union Jack butterfly	ベニヘリシロチョウ	*Delias mysis* (Fabricius)	（チョウ目、シロチョウ科）豪州区
union rustic	ウスクモヨトウ	*Apamea pabulatricula* (Brahm)	（チョウ目、ヤガ科）日本、旧北区

英名	和名	学名	所属、分布、ほか
unipunctate broad bug	ホシハラビロヘリカメムシ	*Homoeocerus unipunctatus* (Thunberg)	(カメムシ目、ヘリカメムシ科) 日本、東洋区
unique-headed bugs			gnat bugs を見よ
unique ranger		*Kedestes lenis* Riley	(チョウ目、セセリチョウ科) エチオピア区
univoltine gall midge		*Lasioptera hungarica* Mohn	(ハエ目、タマバエ科) 全北区
unlined giant chafer			broken-lined giant chafer を見よ
unmargined rove beetles	ツツハネカクシ亜科	Osoriinae	(コウチュウ目、ハネカクシ科) の昆虫の総称
unmarked dagger moth		*Acronicta innotata* Guenée	(チョウ目、ヤガ科) 新北区
unsilvered fritillary			Adiaste fritillary を見よ
unspotted aspen leaf-beetle		*Melasoma tremulae* Fabricius	(コウチュウ目、ハムシ科) 旧北区
unspotted looper moth		*Allagrapha aerea* (Hübner)	(チョウ目、ヤガ科) 新北区
unspotted tentiform apple leaf-miner		*Parornix geminatella* Packard	(チョウ目、ホソガ科) 新北区。unspotted tentiform leafminer moth ともいう
untailed playboy		*Deudorix ecaudata* Gifford	(チョウ目、シジミチョウ科) エチオピア区
Unxia metalmark		*Eurybia unxia* Godman et Salvin	(チョウ目、シジミタテハ科) 新熱帯区
up-winged flies			mayflies を見よ
upholsterer bee			poppy bee を見よ 木の葉の断片を切り取り、地中の巣の内側をおおう習性あり
upland click beetle (1)		*Corymbites cupreus* Fabricius	(コウチュウ目、コメツキムシ科) 旧北区
upland click beetle (2)		*Corymbites purpureus* Poda	(コウチュウ目、コメツキムシ科) 旧北区
upland click beetle (3)		*Corymbites tessellatus* (Linnaeus)	(コウチュウ目、コメツキムシ科) 旧北区
upland click beetle (4)		*Corymbites castaneus* Linnaeus	(コウチュウ目、コメツキムシ科) 旧北区
upland click beetles		*Corymbites*	(コウチュウ目、コメツキムシ科) の昆虫の総称
upland field grasshopper		*Chorthippus apricarius* (Linnaeus)	(バッタ目、バッタ科) 旧北区
upland green bush cricket		*Tettigonia cantans* (Fuessly)	(バッタ目、キリギリス科) 旧北区
upland midget		*Phyllonorycter junoniella* (Zeller)	(チョウ目、ホソガ科) 旧北区
upland summer mayfly		*Ameletus inopinatus* Eaton	(カゲロウ目、フタオカゲロウ科) 旧北区
upland wireworm			upland click beetle (1)(2)(3)(4) を見よ 幼虫の英名
upland wireworms			upland click beetles を見よ
upright Acoloithus moth		*Acoloithus rectarius* Dyar	(チョウ目、マダラガ科) 新北区
upside down flies		Neurochaetidae	(ハエ目) の昆虫の総称 豪州区
Upupa hairstreak		*Apuecla upupa* (Druce)	(チョウ目、シジミチョウ科) 新熱帯区

英名	和名	学名	所属、分布、ほか
Urakawa leafhopper	ウラカワズキンヨコバイ	*Idiocerus urakawensis* Matsumura	（カメムシ目、オオヨコバイ科）日本
Urania skipper		*Phocides urania urania* (Westwood)	（チョウ目、セセリチョウ科）新熱帯区
Urania swallowtail			green page moth を見よ
uranid moths			swallowtail moths を見よ
Uranus opal		*Chrysoritis uranus* (Pennington)	（チョウ目、シジミチョウ科）エチオピア区
urban Anthophora		*Anthophora urbana* Cresson	（ハチ目、コシブトハナバチ科）新北区
urban silverfish		*Ctenolepisma urbana* Slabaugh	（シミ目、シミ科）大洋区
ursine giant skipper		*Megathymus ursus* Poling	（チョウ目、セセリチョウ科）新北区、新熱帯区
urtica false looper			snout (1) を見よ
urtica leaf roller	イラクサヒメハマキ	*Orthotaenia undulana* (Denis et Schiffermüller)	（チョウ目、ハマキガ科）日本、旧北区
urticating anthelid		*Anthela nicothoe* (Boisduval)	（チョウ目、Anthelidae）豪州区
Usambara diadem			red spot diadem を見よ
Utina sister		*Adelpha delinita utina* Hall	（チョウ目、タテハチョウ科）新熱帯区
uzi fly	カイコノウジバエ	*Blepharipa zebina* (Walker)	（ハエ目、ヤドリバエ科）日本、旧北区、東洋区。カイコの寄生ハエとして著名。日本語のウジに由来
uzi fly			multivoltine tachina fly を見よ
Uzza satyr		*Taygetis uzza* Butler	（チョウ目、タテハチョウ科）新熱帯区

英名	和名	学名	所属、分布、ほか
V			
V-lined quaker moth		*Zosteropoda hirtipes* Grote	(チョウ目、ヤガ科) 新北区
V moth		*Macaria wauaria* (Linnaeus)	(チョウ目、シャクガ科) 旧北区。幼虫は V looper。V-moth とも記される
V nephele		*Nephele vau* (Walker)	(チョウ目、スズメガ科) エチオピア区
v pug		*Chloroclystis coronata* Hübner	(チョウ目、シャクガ科) 旧北区
v-pug	クロスジアオナミシャク	*Chloroclystis v-ata* (Haworth)	(チョウ目、シャクガ科) 日本。v-pug moth ともいう
Vaal sprite		*Pseudagrion vaalense* Chutter	(トンボ目、イトトンボ科) エチオピア区
vaccinium scale	シャシャンボコノハカイガラムシ	*Fiorinia vacciniae* Kuwana	(カメムシ目、マルカイガラムシ科) 日本
vagabond Crambus		*Agriphila vulgivagella* (Clemens)	(チョウ目、メイガ科) 新北区。vagabond Crambus moth ともいう
vagabond sod webworm			vagabond Crambus を見よ
vagrant			Australian vagrant を見よ
vagrant ant	ウスヒメアリ	*Plagiolepis alluaudi* Emery	(ハチ目、アリ科) 日本、汎亜熱帯、汎熱帯
vagrant darter		*Sympetrum vulgatum* (Linnaeus)	(トンボ目、トンボ科) 旧北区
vagrant emperor	ヒメギンヤンマ	*Hemianax ephippiger* (Burmeister)	(トンボ目、ヤンマ科) 日本、旧北区、東洋区、エチオピア区
vagrant emperor dragonfly			vagrant emperor を見よ
vagrant grasshopper		*Schistocera nitens nitens* Thunberg	(バッタ目、バッタ科) 新北区。体長7cmの米国普通種
vagrant sympetrum			vagrant darter を見よ
vagrants		*Nephronia*	(チョウ目、シロチョウ科) の昆虫の総称
Valda skipper		*Morys valda* Evans	(チョウ目、セセリチョウ科) 新熱帯区
valdivian archaic moths		Heterobathmiidae	(チョウ目) の昆虫の総称
Valentina ringlet		*Pseudodebis valentina* (Cramer)	(チョウ目、タテハチョウ科) 新熱帯区
valerian pug		*Eupithecia valerianata* (Hübner)	(チョウ目、シャクガ科) 旧北区
Valeriana skipper		*Thorybes valeriana* Plötz	(チョウ目、セセリチョウ科) 新北区
Valeriana skipper		*Codatractus mysie* (Dyar)	(チョウ目、セセリチョウ科) 新北区
valley black gnat		*Leptoconops torrens* (Townsend)	(ハエ目、ヌカカ科) 新北区
valley carpenter bee		*Xylocopa varipuncta* Potton	(ハチ目、ミツバチ科) 新北区
valley elderberry longhorn beetle		*Desmocerus californicus dimorphus* Fisher	(コウチュウ目、カミキリムシ科) 新北区
valley grasshopper		*Oedaleonotus enigma* (Scudder)	(バッタ目、バッタ科) 新北区。カンキツ害虫
vampire leafhoppers		*Draeculacephala*	(カメムシ目、オオヨコバイ科) の昆虫の総称

英名	和名	学名	所属、分布、ほか
vampire moth		*Calyptra eustrigata* Hampson	（チョウ目、ヤガ科）東洋区
Van de Poll's birdwing		*Troides vandepolli* (Snellen)	（チョウ目、アゲハチョウ科）東洋区
Van Duzee treehopper		*Vanduzea segmentata* (Fowler)	（カメムシ目、ツノゼミ科）新北区
van Someren's buff		*Cnodontes vansomereni* Stempffer et Bennet	（チョウ目、シジミチョウ科）エチオピア区
van Someren's gem		*Chloroselas vansomereni* Jackson	（チョウ目、シジミチョウ科）エチオピア区
van Someren's green-banded swallowtail	フトオビルリアゲハ	*Papilio interjectana* Vane-Wright	（チョウ目、アゲハチョウ科）エチオピア区
van Someren's playboy		*Deudorix vansomereni* Stempffer	（チョウ目、シジミチョウ科）エチオピア区
van Son's blue		*Lepidochrysops vansoni* (Swanepoel)	（チョウ目、シジミチョウ科）エチオピア区
van Son's brown		*Stygionympha vansoni* (Pennington)	（チョウ目、タテハチョウ科）エチオピア区
van Son's copper		*Aloeides vansoni* Tite et Dickson	（チョウ目、シジミチョウ科）エチオピア区
van Son's emperor		*Charaxes vansoni* van Someren	（チョウ目、タテハチョウ科）エチオピア区
van Son's playboy		*Deudorix vansoni* Pennington	（チョウ目、シジミチョウ科）エチオピア区
van Son's skolly		*Thestor vansoni* Pennington	（チョウ目、シジミチョウ科）エチオピア区
vanda orchid scale	ランクロホシカイガラムシ	*Genaparlatoria pseudaspidiotus* (Lindinger)	（カメムシ目、マルカイガラムシ科）日本、汎世界
vanda thrips		*Dichromothrips corbetti* (Priesner)	（アザミウマ目、アザミウマ科）新北区
vanhorniids		Vanhorniidae	（ハチ目）の昆虫の総称
Vaninka		*Mesotaenia vaninka* Hewitson	（チョウ目、タテハチョウ科）新熱帯区
Vanoort's crow	エウパトルルリマダラ	*Euploea eupator* Hewitson	（チョウ目、タテハチョウ科）東洋区
vapourer			rusty tussock moth を見よ
vapourer moth			rusty tussock moth を見よ
vapourer moth			common vapourer を見よ
Vareriana cloudywing		*Codatractus valeriana* (Plötz)	（チョウ目、セセリチョウ科）新北区
Vareriana skipper			Vareriana cloudywing を見よ
variable Antepione moth		*Antepione thisoaria* (Guenée)	（チョウ目、シャクガ科）新北区、新熱帯区。variable Antepione ともいう
variable banner		*Bolboneura sylphis sylphis* (Bates)	（チョウ目、タテハチョウ科）新熱帯区
variable bell moth		*Pyrgotis semiferana* Walker	（チョウ目、ハマキガ科）豪州区

英名	和名	学名	所属、分布、ほか
variable blue-skipper		*Pythonides jovianus amaryllis* Staudinger	（チョウ目、セセリチョウ科）新熱帯区
variable burnet moth		*Zygaena cynarae* (Esper)	（チョウ目、マダラガ科）旧北区
variable burnet moth (1)		*Zygaena ephialtes* (Linnaeus)	（チョウ目、マダラガ科）旧北区
variable carpet moth		*Anticlea vasiliata* Guenée	（チョウ目、シャクガ科）新北区
variable cattle-heart	オオシロモンジャコウアゲハ	*Parides erithalion* (Boisduval)	（チョウ目、アゲハチョウ科）新熱帯区
variable chafer		*Gnorimus variabilis* (Linnaeus)	（コウチュウ目、コガネムシ科）旧北区
variable chaff scale			cattleya scale を見よ　米国での英名
variable checkerspot			Chalcedon を見よ
variable chicken louse			chicken wing louse を見よ　米国での英名
variable climbing caterpillar			southern variable dart moth を見よ
variable coenagrion			variable damselfly を見よ
variable Colotis			blue-spotted Arab を見よ
variable cracker		*Hamadryas feronia* (Linnaeus)	（チョウ目、タテハチョウ科）新北区、新熱帯区。*H. f. farinulenta* (Fruhstorfer) も同英名
variable currant aphid		*Ampholophora varians* Patch	（カメムシ目、アブラムシ科）新北区
variable cutworm		*Agrotis porphyricollis* Guenée	（チョウ目、ヤガ科）豪州区
variable damselfly		*Coenagrion pulchellum* (van der Linden)	（トンボ目、イトトンボ科）旧北区。variable bluet ともいう
variable dancer			smoky-winged darner を見よ
variable darner		*Aeschna interrupta* (Walker)	（トンボ目、ヤンマ科）新北区
variable Diadem			variable mimic (2) を見よ
variable eggfly			variable mimic (2) を見よ
variable Emesis		*Emesis mandana* (Cramer)	（チョウ目、シジミタテハ科）新熱帯区
variable false Acraea		*Pseudacraea dolomena* (Hewitson)	（チョウ目、タテハチョウ科）エチオピア区
variable field cricket		*Gryllus lineaticeps* Stål	（バッタ目、コオロギ科）新北区
variable flower lepturine		*Stenocorus nubifer* LeConte	（コウチュウ目、カミキリムシ科）新北区
variable forest carpet moth		*Chloroclystis sandycias* Meyrick	（チョウ目、シャクガ科）豪州区
variable girdle moth		*Enypia venata* (Grote)	（チョウ目、シャクガ科）新北区
variable hairstreak		*Parrhasius orgia* (Hewitson)	（チョウ目、シジミチョウ科）新熱帯区
variable hen louse			chicken wing louse を見よ
variable lady beetle		*Exochomus marginipennis* (LeConte)	（コウチュウ目、テントウムシ科）新北区
variable ladybird	セボシテントウ	*Coelophora inaequaris* (Fabricius)	（コウチュウ目、テントウムシ科）日本、新北区、豪州区

英名	和名	学名	所属、分布、ほか
variable ladybird beetle			variable lady beetle を見よ
variable leaf beetle	サクラサルハムシ	*Cleoporus variabilis* (Baly)	（コウチュウ目、ハムシ科）日本、旧北区、東洋区
variable longhorn beetle		*Stenocorus meridianus* (Linnaeus)	（コウチュウ目、カミキリムシ科）旧北区。variable longhorn ともいう
variable louse			chicken wing louse を見よ　米国での英名
variable Metallata moth		*Metallata absumens* (Walker)	（チョウ目、ヤガ科）新北区
variable mimic (1)		*Eresia ithomioides* (Hewitson)	（チョウ目、タテハチョウ科）新熱帯区
variable mimic (2)	アンテドンムラサキ	*Hypolimnas anthedon* (Doubleday)	（チョウ目、タテハチョウ科）エチオピア区。亜種 *H. a. wahlbergi* (Wallengren) も同英名
variable mimic white		*Dismorphia theucharila* (Doubleday)	（チョウ目、シロチョウ科）新熱帯区
variable mottled-skipper		*Codatractus uvydixa* (Dyar)	（チョウ目、セセリチョウ科）新熱帯区
variable nigranum moth		*Olethreutes nigranum* (Heinrich)	（チョウ目、ハマキガ科）新北区
variable oak leaf caterpillar moth		*Heterocampa manteo* (Doubleday)	（チョウ目、シャチホコガ科）新北区。幼虫は variable oak leaf caterpillar
variable oak leaf-roller		*Epinotia emarginana* (Walsingham)	（チョウ目、ハマキガ科）新北区
variable owlet		*Scythris empetrella* Karsholt et Nielsen	（チョウ目、キヌバコガ科）旧北区
variable polyptychus		*Neopolyptychus compar* (Rothschild et Jordan)	（チョウ目、スズメガ科）エチオピア区
variable red bell		*Epinotia tenerana* (Denis et Schiffermüller)	（チョウ目、ハマキガ科）旧北区
variable reddish Pyrausta moth		*Pyrausta rubricalis* (Hübner)	（チョウ目、メイガ科）新北区
variable reed beetle	スゲハムシ	*Plateumaris sericea* (Linnaeus)	（コウチュウ目、ハムシ科）日本、旧北区
variable rose sawfly			rose argid sawfly (3) を見よ
variable satyr		*Pindis squamistriga* R. Felder	（チョウ目、タテハチョウ科）新熱帯区
variable shaft scale			cattleya scale を見よ
variable shield bug		*Kapunda troughtoni* (Distant)	（カメムシ目、カメムシ科）豪州区
variable skipperling		*Piruna gyrans* (Plötz)	（チョウ目、セセリチョウ科）新熱帯区
variable snout moth			bent-winged owlet moth を見よ
variable swallowtail		*Mimoides phaon phaon* (Boisduval)	（チョウ目、アゲハチョウ科）新熱帯区
variable swift		*Borbo holtzii* (Plötz)	（チョウ目、セセリチョウ科）エチオピア区
variable tigerwing			Menapis tigerwing を見よ
variable tropic moth		*Hemeroplanis scopulepes* Haworth	（チョウ目、ヤガ科）新北区
variable tussock moth			grey tussock moth を見よ

英名	和名	学名	所属、分布、ほか
variable yellow underwing		*Hypocala deflorata* (Fabricius)	（チョウ目、ヤガ科）大洋区、エチオピア区
variable Ypsolopha moth			broad-streaked smudge を見よ
variable Zanclognatha moth		*Zanclognatha laevigata* (Grote)	（チョウ目、ヤガ科）新北区。variable Zanclognatha ともいう
varicolored wasps		Actatinae	（ハチ目、アナバチ科）の昆虫の総称
varicolored wasps		Dimorphidae	（ハチ目）の昆虫の総称
varied carpet beetle	ヒメマルカツオブシムシ	*Anthrenus verbasci* (Linnaeus)	（コウチュウ目、カツオブシムシ科）日本、汎世界。貯蔵穀物、動植物標本の害虫
varied coronet	シロオビヨトウ	*Hadena compta* (Denis et Schiffermüller)	（チョウ目、ヤガ科）日本、旧北区
varied dusky-blue			common dusky blue を見よ
varied eggfly			common eggfly を見よ
varied hairstreak			Inous blue を見よ
varied springtail		*Isotoma viridis* Bourlet	（トビムシ目、ツチトビムシ科）新北区
variegated Acraea skipper		*Fresna nyassae* (Hewitson)	（チョウ目、セセリチョウ科）エチオピア区。variegated Acraea ともいう
variegated batwing		*Achlyodes busirus* (Stoll)	（チョウ目、セセリチョウ科）新熱帯区
variegated brindle			spectacled nettle moth を見よ
variegated carpet beetle			varied carpet beetle を見よ
variegated clearwing		*Godyris zavaleta sosunga* (Reakirt)	（チョウ目、タテハチョウ科）新熱帯区。Zavaleta glasswing を参照
variegated coffee bug		*Antestiopsis lineaticollis* Stål	（カメムシ目、カメムシ科）エチオピア区。コーヒー害虫
variegated crepuscular skipper		*Gretna zaremba* (Plötz)	（チョウ目、セセリチョウ科）エチオピア区
variegated cutworm			pearly underwing を見よ　米国での幼虫の英名
variegated cutworm moth			pearly underwing を見よ
variegated five-ring		*Ypthima methora* Hewitson	（チョウ目、タテハチョウ科）東洋区
variegated flutterer			common picture wing を見よ
variegated fritillary	トケイソウヒョウモン	*Euptoieta claudia* (Cramer)	（チョウ目、タテハチョウ科）新北区、新熱帯区
variegated golden tortrix			apple leafroller (1) を見よ
variegated grape leafhopper		*Erythroneura variabilis* Beamer	（カメムシ目、オオヨコバイ科）新北区
variegated ground cricket		*Neonemobius variegatus* (Bruner)	（バッタ目、コオロギ科）新北区
variegated hairstreak		*Michaelus jebus* (Godart)	（チョウ目、シジミチョウ科）新熱帯区
variegated Lasaia		*Lasaia meris* (Stoll)	（チョウ目、シジミタテハ科）新熱帯区
variegated leafroller moth			black-shaped Platynota moth を見よ

英名	和名	学名	所属、分布、ほか
variegated mud-loving beetles	ナガドロムシ科	Heteroceridae	（コウチュウ目）の昆虫の総称
variegated plushblue		Flos adriana (de Nicéville)	（チョウ目、シジミチョウ科）東洋区
variegated ragweed beetle		Zygogramma bicolorata Pallister	（コウチュウ目、ハムシ科）豪州区
variegated Rajah		Charaxes kahruba (Moore)	（チョウ目、タテハチョウ科）旧北区、東洋区
variegated sailer	マルバネキイロミスジ	Neptis antilope Leech	（チョウ目、タテハチョウ科）東洋区
variegated shieldback		Idiostatus variegatus Caudell	（バッタ目、キリギリス科）新北区
variegated skipper		Diaeus variegata (Plötz)	（チョウ目、セセリチョウ科）新熱帯区
variegated skipper (1)		Gorgythion begga (Prittwitz)	（チョウ目、セセリチョウ科）新北区、新熱帯区
variegated snout-moth			mottled Bomolocha moth を見よ
variegated ticlear			Zavaleta glasswing を見よ
Vari's brown		Pseudonympha varii van Son	（チョウ目、タテハチョウ科）エチオピア区
varnished Apollo	チベットタカネウスバアゲハ	Parnassius acco Gray	（チョウ目、アゲハチョウ科）東洋区
Vashti sphinx		Sphinx vashti Strecker	（チョウ目、スズメガ科）新北区。Vashti sphinx moth ともいう
Vaucher's heath		Coenonympha vaucheri Blachier	（チョウ目、タテハチョウ科）旧北区
vedalia	ベダリアテントウ	Rodolia cardinalis (Mulsant)	（コウチュウ目、テントウムシ科）日本、全北区、東洋区、豪州区。豪州からイセリアカイガラムシ防除に米国に、その後各国に導入された。Vedalia は本種の旧属名
vedalia beetle			vedalia を見よ
vedalia ladybird			vedalia を見よ
vegetable beetle		Gonocephalum elderi (Blackburn)	（コウチュウ目、ゴミムシダマシ科）豪州区
vegetable grasshopper	ミカドフキバッタ	Parapodisma mikado (Bolivar)	（バッタ目、バッタ科）日本、旧北区
vegetable leafhopper		Austroasca viridigrisea (Paoli)	（カメムシ目、オオヨコバイ科）豪州区
vegetable leafminer	トマトハモグリバエ	Liriomyza sativae Blanchard	（ハエ目、ハモグリバエ科）日本、新北区。ジャガイモ、ナス、トマト、ダイズ、キクなどの害虫
vegetable leafminer			garden pea leafminer を見よ
vegetable marrow midge			white bryony midge を見よ
vegetable silver phytometra	マガリキンウワバ	Diachrysia leonina (Oberthür)	（チョウ目、ヤガ科）日本、旧北区
vegetable weevil		Listroderes difficilis Germar	（コウチュウ目、ゾウムシ科）新北区、豪州区
vegetable weevil		Listroderes obliquus (Klug)	（コウチュウ目、ゾウムシ科）新北区、新熱帯区
vegetable weevil (1)	ヤサイゾウムシ	Listroderes costirostris Schoenherr	（コウチュウ目、ゾウムシ科）日本、東洋区、新北区、新熱帯区。多くの野菜、タバコ、キクなどの害虫

英名	和名	学名	所属、分布、ほか
Veia glasswing		*Pteronymia veia* (Hewitson)	(チョウ目、タテハチョウ科) 新熱帯区
veiled ear moth		*Loscopia velata* (Walker)	(チョウ目、ヤガ科) 新北区
veined blue		*Plebejus neurona* (Skinner)	(チョウ目、シジミチョウ科) 新北区
veined Ctenucha moth		*Ctenucha venosa* Walker	(チョウ目、カノコガ科) 新北区
veined dart		*Actinor radians* (Moore)	(チョウ目、セセリチョウ科) 東洋区
veined gold			African golden Arab を見よ
veined golden Arab			veined orange を見よ
veined hawk		*Litosphingia corticea* Jordan	(チョウ目、スズメガ科) エチオピア区
veined jay		*Graphium bathycles* (Zinken)	(チョウ目、アゲハチョウ科) 東洋区
veined Jay	ギンスジミカドアゲハ	*Graphium chiron* Wallace	(チョウ目、アゲハチョウ科) 東洋区
veined labyrinth	アリサンキマダラヒカゲ	*Neope pulaha* (Moore)	(チョウ目、タテハチョウ科) 東洋区
veined orange	ベスタツマアカシロチョウ	*Colotis vesta* (Reiche)	(チョウ目、シロチョウ科) エチオピア区
veined paradise skipper			veined skipper を見よ
veined pierrot		*Tarucus venosus* Moore	(チョウ目、シジミチョウ科) 東洋区
veined scrub hopper		*Acromachus stigmata* (Moore)	(チョウ目、セセリチョウ科) 東洋区
veined skipper		*Abantis venosa* Trimen	(チョウ目、セセリチョウ科) エチオピア区
veined swallowtail			veined swordtail を見よ
veined swordtail	アフリカマダラタイマイ	*Graphium leonidas* (Fabricius)	(チョウ目、アゲハチョウ科) エチオピア区
veined tiger		*Teracotona euprepia* Hampson	(チョウ目、ヒトリガ科) エチオピア区
veined white			green-veined white を見よ
veined white			mustard white を見よ
veined white skipper			Arsalte skipper を見よ
veinous smudge		*Orthotaelia sparganella* (Thunberg)	(チョウ目、ホソハマキモドキガ科) 旧北区
Veleda skipper		*Eprius veleda veleda* (Godman)	(チョウ目、セセリチョウ科) 新熱帯区
Velinus skipper		*Urbanus velinus* (Plötz)	(チョウ目、セセリチョウ科) 新熱帯区
velleda lappet moth			large Tolype を見よ
Velutina cracker		*Hamadryas velutina* Bates	(チョウ目、タテハチョウ科) 新熱帯区
velvet ant		*Mutilla europaea* Linnaeus	(ハチ目、アリバチ科) 旧北区。日本亜種は *M. e. mikado* Cameron ミカドアリバチ
velvet ants (1)	アリバチ科	Mutillidae	(ハチ目) の昆虫の総称
velvet ants		*Dasymutilla*	(ハチ目、アリバチ科) の昆虫の総称
velvet ants			wasps (1) を見よ
velvet armyworm moth		*Spodoptera latifascia* Walker	(チョウ目、ヤガ科) 新北区、新熱帯区。velvet armyworm ともいう

英名	和名	学名	所属、分布、ほか
velvet bean caterpillar		*Anticarsia gemmatalis* Hübner	(チョウ目、ヤガ科) 新北区、新熱帯区。ダイズ害虫。成虫は velvetbean caterpillar moth、velvetbean moth
velvet flat		*Celaenorrhinus ficulnea* (Hewitson)	(チョウ目、セセリチョウ科) 東洋区
velvet hairstreak		*Thepytus echelta* (Hewitson)	(チョウ目、シジミチョウ科) 新熱帯区
velvet sicklewing		*Eantis minna* Evans	(チョウ目、セセリチョウ科) 新熱帯区
velvet-spotted bush-blue			desert Babul blue を見よ　velvet-spotted blue ともいう
velvet-striped grasshopper		*Eritettix simplex* (Scudder)	(バッタ目、バッタ科) 新北区
velvet water bugs	ケシミズカメムシ科	Hebridae	(カメムシ目) の昆虫の総称　体長 3 mm 以下
velvety chafer	ビロウドコガネ	*Maladera japonica japonica* (Motschulsky)	(コウチュウ目、コガネムシ科) 日本
velvety chafer			Asiatic garden beetle を見よ
velvety hawk moth	ビロードスズメ	*Rhagastis mongoliana* (Butler)	(チョウ目、スズメガ科) 日本、旧北区、東洋区
velvety huge-comma	ハグルマトモエ	*Spirama helicina* (Hübner)	(チョウ目、ヤガ科) 日本、東洋区
velvety longicorn beetle	ビロウドカミキリ	*Acalolepta fraudatrix* (Bates)	(コウチュウ目、カミキリムシ科) 日本、旧北区。ヤツデ害虫
velvety shore bugs	メミズムシ科	Ochteridae	(カメムシ目) の昆虫の総称
velvety small noctuid	ビロウドコヤガ	*Anterastria atrata* (Butler)	(チョウ目、ヤガ科) 日本、旧北区
velvety tortrix	ビロードハマキ	*Cerace xanthocosma* Diakonoff	(チョウ目、ハマキガ科) 日本、旧北区
velvety tree ant		*Liometopum occidentale* Emery	(ハチ目、アリ科) 新北区
velvety tree nymph		*Sallya occidentalium* (Mabille)	(チョウ目、タテハチョウ科) エチオピア区
venerable dart moth			dusky cutworm moth を見よ
venerable silverfish		*Tricholepidion gertschi* Wygodzinsky	(シミ目、Lepidotrichidae) 新北区
Venezuelan sister			Boeotia sister を見よ
Venus moth		*Leto venus* (Cramer)	(チョウ目、コウモリガ科) エチオピア区
Venusta grass yellow		*Eurema venusta* (Boisduval)	(チョウ目、シロチョウ科) 新熱帯区
Venusta metalmark		*Calydna venusta venusta* Godman et Salvin	(チョウ目、シジミタテハ科) 新熱帯区
Veracruz sister			tawny-striped sister を見よ
Veracruz sister			white-spot sister を見よ
Veracruz skipperling		*Piruna ceracates* (Hewitson)	(チョウ目、セセリチョウ科) 新熱帯区
Veracruzan ministreak		*Ministrymon inoa* (Godman et Salvin)	(チョウ目、シジミチョウ科) 新熱帯区
Veracruzan Sarota		*Sarota craspediodonta* (Dyar)	(チョウ目、シジミタテハ科) 新熱帯区
Veracruzan skipper		*Oeonus pyste* Godman	(チョウ目、セセリチョウ科) 新熱帯区

英名	和名	学名	所属、分布、ほか
Veracruzan sootywing		*Bolla cybele* Evans	（チョウ目、セセリチョウ科）新熱帯区
Veracruzan Theope		*Theope eupolis* Schaus	（チョウ目、シジミタテハ科）新熱帯区
verbena bud moth		*Endothenia hebesana* (Walker)	（チョウ目、ハマキガ科）全北区
Verdant hawk			Verdant sphinx を見よ
Verdant sphinx		*Euchloron megaera* (Linnaeus)	（チョウ目、スズメガ科）エチオピア区
Verhuell's smut		*Psychoides verhuella* Bruand	（チョウ目、ヒロズコガ科）旧北区
vermileonid flies		Vermileonidae	（ハエ目）の昆虫の総称　シギアブ科に入れることあり
vernal bluet		*Enallagma vernale* Gloyd	（トンボ目、イトトンボ科）新北区
vernal skipper			little glassywing を見よ
vernonia borer moth		*Papaipema limpida* (Guenée)	（チョウ目、ヤガ科）新北区
Vernon's half mourner			bath white を見よ
Verona borer moth		*Papaipema verona* (Smith)	（チョウ目、ヤガ科）新北区
vertical skipper		*Vertica verticalis coatepeca* (Schaus)	（チョウ目、セセリチョウ科）新熱帯区
Vertumnus cattle-heart	アオマエモンジャコウアゲハ	*Parides vertumnus* (Cramer)	（チョウ目、アゲハチョウ科）新熱帯区
vespasianus groundstreak		*Camissecla vespasianus* (Butler et Druce)	（チョウ目、シジミチョウ科）新熱帯区
vesper bluet		*Enallagma vesperum* Calvert	（トンボ目、イトトンボ科）新北区
vesperus		*Vesperus*	（コウチュウ目、カミキリムシ科）の昆虫の総称
vespiary			スズメバチの巣
vespid			スズメバチ科のハチ。hornets (2) を見よ
vespid wasps			hornets (2) を見よ
vespiform thrips			predatory thrips を見よ
vespoid digger-wasps			scoliid wasps を見よ
vespoid wasps			wasps (1) を見よ
Vesta crescent			Vesta crescentspot (1) を見よ
Vesta crescentspot		*Phyciodes vesta* Edwards	（チョウ目、タテハチョウ科）新北区
Vesta crescentspot (1)		*Phyciodes graphica* (R. Felder)	（チョウ目、タテハチョウ科）新北区
vestal		*Rhodometra sacraria* (Linnaeus)	（チョウ目、シャクガ科）旧北区
vestal cuckoo bee		*Bombus vestalis* Geoffroy	（ハチ目、ミツバチ科）旧北区
vestal moth		*Antaeotricha vestalis* Zeller	（チョウ目、マルハキバガ科）新北区
vestal moth		*Cabera variolaria* Guenée	（チョウ目、シャクガ科）新北区
vestal moth			vestal を見よ
vestal tiger moth		*Spilosoma vestalis* Packard	（チョウ目、ヒトリガ科）新北区

英名	和名	学名	所属、分布、ほか
Vestalis metalmark		*Leucochimonia vestalis vestalis* (H. Bates)	(チョウ目、シジミタテハ科) 新熱帯区
vetch aphid		*Megoura viciae* Buckton	(カメムシ目、アブラムシ科) 旧北区、エチオピア区
vetch bruchid		*Bruchus brachialis* Fahraseus	(コウチュウ目、ハムシ科) 新北区
vetch bruchid			vetch weevil を見よ
vetch leaf midge		*Dasineura viciae* (Kieffer)	(ハエ目、タマバエ科) 旧北区
vetch looper moth		*Caenurgia chloropha* (Hübner)	(チョウ目、ヤガ科) 新北区、新熱帯区
vetch weevil		*Laria rufipes* Herbst	(コウチュウ目、ハムシ科) 旧北区、東洋区
vetch weevil			vetch bruchid を見よ
vexans mosquito		*Aedes* (*Aedimorphus*) *vexans* (Meigen)	(ハエ目、カ科) 全北区、東洋区、汎熱帯。水田、湿原などに発生。日本亜種は *A. (A.) v. nipponii* (Theobald) キンイロヤブカ
vexatious mosquito			vexans mosquito を見よ
Viardi white	メキシコキスジシロチョウ	*Pieriballia viardi* (Boisduval)	(チョウ目、シロチョウ科) 新北区、新熱帯区
Vibilia longwing		*Eueides vibilia* (Godart)	(チョウ目、タテハチョウ科) 新熱帯区。*E. v. vialis* Stichel も同英名
viburnum aphid		*Aphis viburniphila* Patch	(カメムシ目、アブラムシ科) 新北区
viburnum aphid (1)		*Aphis viburni* Scopoli	(カメムシ目、アブラムシ科) 旧北区
viburnum beetle		*Pyrrhalta viburni* (Paykull)	(コウチュウ目、ハムシ科) 旧北区。viburnum leaf beetle ともいう
viburnum button			Logan's button (1) を見よ
viburnum clearwing moth		*Synanthedon viburni* Engelhardt	(チョウ目、スカシバガ科) 新北区
viburnum cottony scale	ガマズミワタカイガラモドキ	*Phenacoccus viburnae* Kanda	(カメムシ目、コナカイガラムシ科) 日本
viburnum drepanid	クロスジカギバ	*Oreta turpis* Butler	(チョウ目、カギバガ科) 日本、旧北区
viburnum leaf beetle	サンゴジュハムシ	*Pyrrhalta humeralis* (Chen)	(コウチュウ目、ハムシ科) 日本、旧北区。サンゴジュ害虫
viburnum shoot sawfly	モンクキバチ	*Janus japonicus* Sato	(ハチ目、クキバチ科) 日本
viburnum whitefly		*Aleurotrachelus jelinekii* (von Frauenfeld)	(カメムシ目、コナジラミ科) 旧北区
viceroy	カバイロイチモンジ	*Limenitis archippus* (Cramer)	(チョウ目、タテハチョウ科) 新北区、新熱帯区。*L. a. obsoleta* Edwards も同英名
viceroy butterfly			viceroy を見よ
vicia aphid	タニワタシオナガヒゲナガアブラムシ	*Megoura japonica* (Matsumura)	(カメムシ目、アブラムシ科) 日本、旧北区、東洋区
Victor swallowtail			Victorine swallowtail を見よ
victoria dots		*Micropentila victoriae* Stempffer et Bennet	(チョウ目、シジミチョウ科) エチオピア区
Victoria silverline		*Cigaritis victoriae* (Butler)	(チョウ目、シジミチョウ科) エチオピア区
Victoria white		*Belenois victoria* Dixey	(チョウ目、シロチョウ科) エチオピア区

英名	和名	学名	所属、分布、ほか
Victorine swallowtail		*Papilio victorinus* Doubleday	（チョウ目、アゲハチョウ科）新北区、新熱帯区
Victor's blue		*Lepidochrysops victori* Pringle	（チョウ目、シジミチョウ科）エチオピア区
Vidler's alpine			northwest alpine を見よ
Viennese emperor			giant emperor を見よ
Viereck's skipper		*Atrytonopsis viereckii* (Skinner)	（チョウ目、セセリチョウ科）新北区、新熱帯区
Vietnamese stick insect		*Ramulus artemis* (Westwood)	（ナナフシ目、ナナフシ科）東洋区
vigilant mosquito		*Culex pervigilans* Bergroth	（ハエ目、カ科）豪州区
vigorous Cissusa moth		*Cissusa valens* (Hy. Edwards)	（チョウ目、ヤガ科）新北区
Vindhyan bob		*Arnetta vindhiana* (Moore)	（チョウ目、セセリチョウ科）東洋区
vine borer		*Melittia gloriosa* Edwards	（チョウ目、スカシバガ科）新北区
vine bud moth		*Pyroderces amphisaris* Meyrick	（チョウ目、カザリバガ科）東洋区
vine bud moth		*Theresimima ampelophaga* (Bayle-Barelle)	（チョウ目、マダラガ科）旧北区
vine chafer		*Anomala vitis* Fabricius	（コウチュウ目、コガネムシ科）旧北区
vine chrysomelid worm			brown and black beetle を見よ
vine flea beetle		*Haltica ampelophaga* Guérin-Méneville	（コウチュウ目、ハムシ科）旧北区
vine flower gall midge			grapevine midge を見よ
vine hawk-moth (1)		*Hippotion celerio* (Linnaeus)	（チョウ目、スズメガ科）旧北区、東洋区、豪州区、エチオピア区
vine hawk moth (2)	セスジスズメ	*Theretra oldenlandiae* (Fabricius)	（チョウ目、スズメガ科）日本、東洋区、豪州区。ブドウ、サトイモなどの害虫
vine hawk-moth		*Hippotion rosetta* (Swinhoe)	（チョウ目、スズメガ科）日本、東洋区、豪州区
vine hawk moth			balsam striped hawk を見よ
vine leaf roller			pear leaf roller (1) を見よ
vine leafhopper			green leafhopper (1) を見よ
vine leafhoppers		*Erythroneura*	（カメムシ目、オオヨコバイ科）の昆虫の総称
vine leaf-roller moth			leaf roller (1) を見よ
vine leafroller Tortrix moth			leaf roller (1) を見よ
vine-leaf vagrant	クレオドラシロチョウ	*Eronia cleodora* (Hübner)	（チョウ目、シロチョウ科）エチオピア区
vine louse			grapeleaf louse を見よ
vine mealybug			grape mealybug を見よ
vine moth			grapevine moth (1)(2) を見よ
vine moth			grape cochylid を見よ

1161

英名	和名	学名	所属、分布、ほか
vine scale (1)		*Eulecanium coryli* (Linnaeus)	（カメムシ目、カタカイガラムシ科）新北区
vine scale (2)		*Eulecanium persicae* (Fabricius)	（カメムシ目、カタカイガラムシ科）豪州区
vine sphinx		*Eumorpha vitis* (Linnaeus)	（チョウ目、スズメガ科）新北区、新熱帯区。vine sphinx moth ともいう
vine thrips			grape thrips を見よ
vine tortrix			fruit tree tortrix を見よ
vine weevil		*Orthorhinus klugi* Boheman	（コウチュウ目、ゾウムシ科）新北区、豪州区
vine weevil			black vine weevil を見よ
vine-worm			whortleberry bell を見よ
vinegar flies (1)	ショウジョウバエ科	Drosophilidae	（ハエ目）の昆虫の総称　世界に1,500種
vinegar flies (2)		*Drosophila*	（ハエ目、ショウジョウバエ科）の昆虫の総称
vinegar fly			pomace fly を見よ
vinegar fly			fruit fly (1) を見よ
vine's rustic		*Hoplodrina ambigua* (Denis et Schiffermüller)	（チョウ目、ヤガ科）旧北区
vineyard cicada		*Tibicen haematodes* Scopoli	（カメムシ目、セミ科）旧北区
Vinous oakblue		*Arhopala athada* (Staudinger)	（チョウ目、シジミチョウ科）東洋区
viola Charaxes		*Charaxes viola* Butler	（チョウ目、タテハチョウ科）エチオピア区
viola sawfly			violet sawfly (1) を見よ
violaceous bent-skipper			widespread bent-skipper を見よ
violaceous stink bug	ツマジロカメムシ	*Menida violacea* Motschulsky	（カメムシ目、カメムシ科）日本、旧北区
Viola's wood satyr		*Megisto viola* (Maynard)	（チョウ目、タテハチョウ科）新北区
violescent blue		*Orachrysops violescens* Henning et Henning	（チョウ目、シジミチョウ科）エチオピア区
violet aphid		*Micromyzus violae* (Pergande)	（カメムシ目、アブラムシ科）新北区。温室害虫
violet aphid		*Myzus portulacae* Marshall	（カメムシ目、アブラムシ科）旧北区
violet aphid (1)		*Myzus ornatus* Laing	（カメムシ目、アブラムシ科）汎世界
violet awl		*Hasora leucospila* (Mabille)	（チョウ目、セセリチョウ科）東洋区
violet-banded Palla		*Palla violinitens* (Crowley)	（チョウ目、タテハチョウ科）エチオピア区
violet-banded skipper		*Nyctelius nyctelius* (Latreille)	（チョウ目、セセリチョウ科）新北区
violet-black case		*Coleophora albitarsella* Zeller	（チョウ目、ツツミノガ科）旧北区
violet black-legged robberfly		*Dioctria atricapilla* Meigen	（ハエ目、ムシヒキアブ科）旧北区
violet carpenter bee		*Xylocopa violacea* (Linnaeus)	（ハチ目、ミツバチ科）旧北区、東洋区

英名	和名	学名	所属、分布、ほか
violet click beetle		*Limoniscus violaceus* (Müller)	(コウチュウ目、コメツキムシ科) 旧北区
violet-clouded skipper			olive clouded skipper
violet copper		*Lycaena helle* Schiffermüller	(チョウ目、シジミチョウ科) 旧北区
violet dancer		*Argia violacea* (Hagen)	(トンボ目、イトトンボ科) 新北区
violet dart moth		*Euxoa violaris* Grote et Robinson	(チョウ目、ヤガ科) 新北区
violet Diopetes		*Pilodeudorix violetta* (Aurivillius)	(チョウ目、シジミチョウ科) エチオピア区
violet dropwing		*Trithemis annulata* (Palisot de Beauvois)	(トンボ目、トンボ科) エチオピア区
violet-dusted skipperling		*Piruna dampfi* (Bell)	(チョウ目、セセリチョウ科) 新熱帯区
violet 4-lineblue		*Nacaduba subperusia* (Snellen)	(チョウ目、シジミチョウ科) 東洋区
violet fritillary	スミレチビヒョウモン	*Clossiana dia* (Linnaeus)	(チョウ目、タテハチョウ科) 新北区
violet-frosted skipper		*Mnasicles geta* Godman	(チョウ目、セセリチョウ科) 新熱帯区
violet ground beetle		*Carabus violaceus* Linnaeus	(コウチュウ目、オサムシ科) 旧北区
violet juniper argent		*Argyresthia dilectella* Zeller	(チョウ目、メムシガ科) 旧北区
violet lacewing	ホソバハレギチョウ	*Cethosia myrina* C. et R. Felder	(チョウ目、タテハチョウ科) 東洋区
violet leaf midge		*Dasineura affinis* (Kieffer)	(ハエ目、タマバエ科) 旧北区
violet leaf miner			violet leaf midge を見よ
violet leaf-rolling gall-midge			violet leaf midge を見よ
violet line-blue			dark Ceylon 6-lineblue を見よ
violet looper	スミレシロヒメシャク	*Scopula umbelaria majoraria* (Leech)	(チョウ目、シャクガ科) 日本
violet midge		*Contarinia violicola* (Coquillett)	(ハエ目、タマバエ科) 新北区
violet oil beetle	ムラサキオオツチハンミョウ	*Meloe violaceus* Linnaeus	(コウチュウ目、ツチハンミョウ科) 旧北区
violet onyx	ヒメミツオシジミ	*Horaga albimacula* (Wood-Mason et de Nicéville)	(チョウ目、シジミチョウ科) 東洋区
violet opal		*Chrysoritis violescens* (Dickson)	(チョウ目、シジミチョウ科) エチオピア区
violet-patch skipper		*Monca tyrtaeus* Plötz	(チョウ目、セセリチョウ科) 新北区。violet-patched skipper ともいう
violet-patched skipper		*Monca crispinus* (Plötz)	(チョウ目、セセリチョウ科) 新熱帯区
violet rove beetle		*Staphylinus violaceus* Gravenhorst	(コウチュウ目、ハネカクシ科) 新北区。菌、屍肉、樹皮下に生息
violet sawfly	ヒゲナガハバチ	*Lagidina platycerus* (Marlatt)	(ハチ目、ハバチ科) 日本、旧北区

英名	和名	学名	所属、分布、ほか
violet sawfly (1)		*Ametastegia pallipes* (Spinola)	(ハチ目、ハバチ科) 新北区
violet-shaded long-horn		*Nemophora minimella* (Denis et Schiffermüller)	(チョウ目、マガリガ科) 旧北区
violet silverline		*Cigaritis crustaria* (Holland)	(チョウ目、シジミチョウ科) エチオピア区
violet-spotted Charaxes		*Charaxes violetta* Grose-Smith	(チョウ目、タテハチョウ科) エチオピア区
violet-spotted ciliate blue		*Anthene lysicles* (Hewitson)	(チョウ目、シジミチョウ科) エチオピア区
violet spreadwing		*Carrhenes santes* Bell	(チョウ目、セセリチョウ科) 新熱帯区
violet stem borer			pink borer (1) を見よ
violet-studded skipper		*Morys lyde* (Godman)	(チョウ目、セセリチョウ科) 新熱帯区
violet tail			violet dancer を見よ
violet tanbark beetle	ルリヒラタカミキリ	*Callidium violaceum* (Linnaeus)	(コウチュウ目、カミキリムシ科) 日本、全北区
violet tip			question mark を見よ
violet tip			purple tip を見よ
violet-tipped crow			common Indian crow を見よ
violet-tipped Saliana		*Saliana saladin saladin* Evans	(チョウ目、セセリチョウ科) 新熱帯区
violet tree-brown		*Lethe oviolae* Tsukada et Nishiyama	(チョウ目、タテハチョウ科) 東洋区
violet-washed Charaxes	キオビフタオチョウ	*Charaxes lucretius* (Cramer)	(チョウ目、タテハチョウ科) エチオピア区
violet-washed eyed-metalmark		*Mesosemia telegone telegone* (Boisduval)	(チョウ目、シジミタテハ科) 新熱帯区
violet-washed skipper		*Damas clavus* (Herrich-Schäffer)	(チョウ目、セセリチョウ科) 新熱帯区
violet-winged grasshopper			giant South American grasshopper を見よ
Viper's bugloss gothic		*Hadena irregularis* (Hufnagel)	(チョウ目、ヤガ科) 旧北区。Viper's bugloss ともいう
Viper's bugloss moth			Viper's bugloss gothic を見よ
Viper's bugloss moth			bordered echium ermel を見よ
Viper's bugloss spearwing		*Tinagma ocnerostomella* (Stainton)	(チョウ目、ヒロズコガ科) 旧北区
vipionids		Vipionidae	(ハチ目) の昆虫の総称
Viresco hairstreak		*Theritas viresco* (Druce)	(チョウ目、シジミチョウ科) 新熱帯区
Virgilia wood nymph		*Taygetis virgilia* (Cramer)	(チョウ目、タテハチョウ科) 新熱帯区
virgin			単性生殖をする雌の昆虫
virgin mayflies			pale burrower mayflies を見よ
virgin mayfly		*Polimitarcys virgo* (Olivier)	(カゲロウ目、オオシロカゲロウ科) 旧北区
virgin moth		*Protitame virginalis* (Hulst)	(チョウ目、シャクガ科) 新北区
virgin tiger moth		*Apantesis virgo* (Linnaeus)	(チョウ目、ヒトリガ科) 新北区

英名	和名	学名	所属、分布、ほか
Virginia creeper clearwing moth		*Albuna fraxini* (Edwards)	(チョウ目、スカシバガ科) 新北区。Virginia creeper clearwing ともいう
Virginia creeper deathwatch beetle		*Xyletinus peltatus* Harris	(コウチュウ目、シバンムシ科) 新北区
Virginia creeper leafhopper		*Erythroneura ziczac* Walsh	(カメムシ目、オオヨコバイ科) 新北区
Virginia creeper sphinx moth		*Darapsa myron* (Cramer)	(チョウ目、スズメガ科) 新北区、新熱帯区。Virginia creeper sphinx ともいう
Virginia ctenochid moth			Virginia ctenucha moth を見よ
Virginia ctenucha moth		*Ctenucha virginica* (Esper)	(チョウ目、カノコガ科) 新北区。Virginia ctenucha ともいう
Virginia ermine moth			Virginia tiger moth を見よ
Virginia flower fly			yellowjacket hover fly を見よ
Virginia lady			painted beauty を見よ
Virginia lady hunter's butterfly			American painted lady を見よ
Virginia metalmark			little metalmark を見よ
Virginia pine borer			large flat-head pine heartwood borer beetle を見よ
Virginia pine sawfly		*Neodiprion pratti pratti* (Dyar)	(ハチ目、マツハバチ科) 新北区
Virginia pitch-nodule moth		*Petrova virginiana* Busck	(チョウ目、ハマキガ科) 新北区
Virginia tiger moth		*Diacrisia virginica* (Fabricius)	(チョウ目、ヒトリガ科) 全北区、新熱帯区。Virginian tiger moth ともいう
Virginiana clearwing		*Hyposcada virginiana* (Hewitson)	(チョウ目、タテハチョウ科) 新熱帯区
Virginius skipper		*Virga virginius* (Möschler)	(チョウ目、セセリチョウ科) 新熱帯区
visitation moth		*Dyspyralis illocata* (Warren)	(チョウ目、ヤガ科) 新北区
visiting ants			army ants (1) を見よ
Vista skipper		*Niconiades viridis vista* Evans	(チョウ目、セセリチョウ科) 新熱帯区
vitis plume moth	シラホシトリバ	*Deuterocopus albipunctatus* Fletcher	(チョウ目、トリバガ科) 日本、旧北区。幼虫はブドウ害虫
Vitoria's satyr		*Pedaliodes marmelsi* Vitoria	(チョウ目、タテハチョウ科) 新熱帯区
vivid blue			Macchia blue を見よ
vivid cosmet		*Cosmopterix schmidiella* Frey	(チョウ目、カザリバガ科) 旧北区
vivid dancer		*Argia vivida* Hagen	(トンボ目、イトトンボ科) 新北区
vivid green swallowtail			apple-green swallowtail を見よ
vivid metallic ground beetles		Chlaeniini	(コウチュウ目、オサムシ科) の昆虫の総称
vivid painted lady	オビヒメアカタテハ	*Vanessa myrinna* (Doubleday)	(チョウ目、タテハチョウ科) 新熱帯区

英名	和名	学名	所属、分布、ほか
vocal field cricket		*Gryllus vocalis* Scudder	（バッタ目、コオロギ科）新北区
Vogel's blue		*Plebejus vogelii* (Oberthür)	（チョウ目、シジミチョウ科）旧北区
vole louse		*Hoplopleura acanthopus* (Burmeister)	（シラミ目、フトゲシラミ科）旧北区
Vollenhov's crow	エレウシナルリマダラ	*Euploea eleusina* (Cramer)	（チョウ目、タテハチョウ科）東洋区
Volta swallowtail		*Papilio nobicea* Suffert	（チョウ目、アゲハチョウ科）エチオピア区
Volta swallowtail			Zenobia swallowtail を見よ
volupial Pyrausta moth		*Pyrausta volupialis* (Grote)	（チョウ目、メイガ科）新北区
voyaging glider			ferruginous glider を見よ
vulpina dagger moth			Miller dagger moth を見よ
vulture-feather case		*Coleophora gryphipennella* (Hübner)	（チョウ目、ツツミノガ科）旧北区

英名	和名	学名	所属、分布、ほか
W			
w-marked cutworm moth		*Spaelotis clandestina* (Harris)	(チョウ目、ヤガ科) 新北区。w-marked cutworm ともいう
Waaihoek opal		*Chrysoritis blencathrae* (Heath et Ball)	(チョウ目、シジミチョウ科) エチオピア区
Wagner bean pod weevil		*Apion godmani* Wagner	(コウチュウ目、ホソクチゾウムシ科) 新北区
wainscot grass-veneer	カバイロツトガ	*Chilo phragmitellus* (Hübner)	(チョウ目、メイガ科) 日本、旧北区
wainscot grass-veneer moth		*Eoreuma densella* (Zeller)	(チョウ目、メイガ科) 新北区
wainscot smudge		*Ypsolopha scabrella* (Linnaeus)	(チョウ目、スガ科) 旧北区
wainscots		Leucaniinae	(チョウ目、ヤガ科) の昆虫の総称
wainscots		Nonagriinae	(チョウ目、ヤガ科) の昆虫の総称
waiter	ナミスジツルギタテハ	*Marpesia coresia* Weymer	(チョウ目、タテハチョウ科) 新北区。waiter butterfly ともいう
waiter			waiter daggerwing を見よ
waiter daggerwing		*Marpesia zerynthia* Hübner	(チョウ目、タテハチョウ科) 新北区、新熱帯区
Wakayama scale	ワカヤマシロカイガラムシ	*Aulacaspis wakayamaensis* Kuwana	(カメムシ目、マルカイガラムシ科) 日本
Wakkerstroom copper		*Aloeides merces* Henning et Henning	(チョウ目、シジミチョウ科) エチオピア区
Wakkerstroom widow		*Dingana alaedeus* Henning et Henning	(チョウ目、タテハチョウ科) エチオピア区
walker			Fuller を見よ
Walker's Epinotia moth		*Epinotia transmissana* (Walker)	(チョウ目、ハマキガ科) 新北区
Walker's ground cricket		*Allonemobius walkeri* Howard et Furth	(バッタ目、コオロギ科) 新北区
Walker's heath twist		*Philedonides lunana* (Thunberg)	(チョウ目、ハマキガ科) 旧北区
Walker's lanark tortrix			Walker's heath twist を見よ
Walker's metalmark		*Apodemia walkeri* Godman et Sallvin	(チョウ目、シジミタテハ科) 新北区
Walker's tree cricket		*Oecanthus walkeri* Collins et Symes	(バッタ目、カンタン科) 新北区
walking flower mantis			flower mantis を見よ
walking leaf		*Phyllium*	(ナナフシ目、コノハムシ科) の昆虫の総称
walking leaves			leaf insects を見よ
walking stick		*Diapheromera femorata* (Say)	(ナナフシ目、ナナフシ科) 新北区
walking sticks (1)	ナナフシ科	Phasmatidae	(ナナフシ目) の昆虫の総称
walking sticks (2)	ナナフシ目	Phasmida	の昆虫の総称 世界に約 2,500 種

英名	和名	学名	所属、分布、ほか
walking thread-legged katydid		*Arethaea ambulator* Hebard	（バッタ目、キリギリス科）新北区
walkingsticks		*Diapheromera*	（ナナフシ目、ナナフシ科）の昆虫の総称
wall	メゲラツマジロウラジャノメ（キイロウラジャノメ）	*Pararge megera* (Linnaeus)	（チョウ目、タテハチョウ科）旧北区。壁や堤で陽をあびる習性に由来。wall butterfly ともいう
wall bee		*Chalicodoma muraria* Fabricius	（ハチ目、ハキリバチ科）旧北区
wall brown			wall を見よ
wall grey		*Scoparia murana* Curtis	（チョウ目、メイガ科）旧北区
wall louse			bed bug (1) を見よ
wall mason wasp		*Symmorphus murarius* (Linnaeus)	（ハチ目、スズメバチ科）旧北区
wall wasp			wall mason wasp を見よ
wall wasps			mason wasps を見よ
wallaby flies			louse flies を見よ
Wallacea bluebottle		*Graphium anthedon* (C. et R. Felder)	（チョウ目、アゲハチョウ科）東洋区
Wallace's albatross		*Appias galba* (Wallace)	（チョウ目、シロチョウ科）東洋区
Wallace's black prince		*Rohana macar* Wallace	（チョウ目、タテハチョウ科）東洋区
Wallace's courtesan	スダレゴマダラ	*Euripus robustus* Wallace	（チョウ目、タテハチョウ科）東洋区
Wallace's giant bee		*Megachile pluto* Smith	（ハチ目、ハキリバチ科）東洋区。開張 6.3 cm になるインドネシアの大型種
Wallace's golden birdwing	アカメガネトリバネアゲハ	*Ornithoptera croesus* Wallace	（チョウ目、アゲハチョウ科）東洋区
Wallace's longwing		*Heliconius wallacei* (Reakirt)	（チョウ目、タテハチョウ科）新熱帯区
Wallengren's copper		*Trimenia wallengrenii* (Trimen et Bowker)	（チョウ目、シジミチョウ科）エチオピア区
Wallengren's ranger		*Kedestes wallengrenii* (Trimen)	（チョウ目、セセリチョウ科）エチオピア区
Wallengren's silver-spotted copper			Wallengren's copper を見よ
Wall's grappletail		*Heliogomphus walli* Fraser	（トンボ目、サナエトンボ科）東洋区
Wall's shadowdamsel		*Drepanosticta walli* (Fraser)	（トンボ目、Platystictidae）東洋区
Wallula grasshopper		*Conozoa wallula* (Scudder)	（バッタ目、バッタ科）新北区
wallum cicada		*Cicadetta stradbrokensis* (Distant)	（カメムシ目、セミ科）豪州区
wallum darner		*Austroaeschna cooloola* Theishinger	（トンボ目、ヤンマ科）豪州区
wallum vicetail		*Hemigomphus cooloola* Watson	（トンボ目、サナエトンボ科）豪州区
walnut aphid		*Chromaphis juglandicola* (Kaltenbach)	（カメムシ目、アブラムシ科）全北区
walnut bark beetle	クルミノコキクイムシ	*Cryphalus juglans* Niijima	（コウチュウ目、キクイムシ科）日本。クルミ害虫

英名	和名	学名	所属、分布、ほか
walnut blue		*Chaetoprocta odata* (Hewitson)	(チョウ目、シジミチョウ科) 東洋区
walnut borer			shot-hole borer (3) を見よ
walnut broad-headed leafhopper			walnut leafhopper を見よ
walnut caterpillar moth		*Datana integerrima* Grote et Robinson	(チョウ目、シャチホコガ科) 新北区。幼虫は walnut caterpillar
walnut codling moth			codling moth を見よ
walnut Datana			walnut caterpillar moth を見よ
walnut hornworm	エゾスズメ	*Phyllosphingia dissimilis* (Bremer)	(チョウ目、スズメガ科) 日本、旧北区
walnut husk fly		*Rhagoletis completa* Cresson	(ハエ目、ミバエ科) 新北区。クルミ害虫
walnut lace bug	クルミグンバイ	*Uhlerites latius* Takeya	(カメムシ目、グンバイムシ科) 日本、旧北区。クルミ害虫
walnut leaf beetle	クルミハムシ	*Gastrolina depressa* Bay	(コウチュウ目、ハムシ科) 日本、旧北区。クルミ害虫
walnut leafhopper	クルミヒロズヨコバイ	*Oncopsis juglans* (Matsumura)	(カメムシ目、オオヨコバイ科) 日本、旧北区。クルミ害虫
walnut moth			tobacco moth (1) を見よ
walnut pinhole borer		*Diapus pusillimus* Chapuis	(コウチュウ目、ナガキクイムシ科) 豪州区
walnut scale		*Quadraspidiotus juglansregiae* (Comstock)	(カメムシ目、マルカイガラムシ科) 日本、新北区。クルミ害虫
walnut shoot moth		*Acrobasis demotella* Grote	(チョウ目、メイガ科) 新北区
walnut span worm		*Coniodes plumigeraria* (Hulst)	(チョウ目、シャクガ科) 新北区。クルミ害虫
walnut sphinx		*Laothoe juglandis* (Smith)	(チョウ目、スズメガ科) 新北区。walnut sphinx moth ともいう
walnut worm			codling moth を見よ　幼虫の英名
walrus louse		*Antarctophthirus trichechi* (Boheman)	(シラミ目、カイジュウジラミ科) 旧北区
Walsingham's Agonopterix moth		*Agonopterix walsinghamella* (Busck)	(チョウ目、マルハキバガ科) 新北区
Walsingham's darner			giant darner を見よ
Walsingham's grass tubeworm moth		*Acrolophus propinquus* (Walsingham)	(チョウ目、Acrolophidae) 新北区
waltzing midges			moth flies を見よ
wander			harvester を見よ
wanderer		*Acraea aganice* Hewitson	(チョウ目、タテハチョウ科) エチオピア区
wanderer (1)	アサギシロチョウ	*Pareronia valeria* (Cramer)	(チョウ目、シロチョウ科) 東洋区。*P. v. lutescens* (Butler) も同英名
wanderer			monarch butterfly を見よ　豪州で使用
wanderer			common wanderer (1) を見よ
wanderer butterfly			monarch butterfly を見よ
wanderer moth			harvester を見よ
wanderers	アサギシロチョウ属	*Pareronia*	(チョウ目、シロチョウ科) の昆虫の総称

英名	和名	学名	所属、分布、ほか
wandering brocade moth		*Fishia illocata* (Walker)	(チョウ目、ヤガ科) 新北区。wandering brocade ともいう
wandering Diacrisia			wandering tiger moth を見よ
wandering donkey Acraea		*Acraea neobule* Doubleday	(チョウ目、タテハチョウ科) エチオピア区
wandering glider			globe trotter を見よ
wandering grasshopper	ゴウシュウトビバッタ	*Chortoicetes terminifera* (Walker)	(バッタ目、バッタ科) 豪州区
wandering percher		*Diplacodes bipunctata* (Brauer)	(トンボ目、トンボ科) 豪州区
wandering sandman		*Spialia depauperata australis* de Jong	(チョウ目、セセリチョウ科) エチオピア区。deprived grizzled skipper を参照
wandering skipper		*Panoquina errans* (Skinner)	(チョウ目、セセリチョウ科) 新北区、新熱帯区
wandering tiger moth		*Spilosoma vagans* (Boisduval)	(チョウ目、ヒトリガ科) 新北区
wandering wisp			pygmy wisp を見よ
wanstead grey		*Anarsia spartiella* (Schrank)	(チョウ目、キバガ科) 旧北区、大洋区
warble			cattle warble-fly (1) を見よ
warble flies			bot flies (1) を見よ
warble flies (1)	ウシバエ科	Hypodermatidae	(ハエ目) の昆虫の総称
warble flies (2)		*Hypoderma*	(ハエ目、ウシバエ科) の昆虫の総称
warble flies (3)		Hypodermatinae	(ハエ目、ウシバエ科) の昆虫の総称
warble fly			cattle warble fly (2) を見よ
warble-fly			cattle warble-fly (1) を見よ
warbles			blow flies を見よ
warbles			botflies (1) (2) を見よ　幼虫の英名
wardrobe beetle	オビヒメカツオブシムシ	*Attagenus fasciatus* (Thunberg)	(コウチュウ目、カツオブシムシ科) 日本、汎世界。貯蔵食品害虫
Ward's albatross		*Appias wardii* (Moore)	(チョウ目、シロチョウ科) 東洋区
warehouse beetle		*Trogoderma variabile* Ballion	(コウチュウ目、カツオブシムシ科) 新北区
warehouse beetle (1)	ヒメマダラカツオブシムシ	*Trogoderma inclusum* (LeConte)	(コウチュウ目、カツオブシムシ科) 日本、全北区。貯蔵食品害虫で倉庫に多い
warehouse moth			tobacco moth (1) を見よ
warehouse pirate bug		*Xylocoris flavipes* (Reuter)	(カメムシ目、ハナカメムシ科) 新北区
warm-chevroned moth			early button slug moth を見よ
Warner's Metarranthis moth		*Metarranthis warneri* (Harvey)	(チョウ目、シャクガ科) 新北区
Warren's blue		*Orachrysops warreni* Henning et Henning	(チョウ目、シジミチョウ科) エチオピア区
Warren's hooktip	チャイロフトカギバ	*Oreta erminea* (Esper)	(チョウ目、カギバガ科) 旧北区
Warren's shoot		*Blastesthia posticana* (Zetterstedt)	(チョウ目、ハマキガ科) 旧北区
Warren's shoot moth			Warren's shoot を見よ

英名	和名	学名	所属、分布、ほか
Warren's skipper		*Pyrgus warrenensis* (Verity)	(チョウ目、セセリチョウ科) 旧北区
warrior ant			slave-maker ant を見よ
warrior silver-spotted copper		*Argyraspodes argyraspis* (Trimen)	(チョウ目、シジミチョウ科) エチオピア区
wart-biter	カラフトギス	*Decticus verrucivorus* (Linnaeus)	(バッタ目、キリギリス科) 日本、旧北区
wart-biter bush cricket			wart-biter を見よ
wart-biter tettigoniid			wart-biter を見よ
wart hog louse		*Haematomyzus hopkinsi* Clay	(ハジラミ目、ケモノタンカクハジラミ科) 新北区
warted knot-horn		*Acrobasis repandana* Fabricius	(チョウ目、メイガ科) 旧北区
warted knot-horn		*Acrobasis tumidella* Zincken	(チョウ目、メイガ科) 旧北区
warty leaf beetles	コブハムシ亜科	Clamisinae	(コウチュウ目、ハムシ科) の昆虫の総称
washed-out Zale moth		*Zale metatoides* McDunnough	(チョウ目、ヤガ科) 新北区。washed-out Zale ともいう
Washington Udea moth		*Udea washingtonalis* (Grote)	(チョウ目、メイガ科) 新北区
Washington's lady beetle		*Hippodamia washingtoni* Timberlake	(コウチュウ目、テントウムシ科) 新北区
wasp			common wasp を見よ　昆虫ではなく人だが white anglo-saxon protestant の略称の WASP なる語がある
wasp bee			cuckoo bees (1) を見よ
wasp beetle		*Clytus rhamni* Germar	(コウチュウ目、カミキリムシ科) 旧北区
wasp beetle (1)		*Clytus arietis* (Linnaeus)	(コウチュウ目、カミキリムシ科) 旧北区
wasp fan beetle	ヒトスジオオハナノミ	*Metoecus paradoxus* (Linnaeus)	(コウチュウ目、オオナハノミ科) 日本、旧北区
wasp flies		*Volucella*	(ハエ目、ハナアブ科) の昆虫の総称
wasp flies			thick-headed flies を見よ　豪州での英名
wasp fly		*Physocephala burgessi* (Williston)	(ハエ目、メバエ科) 新北区
wasp-like bees			obtuse-tongued bees を見よ
wasplike longhorn		*Tragidion annulatum* LeConte	(コウチュウ目、カミキリムシ科) 新北区
wasp mantidfly			brown mantisfly を見よ
wasp moths	カノコガ科	Ctenuchidae	(チョウ目) の昆虫の総称　ハチに似ることから
wasp moths			clearwing moths を見よ
wasp wood-soldierfly		*Xylomya maculata* (Meigen)	(ハエ目、キアブモドキ科) 旧北区
wasps (1)	スズメバチ上科	Vespoidea	(ハチ目) の昆虫の総称
wasps (2)	ハチ目	Hymenoptera	の昆虫の総称　とくにスズメバチ科、アナバチ科の種。世界に約 130,000 種。bees, wasps, ants, sawflies を含む
wasps			common wasps (2) を見よ

英名	和名	学名	所属、分布、ほか
waste grain beetle	フタオビツヤゴミムシダマシ	*Alphitophagus bifasciatus* (Say)	(コウチュウ目、ゴミムシダマシ科) 日本、全北区。貯穀害虫
waste grass veneer		*Thisanotia contaminella* Hübner	(チョウ目、メイガ科) 旧北区
Watanabe snow flea	ワタナベトビムシモドキ	*Onychiurus sibiricus* (Tullberg)	(トビムシ目、シロトビムシ科) 日本、旧北区
water beetle			ゲンゴロウ、ガムシなど水生甲虫
water beetle (1)		*Dytiscus marginalis* (Linnaeus)	(コウチュウ目、ゲンゴロウ科) 旧北区
water beetles			diving beetles (1) を見よ
water beetles			water scavenger beetles (1) を見よ 豪州での英名
water betony		*Cucullia scrophulariae* Denis et Schiffermüller	(チョウ目、ヤガ科) 旧北区
water betony moth			water betony を見よ
water betony shark			water betony を見よ
water blue			water bronze を見よ
water boatman		*Sigara arguta* (White)	(カメムシ目、ミズムシ科) 豪州区
water boatman (1)		*Notonecta glauca* Linnaeus	(カメムシ目、マツモムシ科) 旧北区
water boatman bugs	ミズムシ科	Corixidae	(カメムシ目) の昆虫の総称
water boatmen			water boatman bugs を見よ
water boatmen			backswimmers を見よ
water bronze		*Cacyreus tespis* (Herbst)	(チョウ目、シジミチョウ科) エチオピア区
water bug			カメムシ目のコオイムシ科、タイコウチ科などの水生昆虫
water bug			German cockroach を見よ
water bugs	水生カメムシ群	Hydrocorisa	(カメムシ目) の昆虫の総称
water bugs			cockroaches (1) を見よ 米国での英名
water bugs			water scorpions (1) を見よ
water butterfly			dragonflies (1) を見よ
water carpet		*Lampropteryx suffumata* (Denis et Schiffermüller)	(チョウ目、シャクガ科) 旧北区
water Charaxes			Manx Charaxes を見よ
water creeper			saucer bug (1) を見よ
water creepers			creeping water bugs を見よ
water cricket		*Velia caprai* Tamanini	(カメムシ目、カタビロアメンボ科) 旧北区
water cricket			small water strider を見よ
water crickets			ripple bugs を見よ
water crickets			water boatman bugs を見よ
water dropwort aphid	セリスナヨセアブラムシ	*Cavariella oeneanthi* (Shinji)	(カメムシ目、アブラムシ科) 日本、旧北区
water dropwort long-tailed aphid			Columbine aphid を見よ
water ermine		*Spilosoma urticae* (Esper)	(チョウ目、ヒトリガ科) 旧北区

英名	和名	学名	所属、分布、ほか
water flea			water springtail を見よ
water Geranium blue		*Cacyreus palemon* (Stoll)	（チョウ目、シジミチョウ科）エチオピア区
water gnats			water measures を見よ
water gnats			winter crane flies を見よ
water hairstreak	タイワンイチモンジシジミ	*Euaspa milionia* (Hewitson)	（チョウ目、シジミチョウ科）東洋区
water hyacinth stalk borer		*Xubida infesellus* (Walker)	（チョウ目、メイガ科）東洋区、豪州区、新北区、エチオピア区
water ladybird		*Anisosticta novemdecimpunctata* (Linnaeus)	（コウチュウ目、テントウムシ科）旧北区
water lily beetle		*Donacia palmata* Olivier	（コウチュウ目、ハムシ科）新北区
water lovers			giant water scavenger beetles を見よ
water-meadow grasshopper		*Chorthippus montanus* (Charpentier)	（バッタ目、バッタ科）旧北区
water measure		*Hydrometra stagnorum* (Linnaeus)	（カメムシ目、イトアメンボ科）旧北区。water-measurer ともいう
water measurer		*Hydrometra risbeci* Hungerford	（カメムシ目、イトアメンボ科）豪州区
water measures	イトアメンボ科	Hydrometridae	（カメムシ目）の昆虫の総称　世界に100種。水生で小型のナナフシに似た昆虫
water midges			カ科など水生の小型ハエ類の俗称
water moths			caddisflies (1) を見よ
water opal		*Chrysoritis palmus* (Stoll)	（チョウ目、シジミチョウ科）エチオピア区
water pennies			water penny beetles を見よ
water penny beetles	ヒラタドロムシ科	Psephenidae	（コウチュウ目）の昆虫の総称　成虫幼虫とも水生
water prince		*Epicordulia princeps* Hagen	（トンボ目、エゾトンボ科）新北区
water ringlet	ウスズミベニヒカゲ	*Erebia pronoe* (Esper)	（チョウ目、タテハチョウ科）旧北区
water scavenger beetles (1)	ガムシ科	Hydrophilidae	（コウチュウ目）の昆虫の総称　世界に2,000種
water scavenger beetles		*Hydrobius*	（コウチュウ目、ガムシ科）の昆虫の総称
water scorpion		*Ranatra fusca* P. Beauvois	（カメムシ目、タイコウチ科）新北区
water scorpion		*Nepa apiculata* Uhler	（カメムシ目、タイコウチ科）新北区
water scorpion		*Nepa cinerea* Linnaeus	（カメムシ目、タイコウチ科）旧北区
water scorpion			タイコウチ (*Nepa* 属)、ミズカマキリ (*Ranatra* 属) などの水生昆虫
water scorpion (1)		*Laccotrephes tristis* (Stål)	（カメムシ目、タイコウチ科）豪州区
water scorpion (2)		*Ranatra quadridentata* Stål	（カメムシ目、タイコウチ科）新北区
water scorpion (3)	ヒメミズカマキリ	*Ranatra unicolor* Scott	（カメムシ目、タイコウチ科）日本、旧北区
water scorpion			water stick insect を見よ
water scorpion toe-biter			water scorpion (1) を見よ
water scorpions (1)	タイコウチ科	Nepidae	（カメムシ目）の昆虫の総称　世界に200種

英名	和名	学名	所属、分布、ほか
water scorpions		*Curicta*	（カメムシ目、タイコウチ科）の昆虫の総称
water skaters	メダカハネカクシ亜科	Steninae	（コウチュウ目、ハネカクシ科）の昆虫の総称
water skaters			water striders (1) を見よ
water skipper			water watchman を見よ
water snow flat		*Tagiades litigiosa* Möschler	（チョウ目、セセリチョウ科）東洋区
water springtail	ミズトビムシ	*Podura aquatica* Linnaeus	（トビムシ目、ミズトビムシ科）日本、汎世界。淡水、汽水の水面に群れて浮かぶ
water springtails			snowfleas を見よ
water springtails			podurid springtails を見よ
water stick insect		*Ranatra linearis* (Linnaeus)	（カメムシ目、タイコウチ科）旧北区
water strider		*Limnoporus notabilis* (Drake et Hottes)	（カメムシ目、アメンボ科）新北区
water strider			common pond skater を見よ
water strider			common water strider を見よ
water strider bugs			water striders (1) を見よ
water striders		Gerroidea	（カメムシ目）の昆虫の総称
water striders (1)	アメンボ科	Gerridae	（カメムシ目）の昆虫の総称 世界に500種
water tiger		*Paracles laboulbeni* (Bar)	（チョウ目、ヒトリガ科）旧北区
water tiger			water beetle (1) を見よ
water tigers			diving beetles (1) を見よ 米国での英名
water treader			pondweed bug を見よ
water treader bugs			water treaders を見よ
water treaders	ミズカメムシ科	Mesoveliidae	（カメムシ目）の昆虫の総称 半水生で小型。世界に120種
water treaders			water measures を見よ
water veneer		*Acentria ephemerella* (Denis et Schiffermüller)	（チョウ目、メイガ科）全北区。water veneer moth ともいう
water watchman		*Parnara monasi* (Trimen)	（チョウ目、セセリチョウ科）エチオピア区
water willow stem borer moth		*Papaipema sulphurata* Bird	（チョウ目、ヤガ科）新北区
Waterberg copper		*Erikssonia edgei* Gardiner et Terblanche	（チョウ目、シジミチョウ科）エチオピア区
waterboatman			common corixa を見よ
waterboatmen		*Corixa*	（カメムシ目、ミズムシ科）の昆虫の総称
waterbug mill-beetle			common cockroach を見よ
watercress beetle			mustard beetle (2) を見よ
watercress caddis fly		*Limnephilus flavicornis* (Fabricius)	（トビケラ目、エグリトビケラ科）旧北区
watercress caddis worm			watercress caddis fly を見よ 幼虫の英名
watercress fly		*Hydrellia nasturtii* Collin	（ハエ目、ミギワバエ科）旧北区
watercress leaf beetle		*Phaedon viridis* (Melsheimer)	（コウチュウ目、ハムシ科）新北区

英名	和名	学名	所属、分布、ほか
watercress leaf-beetle			mustard beetle (1)(2) を見よ
watercress sharpshooter		*Draeculacephala clypeata* Say	（カメムシ目、オオヨコバイ科）新北区
watercress sharpshooter (1)		*Draeculacephala mollipes* (Say)	（カメムシ目、オオヨコバイ科）新北区、新熱帯区
waterfall redspot		*Austropetalia patricia* Tillyard	（トンボ目、Austropetaliidae）豪州区
Waterhouse's hairstreak			Lithochroa blue を見よ
waterhyacinth moth		*Sameodes albiguttalis* (Warren)	（チョウ目、メイガ科）豪州区
waterhyacinth weevil		*Neochetina eichhorniae* Warner	（コウチュウ目、ゾウムシ科）豪州区
waterhyacinth weevil		*Neochetina bruchi* Hustache	（コウチュウ目、ゾウムシ科）東洋区、豪州区、新熱帯区
waterlily aphid	ハスクビレアブラムシ	*Rhopalosiphum nymphaeae* (Linnaeus)	（カメムシ目、アブラムシ科）日本、汎世界。ナシ、モモ、ウメ、イグサ、サクラ害虫
water-lily beetle			waterlily leaf beetle を見よ
waterlily borer moth		*Elophila gyralis* (Hulst)	（チョウ目、メイガ科）新北区
waterlily delphacid			waterlily planthopper を見よ
waterlily leaf beetle			pund leaf beetle を見よ
waterlily leaf beetles			aquatic leaf beetles を見よ
waterlily leaf cutter		*Synclita obliteralis* (Walker)	（チョウ目、メイガ科）新北区、大洋区、エチオピア区。waterlily leafcutter moth ともいう
water-lily midge		*Endochironomus nymphaea* (Willem)	（ハエ目、ユスリカ科）旧北区
waterlily nymphula	マダラミズメイガ	*Elophila interuptalis* (Pryer)	（チョウ目、メイガ科）日本、旧北区
waterlily owlet moth		*Homophoberia cristata* Morrison	（チョウ目、ヤガ科）新北区。waterlily moth ともいう
waterlily planthopper		*Megamelus angulatus* Osborn	（カメムシ目、ビワハゴロモ科）大洋区
watermilfoil leafcutter moth		*Parapoynx allionealis* Walker	（チョウ目、メイガ科）新北区
waterpenny		*Psephenus lecontei* (LeConte)	（コウチュウ目、ヒラタドロムシ科）新北区
waterpenny		*Eubrianax edwardsi* LeConte	（コウチュウ目、ヒラタドロムシ科）新北区
waterrice planthopper	ニホンウンカ	*Zuleica nipponica* Matsumura et Ishihara	（カメムシ目、ウンカ科）日本
waterscorpion bugs			water scorpions (1) を見よ
water-singer		*Micronecta poweri* (Douglas et Scott)	（カメムシ目、ミズムシ科）旧北区
Watkins' brown Morpho		*Antirrhea watkinsi* (Rosenberg et Talbot)	（チョウ目、タテハチョウ科）新熱帯区
Watson's Arugisa moth		*Arugisa latiorella* (Walker)	（チョウ目、ヤガ科）新北区

英名	和名	学名	所属、分布、ほか
Watson's bushbrown		*Mycalesis adamsoni* Watson	（チョウ目、タテハチョウ科）東洋区
Watson's demon		*Stimula swinhoei* (Elwes et Edwards)	（チョウ目、セセリチョウ科）東洋区
Watson's grass-veneer moth		*Crambus watsonellus* Klots	（チョウ目、メイガ科）新北区。Watson's grass-veneer ともいう
Watson's hairstreak	レタミドリシジミ	*Chrysozephyrus letha* (Watson)	（チョウ目、シジミチョウ科）東洋区
Watson's mottle		*Logania watsoniana* de Nicéville	（チョウ目、シジミチョウ科）東洋区
Watson's Tallula moth		*Tallula watsoni* Barnes et McDunnough	（チョウ目、メイガ科）新北区
wattle apple-gall wasp		*Trichilogaster acaciaelongifoliae* (Froggatt)	（ハチ目、コガネコバチ科）豪州区
wattle bagworm		*Kotochalia junodi* (Heylaerts)	（チョウ目、ミノガ科）新北区、エチオピア区
wattle blue		*Theclinesthes miskini* (Lucas)	（チョウ目、シジミチョウ科）豪州区
wattle cicada		*Cicadetta oldfieldi* (Distant)	（カメムシ目、セミ科）豪州区
wattle goat moth		*Xyleutes eucalypti* (Herrich-Schäffer)	（チョウ目、ボクトウガ科）豪州区
wattle leafminer		*Acrocercops plebeia* (Turner)	（チョウ目、ホソガ科）豪州区
wattle mealybug		*Melanococcus albizziae* (Maskell)	（カメムシ目、コナカイガラムシ科）豪州区
wattle moths		*Dasypolia*	（チョウ目、ヤガ科）の昆虫の総称
wattle root longicorn		*Eurynassa australis* (Boisduval)	（コウチュウ目、カミキリムシ科）豪州区
wattle snout moth		*Digglesia australasiae* (Fabricius)	（チョウ目、シャクガ科）豪州区
wattle tick scale		*Cryptes baccatus* (Maskell)	（カメムシ目、カタカイガラムシ科）豪州区
wave moths			geometer moths を見よ
waved black		*Parascotia fuliginaria* (Linnaeus)	（チョウ目、ヤガ科）旧北区。waved black moth ともいう
waved carpet	キスジハイイロナミシャク	*Hydrelia sylvata* (Denis et Schiffermüller)	（チョウ目、シャクガ科）日本、旧北区
waved carpet			clouded magpie を見よ　原名亜種の英名
waved carpet moth			waved carpet を見よ
waved sphinx		*Ceratomia undulosa* (Walker)	（チョウ目、スズメガ科）新北区。waved sphinx moth ともいう
waved tabby		*Idia aemula* Hübner	（チョウ目、ヤガ科）全北区
waved umber		*Menophra abruptaria* (Thunberg)	（チョウ目　シャクガ科）旧北区
waved umber beauty			waved umber を見よ
waved umber moth			waved umber を見よ
waves	ヒメシャク亜科	Sterrhinae	（チョウ目、シャクガ科）の昆虫の総称

英名	和名	学名	所属、分布、ほか
waves			geometer moths を見よ
wavy-barred sable		*Pyrausta nigrata* Scopoli	(チョウ目、メイガ科) 旧北区
wavy chestnut Y moth		*Autographa mappa* (Grote et Robinson)	(チョウ目、ヤガ科) 新北区。wavy chestnut Y ともいう
wavy-edged leafwing		*Memphis neidhoeferi* (Roger, Escalante et Coronado)	(チョウ目、タテハチョウ科) 新熱帯区
wavy huge-comma	オスグロトモエ	*Spirama retoria* (Clerck)	(チョウ目、ヤガ科) 日本、旧北区
wavy-lined emerald moth		*Synchlora aerata* (Fabricius)	(チョウ目、シャクガ科) 新北区
wavy-lined Heterocampa moth		*Heterocampa biundata* Walker	(チョウ目、シャチホコガ科) 新北区
wavy lined mallow moth		*Bagisara repanda* (Fabricius)	(チョウ目、ヤガ科) 新北区、新熱帯区
wavy-lined Zanclognatha moth		*Zanclognatha jacchusalis* (Walker)	(チョウ目、ヤガ科) 新北区。wavy-lined Zanclognatha ともいう
wavy maplet	チビイシガキチョウ	*Chersonesia rahria* (Moore)	(チョウ目、タテハチョウ科) 東洋区
wavy-marked looper moth	ナミガタエダシャク	*Heterarmia charon* (Butler)	(チョウ目、シャクガ科) 日本、旧北区
wavy-striped flea beetle			striped flea beetle (1) を見よ
wavy-striped thick-legged moth	オキナワアシブトクチバ	*Parallelia arcuata* (Moore)	(チョウ目、ヤガ科) 日本、東洋区
wavy-striped white geometrid	ナミガタシロナミシャク	*Callygris compositata* (Guenée)	(チョウ目、シャクガ科) 日本、旧北区
wavy polyptychus		*Rufoclanis numosae* (Wallengren)	(チョウ目、スズメガ科) エチオピア区
wax dart		*Cupitha purreea* (Moore)	(チョウ目、セセリチョウ科) 東洋区。waxy dart ともいう
wax insect			ろうを分泌する昆虫、とくに中国産のイボタロウムシ
wax moth			greater wax moth を見よ
wax moths	ツヅリガ亜科	Galleriinae	(チョウ目、メイガ科) の昆虫の総称　米国での英名
wax moths			grass moths (2) を見よ
wax scale		*Ceroplastes rusci* Linnaeus	(カメムシ目、カタカイガラムシ科) 旧北区。カンキツ、イチジク害虫
wax scale			Indian wax scale を見よ
wax scale			red wax scale を見よ
wax scales			soft scales を見よ
waxmyrtle wave moth		*Cyclophora myrtaria* (Guenée)	(チョウ目、シャクガ科) 新北区
waxtails		*Ceriagrion*	(トンボ目、イトトンボ科) の昆虫の総称
waxworm			greater wax moth を見よ　幼虫の英名
waxworms			wax moths を見よ
wayside anopheline		*Anopheles pseudopunctipennis* Theobald	(ハエ目、カ科) 新北区、新熱帯区

英名	和名	学名	所属、分布、ほか
weak blues			blues (1) を見よ
weak-banded crescent		*Anthanassa drymaea* (Godman et Salvin)	(チョウ目、タテハチョウ科) 新熱帯区
weak-marked gemmed-satyr		*Cyllopsis parvimaculata* Miller	(チョウ目、タテハチョウ科) 新熱帯区
weaver ant		*Camponotus senex textor* Forel	(ハチ目、アリ科) 新熱帯区
weaver ant (1)	クロトゲアリ	*Polyrhachis (Myrmhopla) dives* F. Smith	(ハチ目、アリ科) 日本、東洋区
weaver ant			red ant (1) を見よ
weaver ants		*Oecophylla*	(ハチ目、アリ科) の昆虫の総称　red ant (1) も参照
weaver beetle	エゾカミキリ	*Lamia textor* (Linnaeus)	(コウチュウ目、カミキリムシ科) 日本、旧北区
Weaver's fritillary			violet fritillary を見よ
Weaver's pigmy		*Fomoria weaveri* (Stainton)	(チョウ目、モグリチビガ科) 旧北区
Weaver's wave		*Sterrha eburnata* Wocke	(チョウ目、シャクガ科) 旧北区
Weaver's wave		*Idaea contiguaria* (Hübner)	(チョウ目、シャクガ科) 旧北区
web			カレハガ科幼虫の巣
webbing clothes moth			clothes moth (1) を見よ
webbing coneworm		*Dioryctria disclusa* Heinrich	(チョウ目、メイガ科) 新北区。幼虫は松かさに食入。成虫は　webbing coneworm moth
Webb's wainscot	キスジウスキヨトウ	*Archanara sparganii* (Esper)	(チョウ目、ヤガ科) 日本、旧北区
Webb's wainscot			white-mottled wainscot を見よ
Weber's piercer			cherry-bark moth を見よ
webspinner		*Haploembia solieri* (Rambur)	(シロアリモドキ目、シロアリモドキ科) 全北区
webspinners (1)	シロアリモドキ目	Embioptera	の昆虫の総称　約 300 種が主として熱帯に分布。絹糸の巣にすむ
webspinners		*Haploembia*	(シロアリモドキ目、シロアリモドキ科) の昆虫の総称
web-spinning caddisflies			net-spinning caddisflies を見よ
web-spinning larch sawfly		*Cephalcia lariciphila* Wachtl	(ハチ目、ヒラタハバチ科) 旧北区
web-spinning larch sawfly			larch web-spinning sawfly を見よ
web-spinning sawflies	ヒラタハバチ科	Pamphiliidae	(ハチ目) の昆虫の総称
web-spinning tree crickets			tree crickets (1) を見よ
webworm		*Cyclosia panthona* Cramer	(チョウ目、マダラガ科) 東洋区
webworm			pasture webworm (1) (2)を見よ
webworms		*Hyphantria*	(チョウ目、ヒトリガ科) の昆虫の総称　米国での英名

英名	和名	学名	所属、分布、ほか
webworms			grass moths (1) を見よ
webworms			leaftiers を見よ
wedge grass-skipper			wedge skipper を見よ
wedge-marked groundling		*Gelechia cuneatella* Douglas	(チョウ目、キバガ科) 旧北区
wedge-shaped bark weevils	ツツキクイゾウムシ亜科	Magdalinae	(コウチュウ目、ゾウムシ科) の昆虫の総称
wedge-shaped beetles	オオハナノミ科	Rhipiphoridae	(コウチュウ目) の昆虫の総称
wedge-shaped leaf beetle			spiny leaf beetle を見よ 米国での英名
wedge-shaped planthopper	クサビウンカ	*Sarima amagisana* Melichar	(カメムシ目、マルウンカ科) 日本、東洋区
wedge skipper		*Anisynta sphenosema* (Meyrick et Lower)	(チョウ目、セセリチョウ科) 豪州区
wedge-spotted cattleheart	オオアオマエモンジャコウアゲハ	*Parides panares panares* (Gray)	(チョウ目、アゲハチョウ科) 新北区、新熱帯区。*P. p. lycimenes* (Boisduval) も同英名
wedgling moth			garden miller を見よ
weed-defoliator moth		*Pareuchaetes pseudoinsulata* Rego Barros	(チョウ目、ヒトリガ科) 東洋区、新熱帯区
weed web moth		*Achyra affinitalis* (Lederer)	(チョウ目、メイガ科) 豪州区
weed weevil		*Lixus mastersi* Pascoe	(コウチュウ目、ゾウムシ科) 豪州区
weedfield sable		*Pyrausta subsequalis* (Guenée)	(チョウ目、メイガ科) 新北区
weevil		*Cryptorhynchus corticolis* Boheman	(コウチュウ目、ゾウムシ科) 新熱帯区。カンキツ害虫
weevil		*Diaprepes esuriens* Gyllenhal	(コウチュウ目、ゾウムシ科) 新熱帯区。カンキツ害虫
weevil		*Diaprepes excavatus* Rosenschold	(コウチュウ目、ゾウムシ科) 新熱帯区。カンキツ害虫
weevil		*Diaprepes sprengleri* Linnaeus	(コウチュウ目、ゾウムシ科) 新熱帯区。カンキツ害虫
weevil		*Lachnopus coffeae* Marshall	(コウチュウ目、ゾウムシ科) 新熱帯区。カンキツ害虫
weevil		*Lachnopus curvipes* Fabricius	(コウチュウ目、ゾウムシ科) 新熱帯区。カンキツ害虫
weevil		*Lachnopus hispidis* Gyllenhal	(コウチュウ目、ゾウムシ科) 新熱帯区。カンキツ害虫
weevil		*Lachnopus sparsimguttatus* Perroud	(コウチュウ目、ゾウムシ科) 新熱帯区。カンキツ害虫
weevil		*Pantomorus glaucus* Perty	(コウチュウ目、ゾウムシ科) 新熱帯区。カンキツ害虫
weevil		*Pantomorus parsevali* Costa Lima	(コウチュウ目、ゾウムシ科) 新熱帯区。カンキツ害虫
weevil		*Prepodes quadrivittatus* Olivier	(コウチュウ目、ゾウムシ科) 新熱帯区。カンキツ害虫
weevil		*Prepodes roseipes* Chevrolat	(コウチュウ目、ゾウムシ科) 新熱帯区。カンキツ害虫

英名	和名	学名	所属、分布、ほか
weevil		*Prepodes amabilis* Waterhouse	（コウチュウ目、ゾウムシ科）新熱帯区。カンキツ害虫
weevil		*Prepodes vittatus* Linnaeus	（コウチュウ目、ゾウムシ科）新熱帯区。カンキツ害虫
weevil		*Protostrophus avidus* Marshall	（コウチュウ目、ゾウムシ科）豪州区。カンキツ害虫
weevil			sugarcane root borer を見よ
weevil borer			silky cane weevil を見よ
weevil wasps		*Cerceris*	（ハチ目、アナバチ科）の昆虫の総称
weevillike planthoppers			issid planthoppers を見よ
weevils (1)	ゾウムシ科	Curculionidae	（コウチュウ目）の昆虫の総称　世界に50,000種
weevils (2)	ゾウムシ上科	Curculionoidea	（コウチュウ目、ゾウムシ科）の昆虫の総称
Weidemeyer's admiral	ワイデマイヤーイチモンジ	*Limenitis weidemeyerii* (Edwards)	（チョウ目、タテハチョウ科）新北区。*L. w. siennafascia* Austin et Mullins 新熱帯区も同英名
Weir's piercer		*Strophedra weirana* (Douglas)	（チョウ目、ハマキガ科）旧北区
Weissman's conehead		*Bucrates* sp.	（バッタ目、キリギリス科）新北区
wekiu lygaeid bug		*Nysius wekiuicola* Ashlock et Gagné	（カメムシ目、ナガカメムシ科）新北区、大洋区。wekiu bug ともいう
Welling's calephelis		*Calephelis wellingi* McAlpine	（チョウ目、シジミタテハ科）新熱帯区
Welling's Gaudy checkerspot		*Chlosyne gaudialis wellingi* Miller et Rotger	（チョウ目、タテハチョウ科）新熱帯区
Welling's leafwing		*Memphis wellingi* Miller et Miller	（チョウ目、タテハチョウ科）新熱帯区
Wellington tree weta		*Hemideina crassidens* (Blanchard)	（バッタ目、Stenopelmatidae）豪州区
Welsh chafer		*Hoplia philanthus* (Fuessley)	（コウチュウ目、コガネムシ科）旧北区
Welsh clearwing		*Synanthedon scoliaeformis* (Borkhausen)	（チョウ目、スカシバガ科）旧北区
Welsh oak longhorn beetle		*Pyrrhidium sanguineum* (Linnaeus)	（コウチュウ目、カミキリムシ科）旧北区
Welsh wave	ミヤマナミシャク	*Venusia cambrica* Curtis	（チョウ目、シャクガ科）日本、旧北区。Welsh wave moth, Welsh waved carpet ともいう
Wenzel's pitch-blister moth		*Retinia wenzeli* (Kearfott)	（チョウ目、ハマキガ科）新北区
Werner's dropwing		*Trithemis werneri* Ris	（トンボ目、トンボ科）エチオピア区
West African fig tree blue		*Myrina subornata* Lathy	（チョウ目、シジミチョウ科）エチオピア区
West African pink borer		*Sesamia nonagrioides botanephaga* Tams et Bowden	（チョウ目、ヤガ科）エチオピア区
West African stem borer		*Bixadus sierricola* White	（コウチュウ目、カミキリムシ科）エチオピア区。コーヒー害虫
west coast bush weta		*Hemideina broughi* (Butler)	（バッタ目、Stenopelmatidae）豪州区

英名	和名	学名	所属、分布、ほか
west coast lady	ニシヒメアカタテハ	*Vanessa annabella* (Field)	(チョウ目、タテハチョウ科) 新北区
West India red scale			rufous scale を見よ
West Indian buckeye	カリブタテハモドキ	*Junonia evarete* Cramer	(チョウ目、タテハチョウ科) 新北区、新熱帯区
West Indian cane weevil			silky cane weevil を見よ
West Indian cane weevil			West Indian sugarcane weevil を見よ
West Indian canefly		*Saccharosydne saccharivora* (Westwood)	(カメムシ目、ウンカ科) 新熱帯区
West Indian drywood termite		*Cryptotermes brevis* (Walker)	(シロアリ目、レイビシロアリ科) 全北区、大洋区、豪州区、新熱帯区、エチオピア区
West Indian flatid		*Anormenis antillarum* (Kirkaldy)	(カメムシ目、アオバハゴロモ科) 新北区
West Indian flatid		*Melormensis basalis* (Walker)	(カメムシ目、アオバハゴロモ科) 新北区
West Indian fruit fly			Caribbean fruit fly を見よ
West Indian fruit fly			mango fruit fly (1) を見よ
West Indian leaf cockroach			discoid cockroach を見よ
West Indian mole cricket			changa を見よ
West Indian peach scale			white peach scale を見よ
West Indian red scale		*Sclenaspidus articulatus* Morgan	(カメムシ目、コナジラミ科) 新熱帯区
West Indian scale			white peach scale を見よ
West Indian sugarcane leafhopper			West Indian canefly を見よ
West Indian sugarcane weevil		*Metamasius hemipterus* (Linnaeus)	(コウチュウ目、ゾウムシ科) 新北区、新熱帯区。West Indian cane weevil ともいう
West Indian sweet potato weevil	イモゾウムシ	*Euscepes postfasciatus* (Fairmaire)	(コウチュウ目、ゾウムシ科) 日本、新北区、汎熱帯。サツマイモ害虫
West Indies fruit fly	ニシインドミバエ	*Anatrepha obliqua* (Macquart)	(ハエ目、ミバエ科) 新熱帯区。West Indian fruit fly ともいう
West-Mexican Anastrus		*Anastrus luctuosus* (Godman et Salvin)	(チョウ目、セセリチョウ科) 新熱帯区
West-Mexican banner		*Catonephele cortesi* de la Maza	(チョウ目、タテハチョウ科) 新熱帯区
West-Mexican clearwing		*Pteronymia rufocincta* (Salvin)	(チョウ目、タテハチョウ科) 新熱帯区
West-Mexican Doberes		*Doberes sobrinus* (Godman et Salvin)	(チョウ目、セセリチョウ科) 新熱帯区
West-Mexican eighty-eight			Astala eighty-eight を見よ
West-Mexican Ipidecla		*Ipidecla miadora* Dyar	(チョウ目、シジミチョウ科) 新熱帯区
West-Mexican Phanus		*Phanus rilma* Evans	(チョウ目、セセリチョウ科) 新熱帯区

英名	和名	学名	所属、分布、ほか
West-Mexican satyr		*Pedaliodes* sp.	（チョウ目、タテハチョウ科）新熱帯区
West-Mexican skipperling		*Dalla faula* (Godman)	（チョウ目、セセリチョウ科）新熱帯区
West-Mexican sootywing		*Staphylus tierra* Evans	（チョウ目、セセリチョウ科）新熱帯区
West-Mexican spurwing		*Antigonus funebris* (R. Felder)	（チョウ目、セセリチョウ科）新熱帯区
West-Mexican sulphur		*Prestonia clarki* Schaus	（チョウ目、シロチョウ科）新熱帯区
West-Mexican swallowtail		*Battus eracon* (Godman et Salvin)	（チョウ目、アゲハチョウ科）新熱帯区
West-Mexican Theope		*Theope villai* Beutelspacher	（チョウ目、シジミタテハ科）新熱帯区
West-Mexican thorn-scrub leafwing		*Fountainea halice tehuana* (Hall)	（チョウ目、タテハチョウ科）新熱帯区。ruddy leafwing を参照
West Texas giant skipper		*Agathymus gilberti* Freeman	（チョウ目、セセリチョウ科）新北区
West-Texas streaky-skipper		*Celotes limpia* Burns	（チョウ目、セセリチョウ科）新北区、新熱帯区
West Virginia white	バージニアシロチョウ	*Pieris virginiensis* Edwards	（チョウ目、シロチョウ科）新北区
Westermann's humming bird		*Atemnora westermanni* (Boisduval)	（チョウ目、スズメガ科）エチオピア区
western			Mormon cricket を見よ
western Amazon ant		*Polyergus breviceps* Emery	（ハチ目、アリ科）新北区
western Apamea moth		*Apamea occidens* Grote	（チョウ目、ヤガ科）新北区。western Apamea ともいう
western apple curculio		*Tachypterellus quadrigibbus magnus* List	（コウチュウ目、ゾウムシ科）新北区。リンゴ他害虫
western armyworm		*Euxoa agrestis* Grote	（チョウ目、ヤガ科）新北区
western Australian brown blowfly		*Calliphora varifrons* Malloch	（ハエ目、クロバエ科）豪州区
western avocado leafroller		*Amorbia cuneana* Walsingham	（チョウ目、ハマキガ科）新北区。成虫は western avocado leafroller moth
western balsam bark beetle		*Dryocoetes confusus* Swaine	（コウチュウ目、キクイムシ科）新北区
western banded elfin			western pine elfin を見よ
western bean cutworm		*Loxagrotis albicosta* (Smith)	（チョウ目、ヤガ科）新北区。western bean cutworm moth ともいう
western bee-fly		*Bombylius canescens* Mikan	（ハエ目、ツリアブ科）旧北区
western bentwinged cicada		*Froggatoides pallida* (Ashton)	（カメムシ目、セミ科）豪州区
western bigeyed bug		*Geocoris pallens* Stål	（カメムシ目、ナガカメムシ科）新北区
western bitter-bush blue		*Theclinesthes hesperia* Sibatani et Grund	（チョウ目、シジミチョウ科）豪州区
western black flea beetle		*Phyllotreta pusilla* Horn	（コウチュウ目、ハムシ科）新北区。トウモロコシ、野菜害虫

英名	和名	学名	所属、分布、ほか
western black swallowtail			Baird's swallowtail を見よ
western blackheaded budworm		Acleris gloverana (Walsingham)	（チョウ目、ハマキガ科）新北区。western black-headed budworm moth ともいう
western black-spot ciliate blue		Anthene starki Larsen	（チョウ目、シジミチョウ科）エチオピア区
western blind cave beetle		Glacicavicola bathysciodes Westcott	（コウチュウ目、タマキノコムシ科）新北区
western blood-red lady beetle		Cycloneda munda (Say)	（コウチュウ目、テントウムシ科）新北区
western blood-red lady beetle			Casey's ladybird beetle を見よ
western bloodsucking conenose		Triatoma protracta (Uhler)	（カメムシ目、サシガメ科）新北区
western blotched leopard		Lachnoptera anticlia (Hübner)	（チョウ目、タテハチョウ科）エチオピア区
western blue Charaxes	ルリモンフタオチョウ	Charaxes smaragdalis Butler	（チョウ目、タテハチョウ科）エチオピア区
western blue sapphire		Heliophorus bakeri Evans	（チョウ目、シジミチョウ科）東洋区
western boxelder bug		Boisea rubrolineata Barber	（カメムシ目、ヒメヘリカメムシ科）新北区
western boxelder twig borer moth		Proteoteras arizonae Kearfott	（チョウ目、ハマキガ科）新北区
western branded skipper		Hesperia colorado (Scudder)	（チョウ目、セセリチョウ科）東洋区
western brassy ringlet		Erebia arvernensis Oberthür	（チョウ目、タテハチョウ科）旧北区
western brown		Heteronympha merope duboulayi (Butler)	（チョウ目、タテハチョウ科）豪州区。common brown を参照
western brown stink bug		Euschistus impunctiventris (Stål)	（カメムシ目、カメムシ科）新北区
western bumble bee		Bombus (Bombus) terricola occidentalis Greene	（ハチ目、コシブトハナバチ科）新北区
western bush katydids		Insara	（バッタ目、キリギリス科）の昆虫の総称
western carpet moth			green-striped forest looper を見よ　western carpet ともいう
western Catoptria			Oregon Catoptria moth を見よ
western cedar bark-beetle			cedar bark beetle を見よ
western cedar borer		Trachykele blondeli Marseul	（コウチュウ目、タマムシ科）新北区
western cherry fruit fly	セイブオウトウミバエ	Rhagoletis indifferens Curran	（ハエ目、ミバエ科）新北区
western chicken flea		Ceratophyllus niger Fox	（ノミ目、ナガノミ科）新北区
western chinch bug		Blissus occiduus Barber	（カメムシ目、ナガカメムシ科）新北区
western cicada killer		Sphecius convallis Patton	（ハチ目、アナバチ科）新北区

英名	和名	学名	所属、分布、ほか
western cigar casebearer			cherry casebearer を見よ
western citrus swallowtail			King page を見よ
western clothes moth		*Tinea occidentella* Chambers	(チョウ目、ヒロズコガ科) 新北区
western cloudywing		*Thorybes diversus* Bell	(チョウ目、セセリチョウ科) 新北区
western clubtail		*Gomphus pulchellus* Selys	(トンボ目、サナエトンボ科) 旧北区
western conifer seed bug	マツヘリカメムシ	*Leptoglossus occidentalis* Heidemann	(カメムシ目、ヘリカメムシ科) 全北区、大洋区
western corn borer		*Diatraea grandiosella* (Dyar)	(チョウ目、メイガ科) 新北区
western corn rootworm		*Diabrotica virgifera* LeConte	(コウチュウ目、ハムシ科) 全北区。トウモロコシ害虫。亜種 *D. v. zeae* Krysan et Smith は Mexican corn rootworm という
western corn rootworm		*Diabrotica viridula* (Fabricius)	(コウチュウ目、ハムシ科) 新熱帯区
western corsair		*Rasahus thoracicus* Stål	(カメムシ目、サシガメ科) 新北区
western courtier		*Sephisa dichroa* (Kollar)	(チョウ目、タテハチョウ科) 旧北区、東洋区
Western Covadonga skipper		*Pheraeus covadonga loxicha* Steinhauser	(チョウ目、セセリチョウ科) 新熱帯区
western creeping water bug		*Ambrysus occidentalis* La Rivers	(カメムシ目、コバンムシ科) 新北区
western cricket			Mormon cricket を見よ
western cypress katydid		*Inscudderia taxodii* Caudell	(バッタ目、キリギリス科) 新北区
western damsel bug		*Nabis alternatus* Parshley	(カメムシ目、マキバサシガメ科) 新北区
western dappled white		*Euchloe crameri* Butler	(チョウ目、シロチョウ科) 旧北区
western darner		*Austroaeschna anacantha* Tillyard	(トンボ目、ヤンマ科) 豪州区
western dobsonflies			dobsonflies (3) を見よ
western dotted border			common dotted border を見よ
western drywood termite	アメリカカンザイシロアリ	*Incisitermes minor* (Hagen)	(シロアリ目、レイビシロアリ科) 新北区、新熱帯区
western elfin		*Incisalia iroides* (Boisduval)	(チョウ目、シジミチョウ科) 新北区
western emperor swallowtail	マルバネサカハチアゲハ	*Papilio menestheus* Drury	(チョウ目、アゲハチョウ科) エチオピア区
western encephalitis mosquito		*Culex tarsalis* Coquillett	(ハエ目、カ科) 新北区
western eyed click beetle			eyed elater を見よ
western false hemlock looper		*Nepytia treemani* Munroe	(チョウ目、シャクガ科) 新北区
western field wireworm		*Limonius infuscatus* Motschulsky	(コウチュウ目、コメツキムシ科) 新北区

英名	和名	学名	所属、分布、ほか
western firefly		*Ellychnia californica* Motschulsky	（コウチュウ目、ホタル科）新北区
western five ring		*Ypthima indecora* Moore	（チョウ目、タテハチョウ科）旧北区
western flat		*Exometoeca nycteris* Meyrick	（チョウ目、セセリチョウ科）豪州区
western flower thrips (1)		*Frankliniella moultoni* Hood	（アザミウマ目、アザミウマ科）新北区。カンキツ害虫
western flower thrips (2)	ミカンキイロアザミウマ	*Frankliniella occidentalis* (Pergande)	（アザミウマ目、アザミウマ科）日本、全北区。カンキツ、野菜、花の害虫
western flying adder			yellow-backed biddie を見よ　北米での英名
western forktail		*Ischnura perparva* Selys	（トンボ目、イトトンボ科）新北区
western fruit beetle		*Syneta albida* LeConte	（コウチュウ目、ハムシ科）新北区
western fruit lecanium			nut scale を見よ　カナダでの英名
western Furcula moth		*Furcula occidentalis* (Lintner)	（チョウ目、シャチホコガ科）新北区
western garden webworm		*Achyra occidentalis* (Packard)	（チョウ目、メイガ科）新北区
western giant cupid		*Lepidochrysops parsimon* (Fabricius)	（チョウ目、シジミチョウ科）エチオピア区
western giant ichneumon		*Megarhyssa nortoni* (Cresson)	（ハチ目、ヒメバチ科）新北区
western goldenhaired blowfly		*Calliphora albifrontalis* Malloch	（ハエ目、クロバエ科）豪州区
western grain sawfly			western wheat-stem sawfly を見よ
western grape leaf skeletonizer		*Harrisina brilliana* Barnes et McDunnough	（チョウ目、マダラガ科）新北区。ブドウ害虫
western grape leaf skeletonizer		*Harrisina metallica* Stretch	（チョウ目、マダラガ科）新北区。ブドウ害虫。成虫は western grape leaf skeletonizer moth
western grape leafhopper			grape leafhopper (1) を見よ
western grape rootworm			brown and black beetle を見よ
western grass dart		*Taractrocera papyria agraulia* (Hewitson)	（チョウ目、セセリチョウ科）豪州区
western grayback			black petaltail を見よ
western harvester ant		*Pogonomyrmex occidentalis* (Cresson)	（ハチ目、アリ科）新北区
western hemlock looper		*Lambdina fiscellaria lugubrosa* (Hulst)	（チョウ目、シャクガ科）新北区
western hemlock looper			hemlock looper (1) を見よ
western Hercules beetle			southwestern Hercules beetle を見よ
western honey bee			honey bee を見よ
western horn lerp		*Creis periculosa* (Olliff)	（カメムシ目、キジラミ科）豪州区
western horse lubber grasshopper			horse lubber を見よ

英名	和名	学名	所属、分布、ほか
western horsefly		*Tabanus punctifer* Osten Sacken	（ハエ目、アブ科）新北区
western Indian drywood termite			West Indian drywood termite を見よ
western inland hunter		*Austrogomphus collaris* Hagen	（トンボ目、サナエトンボ科）豪州区
western jewel		*Hypochrysops halyaetus* Hewitson	（チョウ目、シジミチョウ科）豪州区
western lawn moth		*Tehama bonifatella* (Hulst)	（チョウ目、メイガ科）新北区
western leaf butterfly	ツマアカヒメコノハチョウ	*Junonia cymadoce* (Cramer)	（チョウ目、タテハチョウ科）エチオピア区。blue leaf pansy, western leaf ともいう
western leaf-footed bug		*Leptoglossus clypealis* Heidemann	（カメムシ目、ヘリカメムシ科）新北区
western leaf footed bug			chincha を見よ
western lightningbug			western firefly を見よ
western lily aphid		*Macrosiphum scoliopi* (Monell)	（カメムシ目、アブラムシ科）新北区
western long dash			Sonora skipper を見よ
western longwing		*Capnobotes occidentalis* (Thomas)	（バッタ目、キリギリス科）新北区
western malaria mosquito		*Anopheles freeborni* Aitken	（ハエ目、カ科）新北区、新熱帯区
western marbled white		*Melanargia occitanica* (Esper)	（チョウ目、タテハチョウ科）旧北区
western meadow fritillary	マルバネヒメヒョウモン	*Clossiana epithore* (Edwards)	（チョウ目、タテハチョウ科）新北区
western milkweed longhorn		*Tetraopes basalis* LeConte	（コウチュウ目、カミキリムシ科）新北区
western mole cricket		*Gryllotalpa cultiger* Uhler	（バッタ目、ケラ科）新北区
western mountain flower longhorn		*Cortodera subpilosa* (LeConte)	（コウチュウ目、カミキリムシ科）新北区
western oak dusky wing			propertius duskywing を見よ
western oak looper		*Lambdina fiscellaria somniaria* (Hulst)	（チョウ目、シャクガ科）新北区
western oak looper			hemlock looper (1) を見よ
western orussus		*Orussus occidentalis* Cresson	（ハチ目、ヤドリキバチ科）新北区
western painted lady	ルリボシアカタテハ	*Vanessa carye* (Hübner)	（チョウ目、タテハチョウ科）新北区
western Pero		*Pero occidentalis* (Hulst)	（チョウ目、シャクガ科）新北区
western petaltail		*Petalura hesperia* Watson	（トンボ目、ムカシヤンマ科）豪州区
western pie		*Tuxentius hesperis* (Vari)	（チョウ目、シジミチョウ科）エチオピア区
western pine beetle	アメリカマツノコキクイムシ	*Dendroctonus brevicornis* LeConte	（コウチュウ目、キクイムシ科）新北区
western pine borer			large flat-head pine heartwood borer を見よ

英名	和名	学名	所属、分布、ほか
western pine elfin		*Incisalia eryphon* (Boisduval)	（チョウ目、シジミチョウ科）新北区、新熱帯区
western pine moth		*Dioryctria cambiicola* (Dyar)	（チョウ目、メイガ科）新北区
western pine shoot borer		*Eucosma sonomana* Kearfott	（チョウ目、ハマキガ科）新北区。western pine shoot borer moth ともいう
western plant bug		*Rhinacloa forticornis* Reuter	（カメムシ目、カスミカメムシ科）新北区
western policeman		*Coeliades hanno* (Plötz)	（チョウ目、セセリチョウ科）エチオピア区
western Polyphemus moth		*Antheraea oculea* (Neumoegen)	（チョウ目、ヤママユガ科）新北区
western poplar clearwing		*Paranthrene robiniae* (Hy. Edwards)	（チョウ目、スカシバガ科）新北区。western poplar clearwing moth ともいう
western poplar sphinx		*Pachysphinx occidentalis* (Hy. Edwards)	（チョウ目、スズメガ科）新北区。western poplar sphinx moth ともいう
western potato flea beetle		*Epitrix subcrinita* (LeConte)	（コウチュウ目、ハムシ科）新北区。野菜害虫
western potato leafhopper		*Empoasca abrupta* DeLong	（カメムシ目、オオヨコバイ科）新北区
western pygmy blue	ピグミーシジミ	*Brephidium exilis* (Boisduval)	（チョウ目、シジミチョウ科）新北区、大洋区、新熱帯区
western ragged skipper		*Caprona adelica* Karsch	（チョウ目、セセリチョウ科）エチオピア区
western raspberry fruitworm		*Byturus bakeri* Barber	（コウチュウ目、キスイモドキ科）新北区
western Rectilinea slug moth		*Apoda latomia* (Harvey)	（チョウ目、イラガ科）新北区
western red Charaxes		*Charaxes cynthia* Butler	（チョウ目、タテハチョウ科）エチオピア区
western red damsel		*Amphiagrion abbreviatum* (Selys)	（トンボ目、イトトンボ科）新北区
western red hunter		*Austroepigomphus gordoni* (Watson)	（トンボ目、サナエトンボ科）豪州区
western red-shouldered plant bug		*Thyauta pallidovirens* (Stål)	（カメムシ目、カメムシ科）新北区
western river cruiser			magnificent skimmer を見よ
western rose chafer		*Macrodactylus uniformis* Horn	（コウチュウ目、コガネムシ科）新北区。野生のブドウに多い
western round-winged katydid		*Amblycorypha parvipennis* Stål	（バッタ目、キリギリス科）新北区
western sallow		*Sunira decipiens* Grote	（チョウ目、ヤガ科）新北区
western sand wasp		*Bembix cornata* Parker	（ハチ目、アナバチ科）新北区
western shieldback		*Ateloplus hesperus* Hebard	（バッタ目、キリギリス科）新北区
western short-horned walking stick		*Parabacillus hesperus* Hebard	（ナナフシ目、Heteromeniidae）新北区
western skiff moth		*Prolimacodes trigona* (Hy. Edwards)	（チョウ目、イラガ科）新北区
western skipperling			Garita skipperling を見よ
western slave-making ant			western Amazon ant を見よ

英名	和名	学名	所属、分布、ほか
western smoke fly		*Microsania occidentalis* Malloch	(ハエ目、ヒラタアシバエ科) 新北区
western spotted cucumber beetle			spotted cucumber beetle を見よ
western spruce budworm		*Choristoneura occidentalis* Freeman	(チョウ目、ハマキガ科) 新北区。western spruce budworm moth ともいう
western spruce budworm (1)		*Choristoneura biennis* Freeman	(チョウ目、ハマキガ科) 新北区
western striped albatross			striped albatross を見よ
western striped cricket		*Miogryllus lineatus* (Scudder)	(バッタ目、コオロギ科) 新北区
western striped cucumber beetle		*Acalymma trivittata* (Mannerheim)	(コウチュウ目、ハムシ科) 新北区
western striped cucumber beetle			twelve-spotted cucumber beetle を見よ
western striped flea beetle		*Phyllotreta ramosa* (Crotch)	(コウチュウ目、ハムシ科) 新北区。野菜害虫
western stutter-trilling cricket		*Gryllus integer* Scudder	(バッタ目、コオロギ科) 新北区
western subterranean termite		*Reticulitermes hesperus* Banks	(シロアリ目、ミゾガシラシロアリ科) 新北区
western sulphur	ロッキーモンキチョウ	*Colias occidentalis* Scudder	(チョウ目、シロチョウ科) 新北区
western swallowtail			anise swallowtail を見よ
western tailed blue	メスルリツバメシジミ	*Cupido amyntula* (Boisduval)	(チョウ目、シジミチョウ科) 新北区
western tailed blue	ニシツバメシジミ	*Everes amyntas albrighti* Clench	(チョウ目、シジミチョウ科) 新北区
western tent caterpillar			California tent caterpillar を見よ
western tent caterpillar			Pacific tent caterpillar を見よ
western tent caterpillar moth			California tent caterpillar を見よ
western tentiform leafminer		*Phyllonorycter elmaella* Doganlar et Mutuura	(チョウ目、ホソガ科) 新北区
western Tetanolita moth		*Tetanolita palligera* Smith	(チョウ目、ヤガ科) 新北区
western thatching ant			thatching ant (1) を見よ　西部の屋根ふきアリの意
western tiger			western tiger swallowtail を見よ
western tiger crane fly		*Nephrotoma suturalis wulpiana* (Bergroth)	(ハエ目、ガガンボ科) 新北区
western tiger flat		*Eagris tigris* Evans	(チョウ目、セセリチョウ科) エチオピア区
western tiger swallowtail		*Papilio rutulus* Lucas	(チョウ目、アゲハチョウ科) 新北区
western tree cricket		*Oecanthus californicus* (Saussure)	(バッタ目、カンタン科) 新北区

英名	和名	学名	所属、分布、ほか
western treehole mosquito		*Aedes sierrensis* (Indlow)	(ハエ目、カ科) 新北区
western treehole mosquito		*Ochlerotatus deserticola* Zavortink	(ハエ目、カ科) 新北区
western true katydids		*Paracyrtophyllus*	(バッタ目、キリギリス科) の昆虫の総称
western tumblebug		*Canthon imitator* Brown	(コウチュウ目、コガネムシ科) 新北区。糞で球を作り、産卵して地中にうめる
western tussock moth		*Orgyia vetusta* (Boisduval)	(チョウ目、ドクガ科) 新北区。雌は無翅
western twig borer			powderpost bostrichid を見よ
western velvet ant		*Dasymutilla sackenii* (Cresson)	(ハチ目、アリバチ科) 新北区。ハナダカバチに寄生
western water scorpion			common water scorpion を見よ
western wheat aphid		*Diuraphis (Holcaphis) tritici* (Gillette)	(カメムシ目、アブラムシ科) 新北区
western wheat aphid		*Brachycolus frequens* (Walker)	(カメムシ目、アブラムシ科) 旧北区
western wheat stem maggot		*Hylemya cerealis* (Gillette)	(ハエ目、ハナバエ科) 新北区
western wheat-stem sawfly		*Syrista orientalis* Tischbein	(ハチ目、クキバチ科) 旧北区
western white		*Pontiia occidentalis* (Reakirt)	(チョウ目、シロチョウ科) 新北区
western white-ribboned carpet moth		*Mesoleuca gratulata* (Walker)	(チョウ目、シャクガ科) 新北区
western widow		*Libellula forensis* Hagen	(トンボ目、トンボ科) 新北区
western willow spreadwing			willow emerald damselfly を見よ
western willow clearwing moth		*Synanthedon albicornis* (Edwards)	(チョウ目、スカシバガ科) 新北区
western w-marked cutworm		*Spaelotis havilae* (Grote)	(チョウ目、ヤガ科) 旧北区
western wood cockroach		*Parcoblatta americana* Scudder	(ゴキブリ目、チャバネゴキブリ科) 新北区
western woolly legs		*Lachnocnema brimo* Karsch	(チョウ目、シジミチョウ科) エチオピア区
western woolly legs		*Lachnocnema vuattouxi* Libert	(チョウ目、シジミチョウ科) エチオピア区
western woolly legs (1)		*Lachnocnema emperamus* (Snellen)	(チョウ目、シジミチョウ科) エチオピア区
western Xenica		*Geitoneura minyas* (Waterhouse et Lyell)	(チョウ目、タテハチョウ科) 豪州区
western yellow-faced			yellow-faced bumble bee を見よ
western yellowjacket		*Vespula pennsylvanica* (Saussure)	(ハチ目、スズメバチ科) 新北区
western yellow-striped armyworm		*Spodoptera praefica* (Grote)	(チョウ目、ヤガ科) 新北区。アルファルファ害虫。western yellowstriped armyworm moth ともいう

英名	和名	学名	所属、分布、ほか
western Zobera		*Zobera marginata* Freeman	（チョウ目、セセリチョウ科）新熱帯区
Westfall's clubtail		*Gomphus westfalli* Carle et May	（トンボ目、サナエトンボ科）新北区
Westfall's dancer		*Argia westfalli* Garrison	（トンボ目、イトトンボ科）新熱帯区
Westfall's snaketail		*Ophiogomphus westfalli* Cook et Daigle	（トンボ目、サナエトンボ科）新北区
Westwood's king crow		*Euploea westwoodii* C. et R. Felder	（チョウ目、タテハチョウ科）東洋区
Westwood's leafwing		*Memphis xenocles* Westwood	（チョウ目、タテハチョウ科）新熱帯区
Westwood's satyr		*Euptychia westwoodi* Butler	（チョウ目、タテハチョウ科）新熱帯区
Westwood's white-lady	アガメデスタイマイ	*Graphium agamedes* (Westwood)	（チョウ目、アゲハチョウ科）エチオピア区
wet-wood termite			Formosan subterranean termite を見よ
wet-wood termites			subterranean termites (1) を見よ
weta		*Deinacrida*	（バッタ目、Stenopelmatidae）の昆虫の総称 weta はニュージーランドのマオリ語に由来
weta			cave crickets (2) を見よ
weta-punga		*Deinacrida heteracantha* White	（バッタ目、Stenopelmatidae）豪州区。巨大な絶滅危惧種
Weymer's banded nymph		*Manerebia satura* (Weymer)	（チョウ目、タテハチョウ科）新熱帯区
Weymer's bark		*Rareuptychia clio* Staudinger	（チョウ目、タテハチョウ科）新熱帯区
Weymer's crow			broad-banded crow を見よ
Weymer's flasher		*Astraptes weymeri* (Plötz)	（チョウ目、セセリチョウ科）新熱帯区
Weymer's glasswing	ヘリボシトンボマダラ	*Hyalyris latilimbata* (Weymer)	（チョウ目、タテハチョウ科）新熱帯区
Weymer's glider	ウエーマーシロタテハ	*Cymothoe weymeri* Suffert	（チョウ、タテハチョウ科）エチオピア区
Weymer's mountain satyr		*Lymanopoda rana* Thieme	（チョウ目、タテハチョウ科）新熱帯区
Weymer's ringlet		*Cissia proba* (Weymer)	（チョウ目、タテハチョウ科）新熱帯区
Weymouth fir seed chalcid		*Megastigmus atedius* Walker	（ハチ目、オナガコバチ科）旧北区。イングランド南部の地名に由来
Weymouth pine adelges			pine bark adelgid を見よ
whame-flies			clegs を見よ
whame-flies			horse fies (2) を見よ
wharf borer	ツマグロカミキリモドキ	*Nacerdes melanura* (Linnaeus)	（コウチュウ目、カミキリモドキ科）日本、汎世界
wheat and barley Anoecia		*Anoecia vagans* (Koch)	（カメムシ目、アブラムシ科）旧北区
wheat and barley Rhizobius		*Aploneura graminis* Buckton	（カメムシ目、アブラムシ科）旧北区

英名	和名	学名	所属、分布、ほか
wheat and barley stem fly		*Phorbia flavibasis* (Stein)	(ハエ目、ハナバエ科) 旧北区
wheat aphid			oat bird-cherry aphid を見よ
wheat aphid			greenbug を見よ
wheat armyworm moth		*Mythimna sequax* (Franclemont)	(チョウ目、ヤガ科) 新北区。wheat armyworm ともいう
wheat beetle			cadelle を見よ
wheat blossom midge		*Contarinia tritici* (Kirby)	(ハエ目、タマバエ科) 旧北区
wheat blossom midge			wheat midge を見よ
wheat bulb fly		*Delia montana* Malloch	(ハエ目、ハナバエ科) 新北区
wheat bulb fly		*Hylemya coarctata* (Fallén)	(ハエ目、ハナバエ科) 旧北区
wheat cutworm			wheat earworm を見よ
wheat cutworm			ear miner moth を見よ
wheat earworm		*Apamea sordens* (Hufnagel)	(チョウ目、ヤガ科) 全北区。日本亜種は *A. s. basistriga* (Staudinger) シロミミハイイロヨトウ
wheat elongate flea beetle			indigo flea beetle を見よ
wheat flea beetle		*Crepidodera ferruginea* (Scopoli)	(コウチュウ目、ハムシ科) 旧北区
wheat flour beetle			red flour beetle を見よ
wheat grain thrips			wheat thrips を見よ
wheat head armyworm		*Faronta albilinea* Hübner	(チョウ目、ヤガ科) 新北区、新熱帯区
wheat head armyworm			wheat head armyworm moth を見よ
wheat head armyworm moth		*Faronta diffusa* (Walker)	(チョウ目、ヤガ科) 新北区
wheat jointworm		*Harmolita tritici* (Fitch)	(ハチ目、カタビロコバチ科) 新北区
wheat leaf beetle	ムギクビボソハムシ	*Oulema erichsoni* (Suffrian)	(コウチュウ目、ハムシ科) 日本、旧北区。ムギ害虫
wheat leaf beetle			cereal leaf beetle (2) を見よ
wheat leaf bug	ムギカスミカメムシ	*Stenodema calcaratum* (Fallén)	(カメムシ目、カスミカメムシ科) 日本、旧北区。ムギ、イネ害虫
wheat leaf miner		*Syringopais temperatella* (Lederer)	(チョウ目、キヌバコガ科) 旧北区
wheat leafhopper	ムギトガリヨコバイ	*Sorhoanus tritici* (Matsumura)	(カメムシ目、オオヨコバイ科) 日本、旧北区。ムギ害虫
wheat leafminer	ムギスジハモグリバエ	*Chromatomyia nigra* (Meigen)	(ハエ目、ハモグリバエ科) 日本、全北区。ムギ害虫
wheat louse			greenbug を見よ
wheat louse			grain aphid (1) を見よ
wheat maggot			wheat blossom midge を見よ
wheat midge	ムギアカタマバエ	*Sitodiplosis mosellana* (Gehin)	(ハエ目、タマバエ科) 日本、全北区。ムギ害虫
wheat midge			Hessian fly を見よ

英名	和名	学名	所属、分布、ほか
wheat moth			European grain moth を見よ
wheat mud beetle		*Helophorus nubilus* Fabricius	（コウチュウ目、ガムシ科）旧北区
wheat planthopper	キタウンカ	*Javesella pellucida* (Fabricius)	（カメムシ目、ウンカ科）日本、旧北区。ムギ害虫
wheat plant-louse			grain aphid (1) を見よ
wheat root scarab		*Sericesthis consanguinea* (Blackburn)	（コウチュウ目、コガネムシ科）豪州区
wheat sapsucker			wheat midge を見よ
wheat sawfly (1)	オオムネアカハバチ	*Dolerus ephippiatus* Smith	（ハチ目、ハバチ科）日本、旧北区。ムギ害虫
wheat sawfly (2)	ムギハバチ	*Dolerus lewisi* Cameron	（ハチ目、ハバチ科）日本、旧北区。ムギ害虫
wheat sawfly borer			wheat stem sawfly (1) を見よ
wheat shoot beetle			ivy and hop weevil を見よ
wheat shoot beetle			wheat mud beetle を見よ
wheat sirividhi			cereal leaf-miner を見よ
wheat spotted noctuid	ホシミミヨトウ	*Mesapamea concinnata* Heinicke	（チョウ目、ヤガ科）日本、旧北区
wheat stem borer			wheat stem sawfly (1) を見よ
wheat stem flea beetle			wheat flea beetle を見よ
wheat stem maggot	ムギキカラバエ	*Chlorops mugivorus* Nishijima et Kanmiya	（ハエ目、キモグリバエ科）日本。ムギ害虫
wheat stem maggot	ムギキモグリバエ	*Mereomyza nigriventris* Macquart	（ハエ目、キモグリバエ科）日本、旧北区。ムギ害虫
wheat stem maggot	ニセムギキモグリバエ	*Meromyza grandifemoris* Kanmiya	（ハエ目、キモグリバエ科）日本。ムギ害虫
wheat stem maggot	コムギクキハナバエ	*Atherigona falcata* Thomson	（ハエ目、イエバエ科）日本。ムギ害虫
wheat stem maggot		*Meromyza americana* Fitch	（ハエ目、キモグリバエ科）新北区。ムギ害虫
wheat stem sawfly		*Cephus cinctus* Norton	（ハチ目、クキバチ科）新北区
wheat stem sawfly (1)		*Cephus pygmaeus* (Linnaeus)	（ハチ目、クキバチ科）全北区
wheat strawworm		*Harmolita grandis* (Riley)	（ハチ目、カタビロコバチ科）新北区
wheat thrips		*Haplothrips tritici* Kurd	（アザミウマ目、クダアザミウマ科）旧北区
wheat thrips			flower thrips (2) を見よ
wheat wireworm		*Agriotes maneus* (Say)	（コウチュウ目、コメツキムシ科）新北区
wheat wireworm	カバイロコメツキ	*Ectinus sericeus* (Candèze)	（コウチュウ目、コメツキムシ科）日本、旧北区。ムギ、トウモロコシ、ジャガイモ、カンキツなどの害虫
wheel bug		*Arilus cristatus* (Linnaeus)	（カメムシ目、サシガメ科）新北区。体長28〜36 mm。捕食性
wheeling glider			keyhole glider を見よ
wherrymen			water striders (1) を見よ
whip-marked snout moth		*Microtheoris vibicalis* Zeller	（チョウ目、メイガ科）新北区

英名	和名	学名	所属、分布、ほか
whiplash rove beetle		*Paederus australis* Guérin-Méneville	（コウチュウ目、ハネカクシ科）豪州区
whiplash rove beetle		*Paederus cruenticollis* Germar	（コウチュウ目、ハネカクシ科）豪州区
whirlabout		*Polites vibex vibex* (Geyer)	（チョウ目、セセリチョウ科）新北区
whirligig			whirligig beetle (1) を見よ
whirligig beetle		*Gyrinus limbatus* Say	（コウチュウ目、ミズスマシ科）新北区
whirligig beetle (1)		*Gyrinus natator* (Linnaeus)	（コウチュウ目、ミズスマシ科）旧北区
whirligig beetles	ミズスマシ科	Gyrinidae	（コウチュウ目）の昆虫の総称
whirligigs			whirligig beetles を見よ
whistling moth			common whistling moth を見よ
whistling moths		Hecatera	（チョウ目、ヤガ科）の昆虫の総称
white			common white を見よ
white admirable			blue admiral を見よ
white admiral		*Ladoga camilla* (Linnaeus)	（チョウ目、タテハチョウ科）旧北区。日本亜種は *L. c. japonica* (Ménétriès) イチモンジチョウ。white admiral butterfly ともいう
white admiral (1)	アメリカイチモンジ（アメリカアオイチモンジ）	*Limenitis arthemis* (Drury)	（チョウ目、タテハチョウ科）新北区
white albatross			common albatross を見よ
white and orange Halimede			yellow patch white を見よ
white angled sulphur			ghost brimstone を見よ
white Annaphila		*Annaphila diva* Grote	（チョウ目、ヤガ科）新北区
white ant			eastern subterranean termite を見よ
white ants			termites を見よ
white-apex-hindwinged noctuid	エゾシロシタバ	*Catocala dissimilis* Bremer	（チョウ目、ヤガ科）日本、旧北区
white apple leafhopper		*Typhlocyba pomaria* McAtee	（カメムシ目、オオヨコバイ科）新北区。リンゴ他害虫
white Arab		*Colotis vestalis* (Butler)	（チョウ目、シロチョウ科）東洋区、エチオピア区
white-backed longicorn beetle	ネジロカミキリ	*Pogonacherus seminiveus* Bates	（コウチュウ目、カミキリムシ科）日本、旧北区
whitebacked planthopper			white-backed rice planthopper を見よ
white-backed rice planthopper	セジロウンカ	*Sogatella furcifera* (Horvath)	（カメムシ目、ウンカ科）日本、旧北区、東洋区。イネ害虫
white-backed wisp			pinhead wisp を見よ
white bamboo scale			bamboo diaspidid scale を見よ
white-banded awl		*Hasora taminatus* (Hübner)	（チョウ目、セセリチョウ科）東洋区。*H. t. malayana* (Felder et Felder) も同英名
white-banded bark beetle	シラオビキクイムシ	*Hylesinus cingulatus* Blandford	（コウチュウ目、キクイムシ科）日本、旧北区
white-banded black moth		*Rheumaptera subhastata* (Nolcken)	（チョウ目、シャクガ科）全北区

英名	和名	学名	所属、分布、ほか
white-banded black zygaenid	ホタルガ	*Pidorus atratus* Butler	(チョウ目、マダラガ科) 日本、旧北区
white-banded carpet		*Spargania luctuata* (Denis et Schiffermüller)	(チョウ目、シャクガ科) 旧北区
white-banded castor		*Ariadne albifascia* (Joicey et Talbot)	(チョウ、タテハチョウ科) エチオピア区
white-banded clerid		*Paratillus carus* (Newman)	(コウチュウ目、カッコウムシ科) 豪州区
white-banded daggerwing		*Marpesia crethon* (Fabricius)	(チョウ目、タテハチョウ科) 新熱帯区
white-banded day sphinx			titan sphinx moth を見よ
white-banded delicate geometrid			clouded border を見よ
white-banded elm leafhopper		*Scaphoideus luteolus* Van Duzee	(カメムシ目、オオヨコバイ科) 新北区
white-banded eucosmid			double-arched roller を見よ
white-banded firetip		*Pyrrhopyge crida* Hewitson	(チョウ目、セセリチョウ科) 新熱帯区
white-banded flat	ダイミョウセセリ	*Daimio tethys* (Ménétriès)	(チョウ目、セセリチョウ科) 日本、旧北区
white-banded flat			large streaked flat を見よ
white-banded flat			dusky yellow-breast flat を見よ
white-banded froghopper			common spittlebug を見よ
white-banded grayling		*Pseudochazara anthelea* (Hübner)	(チョウ目、タテハチョウ科) 旧北区
white-banded greenish geometrid	シロオビアオシャク	*Geometra sponsaria* (Bremer)	(チョウ目、シャクガ科) 日本、旧北区
white-banded hairstreak			bright Cerulean を見よ
white-banded hedge blue		*Lestranicus transpectus* (Moore)	(チョウ目、シジミチョウ科) 東洋区
white-banded line-blue			white-line blue (1) を見よ
white banded longhorn beetle		*Poecilium alni* (Linnaeus)	(コウチュウ目、カミキリムシ科) 旧北区
white-banded longicorn beetle	シロオビカミキリ	*Phymatodes albicinctus* Bates	(コウチュウ目、カミキリムシ科) 日本。ブドウ害虫
white-banded metalmark		*Hypophylla sudias sudias* (Hewitson)	(チョウ目、シジミタテハ科) 新熱帯区
white-banded mountain satyr		*Lymanopoda albocincta* Hewitson	(チョウ目、タテハチョウ科) 新熱帯区
white banded Nephele		*Nephele rosae* Butler	(チョウ目、スズメガ科) エチオピア区
white-banded noctuid	シラオビキリガ	*Cosmia camptostigma* (Ménétriès)	(チョウ目、ヤガ科) 日本、旧北区
white-banded pallid ringlet		*Euptychoides albofasciata* (Hewitson)	(チョウ目、タテハチョウ科) 新熱帯区
white-banded plane			common aeroplane を見よ

英名	和名	学名	所属、分布、ほか
white-banded pointed longicorn	トガリシロオビサビカミキリ	*Pterolophia caudata caudata* (Bates)	(コウチュウ目、カミキリムシ科) 日本
white-banded red-eye		*Pteroteinon caenira* (Hewitson)	(チョウ目、セセリチョウ科) エチオピア区
white-banded ringlet			yellow-banded ringlet を見よ
white-banded royal		*Ancema cotys* (Hewitson)	(チョウ目、シジミチョウ科) 東洋区。
white-banded salt marsh mosquito			golden saltmarsh mosquito を見よ
white-banded satyr			Metaleuca ringlet を見よ
white-banded skipper		*Heliopetes domicella* (Erichson)	(チョウ目、セセリチョウ科) 新北区、新熱帯区
white-banded swallowtail	トガリイチモンジアゲハ	*Papilio echeriodes* Trimen	(チョウ目、アゲハチョウ科) エチオピア区
white-banded Telphusa moth		*Telphusa latifasciella* (Chambers)	(チョウ目、キバガ科) 新北区
white-banded toothed carpet moth			common carpet を見よ　white-banded toothed carpet ともいう
white-banded tussock moth	シロオビドクガ	*Numenes disparilis* Staudinger	(チョウ目、ドクガ科) 日本、旧北区
white-bar bushbrown		*Mycalesis anaxias* Hewitson	(チョウ目、タテハチョウ科) 東洋区
white-bar mountain satyr		*Pedaliodes palaepolis* Hewitson	(チョウ目、タテハチョウ科) 新熱帯区
white-barred Acraea		*Acraea encedon* (Linnaeus)	(チョウ目、タテハチョウ科) エチオピア区
white-barred beech pigmy			gold-barred beech pigmy (1) を見よ
white-barred Charaxes	シロオビフタオチョウ	*Charaxes brutus* (Cramer)	(チョウ目、タテハチョウ科) エチオピア区。亜種 *C. b. natalensis* Staudinger も同英名
white-barred clearwing		*Synanthedon spheciformis* (Denis et Schiffermüller)	(チョウ目、スカシバガ科) 旧北区
white-barred emerald moth		*Nemoria bifilata* (Walker)	(チョウ目、シャクガ科) 新北区
white-barred emperor			white-barred Charaxes を見よ　white barred emperor とも記す
white-barred groundling		*Recurvaria leucatella* (Clerck)	(チョウ目、キバガ科) 旧北区
white-barred gypsy		*Palasea albimacula* Wallengren	(チョウ目、ドクガ科) 旧北区
white barred hawk		*Leucostrophus alterhirundo* d'Abrera	(チョウ目、スズメガ科) エチオピア区
white-barred lady slipper		*Pierella hortona* (Hewitson)	(チョウ目、タテハチョウ科) 新熱帯区
white-barred longwing			Cydno longwing を見よ
white-barred satyr		*Parataygetis albinotata* (Butler)	(チョウ目、タテハチョウ科) 新熱帯区
white-barred sister	シロモンイチモンジ	*Adelpha epione* (Godart)	(チョウ目、タテハチョウ科) 新熱帯区
white-barred skipper			parchment skipper を見よ

英名	和名	学名	所属、分布、ほか
white-barred soldier		*Oxycera morrisii* Curtis	(ハエ目、ミズアブ科) 旧北区
white beach tiger beetle		*Cicindela (Habroscelimorpha) dorsalis* Say	(コウチュウ目、ハンミョウ科) 新北区
white beauty			Camberwell beauty を見よ
white-belted ringtail		*Erpetogomphus compositus* (Hagen)	(トンボ目、サナエトンボ科) 新北区
white blind springtails		*Onychiurus*	(トビムシ目、シロトビムシ科) の昆虫の総称
white blotch oak leaf miner		*Phyllonorycter hamadryadella* Clemens	(チョウ目、ホソガ科) 新北区
white-blotched Heterocampa moth		*Heterocampa umbrata* Walker	(チョウ目、シャチホコガ科) 新北区
white-bodied Estigmene			agreeable tiger moth を見よ
white-bordered copper		*Lycaena pavana* (Kollar)	(チョウ目、シジミチョウ科) 東洋区
white-bordered crest		*Dichomeris marginella* (Fabricius)	(チョウ目、キバガ科) 全北区
white borer			sugarcane shoot borer を見よ
white borer			white rice stemborer を見よ
white-brand skipper		*Toxidia rietmanni* (Semper)	(チョウ目、セセリチョウ科) 豪州区
white branded swift			banana skipper (1) を見よ
white branded swift			Caribbean ruby-eye をみよ
white bryony midge		*Jaapiella bryoniae* Bouché	(ハエ目、タマバエ科) 旧北区
white buff	ツマグロシロシジミ	*Teriomima subpunctata* Kirby	(チョウ目、シジミチョウ科) エチオピア区
white butterflies			cabbage whites を見よ
white butterfly			black-veined white (1) を見よ
white butterfly			common white を見よ　*P. rapae* Linnaeus の豪州での英名
white cabbage butterfly			common white を見よ
white-caudate longicorn	アトジロサビカミキリ	*Pterolophia zonata* (Bates)	(コウチュウ目、カミキリムシ科) 日本
white cedar moth			cedar tussock を見よ
white-centered bent-skipper		*Theagenes aegides* (Herrich-Schäffer)	(チョウ目、セセリチョウ科) 新熱帯区
white-centered ruby-eye		*Cobalus virbius fidicula* Hewitson	(チョウ目、セセリチョウ科) 新熱帯区
white Cerulean	センペルウラナミシジミ	*Jamides cleodus* (C. et R. Felder)	(チョウ目、シジミチョウ科) 東洋区
white Cerulean		*Jamides pura* (Moore)	(チョウ目、シジミチョウ科) 東洋区
white checkered skipper			large grizzled skipper (1) を見よ
white-cheeked scarlet-eye			Midas skipper を見よ

英名	和名	学名	所属、分布、ほか
white-cloack bell		*Gypsonoma sociana* Haworth	（チョウ目、ハマキガ科）旧北区
white-cloaked skipper		*Leucochitonea levubu* Wallengren	（チョウ目、セセリチョウ科）エチオピア区
white-clouded longhorn beetle		*Mesosa nebulosa* (Fabricius)	（コウチュウ目、カミキリムシ科）旧北区
white-clubbed grass-veneer moth		*Crambus alboclavellus* Zeller	（チョウ目、メイガ科）新北区
white clover-head weevil		*Apion flavipes* (Paykull)	（コウチュウ目、ホソクチゾウムシ科）旧北区
white clover seed weevil		*Apion dichroum* Bedel	（コウチュウ目、ホソクチゾウムシ科）旧北区
white clover weevil			clover leaf weevil (2) を見よ
white coffee borer		*Anthores leuconotus* Pascoe	（コウチュウ目、カミキリムシ科）エチオピア区。コーヒー害虫
white coffee leaf-miner			coffee leaf miner (1) を見よ
white colon		*Sideridis albicolon* (Hübner)	（チョウ目、ヤガ科）旧北区
white commodore	ムラサキイチモンジ	*Parasarpa dudu* (Doubleday)	（チョウ目、タテハチョウ科）東洋区
white-costate noctuid	マエジロヤガ	*Ochropleura plecta glaucimacula* (Graeser)	（チョウ目、ヤガ科）日本
white-crescent longtail		*Codatractus alcaeus* (Hewitson)	（チョウ目、セセリチョウ科）新北区
white-crescent mottled-skipper			white-crescent longtail を見よ
white-crescent swallowtail		*Mimoides thymbraeus thymbraeus* (Boisduval)	（チョウ目、アゲハチョウ科）新熱帯区。*M. t. aconophos* (Gray) も同英名
white-crossed grasshopper		*Auclocara femoratum* Scudder	（バッタ目、バッタ科）新北区
white-crossed seed bug		*Neacoryphus bicrucis* (Say)	（カメムシ目、ナガカメムシ科）新北区
white cutworm			white cutworm moth を見よ
white cutworm moth		*Euxoa (Pleonectopoda) scandens* (Riley)	（チョウ目、ヤガ科）新北区
white cypress longicorn		*Uracanthus pallens* Hope	（コウチュウ目、カミキリムシ科）豪州区
white dagger moth	シロケンモン	*Acronicta leporina leporella* Staudinger	（チョウ目、ヤガ科）日本
white-dappled swallowtail	シロマダラタイマイ	*Graphium philonoe* (Ward)	（チョウ目、アゲハチョウ科）エチオピア区
white dart		*Andronymus caesar* (Fabricius)	（チョウ目、セセリチョウ科）エチオピア区
white dawnfly		*Capila pieridoides* (Moore)	（チョウ目、セセリチョウ科）東洋区
whitedisc hedge blue		*Celatoxia albidisca* (Moore)	（チョウ目、シジミチョウ科）東洋区
white dot	シラホシヨトウ	*Melanchra persicariae* (Linnaeus)	（チョウ目、ヤガ科）日本、旧北区。ダイズ、アブラナ科野菜、テンサイなどの害虫

英名	和名	学名	所属、分布、ほか
white-dot oakblue		*Arhopala democritus democritus* Fabricius	（チョウ目、シジミチョウ科）東洋区
white-dotted brown-striped noctuid	シロモンオビヨトウ	*Athetis lineosa* (Moore)	（チョウ目、ヤガ科）日本、旧北区、東洋区
white-dotted cattleheart		*Parides alopius* Godman et Salvin	（チョウ目、アゲハチョウ科）新北区
white-dotted crescent		*Castilia ofella* (Hewitson)	（チョウ目、タテハチョウ科）新熱帯区
white-dotted flower chafer	シロテンハナムグリ	*Protaetia orientalis submarmorea* (Burmeister)	（コウチュウ目、コガネムシ科）日本
white-dotted little noctuid	シラホシコヤガ	*Enispa leucosticta* Hampson	（チョウ目、ヤガ科）日本
white-dotted prominent			tawny prominent (1) を見よ
white-dotted ringlet		*Forsterinaria neonympha* (C. et R. Felder)	（チョウ目、タテハチョウ科）新熱帯区
white-dotted satyr		*Forsterinaria neonympha umbracea* Butler et Druce	（チョウ目、タテハチョウ科）新熱帯区。white-dotted ringlet を参照
white dragontail	シロスソビキアゲハ	*Lamproptera curius* (Fabricius)	（チョウ目、アゲハチョウ科）旧北区、東洋区
white drummer		*Arunia perulata* (Guérin-Méneville)	（カメムシ目、セミ科）豪州区
white-dusted mountain satyr		*Lymanopoda obsoleta* (Westwood)	（チョウ目、タテハチョウ科）新熱帯区
white-eared gothic			feathered ear を見よ
white edge moth		*Oruza albocostaliata* (Packard)	（チョウ目、ヤガ科）新北区
white-edged Ancylis moth		*Ancylis albacostana* Kearfott	（チョウ目、ハマキガ科）新北区
white-edged blue baron		*Euthalia phemius* (Doubleday)	（チョウ目、タテハチョウ科）東洋区
white-edged bushbrown		*Mycalesis mestra* Hewitson	（チョウ目、タテハチョウ科）東洋区
white-edged cloudywing			Skinner's cloudywing を見よ
white-edged Coleotechnites moth		*Coleotechnites albicostatus* (Freeman)	（チョウ目、キバガ科）新北区
white-edged furze case		*Coleophora albicosta* (Haworth)	（チョウ目、ツツミノガ科）旧北区
white-edged longtail		*Urbanus albimargo* (Mabille)	（チョウ目、セセリチョウ科）新熱帯区
white-edged Pima moth		*Pima albiplagiatella* (Packard)	（チョウ目、メイガ科）新北区
white-edged roadside-skipper		*Amblyscirtes fimbriata pallida* Freeman	（チョウ目、セセリチョウ科）新熱帯区
white-edged woodbrown		*Lethe visrava* (Moore)	（チョウ目、タテハチョウ科）東洋区
white emperor		*Helcyra hemina* Hewitson	（チョウ目、タテハチョウ科）東洋区
white ermine		*Spilosoma menthastri* Denis et Schiffermüller	（チョウ目、ヒトリガ科）旧北区

英名	和名	学名	所属、分布、ほか
white ermine moth	キハラゴマダラヒトリ	*Spilosoma lubricipeda* (Linnaeus)	(チョウ目、ヒトリガ科) 日本、旧北区、新熱帯区。ダイズ、カンキツなど果樹、サクラ、花卉の害虫。white ermine ともいう
white-etched hairstreak		*Contrafacia bassania* (Hewitson)	(チョウ目、シジミチョウ科) 新熱帯区
white Eulithis moth		*Eulithis explanata* (Walker)	(チョウ目、シャクガ科) 新北区
white-eyed borer moth		*Iodopepla u-album* (Guenée)	(チョウ目、ヤガ科) 新北区
white-faced darter (1)		*Leucorrhinia dubia* (Van der Linden)	(トンボ目、トンボ科) 旧北区。日本亜種は *L. d. orientalis* Selys カオジロトンボ
white-faced darter (2)		*Leucorrhinia albifrons* (Burmeister)	(トンボ目、トンボ科) 旧北区
white-faced dragonfly			white-faced darter (1) を見よ
white-faced hornet			baldfaced hornet を見よ
white-faced meadowhawk		*Sympetrum obtrusum* (Hagen)	(トンボ目、トンボ科) 新北区
white-faced skimmers		*Leucorrhinia*	(トンボ目、トンボ科) の昆虫の総称
white fall geometrid	シロトゲエダシャク	*Phigalia verecundaria* (Leech)	(チョウ目、シャクガ科) 日本、旧北区
white fir cone maggot		*Spermatolonchaea viridana* Meigen	(ハエ目、クロツヤバエ科) 旧北区。幼虫の英名
white-fir needle miner		*Epinotia meritana* Heinrich	(チョウ目、ハマキガ科) 新北区。white-fir needle miner moth ともいう
white flag		*Melete leucanthe* (C. et R. Felder)	(チョウ目、シロチョウ科) 新熱帯区。
white flannel moth		*Lagoa erispata* Packard	(チョウ目、Megalopygidae) 新北区
white flannel moth		*Norape ovina* (Sepp)	(チョウ目、Megalopygidae) 新北区、新熱帯区
white flannel moth			yellow flannel moth (1) を見よ
white flats		*Seseria*	(チョウ目、セセリチョウ科) の昆虫の総称
white flax moth		*Orthoclydon praefectata* Walker	(チョウ目、シャクガ科) 豪州区
white flies		*Aleyrodina*	(カメムシ目) の昆虫の総称
whiteflies	コナジラミ科	*Aleyrodidae*	(カメムシ目) の昆虫の総称　世界に約1,200種
whitefly			citrus whitefly (1) を見よ
whitefly predator		*Delphastus dejavu* Gordon	(コウチュウ目、テントウムシ科) 新北区
white-foot bell			mugwort leaf caterpillar を見よ
white-footed house ant		*Technomyrmex albipes* (F. Smith)	(ハチ目、アリ科) 日本、東洋区、豪州区
white four-line blue		*Nacaduba angusta kerriana* Distant	(チョウ目、シジミチョウ科) 東洋区。white lineblue を参照
white fourring	セイロンウラナミジャノメ	*Ypthima ceylonica* Hewitson	(チョウ目、タテハチョウ科) 東洋区
white-fringed beetle	シロヘリクチブトゾウムシ	*Graphognathus leucoloma* (Boheman)	(コウチュウ目、ゾウムシ科) 新熱帯区、エチオピア区。南米原産の栽培植物害虫

英名	和名	学名	所属、分布、ほか
white-fringed beetles			white-fringed weevils を見よ
white-fringed cloudywing			drusius cloudywing を見よ
white-fringed emerald moth		*Nemoria mimosaria* (Guenée)	（チョウ目、シャクガ科）新北区
white-fringed longhorn		*Neoptychodes trilineatus* (Linnaeus)	（コウチュウ目、カミキリムシ科）新北区
white-fringed meridian duskywing		*Erynnis meridianus fieldi* Burns	（チョウ目、セセリチョウ科）新熱帯区。southern duskywing を参照
white-fringed Pyrausta moth		*Pyrausta niveicilialis* (Grote)	（チョウ目、メイガ科）新北区
white-fringed recluse		*Caenides dacena* (Hewitson)	（チョウ目、セセリチョウ科）エチオピア区
white-fringed sleepy duskywing		*Erynnis brizo mulleri* (Draudt)	（チョウ目、セセリチョウ科）新熱帯区。sleepy duskywing を参照
white-fringed swift		*Sabera fuliginosa fuliginosa* (Miskin)	（チョウ目、コウモリガ科）豪州区
white-fringed swift		*Sabera fuligenosa* (Miskin)	（チョウ目、セセリチョウ科）豪州区
whitefringed weevil			white-fringed beetle を見よ
white-fringed weevils		*Graphognathus*	（コウチュウ目、ゾウムシ科）の昆虫の総称
white-frosted hairstreak		*Brevianta tolmides* (C. et R. Felder)	（チョウ目、シジミチョウ科）新熱帯区
white-fumated burdock weevil	シラクモゴボウゾウムシ	*Larinus griseopilosus* Roelofs	（コウチュウ目、ゾウムシ科）日本
white Furcula moth		*Furcula borealis* (Guérin-Méneville)	（チョウ目、シャチホコガ科）新北区
white grass dart		*Taractrocera papyria* (Boisduval)	（チョウ目、セセリチョウ科）豪州区
white grass yellow			ghost yellow を見よ
white ground mealybug		*Rhizoecus leucosomus* (Cockerell)	（カメムシ目、コナカイガラムシ科）新北区
white ground mealybug			cacticans mealybug を見よ
white ground pearl		*Promargarodes australis* Jakubski	（カメムシ目、ワタフキカイガラムシ科）豪州区
white grub		*Clemora smithi* Arrow	（コウチュウ目、コガネムシ科）東洋区、エチオピア区
white grub (1)		*Anomala antigua* Piper	（コウチュウ目、コガネムシ科）東洋区
white grub			June beetle (1) を見よ
white grub beetle	ケブカアカチャコガネ	*Dasylepida ishigakiensis* (Niijima et Kinoshita)	（コウチュウ目、コガネムシ科）日本（琉球）サトウキビ害虫
white-grub cockchafer			June beetle (1) を見よ
white grub parasites			tiphiid wasps を見よ
white grub wasp		*Triscolia ardens* (Smith)	（ハチ目、ツチバチ科）新北区
white grubs		*Phyllopertha*	（コウチュウ目、コガネムシ科）の昆虫の総称
white grubs			May beetles (2) を見よ

英名	和名	学名	所属、分布、ほか
white hairstreak		*Strymon albata* (C. et R. Felder)	(チョウ目、シジミチョウ科) 新熱帯区
white hairy caterpillar			ber moth を見よ
white-headed leafhopper	シロズキンヨコバイ	*Idiocerus ishiyamae* Matsumura	(カメムシ目、オオヨコバイ科) 日本、旧北区
white-headed Monopis moth		*Monopis monachella* (Hübner)	(チョウ目、ヒロズコガ科) 日本、全北区、東洋区、豪州区、新熱帯区
white-headed paddy planthopper			white-backed rice planthopper を見よ
white-headed prominent moth		*Symmerista albifrons* (Smith)	(チョウ目、シャチホコガ科) 新北区
white-headed thorn		*Xanthisthisa niveifrons* Prout	(チョウ目、シャクガ科) エチオピア区
white hedge blue		*Udara akasa* (Horsfield)	(チョウ目、シジミチョウ科) 東洋区
white-hindwinged Catocala	シロシタバ	*Catocala nivea* Butler	(チョウ目、ヤガ科) 日本、東洋区
white-hindwinged dagger moth	シロシタケンモン	*Hylonycta hercules* (Felder et Rogenhofer)	(チョウ目、ヤガ科) 日本、旧北区
white-hindwinged geometrid	アトジロエダシャク	*Pachyligia dolosa* Butler	(チョウ目、シャクガ科) 日本、旧北区
white-hindwinged zygaenid	シロシタホタルガ	*Chalcosia remota yaeyamana* Matsumura	(チョウ目、マダラガ科) 日本、旧北区
white imperial	テイオウシジミ	*Neomyrina nivea periculosa* Fruhstorfer	(チョウ目、シジミチョウ科) 東洋区
white lace Epinotia moth		*Epinotia madderana* (Kearfott)	(チョウ目、ハマキガ科) 新北区
white lace lerp		*Cardiaspina albitextura* Taylor	(カメムシ目、キジラミ科) 豪州区
white lacewings			dustywings を見よ 英国での英名
white lady			Angola white-lady swordtail を見よ
white-lady swallowtail			white lady swordtail を見よ
white lady swordtail		*Graphium morania* (Angas)	(チョウ目、アゲハチョウ科) エチオピア区
white leafhopper			rice leafhopper (1) を見よ
white-legged cherry sawfly			gooseberry sawfly (2) を見よ
white-legged damselfly		*Platycnemis pennipes* (Pallas)	(トンボ目、モノサシトンボ科) 旧北区
white-legged mayflies		*Isonychia*	(カゲロウ目、ヒトリカゲロウ科) の昆虫の総称
white-legged metalmark		*Mesene leucopus* Godman et Salvin	(チョウ目、シジミタテハ科) 新熱帯区
white-letter hairstreak		*Strymonidia w-album* (Knoch)	(チョウ目、シジミチョウ科) 旧北区。日本亜種は *S. w-album fentoni* (Butler) カラスシジミ
white-letter hairstreak	ミヤマカラスシジミ	*Strymonidia mera* (Janson)	(チョウ目、シジミチョウ科) 日本

英名	和名	学名	所属、分布、ほか
white line Temnora		*Temnora albilinea* Rothschild	（チョウ目、スズメガ科）エチオピア区
white lineblue		*Nacaduba angusta* (Druce)	（チョウ目、シジミチョウ科）東洋区
white-line blue (1)	アマミウラナミシジミ	*Nacaduba kurava* (Moore)	（チョウ目、シジミチョウ科）豪州区。*N. k. nemana* Fruhstorfer も同英名
white-line bush brown		*Bicyclus medontias* (Hewitson)	（チョウ目、タテハチョウ科）エチオピア区
white-line bushbrown		*Mycalesis malsara* Moore	（チョウ目、タテハチョウ科）東洋区
white-line dart moth		*Euxoa tritici* (Linnaeus)	（チョウ目、ヤガ科）旧北区。white-line dart ともいう
white-line dart moth		*Scotia segetum* Schiffermüller	（チョウ目、ヤガ科）旧北区
whiteline hairstreak		*Satyrium sassanides* (Kollar)	（チョウ目、シジミチョウ科）旧北区、東洋区
white-line leafroller moth		*Amorbia humerosana* Clemens	（チョウ目、ハマキガ科）新北区
white-line snout		*Schrankia taenialis* (Hübner)	（チョウ目、ヤガ科）旧北区
white-lined Bomolocha moth		*Hypena abalienalis* Walker	（チョウ目、ヤガ科）新北区。white-lined Hypena ともいう
white-lined grape vine hawk moth			vine hawk-moth (1) を見よ　南アフリカでの英名
white-lined grasshopper		*Trimerotropis albolineata* (Brunner)	（バッタ目、バッタ科）新北区
white-lined graylet moth		*Hyperstrotia villificans* (Barnes et McDunnough)	（チョウ目、ヤガ科）新北区。white-lined graylet ともいう
white-lined green hairstreak	ロッキーウラアオシジミ	*Callophrys sheridanii* (Edwards)	（チョウ目、シジミチョウ科）新北区
white-lined greenish small geometrid	コウスアオシャク	*Chloriasa obliterata* (Walker)	（チョウ目、シャクガ科）日本、旧北区
whitelined hawk moth			striped hawk を見よ
white-lined hawkmoth			striped hawk を見よ
whitelined looper moth		*Epirrita pulchraria* (Taylor)	（チョウ目、シャクガ科）新北区
white-lined satyr		*Parataygetis lineata* Godman et Salvin	（チョウ目、タテハチョウ科）新熱帯区
white-lined skimmer			two-striped skimmer を見よ
white-lined small noctuid	シロスジシマコヤガ	*Corgatha dictaria* (Walker)	（チョウ目、ヤガ科）日本、旧北区
white-lined sphinx			striped hawk を見よ
white-lined sphinx moth			striped hawk を見よ
white litchi scale		*Pseudaulacaspis major* (Cockerell)	（カメムシ目、マルカイガラムシ科）大洋区
white longicorn beetle	ムネホシシロカミキリ	*Olenecamptus clarus* Pascoe	（コウチュウ目、カミキリムシ科）日本、旧北区

英名	和名	学名	所属、分布、ほか
white louse scale			citrus snow scale を見よ
white louse scale predator		*Chilocorus circumdatus* Gyllenhal	（コウチュウ目、テントウムシ科）豪州区
white-M hairstreak		*Parrhasius m-album* (Boisduval et LeConte)	（チョウ目、シジミチョウ科）新北、新熱帯区
white Malachite		*Chlorolestes umbratus* Hagen	（トンボ目、Synlestidae）エチオピア区
white mango scale			mango scale (1) を見よ
white-mantled wainscot			white-mottled wainscot を見よ
white-margined cockroach		*Cutilia soror* (Brunner)	（ゴキブリ目、ゴキブリ科）新北区
white-margined leafhopper	マエジロオオヨコバイ	*Kolla atramentaria* (Motschulsky)	（カメムシ目、オオヨコバイ科）日本、旧北区、東洋区
white-margined moonbeam		*Philiris ziska* (Grose-Smith)	（チョウ目、シジミチョウ科）豪州区
white-margined stink bug	シロヘリカメムシ	*Aenalis lewisi* (Scott)	（カメムシ目、カメムシ科）日本、旧北区
white-marked brown moth		*Selidosema leucelaea* Meyrick	（チョウ目、シャクガ科）豪州区
white-marked Cydia moth		*Cydia albimaculana* (Fernald)	（チョウ目、ハマキガ科）新北区
white-marked Emesis		*Emesis fastidiosa* Ménétriès	（チョウ目、シジミタテハ科）新熱帯区
white-marked fleahopper		*Spanagonicus albofasciatus* (Reuter)	（カメムシ目、カスミカメムシ科）新北区、新熱帯区、大洋区
white-marked gourd-shaped weevil	オビモンヒョウタンゾウムシ	*Amystax fasciatus* Roelofs	（コウチュウ目、ゾウムシ科）日本
white-marked moth	ムラサキウスモンヤガ	*Cerastis leucographa* (Denis et Schiffermüller)	（チョウ目、ヤガ科）日本、旧北区。white-marked ともいう
white-marked shieldback		*Plagiostira albonotata* Scudder	（バッタ目、キリギリス科）新北区
white-marked spider beetle	ヒョウホンムシ	*Ptinus fur* (Linnaeus)	（コウチュウ目、ヒョウホンムシ科）日本、汎世界
white-marked tentmaker			rusty-lined leaf tier を見よ
white-marked treehopper		*Tricentrus albomaculatus* Distant	（カメムシ目、ツノゼミ科）新北区、大洋区
white-marked tussock moth		*Orgyia leucostigma* (J. E. Smith)	（チョウ目、ドクガ科）新北区。雌は無翅。亜種 *O. l. intermedia* Fitch も同英名
white-mark-hindwinged noctuid	コシロシタバ	*Catocala actaea* Felder et Rogenhofer	（チョウ目、ヤガ科）日本、旧北区
white-marmorated broad longicorn	ニホンゴマフカミキリ	*Mesosa myops japonica* Bates	（コウチュウ目、カミキリムシ科）日本
white marmorated longicorn			four-spotted conifer longicorn beetle を見よ
white-marmorated small noctuid			rice false looper (1) を見よ
white-masked whisp		*Agriocnemis falcifera* Pinhey	（トンボ目、イトトンボ科）エチオピア区

英名	和名	学名	所属、分布、ほか
white May frit			heath fritillary を見よ
white metalmark		*Hermathena oweni* Schaus	（チョウ目、シジミタテハ科）新熱帯区
white-middled longicorn	ナカジロサビカミキリ	*Pterolophia jugosa jugosa* (Bates)	（コウチュウ目、カミキリムシ科）日本
white migrant			mottled migrant を見よ
white mimic			large glasswing を見よ
white Morpho		*Morpho polyphemus* Westwood	（チョウ目、タテハチョウ科）新北区、新熱帯区
whitemoth		*Asmicridea edwardsii* (McLachlan)	（トビケラ目、シマトビケラ科）豪州区
white-mottled greenish geometrid	シロフアオシャク	*Ochrognesia difficta* (Walker)	（チョウ目、シャクガ科）日本、旧北区
white-mottled wainscot		*Archanara neurica* (Hübner)	（チョウ目、ヤガ科）旧北区
white mountain copper		*Lycaena rubidus ferrisi* Johnson et Balogh	（チョウ目、シジミチョウ科）新北区。Ruddy copper を参照
white muscardine fungus beetle			white grub (1) を見よ
white mussel scale			cassava scale を見よ
white-necked noctuid	シロクビキリガ	*Lithophane consocia* (Borkhausen)	（チョウ目、ヤガ科）日本
white nymph			jezebel nymph を見よ
white oak borer		*Goes tigrinus* (De Geer)	（コウチュウ目、カミキリムシ科）新北区。white oak borer beetle ともいう
white oak leaf miner		*Lithocolletis hamadryadella* Clemens	（チョウ目、ホソガ科）新北区
white oak-blue			small oak blue を見よ
white orange tip		*Ixias marianne* (Cramer)	（チョウ目、シロチョウ科）東洋区
white owl	アッサムムカシヒカゲ	*Neorina patria* Leech	（チョウ目、タテハチョウ科）東洋区
white paddy stem borer			white rice stemborer を見よ
white padi cicadellid			rice leafhopper (1) を見よ
white palm scale		*Phenacaspis eugeniae* (Maskell)	（カメムシ目、マルカイガラムシ科）豪州区
white patch			white-patched skipper (1) を見よ
white-patched eighty-eight		*Diaethria bacchis* (Doubleday)	（チョウ目、タテハチョウ科）新熱帯区
white-patched leafwing		*Memphis artacaena* (Hewitson)	（チョウ目、タテハチョウ科）新熱帯区
white-patched ruby-eye		*Carystus phorcus phorcus* (Cramer)	（チョウ目、セセリチョウ科）新熱帯区
white-patched skipper		*Chiomara georgina* (Reakirt)	（チョウ目、セセリチョウ科）新北区
white-patched skipper (1)		*Chiomara asychis* (Stoll)	（チョウ目、セセリチョウ科）新北区、新熱帯区
white peach scale	クワシロカイガラムシ	*Pseudaulacaspis pentagona* (Targioni-Tozzetti)	（カメムシ目、マルカイガラムシ科）日本、汎世界。クワ、チャ、モモ、ツツジ、ポプラなどの害虫。米国、豪州での英名

英名	和名	学名	所属、分布、ほか
white peacock	ウスベニタテハ	*Anartia jatrophae* (Linnaeus)	（チョウ目、タテハチョウ科）新北区、新熱帯区
white petticoat			Camberwell beauty を見よ
white pie		*Tuxentius calice* (Hopffer)	（チョウ目、シジミチョウ科）エチオピア区
white pierrot			white pie を見よ
white pine angle moth		*Macaria pinistrobata* (Ferguson)	（チョウ目、シャクガ科）新北区
white pine aphid			pine bark adelgid を見よ　white-pine aphid とも記す
white pine barkminer moth		*Marmara fasciella* (Chambers)	（チョウ目、ホソガ科）新北区
white pine cone beetle		*Conophthorus coniperda* (Schwarz)	（コウチュウ目、キクイムシ科）新北区
white pine cone borer		*Eucosma tocullionana* Heinrich	（チョウ目、ハマキガ科）新北区。white pinecone borer moth ともいう
white pine cutworm			northern variable dart moth を見よ
white pine louse			pine bark adelgid を見よ
white pine sawfly		*Neodiprion pinetum* (Norton)	（ハチ目、マツハバチ科）新北区
white pine weevil		*Pissodes strobi* (Peck)	（コウチュウ目、ゾウムシ科）新北区、マツ害虫
white pine weevil			minor pine weevil を見よ
white pinion spotted		*Lomographa bimaculata* (Fabricius)	（チョウ目、シャクガ科）旧北区。日本亜種は *L. b. subnotata* (Warren) フタホシシロエダシャク。white-pinion spotted moth ともいう
white-plagued sphinx		*Manduca albiplaga* (Walker)	（チョウ目、スズメガ科）新北区、新熱帯区
white plume moth		*Alucita monospilalis* Walker	（チョウ目、ニジュウシトリバガ科）豪州区
white plume moth (1)		*Pterophorus pentadactyla* (Linnaeus)	（チョウ目、トリバガ科）旧北区
white-point		*Mythimna albipuncta* (Denis et Schiffermüller)	（チョウ目、ヤガ科）旧北区
white point		*Sideridis albipuncta* (Leech)	（チョウ目、ヤガ科）日本
white-point wainscot			white-point を見よ
white poplar leafhopper		*Idiocerus cognatus* Fieber	（カメムシ目、オオヨコバイ科）旧北区
white-posted metalmark		*Calicosama lilina* (Butler)	（チョウ目、シジミタテハ科）新熱帯区
white prominent	モンキシロシャチホコ	*Leucodonta bicoloria* (Denis et Schiffermüller)	（チョウ目、シャチホコガ科）日本、旧北区
white prunicola scale		*Pseudaulacaspis prunicola* (Maskell)	（カメムシ目、マルカイガラムシ科）日本
white psychid	コンドウシロミノガ	*Chalioides kondonis* Kondo	（チョウ目、ミノガ科）日本
white punch		*Dodona henrici* Holland	（チョウ目、シジミタテハ科）東洋区
white-pupiled scallop moth			cotton semi-looper を見よ

英名	和名	学名	所属、分布、ほか
white-rayed checkerspot		*Chlosyne ehrenbergii* (Behr)	(チョウ目、タテハチョウ科) 新熱帯区
white-rayed metalmark		*Brachyglenis esthema* C. et R. Felder	(チョウ目、シジミタテハ科) 新熱帯区
white-rayed metalmark		*Hades noctula* Westwood	(チョウ目、シジミタテハ科) 新熱帯区
white-rayed metalmark		*Melanis acroleuca acroleuca* (R. Felder)	(チョウ目、シジミタテハ科) 新熱帯区 *M. a. huasteca* J. White et A. White も同英名
white-rayed patch			white-rayed checkerspot を見よ
white-ribboned carpet moth		*Mesoleuca ruficillata* (Guenée)	(チョウ目、シャクガ科) 新北区
white rice borer		*Maliarpha separatella* Ragonot	(チョウ目、メイガ科) 東洋区、エチオピア区。イネ害虫
white rice borer			white rice stemborer を見よ
white rice borer			yellow rice borer を見よ
white rice leafhopper		*Cofana unimaculatus* (Distant)	(カメムシ目、オオヨコバイ科) 東洋区
white rice leafhopper		*Yasumatsuus mimicus* (Distant)	(カメムシ目、オオヨコバイ科) 東洋区。イネ害虫
white rice leafhopper			rice leafhopper (1) を見よ
white rice stemborer		*Scirpophaga innotata* (Walker)	(チョウ目、メイガ科) 東洋区、豪州区。イネ害虫
white-ringed meadowbrown		*Hyponephele davendra* (Moore)	(チョウ目、タテハチョウ科) 旧北区
white-roped Glaphyria moth		*Glaphyria sequistrialis* Hübner	(チョウ目、メイガ科) 新北区
white rose looper moth			clouded silver を見よ
white round bamboo scale	タケシロマルカイガラムシ	*Odonaspis secreta* (Cockerell)	(カメムシ目、マルカイガラムシ科) 日本、汎世界
white royal		*Pratapa deva* (Moore)	(チョウ目、シジミチョウ科) 東洋区
white royal		*Tajuria illurgis* (Hewitson)	(チョウ目、シジミチョウ科) 東洋区
white sailor		*Dynamine theseus* (C. et R. Felder)	(チョウ目、タテハチョウ科) 新熱帯区
white satin			satin moth (1) (2) を見よ
white satin moth			satin moth (1) を見よ
white satyr		*Pareuptychia ocirrhoe* (Fabricius)	(チョウ目、タテハチョウ科) 新北区、新熱帯区
white scale			lesser snow scale を見よ
white scale			white peach scale を見よ
white-scaled flower weevil		*Odontocorynus denticornis* Casey	(コウチュウ目、ゾウムシ科) 新北区
white-scalloped hairstreak		*Brevianta busa* (Godman et Salvin)	(チョウ目、シジミチョウ科) 新熱帯区
white scrub-hairstreak			white hairstreak を見よ
white-shafted plume		*Pterophorus tridactyla* Linnaeus	(チョウ目、トリバガ科) 旧北区

英名	和名	学名	所属、分布、ほか
white-shouldered house moth		*Endrosis sarcitrella* (Linnaeus)	（チョウ目、マルハキバガ科）全北区、大洋区、豪州区。white-shouldered moth ともいう
white-shouldered longicorn	カタシロゴマフカミキリ	*Mesosa hirsuta hirsuta* Bates	（コウチュウ目、カミキリムシ科）日本
white-shouldered smudge			oak moth を見よ
white-shouldered tubic			white-shouldered house moth を見よ
white slant-line moth		*Tetracis cachexiata* Guenée	（チョウ目、シャクガ科）新北区。white slant ともいう
white small argent		*Argyresthia abdominalis* Zeller	（チョウ目、メムシガ科）旧北区
white snow flea	シロトビムシ	*Onychiurus folsomi* (Schäffer)	（トビムシ目、シロトビムシ科）日本、全北区、豪州区
white speck moth			armyworm (1) を見よ　white speck ともいう
white-speck			armyworm (1) を見よ
white-speck wainscot			armyworm (1) を見よ
white-speckled smoke		*Narycia moniliferella* Villers	（チョウ目、ミノガ科）旧北区
white speck ringlet	シロコモンベニヒカゲ	*Erebia claudina* (Borkhausen)	（チョウ目、タテハチョウ科）旧北区
white spot		*Hadena albimacula* (Borkhausen)	（チョウ目、ヤガ科）旧北区
white spot			one-spotted variant moth を見よ
white-spot coronet			white spot を見よ
white-spot falcon		*Corades medeba* Hewitson	（チョウ目、タテハチョウ科）新熱帯区
white-spot forester		*Bebearia phranza* (Hewitson)	（チョウ目、タテハチョウ科）エチオピア区
white-spot marbled		*Jaspidia pygarga* Hufnagel	（チョウ目、ヤガ科）旧北区
white spot palmer		*Eetion elia* (Hewitson)	（チョウ目、セセリチョウ科）東洋区
white-spot purple		*Eriocrania unimaculella* (Zetterstedt)	（チョウ目、スイコバネガ科）旧北区
white-spot sister		*Adelpha leucophthalma* (Latreille)	（チョウ目、タテハチョウ科）新熱帯区
white spotted sapphire		*Iolaus lulua* (Riley)	（チョウ目、シジミチョウ科）エチオピア区
white-spotted leaf beetle		*Monolepta signata* Olivier	（コウチュウ目、ハムシ科）東洋区
white-spotted Agrias	アミドンミイロタテハ	*Agrias amydon* Hewitson	（チョウ目、タテハチョウ科）新熱帯区
white-spotted beak		*Libythea narina* Godart	（チョウ目、タテハチョウ科）東洋区
white-spotted brown moth		*Diastictis ventralis* (Grote et Robinson)	（チョウ目、メイガ科）新北区
white-spotted bug	シラホシカメムシ	*Eysarcoris ventralis* (Westwood)	（カメムシ目、カメムシ科）日本、東洋区。イネ、ダイズ、カキなどの害虫
white-spotted cankerworm moth		*Paleacrita merriccata* Dyar	（チョウ目、シャクガ科）新北区

英名	和名	学名	所属、分布、ほか
white-spotted clearwing		*Greta annette annette* (Guérin-Méneville)	（チョウ目、タテハチョウ科）新熱帯区
white-spotted commodore		*Precis limnoria* (Klug)	（チョウ目、タテハチョウ科）エチオピア区
white-spotted Emesis		*Emesis aurimna* (Boisduval)	（チョウ目、シジミタテハ科）新熱帯区。
white-spotted eucosmid			dotted cloaked marble (1) を見よ
white-spotted eyed-metalmark		*Mesosemia albipuncta* Schaus	（チョウ目、シジミタテハ科）新熱帯区
white-spotted flash		*Deudorix democles* Miskin	（チョウ目、シジミチョウ科）豪州区
white-spotted flasher		*Astraptes enotrus* (Stoll)	（チョウ目、セセリチョウ科）新熱帯区
white-spotted flower chafer	シラホシハナムグリ	*Protactia brevitarsis brevitarsis* (Lewis)	（コウチュウ目、コガネムシ科）日本
white-spotted globular stink bug	マルシラホシカメムシ	*Eysarcoris guttiger* (Thunberg)	（カメムシ目、カメムシ科）日本、旧北区、東洋区
white-spotted hairstreak		*Euaspa ziha* (Hewitson)	（チョウ目、シジミチョウ科）東洋区
white-spotted hairstreak		*Siderus leucophaeus* (Hübner)	（チョウ目、シジミチョウ科）新熱帯区
white-spotted hairstreak		*Thecla ziha* (Hewitson)	（チョウ目、シジミチョウ科）東洋区
white-spotted Hedya moth			twinspotted budworm を見よ
white-spotted leafroller moth		*Argyrotaenia alisellana* (Robinson)	（チョウ目、ハマキガ科）新北区。幼虫は white-spotted leafroller
white-spotted longhorn beetle			fig longicorn beetle を見よ
white-spotted longicorn beetle	シラホシカミキリ	*Glenea relicta relicta* Pascoe	（コウチュウ目、カミキリムシ科）日本、旧北区、東洋区
white-spotted longicorn beetle (1)	ゴマダラカミキリ	*Anoplophora malasiaca* (Thomson)	（コウチュウ目、カミキリムシ科）日本、旧北区、東洋区。カンキツ、イチジク、チャ、カエデ、ポプラなど多くの果樹、樹木害虫
white-spotted metalmark		*Napaea theages theages* (Godman et Salvin)	（チョウ目、シジミタテハ科）新熱帯区
white-spotted orange moth		*Diastictis argyralis* Hübner	（チョウ目、メイガ科）新北区
white-spotted pinion		*Cosmia diffinis* (Linnaeus)	（チョウ目、ヤガ科）旧北区
white-spotted Prepona			turquoise-banded shoemaker を見よ
white-spotted Properigea moth		*Properigea albimacula* (Barnes et McDunnough)	（チョウ目、ヤガ科）新北区
white-spotted pug		*Eupithecia albipunctata* Hufnagel	（チョウ目、シャクガ科）旧北区
white-spotted pug (1)	シロテンカバナミシャク	*Eupithecia tripunctaria* Herrich-Schäffer	（チョウ目、シャクガ科）日本、全北区
white-spotted Redectis moth		*Redectis vitrea* (Grote)	（チョウ目、ヤガ科）新北区。white-spotted Redectis ともいう

英名	和名	学名	所属、分布、ほか
white-spotted sable	シロモンクロノメイガ	*Anania funebris* (Butler)	(チョウ目、メイガ科) 旧北区。成虫は white-spotted sable moth。日本亜種は *A. f. assimilis* (Butler) と *A. f. astrifera* (Butler)
white-spotted satyr		*Manataria hercyna* (Hübner)	(チョウ目、タテハチョウ科) 新熱帯区
white-spotted sawyer			black pine sawyer を見よ white-spotted sawyer beetle ともいう
white-spotted sister		*Adelpha demialba* (Butler)	(チョウ目、タテハチョウ科) 新熱帯区
white-spotted slender		*Calliste denticulella* (Thunberg)	(チョウ目、ホソガ科) 旧北区
white-spotted small noctuid			marbled white-spot を見よ
white-spotted spined bug	トゲシラホシカメムシ	*Eysarcoris aeneus* Scopoli	(カメムシ目、カメムシ科) 日本、旧北区。イネ、ダイズ、ウリ類などの害虫
white-spotted spined stink bug			white-spotted spined bug を見よ
white-spotted tadpole		*Syrmatia dorilas* (Cramer)	(チョウ目、シジミタテハ科) 新熱帯区。white-spotted tadpole butterfly ともいう
white-spotted truncate-tipped geometrid	モンシロツマキリエダシャク	*Zethenia albonotaria* (Bremer)	(チョウ目、シャクガ科) 日本
white-spotted tussock moth			Japanese tussock moth を見よ
white-spotted woodrush dwarf		*Biselachista trapeziella* Stainton	(チョウ目、クサモグリガ科) 旧北区
white spring moth		*Lomographa vestaliana* (Guenée)	(チョウ目、シャクガ科) 新北区
white springtail		*Folsomia candida* Willem	(トビムシ目、ツチトビムシ科) 日本
white-sprinkled groundling		*Caryocolum fraternella* (Douglas)	(チョウ目、キバガ科) 旧北区
white spruce beetle		*Cryphalus piceae* Ratzeburg	(コウチュウ目、キクイムシ科) 旧北区
white spurwing		*Antigonus emorsus* (Felder)	(チョウ目、セセリチョウ科) 新北区
white-stained oakblue		*Arhopala aida aida* (Hewitson)	(チョウ目、シジミチョウ科) 東洋区
white stem borer			white rice stemborer を見よ
white stem borer			white coffee borer を見よ
white stem borer			South American white borer を見よ
whitestemmed gum moth			giant anthelid を見よ
white-stiched metalmark		*Napaea eucharila picina* Stichel	(チョウ目、シジミタテハ科) 新熱帯区
white-stockinged black fly			buffalo gnat (2) を見よ
white-streaked brown midget			apple leafminer (3) を見よ
white-streaked looper moth		*Plusia venusta* Walker	(チョウ目、ヤガ科) 新北区。white-streaked looper ともいう

英名	和名	学名	所属、分布、ほか
white-streaked prominent moth		*Oligocentria lignicolor* (Walker)	（チョウ目、シャチホコガ科）新北区
white-streaked Saturnia moth		*Saturnia albofasciata* (Johnson)	（チョウ目、ヤママユガ科）新北区
white-streaked sober		*Syncopacma cinctella* (Clerck)	（チョウ目、キバガ科）旧北区
white-striated planthopper		*Nisia atrovenosa* Lethierry	（カメムシ目、シマウンカ科）東洋区。イネ害虫
white-striped Aguna		*Aguna albistria leucogramma* (Mabille)	（チョウ目、セセリチョウ科）新熱帯区
white-striped black moth		*Trichodezia albovittata* (Guenée)	（チョウ目、シャクガ科）新北区
white-striped brown Scoparia moth		*Scoparia dinodes* Meyrick	（チョウ目、メイガ科）豪州区。white-striped brown Scoparia ともいう
white-striped chafer	シロスジコガネ	*Pelyphylla albolineata* (Motschulsky)	（コウチュウ目、コガネムシ科）日本
white-striped groundstreak		*Calycopis clarina* (Hewitson)	（チョウ目、シジミチョウ科）新熱帯区
white-striped huge comma	シロスジトモエ	*Metopia rectifasciata* (Ménétriès)	（チョウ目、ヤガ科）日本、旧北区
white-striped leafhopper	シロセスジヨコバイ	*Scaphoideus albovittatus* Matsumura	（カメムシ目、オオヨコバイ科）日本、旧北区
white-striped longicorn beetle	シロスジカミキリ	*Batocera lineolata* Chevrolat	（コウチュウ目、カミキリムシ科）日本、旧北区、東洋区。ナシ、イチジク、エノキ、ポプラ、カエデなどの害虫。white-striped longicorn ともいう
white-striped longtail		*Chioides catillus* (Cramer)	（チョウ目、セセリチョウ科）新北区
white-striped longtail		*Chioides albofasciatus* (Hewitson)	（チョウ目、セセリチョウ科）新熱帯区
white-striped phytometra	エゾギクキンウワバ	*Ctenoplusia albostriata* (Bremer et Grey)	（チョウ目、ヤガ科）日本、東洋区、豪州区
white-striped prominent	シロジマシャチホコ	*Pheosia fusiformis* Matsumura	（チョウ目、シャチホコガ科）日本
white-striped weevil		*Perperus lateralis* Boisduval	（コウチュウ目、ゾウムシ科）豪州区
white striped planthopper	セスジウンカ	*Terthron albovittatum* (Matsumura)	（カメムシ目、ウンカ科）日本
white striped weevil		*Perperus insularis* Boheman	（コウチュウ目、ゾウムシ科）豪州区。カンキツ害虫
white sugarcane borer		*Scirpophaga auriflua* Zeller	（チョウ目、メイガ科）東洋区
white sugarcane scale			sugarcane scale (1) を見よ
white swan louse		*Ornithobius cygni* (Linnaeus)	（ハジラミ目、チョウカクハジラミ科）新北区
white tail		*Libellula lydia* Drury	（トンボ目、トンボ科）新北区、大洋区。white-tailed skimmer ともいう
white-tail hopper		*Platylesches galesa* (Hewitson)	（チョウ目、セセリチョウ科）エチオピア区
white tail skimmers			red skimmers を見よ

英名	和名	学名	所属、分布、ほか
white-tailed bamboo scale			bamboo scale (2) を見よ
white-tailed bumble bee		*Bombus lucorum* Linnaeus	（ハチ目、ミツバチ科）旧北区
white-tailed diver moth		*Bellura gortynoides* Walker	（チョウ目、ヤガ科）新北区。white-tailed diver ともいう
white-tailed fruitworm moth		*Cenopis albicaudana* (Busck)	（チョウ目、ハマキガ科）新北区
white-tailed hornet			baldfaced hornet を見よ
white-tailed longtail		*Urbanus doryssus chales* (Godman et Salvin)	（チョウ目、セセリチョウ科）新熱帯区
white-tailed longtail			white-tailed skipper を見よ
white-tailed Ridens		*Ridens mercedes* Steinhauser	（チョウ目、セセリチョウ科）新熱帯区
white-tailed skimmer		*Orthetrum albistylum* Selys	（トンボ目、トンボ科）日本、旧北区
white-tailed skipper	シロフチオナガセセリ	*Urbanus doryssus* (Swainson)	（チョウ目、セセリチョウ科）新北区
white-tailed sylph		*Macrothemis pseudimitans* (Calvert)	（トンボ目、トンボ科）新熱帯区
white tarata moth		*Nymphostola galactina* Felder	（チョウ目、マルハキバガ科）豪州区
white Theope		*Theope pieridoides* C. et R. Felder	（チョウ目、シジミタテハ科）新熱帯区
whitethorn grey		*Dipleurina lacustrata* (Panzer)	（チョウ目、メイガ科）旧北区
white tiger			common tiger (1) を見よ
white tiger moth	オビヒトリ	*Spilarctia subcarnea* Walker	（チョウ目、ヒトリガ科）日本、旧北区、東洋区
white tiger swallowtail			pale swallowtail を見よ
white tip borer			top borer (1) を見よ
white-tip clothes moth			carpet moth を見よ
white-tip moth			carpet moth を見よ
white-tipped baron		*Euthalia merta* (Moore)	（チョウ目、タテハチョウ科）東洋区
white-tipped black moth		*Melanochroia chephise* (Cramer)	（チョウ目、シャクガ科）新北区
white-tipped blue		*Eicochrysops hippocrates* (Fabricius)	（チョウ目、シジミチョウ科）エチオピア区
white-tipped eucosmid	オオウスツヤヒメハマキ	*Hedya semiassana* (Kennel)	（チョウ目、ハマキガ科）日本、旧北区
white-tipped hairstreak		*Atlides gaumeri* (Godman)	（チョウ目、シジミチョウ科）新熱帯区
white-tipped lineblue		*Prosotas noreia* (Felder)	（チョウ目、シジミチョウ科）東洋区
white-tipped long-leg weevil	オジロアシナガゾウムシ	*Mesalcidodes trifidus* (Pascoe)	（コウチュウ目、ゾウムシ科）日本、東洋区
white-tipped looper moth	ツマジロエダシャク	*Trigonoptila latimarginaria* (Leech)	（チョウ目、シャクガ科）日本、旧北区、東洋区

英名	和名	学名	所属、分布、ほか
white-tipped metalmark		*Melanis acroleuca huasteca* J. White et A. White	(チョウ目、シジミタテハ科) 新熱帯区
white-tipped metalmark		*Melanis cephise* (Ménétriès)	(チョウ目、シジミタテハ科) 新熱帯区
white-tipped Phanus		*Phanus albiapicalis* Austin	(チョウ目、セセリチョウ科) 新熱帯区
white-tipped spreadwing		*Lestes elatus* Hagen	(トンボ目、アオイトトンボ科) 東洋区
white-trailed skipper		*Gindanes brebisson panaetius* Godman et Salvin	(チョウ目、セセリチョウ科) 新熱帯区
white tree crickets	カンタン科	Oecanthidae	(バッタ目) の昆虫の総称
white tree crickets			long-horned grasshoppers (3) を見よ
white-triangle Tortrix moth		*Clepsis persicana* (Fitch)	(チョウ目、ハマキガ科) 新北区。white-triangle Tortrix ともいう
white-trimmed brown pyralid moth		*Abegesta remellalis* (Druce)	(チョウ目、メイガ科) 新北区
white-tufted button	トサカハマキ	*Acleris cristana* (Denis et Schiffermüller)	(チョウ目、ハマキガ科) 日本、旧北区。リンゴ、バラ、カシ類などの害虫
white-tufted leaf sitter		*Gorgyra heterochrus* (Mabille)	(チョウ目、セセリチョウ科) エチオピア区
white tussock moth		*Laelia monoscola* Collenette	(チョウ目、ドクガ科) 旧北区
white underwing moth		*Catocala relicta* Walker	(チョウ目、ヤガ科) 新北区。white underwing ともいう
white-unibanded geometrid	シロホソオビクロナミシャク	*Baptria tibiale aterimma* (Butler)	(チョウ目、シャクガ科) 日本
white-unispotted yellowish geometrid			brindled white-spot (1) を見よ
white-vein skipper			Uncas skipper を見よ
white-veined arctic		*Oeneis taygete* Geyer	(チョウ目、タテハチョウ科) 新北区
white-veined arctic			arctic grayling を見よ
white-veined dagger		*Simyra henrici* (Grote)	(チョウ目、ヤガ科) 新北区
white-veined leafhopper	シロミャクイチモンジヨコバイ	*Paramesodes albinervosus* (Matsumura)	(カメムシ目、オオヨコバイ科) 日本、東洋区
white-veined sand-skipper			white-veined skipper を見よ
white-veined skipper		*Anisynta albovenata* Waterhouse	(チョウ目、セセリチョウ科) 豪州区
white wainscot		*Leucania l-album* Linnaeus	(チョウ目、ヤガ科) 旧北区
white walnut longicorn beetle	タカサゴシロカミキリ	*Olenecamptus formosanus* Pic	(コウチュウ目、カミキリムシ科) 日本、東洋区。white walnut longicorn ともいう
whitewater darner		*Austroaeschna inermis* Martin	(トンボ目、ヤンマ科) 豪州区
white waved carpet			small white wave (1) を見よ
white waved silver			common white wave を見よ
white wax parasite		*Anicetus communis* (Annecke)	(ハチ目、トビコバチ科) 豪州区

英名	和名	学名	所属、分布、ほか
white wax parasite		*Paraceraptrocerus nyasicus* (Compere)	(ハチ目、トビコバチ科) 豪州区
white wax parasite		*Tetrastichus ceroplastae* (Girault)	(ハチ目、ヒメコバチ科) 豪州区
white wax scale		*Ceroplastes destructor* Newstead	(カメムシ目、カタカイガラムシ科) 豪州区、エチオピア区
white wax scale			Chinese wax scale (1) を見よ
white waxy scale		*Gascardia brevicauda* (Hall)	(カメムシ目、カタカイガラムシ科) エチオピア区
white waxy scale			white wax scale を見よ
white-whiskered grasshopper		*Ageneotettix deorum* (Scudder)	(バッタ目、バッタ科) 新北区
white-winged forest sylph		*Ceratrichia nothus* (Fabricius)	(チョウ目、セセリチョウ科) エチオピア区
white-winged ghost moth			ghost moth を見よ
white-winged march fly		*Bibio albipennis* Say	(ハエ目、ケバエ科) 新北区
white witch			giant agrippa を見よ
white witch moth			giant agrippa を見よ
white yellow		*Eurema albula celata* (R. Felder)	(チョウ目、シロチョウ科) 新熱帯区。white grass yellow を参照
whitened bluewing	キアニスルリツヤタテハ	*Myscelia cyaniris* Doubleday	(チョウ目、タテハチョウ科) 新熱帯区
whitened crescent		*Janatella leucodesma* (Felder et Felder)	(チョウ目、タテハチョウ科) 新熱帯区
whitened eyed-metalmark		*Mesosemia zonalis* Godman et Salvin	(チョウ目、シジミタテハ科) 新熱帯区
whitened flasher		*Astraptes creteus crana* Evans	(チョウ目、セセリチョウ科) 新熱帯区
whites	モンシロチョウ亜科	Pierinae	(チョウ目、シロチョウ科) の昆虫の総称
whites	モンシロチョウ属	*Pieris*	(チョウ目、シロチョウ科) の昆虫の総称
White's Sicilian stick-insect		*Bacillus whitei* Nascetti et Bullini	(ナナフシ目、コブナナフシ科) 旧北区
whitish-striped looper moth	ギンスジエダシャク	*Chariaspilates formosaria* (Eversmann)	(チョウ目、シャクガ科) 日本、旧北区
whitish tailed geometrid			tailed geometrid (1) を見よ
Whitmer's grass-veneer moth		*Crambus whitmerellus* Klots	(チョウ目、メイガ科) 新北区。Whitmer's grass-veneer ともいう
Whitney's underwing moth		*Catocala whitneyi* Dodge	(チョウ目、ヤガ科) 新北区。Whitney's underwing ともいう
whorl maggot			rice whorl maggot (1) を見よ
whortleberry bell	ミヤマカギバヒメハマキ	*Ancylis myrtillana* (Treitschke)	(チョウ目、ハマキガ科) 日本、旧北区
whortleberry pistol case			scar large case を見よ

英名	和名	学名	所属、分布、ほか
Wichgraf's brown		*Stygionympha wichgrafi* van Son	（チョウ目、タテハチョウ科）エチオピア区
Wicken loosestrife neb		*Monochroa conspersella* (Herrich-Schäffer)	（チョウ目、キバガ科）旧北区
wicker-work obscure		*Brachmia blandella* (Fabricius)	（チョウ目、キバガ科）旧北区
Wickham's calligrapha		*Calligrapha wickhami* Bowditch	（コウチュウ目、ハムシ科）新北区
wide-armed mantis		*Cilnia humeralis* Saussure	（カマキリ目、カマキリ科）エチオピア区
wide-banded commodore	ベニオビタテハモドキ	*Precis sinuata* Plötz	（チョウ目、タテハチョウ科）エチオピア区
wide-bordered satyr		*Satyrotaygetis satyrina* (Bates)	（チョウ目、タテハチョウ科）新北区、新熱帯区
wide-brand crow		*Euploea netscheri* Snellen	（チョウ目、タテハチョウ科）豪州区
wide-brand grass dart		*Suniana sunias* Felder	（チョウ目、セセリチョウ科）東洋区
widespread bent-skipper		*Cycloglypha thrasibulus thrasibulus* (Fabricius)	（チョウ目、セセリチョウ科）新熱帯区
widespread glory		*Myscelus amystis* (Hewitson)	（チョウ目、セセリチョウ科）新熱帯区
widespread Myscelus		*Myscelus amystis hages* Godman et Salvin	（チョウ目、セセリチョウ科）新熱帯区
widespread Phanus			ant-bird skipper を見よ
wide-stripe grass-veneer moth		*Crambus unistrialellus* Packard	（チョウ目、メイガ科）新北区
wide-striped leafroller			Allen's Tortrix moth を見よ
widow		*Libellula luctuosa* Burmeister	（トンボ目、トンボ科）新北区
widow skimmer			widow を見よ
widow sphinx		*Acanthosphinx guessfeldti* (Dewitz)	（チョウ目、スズメガ科）エチオピア区
widow underwing moth		*Catocala vidua* (J. E. Smith)	（チョウ目、ヤガ科）新北区
widows		*Palpopleura*	（トンボ目、トンボ科）の昆虫の総称
wiggler			mosquitoes (1) を見よ
Wijaya's scissortail		*Microgomphus wijaya* Lieftinck	（トンボ目、サナエトンボ科）東洋区
wild carrot wasp		*Gasteruption assectator* (Linnaeus)	（ハチ目、コンボウヤセバチ科）新北区
wild cherry moth			spotted Apatelodes moth を見よ
wild cherry sphinx		*Sphinx drupiferarum* (J. E. Smith)	（チョウ目、スズメガ科）全北区。wild cherry sphinx moth ともいう
wild-flying piercer		*Cydia orobana* Treitschke	（チョウ目　ハマキガ科）旧北区
wild forget-me not moth			Sierran pericopid を見よ
wild indigo duskywing		*Erynnis baptisiae* Forbes	（チョウ目、セセリチョウ科）新北区

英名	和名	学名	所属、分布、ほか
wild mulberry silkmoth	クワコ	*Bombyx mandarina* (Moore)	（チョウ目、カイコガ科）日本、旧北区
wild parsnip aphid		*Cavariella pastinacae* (Linnaeus)	（カメムシ目、アブラムシ科）旧北区
wild service aphid		*Dysaphis aucupariae* (Buckton)	（カメムシ目、アブラムシ科）豪州区
wild silk moth			giant silk moth を見よ
wild silkworm		*Epanaphe moloneyi* Druce	（チョウ目、シャチホコガ科）エチオピア区
wild silkworm		*Epanaphe vuilleti* de Joanis	（チョウ目、シャチホコガ科）エチオピア区
wild silkworm			wild mulberry silkmoth を見よ
wild silkworms			giant silkworm moths を見よ
wild walnut fly		*Rhagoletis suavis* (Loew)	（ハエ目、ミバエ科）新北区
Wilkinson's case		*Coleophora orbitella* Zeller	（チョウ目、ツツミノガ科）旧北区
Wilkinson's drill		*Dichrorampha consortana* (Stephens)	（チョウ目、ハマキガ科）旧北区
Williams' plume moth		*Platyptilia williamsi* Grinnell	（チョウ目、トリバガ科）新北区
Williams' tiger moth		*Apantesis williamsii* (Dudges)	（チョウ目、ヒトリガ科）新北区
Williamson's Hawaiian damselfly		*Megalagrion williamsoni* (Perkins)	（トンボ目、イトトンボ科）大洋区
Williston's bee fly		*Poecilanthrax willistoni* (Coquillett)	（ハエ目、ツリアブ科）新北区、新熱帯区
Willmott's admiral		*Hypanartia trimaculata* (Lamas, Willmott et Hall)	（チョウ目、タテハチョウ科）新熱帯区
willow aphid		*Aphis salicariae* Koch	（カメムシ目、アブラムシ科）新北区
willow aphid			small willow aphid を見よ
willow aphid			celery aphid (1) を見よ
willow aphid			large willow aphid を見よ
willow aphid			carrot aphid (1) を見よ
willow aphids		*Cavariella*	（カメムシ目、アブラムシ科）の昆虫の総称
willow barkhopper	ヤナギカワウンカ	*Andes marmoratus* (Uhler)	（カメムシ目、ヒシウンカ科）日本、旧北区
willow beaked-gall midge	ヤナギマルタマバエ	*Rhabdophaga rigidae* (Osten Sacken)	（ハエ目、タマバエ科）日本、新北区。ヤナギ害虫
willow-beaker gall midge			willow beaked-gall midge を見よ
willow bean-gall sawfly		*Pontania proxima* (Lepeletier)	（ハチ目、ハバチ科）全北区
willow beauty			Camberwell beauty を見よ
willow beauty moth		*Peribatodes rhomboidaria* (Denis et Schiffermüller)	（チョウ目、シャクガ科）旧北区。willow beauty ともいう
willow beetle	ヤナギシリジロゾウムシ	*Cryptorhynchus lapathi* (Linnaeus)	（コウチュウ目、ゾウムシ科）日本、全北区ヤナギ害虫
willow beetle			yellow willow leaf beetle を見よ
willow beetle			brassy willow leaf beetle を見よ
willow black minute weevil	クロノミゾウ	*Rhynchaenus stigma* (Germar)	（コウチュウ目、ゾウムシ科）日本、旧北区

英名	和名	学名	所属、分布、ほか
willow blue leaf beetle			willow leaf beetle (1) を見よ
willow-bog fritillary			Freya's fritillary を見よ
willow borer			willow beetle を見よ
willow borer			willow longhorn を見よ
willow-boring sawfly			willow wood wasp を見よ
willow bud sawfly	ヤナギメフシハバチ	*Euura mucronata* (Hartig)	(ハチ目、ハバチ科) 日本、旧北区
willow butterfly			Camberwell beauty を見よ
willow button-top midge		*Rhabdophaga heterobia* (Low)	(ハエ目、タマバエ科) 旧北区
willow cambium miner		*Phytobia barnesi* Hendel	(ハエ目、ハモグリバエ科) 旧北区
willow carrot aphid			carrot aphid (1) を見よ
willow clearwing moth		*Synanthedon sigmoidea* (Beutenmüller)	(チョウ目、スカシバガ科) 新北区
willow cone gall			willow stem gall を見よ
willow cottony scale	ヤナギワタカイガラムシ	*Pulvinaria oyamae* Kuwana	(カメムシ目、カタカイガラムシ科) 日本
willow curculio	ヤナギルリチョッキリ	*Neocoenorrhinus interruptus* (Voss)	(コウチュウ目、ゾウムシ科) 日本
willow cutworm			double kidney を見よ
willow dagger moth			impressed dagger moth を見よ
willow dart moth		*Cerastis salicarum* (Walker)	(チョウ目、ヤガ科) 新北区
willow emerald damselfly		*Lestes viridis* (Van der Linden)	(トンボ目、アオイトトンボ科) 旧北区
willow ermine moth		*Yponomeuta rorellus* (Hübner)	(チョウ目、スガ科) 旧北区。willow ermine ともいう
willow eucosmid	ヤナギコハマキ	*Gypsonoma bifasciata* Kuznetzov	(チョウ目、ハマキガ科) 日本、旧北区。ヤナギ害虫
willow fairy moth		*Adela purpurea* Walker	(チョウ目、マガリガ科) 新北区
willow flea beetle		*Chalcoides aurata* (Marsham)	(コウチュウ目、ハムシ科) 旧北区
willow flea beetle (1)	ヤナギノミゾウムシ	*Orchestes salicis* (Linnaeus)	(コウチュウ目、ゾウムシ科) 日本、旧北区
willow flea weevil		*Rhynchaenus rufipes* (LeConte)	(コウチュウ目、ゾウムシ科) 新北区
willow flea weevil			willow flea beetle (1) を見よ
willow flies			rolled-winged stoneflies を見よ
willow fly		*Nemoura cinerea* (Retzius)	(カワゲラ目、オナシカワゲラ科) 旧北区
willow fly			yellow sally を見よ
willow froghopper		*Aphrophora salicina* (Goeze)	(カメムシ目、アワフキムシ科) 旧北区
willow froghopper (1)	マエキアワフキ	*Aphrophora pectoralis* Matsumura	(カメムシ目、アワフキムシ科) 日本、旧北区。ヤナギ害虫

英名	和名	学名	所属、分布、ほか
willow gall inquiline moth		*Sciota carneella* (Hulst)	（チョウ目、メイガ科）新北区
willow gall moth		*Cydia gallaesaliciana* (Riley)	（チョウ目、ハマキガ科）新北区
willow gall sawflies		*Pontania*	（ハチ目、ハバチ科）の昆虫の総称
willow gall sawfly	シバヤナギコブハバチ	*Pontania shibayanagii* (Togashi et Usuba)	（ハチ目、ハバチ科）日本。ヤナギ害虫
willow gall weevil		*Archarius salicivorus* (Paykull)	（コウチュウ目、ゾウムシ科）旧北区
willow geometrid	アトボシエダシャク	*Cephis advenaria* (Hübner)	（チョウ目、シャクガ科）日本、旧北区
willow ghost moth		*Sthenopis thule* (Strecker)	（チョウ目、コウモリガ科）新北区
willow Gypsonoma moth		*Gypsonoma salicicolana* (Clemens)	（チョウ目、ハマキガ科）新北区
willow hawk moth (1)		*Smerinthus ocellata* (Linnaeus)	（チョウ目、スズメガ科）旧北区
willow hawk moth (2)		*Smerinthus cerisyi* (Kirby)	（チョウ目、スズメガ科）新北区
willow hawk moth			willow hawk moth (1) を見よ
willowherb hawk-moth		*Proserpinus proserpina* Pallas	（チョウ目、スズメガ科）旧北区
willow herb looper			small phoenix を見よ
willow hornworm	ヒメウチスズメ	*Smerinthus caecus* Ménétriès	（チョウ目、スズメガ科）日本、旧北区
willow kitten			western Furcula moth を見よ
willow lace bug	ヤナギグンバイ	*Metasalis populi* Takeya	（カメムシ目、グンバイムシ科）日本、旧北区、東洋区。ヤナギ害虫
willow leaf beetle		*Chrysomela lapponica* Linnaeus	（コウチュウ目、ハムシ科）新北区
willow leaf beetle (1)	ヤナギルリハムシ	*Plagiodera versicolora* (Laicharting)	（コウチュウ目、ハムシ科）日本、全北区。ヤナギ害虫
willow leaf beetle			elm tree beetle を見よ
willow leafblotch miner moth		*Micrurapteryx salicifoliella* (Chambers)	（チョウ目、ホソガ科）新北区
willow leaf blotch miner moth			long-streaked sallow midget を見よ
willow leaf-folding midge		*Rhabdophaga clausilia* (Bremi)	（ハエ目、タマバエ科）旧北区
willow leaf-folding sawfly	カワラヤナギハバチ	*Pontania vesicator* (Bremi)	（ハチ目、ハバチ科）日本、旧北区
willow leafgall sawfly		*Euura pacifica* Marlatt	（ハチ目、ハバチ科）新北区
willow leafhopper	ヤナギヒメヨコバイ	*Empoasca smaragdulus* (Fallén)	（カメムシ目、オオヨコバイ科）日本、旧北区
willow leafminer	ヤナギハモグリバエ	*Paraphytomyza populi* (Kaltenbach)	（ハエ目、ハモグリバエ科）日本、旧北区。ヤナギ害虫
willow longhorn		*Xylotrechus insignis* LeConte	（コウチュウ目、カミキリムシ科）新北区
willow looper	ウチジロナミシャク	*Dysstroma truncata fusconebulosa* Inoue	（チョウ目、シャクガ科）日本

英名	和名	学名	所属、分布、ほか
willow minute weevil	ヤナギノミゾウムシ	*Rhynchaenus salicis* (Linnaeus)	(コウチュウ目、ゾウムシ科) 日本、旧北区
willow moth			satin moth (1) を見よ
willow nestmaker		*Ichthyura apicalis* Walker	(チョウ目、シャチホコガ科) 新北区
willow noctuid			angle-striped sallow を見よ
willow oystershell scale	ヤナギカキカイガラムシ	*Lepidosaphes yanagicola* Kuwana	(カメムシ目、マルカイガラムシ科) 日本、旧北区
willow pea-gall sawfly		*Pontania viminalis* (Linnaeus)	(ハチ目、ハバチ科) 旧北区。本種の虫えいを willow pea-gall という
willow phylloxera			black willow aphid を見よ
willow pinecone gall midge		*Rhabdophaga strobiloides* (Osten Sacken)	(ハエ目、タマバエ科) 新北区
willow plant-louse			black willow aphid を見よ
willow prominent	キエグリシャチホコ	*Himeropteryx miraculosa* Staudinger	(チョウ目、シャチホコガ科) 日本、旧北区、東洋区
willow prominent			puss moth を見よ
willow psylla	ヤナギキジラミ	*Psylla ambigua* Foerster	(カメムシ目、キジラミ科) 日本、旧北区
willow red-gall sawfly			willow bean-gall sawfly を見よ
willow red-spotted sawfly	ヤナギアカモンハバチ	*Pristiphora salicivora* (Takeuchi)	(ハチ目、ハバチ科) 日本
willow rosette-gall midge			European rosette willow gall midge を見よ
willow sawfly	セグロヤナギハバチ	*Amauronematus fallax* (Lepeletier)	(ハチ目、ハバチ科) 日本、旧北区。ヤナギ害虫
willow sawfly		*Pteronidea salicis* (Linnaeus)	(ハチ目、ハバチ科) 旧北区
willow sawfly	ヤナギチュウレンジバチ	*Arge enodis* Linnaeus	(ハチ目、ミフシハバチ科) 日本、旧北区
willow sawfly		*Nematus salicis* (Linnaeus)	(ハチ目、ハバチ科) 旧北区
willow sawfly (1)		*Nematus ventralis* Say	(ハチ目、ハバチ科) 旧北区、新北区
willow sawfly (2)	キアシヒゲナガハバチ	*Nematus crassus* (Fallén)	(ハチ目、ハバチ科) 日本、旧北区。ヤナギ害虫
willow scale (1)		*Chionaspis salicis* (Linnaeus)	(カメムシ目、マルカイガラムシ科) 全北区
willow scale (2)	ヤナギシロカイガラムシ	*Chionaspis salicisnigrae* (Walsh)	(カメムシ目、マルカイガラムシ科) 日本、新北区。ヤナギ害虫
willow shoot moth		*Agonopterix conterminella* (Zeller)	(チョウ目、マルハキバガ科) 旧北区
willow shoot sawfly		*Janus abbreviatus* (Say)	(ハチ目、クキバチ科) 新北区
willow shot-hole midge		*Rhabdophaga justini* Barnes	(ハエ目、タマバエ科) 旧北区
willow shot-hole midge		*Rhabdophaga triandraperda* Barnes	(ハエ目、タマバエ科) 旧北区
willow shot-hole midge (1)		*Rhabdophaga saliciperda* (Dufour)	(ハエ目、タマバエ科) 旧北区
willow shot-hole midge			shot hole midge を見よ

英名	和名	学名	所属、分布、ほか
willow sphinx			willow hawk moth (2) を見よ
willow spittlebug			willow froghopper (1) を見よ
willow stem aphid			large willow aphid を見よ
willow stem gall		*Rhabdophaga salicisbatatas* (Osten Sacken)	（ハエ目、タマバエ科）の虫えい。新北区
willow stem sawfly		*Janus luteipes* (Lepeletier)	（ハチ目、クキバチ科）旧北区
willow sucker	ヤナギトガリキジラミ	*Bactericera salicivora* (Reuter)	（カメムシ目、トガリキジラミ科）日本、旧北区。ヤナギ害虫
willow sulphur			Scudder's willow sulphur を見よ
willow tail prominent	ナカグロモクメシャチホコ	*Furcula furcula lanigera* (Butler)	（チョウ目、シャチホコガ科）日本、旧北区。sallow kitten (1) も参照
willow terminal leaf midge		*Rhabdophaga terminalis* (Loew)	（ハエ目、タマバエ科）旧北区
willow tortrix moth	ミヤマヤナギヒメハマキ	*Epinotia cruciana* (Linnaeus)	（チョウ目、ハマキガ科）日本、全北区。willow tortrix ともいう
willow-tree midge			Japanese beech gall midge を見よ
willow two-tailed aphid			salix black hairy aphid を見よ
willow weevil			willow beetle を見よ
willow white scale			willow scale (2) を見よ
willow white-tipped eucosmid			sallow marble を見よ
willow wood midge			willow shot-hole midge (1) を見よ
willow wood wasp		*Xiphydria prolongata* (Geoffroy)	（ハチ目、クビナガキバチ科）旧北区
willowherb hawk-moth		*Proserpinus proserpina* Pallas	（チョウ目、スズメガ科）旧北区
Wilson sphinx		*Hyles wilsoni* (Rothschild)	（チョウ目、スズメガ科）新北区、大洋区。Wilson's sphinx ともいう
Wilson's ciliate blue		*Anthene wilsoni* (Talbot)	（チョウ目、シジミチョウ科）エチオピア区
Wilson's groundling		*Brachythemis wilsoni* Pinhey	（トンボ目、トンボ科）エチオピア区
Wilson's wide-headed ambrosia beetle		*Platypus wilsoni* Swaine	（コウチュウ目、ナガキクイムシ科）新北区
Wilson's wood-nymph moth		*Xerociris wilsonii* Grote	（チョウ目、ヤガ科）新北区
Wimmer's bell		*Eucosma lacteana* (Treitschke)	（チョウ目、ハマキガ科）旧北区
wind river fritillary		*Clossiana improba harryi* Ferris	（チョウ目、タテハチョウ科）新北区
window Acraea			Roibok Acraea を見よ
window flies	マドギワアブ科	Scenopinidae	（ハエ目）の昆虫の総称
window flies			wood gnats を見よ　豪州で使用
window fly			house windowfly を見よ
window midge			wood gnat を見よ

英名	和名	学名	所属、分布、ほか
window midges			wood gnats を見よ 英国で使用
window-midge			house windowfly を見よ
window-winged moth		*Meskea dyspteraria* Grote	(チョウ目、マドガ科) 新北区
window-winged moths	マドガ科	Thyrididae	(チョウ目) の昆虫の総称
window-winged skipper			glassy-winged skipper を見よ
Wind's gemmed-satyr		*Cyllopsis windi* Miller	(チョウ目、タテハチョウ科) 新熱帯区
Wind's silverdrop		*Epargyreus windi* Freeman	(チョウ目、セセリチョウ科) 新熱帯区
Wind's skipper		*Windia windi* Freeman	(チョウ目、セセリチョウ科) 新熱帯区
wine-bottle corks			yellow V carl を見よ
wine cork moth			European grain moth を見よ
wine flies			vinegar flies (1) を見よ
wine fly			pomace fly を見よ
wineland blue		*Lepidochrysops bacchus* Riley	(チョウ目、シジミチョウ科) エチオピア区
wine-tinted Oenobotys moth		*Oenobotys vinotinctalis* (Hampson)	(チョウ目、メイガ科) 新北区
wing louse			chicken wing louse を見よ
wing louse of poultry			chicken wing louse を見よ
winged bush crickets			bush crickets (3) を見よ winged-bush crickets とも記す
winged euonymous scale			willow oystershell scale を見よ
winged flies			flies を見よ
winged insects	有翅亜綱	Pterygota	の昆虫の総称
winged walking sticks			walking sticks (1) を見よ 有翅型の英名
wingless bat-flies			bat flies (2) を見よ
wingless camel crickets			camel crickets (1) を見よ
wingless cockroach		*Calolampra elegans* Roth et Princis	(ゴキブリ目、ブラベルスゴキブリ科) 豪州区
wingless cockroach		*Calolampra solida* Roth et Princis	(ゴキブリ目、ブラベルスゴキブリ科) 豪州区
wingless flies			bat flies (1) を見よ
wingless grasshopper		*Phaulacridum vittatum* (Sjöstedt)	(バッタ目、バッタ科) 豪州区
wingless grasshopper		*Paraidemona mimica* Scudder	(バッタ目、バッタ科) 大洋区
wingless hangingfly		*Apterobittacus apterus* (McLachlan)	(シリアゲムシ目、ガガンボモドキ科) 新北区
wingless long-horned grasshoppers			cave crickets (1) を見よ

英名	和名	学名	所属、分布、ほか
wingless meadow katydid		*Odontoxiphidium apterum* Morse	(バッタ目、キリギリス科) 新北区
wingless mountain grasshopper		*Zubouskya glacialis* (Scudder)	(バッタ目、バッタ科) 新北区。山地の草地や樹上でみつかる
wingless prickly pear longicorn		*Moneilema ulkei* Horn	(コウチュウ目、カミキリムシ科) 豪州区
wingless scorpionfly			wingless hangingfly を見よ
wingless short-horned grasshoppers			monkey grasshoppers (1)(2) を見よ
wingless soldier fly		*Boreoides subulatus* Hardy	(ハエ目、ミズアブ科) 豪州区
wingless weevil		*Otiorhynchus dieckmanni* Magnano	(コウチュウ目、ゾウムシ科) 旧北区
winter cherry bug	ホオズキカメムシ	*Acanthocoris sordidus* (Thunberg)	(カメムシ目、ヘリカメムシ科) 日本、東洋区。ジャガイモ、ナス、トマト、ナシ、タバコなどの害虫
winter corbie			underground grassgrub を見よ
winter crane flies	ガガンボダマシ科	Trichoceridae	(ハエ目) の昆虫の総称
winter crane fly			winter gnat (1) を見よ 米国での英名
winter cutworm			cutworm (2) を見よ 米国での幼虫の英名
winter damselfly		*Sympycna fusca* (Van der Linden)	(トンボ目、アオイトトンボ科) 旧北区
winter egg			bark aphids を見よ 卵の英名
winter form pear psylla			pear sucker を見よ
winter gnat	オナガガガンボダマシ	*Trichocera regelarionis* (Linnaeus)	(ハエ目、ガガンボダマシ科) 日本、全北区
winter gnat		*Trichocera annulata* Meigen	(ハエ目、ガガンボダマシ科) 全北区
winter gnat (1)	フユガガンボダマシ	*Trichocera hiemalis* (De Geer)	(ハエ目、ガガンボダマシ科) 日本、旧北区
winter gnats		*Petaurista*	(ハエ目、ガガンボダマシ科) の昆虫の総称
winter gnats			winter crane flies を見よ
winter midges			winter crane flies を見よ
winter moth	アミスジフユナミシャク	*Operophtera brumata* (Linnaeus)	(チョウ目、シャクガ科) 日本、全北区。成虫が冬に出現
winter moth			lucerne looper を見よ
winter stem weevil			black winter weevil を見よ
winter stoneflies	ミジカオカワゲラ科	Taeniopterygidae	(カワゲラ目) の昆虫の総称
winter stoneflies			smoky stoneflies を見よ
winter turnip gnat			winter gnat (1) を見よ
winter webworm		*Ocnogyna baetica* Rambur	(チョウ目、ヒトリガ科) 旧北区
Winton's ladybug beetle		*Allenius iviei* Escalona et Slipinski	(コウチュウ目、テントウムシ科) 新北区
wireworm			common click beetle (1) を見よ
wireworm			obscure click beetle を見よ
wireworm			click beetle (2) を見よ

英名	和名	学名	所属、分布、ほか
wireworms			click beetles (1) を見よ
Wirth hazel dagger moth		*Acronicta hamamelis* Guenée	（チョウ目、ヤガ科）新北区
wise Rajah	ホソオビフタオチョウ	*Charaxes solon* (Fabricius)	（チョウ目、タテハチョウ科）東洋区
wisps		*Agriocnemis*	（トンボ目、イトトンボ科）の昆虫の総称
wisteria bud miner	フジタマモグリバエ	*Hexomyza websteri* (Malloch)	（ハエ目、ハモグリバエ科）日本、新北区。フジ害虫
wisteria flower-bud midge	フジツボミタマバエ	*Dasineura wistariae* (Mani)	（ハエ目、タマバエ科）日本。フジ害虫
wisteria leafminer	フジハモグリバエ	*Agromyza wistariae* Sasakawa	（ハエ目、ハモグリバエ科）日本。フジ害虫
wisteria longicorn beetle	ホソキリンゴカミキリ	*Oberea vittata* Blessig	（コウチュウ目、カミキリムシ科）日本、旧北区。フジ害虫。wisteria longicorn ともいう
wisteria scurfy scale	フジシロナガカイガラムシ	*Chionaspis wistariae* Cooley	（カメムシ目、マルカイガラムシ科）日本、新北区。フジ害虫
witch		*Araotes lapithis* (Moore)	（チョウ目、シジミチョウ科）東洋区。*A. l. uruwela* Fruhstorfer も同英名
witch			clothes moth (1) を見よ
witch hazel cone gall aphid		*Harnaphis hamamelidis* (Fitch)	（カメムシ目、アブラムシ科）新北区
witch hazel leaf-gall aphid			witch hazel cone gall aphid を見よ
witch moth			black witch を見よ
witchelly grub		*Endoxyla leucomochla* (Turner)	（チョウ目、ボクトウガ科）豪州区
witches			moths を見よ
witchetty grub			witjuti grub を見よ
withered Mocis moth		*Mocis marcida* (Guenée)	（チョウ目、ヤガ科）新北区。withered Mocis ともいう
witjuti grub		*Xyleutes cinereus* (Tepper)	（チョウ目、ボクトウガ科）豪州区。アボリジニが witjuti tree (*Acacia kempeana*) につく幼虫を食べる
Wittfeld's forester moth		*Alypia wittfeldii* H. Edwards	（チョウ目、ヤガ科）新北区。Wittfeldt's forester ともいう
wizard	ソトグロカバタテハ	*Rhinopalpa polynice* (Cramer)	（チョウ目、タテハチョウ科）東洋区。*R. p. eudoxia* Guérin-Méneville も同英名
woad aphid		*Acyrthosiphon genistae* Mordvilko	（カメムシ目、アブラムシ科）旧北区
wolf moth			European grain moth を見よ
Wolkberg sandman		*Spialia secessus* (Trimen)	（チョウ目、セセリチョウ科）エチオピア区
Wolkberg widow		*Dingana clara* (van Son)	（チョウ目、タテハチョウ科）エチオピア区
Wolkberg zulu		*Alaena margaritacea* Eltringham	（チョウ目、シジミチョウ科）エチオピア区
Wollaston's forest sylph		*Ceratrichia wollastoni* Heron	（チョウ目、セセリチョウ科）エチオピア区
wonder brown		*Heteronympha mirifica* (Butler)	（チョウ目、タテハチョウ科）豪州区。wonder brown butterfly ともいう

英名	和名	学名	所属、分布、ほか
wonderful hairstreak	ヒマラヤミドリシジミ	*Thermozephyrus ataxus* (Westwood)	(チョウ目、シジミチョウ科) 東洋区。日本亜種は *T. ataxus kirishimaensis* (Okajima) キリシマミドリシジミ
wonderful underwing moth		*Catocala mira* Grote	(チョウ目、ヤガ科) 新北区。wonderful underwing ともいう
wondrous Epitola		*Cerautola miranda* Staudinger	(チョウ目、シジミチョウ科) エチオピア区
wood and pool mosquito		*Ochlerotatus canadensis* (Theobald)	(ハエ目、カ科) 新北区
wood ant			common red ant を見よ
wood ant			herculean ant (1) を見よ
wood-ant clothes		*Myrmecozela ochraceella* (Tengstrom)	(チョウ目、ヒロズコガ科) 旧北区
wood-ant clothes moth			wood-ant clothes を見よ
wood ants			carpenter ants を見よ
wood ants			termites を見よ
wood bees			carpenter bees を見よ
wood blue			holly blue を見よ　米国での英名
wood borer			樹木の木質部に穴をあける各種昆虫の幼虫。テッポウムシ (カミキリムシの幼虫) など
wood borer beetles			long-horned beetles を見よ
wood-borer moths		*Indarbela*	(チョウ目、Metarbelidae) の昆虫の総称
wood-borer moths			tropical carpenterworm moths を見よ
wood borer parasites			stephanid wasps を見よ　南アフリカでの英名
wood borers			carpenter moths (1) を見よ
wood borers			false powderpost beetles を見よ
wood borers			snow scorpion flies を見よ
wood-boring beetles		*Teredilia*	(コウチュウ目) の昆虫の総称
wood-boring fly		*Tanyptera atrata* (Linnaeus)	(ハエ目、ガガンボ科) 旧北区
wood-boring weevil		*Pentarthrum huttoni* Wollaston	(コウチュウ目、ゾウムシ科) 旧北区
wood-boring weevils		*Euophryum*	(コウチュウ目、ゾウムシ科) の昆虫の総称
wood carder bee		*Anthidium manicatum* (Linnaeus)	(ハチ目、ハキリバチ科) 旧北区
wood carpet		*Epirrhoe rivata* (Hübner)	(チョウ目、シャクガ科) 旧北区
wood cockroach		*Parcoblatta pennsylvanica* (De Geer)	(ゴキブリ目、チャバネゴキブリ科) 新北区
wood cockroaches		*Parcoblatta*	(ゴキブリ目、チャバネゴキブリ科) の昆虫の総称
wood-colored Apamea moth		*Apamea lignicolora* Guenée	(チョウ目、ヤガ科) 新北区。wood-colored Apamea, wood-colored quaker ともいう
wood cricket		*Nemobius sylvestris* (Bosc)	(バッタ目、コオロギ科) 旧北区
wood crickets		Nemobiinae	(バッタ目、コオロギ科) の昆虫の総称
wood crickets			tree crickets (1) を見よ

英名	和名	学名	所属、分布、ほか
wood-destroying termite			crater termite を見よ
wood frit			heath fritillary を見よ
wood gnat		*Sylvicola fenestralis* (Scopoli)	(ハエ目、カバエ科) 新北区
wood gnat			snow mosquito を見よ
wood gnats	カバエ科	Anisopodidae	(ハエ目) の昆虫の総称　米国で使用
woodgrain leafroller moth		*Pandemis lamprosana* (Robinson)	(チョウ目、ハマキガ科) 新北区
wood grass-veneer		*Crambus silvella* Hübner	(チョウ目、メイガ科) 旧北区
wood leopard			leopard moth を見よ
wood leopard moth			leopard moth を見よ
wood lice			common barklice を見よ　ワラジムシ綱の動物の総称　でもある
wood-louse			booklouse (2) を見よ
woodlouse ant		*Myrmecina graminicola* (Latreille)	(ハチ目、アリ科) 旧北区。日本亜種は *M. g. nipponica* Wheeler カドフシアリ
woodlouse flies	タンカクヤドリバエ科	Rhinophoridae	(ハエ目) の昆虫の総称
Wood-Mason's bushbrown	カギコジャノメ	*Mycalesis suaveolens* Wood-Mason et de Nicéville	(チョウ目、タテハチョウ科) 東洋区
wood moths			carpenter moths (1) を見よ
wood nymph		*Cercyonis pegala nephele* (Kirby)	(チョウ目、タテハチョウ科) 新北区
wood rat flea		*Hystrichopsylla dippiei* Rothschild	(ノミ目、ケブカノミ科) 新北区
wood ringlet			ringlet (1) を見よ
woodrose bug		*Graptostethus manillensis* (Stål)	(カメムシ目、ナガカメムシ科) 新北区
woodrose bugs		*Graptostethus*	(カメムシ目、ナガカメムシ科) の昆虫の総称
wood-rush dwarf		*Elachista regificella* Sircom	(チョウ目、クサモグリガ科) 旧北区
wood satyr		*Euptychia cymela* (Cramer)	(チョウ目、タテハチョウ科) 新北区
wood satyrs		*Euptychia*	(チョウ目、タテハチョウ科) の昆虫の総称
wood snipefly		*Rhagio annulatus* (De Geer)	(ハエ目、シギアブ科) 日本、旧北区
wood swift		*Triodia sylvina* Linnaeus	(チョウ目、コウモリガ科) 旧北区。成虫は wood swift moth
wood tiger		*Parasemia plantaginis* (Linnaeus)	(チョウ目、ヒトリガ科) 旧北区。日本に 4 亜種 *P. p. jezoensis* Inoue ヒメキシタヒトリ 北海道、*P. p. macromera* (Butler) 本州、*P. p. melanissima* Inoue 本州、*P. p. melanomera* (Butler) 本州。wood tiger moth ともいう。ヒメキシタヒトリは plantain arctiid ともいう
wood tiger beetle		*Cicindela sylvatica* Linnaeus	(コウチュウ目、ハンミョウ科) 旧北区
wood tortoiseshell			large tortoiseshell を見よ

英名	和名	学名	所属、分布、ほか
wood wasp		*Pamphilius sylvaticus* (Linnaeus)	（ハチ目、ヒラタハバチ科）旧北区
wood wasp			tree wasp を見よ
wood wasp			horntail (1) を見よ
wood wasps	キバチ上科	Siricoidea	（ハチ目）の昆虫の総称　クキバチ上科 Cephoidea にも使われる
wood wasps	クビナガキバチ科	Xiphydriidae	（ハチ目）の昆虫の総称
wood wasps			horntail wasps を見よ　豪州での英名
wood wasps			horntails (1) を見よ
wood white	セイヨウヒメシロチョウ	*Leptidea sinapis* (Linnaeus)	（チョウ目、シロチョウ科）旧北区
wood white (1)	ベニモンカザリシロチョウ	*Delias aganippe* (Donovan)	（チョウ目、シロチョウ科）豪州区。wood white butterfly ともいう
wood worm	ツマグロツツシンクイ	*Hylecoetus dermestoides cossis* Lewis	（コウチュウ目、ツツシンクイ科）日本
wood worm			furniture beetle (2) を見よ
woodland brown		*Lopinga achine* (Scopoli)	（チョウ目、タテハメチョウ科）旧北区。日本亜種は *L. a. achinoides* (Butler) ウラジャノメ
woodland chocolate moth		*Argyrostrotis sylvarum* (Guenée)	（チョウ目、ヤガ科）新北区
woodland cicadas		*Platypedia*	（カメムシ目、セミ科）の昆虫の総称
woodland floodwater mosquito		*Ochlerotatus abserratus* (Felt et Young)	（ハエ目、カ科）新北区
woodland grasshopper		*Omocestus ventralis* (Zetterstedt)	（バッタ目、バッタ科）旧北区
woodland grasshopper		*Omocestus rufipes* (Zetterstedt)	（バッタ目、バッタ科）旧北区
woodland grayling		*Hipparchia fagi* Scopoli	（チョウ目、タテハチョウ科）旧北区
woodland ground beetles		Pterostichini	（コウチュウ目、オサムシ科）の昆虫の総称
woodland malaria mosquito		*Anopheles punctipennis* (Say)	（ハエ目、カ科）新北区
woodland meadow katydid		*Conocephalus nemoralis* (Scudder)	（バッタ目、キリギリス科）新北区
woodland mosquito		*Ochlerotatus excrucians* (Walker)	（ハエ目、カ科）新北区
woodland ringlet	ステップベニヒカゲ	*Erebia medusa* Fabricius	（チョウ目、タテハチョウ科）旧北区
woodland skipper		*Ochlodes sylvanoides* (Boisduval)	（チョウ目、セセリチョウ科）新北区、新熱帯区
woodland white		*Appias sylvia* (Fabricius)	（チョウ目、シロチョウ科）エチオピア区
Wood's bell			pine cone moth を見よ
Wood's birch pigmy		*Ectoedemia minimella* Rebel	（チョウ目、モグリチビガ科）旧北区
Wood's dart		*Agrotis graslini* (Rambur)	（チョウ目、ヤガ科）旧北区
woods eyed brown			Appalachian brown を見よ
woods firefly		*Photuris pennsylvanicus* De Geer	（コウチュウ目、ホタル科）新北区。野原より林に多い

英名	和名	学名	所属、分布、ほか
woods weevil		*Nemocestes incomptus* (Horn)	(コウチュウ目、ゾウムシ科) 新北区
Woodward's sailer		*Neptis woodwardi* Sharpe	(チョウ目、タテハチョウ科) エチオピア区
woodwig			dragonflies (1) を見よ
woodworm			common furniture beetle を見よ
woody angle moth		*Macaria aequiferaria* Walker	(チョウ目、シャクガ科) 新北区
woody underwing moth		*Catocala grynea* Cramer	(チョウ目、ヤガ科) 新北区。woody underwing ともいう
wool-bearing gall wasp		*Andricus quercuslanigera* (Ashmead)	(ハチ目、タマバチ科) 新北区
wool-carder bee	トモンハナバチ	*Anthidium septemspinosum* Lepeletier	(ハチ目、ハキリバチ科) 日本、旧北区
wool maggot			green botfly を見よ　幼虫の英名
wool maggot			black blow fly を見よ　幼虫の英名
woolly alder aphid		*Prociphilus tessallatus* (Fitch)	(カメムシ目、アブラムシ科) 新北区
woolly and gall aphids			woolly aphids を見よ
woolly ants			velvet ants (1) を見よ　米国での英名
woolly aphid			woolly apple aphid を見よ
woolly aphid parasite	ワタムシヤドリコバチ	*Aphelinus mali* (Haldeman)	(ハチ目、ツヤコバチ科) 日本、汎世界
woolly aphids	ワタアブラムシ科	Eriosomatidae	(カメムシ目) の昆虫の総称　アブラムシ科に入れられる
woolly aphis parasite			woolly aphid parasite を見よ
woolly aphis parasite wasp			woolly aphid parasite を見よ
woolly apple aphid	リンゴワタムシ	*Eriosoma lanigerum* (Hausmann)	(カメムシ目、アブラムシ科) 日本、全北区。リンゴ害虫
woolly apple aphid parasite			woolly aphid parasite を見よ
woolly apple aphis			woolly apple aphid を見よ
woolly-backed moth		*Olethreutes furfuranum* (McDunnough)	(チョウ目、ハマキガ科) 新北区
woolly balsam aphid			silver fir adelgid を見よ
woolly balsam aphid			balsam woolly adelgid を見よ
woolly bear			banded woollybear を見よ
woolly bear			ヒトリガの幼虫など毛の多い各種の幼虫
woolly bear			black woolly-bear を見よ
woolly bear caterpillar			woollybear tiger moth を見よ
woolly bear caterpillar moth			banded woollybear を見よ
woolly bear moth			banded woollybear を見よ
woolly bears			magpie moths を見よ
woolly bears			giant lappet moths を見よ

英名	和名	学名	所属、分布、ほか
woolly bears			tiger moths (1) を見よ
woolly bears			carpet beetles を見よ
woolly bee flies		Systoechus	(ハエ目、ツリアブ科) の昆虫の総称
woolly beech aphid	ブナハアブラムシ	Phyllaphis fagi (Linnaeus)	(カメムシ目、アブラムシ科) 日本、全北区、豪州区。ブナノキ害虫
woolly beech-leaf aphid			woolly beech aphid を見よ
woolly currant scale (1)		Pulvinaria betulae (Linnaeus)	(カメムシ目、カタカイガラムシ科) 旧北区
woolly currant scale (2)		Pulvinaria ribesiae Signoret	(カメムシ目、カタカイガラムシ科) 旧北区
woolly currant scale			cottony maple scale (1) を見よ
woolly darkling beetle		Eleodes osculans LeConte	(コウチュウ目、ゴミムシダマシ科) 新北区
woolly elm aphid		Eriosoma americanum (Riley)	(カメムシ目、アブラムシ科) 新北区
woolly elm aphid			woolly pear aphid を見よ
woolly fir aphid		Prociphilus pini (Koch)	(カメムシ目、アブラムシ科) 旧北区
woolly grass-veneer moth		Thaumatopsis pexellus (Zeller)	(チョウ目、メイガ科) 新北区。woolly grass-veneer ともいう
woolly gray moth		Lycia ypsilon (Forbes)	(チョウ目、シャクガ科) 新北区
woolly ground beetle		Cratidus osculans LeConte	(コウチュウ目、ゴミムシダマシ科) 新北区
woolly larch aphid			larch aphid (1) を見よ
woolly legs		Lachnocnema	(チョウ目、シジミチョウ科) の昆虫の総称
woolly legs (1)	アフリカヨコバイシジミ	Lachnocnema bibulus (Fabricius)	(チョウ目、シジミチョウ科) エチオピア区
woolly meadow moth		Penthophera morio Linnaeus	(チョウ目、ドクガ科) 旧北区
woolly pear aphid		Eriosoma lanuginosum Hartig	(カメムシ目、アブラムシ科) 旧北区
woolly pear aphid			pear root aphid を見よ
woolly pine aphid			pine adelgid を見よ　豪州での英名
woolly pine scale		Pseudophilippia quaintancii Cockerell	(カメムシ目、カタカイガラムシ科) 新北区
woolly psyllid		Psylla floccosa (Patch)	(カメムシ目、キジラミ科) 新北区
woolly pyrol moth			velvet bean caterpillar を見よ
woolly vine scale			cottony maple scale (1) を見よ
woolly white fly			citrus whitefly (1) を見よ
woolly whitefly	ミカンワタコナジラミ	Aleurothrixus floccosus (Maskell)	(カメムシ目、コナジラミ科) 日本、新北区、大洋区、新熱帯区。カンキツ害虫
woollybear tiger moth		Spilosoma glatignyi (Le Guillou)	(チョウ目、ヒトリガ科) 豪州区
Worcester copper		Aloeides lutescens Tite et Dickson	(チョウ目、シジミチョウ科) エチオピア区
workerless inquiline ant		Pogonomyrmex colei (Rissing)	(ハチ目、アリ科) 新北区

英名	和名	学名	所属、分布、ほか
workers	職蟻		社会性のハチとシロアリで性的に不完全な雌虫。造巣、育虫などに従事し、働バチと俗称シロアリでは兵アリ以外の性的不発達の型
worm			一般に柔らかく細長い虫を指す
worm-lions		*Vermileo*	(ハエ目、シギアブ科) の昆虫の総称
wormwood		*Cucullia absinthii* (Linnaeus)	(チョウ目、ヤガ科) 旧北区
wormwood			wormwood knot-horn を見よ
wormwood knot-horn		*Euzophera cinerosella* Zeller	(チョウ目、メイガ科) 旧北区
wormwood leaf-tier		*Phaneta artemisiana* Walsingham	(チョウ目、ハマキガ科) 新北区
wormwood moth			wormwood を見よ
wormwood pug	ホソチビナミシャク	*Eupithecia absinthiata* (Clerck)	(チョウ目、シャクガ科) 日本、旧北区
wormwood webworm			knot-horn を見よ
wornil			cattle warble fly (2) を見よ　幼虫の英名
wounded hawk		*Neoclanis basalis* (Walker)	(チョウ目、スズメガ科) エチオピア区
wounded lady beetle		*Axion plagiatum* (Olivier)	(コウチュウ目、テントウムシ科) 新北区
wounded-tree beetles	ヒメトゲムシ科	Nosodendridae	(コウチュウ目) の昆虫の総称　マルトゲムシ科に含まれる
woundwort pearl		*Phlyctaenia stachydalis* (Zincken)	(チョウ目、メイガ科) 旧北区
woundwort pearl	ヒメトガリノメイガ	*Anania verbascalis* (Denis et Schiffermüller)	(チョウ目、メイガ科) 日本、旧北区。ゴボウ、キク害虫
Wreford's grizzled skipper		*Spialia wrefordi* Evans	(チョウ目、セセリチョウ科) エチオピア区
wretched Olethreutes moth		*Olethreutes exoletum* (Zeller)	(チョウ目、ハマキガ科) 新北区
wrigglers			mosquitoes (1) を見よ　カの幼虫ボウフラ
Wright's Calephelis			Wright's metalmark を見よ
Wright's Hulstina moth		*Hulstina wrightiaria* (Hulst)	(チョウ目、シャクガ科) 新北区
Wright's metalmark		*Calephelis wrighti* Holland	(チョウ目、シジミタテハ科) 新北区
Wright's thread-waisted wasp		*Ammophila wrightii* (Cresson)	(ハチ目、アナバチ科) 新北区
wrinckled grasshopper		*Hippiscus rugosus* Scudder	(バッタ目、バッタ科) 新北区
wrinkled bark beetles	セスジムシ科	Rhysodidae	(コウチュウ目) の昆虫の総称
wrinkled sucking lice	ケモノジラミ科	Haematopinidae	(シラミ目) の昆虫の総称
Wyatt's Stagmatophora moth		*Stagmatophora wyattella* Barnes et Busck	(チョウ目、カザリバガ科) 新北区
Wykeham's blue		*Lepidochrysops wykehami* Tite	(チョウ目、シジミチョウ科) エチオピア区
Wykeham's grey		*Crudaria wykehami* Dickson	(チョウ目、シジミチョウ科) エチオピア区

英名	和名	学名	所属、分布、ほか
Wykeham's silver-spotted copper		*Trimenia wykehami* (Dickson)	(チョウ目、シジミチョウ科) エチオピア区
Wymore's shieldback		*Idiostatus wymorei* Caudell	(バッタ目、キリギリス科) 新北区
Wyoming ringlet			Hayden's ringlet を見よ

英名	和名	学名	所属、分布、ほか

X

英名	和名	学名	所属、分布、ほか
X-linear grass-veneer moth		*Xubida linearella* (Zeller)	（チョウ目、メイガ科）新北区
Xami hairstreak		*Callophrys xami* (Reakirt)	（チョウ目、シジミチョウ科）新北区
xanthic case		*Coleophora chalcogrammella* Zeller	（チョウ目、ツツミノガ科）旧北区
xanthic cosmet		*Mompha ochraceella* (Curtis)	（チョウ目、カザリバガ科）旧北区
Xanthochlora grass yellow		*Eurema xanthochlora* (Kollar)	（チョウ目、シロチョウ科）新熱帯区
Xanthocles longwing		*Heliconius xanthocles* Bates	（チョウ目、タテハチョウ科）新熱帯区
Xanthomera skipper		*Neohesperilla xanthomera* (Meyrick et Lower)	（チョウ目、セセリチョウ科）豪州区
Xanthophysa moth		*Xanthophysa psychialis* (Hulst)	（チョウ目、メイガ科）新北区
xanthoxalis whitefly			Oxalis whitefly を見よ
Xerces blue	クセルスシジミ	*Glaucopsyche xerces* (Boisduval)	（チョウ目、シジミチョウ科）新北区
Xicaque satyr		*Paramacera xicaque* (Reakirt)	（チョウ目、タテハチョウ科）新北区
Ximena sister		*Adelpha ximena* (C. et R. Felder)	（チョウ目、タテハチョウ科）新熱帯区
Xiphiphora skipper		*Neohesperilla xiphiphora* (Lower)	（チョウ目、セセリチョウ科）豪州区
xuthus swallowtail			citrus swallowtail (1) を見よ
xyelid sawflies	ナギナタハバチ科	Xyelidae	（ハチ目）の昆虫の総称
xyelids			xyelid sawflies を見よ
xylocopids			carpenter bees を見よ
xylomyid flies	キアブモドキ科	Xylomyidae	（ハエ目）の昆虫の総称
xylomyid flies		*Solva*	（ハエ目、キアブモドキ科）の昆虫の総称
xylophage			木食い虫
xylophagid flies	キアブ科	Xylophagidae	（ハエ目）の昆虫の総称
xyloryctid moths	ヒロバキバガ科	Xyloryctidae	（チョウ目）の昆虫の総称

英名	和名	学名	所属、分布、ほか
Y			
Y-moths			gems (1) を見よ
Yagi snow flea	ヤギシロトビムシ	*Onychiurus pseudarmatus yagii* Kinoshita	(トビムシ目、シロトビムシ科) 日本
yam beetle		*Prionoryctes caniculus* Arrow	(コウチュウ目、コガネムシ科) エチオピア区
yam hawk moth			yellow hawk moth を見よ
yam leafminer	ヤマノイモコガ	*Acrolepiopsis suzukiella* (Matsumura)	(チョウ目、アトヒゲコガ科) 日本。ヤマノイモ害虫
yam scale		*Aspidiotus hartii* Cockerell	(カメムシ目、マルカイガラムシ科) 豪州区、大洋区。ヤム害虫
yam tuber beetle		*Heteroligus melas* Billberg	(コウチュウ目、コガネムシ科) エチオピア区。ヤム害虫
yam tuber beetle		*Heteroligus appius* Burmeister	(コウチュウ目、コガネムシ科) エチオピア区。ヤム害虫
yam tuber beetle		*Prionoryctes canaliculus* Arrow	(コウチュウ目、コガネムシ科) エチオピア区。ヤム害虫
yam tuber beetle		*Prionoryctes rufopiceus* Arrow	(コウチュウ目、コガネムシ科) エチオピア区。ヤム害虫
yam weevil			sweetpotato weevil (1) を見よ
Yamada lasiocampid			quercus lasiocampid (1) を見よ
yamamai			Japanese silkmoth を見よ
Yamato orneodid	ヤマトニジュウシトリバ	*Alucita japonica* (Matsumura)	(チョウ目、ニジュウシトリバガ科) 日本
yamflies	オナガアカシジミ属	*Loxura*	(チョウ目、シジミチョウ科) の昆虫の総称　幼虫が yam ヤマノイモ を食べる
yamfly	オナガアカシジミ	*Loxura atymnus* Stoll	(チョウ目、シジミチョウ科) 東洋区。*L. a. fuconius* Fruhstorfer も同英名
yanagicola oystershell scale			willow oystershell scale を見よ
yanagicola scale			willow oystershell scale を見よ
Yankee settler			German cockroach を見よ
Yano aphid	ヤノクチナガオオアブラムシ	*Stomaphis yanonis* Takahashi	(カメムシ目、アブラムシ科) 日本
Yano distylium gall aphid	ヤノイスアブラムシ	*Neothoracaphis yanonis* (Matsumura)	(カメムシ目、アブラムシ科) 日本
Yano leafminer			barley leafminer (2) を見よ
yanone scale			arrowhead scale を見よ
Yapapai giant skipper		*Aryxna baueri* (Stallings et Turner)	(チョウ目、セセリチョウ科) 新北区
Yaqui dancer		*Argia carlcooki* Daigle	(トンボ目、イトトンボ科) 新北区、新熱帯区
yarn moths		*Tolype*	(チョウ目、カレハガ科) の昆虫の総称
yarrow biston moth			belted beauty を見よ　米国での英名
yarrow leaf beetle		*Cassida denticollis* Suffrian	(コウチュウ目、ハムシ科) 旧北区
yarrow pug			milfoil pug を見よ
Yarrow's tiger moth		*Pararctia yarrowii* (Stretch)	(チョウ目、ヒトリガ科) 新北区

英名	和名	学名	所属、分布、ほか
Yasushi ensign scale	ヤスシハカマカイガラムシ	*Orthezia yasushii* Kuwana	(カメムシ目、ハカマカイガラムシ科) 日本、旧北区、東洋区。アケビ害虫
Yasushi orthezia			Yasushi ensign scale を見よ
Yatta three-ring		*Ypthima yatta* Kielland	(チョウ目、タテハチョウ科) エチオピア区
Yavapai giant skipper		*Agathymus baueri* (Stallings et Turner)	(チョウ目、セセリチョウ科) 新北区
Yeates's flat-body		*Agonopterix yeatiana* (Fabricius)	(チョウ目、マルハキバガ科) 旧北区
Yehl skipper		*Poanes yehl* (Skinner)	(チョウ目、セセリチョウ科) 新北区
yellow turf ant			yellow meadow ant を見よ
yellow Acraea skipper			common Acraea skipper を見よ
yellow admiral		*Bassaris itea* (Fabricius)	(チョウ目、タテハチョウ科) 豪州区
yellow admiral			Australian admiral を見よ
yellow admiral			short-tailed admiral を見よ
yellow albatross			Ceylon lesser albatross を見よ
yellow alfalfa geometrid		*Tephrina arenacearia* Denis et Schiffermüller	(チョウ目、シャクガ科) 旧北区
yellow and black mud dauber			mud dauber を見よ
yellow and black potter wasp		*Delta campaniformis campaniformis* (Fabricius)	(ハチ目、スズメバチ科) 新北区
yellow and black tachinid		*Xanthoepalpus bicolor* (Williston)	(ハエ目、ヤドリバエ科) 新北区。yellow-and-black tachinid とも記される
yellow and brown looper moth		*Xanthorhoe prasinias* Meyrick	(チョウ目、シャクガ科) 豪州区
yellow angled sulphur			yellow brimstone を見よ
yellow ant			yellow meadow ant を見よ
yellow Apollo			Eversmann's Parnassian を見よ
yellow apple leafhopper			apple leafhopper (4) を見よ　カナダでの英名
yellow apricot			orange-barred sulphur を見よ
yellow auger beetle		*Xylotillus lindi* (Blackburn)	(コウチュウ目、ナガシンクイムシ科) 豪州区
yellow-backed biddie		*Cordulegaster dorsalis* (Hagen)	(トンボ目、オニヤンマ科) 新北区
yellow band dart		*Potanthus pava* (Fruhstorfer)	(チョウ目、セセリチョウ科) 東洋区
yellow-banded Acraea		*Acraea cabira* Hopffer	(チョウ目、タテハチョウ科) エチオピア区
yellow banded awl		*Hasora schoenherr* (Latreille)	(チョウ目、セセリチョウ科) 東洋区。*H. s. chuza* Hewitson も同英名
yellow-banded Bematistes		*Acraea obliqua* (Aurivillius)	(チョウ目、タテハチョウ科) エチオピア区
yellow-banded black geometrid	キオビエダシャク	*Milionia basalis pryeri* Druce	(チョウ目、シャクガ科) 日本、東洋区
yellow-banded dart		*Ocybadistes walkeri* Heron	(チョウ目、セセリチョウ科) 豪州区

英名	和名	学名	所属、分布、ほか
yellow-banded day sphinx		*Proserpinus flavofasciata* (Walker)	(チョウ目、スズメガ科) 新北区。yellow-banded day sphinx moth ともいう
yellow-banded eighty-eight		*Callicore atacama* (Hewitson)	(チョウ目、タテハチョウ科) 新熱帯区
yellow-banded evening brown		*Gnophodes betsimena* (Boisduval)	(チョウ目、タテハチョウ科) エチオピア区。banded evening brown ともいう
yellow-banded Jezebel		*Delias ennia* (Wallace)	(チョウ目、シロチョウ科) 豪州区
yellow-banded ringlet	キオビチビベニヒカゲ	*Erebia flavofasciata* Heyne	(チョウ目、タテハチョウ科) 旧北区
yellow-banded sapphire		*Iolaus aphanaeoides* Trimen	(チョウ目、シジミチョウ科) エチオピア区
yellow-banded sapphire (1)		*Iolaus diametra* Karsch	(チョウ目、シジミチョウ科) エチオピア区
yellow-banded skipper		*Pyrgus sidae* Esper	(チョウ目、セセリチョウ科) 旧北区
yellow-banded stonefly		*Calineuria californica* (Banks)	(カワゲラ目、カワゲラ科) 新北区
yellow-banded underwing moth		*Catocala cerogama* Guenée	(チョウ目、ヤガ科) 新北区。yellow-banded underwing ともいう
yellow-banded wasp moth		*Syntomeida ipomoeae* Harris	(チョウ目、ヒトリガ科) 新北区。幼虫はアサガオを食う
yellow barred		*Xanthotaenia busiris busiris* Westwood	(チョウ目、タテハチョウ科) 東洋区
yellow-barred brindle	ルリオビナミシャク	*Acasis viretata viretata* (Hübner)	(チョウ目、シャクガ科) 日本、旧北区
yellow-barred dwarf			banded false-brome dwarf を見よ
yellow-barred heliconian			zebra longwing を見よ
yellow barred Nephele		*Nephele oenopion* (Hübner)	(チョウ目、スズメガ科) エチオピア区
yellow-base flitter		*Quedara basiflava* (de Nicéville)	(チョウ目、セセリチョウ科) 東洋区
yellow-based Cacozelia moth		*Cacozelia basiochrealis* Grote	(チョウ目、メイガ科) 新北区。yellow-based Cacozelia ともいう
yellow-based metalmark		*Isapis agyrtus hera* Godman et Salvin	(チョウ目、シジミタテハ科) 新熱帯区
yellow bat's flea		*Ceratophyllus elongatus* Curtis	(ノミ目、ナガノミ科) 全北区
yellow bear			Virginia tiger moth を見よ　幼虫の英名
yellow bear caterpillar			Virginia tiger moth を見よ　幼虫の英名
yellow bear moth			Virginia tiger moth を見よ
yellow belle		*Aspilates ochrearia* (Rossi)	(チョウ目、シャクガ科) 旧北区
yellowbelly		*Psaltoda harrisii* (Leach)	(カメムシ目、セミ科) 豪州区
yellow-belly blue lema	キバラルリクビボソハムシ	*Lema concinipennis* Baly	(コウチュウ目、ハムシ科) 日本、旧北区、東洋区
yellow-belly dagger moth	キバラケンモン	*Trichosea champa* (Moore)	(チョウ目、ヤガ科) 日本、東洋区

英名	和名	学名	所属、分布、ほか
yellow-belly leaf-cutting bee	キバラハキリバチ	*Megachile xanthothris* Yasumatsu et Hirashima	（ハチ目、ハキリバチ科）日本、旧北区
yellow-belly small leaf beetle	キバラヒメハムシ	*Exosoma flaviventris* (Motschulsky)	（コウチュウ目、ハムシ科）日本、旧北区、東洋区
yellow-belly tiger moth	キバラヒトリ	*Epatolmis caesaria* (Goeze)	（チョウ目、ヒトリガ科）日本、旧北区
yellow-bellyblack-dotted arctiid			white ermine を見よ
yellow-belted burnet		*Syntomis phegea* (Linnaeus)	（チョウ目、マダラガ科）旧北区
yellow birch leafroller moth		*Ancylis discigerana* (Walker)	（チョウ目、ハマキガ科）新北区
yellow-bodied clubtail	ネプトゥヌスホソバジャオウアゲハ	*Atrophaneura neptunus* (Guérin-Méneville)	（チョウ目、アゲハチョウ科）東洋区。後翅の尾状突起が棍棒状のため
yellow bodied clubtail	アカモンホソバベニモンアゲハ	*Pachliopta neptunus* (Guérin)	（チョウ目、アゲハチョウ科）東洋区。yellow clubtail ともいう
yellow-bordered cutworm moth		*Melanchra tartarea* Butler	（チョウ目、ヤガ科）豪州区
yellow-bordered owl			owl butterfly (2) を見よ
yellow borer			yellow rice borer を見よ
yellow-bottomed Theope			azure Theope を見よ
yellow bowed longhorn beetle			oak wasp beetle を見よ
yellowbox lerp		*Lasiopsylla rotundipennis* Froggatt	（カメムシ目、カイガラキジラミ科）豪州区
yellow brimstone	オオヤマキチョウ	*Anteos maerula* (Fabricius)	（チョウ目、シロチョウ科）新北区、新熱帯区
yellow-brown leaf-cut weevil	チャイロチョッキリ	*Aderorhinus crioceroides* (Roelofs)	（コウチュウ目、オトシブミ科）日本、東洋区
yellow-brown minute leaf beetle			sugi leaf beetle を見よ
yellow-brown stink bug			brown marmorated stink bug を見よ
yellow bumble bee			golden northern bumble bee を見よ
yellow caper white	キイロヘリグロシロチョウ	*Belenois solilucis* Butler	（チョウ目、シロチョウ科）エチオピア区
yellow chafer	オオキイロコガネ	*Pollaplonyx flavidus* Waterhouse	（コウチュウ目、コガネムシ科）日本
yellow chequered lancer		*Plastingia pellonia* Fruhstorfer	（チョウ目、セセリチョウ科）東洋区
yellow-cloaked midget			half-yellow moth を見よ
yellow clover aphid		*Therioaphis trifolii* (Monell)	（カメムシ目、アブラムシ科）全北区。spotted alfalfa-like aphid を参照
yellow clubtail		*Gomphus simillimus* Selys	（トンボ目、サナエトンボ科）旧北区
yellow-collared scape moth		*Cisseps fulvicollis* (Hübner)	（チョウ目、カノコガ科）新北区
yellow-collared slug moth		*Apoda y-inversum* (Packard)	（チョウ目、イラガ科）新北区

英名	和名	学名	所属、分布、ほか
yellow-costate banded geometrid	マエキオエダシャク	*Plesiomorpha flaviceps* (Butler)	(チョウ目、シャクガ科) 日本、東洋区
yellow-costate greenish geometrid	キマエアオシャク	*Neohipparchus vallata* (Butler)	(チョウ目、シャクガ科) 日本、旧北区、東洋区
yellow-costate leaf like moth			fruit piercing moth (3) を見よ
yellow coster	ホソチョウ	*Acraea issoria* (Hübner)	(チョウ目、タテハチョウ科) 旧北区、東洋区
yellow coster		*Acraea vesta* Fabricius	(チョウ目、タテハチョウ科) 東洋区
yellow cottony-cushion scale	キイロワタフキカイガラムシ	*Icerya seychellarum* (Westwood)	(カメムシ目、ワタフキカイガラムシ科) 日本、東洋区、豪州区、エチオピア区、大洋区。カンキツ、チャ、ツバキなどの害虫
yellow cracker	ウラキカスリタテハ	*Hamadryas fornax* Hübner	(チョウ目、タテハチョウ科) 新熱帯区。*H. f. fornacalia* (Fruhstorfer) は orange cracker という
yellow crazy ant		*Anopholepis gracilipes* (F. Smith)	(ハエ目、アリ科) エチオピア区、東洋区、豪州区
yellow-crested spangle		*Papilio elephenor* Doubleday	(チョウ目、アゲハチョウ科) 東洋区
yellow-dappled longicorn beetle	キマダラカミキリ	*Aeolesthes chrysothrix* (Bates)	(コウチュウ目、カミキリムシ科) 日本。カシ類、ネムノキ害虫。yellow-dappled longicorn ともいう
yellow diacrisia moth			clouded buff を見よ
yellow disc oakblue		*Arhopala perimuta* (Moore)	(チョウ目、シジミチョウ科) 東洋区
yellow-disc oakblue		*Arhopala singla* (de Nicéville)	(チョウ目、シジミチョウ科) 東洋区
yellowdisc tailless oakblue		*Darasana perimuta* (Moore)	(チョウ目、シジミチョウ科) 東洋区
yellowdisc tailless oakblue			yelllow disc oakblue を見よ
yellow dryad		*Aemona amathusia* (Hewitson)	(チョウ目、タテハチョウ科) 東洋区
yellow dung fly	ヒメフンバエ	*Scatophaga stercoraria* (Linnaeus)	(ハエ目、フンバエ科) 日本、全北区、新熱帯区
yellow dung-flies	フンバエモドキ科	Cordyluridae	(ハエ目) の昆虫の総称
yellow-dusted cream moth		*Cabera erythemaria* Guenée	(チョウ目、シャクガ科) 新北区
yellow-edged flasher		*Astraptes phalaecus* (Godman et Salvin)	(チョウ目、セセリチョウ科) 新熱帯区
yellow-edged footman			bicolored moth を見よ
yellow-edged giant owl			gold-edged giant owl を見よ
yellow-edged Pygarctia		*Pygarctia abdominalis* Grote	(チョウ目、ヒトリガ科) 新北区。yellow-edged Pygarctia moth ともいう
yellow-edged ruby-eye		*Orses cynisca* (Swainson)	(チョウ目、セセリチョウ科) 新熱帯区
yellow emperor dragonfly			Australian emperor dragonfly を見よ
yellow-eyed plane		*Neptis praslini* Boisduval	(チョウ目、タテハチョウ科) 豪州区

英名	和名	学名	所属、分布、ほか
yellow face			inspector を見よ
yellowface Kauai damselfly		*Megalagrion orobates* (Perkins)	(トンボ目、イトトンボ科) 大洋区。yellowfaced Kauai damselfly ともいう
yellow-faced bees (1)	ムカシハナバチ科	Colletidae	(ハチ目) の昆虫の総称
yellow-faced bees	チビムカシハナバチ亜科	Hylaeinae	(ハチ目、ムカシハナバチ科) の昆虫の総称
yellow-faced bees		*Hylaeus*	(ハチ目、ムカシハナバチ科) の昆虫の総称
yellow-faced bees			obtuse-tongued bees を見よ
yellow-faced bumble bee		*Bombus vosnesenskii* (Radokowsky)	(ハチ目、ミツバチ科) 新北区
yellow-faced leafhopper		*Scaphytopius loricatus* (Van Duzee)	(カメムシ目、オオヨコバイ科) 新北区
yellow-faced sprite		*Pseudagrion citricola* Barnard	(トンボ目、イトトンボ科) エチオピア区
yellow false lady beetle	キイロテントウダマシ	*Saula japonica* Gorham	(コウチュウ目、テントウムシダマシ科) 日本
yellow featherleg		*Copera marginipes* (Rambur)	(トンボ目、モノサシトンボ科) 東洋区
yellow fever mosquito	ネッタイシマカ	*Aedes aegypti* (Linnaeus)	(ハエ目、カ科) 日本、汎熱帯。黄熱、デング熱を媒介
yellow flannel moth (1)		*Lagoa crispata* Packard	(チョウ目、Megalopygidae) 新北区
yellow flannel moth (2)		*Lagoa pyxidifera* (Smith)	(チョウ目、Megalopygidae) 新北区
yellow flash		*Rapala domitia domitia* (Hewitson)	(チョウ目、シジミチョウ科) 東洋区
yellow flat		*Mooreana trichoneura* (Moore)	(チョウ目、セセリチョウ科) 東洋区
yellow-flattened leaf beetle	ズグロキハムシ	*Gastrolinoides japonicus* (Harold)	(コウチュウ目、ハムシ科) 日本
yellow flower thrips		*Frankliniella schultzei* (Trybom)	(アザミウマ目、アザミウマ科) 大洋区、豪州区、新熱帯区、エチオピア区
yellow flower wasp		*Campsomeris tasmaniensis* Saussure	(ハチ目、ツチバチ科) 豪州区
yellow fly		*Diachlorus ferrugatus* (Fabricius)	(ハエ目、アブ科) 新北区
yellow fly			yellow rice leafhopper を見よ
yellow forest sylph		*Ceratrichia flava* Hewitson	(チョウ目、セセリチョウ科) エチオピア区
yellow-fringed conch		*Cochylis flaviciliana* (Westwood)	(チョウ目、ホソハマキガ科) 旧北区
yellow-fringed Dolichomia moth		*Hypsopygia olinalis* (Guenée)	(チョウ目、メイガ科) 新北区。yellow-fringed Dolichomia ともいう
yellow-fringed scale			green shield scale を見よ
yellow-fringed swift		*Caltoris aurociliata* (Elwes et Edwards)	(チョウ目、セセリチョウ科) 東洋区
yellow-fringed underwing			Hulst's underwing moth を見よ
yellow-fronted owl		*Caligo telamonius* (C. et R. Felder)	(チョウ目、タテハチョウ科) 新熱帯区

英名	和名	学名	所属、分布、ほか
yellow fruit fly	キイロケブカミバエ	*Xyphosia punctigera* (Coquillett)	(ハエ目、ミバエ科) 日本、旧北区
yellow glassy tiger		*Parantica aspasia aspasia* (Fabricius)	(チョウ目、タテハチョウ科) 東洋区
yellow glassy tiger			chestnut tiger を見よ
yellow gnat			buffalo gnat (3) を見よ
yellow gorgon			outlet sword を見よ
yellow-gray underwing moth		*Catocala retecta* Grote	(チョウ目、ヤガ科) 新北区。yellow-gray underwing ともいう
yellow-hair-dappled longicorn beetle	ケマダラカミキリ	*Agapanthia daurica* Ganglebauer	(コウチュウ目、カミキリムシ科) 日本、旧北区。yellow-hair-dappled longicorn ともいう
yellow-haired dagger moth		*Acronicta impleta* Walker	(チョウ目、ヤガ科) 新北区
yellow-haired dung-fly			yellow dung fly を見よ
yellow-haired skipper		*Cogia cajeta cajeta* (Herrich-Schäffer)	(チョウ目、セセリチョウ科) 新熱帯区
yellow hawk moth	キイロスズメ	*Theretra nessus* (Drury)	(チョウ目、スズメガ科) 日本、旧北区、東洋区、大洋区、豪州区
yellow-head tubic		*Pseudatemelia flavifrontella* (Denis et Schiffermüller)	(チョウ目、マルハキバガ科) 旧北区
yellowhead vineworm			yellow-headed fireworm を見よ
yellow-headed anthelid		*Nataxa flavescens* (Walker)	(チョウ目、Anthelidae) 豪州区
yellow-headed borer		*Diatraea centrella* (Möschler)	(チョウ目、メイガ科) 新北区、新熱帯区。イネ害虫
yellow-headed borer			orange stem borer (2) を見よ
yellowheaded cockchafer		*Sericesthis harti* (Sharp)	(コウチュウ目、コガネムシ科) 豪州区
yellow-headed coffee borer		*Dirphya princeps* Jordan	(コウチュウ目、カミキリムシ科) エチオピア区。コーヒー害虫
yellow-headed cosmet		*Spulerina flavicaput* (Haworth)	(チョウ目、ホソガ科) 旧北区
yellow-headed cosmet		*Heinemannia aurifrontella* (Geyer)	(チョウ目、カザリバガ科) 旧北区
yellow-headed cranberry worm			yellow-headed fireworm を見よ
yellowheaded cutworm		*Apamea arctica* Freyer	(チョウ目、ヤガ科) 新北区
yellow-headed cutworm moth		*Apamea amputatrix* (Fitch)	(チョウ目、ヤガ科) 新北区。yellow-headed cutworm ともいう
yellow-headed fireworm		*Acleris minuta* (Robinson)	(チョウ目、ハマキガ科) 新北区。yellow-headed fireworm moth ともいう
yellow headed leafhopper		*Carneocephala flaviceps* (Riley)	(カメムシ目、オオヨコバイ科) 新北区
yellow-headed lichen moth		*Crambidia cephalica* (Grote et Robinson)	(チョウ目、ヒトリガ科) 新北区

英名	和名	学名	所属、分布、ほか
yellow-headed looper			eastern pine looper を見よ　yellow-headed looper moth ともいう
yellow headed rhombic planthopper	キガシラヒシウンカ	*Kuvera flaviceps* (Matsumura)	（カメムシ目、ヒシウンカ科）日本、旧北区
yellow-headed snail parasitic blowfly			snail parasitic blowfly を見よ
yellowheaded spruce sawfly		*Pikonema alaskensis* (Rohwer)	（ハチ目、ハバチ科）新北区
yellow-headed stem borer			orange stem borer (2) を見よ
yellow-headed sugarcane moth borer			yellow-headed borer を見よ
yellow Helen		*Papilio nephelus* Boisduval	（チョウ目、アゲハチョウ科）旧北区。東洋区の亜種　*P. n. chaon* Westwood タイワンモンキアゲハ
yellow Helen (1)	フスクスアゲハ（ネッタイモンキアゲハ）	*Papilio fuscus* Goeze	（チョウ目、アゲハチョウ科）東洋区、豪州区。ギリシャ神話の女神ヘレネーに因む
yellow-hindwinged arctiid	キシタホソバ	*Eilema griseola aegrota* (Butler)	（チョウ目、ヒトリガ科）日本
yellow-hindwinged catocala	キシタバ	*Catocala patala* Felder et Rogenhofer	（チョウ目、ヤガ科）日本、旧北区、東洋区
yellow-hindwinged geometrid	キシタエダシャク	*Arichanna melanaria fraterna* (Butler)	（チョウ目、シャクガ科）日本
yellow-hindwinged noctuid	キシタアツバ	*Hypena claripennis* (Butler)	（チョウ目、ヤガ科）日本、旧北区
yellow-hindwinged tussock moth	シタキドクガ	*Calliteara taiwana aurifera* (Scriba)	（チョウ目、ドクガ科）日本
yellowhorn moth		*Colocasia flavicornis* (Smith)	（チョウ目、ヤガ科）新北区。yellowhorn ともいう
yellow-horned			yellowhorn moth を見よ
yellowhorned clerid		*Trogodendron fasciculatum* (Schreibers)	（コウチュウ目、カッコウムシ科）豪州区
yellow-horned horntail			horntail (1) を見よ
yellow-horned lutestring			yellowhorn moth を見よ
yellow horned moth	ミスジトガリバ	*Achlya flavicornis* (Linnaeus)	（チョウ目、トガリバガ科）日本、旧北区。yellow horned ともいう
yellow hornet	ケブカスズメバチ	*Vespa simillima* Smith	（ハチ目、スズメバチ科）日本
yellow hornet			Japanese yellow hornet を見よ
yellow house ant			Pharaoh's ant を見よ
yellow Jack pine tip moth		*Rhyacionia sonia* Miller	（チョウ目、ハマキガ科）新北区
yellow jacket			giant hornet を見よ
yellow jackets			hornets (1) (2) を見よ
yellow Jezebel	アゴスチーナカザリシロチョウ	*Delias agostina* (Hewitson)	（チョウ目、シロチョウ科）東洋区

英名	和名	学名	所属、分布、ほか
yellow kaiser		*Penthema lisarda* (Doubleday)	(チョウ目、タテハチョウ科) 東洋区
yellow kite		*Eurytides iphitas* Hübner	(チョウ目、アゲハチョウ科) 新熱帯区
yellow kite-swallowtail		*Eurytides calliste calliste* (Bates)	(チョウ目、アゲハチョウ科) 新熱帯区
yellow labyrinth			Chinese labyrinth (1) を見よ
yellow largest dart		*Paronymus xanthias* (Mabille)	(チョウ目、セセリチョウ科) エチオピア区
yellow leaf-cut weevil	ヒゲナガオトシブミ	*Paratrachelophorus longicornis* (Roelofs)	(コウチュウ目、オトシブミ科) 日本
yellow leafhopper		*Zygina zealandica* (Meyers)	(カメムシ目、オオヨコバイ科) 豪州区
yellow leafhopper			yellow rice leafhopper を見よ
yellow leafhopper			West Indian canefly を見よ
yellow-leg long-horned green leaf beetle	キアシヒゲナガアオハムシ	*Clerotilia flavomarginata* Jacoby	(コウチュウ目、ハムシ科) 日本、旧北区
yellow-leg long-horned sawfly			willow sawfly (2) を見よ
yellow-leg three-striped miner fly	キアシミスジキモグリバエ	*Chlorops hypostigma* Meigen	(ハエ目、キモグリバエ科) 旧北区
yellow-legged paper wasp		*Mischocyttarus flavitarsis* (Saussure)	(ハチ目、スズメバチ科) 新北区
yellow-legged black		*Pachygaster leachii* Stephens	(ハエ目、ミズアブ科) 旧北区
yellow-legged black leafcut weevil	ヒメクロオトシブミ	*Apoderus erythrogaster* Vollenhoven	(コウチュウ目、オトシブミ科) 日本
yellow-legged black legionnaire		*Beris morrisii* Dale	(ハエ目、ミズアブ科) 旧北区
yellow-legged centurion		*Sargus flavipes* Meigen	(ハエ目、ミズアブ科) 旧北区
yellow-legged clearwing		*Synanthedon vespiformis* (Linnaeus)	(チョウ目、スカシバガ科) 旧北区
yellow-legged dragonfly		*Gomphus flavipes* (Charpentier)	(トンボ目、サナエトンボ科) 旧北区
yellow-legged glasswing		*Episcada clausina* (Hewitson)	(チョウ目、タテハチョウ科) 新熱帯区
yellow-legged hornet			Asian predatory wasp を見よ
yellow-legged lema	キアシクビボソハムシ	*Oulema tristis* (Herbst)	(コウチュウ目、ハムシ科) 日本、旧北区
yellow-legged meadowhawk		*Sympetrum vicinum* (Hagen)	(トンボ目、トンボ科) 新北区
yellow-legged mining bee		*Andrena flavipes* Panzer	(ハチ目、ヒメハナバチ科) 旧北区
yellow-legged ringtail		*Erpetogomphus crotalinus* (Hagen)	(トンボ目、サナエトンボ科) 新北区、新熱帯区
yellow-legged rose leaf beetle	キアシルリツツハムシ	*Cryptocephalus fortunatus* Baly	(コウチュウ目、ハムシ科) 日本

英名	和名	学名	所属、分布、ほか
yellow-legged tortoiseshell			scarce tortoiseshell を見よ
yellow-legged tussock moth	キアシドクガ	*Ivela auripes* (Butler)	(チョウ目、ドクガ科) 日本、旧北区
yellow-legged water-snipefly	ハマダラシギアブ	*Atherix ibis* (Fabricius)	(ハエ目、ナガレアブ科) 旧北区。日本亜種は *A. i. japonica* Nagatomi ハマダラナガレアブ
yellow-line quaker		*Agrochola macilenta* (Hübner)	(チョウ目、ヤガ科) 旧北区
yellow-lined angle moth		*Digrammia mellistrigata* (Grote)	(チョウ目、シャクガ科) 新北区
yellow-lined caterpillar			tawny prominent (1) を見よ
yellow-lined chocolate moth		*Argyrostrotis flavistriaria* (Hübner)	(チョウ目、ヤガ科) 新北区
yellow-lined owlet moth		*Colobochyla interpuncta* (Grote)	(チョウ目、ヤガ科) 新北区。yellow-lined owlet ともいう
yellow-lined skimmer		*Orthemis biolleyi* Calvert	(トンボ目、トンボ科) 新熱帯区
yellow-lined thorn			pale Metanema moth を見よ
yellow Liptena		*Liptena homeyeri* Dewitz	(チョウ目、シジミチョウ科) エチオピア区
yellow longicorn		*Phoracantha recurva* Newman	(コウチュウ目、カミキリムシ科) 豪州区
yellow long-legged wasp	コアシナガバチ	*Polistes snelleni* Saussure	(ハチ目、スズメバチ科) 日本、旧北区
yellow long-necked leaf beetle	キイロクビナガハムシ	*Lilioceris rugata* (Baly)	(コウチュウ目、ハムシ科) 日本
yellow-margined buprestid	キンヘリタマムシ	*Scintillatryx pretiosa bellula* (Lewis)	(コウチュウ目、タマムシ科) 日本
yellow-margined leaf beetle		*Microtheca ochroloma* Stål	(コウチュウ目、ハムシ科) 新北区
yellow-margined shieldback		*Clinopleura flavomarginata* Scudder	(バッタ目、キリギリス科) 新北区
yellow-margined spined beetle	ヒメキベリトゲハムシ	*Dactylispa angulosa* (Solsky)	(コウチュウ目、ハムシ科) 日本、旧北区
yellow-marked blue leaf beetle	キボシルリハムシ	*Smaragdina aurita* (Linnaeus)	(コウチュウ目、ハムシ科) 日本、旧北区
yellow-marked buprestids	フナガタタマムシ亜科	Acmaeoderinae	(コウチュウ目、タマムシ科) の昆虫の総称
yellow-marked cotton green moth			spotted bollworm (2) を見よ
yellow-marmorated drepanid			red-legged drepanid を見よ
yellow marmorated stink bug	キマダラカメムシ	*Erthesina fullo* (Thunberg)	(カメムシ目、カメムシ科) 日本、東洋区
yellow May		*Heptagenia sulphurea* (Müller)	(カゲロウ目、ヒラタカゲロウ科) 旧北区
yellow mayfly		*Potamanthus luteus* (Linnaeus)	(カゲロウ目、カワカゲロウ科) 日本、旧北区
yellow meadow ant	キイロケアリ	*Lasius flavus* (Fabricius)	(ハチ目、アリ科) 日本、全北区

英名	和名	学名	所属、分布、ほか
yellow mealworm	チャイロコメノゴミムシダマシ	*Tenebrio molitor* Linnaeus	(コウチュウ目、ゴミムシダマシ科) 日本、汎世界。貯穀害虫
yellow mealworm beetle			yellow mealworm を見よ
yellow metalmark		*Mesene silaris* Godman et Salvin	(チョウ目、シジミタテハ科) 新熱帯区
yellow migrant		*Catopsilia gorgophone* (Boisduval)	(チョウ目、シロチョウ科) 豪州区
yellow migrant			orange migrant を見よ
yellow minute bark beetle			pine bark beetle (1) を見よ
yellow minute flea beetle	オオバコトビハムシ	*Longitarsus scutellaris* (Rey)	(コウチュウ目、ハムシ科) 日本、旧北区、東洋区
yellow mirid			apple dumpling bug を見よ
yellow Mocis moth		*Mocis disseverans* (Walker)	(チョウ目、ヤガ科) 新北区
yellow Monday			greengrocer を見よ
yellow Monolepta beetle			red-shouldered leaf beetle (1) を見よ
yellow mulberry leaf beetle	キイロクワハムシ	*Monolepta pallidula* (Baly)	(コウチュウ目、ハムシ科) 日本、旧北区、東洋区
yellow mulberry leafhopper	クワキヨコバイ	*Pagaronia guttigera* (Uhler)	(カメムシ目、オオヨコバイ科) 日本、旧北区。クワ、カキ、ムギなどの害虫
yellow-necked apple caterpillar			yellow-necked caterpillar を見よ
yellow necked apple worm			yellow-necked caterpillar を見よ
yellow-necked caterpillar		*Datana ministra* (Drury)	(チョウ目、シャチホコガ科) 新北区。リンゴ他害虫。yellow-neck caterpillar ともいう
yellow-necked caterpillar moth			yellow-necked caterpillar を見よ
yellow-necked drywood termite		*Kalotermes flavicollis* (Fabricius)	(シロアリ目、レイビシロアリ科) 全北区、東洋区。yellownecked dry-wood termite ともいう
yellow nutsedge moth		*Diploschizia impigritella* (Clemens)	(チョウ目、ホソハマキモドキガ科) 新北区
yellow ochre			rare white-spot skipper を見よ
yellow onyx		*Horaga moulmeina* Moore	(チョウ目、シジミチョウ科) 東洋区
yellow onyx (1)	トガリミツオシジミ	*Horaga syrinx* (Felder)	(チョウ目、シジミチョウ科) 東洋区。*H. s. maenala* Hewitson も同英名
yellow Ophion	オオアメバチ	*Ophion luteus* (Linnaeus)	(ハチ目、ヒメバチ科) 日本、全北区、エチオピア区
yellow orange tip		*Colotis auxo* (Lucas)	(チョウ目、シロチョウ科) エチオピア区
yellow orange tip	メスシロキチョウ	*Ixias pyrene* (Linnaeus)	(チョウ目、シロチョウ科) 東洋区。*I. p. verna* Druce, *I. p. insignis* Butler も同英名
yellow orchid aphid			orchid aphid (2) を見よ
yellow orchid thrips		*Anaphrothrips orchidaceus* Bagnall	(アザミウマ目、アザミウマ科) 旧北区
yellow orchid thrips			orchid thrips を見よ
yellow owl	キオビムカシヒカゲ	*Neorina hilda* Westwood	(チョウ目、タテハチョウ科) 東洋区

英名	和名	学名	所属、分布、ほか
yellow paddy stem borer			yellow rice borer を見よ
yellow palm dart		*Cephrenes trichopepla* (Lower)	（チョウ目、セセリチョウ科）東洋区、豪州区
yellow pansy	ルリボシタテハモドキ	*Junonia hierta* (Fabricius)	（チョウ目、タテハチョウ科）東洋区、エチオピア区
yellow paper wasp		*Polistes dominulus* (Christ)	（ハチ目、スズメバチ科）豪州区
yellow pasha		*Herona marathus angustata* Moore	（チョウ目、タテハチョウ科）東洋区。Pasha を参照
yellowpatch skipper			Peck's skipper (1) (2) を見よ
yellow patch white		*Colotis halimede* (Klug)	（チョウ目、シロチョウ科）エチオピア区
yellow-patched bent-skipper		*Ebrietas osyris* (Staudinger)	（チョウ目、セセリチョウ科）新熱帯区
yellow-patched leafhopper	モンキヒロズヨコバイ	*Oncopsis mali* (Matsumura)	（カメムシ目、オオヨコバイ科）日本
yellow-patched satyr		*Oxeoschistus tauropolis tauropolis* (Westwood)	（チョウ目、タテハチョウ科）新北区、新熱帯区
yellow peach moth			peach moth を見よ
yellow peach moth parasite		*Argyrophylax proclinata* Crosskey	（ハエ目、ヤドリバエ科）豪州区
yellow pecan aphid		*Monelliopsis pecanis* Bissell	（カメムシ目、アブラムシ科）新北区、新熱帯区、エチオピア区
yellow pine engraver			California five-spined Ips を見よ
yellow plant bug		*Metriorhynchomiris dislocatus* (Say)	（カメムシ目、カスミカメムシ科）新北区
yellow Presba		*Syncordulia gracilis* Burmeister	（トンボ目、エゾトンボ科）エチオピア区
yellow Rajah	ヘリグロフタオチョウ	*Charaxes marmax* Westwood	（チョウ目、タテハチョウ科）東洋区
yellow rice borer	サンカメイガ	*Scirpophaga incertulas* (Walker)	（チョウ目、メイガ科）日本、東洋区、エチオピア区。イネ害虫
yellow rice leafhopper	キイロヒメヨコバイ	*Thaia subrufa* (Motschulsky)	（カメムシ目、オオヨコバイ科）日本、東洋区。サトウキビ害虫
yellow-rimmed eighty-eight		*Callicore texa heroica* (Fruhstorfer)	（チョウ、タテハチョウ科）新熱帯区。*C. t. loxicha* Maza et Maza, *C. t. tacana* Maza et Maza, *C. t. titania* (Salvin) も同英名。Texa numberwing を参照
yellow-rimmed flasher			dull Astraptes を見よ
yellow-rimmed scarlet-eye		*Ocyba calathana calanus* (Godman et Salsin)	（チョウ目、セセリチョウ科）新熱帯区
yellow-rimmed skipper		*Aethilla lavochrea* Butler	（チョウ目、セセリチョウ科）新熱帯区
yellow-ringed carpet		*Entephria flavicinctata* (Hübner)	（チョウ目、シャクガ科）旧北区
yellow-ringed cosmet		*Mompha miscella* (Denis et Schiffermüller)	（チョウ目、カザリバガ科）旧北区
yellow-ringed sympetrum			yellow-winged darter を見よ

英名	和名	学名	所属、分布、ほか
yellow-rippled white looper	キナミシロヒメシャク	*Scopula superior* (Butler)	（チョウ目、シャクガ科）日本、旧北区
yellow rose aphid			green rose aphid を見よ
yellow sailer	ホリシャミスジ	*Neptis ananta* Moore	（チョウ目、タテハチョウ科）東洋区
yellow sally		*Chloroperla viridis* Fabricius	（カワゲラ目、ミドリカワゲラ科）旧北区
yellow satin grass-veneer	ウスギンツトガ	*Crambus perlellus* (Scopoli)	（チョウ目、メイガ科）旧北区。トウモロコシ害虫
yellow satin-flower case		*Coleophora solitariella* (Zeller)	（チョウ目、ツツミノガ科）旧北区
yellow scale	キマルカイガラムシ	*Aonidiella citrina* (Coquillett)	（カメムシ目、マルカイガラムシ科）日本、新北区、豪州区、汎熱帯。カンキツ害虫
yellow scallop moth			cotton leaf caterpillar (1) を見よ
yellow sedge borer			subflava sedge borer moth を見よ
yellow shell		*Camptogramma bilineata* (Linnaeus)	（チョウ目、シャクガ科）旧北区
yellow shieldback		*Ateloplus luteus* Caudell	（バッタ目、キリギリス科）新北区
yellow short-barred conch		*Aethes dilucidana* (Stephens)	（チョウ目、ホソハマキガ科）旧北区
yellow-shouldered dung beetle		*Liatongus militaris* (Castelnau)	（コウチュウ目、コガネムシ科）豪州区
yellow-shouldered lady beetle		*Scymnodes lividigaster* (Mulsant)	（コウチュウ目、テントウムシ科）新北区
yellow-shouldered ladybird			yellow-shouldered lady beetle を見よ
yellow-shouldered slug moth		*Lithacodes fasciola* (Herrich-Schäffer)	（チョウ目、イラガ科）新北区
yellow-shouldered souring beetle		*Urophorus humeralis* (Fabricius)	（コウチュウ目、ケシキスイ科）大洋区
yellow-sided clubtail		*Stylurus potulentus* (Needham)	（トンボ目、サナエトンボ科）新北区
yellow silverline		*Cigaritis epargyros* (Eversmann)	（チョウ目、シジミチョウ科）旧北区
yellow skipper		*Hesperilla flavescens* Waterhouse	（チョウ目、セセリチョウ科）豪州区
yellow slant-line moth		*Tetracis crocallata* Guenée	（チョウ目、シャクガ科）新北区
yellow snout-moth			spotted grass moth を見よ
yellow soldier fly		*Pthecicus trivitatus* (Say)	（ハエ目、ミズアブ科）新北区
yellow soldier fly		*Inopus flavus* (James)	（ハエ目、ミズアブ科）豪州区
yellow spear-winged fly		*Lonchoptera lutea* Panzer	（ハエ目、ヤリバエ科）旧北区、東洋区
yellow spider beetle			golden spider beetle を見よ
yellow-spiked wood nymph		*Posttaygetis penelea* (Cramer)	（チョウ目、タテハチョウ科）新熱帯区
yellow splendour		*Colotis protomedia* (Klug)	（チョウ目、シロチョウ科）エチオピア区

英名	和名	学名	所属、分布、ほか
yellow spot blue	ウラギンミナミシジミ	*Candalides xanthospilos* (Hübner)	(チョウ目、シジミチョウ科) 豪州区。yellow spot blue butterfly, yellow-spotted blue ともいう
yellow spot jewel		*Hypochrysops byzos* (Boisduval)	(チョウ目、シジミチョウ科) 豪州区。yellow jewel ともいう
yellow-spot swift		*Polytremis eltola* (Hewitson)	(チョウ目、セセリチョウ科) 東洋区
yellow-spotted black weevil	クロキボシゾウムシ	*Pissodes obscurus* Roelofs	(コウチュウ目、ゾウムシ科) 日本
yellow-spotted chafer		*Odontria xanthosticta* White	(コウチュウ目、コガネムシ科) 豪州区
yellow-spotted cranefly			spotted cranefly (1) を見よ
yellow-spotted emerald		*Somatochlora flavomaculata* (Van der Linden)	(トンボ目、エゾトンボ科) 旧北区
yellow-spotted fir weevil	トドマツキボシゾウムシ	*Pissodes cembrae* Motschulsky	(コウチュウ目、ゾウムシ科) 日本、旧北区。トドマツ、エゾマツ類などの害虫
yellow-spotted froghopper			paler froghopper を見よ
yellow spotted Gonatryx			ghost brimstone を見よ
yellow-spotted graylet moth		*Hyperstrotia flaviguttata* (Grote)	(チョウ目、ヤガ科) 新北区。yellow-spotted graylet ともいう
yellow-spotted greyish geometrid			Blomer's rivulet を見よ
yellow-spotted Jezebel			Nysa Jezebel を見よ
yellow-spotted lady beetles		Hyperaspini	(コウチュウ目、テントウムシ科) の昆虫の総称
yellow-spotted ladybird beetles			yellow-spotted lady beetles を見よ
yellow-spotted lancewing	キモンクロササベリガ	*Phaulernis fulviguttella* (Zeller)	(チョウ目、ササベリガ科) 旧北区
yellow-spotted lift		*Antispila metallella* (Denis et Schiffermüller)	(チョウ目、ツヤコガ科) 旧北区
yellow-spotted longicorn beetle	キボシカミキリ	*Psacothea hilaris hilaris* (Pascoe)	(コウチュウ目、カミキリムシ科) 日本、旧北区、東洋区。イチジク、クワ害虫。yellow-spotted longicorn ともいう
yellow-spotted pine weevil	マツキボシゾウムシ	*Pissodes nitidus* Roelofs	(コウチュウ目、ゾウムシ科) 日本、旧北区。マツ類、ヒマラヤスギ害虫
yellow-spotted Renia moth		*Renia flavipunctalis* (Geyer)	(チョウ目、ヤガ科) 新北区。yellow-spotted Renia ともいう
yellow-spotted ringlet	マントベニヒカゲ	*Erebia manto* (Schiffermüller)	(チョウ目、タテハチョウ科) 旧北区
yellow-spotted ruby-eye		*Carystoides sicania orbius* (Godman)	(チョウ目、セセリチョウ科) 新熱帯区
yellow-spotted ruby-eye		*Telles arcalaus* (Stoll)	(チョウ目、セセリチョウ科) 新熱帯区
yellow-spotted small noctuid	キマダラコヤガ	*Emmelia trabealis* (Scopoli)	(チョウ目、ヤガ科) 日本、旧北区

英名	和名	学名	所属、分布、ほか
yellow-spotted Telemiades		*Telemiades avitus* (Stoll)	（チョウ目、セセリチョウ科）新熱帯区
yellow spotted Temnora		*Temnora acitula* (Holland)	（チョウ目、スズメガ科）エチオピア区
yellow-spotted tiger moth		*Halisidota maculata* (Harris)	（チョウ目、ヒトリガ科）新北区。幼虫はポプラ、カエデなどの害虫
yellow spotted water beetle			marbled diving beetle を見よ
yellow-spotted water beetles		*Thermonectus*	（コウチュウ目、ゲンゴロウ科）の昆虫の総称
yellow-spotted webworm moth		*Anageshna primordialis* (Dyar)	（チョウ目、メイガ科）新北区
yellow-spotted white prominent			white prominent を見よ
yellow-spotted whiteface		*Leucorrhinia pectoralis* (Charpentier)	（トンボ目、トンボ科）旧北区
yellow-spotted willow-slug			willow sawfly (1) を見よ　幼虫の英名
yellow Spragueia moth		*Spragueia apicalis* Herrich-Schäffer	（チョウ目、ヤガ科）新北区、新熱帯区。yellow Spragueia ともいう
yellow springtail		*Bourletiella lutea* (Lubbock)	（トビムシ目、マルトビムシ科）旧北区
yellow springtail			sugarcane springtail を見よ
yellow spruce budworm		*Zeiraphera fortunana* Kearfott	（チョウ目、ハマキガ科）新北区。成虫は yellow spruce budworm moth
yellow-stained skipper		*Poanes inimica* (Butler et Druce)	（チョウ目、セセリチョウ科）新北区、新熱帯区
yellow stem borer			yellow rice borer を見よ
yellow-striped armyworm			western yellow-striped armyworm を見よ
yellow-striped armyworm moth		*Spodoptera ornithogalli* (Guenée)	（チョウ目、ヤガ科）新北区、新熱帯区。幼虫はワタその他多くの植物につく。幼虫は yellow striped armyworm
yellowstriped caterpillar			spotted Datana moth を見よ
yellowstriped cauliflower army worm			common cutworm を見よ
yellow striped chafer	キスジコガネ	*Phyllopertha irregularis* Waterhouse	（コウチュウ目、コガネムシ科）日本
yellow striped-edge piercer		*Lathronympha strigana* (Fabricius)	（チョウ目、ハマキガ科）旧北区
yellow-striped flea beetle			striped flea beetle (1) を見よ
yellow-striped flutterer		*Rhyothemis phyllis* (Sulzer)	（トンボ目、トンボ科）豪州区。yellow-barred flutterer ともいう
yellow-striped hunter		*Austrogomphus guerini* (Rambur)	（トンボ目、サナエトンボ科）豪州区
yellow-striped leafhopper			hop jumper を見よ

英名	和名	学名	所属、分布、ほか
yellowstriped oakworm			Peigler's oakworm moth を見よ
yellow-striped ruby-eye		*Dubiella fiscella belpa* Evans	（チョウ目、セセリチョウ科）新熱帯区
yellow-striped sister		*Adelpha leuceria leuceria* (Druce)	（チョウ目、タテハチョウ科）新熱帯区
yellow-striped zygaenid	キスジホソマダラ	*Balataea gracilis* (Walker)	（チョウ目、マダラガ科）日本、旧北区
yellow sugarcane aphid		*Sipha flava* (Forbes)	（カメムシ目、アブラムシ科）新北区、新熱帯区
yellow sugarcane cicada		*Parnkalla muelleri* (Distant)	（カメムシ目、セミ科）豪州区
yellow sunflower moth		*Stiria rugifrons* Grote	（チョウ目、ヤガ科）新北区
yellow swarming fly			small cluster fly を見よ
yellow swift		*Borbo impar* Mabille	（チョウ目、セセリチョウ科）豪州区
yellow sylph		*Ceratrichia semlikensis* Joicey et Talbot	（チョウ目、セセリチョウ科）エチオピア区
yellow tail			brown-tail moth (1)(2) を見よ
yellow-tail moth			browntail moth (1)(2) を見よ　yellow-tail, yellow tail moth ともいう。英国での英名
yellow-tailed swallowtail			Cramer's swallowtail を見よ
yellow tea thrips		*Scirtothrips kenyensis* Mound	（アザミウマ目、アザミウマ科）エチオピア区
yellow tea thrips (1)	チャノキイロアザミウマ	*Scirtothrips dorsalis* Hood	（アザミウマ目、アザミウマ科）日本、東洋区。チャ、果樹など多くの作物害虫
yellow tent caterpillar		*Euschausia ingens* Edwards	（チョウ目、ヒトリガ科）新北区
yellow three-spot moth		*Apamea helva* Grote	（チョウ目、ヤガ科）新北区。yellow three-spot ともいう
yellow thrips			yellow orchid thrips を見よ
yellow tiger longicorn	キイロトラカミキリ	*Demonax notabilis* (Pascoe)	（コウチュウ目、カミキリムシ科）日本、旧北区、東洋区
yellow tinsel		*Catapaecilma subochrea* (Elwes)	（チョウ目、シジミチョウ科）東洋区
yellow tinsel		*Pentila subochracea* Hawker-Smith	（チョウ目、シジミチョウ科）東洋区
yellow tip	ツマキチョウ	*Anthocharis scolymus* Butler	（チョウ目、シロチョウ科）日本、旧北区
yellow-tipped bug	ツマキヘリカメムシ	*Hygia opaca* (Uhler)	（カメムシ目、ヘリカメムシ科）日本、旧北区、東洋区
yellow-tipped flasher			dull Astraptes を見よ
yellow-tipped lema	トゲアシクビボソハムシ	*Lema coronata* Baly	（コウチュウ目、ハムシ科）日本、東洋区
yellow-tipped malthine		*Malthinus flaveolus* (Herbst)	（コウチュウ目、ジョウカイボン科）旧北区
yellow-tipped melyrid	ツマキアオジョウカイモドキ	*Malachius prolongatus* Motschulsky	（コウチュウ目、ジョウカイモドキ科）日本

英名	和名	学名	所属、分布、ほか
yellow-tipped prominent			quercus caterpillar を見よ
yellow-tipped pyralid			top borer (1) を見よ
yellow-tipped soldier		*Oxycera terminata* Wiedemann	(ハエ目、ミズアブ科) 旧北区
yellow-tipped ticlear			Tutia clearwing を見よ
yellow top borer			early shoot borer を見よ　yellow top-borer とも記す
yellow-tufted Prepona		*Prepona laertes octavia* Fruhstorfer	(チョウ目、タテハチョウ科) 新熱帯区
yellow turf ant			yellow meadow ant を見よ
yellow tussock			pale tussock を見よ
yellow tussock moth	リンゴドクガ	*Calliteara pseudabietis* Butler	(チョウ目、ドクガ科) 日本。リンゴ、サクラ、カエデなどの害虫
yellow tussock moth	キドクガ	*Euproctis piperita* Oberthür	(チョウ目、ドクガ科) 日本、旧北区。リンゴ、マメ科牧草などの害虫
yellow underwing			common yellow underwing moth を見よ
yellow underwing moth			common yellow underwing moth を見よ
yellow-underwinged pearl		*Uresiphita gilvata* (Fabricius)	(チョウ目、メイガ科) 旧北区
yellow-underwings	モンヤガ亜科	Noctuinae	(チョウ目、ヤガ科) の昆虫の総称
yellow V carl		*Oinophila v-flava* (Haworth)	(チョウ目、ヒロズコガ科) 旧北区
yellow v moth			yellow V carl を見よ
yellow vein lancer		*Pyroneura latoia latoia* (Hewitson)	(チョウ目、セセリチョウ科) 東洋区
yellow-veined Acraea	キシタホソチョウ	*Acraea parrhasia* (Fabricius)	(チョウ目、タテハチョウ科) エチオピア区
yellow-veined geometer moth		*Orthofidonia flavivenata* (Hulst)	(チョウ目、シャクガ科) 新北区
yellow-veined moth		*Microtheoris ophionalis* (Walker)	(チョウ目、メイガ科) 新北区
yellow-veined noctuid	キミャクヨトウ	*Dictyestra dissecta* (Walker)	(チョウ目、ヤガ科) 日本、東洋区
yellow-veined skipper		*Parphorus decora* (Herrich-Schäffer)	(チョウ目、セセリチョウ科) 新熱帯区
yellow-veined widow		*Palpopleura jucunda* Rambur	(トンボ目、トンボ科) エチオピア区
yellow velvet beetle		*Cosmosalia chrysocoma* (Kirby)	(コウチュウ目、カミキリムシ科) 新北区
yellow-vested moth		*Rectiostoma xanthobasis* Zeller	(チョウ目、マルハキバガ科) 新北区
yellow-washed Metarranthis moth		*Metarranthis obfirmaria* (Hübner)	(チョウ目、シャクガ科) 新北区
yellow-washed ruby-eye		*Tromba xanthura* (Godman)	(チョウ目、セセリチョウ科) 新熱帯区
yellow wave moth		*Hybroma servulella* Clemens	(チョウ目、ヒロズコガ科) 新北区

英名	和名	学名	所属、分布、ほか
yellow waved carpet			small yellow wave を見よ
yellow waxtail		*Ceriagrion coromandelianum* (Fabricius)	（トンボ目、イトトンボ科）東洋区
yellow willow aphid			celery aphid (1) を見よ
yellow willow beetle			yellow willow leaf beetle を見よ
yellow willow leaf beetle		*Galerucella lineola* (Fabricius)	（コウチュウ目、ハムシ科）旧北区
yellow-winged bug	キベリヘリカメムシ	*Megalotomus costalis* Stål	（カメムシ目、ホソヘリカメムシ科）日本、旧北区
yellow-winged darter	エゾアカネ	*Sympetrum flaveolum flaveolum* (Linnaeus)	（トンボ目、トンボ科）日本、旧北区
yellow-winged grasshopper			orange-winged grasshopper を見よ
yellow-winged locust		*Gastrimargus musicus* (Fabricius)	（バッタ目、バッタ科）豪州区
yellow-winged oak leafroller moth		*Argyrotaenia quercifoliana* Fitch	（チョウ目、ハマキガ科）新北区
yellow-winged Pareuchaetes		*Pareuchaetes insulata* (Walker)	（チョウ目、ヒトリガ科）新熱帯区。yellow-winged Pareuchaetes moth ともいう
yellow-winged sympetrum			yellow-winged darter を見よ
yellow winged Temnora		*Temnora pylades* Rothschild et Jordan	（チョウ目、スズメガ科）エチオピア区
yellow woodbrown		*Lethe nicetas* (Hewitson)	（チョウ目、タテハチョウ科）東洋区
yellow woolly-bear			Virginia tiger moth を見よ
yellow woollybear moth			Virginia tiger moth を見よ
yellow zebra	デウカリオンタイマイ	*Graphium deucalion* (Boisduval)	（チョウ目、アゲハチョウ科）東洋区
yellow Zulu		*Alaena amazoula ochroma* Vári	（チョウ目、シジミチョウ科）エチオピア区。
yellowish asura	ヒメキホソガ	*Asura dharma* (Moore)	（チョウ目、ヒトリガ科）日本、東洋区
yellowish carpenter ant	キイロシリアゲアリ	*Crematogaster osakensis* Forel	（ハチ目、アリ科）日本、旧北区
yellowish chewing louse of sheep and goat			sheep biting louse を見よ
yellowish chewing louse of sheep and goats		*Bovicola penicillata* Cuvier	（ハジラミ目、ケモノハジラミ科）エチオピア区、豪州区
yellowish darter moth	キハダケンモン	*Triaena leucocuspis* (Butler)	（チョウ目、ヤガ科）日本
yellowish elongate chafer	ナガチャコガネ	*Heptophylla picea picea* Motschulsky	（コウチュウ目、コガネムシ科）日本。リンゴ、チャ、トドマツ類などの害虫
yellowish horn prominent	トビギンボシシャチホコ	*Eguria ornata* (Oberthür)	（チョウ目、シャチホコガ科）日本、旧北区
yellowish lady beetle	チャイロテントウ	*Micraspis discolor* (Fabricius)	（コウチュウ目、テントウムシ科）日本、東洋区

英名	和名	学名	所属、分布、ほか
yellowish Miltochrista	ハガタキコケガ	*Miltochrista pallida* (Bremer)	（チョウ目、ヒトリガ科）日本、旧北区
yellowish prominent	キシャチホコ	*Torigea straminea* (Moore)	（チョウ目、シャチホコガ科）日本、旧北区
yellowish skipper			yellow skipper を見よ
yellowish tree cricket			tree cricket (2) を見よ
yellowish white tailed geometrid			tailed geometrid (2) を見よ
yellowish Zanclognatha moth		*Zanclognatha marcidilinea* (Walker)	（チョウ目、ヤガ科）新北区。yellowish Zanclognatha ともいう
yellowjack			southern yellowjack を見よ
yellowjack sailer		*Lasippa viraja* (Moore)	（チョウ目、タテハチョウ科）東洋区
yellowjacket fly		*Chrysotoxum integre* Williston	（ハエ目、ハナアブ科）新北区
yellowjacket hover fly		*Milesia virginiensis* (Drury)	（ハエ目、ハナアブ科）新北区
yellowjackets		*Dolichovespula*, *Vespula*	（ハチ目、スズメバチ科）の昆虫の総称　ヨーロッパでは両属に対し wasps や true wasps を使うこと多し。
yellowjackets			wasps (1) を見よ
yellowjackets			common wasps (1)(2) を見よ
yellowjackets			hornets (1) を見よ
Yellowstone ringlet			Hayden's ringlet を見よ
yellow stoneroot borer moth		*Papaipema astuta* Bird	（チョウ目、ヤガ科）新北区
yellows	モンキチョウ亜科	Coliadinae	（チョウ目、シロチョウ科）の昆虫の総称
Yeomans	ミナミヒョウモン属	*Cirrochroa*	（チョウ目、タテハチョウ科）の昆虫の総称
Yerba Santa bird-dropping moth		*Ethmia arctostaphylella* Walsingham	（チョウ目、スエヒロキバガ科）新北区。Yerba Santa の葉上の鳥の糞に似ることに由来
Yerbury's elf		*Tetrathemis yerburii* Kirby	（トンボ目、トンボ科）東洋区
Yerbury's sailer		*Neptis nata yerburii* Butler	（チョウ目、タテハチョウ科）東洋区。dirty sailer を参照
Yersin's ground mantis		*Yersiniops sophronica* Rehn et Hebart	（カマキリ目、カマキリ科）新北区
Yesso spruce aphid	エゾマツアブラムシ	*Cinara ezoana* Inouye	（カメムシ目、アブラムシ科）日本、旧北区
Yesso spruce bark beetle	アラキサエダキクイムシ	*Pityophthorus arakii* Sawamoto	（コウチュウ目、キクイムシ科）日本
Yesso spruce lasiocampid			hemlock caterpillar を見よ
Yesso spruce longicorn			black spruce beetle を見よ
Yesso spruce minute scolytid	トウヒノヒメキクイムシ	*Pityophthorus jucundus* Blandford	（コウチュウ目、キクイムシ科）日本、旧北区
Yesso spruce narrow bark beetle	カバイロホソキクイムシ	*Crypturgus tuberosus* Niijima	（コウチュウ目、キクイムシ科）日本、旧北区
yew gall midge		*Taxomyia taxi* (Inchbald)	（ハエ目、タマバエ科）旧北区。本種が yew イチイの新梢につくるアザミ状の虫えいは 10〜20 mm で yew artichoke gall という

英名	和名	学名	所属、分布、ほか
yew scale		*Parthenolecanium pomeranicum* (Kawlecki)	（カメムシ目、カタカイガラムシ科）旧北区
yew scale		*Eulecanium crudum* Green	（カメムシ目、カタカイガラムシ科）旧北区
Yezo bark beetle	エゾキクイムシ	*Polygraphus jezoensis* Niijima	（コウチュウ目、キクイムシ科）日本、旧北区
Yezo blue leaf-cut weevil			maple broad-mouth weevil を見よ
Yezo cicada	エゾゼミ	*Lyristes japonicus* (Kato)	（カメムシ目、セミ科）日本、旧北区
Yezo daimyo oak bug	ヘラクヌギカメムシ	*Urostylis annulicornis* Scott	（カメムシ目、クヌギカメムシ科）日本、旧北区
Yezo forested geometrid	ハイイロオオエダシャク	*Biston regalis comitata* (Warren)	（チョウ目、シャクガ科）日本、旧北区、東洋区
Yezo four-eyed silk moth	エゾヨツメ	*Aglia tau microtau* Inoue	（チョウ目、ヤママユガ科）日本
Yezo horntail	ニセアカアシクビナガキバチ	*Xiphydria palaeanartica* Semonov-Tian-Shanskij	（ハチ目、クビナガキバチ科）日本、旧北区
Yezo longicorn			weaver beetle を見よ
Yezo longicorn beetle			weaver beetle を見よ
Yezo pectinate-horned prominent	クシヒゲシャチホコ	*Ptilophora jezoensis* (Matsumura)	（チョウ目、シャチホコガ科）日本、旧北区
Yezo red-hindwinged catocala			red underwing を見よ
Yezo rice grasshopper			rice grasshopper (5) を見よ
Yezo small weevil	エゾヒメゾウ	*Baris ezoana* Kono	（コウチュウ目、ゾウムシ科）日本
Yezo spring cicada	エゾハルゼミ	*Terpnosia nigricosta* (Motschulsky)	（カメムシ目、セミ科）日本、旧北区
Yojoa hairstreak		*Strymon yojoa* (Reakirt)	（チョウ目、シジミチョウ科）新熱帯区
Yojoa scrub-hairstreak			Yojoa hairstreak を見よ
Yokoyama marmorated aphid	カシワトゲマダラアブラムシ	*Tuberculatus yokoyamai* (Takahashi)	（カメムシ目、アブラムシ科）日本、旧北区
yomena aphid	ヨメナヒメヒゲナガアブラムシ	*Macrosiphoniella yomenae* (Shinji)	（カメムシ目、アブラムシ科）日本、旧北区、東洋区
Yosemite bark weevil		*Pissodes schwarzi* Hopkins	（コウチュウ目、ゾウムシ科）新北区
Yosemite shieldback		*Decticita yosemite* Rentz et Birchim	（バッタ目、キリギリス科）新北区
yoshibue elongate scolytid	ヨシブエナガキクイムシ	*Platypus calamus* Blandford	（コウチュウ目、ナガキクイムシ科）日本、旧北区、東洋区
Young's alpine			Yukon alpine を見よ
Young's grass-veneer moth		*Crambus youngellus* Kearfott	（チョウ目、メイガ科）新北区。Young's grass-veneer ともいう
youthful underwing moth		*Catocala subnata* Grote	（チョウ目、ヤガ科）新北区。youthful underwing ともいう
yponomeutid moths			ermine moths (1) を見よ
yponomeutids			ermine moths (1) を見よ
Yuba skipper			Juba skipper を見よ
Yucatan Calephelis		*Calephelis yucatana* McAlpine	（チョウ目、シジミタテハ科）新熱帯区

英名	和名	学名	所属、分布、ほか
Yucatan katydid		*Phrixa maya* Saussure et Pictet	(バッタ目、キリギリス科) 新北区
Yucatan ministreak		*Ministrymon* sp.	(チョウ目、シジミチョウ科) 新熱帯区
Yucatan mottled-skipper		*Codatractus yucatanus* Freeman	(チョウ目、セセリチョウ科) 新熱帯区
Yucatan pipevine swallowtail	アオジャコウ	*Battus philenor acauda* (Oberthür)	(チョウ目、アゲハチョウ科) 新熱帯区。pipevine swallowtail を参照
Yucatan thorn-scrub leafwing		*Fountainea halice maya* (Witt)	(チョウ目、タテハチョウ科) 新熱帯区。ruddy leafwing を参照
yucca giant-skipper		*Megathymus yuccae* (Boisduval et LeConte)	(チョウ目、セセリチョウ科) 新北区。*M. y. reinthali* Freeman も同英名
yucca moth		*Tegeticula yuccasela* (Riley)	(チョウ目、マガリガ科) 新北区。開帳 20～25 mm。幼虫はユッカの実を食う
yucca moths		Prodoxinae	(チョウ目、マガリガ科) の昆虫の総称
yucca moths		*Tegeticula*	(チョウ目、マガリガ科) の昆虫の総称
yucca moths (1)	マガリガ科	Incurvariidae	(チョウ目) の昆虫の総称
yucca moths (2)		Prodoxidae	(チョウ目) の昆虫の総称 マガリガ科の亜科とされる
yucca moths (3)		*Prodoxus*	(チョウ目、マガリガ科) の昆虫の総称
yucca plant bug		*Halticotoma valida* Townsend	(カメムシ目、カスミカメムシ科) 新北区。ユッカにつく
yucca skipper			yucca giant-skipper を見よ
yucca weevil		*Scyphophorus yuccae* Horn	(コウチュウ目、ゾウムシ科) 新北区
Yugoslavian green-veined white		*Pieris balcana* (Lorkovic)	(チョウ目、シロチョウ科) 旧北区
Yukon alpine	アラスカベニヒカゲ	*Erebia youngi* Holland	(チョウ目、タテハチョウ科) 新北区
Yukon blue	カラフトルリシジミ	*Vacciniina optilete* Knoch	(チョウ目、シジミチョウ科) 全北区
Yukon blue			cranberry blue を見よ
Yule's dotted border		*Mylothris yulei* Butler	(チョウ目、シロチョウ科) エチオピア区
Yuma skipper		*Ochlodes yuma* (Edwards)	(チョウ目、セセリチョウ科) 新北区
Yunnan admiral		*Limenitis homeyeri* Tancre	(チョウ目、タテハチョウ科) 旧北区
Yunnan black-veined white	シナミヤマシロチョウ	*Aporia goutellei* (Oberthür)	(チョウ目、シロチョウ科) 旧北区
Yunnan chequred blue		*Sinia lanty* Oberthür	(チョウ目、シジミチョウ科) 旧北区
Yunnan peacock		*Papilio syfanius* Oberthür	(チョウ目、アゲハチョウ科) 旧北区
Yunnan sailor	シナオオミスジ	*Neptis dejeani* Oberthür	(チョウ目、タテハチョウ科) 旧北区

英名	和名	学名	所属、分布、ほか
Z			
Zabulon skipper		*Poanes zabulon* (Boisduval et LeConte)	（チョウ目、セセリチョウ科）新北区、新熱帯区
Zaddach's forester	ザダックボカシタテハ	*Euphaedra zaddachi* Dewitz	（チョウ目、タテハチョウ科）エチオピア区。亜種 *E. z. elephantina* Staudinger も同英名
Zaddach's mimic forester			Zaddach's forester を見よ
Zaela mimic white		*Dismorphia zaela* (Hewitson)	（チョウ目、シロチョウ科）新熱帯区
Zambezi siphontail		*Neurogomphus zambeziensis* Cammaerts	（トンボ目、サナエトンボ科）エチオピア区
Zambezi skipper		*Abantis zambesiaca* (Westwood)	（チョウ目、セセリチョウ科）エチオピア区
zanolid moths		Zanolidae	（チョウ目）の昆虫の総称　オビガ科に入れられることあり
zanolid moths		*Apatelodes*	（チョウ目、Zanolidae）の昆虫の総称
Zapater's ringlet		*Erebia zapateri* Oberthür	（チョウ目、タテハチョウ科）旧北区
zapulata moth		*Choristoneura zapulata* (Robinson)	（チョウ目、ハマキガ科）新北区
Zarucco duskywing		*Erynnis zarucco* (Lucas)	（チョウ目、セセリチョウ科）新北区
Zathoe mimic white		*Dismorphia zathoe* (Hewitson)	（チョウ目、シロチョウ科）新熱帯区
Zavaleta clearwing			Zavaleta glasswing を見よ
Zavaleta glasswing	シラホシオオスカシマダラ	*Godyris zavaleta* (Hewitson)	（チョウ目、タテハチョウ科）新熱帯区
zea clearwing		*Oleria zea zea* (Hewitson)	（チョウ目、タテハチョウ科）新熱帯区
zea sister		*Adelpha zea* (Hewitson)	（チョウ目、タテハチョウ科）新熱帯区
Zebina hairstreak		*Rekoa zebina* (Hewitson)	（チョウ目、シジミチョウ科）新北区
zebra (1)	ミスジシロチョウ	*Pinacopteryx eriphia* (Godart)	（チョウ目、シロチョウ科）エチオピア区
zebra (2)		*Colobura dirce* (Linnaeus)	（チョウ目、タテハチョウ科）新熱帯区
zebra			Malayan zebra を見よ
zebra			zebra longwing を見よ
zebra beauty			zebra sapseeker を見よ
zebra blue (1)		*Leptotes plinius pseudocassius* (Murray)	（チョウ目、シジミチョウ科）豪州区
zebra blue (2)	カクモンシジミ	*Leptotes plinius* (Fabricius)	（チョウ目、シジミチョウ科）日本、旧北区、東洋区、豪州区
zebra blues	カクモンシジミ属	*Leptotes*	（チョウ目、シジミチョウ科）の昆虫の総称
zebra butterfly			zebra (2) を見よ
zebra butterfly			zebra longwing を見よ
zebra caterpillar			zebra caterpillar moth を見よ　北米での英名
zebra caterpillar moth		*Melanchra picta* (Harris)	（チョウ目、ヤガ科）新北区。キャベツ害虫
zebra clubtail		*Stylurus scudderi* (Selys)	（トンボ目、サナエトンボ科）新北区
zebra Conchylodes moth		*Conchylodes ovulalis* (Guenée)	（チョウ目、メイガ科）新北区

英名	和名	学名	所属、分布、ほか
zebra cross-streak			zebra-striped hairstreak を見よ
zebra-crossing hairstreak			zebra-striped hairstreak を見よ
zebra cutworm		*Lecanobia nevadae* Grote	(チョウ目、ヤガ科) 新北区
zebra fantasy		*Pseudaletis zebra* Holland	(チョウ目、シジミチョウ科) エチオピア区
zebra grizzled skipper		*Spialia zebra* (Butler)	(チョウ目、セセリチョウ科) 東洋区
zebra hairstreak			zebra-striped hairstreak を見よ
zebra heliconian			zebra longwing を見よ
zebra longwing	キジマドクチョウ (チャリトニアドクチョウ)	*Heliconius charithonius* (Linnaeus)	(チョウ目、タテハチョウ科) 新北区、新熱帯区
zebra sapseeker	チグリスタテハ	*Tigridia acesta* (Linnaeus)	(チョウ目、タテハチョウ科) 新熱帯区
zebra scaly cricket		*Cycloptilum zebra* (Rehn et Hebard)	(バッタ目、コオロギ科) 新北区
zebra-striped hairstreak		*Panthiades bathildis* (C. et R. Felder)	(チョウ目、シジミチョウ科) 新熱帯区
zebra swallowtail	トラフタイマイ	*Protographium (Neographium) marcellus* (Cramer)	(チョウ目、アゲハチョウ科) 新熱帯区
zebra teaser		*Arawacus separata* (Lathy)	(チョウ目、シジミチョウ科) 新熱帯区
zebra-tipped metalmark	マーガレットシジミタテハ	*Mesene margaretta margaretta* (White)	(チョウ目、シジミタテハ科) 新熱帯区
zebra vicetail		*Hemigomphus comitatus* (Tillyard)	(トンボ目、サナエトンボ科) 豪州区
zebra webworm moth		*Atteva zebra* Duckworth	(チョウ目、スガ科) 新熱帯区
zebra white			zebra (1) を見よ
Zela Emesis			Zela metalmark を見よ
Zela metalmark		*Emesis zela* Butler	(チョウ目、シジミタテハ科) 新北区、新熱帯区
Zelica clearwing		*Oleria zelica* (Hewitson)	(チョウ目、タテハチョウ科) 新熱帯区。*O. z. pagasa* (Druce) も同英名
Zelica ticlear		*Oleria zelica pagasa* (Druce)	(チョウ目、タテハチョウ科) 新熱帯区
Zelica untailed Charaxes		*Charaxes zelica* Butler	(チョウ目、タテハチョウ科) エチオピア区
zelkova aphid	ニレクロマダラアブラムシ	*Chromocallis nirecola* (Shinji)	(カメムシ目、アブラムシ科) 日本、旧北区
zelkova horn worm	ウンモンスズメ	*Callambulyx tatarinovii gabyae* Bryk	(チョウ目、スズメガ科) 日本
zelkova jumping weevil	アカアシノミゾウムシ	*Orchestes sanguinipes* (Roelofs)	(コウチュウ目、ゾウムシ科) 日本。ケヤキ害虫
zelkova leaf buprestid	ヤノナミガタチビタマムシ	*Trachys yanoi* Y. Kurosawa	(コウチュウ目、タマムシ科) 日本、旧北区。ケヤキ害虫
zelkova sawfly	ケヤキナメクジハバチ	*Caliroa zelkovae* Oishi	(ハチ目、ハバチ科) 日本。ケヤキ害虫
zelkova scale	ケヤキフクロカイガラムシ	*Eriococcus abeliceae* Kuwana	(カメムシ目、フクロカイガラムシ科) 日本

英名	和名	学名	所属、分布、ほか
Zeller's Epipaschia moth		*Macalla zelleri* Grote	（チョウ目、メイガ科）新北区
Zeller's Ethmia moth		*Ethmia zelleriella* (Chambers)	（チョウ目、スエヒロキバガ科）新北区
Zeller's midget moth		*Phyllonorycter messaniella* (Zeller)	（チョウ目、ホソガ科）旧北区、豪州区
Zeller's skipper			Borbo skipper を見よ
Zelphanta numberwing		*Callicore hystaspes* (Fabricius)	（チョウ目、タテハチョウ科）新熱帯区
Zenker's glider		*Cymothoe zenkeri* Richelmann	（チョウ、タテハチョウ科）エチオピア区
Zenobia swallowtail		*Papilio zenobia* Fabricius	（チョウ目、アゲハチョウ科）エチオピア区
zephyr	ゼフィルスシータテハ	*Polygonia zephyrus* (Edwards)	（チョウ目、タテハチョウ科）新北区
zephyr anglewing			zephyr を見よ
zephyr blue		*Plebejus pylaon* (Fischer de Waldheim)	（チョウ目、シジミチョウ科）旧北区
zephyr comma			zephyr を見よ
zephyr-eyed silkmoth		*Automeris zephyria* Grote	（チョウ目、ヤママユガ科）新北区
Zephyritis Morpho		*Morpho zephyritis* Butler	（チョウ目、タテハチョウ科）新熱帯区
Zereda skipper		*Chalypyge zereda* Hewitson	（チョウ目、セセリチョウ科）新熱帯区
Zerene fritillary		*Speyeria zerene* (Boisduval)	（チョウ目、タテハチョウ科）新北区
Zerlina glasswing		*Pteronymia zerlina* Hewitson	（チョウ目、タテハチョウ科）新熱帯区
Zestos skipper		*Epargyreus zestos* Geyer	（チョウ目、セセリチョウ科）新北区
Zeurippa metalmark		*Hypophylla zeurippa* Boisduval	（チョウ目、シジミタテハ科）新熱帯区
Zeutus banded-skipper		*Calliades zeutus* (Möschler)	（チョウ目、セセリチョウ科）新熱帯区
Ziba groundstreak		*Strymon ziba* (Hewitson)	（チョウ目、シジミチョウ科）新熱帯区
Ziba scrub-hairstreak			Ziba groundstreak を見よ
zigzag flat		*Odina decoratus* Hewitson	（チョウ目、セセリチョウ科）東洋区
zigzag fritillary			Freya's fritillary を見よ
zigzag Furcula moth		*Furcula scolopendrina* (Boisduval)	（チョウ目、シャチホコガ科）新北区
zigzag Herpetogramma moth			Herpetogramma moth を見よ
zigzag leafhopper			zigzag rice leafhopper を見よ
zigzag rice leafhopper	イナズマヨコバイ	*Recilia dorsalis* (Motschulsky)	（カメムシ目、オオヨコバイ科）日本、東洋区。イネ害虫
zigzag striped leafhopper			zigzag rice leafhopper を見よ
zig-zag winged rice leafhopper			zigzag rice leafhopper を見よ

英名	和名	学名	所属、分布、ほか
zilpa longtail			zilpa skipper を見よ
zilpa skipper		*Chioides zilpa* (Butler)	（チョウ目、セセリチョウ科）新北区
Zimmerman pine moth		*Dioryctria zimmermani* (Denis et Schiffermüller)	（チョウ目、メイガ科）新北区。幼虫は松枝の樹皮下につく
Zina sister			zea sister を見よ
Zincken's tiger		*Parantica albata* (Zincken)	（チョウ目、タテハチョウ科）東洋区
zizyphus scale			citrus scale を見よ
zodiac moth		*Alcides zodiaca* (Butler)	（チョウ目、ツバメガ科）豪州区
Zoega's conch		*Agapeta zoegana* (Linnaeus)	（チョウ目、ハマキガ科）旧北区
zorapteran		*Zorotypus snyderi* Caudell	（ジュズヒゲムシ目、ジュズヒゲムシ科）新北区
zorapterans			angel insects (1) を見よ
Zorea eyemark		*Mesosemia zorea* Hewitson	（チョウ目、シジミタテハ科）新熱帯区
Zorilla sootywing		*Bolla zorilla* Plötz	（チョウ目、セセリチョウ科）新熱帯区
Zorion longhorn beetle		*Zorion minutum* Fabricius	（コウチュウ目、カミキリムシ科）豪州区
Zullich's blue		*Agriades zullichi* Hemming	（チョウ目、シジミチョウ科）旧北区。
Zulu		*Alaena picata* Sharpe	（チョウ目、シジミチョウ科）エチオピア区
Zulu blue		*Lepidochrysops ignota* (Trimen)	（チョウ目、シジミチョウ科）エチオピア区
Zulu buff		*Terionima zuluana* van Son	（チョウ目、シジミチョウ科）エチオピア区
Zulu shadefly		*Coenyra hebe* Trimen	（チョウ目、タテハチョウ科）エチオピア区
Zulus		*Alaena*	（チョウ目、シジミチョウ科）の昆虫の総称　ズールー族に因む
Zuni tiger-moth		*Arachnis zuni* Neumoegen	（チョウ目、ヒトリガ科）新北区
Zweifel's skipper		*Atrytonopsis zweifeli* Freeman	（チョウ目、セセリチョウ科）新熱帯区
zygaenids			leaf skeletonizer moths (2) を見よ

世界の昆虫英名辞典
vol. 2 M-Z

発行日　2018年5月12日　初版　第1刷

編　集
やのこうじ
矢野宏二

発　行
とうかしょぼう
櫂歌書房

ISBN 978-4-434-24028-7

有限会社 櫂歌書房
〒811-1365 福岡市南区皿山4丁目14-2
TEL: 092-511-8111　FAX: 092-511-6641
E-mail: e@touka.com　http://www.touka.com

発売所　星雲社